Analytical and Experimental Modal Analysis

This book covers the fundamentals and basic concepts of analytical and experimental approaches to modal analysis. In practice, the analytical approach based on lumped parameter and finite element models is widely used for modal analysis and simulation, and experimental modal analysis is widely used for modal identification and model validation. This book is inspired by this consideration and is written to give a complete picture of modal analysis.

Features:

- Presents a systematic development of the relevant concepts and methods of the analytical and experimental modal analyses.
- Covers phase resonance testing and operational modal analysis.
- Provides the relevant signal processing concepts.
- Includes applications like model validation and updating, force identification, and structural modification.
- Contains simulations, examples, and MATLAB® programs to enhance understanding.

This book is aimed at senior undergraduates/graduates, researchers, and engineers from mechanical, aerospace, automotive, civil, and structural engineering disciplines.

Analytical and Experimental Modal Analysis

Subodh V. Modak

CRC Press
Taylor & Francis Group
Boca Raton London New York

CRC Press is an imprint of the
Taylor & Francis Group, an **informa** business

MATLAB® is a trademark of The MathWorks, Inc. and is used with permission. The MathWorks does not warrant the accuracy of the text or exercises in this book. This book's use or discussion of MATLAB® software or related products does not constitute endorsement or sponsorship by The MathWorks of a particular pedagogical approach or particular use of the MATLAB® software

Designed cover image: © Shutterstock

First edition published 2024
by CRC Press
6000 Broken Sound Parkway NW, Suite 300, Boca Raton, FL 33487-2742

and by CRC Press
4 Park Square, Milton Park, Abingdon, Oxon, OX14 4RN

CRC Press is an imprint of Taylor & Francis Group, LLC

© 2024 Subodh V. Modak

ISBN: 9781138318151 (hbk)
ISBN: 9781032551029 (pbk)
ISBN: 9780429454783 (ebk)

DOI: 10.1201/9780429454783

Typeset in Times
by codeMantra

To my wife, Anamika; daughter, Priyamvada; son, Sadanand and to the memory of my parents, Vandana and Vasant Modak

Contents

Preface

This book is about modal analysis and covers the analytical and experimental approaches. The reason to cover both approaches stems from the way the modal analysis is practiced today.

MOTIVATION

The field of modal analysis today depends on analytical and experimental approaches. A few decades back, when the numerical modeling and analysis methods were not developed, the experimental approach was primarily used and saw rapid developments. However, with developments in the finite element method (FEM), numerical modeling of the vibrating systems or structures has emerged as a powerful approach for dynamic and modal analysis. The analysis using numerical models is generally faster and more economical and allows easy exploration of optimal designs and solutions. However, despite the sophistication of the numerical models, the accuracy of their predictions needs to be validated. The model validation demands accurate reference data, and the experimental modal analysis or modal testing plays a vital role in this regard as it can provide such data. In addition, the output of the experimental modal analysis provides a system model that can be used independently for analysis and design. This way, the analytical and experimental approaches have become indispensable tools for modal analysis today.

The individuals concerned with modal analysis benefit by understanding both approaches. It allows chalking out a dynamic analysis and design strategy that is effective, robust, and optimal in terms of the goals attained and the utilization of the available resources.

The inspiration and need to cover the analytical and experimental approaches to modal analysis in a single book is driven by the above considerations. In addition, the author believes that a good understanding of the underlying theory is essential for carrying out the experimental modal analysis correctly. Hence, this book tries to present a systematic development of the relevant background concepts, theories, and formulation of the modal analysis methods.

This book has evolved from the author's experience of teaching a course on modal analysis at the postgraduate level for several years at IIT Delhi. The author's study and research for over two decades further strengthened his experience in this field.

AUDIENCE

This book is relevant to individuals from mechanical, aerospace, automotive, and civil and structural engineering disciplines interested in modal analysis. Specifically, it may be of interest to

- students taking courses in modal analysis, modal testing, vibration, structural dynamics, and modal analysis applications;
- engineers working on projects related to modal testing, NVH, or dynamic testing and wanting to develop a thorough understanding of the experimental approaches to modal analysis and its theory;
- researchers in the fields of analytical and experimental modal analyses and the related areas such as model updating, coupled structural analysis, and operational modal analysis;
- teachers teaching courses or guiding projects and research related to modal analysis, vibrations, FEM for dynamics, FE model updating, and modal analysis applications.

This book can also be read by anyone passionate about studying modal analysis but familiar with the basics of mechanics, calculus, and linear algebra.

KEY FEATURES OF THE BOOK

This book comprehensively covers various methods and techniques in modal analysis and can be used as a textbook for analytical and experimental modal analysis courses. It emphasizes the concepts and fundamentals, and each topic is dealt with from this perspective. The idea is to provide a conceptual clarity of various topics this book covers.

Some key features include:

- Emphasis on concepts with a clear explanation
- Presentation of the material in a step-by-step and logical manner
- Coverage of the detailed mathematical formulation and equations

- Simulations to demonstrate the modal testing techniques and methods
- Solved examples to provide insight into the concepts learned
- Examples based on MATLAB[1] programs to demonstrate computer implementation of the modal analysis concepts
- Unsolved problems to strengthen the understanding of the modal analysis concepts
- Review questions at the end of each chapter
- A large number of carefully made figures and plots for easier understanding

This book is written as a self-contained piece of material. It doesn't assume any prior knowledge of the vibration theory, and the necessary and relevant concepts and details are covered in this book. The experimental modal analysis methods based on the hammer and shaker testing are covered in detail, with separate chapters devoted to phase resonance testing, operational modal analysis, model updating, and modal testing applications. Since the FEM is widely used for analytical modal analysis, a chapter on FEM for dynamics is included.[1]

OVERVIEW OF THE BOOK

Chapter 1 gives a basic introduction to the topics covered in various chapters. It highlights the critical concepts and methods addressed in each of them. Chapters 2–15 can be broadly classified as follows.

- Chapters 2–6: Analytical modal analysis
- Chapters 7–12: Experimental modal analysis
- Chapter 13: Operational modal analysis
- Chapters 14 and 15: Applications of experimental modal analysis

Chapter 2 covers the lumped parameter approach to develop single and multi-degree freedom models of vibrating systems. The fundamental dynamical principles, including the principle of virtual work, Lagrange's equations, and Hamilton's principle, are covered to derive the equations of motion. The reciprocity theorem for flexibility and stiffness coefficients is derived. Examples of the lumped parameter modeling of longitudinal, torsional, and transverse vibrations are presented. Most literature considers flexibility influence coefficients to derive the stiffness matrices for the transverse vibration problems. This chapter presents the derivation of the stiffness matrix coefficients by direct solution of the beam differential equation.

Chapter 3 presents the finite element method (FEM) for the dynamics problems. The finite element matrices for the bar, beam, and frame elements are formulated using Hamilton's principle. The chapter then describes the basic ideas of the assembly of the element matrices and the imposition of boundary conditions. A few examples of FE modeling of dynamic systems are presented at the end.

Chapter 4 is concerned with the analytical modal analysis of the single degree of freedom (SDOF) systems. The free and forced vibration of undamped and damped systems with viscous and hysteretic damping is considered. The concepts of frequency response function (FRF) and impulse response function (IRF) are introduced. Response analysis by the Laplace transform approach is also covered.

Chapter 5 presents the analytical modal analysis of undamped multi-degree of freedom (MDOF) systems covering their free vibration, eigenvalue problem, and orthogonality properties. The expansion theorem and concepts of physical and modal spaces are introduced. The free and forced responses using modal analysis are obtained. The ideas of FRF and IRFs are extended to the undamped MDOF systems.

Chapter 6 presents the analytical modal analysis of damped MDOF systems. Free and forced vibration of systems with proportional and nonproportional structural and viscous damping are covered. Eigenvalue problems are formulated, and the orthogonality properties are derived. The modal analysis is used to find the free and forced response to harmonic and transient forces. FRF and IRFs for damped systems are derived along with their relationships to the modal models, which form the basis for modal parameter estimation in experimental modal analysis.

Chapter 7 is concerned with the characteristics and graphical representation of the FRFs for SDOF and MDOF systems. Graphical representation of the FRFs using Bode, Nyquist, and real and imaginary plots is presented. The concepts of stiffness and mass lines for SDOF systems and antiresonances for undamped and damped MDOF systems are introduced.

Chapter 8 presents comprehensive coverage of signal processing for modal analysis. Fourier analysis of continuous and discrete periodic and aperiodic signals is presented. The concepts of discrete Fourier series (DFS), discrete-time Fourier transform (DTFT), and discrete Fourier transform (DFT) are developed. The idea of the fast Fourier transform (FFT) is

---

[1] MATLAB® is a registered trademark of The MathWorks, Inc.

discussed. This chapter presents the relationships between the parameters of a discrete signal and its DFT. Some critical issues in digital signal processing, such as sampling, aliasing, leakage, windowing, and quantization, are discussed. It is followed by the Fourier analysis of random signals covering the basic concepts of auto and cross-correlations, power and cross-spectral densities, and white noise.

Chapter 9 presents the FRF measurement using an impact hammer. The basic constructions of the impact hammer and accelerometer are described. This chapter covers the theory of seismic pickup and piezoelectric accelerometers. The role of a charge amplifier is analyzed, and IEPE/ICP transducers are described. The accelerometer selection for modal testing and various accelerometer mounting methods are discussed. The H1 and H2 estimates of FRFs are presented. FRF measurement using an impact hammer is simulated on a cantilever structure to clearly demonstrate the various issues involved.

FRF measurement using a vibration shaker is dealt with in **Chapter 10**. The construction of an electromagnetic vibration exciter is first explained. A lumped parameter model of the shaker- structure system is developed, and the dynamic interaction between the two is analyzed. The use of force transducer, stinger, and impedance head is discussed. Various excitation signals used in shaker testing are described. FRF measurement using a shaker is simulated on a cantilever structure. The simulation demonstrates the application of various excitation signals, such as random, pseudo-random, periodic random, burst random, and sine chirp, for modal testing.

Chapter 11 presents modal parameter estimation methods, also called curve fitting methods. The SDOF methods such as peak picking, circle fit, line fit, and modified line fit, are presented with all relevant mathematical details. The concept of residual is described. The MDOF frequency-domain methods like rational fraction polynomial (RFP) and global RFP and the time-domain methods such as complex exponential method (CEM), least-square CEM, the Ibrahim time-domain (ITD) method, and eigensystem realization algorithm (ERA) are described in detail. The modal parameter estimation with each method is carried out using the simulated data on the cantilever structure to develop further insight into the working of the methods.

Chapter 12 deals with phase resonance testing. The primary objective and motivation of phase resonance testing are first discussed, and the phase resonance condition is derived. The basic principle of estimating force appropriation vector to excite normal modes is then described. Methods such as the extended Asher's method, real mode indicator function (RMIF), and multivariate mode indicator function (MMIF) to excite normal modes are presented. The phase resonance testing is then simulated on a 3-DOF system with proportional and nonproportional damping and the cantilever structure.

The basic concept and approach of operational modal analysis (OMA), also referred to as the output-only modal analysis, are presented in **Chapter 13**. The estimation of free decays using the correlation functions and random decrement is described. OMA in the time domain using ERA is simulated on the cantilever structure, highlighting the basic steps involved. In the end, OMA in the frequency domain and the elimination of harmonics in OMA are discussed.

Chapter 14 presents the applications of EMA. The incomplete nature of the EMA data is highlighted. This chapter covers response simulation, dynamic design, local structural modification, coupled structural analysis using fixed and free interface methods, and force identification using experimental modal model.

Finite element model validation and updating, another application of EMA, is presented in **Chapter 15**. The need for FE model validation is initially described, followed by a discussion of the possible sources of modeling inaccuracies in an FE model. Some model correlation indices are then presented. This chapter then covers the methods of model reduction and mode shape expansion. The issue of updating parameter selection is discussed. The mathematical formulation of one model updating method from each class is described in detail. Model updating with these methods is carried out using the simulated data on the cantilever structure to bring out the basic steps and issues involved. The ill-conditioning and regularization in model updating is discussed. This chapter concludes by explaining briefly the need and concept of stochastic FE model updating.

SUGGESTED COURSE PLANS

The material covered in this book is primarily meant for a course on experimental modal analysis/modal testing/modal analysis but is also valuable for courses on vibration and structural dynamics. Many universities offer a course on experimental modal analysis at the postgraduate level. Courses on this subject can be introduced at other universities wherever relevant, and three courses with course objectives and broad contents are suggested below to meet different learning requirements. The course plans are only indicative, and the contents and the coverage from the suggested chapters can be reduced or increased based on the course credits.

Course 1

Course name: Experimental modal analysis

Course objective: To cover the theoretical basis of experimental modal analysis, signal processing for modal analysis, FRF measurement using hammer and shaker, curve fitting, and a brief idea of phase resonance testing and operational modal analysis

Course level: Postgraduate

Relevant chapters	Topics
Chapters 4, 5 and 6	Selected topics from these chapters to lay the theoretical basis of modal analysis: eigenvalue problem for systems with and without damping, FRFs, IRFs, and solution by modal analysis
Chapter 7	Characteristics of FRFs, various FRF plots, stiffness, and mass lines
Chapter 8	Fourier analysis and Fourier transform; discrete Fourier series, DFT, FFT, aliasing, sampling theorem, leakage, windowing, quantization noise; Fourier analysis of random signals, auto and cross-correlation, PSD and CPSD, and white noise
Chapter 9	FRF measurement using an impact hammer
Chapter 10	FRF measurement using a shaker
Chapter 11	Curve fitting (a few select methods)
Chapters 12, 13	A brief idea of phase resonance testing and OMA
Chapters 14 and 15	A brief overview of EMA applications

Course 2

Course name: Analytical modal analysis and model validation

Course objectives: To cover the principles of dynamics, analytical modeling using lumped parameter models, FEM for dynamics, modal analysis using analytical models, brief introduction of EMA using hammer excitation, FE model correlation and validation, and FE model updating

This course has more emphasis on analytical modal analysis. It gives the necessary exposure to EMA to validate the analytical models, which has become an essential requirement in industries today.

Course level: Postgraduate

Relevant chapters	Topics
Chapter 2	Principles of dynamics and lumped parameter models
Chapter 3	FEM for dynamics
Chapter 4	Modal analysis of undamped and damped SDOF systems
Chapter 5	Modal analysis of undamped MDOF systems
Chapter 6	Modal analysis of damped MDOF systems
Chapters 8, 9 and 11	A brief idea of signal processing; a brief idea of EMA using hammer excitation;
Chapter 15	FE model validation and updating

Course 3

Course name: Vibration and modal analysis

Course objectives: To cover the principles of dynamics, lumped parameter modeling of vibrating systems, modal analysis of MDOF systems, basic signal processing and introduction to EMA using hammer excitation

This course combines the coverage of basic vibration theory with an introduction to EMA. The theoretical studies in vibration are complemented with modal testing to extend the vibration theory to practice.

Course level: Undergraduate/Postgraduate

Relevant chapters	Remarks
Chapter 2	Dynamics principles, lumped parameter models
Chapter 4	Modal analysis of undamped and damped SDOF systems
Chapter 5	Modal analysis of undamped MDOF system
Chapter 6	Modal analysis of damped MDOF system
Chapter 7	Characteristics of FRFs
Chapter 8	Fourier analysis and Fourier transform; discrete Fourier series, DFT, FFT, aliasing, windowing, quantization noise
Chapter 9	Introduction to EMA with an impact hammer
Chapter 11	The basic idea of curve fitting; Modal parameter estimation with a few selected methods

ACKNOWLEDGMENTS

I would like to express my deepest gratitude to Prof. T. K. Kundra and the late Prof. B. C. Nakra of IIT Delhi for their supervision and encouragement to work on modal analysis and model updating during my doctoral studies.

Fruitful interactions with all the students whom I supervised for their doctoral, master's, or bachelor's theses are gratefully acknowledged.

I want to thank IIT Delhi for granting me the sabbatical leave to write this book.

I also thankfully acknowledge the cooperation and support provided by Dr. Gagandeep Singh, Ms. Aditi Mittal, Ms. Aimée Crickmore of Taylor and Francis, and Ms. Assunta Petrone of Codemantra during the writing, copy editing, and production stages of the book and CRC Press for its publication.

Work of this kind often takes a significant toll on family time. I am indebted to my wife, daughter, and son for their cooperation and encouragement and for providing many small pieces of helpful advice.

Subodh V. Modak

Author

Subodh V. Modak is Professor in the Department of Mechanical Engineering at the Indian Institute of Technology, Delhi. He received his bachelor's degree in Mechanical Engineering from Shri G.S. Institute of Technology and Science, Indore, master's degree in Design Engineering from the Indian Institute of Technology, Bombay, and PhD from the Indian Institute of Technology, Delhi. His research and teaching interests include experimental modal analysis, model updating, vibration and acoustics, control engineering, and active noise control. His research is published in many leading and reputed journals.

1 Introduction

1.1 WHAT IS MODAL ANALYSIS?

A physical system or a structure possesses natural modes of vibrations. The natural modes are the distinct or principal ways a system can vibrate naturally without external excitation. These modes also determine the system's dynamic response to dynamic forces. Each natural mode is associated with natural frequency, mode shape, and damping factor, and the modal analysis is the process of determining these characteristics of the natural modes. Another objective of the modal analysis is to analyze the system's dynamic response in terms of the responses in its modes of vibration. It is based on the principle that the dynamic response can be expressed as the superposition of the responses in natural modes. The modal analysis provides valuable insight into how the structure responds to external forces, which can be used to reduce the resulting responses.

The natural modes of vibrations are central to the modal analysis. Since the modes are defined for a linear system, the theory of modal analysis is based on the assumption that the system under consideration is linear. Many systems in practice have some degree of nonlinearity. However, if the amplitudes of motions are small, these systems can be approximated as linear systems, and the modal analysis can be used.

Developments in the modal analysis were motivated by the need to address the problems of excessive vibrations. It has helped analyze the cause of the vibration problems and find solutions. Today, modal analysis is applied to analyze and design various dynamic systems in practice. Notable applications include automotive vehicles, rail vehicles, airplanes, helicopters, ships, submarines, machine tools, power plant equipment, civil engineering structures, consumer products, and sports goods.

1.2 VIBRATION, ITS CAUSES, AND DETRIMENTAL EFFECTS

1.2.1 WHAT IS VIBRATION?

The principal motivation behind the modal analysis is reducing the unacceptable levels of vibrations. Vibration or vibratory motion refers to the oscillatory motion of a body or system. Most engineering systems in practice are subjected to dynamic forces and hence undergo vibrations. The motion of a simple pendulum is one of the simplest oscillatory motions. We all have experienced upward and downward motion while sitting in a moving car or train, which is a result of the vibratory motion of those systems. The noise emanating from many machines during their operation is also in many cases due to vibratory motion.

Figure 1.1 depicts a system through a block diagram. It shows that when an input force acts on the system, the output, such as displacement, velocity, and acceleration, may be produced. If the input force is dynamic, it varies with time and results in a time-varying/dynamic output, which is nothing but vibratory motion.

1.2.2 CAUSES OF VIBRATION

The dynamic forces are responsible for vibration and may originate from different sources. Physical systems with rotating parts, such as fans, pumps, turbines, gear and belt drives, electric motors, and generators, often have some mass unbalance, leading to centrifugal forces during operation. The centrifugal forces are rotating and generate vibrations. Systems with sliding parts, like IC engines and reciprocating compressors, have reciprocating unbalance, giving rise to unbalance forces, causing vibrations.

Operational forces may also cause vibrations. Machine tool structures are subjected to dynamic forces during operation causing vibration. Impacts in a forging press generate transient vibrations in the press and surrounding structures. An aircraft

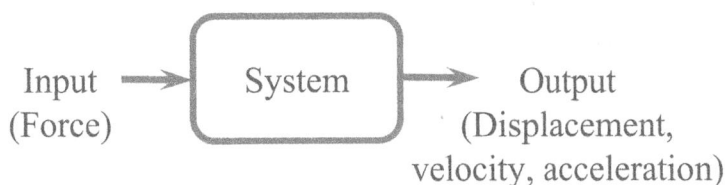

Input → System → Output
(Force)   (Displacement, velocity, acceleration)

FIGURE 1.1 Block diagram of a system.

DOI: 10.1201/9780429454783-1

is subjected to excitations from the engine and turbulence. Onboard machinery on a ship or a submarine applies vibratory forces, causing their vibration.

Dynamic forces and disturbances may also arise from ambient conditions. Structures like buildings may be subjected to loads due to wind and/or ground motion. Waves in the sea excite a ship's hull, causing its vibrations.

The dynamic forces may also arise due to fluid flow in pipes and ducts (causing flow-induced vibrations) or due to the flow of air or fluid past a body (causing self-excited vibrations). Friction can also be a source of excitation leading to vibrations (like in stick-slip motion).

1.2.3 DETRIMENTAL EFFECTS OF VIBRATION

The vibrations are generally undesirable. Vibrations result in oscillatory displacements, strains, and stresses, causing fatigue. Take the example of a simple beam supporting a motor. The beam undergoes transverse vibration due to dynamic forces from the motor. The transverse vibration causes a cyclic contraction and stretching of the longitudinal fibers of the beam, causing fatigue due to oscillatory strains and stresses. The possibility of resonance further compounds the problem because operation at or near the system's natural frequency amplifies the fatigue strains and stresses. Since higher fatigue stresses are associated with lower fatigue life, vibrations reduce the life of a system.

Vibrations may also increase friction and wear and reduce the life of bearings. Vibrations in automotive and other vehicles decrease ride comfort. In a satellite tracking system, the antenna's orientation may be affected by its vibrations. Similarly, the positioning accuracy of a robot may be affected by its vibrations. Vibrations are often accompanied by noise, which is also undesirable.

Some applications exist where the vibrations may be intentionally generated to serve some useful purpose, like in some musical instruments, vibratory conveyors, and vibration exciters. However, for most engineering systems, the vibrations are undesirable, and the objective is to ensure that the vibrations are within the permissible limits. This requires designing the system with favorable dynamics, and the modal analysis provides an efficient tool to achieve this objective.

1.3 NEED FOR MODAL ANALYSIS

A static force on a system causes a static response. But an equal amplitude dynamic force on the system may cause a response much more than the static. This is due to the dynamic or time-varying nature of the force, which brings into play the inertia forces and the system's natural frequencies that may amplify the response beyond the static response. If the frequency of the dynamic force coincides with any of the system's natural frequencies, then resonance occurs causing a large response, and hence the resonant operation of a system needs to be avoided. The resonant operation can be avoided by changing either the excitation frequencies, if that is possible, or the natural frequencies by suitable modifications in the system. However, exercising these remedies requires knowledge of the natural frequencies of the system, and experimental modal analysis (EMA) or analytical modal analysis serves this purpose. Thus, avoidance of resonant operation of systems is one primary motivation for modal analysis.

The vibration response of a system depends on the input forces and the system characteristics. If the response level is unacceptable, then the question arises why it is high. Is it due to a higher input, or is it that the system characteristics amplify the effect of the input?

A simple measurement of the output and its analysis may not help diagnose the cause precisely since the origin of the problem could be from the input and/or the system. To diagnose the cause, one must look at how the input forces interact with the system to produce the output.

Through modal analysis, the vibration modes of the system can be obtained, and this knowledge, along with the information about the dynamic forces, enables a systematic study of the interaction between the dynamic forces and the system characteristics and helps answer the above questions. Thus, troubleshooting vibration (and noise) problems form another motivation for modal analysis. The modal analysis can also be used in many other applications, as discussed in Section 1.7.

1.4 ANALYTICAL MODAL ANALYSIS

Analytical modal analysis is the modal analysis of the system using a dynamic mathematical model of the system. The mathematical model is a spatial model, which describes the system through mass, stiffness, and damping matrices.

The mathematical model can be a lumped parameter model or a numerical model derived from the finite element method or some other numerical technique. The lumped parameter model is based on the idealization of the structure using lumped masses, springs, and dampers. The solution of a lumped parameter model requires less computation, but the predictions may not be accurate enough due to the modeling approximations. These models may be appropriate where the system's elements or parts can be approximated as lumped masses or stiffnesses. These models can also be used to obtain a quick initial estimate of the dynamic characteristics before a more sophisticated model is made. The equations of motion from the lumped parameter models can be derived by applying physical laws such as Newton's law, the principle of virtual work, Lagrange's equation, or Hamilton's principle. Irrespective of the physical law used, the final equation is in the form of simultaneous second-order ordinary differential equations. We consider the lumped parameter modeling of dynamic systems in Chapter 2.

The finite element method (FEM) is a numerical approach to build a discrete mathematical model of a system and is widely used for analytical modal analysis. FEM is a powerful and versatile approach for modeling systems with complex geometries, boundary conditions, and material models. FEM represents a continuous system as an assemblage of finite-size elements and approximates the displacement field over a finite element by a simple function/polynomial. Since the element is small compared to the size of the physical domain, the approximation over an element can be carried out with a simple function, and the displacement field over the complete domain can be approximated by the assembly of these simple but piecewise approximations. We study, in Chapter 3, the basics of FEM for dynamics.

The spatial dynamic multi-degree-of-freedom mathematical model, whether obtained by a lumped parameter approach or by FEM, is in the form of mass, stiffness, and damping matrices along with a force vector. The damping may be either viscous or hysteretic and modeled as proportional or nonproportional.

The spatial model is used to set up and solve the eigenvalue problem to find natural frequencies, mode shapes, and damping factors, which constitute the modal model. For a system with nonproportional viscous damping, the equations of motion are transformed to state space to set up the eigenvalue problem. The solution leads to a set of 2N complex conjugate eigenvalues and corresponding complex conjugate eigenvectors.

The equations of motion of a system in physical space are coupled, meaning they need to be solved simultaneously. The coordinate transformation using the system eigenvectors as the basis leads to a set of equivalent but decoupled equations. The decoupled equations describe the vibration of the system in modal space as each equation corresponds to vibration in a natural mode of the system. All the properties in the physical space have a corresponding reflection in the modal space; hence, we have modal mass, modal stiffness, modal damping, and modal force. The advantage of the modal domain description is that the effect of the external force on individual modes can be studied, and their contributions to the total response can be analyzed.

The frequency response functions (FRFs) and impulse response functions (IRFs) are referred to as response models of the system in frequency and time domains, respectively. An FRF is the system response to a unit amplitude harmonic force, while an IRF is the system response to a unit impulse. The system's dynamic response can be determined to a given input using any of the four analytical models, i.e., spatial model, modal model, FRFs, and IRFs.

The analytical modal analysis of single-degree-of-freedom (SDOF) systems is presented in Chapter 4, while that of undamped and damped multi-degree-of-freedom (MDOF) systems is presented in Chapters 5 and 6, respectively. The analytical modal analysis also provides the theoretical basis for the experimental route to modal analysis, i.e., the EMA, covered in later chapters.

1.5 EXPERIMENTAL MODAL ANALYSIS

Experimental modal analysis (EMA) identifies the modal model, i.e., the natural frequencies, mode shapes, and damping factors, experimentally. However, EMA, also known as modal testing, follows an inverse route and starts from the relationship between the input and output in the response domain. This relationship is characterized by the FRFs in the frequency domain and the IRFs in the time domain.

In EMA, the system is excited by applying input forces. The inputs and outputs are measured and analyzed to estimate the FRFs of the system. The IRFs, if required, can be obtained by Fourier transform of the estimated FRFs. The modal model of the system is then identified by curve fitting the estimated FRFs or IRFs.

1.5.1 A BRIEF HISTORICAL BACKGROUND OF EMA

The developments to experimentally identify the modal characteristics started in the aircraft industry in the 1960s to address the problem of self-excited vibrations called flutter. At that time, numerical methods, like the FEM, were not developed. Analytical solutions were limited to problems of idealized geometries and could not be used to predict the dynamics of a complex structure like an aircraft. It led to the need to identify the dynamic characteristics experimentally by making the necessary measurements on the structure. The development of an experimental technique was also motivated by the needs of the machine tool industry to address the problems of machine tool vibrations, including the problem of chatter (also a self-excited vibration phenomenon). Later, the developments in EMA received a great impetus with advances in (a) digital signal processing and the related hardware, which made possible real-time computation of discrete Fourier transform of a signal, (b) sensors, like accelerometers and force transducers, which allowed an accurate measurement of the dynamic response and force, (c) excitation systems, like impact hammers and vibration exciters, which allowed artificial excitation of a structure, and (d) the theoretical developments of methods to estimate modal parameters from the measured data.

Though the developments in EMA started with the sole goal of identifying the dynamic characteristics experimentally, another application of this technique emerged over the years for validation and, if necessary, updating of the numerical models. With the numerical technique like FEM, and enormous computational power available, the analytical modal analysis can be performed even for complex structures. However, the question remains about the accuracy of the analytical predictions and requires validation of the model before it can be used reliably. In this regard, the EMA provides independent reference data for validation and plays a vital role in structural dynamic analysis and design.

1.5.2 Frequency Response Functions (FRFs)

In EMA, the measured FRFs (or IRFs) provide basic information about the system from which the system's dynamic characteristics can be identified. An FRF is defined for a pair of the response and force degrees of freedom and quantifies the response to a unit amplitude harmonic force. The FRFs can be presented graphically in multiple ways such as the Bode plots (which are the plots of the magnitude and phase of the FRF with frequency), Nyquist plot (which is the plot of the real versus imaginary parts of the FRF), or the plot of real and imaginary parts with frequency. An FRF shows an asymptotic behavior characterized by stiffness and mass lines. The FRF has large magnitude at the resonant frequencies and may also have antiresonances, which are like inverted peaks, where the magnitude is close to zero. We study the characteristics of the FRFs in Chapter 7.

1.5.3 Signal Processing for Modal Analysis

The estimation of the FRFs involves two steps. The first step is the measurement and acquisition of the force and response signals, and the second step is processing the measured signals. The force transducers and accelerometers are generally used for force and response measurement, respectively. The outputs of these transducers are discretized or sampled, and the discrete signals are recorded. The discrete signals are then further processed to obtain the FRFs/IRFs and other quantities of interest. The continuous signals can be transformed to the frequency domain using the Fourier transform. Thus, we see that the estimation of FRFs and IRFs involves signal processing, and understanding the same is necessary to choose appropriate measurement parameters and avoid errors.

In Chapter 8, we first look at the Fourier analysis of the continuous signals covering the Fourier series expansion and Fourier transform, and the concepts are then extended to the discrete signals. This leads to discrete Fourier series (DFS), which is the Fourier series expansion of discrete periodic signals. If the discrete signal is aperiodic, it needs to be Fourier transformed, leading to discrete-time Fourier transform (DTFT). In practice, the measured record length is treated as the period of a periodic signal, and the DFS is computed. The one period of DFS is called the discrete Fourier transform (DFT). The computation of DFT requires significant calculations, which can be done efficiently using the Fast Fourier transform (FFT).

The sampling of the continuous signal in the time domain leads to periodicity in the frequency domain. It means that the frequency components beyond a specific frequency, fs/2 if fs is the sampling frequency, can't be evaluated. To obtain a correct frequency domain representation of a signal, the Nyquist sampling theorem states that the sampling frequency must be equal to or more than twice the highest frequency component in the signal. Violation of this condition leads to the aliasing of the signal and should be avoided. Another critical issue in signal processing is the truncation of a signal in the time domain, which leads to 'leakage' in the frequency domain. It distorts the frequency domain representation. The leakage can be reduced by using window functions. The quantization error is an error that occurs due to the quantization of the continuous distribution of the signal amplitudes into discrete levels. Random signals are also used in modal testing, and hence, the frequency domain representation of random signals is required for FRF estimation. If the signal is random, then the autocorrelation of the signal is computed, which is then Fourier transformed to obtain power spectral density (PSD). Chapter 8 covers all these issues presenting the fundamental concepts necessary to understand the processing of signals for modal analysis.

1.5.4 FRF Measurement Using an Impact Hammer

One of the methods to apply the excitation force for EMA is using an impact hammer. An impact hammer has a force transducer at its front end and allows measurement of the force applied. Generally, there is a provision to attach a hammer tip at the front end and it can be a rubber, plastic, or metal tip and can be selected depending on the frequency range to be excited. The softer the tip, the shorter the frequency range that is excited. A piezoelectric accelerometer is often used to measure the acceleration response. The IEPE (or ICP) accelerometers are popular and convenient to use because they have a built-in charge amplifier, thus not requiring an external charge amplifier. A laser Doppler vibrometer can also be used for response measurements. The advantage is that the mass loading of the structure is eliminated as it is a non-contact transducer. It also enables measurement from a distance and at locations physically inaccessible, provided there is a line of sight.

Once the force and response signals are acquired, the FRF can be estimated as the ratio of the Fourier transforms of the response and force signals. H1 and H2 are FRF estimates based on auto and cross spectrums of the force and response signals and help reduce the influence of the measurement noise on the estimated FRF. The quality of a measured FRF can be judged using the coherence function, with its value close to 1.0, indicating a good measurement. The coherence function quantifies the degree of the linear relationship between the input force and the output response.

It is seen that any one row or column of the FRF matrix is enough to identify the modal model. In hammer testing, it is generally convenient to move the hammer from one location to another to apply the impact and fix the response transducer at one location. This process of FRF measurement is referred to as the roving hammer test. It leads to the measurement of one row of the FRF matrix. The number of measurement points should be enough to prevent spatial aliasing of the mode shapes.

FFT parameters should be appropriately chosen for FRF measurement. The relevant parameters are the frequency range and resolution, record length, sampling frequency, and time samples. These are interrelated, and two of the variables can be independently selected. The sampling frequency is related to the frequency range and should satisfy the sampling theorem to avoid aliasing. The force and exponential windows can be used for force and response signals, respectively, if necessary. The measured FRF can be checked for reciprocity, linearity, and repeatability. Another decision to be made is regarding the boundary conditions for testing. It could be free-free or fixed boundary conditions or the structure's prevailing boundary conditions, i.e., in-situ conditions. The whole measurement system can be calibrated for FRF measurement against a lumped rigid mass instead of calibrating individual sensors. EMA using hammer excitation is covered in Chapter 9.

1.5.5 FRF Measurement Using a Shaker

When the structure to be excited is massive, an electromagnetic vibration exciter can be a better option to ensure an adequate response signal and signal-to-noise ratio. The electromagnetic exciter is driven by a power amplifier and a signal generator. The advantage is that with the signal generator, any desired excitation signal, such as harmonic, periodic, sine chirp, random, or burst random, can be applied. A force transducer is attached to the structure where the excitation is to be applied, and the shaker is attached to the force transducer through a thin rod called a stinger. The stinger is stiff in the axial direction but flexible laterally. Force dropout at the resonant frequencies of the structure occurs due to interaction between the structure and the exciter. In shaker testing, it is generally convenient to fix the exciter at one location and change the location of the response transducers from one measurement to another. This process of FRF measurement is referred to as the roving accelerometer test. It leads to the measurement of one column of the FRF matrix. EMA using shaker excitation is covered in Chapter 10.

1.5.6 Modal Parameter Estimation Methods

Once the FRFs are measured, the last step in EMA is to estimate modal parameters from them. The idea is to fit the known theoretical relationship between the FRFs and the modal model, studied in analytical modal analysis, to the measured data. Hence, the estimation process is also called curve fitting. One main classification of the curve-fitting methods is the SDOF and MDOF methods. The SDOF methods are based on the assumption that the frequency response around a resonant peak in an FRF is dominated by the corresponding mode. Thus, an SDOF system FRF is fitted to the data in this range, simplifying the curve-fitting process. The peak-picking, circle-fit, and line-fit methods are the main SDOF methods. The Nyquist plot of FRF for an SDOF system traces a circle, and the circle-fit method utilizes this property to fit the FRF. The line-fit method utilizes the property that the real and imaginary parts of the inverse FRF of an SDOF system trace straight lines in the resonance frequency region when plotted against the frequency square. Contributions of the modes outside the range or band of analysis are called residuals. Many methods try to account for them in some form to improve the accuracy of the identified modal parameters.

The rational fraction polynomial (RFP) method is an MDOF curve-fitting method. It fits multiple modes simultaneously and is expected to yield more accurate modal parameter estimates as it tries to account for the contribution of all the modes in the chosen band. The method is based on the rational fraction form of the FRF. Analysis of one FRF at a time is often referred to as a local curve-fitting approach. But since the natural frequencies and damping factors should be the same irrespective of the FRF, the global curve-fitting methods seek to identify these modal parameters by analyzing all the measured FRFs simultaneously. This is expected to yield more consistent estimates of these global modal parameters.

The curve-fitting methods in the time domain operate on the IRFs identified by the inverse DFT of the FRFs. The complex exponential method is one of the earliest methods for modal analysis. The eigensystem realization algorithm identifies the system model in state space, from which the modal parameters are identified by eigenvalue analysis. One of the issues in the time-domain methods is the choice of the model order for fitting the data. The stabilization diagram is often used to decide the correct model order. The Ibrahim time-domain method can be used to identify the modal parameters from the free decay responses; hence, this method can also be used with ambient response data. Chapter 11 covers the modal parameter estimation methods.

1.5.7 PHASE RESONANCE TESTING

The curve-fitting methods to estimate modal parameters from the measured FRFs are called phase separation techniques. In these techniques, the contributions of different modes need to be separated to identify the modal parameters of the individual modes. Moreover, the modes obtained in the process are complex as they are influenced by damping in the structure.

Phase resonance testing is a method of EMA where the normal modes of the structure, that is, the undamped modes, are sought. The normal modes are excited when the responses are in quadrature with the applied force, i.e. have a phase of 90^0, a state of vibration referred to as the phase resonance condition. The knowledge of the normal modes is useful to validate the numerical models, which are generally undamped or are based on proportional damping. For exciting a normal mode, the force vector needs to be tuned to ensure that the phase resonance condition is satisfied.

Mode indicator functions (like real mode indicator function (RMIF) and multivariate mode indicator function (MMIF)) allow the detection of undamped natural frequencies and determination of the force vectors (called force appropriation vectors) needed to excite normal modes. Chapter 12 covers phase resonance testing.

1.6 OPERATIONAL MODAL ANALYSIS

Operational Modal Analysis (OMA) is a technique in which the modal analysis of a structure is performed based on the analysis of its ambient response to operational excitation. The modal analysis is carried out based on only measured outputs with no knowledge or measurement of the excitation causing the outputs. The artificial excitation needed for EMA is difficult to apply for huge structures, such as bridges and tall buildings. However, these structures vibrate under the natural excitations present due to wind, traffic, and other sources. These excitations are generally random, due to which the ambient response contain the modes in the frequency range of the ambient excitation. The OMA assumes the excitation as white noise and performs modal parameter identification on the ambient random response data. In OMA, the responses are measured at the selected locations on the structure. Some reference responses at the locations common to all sets of measured responses are also acquired. The reference responses allow the scaling and normalization of the responses measured at different times. In OMA, the correlation functions or free decays are computed from the measured responses and are analyzed to find the modal parameters. Random decrement is another concept that can be used to extract the free decays from the ambient responses. Some EMA techniques, like the Ibrahim time-domain method and the eigensystem realization algorithm can work with the free decays and can be used for OMA. Frequency domain decomposition and stochastic subspace identification are other OMA techniques. Chapter 13 covers the OMA.

1.7 APPLICATIONS OF EMA

The output of EMA is a modal model comprising the natural frequencies, mode shapes, and damping factors. It forms an alternative dynamic mathematical model of the system. The modal model obtained from EMA is said to be incomplete compared to an FE model of the system because it has information about a limited number of modes and a limited number of coordinates. There are many applications where the experimental modal model can be used, as listed below.

- Troubleshooting and avoidance of resonant operation
- Validation of the FE model and its updating
- Simulating the response of the system to unknown forces
- Identification of dynamic forces
- Predicting the effects of local structural modifications
- For constructing the model of an assembled structure by coupled structural analysis
- For damage detection and structural health monitoring
- System model development for structural and active control applications

1.8 FINITE ELEMENT MODEL VALIDATION AND UPDATING

FEM is a well-established technique for the modeling and simulation of dynamic systems. However, an FE model may have inaccuracies associated with modeling material properties, boundary conditions, joints, and damping and due to idealization and simplifications of the geometric details. Thus, it is necessary to validate the accuracy of the FE model predictions before the model can be used for its intended application. The modal model obtained from EMA serves as the reference to compare and validate the FE model predictions.

The first step in comparison is to pair the modes from the FE model and test. It is done by finding the correlation between the FE and test mode shapes, and the modal assurance criterion (MAC) is widely used for this purpose. Cross-orthogonality check is another measure that checks the orthogonality between the measured and analytical mode shapes. One issue in comparing FE and test mode shapes is the inconsistency between the size of these vectors. It can be resolved by either reducing the FE model to the size of the measured mode shapes or by expanding the measured mode shapes to the size of the FE model. Guyan reduction, dynamic reduction, and system equivalent reduction/expansion process (SEREP) are some of the techniques used in this regard.

At the end of the comparison, if it is determined that the correlation with the test data is satisfactory, then the FE model is considered validated and can be used confidently. Otherwise, the next step is to see if any adjustment in the FE model parameters can be made to improve its correlation with the test data. This process of adjusting the model parameters is called model updating.

In model updating, the idea is to update parameters related to the modeling accuracies in the model. The choices of the updating parameters are mainly guided by engineering judgment about the sources of inaccuracies in the model. But the challenge lies in the fact that often there may be more parameters to be updated than the available data. Model updating being an inverse process, there are also challenges with having a well-conditioned problem; else, the unknown parameters cannot be accurately estimated.

The model updating methods can be classified as direct and iterative methods. The direct methods are one-step methods that seek minimum changes in the mass and stiffness matrices to match the chosen FE model natural frequencies and modes shapes with their test counterparts. These methods don't provide a choice of what is changed during updating and hence generally don't yield physically meaningful changes in the model. The iterative methods iteratively change the chosen parameters to improve the FE model correlation with the test data. The iterative methods may be based on modal or FRF data. Regularization techniques can be used whenever the updating problem is not well-conditioned.

Conventionally, the model updating is carried out deterministically. It means that the updating is performed with the test data from a sample of the test structure. But in practice, the test data obtained on nominally identical samples corresponding to the same design show the variability of the dynamic characteristics. Therefore, the recent interest in model updating has been in developing updating methods that also identify the statistical variabilities of the updating parameters from the measured variability of the dynamic characteristics. This process is called stochastic FE model updating. Chapter 12 considers FE model validation and updating.

1.9 OVERVIEW OF THE FIELD OF MODAL ANALYSIS

The field of modal analysis encompasses both analytical and experimental techniques. Figure 1.2 shows the scope and techniques of modal analysis and its applications. The content of each block in the figure is briefly discussed in the previous sections. This book addresses most of these aspects in Chapters 2–15.

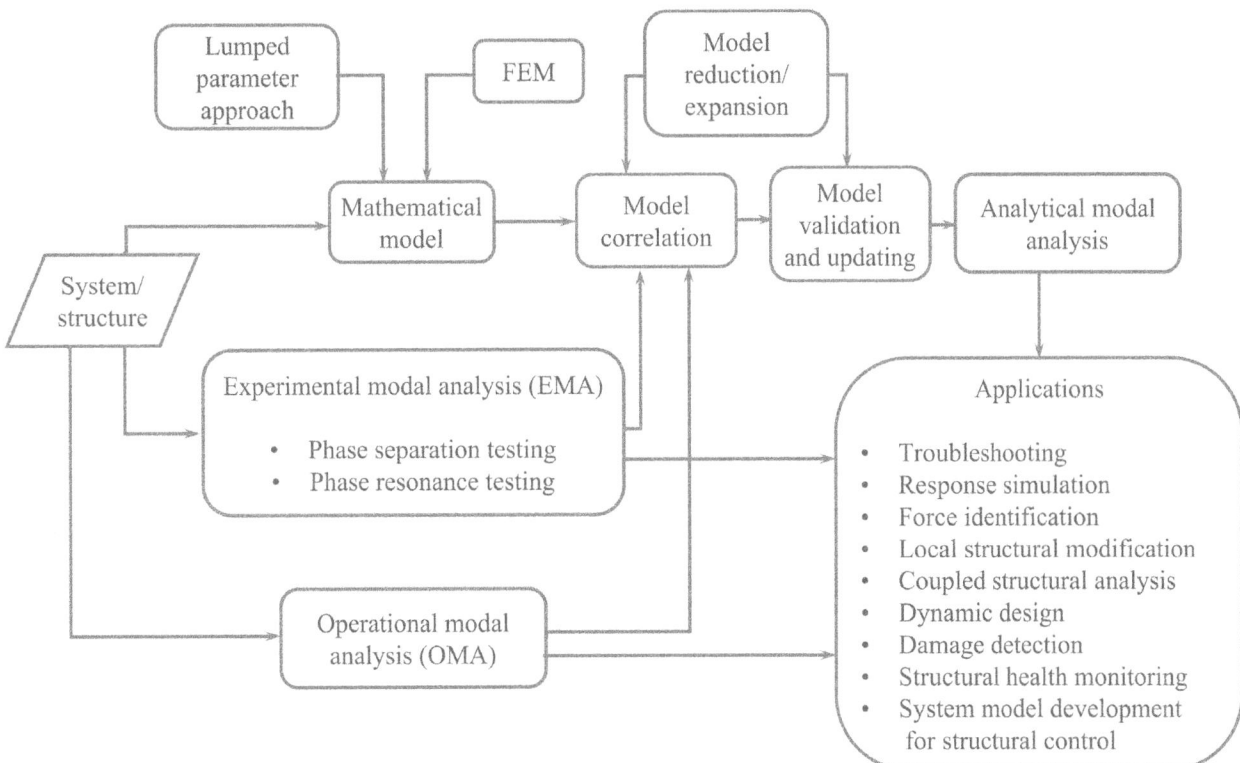

FIGURE 1.2 Scope and techniques of modal analysis.

REVIEW QUESTIONS

1. What is modal analysis?
2. What is the need for modal analysis?
3. What is the difference between the analytical and experimental approaches to modal analyses?
4. List the various applications where a model derived through experimental modal analysis can be used.

2 Lumped Parameter Modeling of Vibrating Systems

2.1 INTRODUCTION

Modal analysis is defined in Chapter 1 as determining a system's natural frequencies, mode shapes, and damping factors. It also means analyzing the system's dynamic response in terms of the responses in individual vibration modes. The process of achieving these goals using a mathematical model of the system is called analytical modal analysis. The analytical modal analysis consists of two broad steps. The first step is to build a mathematical model of the system's dynamic behavior. The second step involves using this model to perform the modal analysis. The present and the next chapter deal with the first step, and the goal is to show how a mathematical model for a given vibrating system can be built. Chapters 4–6 deal with the second step.

The present chapter covers modeling a vibrating system using a lumped parameter approach. This approach uses engineering judgment about the roles of various system elements in vibratory motion to obtain lumped parameter approximation of the system and the principles of solid mechanics and dynamics to obtain governing mathematical equations. This chapter describes dynamics principles like Newton's law, energy principle, virtual work principle, Lagrange's equation, and Hamilton's principle to derive the equations of motion. The principles of reciprocity of flexibility and stiffness influence coefficients are described. The development of lumped parameter models of systems made up of beam-like members is difficult due to the coupling of transverse and rotational degrees of freedom. For these systems, most literature follow an approach based on the flexibility influence coefficients to derive the flexibility matrix first, which is inverted to obtain the stiffness matrix. An alternative approach is described in this chapter, in which the stiffness matrix coefficients are directly obtained from the fundamental principles using the theory of bending, thus averting the need for matrix inversion. Several examples of lumped parameter models and the application of the dynamical principles to obtain the equations of motion are given.

2.2 MATHEMATICAL MODELS OF VIBRATING SYSTEMS

A mathematical model expresses the relationship between the input, output, and system parameters through mathematical equations. The system is assumed to be linear and time-invariant for modal analysis. The vibratory motion is a dynamic phenomenon since the system's output at the present instant depends not only on the inputs at the present instant but also on the past inputs. As a result, the mathematical model of a vibrating system is in the form of differential (or partial differential) equations instead of simple algebraic equations.

All systems, in practice, are continuous since the mass, stiffness, and damping characteristics are distributed continuously throughout the system. The dynamics of a continuous system is described by partial differential equations, which are solvable for systems with simple geometry, boundary conditions, and material properties. However, the practical systems are often much more complex, and the solution of partial differential equations is generally impossible. For practical systems, the lumped parameter and numerical approaches like FEM are viable and widely used for mathematical modeling. In the next section, we consider the lumped parameter approach to developing models of vibratory systems.

2.3 LUMPED PARAMETER MODELING

A lumped parameter model of a vibratory system represents the system by an assemblage of discrete idealized physical elements denoting those physical properties that affect the vibration of the system. In the next section, we look at the properties that need to be represented in such a model.

2.3.1 ELEMENTS OF A LUMPED PARAMETER MODEL

Vibration is an oscillatory motion. Since the system undergoing oscillations possesses mass, which affects the oscillations, the system's mass is one of the fundamental physical properties that need to be represented in the model used for studying vibratory motion. Since, during oscillations, the body is periodically restored to its mean position, there must be a restoring force. In structural systems, the restoring force is due to the system's elasticity and arises due to the deformation of the system as a result of the system displacement from its mean position. Therefore, the second fundamental property which needs to be accounted

DOI: 10.1201/9780429454783-2

for in the model is the stiffness of the system, defined as the elastic force per unit displacement. Thus, mass and stiffness are two fundamental properties that must be included in a model to study the oscillatory motion of a dynamic system. The mass and stiffness properties are represented in a lumped parameter model through discrete rigid masses/inertias and massless springs, respectively.

In energy terms, the discrete masses in the model allow modeling the ability of the system to store kinetic energy due to its motion, and the discrete springs allow modeling the ability of the system to store potential/strain energy due to its deformation. A lumped parameter model representing the mass and stiffness characteristics of the system is called an **undamped model**.

In practice, the system's energy is dissipated during the vibratory motion due to work done against the forces that resist the system's motion. These forces may arise due to the presence of viscous friction (referred to as viscous damping), dry friction (referred to as Coulomb damping), or due to internal friction in the material (referred to as structural or hysteresis damping) or a combination of these mechanisms. There may also be other physical mechanisms inside the material that may cause energy dissipation. The vibratory motion also dissipates energy due to the generation of acoustic waves in the surrounding medium (referred to as acoustic damping). The forces causing the dissipation of the system's energy are called damping forces, and dampers are used in the model to represent the energy-dissipating characteristics of the system. Often a viscous damper, in the form of a viscous dashpot, is used to model damping due to ease in the mathematical analysis. The viscous dashpot is characterized by the viscous damping coefficient, defined as the resistive force of the dashpot per unit velocity. Thus, the mass, stiffness, and damping properties are needed to model the vibratory motion of a dynamic system. The resulting model is called a **damped model** of the vibratory system.

2.3.2 MEANING OF LUMPED PARAMETER REPRESENTATION

In a practical system, the different parts of the system have mass, elasticity, and energy dissipation; therefore, they possess mass, stiffness, and damping properties. Thus, the mass, stiffness, and damping properties are distributed throughout the volume of a physical system. The lumped parameter model represents the distributed properties by lumped or discrete, masses, springs, and damping elements. The three properties are said to be 'lumped' because the properties of these discrete elements are mutually exclusive. A mass element has only mass and no stiffness and damping, a spring element has only stiffness and no mass and damping, while a damper has only damping and no mass and stiffness.

Example 2.1: Lumped Parameter Model, Degree of Freedom, Non-Uniqueness of the Model

In this example, we see how the lumping is done, the concept of degree of freedom, and how the intended application or purpose of the model needs to be considered while constructing a lumped parameter representation.

Let us consider lumped parameter modeling for studying the dynamic behavior of a car moving on a road, as shown in Figure 2.1.

FIGURE 2.1 Car moving on a road.

One application of this model is that it can be used to study the effect of engine and transmission line excitations and road irregularities on the passengers' ride comfort. The ride comfort is influenced by the excitation forces transmitted to the passenger seats. The transmitted forces depend on the suspension design, seat mountings, the mass/inertia of the body and various parts, tire properties, and everything else that affects the mass, stiffness, and damping characteristics. The lumped parameter model can be used to fine-tune the design to maximize ride comfort.

SINGLE-DEGREE-OF-FREEDOM (SDOF) MODEL

Let us suppose the objective is to study the bouncing motion of the car (i.e., the displacement in the vertical direction). The simplest approach could be to represent the mass of the whole car (M_c) along with passengers (M_p) by a rigid mass $M = M_c + M_p$. The restoring forces for the bouncing motion are mainly contributed by the suspension and the tire stiffness at the four wheels. Let the stiffnesses of the suspension and tire at each wheel are K_s and K_t N/m, respectively. Since the two stiffnesses are in series, a spring element with an equivalent stiffness $K = 4K_s K_t / (K_s + K_t)$ can be used to model the stiffness of the system. The values of K_s and K_t can be theoretically estimated using geometry and material properties and mechanics principles or obtained by laboratory tests measuring force-deflection characteristics. Energy dissipation may occur due to viscous damping/hysteresis/friction in the suspension, tires, and other parts. All these damping sources can be represented by an equivalent viscous damping coefficient of C N-sec/m. In the absence of any reliable model to estimate the damping in the system analytically, it can be estimated experimentally, for example, through a rap test or a modal test.

Figure 2.2 shows the model conceived in this manner, where M, C, and K show the car's lumped mass, damping, and stiffness. F(t) represents the excitation forces acting on the car, while y(t) is the displacement base excitation due to road irregularities. x(t) is the resulting vertical displacement of the car measured from its resting position. It is also assumed for the model in Figure 2.2 that the lumped mass is a point mass and cannot have sideways motion.

In this model, one coordinate, the coordinate x(t), is required to specify the motion of the mass at any given time, and hence we say that the 'degree of freedom' of this model is one. Therefore, the **degree of freedom** is the number of independent coordinates needed to specify the displaced configuration of a system at any time. With this specification, we can determine the absolute position of the mass or masses in the system.

2 DOF Model

The accuracy of the model can be improved by lumping the mass, stiffness, and damping that more closely represent the system. It is noted that the vertical movement of the wheel and car body can be different due to the flexible suspension between them. The model shown in Figure 2.3 accounts for this, which represents the tire and axle mass (M_t) and the body mass over the suspension (M_b) separately. K_s and K_t represent the equivalent stiffnesses of the suspension and tire, respectively. Now two coordinates, $x_1(t)$ and $x_2(t)$, are required to specify the displaced configuration of the two masses from their reference positions; therefore, it is a two-degree-of-freedom model. The models with more than one degree of freedom are called **multi-degrees-of-freedom (MDOF) models**. The model in Figure 2.3 is relatively a closer representation of the system than the model in Figure 2.3 and hence is expected to yield a more accurate prediction of the motion $x_1(t)$ of the car body.

4 DOF Model

Let's say we are also interested in predicting the pitching motion of the car, which is nothing but the slope of the car body, in Figure 2.1, in the plane of the paper. The pitching motion occurs due to different vertical displacements at the front and the rear wheels.

It necessitates a separate representation of the parameters of the front and rear suspensions and tires, as shown in Figure 2.4. We use the following symbols. The subscripts 'f' and 'r' denote the front and rear properties. Let M_b be the mass of the car body and I_b be its moment of inertia about an axis through its center of gravity and perpendicular to the paper.

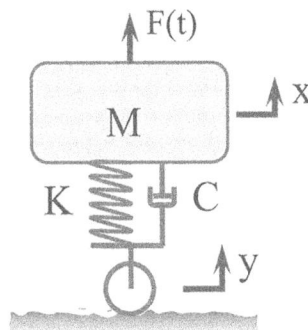

FIGURE 2.2 SDOF model.

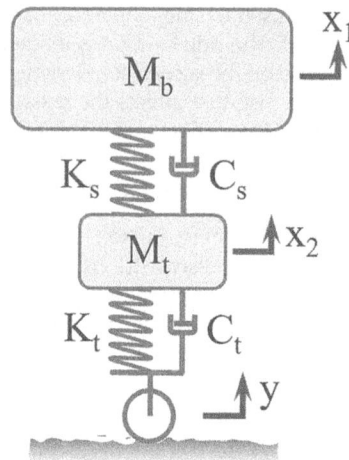

FIGURE 2.3 2-DOF model.

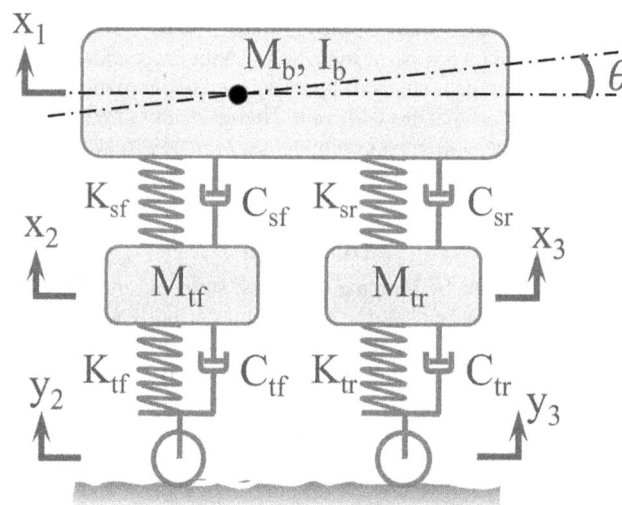

FIGURE 2.4 4-DOF model.

This model has four degrees of freedom (x_1, θ, x_2 and x_3). It allows for predicting both the bouncing (x_1) and pitching (θ) motions of the car body. It also predicts the vertical displacements of the front (x_2) and real wheels (x_3).

7 DOF MODEL

If the rolling motion is also to be predicted, then it requires incorporating the properties of all four wheels separately into the model, as shown in Figure 2.5. This model has seven degrees of freedom and can predict the bouncing (x_1), pitching (θ), and rolling (ϕ) motions of the car body.

More refinements can be done to further improve the model by representing the passenger seat properties distinctly in the model. From the above, we draw the following conclusions:

- As the distribution of the mass, stiffness, and damping is represented more accurately in the lumped parameter representation, the accuracy of the predictions increases, but the computational effort also increases due to the larger model size.
- A mathematical model of a given system is not unique. There can be multiple models depending on the purpose of the model and the accuracy of the predictions required.

2.3.3 MEANING OF THE TERMS SDOF SYSTEM AND MDOF SYSTEM

The terms SDOF system and MDOF system are often used in vibration and modal analysis. Strictly speaking, all practical systems are continuous, even those regarded as lumped systems. For example, an experimental spring-mass system with a rigid lumped mass is also continuous due to distributed mass of its spring. This system is often said to have one degree of freedom,

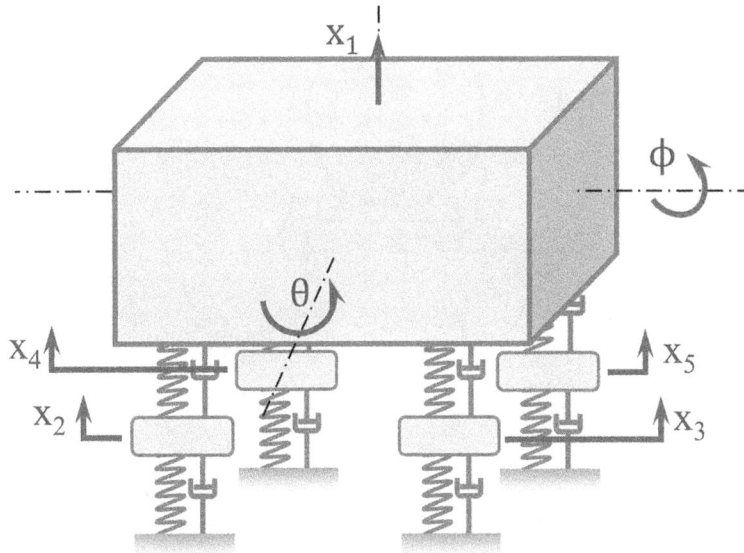

FIGURE 2.5 7-DOF model.

but this conclusion is based on our interest in studying the motion of the mass alone, with no regard to the motion of the infinite number of points on the spring. If the displacements of the points on the spring are also to be prescribed, then the system's degree of freedom would be infinite, and it can no longer be considered as an SDOF system.

Therefore, an SDOF system refers to a physical system that is continuous but whose dynamics for a given objective can be approximated by an SDOF model. Similarly, a continuous system whose dynamics for a given objective can be approximated by an MDOF model may be called an MDOF system.

From a modal perspective, the dynamic behavior of a continuous system can be approximated by an SDOF model if the frequency range of interest covers only one mode of vibration of the system and all the other modes are located far away from this range. In that case, the continuous system effectively behaves like an SDOF system for the frequency range of interest. If the frequency range of interest covers more than one mode, then the continuous system can't be approximated by an SDOF model but requires an MDOF model. In that case, the system behaves like an MDOF system for the frequency range of interest.

Example 2.2: Lumped Parameter Model of a Large Motor

In this example, we consider lumped parameter model of a large electric motor shown in Figure 2.6. The dynamic forces generated during (like due to unbalance) operation produce vibrations. If the motor is not properly isolated, a large force may be transmitted to the floor, which may shake up the floor and adversely affect the operation of the nearby machines and human safety. The motor is typically mounted on a heavy inertia block, isolated from the surroundings via isolators to reduce the force transmitted to the ground. The isolators are chosen to make the natural frequency of the combined system much less than the frequency of the dynamic forces. This choice ensures that the transmitted force is lesser than the dynamic force acting on the machine. The inertia block adds more mass to the system and helps reduce the vibration levels. It also provides a rigid foundation and lowers the CG of the system, thereby improving stability.

Let the objective is to make a lumped parameter model for analyzing the motor's vibration in the vertical direction due to mass unbalance in the motor. It is observed that the motor mass (M_m) and block mass (M_b) are much more than the isolator mass and are relatively rigid. On the other hand, the isolator deforms and dissipates energy, thus contributing to stiffness and damping.

Therefore, a lumped mass $M = M_m + M_b$ is used to represent the motor and block, while the spring and dashpot of stiffness K and damping coefficient C are used to represent the isolator. Figure 2.7 shows the lumped parameter model. $F(t) = M_m.e.\omega^2 \sin\omega t$ is the force on the motor in the vertical direction, $M_m.e$ is the unbalance in the motor, and ω is the angular velocity of rotation. The displacement of the mass in the vertical direction is the only degree of freedom.

Example 2.3: Lumped Parameter Model for Torsional Vibrations in a Diesel Generator

In this example, we consider a lumped parameter model of a diesel generator to study the torsional vibrations of its crankshaft. The system consists mainly of an internal combustion engine coupled to an electric generator and converts the heat energy of the fuel to mechanical energy and then to electrical energy. The torque generated by the engine fluctuates since

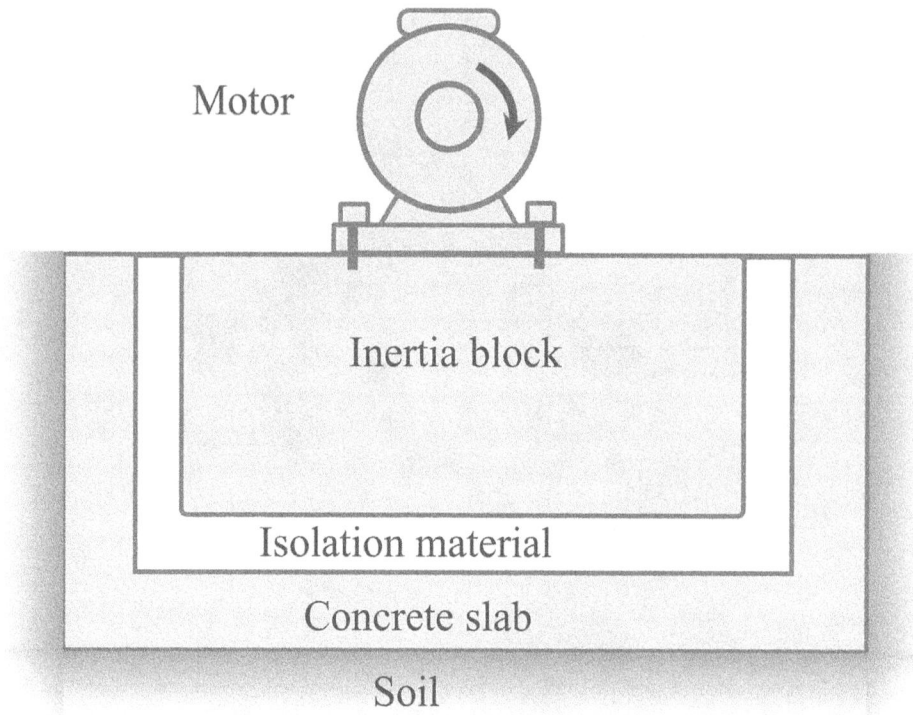

FIGURE 2.6 A large electric motor.

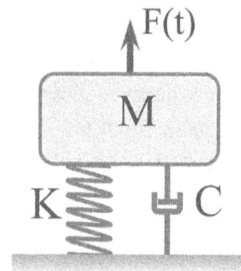

FIGURE 2.7 Lumped parameter model.

the torque is developed only during the power stroke. Consequently, the torsion produced in the crankshaft also fluctuates, leading to torsional vibrations. Figure 2.8 shows a schematic of a four-cylinder diesel generator. The flywheel is used to reduce the fluctuation of speed, while a torsional damper is used to dampen the torsional vibrations.

A lumped parameter model of the system to study its torsional vibration is shown in Figure 2.9. The physical properties relevant to the torsional vibration are the polar moment of inertia about the axis of rotation, torsional stiffness, and rotational damping coefficient. The equivalent rotational inertia can be calculated for each cylinder, considering the inertia of its rotating and reciprocating elements. The contribution of the reciprocating elements varies with rotation, but an average value can be taken as an approximation. The distributed rotational inertia of the crankshaft can be lumped at four cylinders based on kinetic energy equivalence.

Thus, the rotational inertias corresponding to the four cylinders are modeled as inertia elements with the polar moment of inertias J_1, J_2, J_3 and J_4 kg $-$ m^2. Since the four cylinders are identical, the inertias are equal. Similarly, the torsional damper, the flywheel, and the rotor of the generator are the other elements that have significant rotational inertia and are modeled as inertia elements with inertias J_5, J_6, and J_7 kg $-$ m^2 respectively.

The torsional stiffness between two inertia elements is mainly contributed by the portion of the crankshaft between the corresponding elements. Let these stiffnesses be K_{t1}, K_{t2}, K_{t3}, K_{t4}, K_{t5} and K_{t6} N-m/rad. For a shaft of length L, modulus of rigidity G, and polar second moment of area J, the torsional stiffness, i.e., the torque to produce a unit angle of twist, is GJ/L. However, the crankshaft is not a simple straight shaft but has crank arms and crankpins in between; therefore, this formula is only an approximation of the stiffness. A more accurate estimate can be obtained through a finite element model.

The energy dissipation arises for each cylinder due to the friction at the piston ring, piston pin, and crankpin. The resulting damping is modeled by a viscous dashpot resisting the rotational motion of the inertia

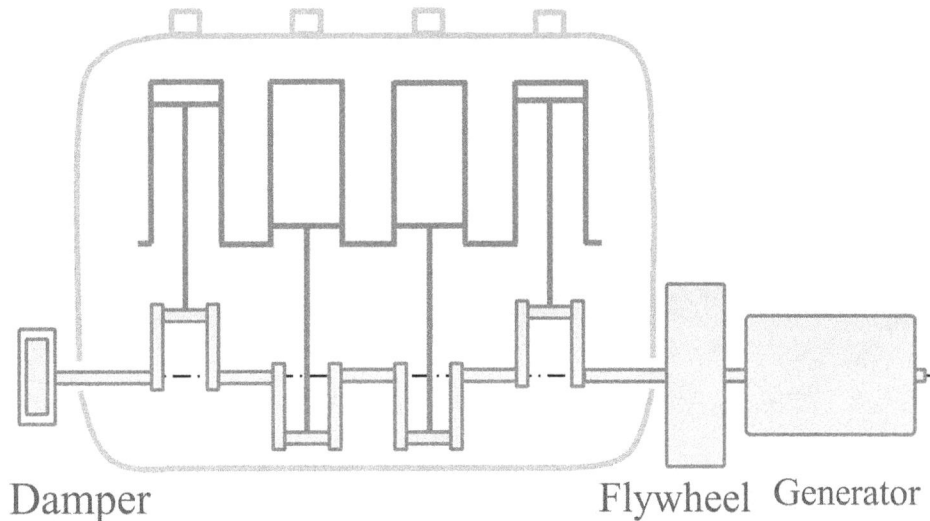

FIGURE 2.8 A schematic of a four-cylinder diesel generator.

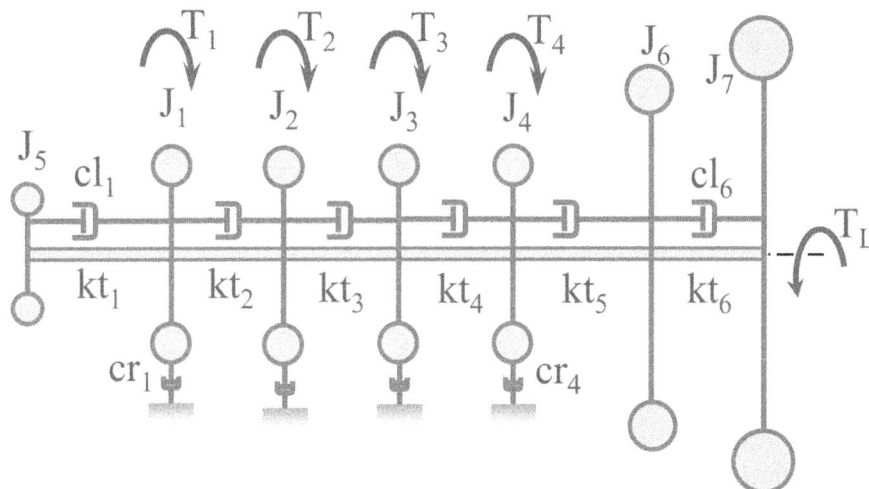

FIGURE 2.9 A lumped parameter model for the torsional vibration in the diesel generator system.

element of that cylinder. These are represented in the figure by dashpots with viscous damping coefficients C_{r1}, C_{r2}, C_{r3} and C_{r4} N – m/rad/sec. Energy dissipation also occurs in the bearings supporting the crankshaft and the generator shaft and due to internal damping in the crankshaft. These are modeled by viscous dashpots between two consecutive inertia elements with viscous damping coefficients C_{l1}, C_{l2}, C_{l3}, C_{l4}, C_{l5} and C_{l6} N – m/rad/sec. The model has seven degrees of freedom corresponding to the angular rotations of each of the seven inertia elements.

$T_1(t)$, $T_2(t)$, $T_3(t)$ and $T_4(t)$ N – m are the toques developed by the four cylinders, acting on the corresponding inertia elements, while $T_L(t)$ denotes the load torque.

Example 2.4: Lumped Parameter Model of Longitudinal Vibrations of a Propeller Shaft

The propeller shaft in a ship or submarine may undergo longitudinal vibrations. The propeller generates the thrust required for propulsion and is driven by an engine or a motor. Figure 2.10 shows a schematic of the propulsion system of a submarine. The shaft supported by bearings at several locations along its length, is coupled to other elements by couplings and is supported by a thrust bearing at the end.

The turbulence due to the asymmetry of the hull, fins, and rudders produces a non-uniform flow field around the propeller causing periodic variation in the axial thrust. The frequency of the variation of the axial thrust is equal to the blade pass frequency given by the product of the rotational frequency multiplied by the number of propeller blades. The varying axial thrust causes the longitudinal vibrations of the propeller shaft.

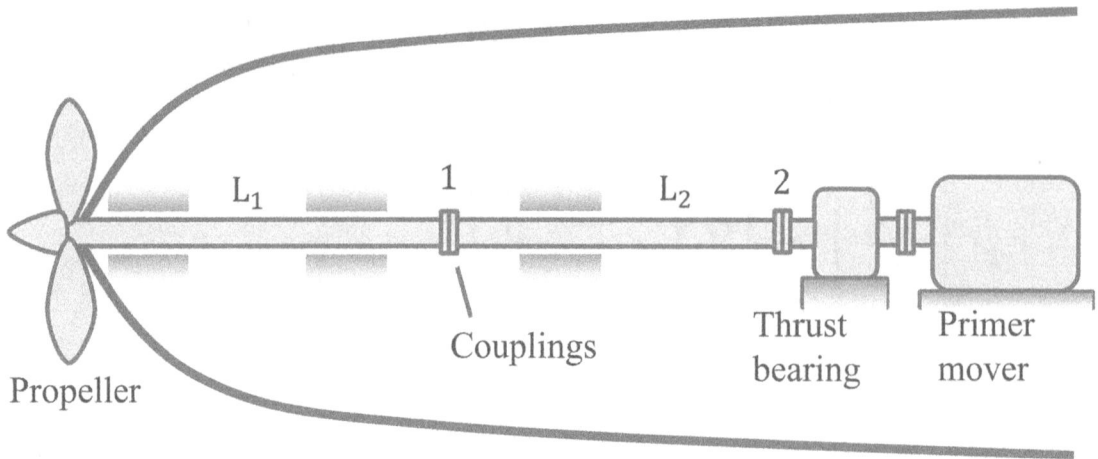

FIGURE 2.10   A schematic of the propulsion system of a submarine.

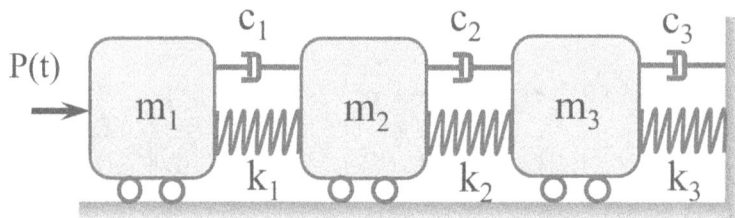

FIGURE 2.11   A lumped parameter model for longitudinal vibrations of the propeller shaft

Figure 2.11 shows a three-DOF lumped parameter model to study the longitudinal vibrations of the propeller shaft. The distributed mass can be lumped based on kinetic energy equivalence. The various lumped elements can be approximated as follows.

- Lumped mass m_1 represents the mass of the propeller plus half the mass of the shaft length L_1.
- Lumped mass m_2 represents the mass of the coupling-1 plus half the mass of the shaft lengths L_1 and L_2.
- Lumped mass m_3 represents the mass of the coupling-2 plus half the mass of the shaft length L_2 and the mass of the thrust collar.
- Lumped stiffness k_1 is the longitudinal stiffness of the shaft L_1. It is the axial force needed to produce a unit axial deformation and is equal to A_1E_1/L_1, where A_1,E_1 and L_1 are the area of cross-section, modulus of elasticity, and length of the shaft, respectively.
- Similarly, stiffness k_2 is the longitudinal stiffness of the shaft L_2.
- Lumped stiffness k_3 is the equivalent stiffness of the thrust bearing support and the shaft length between coupling 2 and the thrust bearing. Spring k_3 is assumed to be fixed at the thrust bearing end.
- The damping in the form of viscous dashpots with viscous damping coefficients c_1, c_2 and c_3 N − sec/m can also be incorporated into the model.
- $P(t)$ is the axial thrust acting on the propeller.

The study of longitudinal vibrations also enables estimating the forces transmitted to the hull and predicting the resulting hull vibration and underwater noise.

2.4   DERIVATION OF THE GOVERNING EQUATIONS

In Section 2.3 and through various examples, we considered lumped parameter modeling of dynamic systems, which gave us the physical models of the systems. The next question is how the governing equations of motion for the model relating the input, output, and model parameters can be obtained. The governing equations can be derived using any available dynamical laws or principles. All these laws are equivalent and yield the same set of equations, though the ease of applying the various laws to a given problem may differ. These laws are presented in the following sections, with examples demonstrating their application.

2.5 NEWTON'S LAW

2.5.1 SDOF System

Consider an SDOF system with mass, spring, and viscous dashpot. Figure 2.12a shows the spring and the dashpot in unde-formed states. Due to the gravitational force on the lumped mass, the spring undergoes a deflection δ, as shown in Figure 2.12b. Figure 2.12c shows the free body diagram of the mass, showing the gravitational force mg and the force $k\delta$ due to the static deflection of the spring. Since the mass is under the static equilibrium, it follows that

$$k\delta = mg \tag{2.1}$$

Most physical systems in practice undergo static deflection due to their weight, like the system in Figure 2.12b, and attain static equilibrium, which serves as the mean position about which the system vibrates when subjected to excitation forces.

Under the action of the external force $f(t)$, let the displacement of the mass at an instant be x from the static equilibrium position, as shown in Figure 2.12d. The displacement x in the upward direction creates extension in the spring and is treated as positive displacement. We use the convention that a dot over a variable represents differentiation with time. Two such dots represent double differentiation. Figure 2.12e shows the forces on the mass under the dynamic condition. Applying Newton's second law,

$$m\frac{d^2x}{dt^2} = m\ddot{x} = \sum_i F_i \tag{2.2}$$

$$m\ddot{x} = f(t) - mg - k(x - \delta) - c\dot{x} \tag{2.3}$$

In light of Eq. (2.1), Eq. (2.3) simplifies to

$$m\ddot{x} + c\dot{x} + kx = f(t) \tag{2.4}$$

Equation (2.4) is the equation governing the motion of the mass under the action of the external force. The governing equation is a linear second-order differential equation with constant coefficients. We see that the static deflection and gravity force don't appear in the final equation. It is because the dynamic equilibrium equations are written in terms of the variable x measured from the static equilibrium position, where the static spring force balances the gravity force.

2.5.2 MDOF System

The procedure followed in Section 2.5.1 can be easily extended to an MDOF system. Let us look at the basic procedure by deriving the governing equations for a two-DOF system (Figure 2.13a). It is assumed that the system is under static

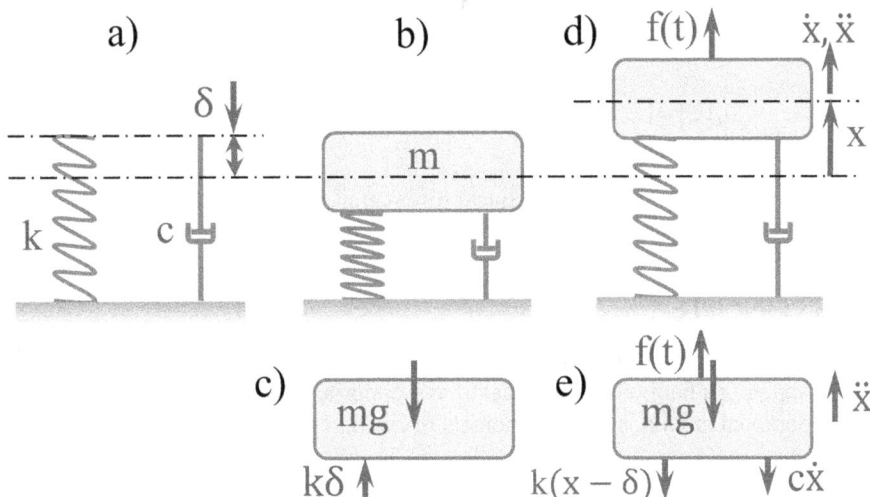

FIGURE 2.12   SDOF system (a) Undeformed state, (b) static equilibrium, (c) forces under static equilibrium, (d) dynamic condition, (e) forces under the dynamic condition.

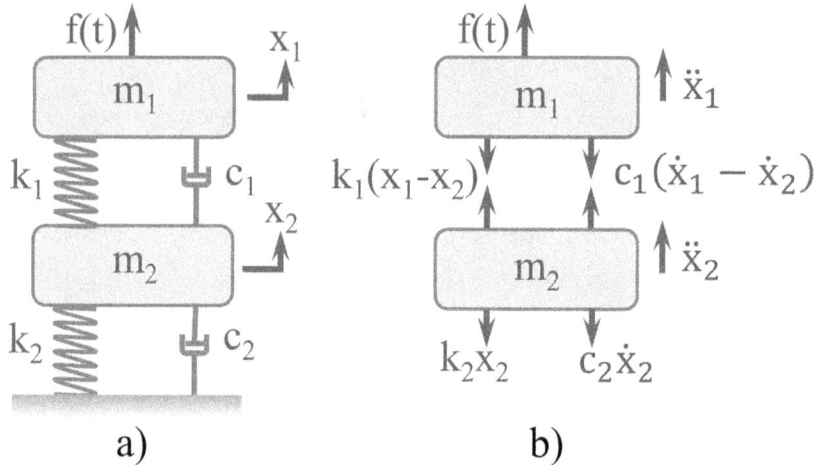

FIGURE 2.13 (a) MDOF system, (b) free body diagram.

equilibrium; therefore, the static forces in springs balance the gravity forces, as we saw for the SDOF system. Hence, these forces are not considered in the following equations.

The displacements x_1 and x_2 of the masses measured from their respective static equilibrium positions are the two DOFs. Figure 2.13b shows the forces on the two masses under the dynamic condition. Applying Newton's law individually to the two masses, we get

$$m_1\ddot{x}_1 = f(t) - k_1(x_1 - x_2) - c_1(\dot{x}_1 - \dot{x}_2) \tag{2.5}$$

$$m_2\ddot{x}_2 = k_1(x_1 - x_2) + c_1(\dot{x}_1 - \dot{x}_2) - k_2 x_2 - c_2 \dot{x}_2 \tag{2.6}$$

The two equations can be combined into a matrix equation as follows:

$$\begin{bmatrix} m_1 & 0 \\ 0 & m_2 \end{bmatrix} \begin{Bmatrix} \ddot{x}_1 \\ \ddot{x}_2 \end{Bmatrix} + \begin{bmatrix} c_1 & -c_1 \\ -c_1 & c_1 + c_2 \end{bmatrix} \begin{Bmatrix} \dot{x}_1 \\ \dot{x}_2 \end{Bmatrix} + \begin{bmatrix} k_1 & -k_1 \\ -k_1 & k_1 + k_2 \end{bmatrix} \begin{Bmatrix} x_1 \\ x_2 \end{Bmatrix} = \begin{Bmatrix} f(t) \\ 0 \end{Bmatrix} \tag{2.7}$$

Equation (2.7) can be represented as

$$[M]\{\ddot{x}\} + [C]\{\dot{x}\} + [K]\{x\} = \{F\} \tag{2.8}$$

where $[M], [C]$ and $[K]$ are mass, damping, and stiffness matrices of order 2×2, while $\{F\}$ is force/excitation vector of order 2×1. They are given by

$$[M] = \begin{bmatrix} m_1 & 0 \\ 0 & m_2 \end{bmatrix}, [C] = \begin{bmatrix} c_1 & -c_1 \\ -c_1 & c_1 + c_2 \end{bmatrix}, [K] = \begin{bmatrix} k_1 & -k_1 \\ -k_1 & k_1 + k_2 \end{bmatrix}, \text{ and } \{F\} = \begin{Bmatrix} f(t) \\ 0 \end{Bmatrix} \tag{2.9}$$

Equation (2.8) is a set of two simultaneous, ordinary, linear differential equations with constant coefficients. There would be N such equations for an MDOF system with N DOFs, and the order of the system matrices would be $N \times N$

2.6   D'ALEMBERT'S PRINCIPLE

As per D'Alembert's principle a body of mass m undergoing an acceleration $\ddot{x}$ can be considered to be under a state of dynamic equilibrium by applying a fictitious force equal to $m\ddot{x}$ opposite to the direction of $\ddot{x}$. This force is called inertia force. Similarly, for the rotational motion, a body of moment of inertia J about the axis of rotation undergoing an angular acceleration $\ddot{\theta}$ can be considered to be under a state of rotational dynamic equilibrium by applying a fictitious toque equal to $J\ddot{\theta}$ opposite to the direction of $\ddot{\theta}$. This toque is called inertia toque.

Thus, by including inertia forces/inertia toques, the system would be in dynamic equilibrium. For the SDOF system considered in Section 2.5.1, we can write the equilibrium equation using D'Alembert's principle as

$$m\ddot{x} - \sum_i F_i = 0 \qquad (2.10)$$

2.7 LAW OF CONSERVATION OF ENERGY

A physical system possesses mechanical energy in the form of kinetic and potential energies. If there are any dissipative/damping forces, the system has to work against these forces, and the system's total energy reduces with time. External forces acting on the system also alter the energy level of the system.

If the system is undamped and there are no external forces, that is, the system is vibrating freely, then the system's total energy does not change with time. This statement is the law of conservation of energy and can be used to study the free undamped vibration of a system.

Example 2.5: Law of Conservation of Energy Applied to an SDOF System

Consider an undamped SDOF system shown in Figure 2.14a. At the static equilibrium position, the spring compresses by δ. At the time 't', the mass has displacement x measured from the static equilibrium position. The total energy E of the mass equals the sum of kinetic (T) and potential energy (U). That is,

$$E = T + U$$

U includes the strain energy of the spring and the work done against the gravity force, also a conservative force. Figure 2.14 shows that the spring force changes from $-k\delta$ to $-k\delta + kx$ when the mass displaces from 0 to x. The strain energy of the spring is equal to the product of the average spring force and the displacement x, and the potential energy of mass due to gravity is mgx. Therefore, we get

$$E = \frac{1}{2}m\dot{x}^2 + mgx + \frac{-k\delta - k\delta + kx}{2}x$$

Since, at the static equilibrium position, we have $k\delta = mg$, we get

$$E = \frac{1}{2}m\dot{x}^2 + \frac{1}{2}kx^2$$

As the total energy is constant for an undamped and unforced system, the time rate of change of E must be zero. In the equation form,

$$\frac{dE}{dt} = 0$$

Substituting for E and simplifying, we obtain

$$m\ddot{x} + kx = 0$$

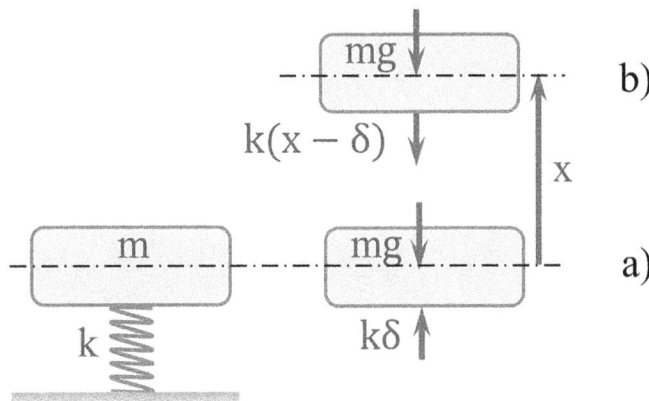

FIGURE 2.14 Undamped SDOF system.

which is the equation of motion of the system.

An equivalent statement of the energy conservation principle can be stated as

$$T_{max} = U_{max}$$

For the system in Figure 2.14a, when the mass is at an extreme position, the kinetic energy is zero while the potential energy is maximum. On the other hand, when the mass is at its mean position, the potential energy is zero, while the kinetic energy is maximum. Since the total energy is constant, the maximum kinetic energy (T_{max}) must be equal to the maximum potential energy (U_{max}).

For MDOF systems also, the energy conservation principle is true for undamped and unforced cases. However, it yields only one equation and can't be used to obtain the complete equations of motion.

2.8   PRINCIPLE OF VIRTUAL WORK

Newton's law and D'Alembert's principle deal directly with the forces. Since the forces are vector quantities, the vector sum needs to be considered. The process becomes more involved for the MDOF systems. The principle of virtual work is a scalar approach based on the work done by the forces when virtual displacements are given to the system.

We first introduce the generalized coordinates, present the virtual work principle for statics, and then extend the principle to dynamics.

2.8.1   GENERALIZED COORDINATES

The concept of generalized coordinates is essential for the virtual work principle and for Lagrange's equations and Hamilton's principle presented in subsequent sections. The generalized coordinates are concerned with choosing coordinates to describe a given system. The coordinates selected in describing a system may not necessarily be independent of each other. For example, for the simple oscillating pendulum shown in Figure 2.15, the position of the point mass of the pendulum during its oscillation might be of interest to us. Therefore, we may choose the coordinates x and y of the point mass to describe its position. However, since the length l of the pendulum is constant, they are related by the following relationship or constraint:

$$x^2 + y^2 = l^2 \tag{2.11}$$

Because of the relationship (2.11), x and y are dependent coordinates. Alternatively, we may select the coordinate θ shown in the figure. The coordinate θ completely specifies the position of the point mass of the pendulum by the following relationships:

$$x = l\sin\theta \quad \text{and} \quad y = l\cos\theta \tag{2.12}$$

Therefore, the coordinate θ not only describes the displaced configuration of the system but is also independent, and such a coordinate is called a **generalized coordinate**. In this example, the coordinate x or y also qualifies as a generalized coordinate. In general, the number of generalized coordinates (n) is equal to the number of chosen coordinates (m) minus the number of constraint equations (r), if any. The constraints are called holonomic if we can eliminate the 'r' number of

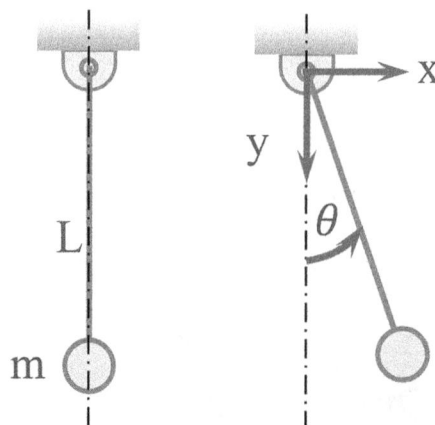

FIGURE 2.15   Simple oscillating pendulum.

coordinates from the 'm' coordinates using the constraint equations. In the pendulum example, the constraint is holonomic since one of the coordinates, say y, can be eliminated using the expression $y = \sqrt{1^2 - x^2}$, leaving only the coordinate x as the independent coordinate.

Therefore, the generalized coordinates form a set of independent coordinates to describe the configuration of a given system. As mentioned in Example 2.1, the degree of freedom of a system is also the number of independent coordinates to describe the configuration of a system, and hence the number of generalized coordinates of a system is equal to the degree of freedom of the system. Generally, the degrees of freedom are the physical coordinates like displacements and angular positions, which often are of direct interest to us. In contrast, the generalized coordinates need not necessarily be the physical coordinates.

Since the generalized coordinates are independent, any coordinate can be varied independently while the other coordinates are held constant. The dependent coordinates, however, don't have this property. We would represent the 'n' generalized coordinates by the notations $q_1, q_2,.. q_n$.

2.8.2 Principle of Virtual Work for Statics

The principle of virtual work states that if a system in equilibrium under the action of a set of forces is given a virtual displacement consistent with the geometric constraints on the system, then the total virtual work done by the forces is zero.

The displacement given to the system from the equilibrium position is only hypothetical and is called 'virtual displacement.' But it should be such that it does not violate the system's constraints (i.e., the geometric boundary conditions). The virtual displacement is infinitesimal and arbitrary, and it is assumed that the forces don't change when the system moves through the virtual displacement. The virtual displacement is also considered to occur instantaneously, i.e., with no passage of time.

We first look at this principle in the case of statics and then extend it to dynamics. We take a simple example to understand the principle of virtual work. Consider the static equilibrium of a particle of negligible mass with forces $\overrightarrow{P_1}$ and $\overrightarrow{P_2}$ acting on it. The particle is also constrained to remain on a frictionless surface, as shown in Figure 2.16a. Figure 2.16b shows the free-body diagram of the particle with $\vec{N}$ being the normal reaction. The vector sum of all the forces must be zero for the static equilibrium. Hence,

$$\overrightarrow{P_1} + \overrightarrow{P_2} + \vec{N} = 0 \tag{2.13}$$

Let the position vector of the particle is $\vec{r}$ and the symbol δ is used to represent an infinitesimal and arbitrary change. Let a virtual displacement $\delta\vec{r}$ is given to the particle satisfying the constraint on the particle. Because of this, $\delta\vec{r}$ needs to be parallel to the constraint plane as shown in Figure 2.16c. If we multiply Eq. (2.13) by $\delta\vec{r}$ and note that $\vec{N}.\delta\vec{r} = 0$, we get

$$\overrightarrow{P_1}.\delta\vec{r} + \overrightarrow{P_2}.\delta\vec{r} = 0 \tag{2.14}$$

The quantities appearing on the LHS of Eq. (2.14) are the virtual works by the corresponding forces because they arise due to the virtual displacement $\delta\vec{r}$. Since we started from the condition of static equilibrium (Eq. 2.13) to reach Eq. (2.14), Eq. (2.14) must be an equivalent statement of the static equilibrium. Equation (2.14) essentially says that the total virtual work of the forces due to the virtual displacement of the body under the static equilibrium must be zero. It is seen from Eq. (2.14) that the

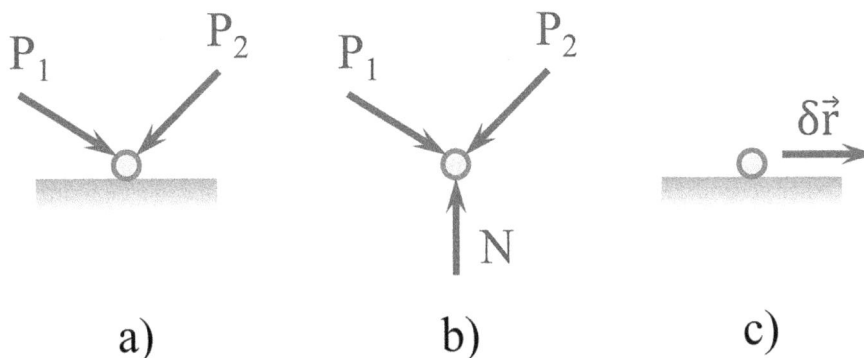

FIGURE 2.16 (a) A particle constrained to remain on a frictionless surface, (b) static equilibrium, and (c) virtual displacement.

forces of constraints don't do virtual work when the virtual displacements are consistent with the constraints, and therefore, they don't appear in the virtual work equation. Equation (2.14) can be written in a general form as

$$\delta W = \sum_i \vec{P_i}.\delta\vec{r_i} = 0 \qquad (2.15)$$

where the force $\vec{P_i}$ undergoes a virtual displacement $\delta\vec{r_i}$. Equation (2.15) can be written with the forces and displacements represented as vector quantities. Expressing them in terms of their components along the three axes, we obtain

$$\delta W = \sum_i \left(P_{ix}\hat{i} + P_{iy}\hat{j} + P_{iz}\hat{k} \right).\left(\delta r_{ix}\hat{i} + \delta r_{iy}\hat{j} + \delta r_{iz}\hat{k} \right) = 0 \qquad (2.16)$$

$$\delta W = \sum_i \left(P_{ix}.\delta r_{ix} + P_{iy}.\delta r_{iy} + P_{iz}.\delta r_{iz} \right) = 0 \qquad (2.17)$$

Let us denote all the forces in Eq. (2.17) by symbols $\{F_1, F_2,.. F_m\}$ and the corresponding virtual displacements by $\{\delta u_1, \delta u_2,.. \delta u_m\}$. With this notation, Eq. (2.17) can be rewritten as

$$\delta W = \sum_{i=1}^m F_i.\delta u_i = 0 \qquad (2.18)$$

Eq. (2.18) is the statement of the principle of virtual work for a system represented by m coordinates.

Let us consider a system of rigid bodies interconnected together where the bodies can have relative motion between them and see how the principle expressed by Eq. (2.18) can be applied. Let the system be under static equilibrium with the external forces acting. We now give a virtual displacement to the system and determine the virtual work due to each force. The forces that do non-zero virtual work are called active forces, and only such forces need to be considered while applying Eq. (2.18).

The active forces are the external forces, forces due to the deformation of a flexible element, and forces due to gravity not balanced by static deflection. However, the reaction forces at fixed support do no virtual work as the displacement at the point of their action is zero. Similarly, the forces between the connecting bodies, assuming no friction, do no net virtual work. It is because the forces on the two bodies would be equal and opposite, making the net virtual work zero. The virtual work is also zero due to internal forces in a rigid body. It is because the internal force is a force of attraction between any two particles of a rigid body. These forces occur in pairs of equal, opposite, and collinear forces undergoing equal virtual displacement, making the net virtual work zero.

The virtual work due to the friction forces needs to be accounted for if relative sliding between the contacting surfaces occurs. But if relative sliding is not there, then the virtual work of the friction force is zero. For example, a wheel having a pure rolling motion on a surface has no slip with the surface resulting in zero virtual work due to friction force.

Thus, we see that in applying the principle of virtual work to the interconnected rigid bodies, the reaction forces at fixed supports and the interconnection forces at connections (neglecting friction) between the bodies do not play any role. It simplifies analyzing the equilibrium position compared to a vector-based approach.

Given the above discussion, the forces F_i^s in Eq. (2.18) represent the active forces and δu_i represent the corresponding virtual displacements in their directions. What are δu_i? Let the configuration of the system is described by n number of generalized coordinates $\{q_1, q_2,.. q_n\}$. Since the generalized coordinates are independent, we first consider a virtual displacement δq_1 along q_1 with no virtual displacement in other generalized coordinates. For this configuration of the system, δu_i are expressed in terms of δq_1 and then the virtual work of all the active forces, represented by δW_1, are found. Repeating this for the other generalized coordinates gives the total virtual work and can be written as

$$\delta W = \delta W_1 + \delta W_2 +.. + \delta W_n \qquad (2.19)$$

Since the virtual displacements $\{\delta q_1, \delta q_2,.. \delta q_n\}$ are independent and arbitrary, the coefficient of each virtual displacement on the RHS of Eq. (2.19) must be zero for the total virtual work to be zero. This yields as many equations as the number of generalized coordinates. These are the equations that govern the equilibrium of the system.

Example 2.6: Principle of Virtual Work for Statics Applied to an Interconnected System

Consider an interconnected system of two rigid links, OA and AB, each of length L (Figure 2.17a). End O is hinged, while end B is connected to a small roller that can roll between the guides which is connected to a spring of stiffness K. Neglecting the mass of the links and friction, find the force P necessary to maintain the static equilibrium of links at an angle θ, if the spring is unstretched at $\theta = 0°$.

Solution:

Figure 2.17b shows the system's configuration at the equilibrium position with the active forces, the applied force P, and the spring force Ku due to the elongation u of the spring. As explained previously, the reaction forces at the supports O and B and the interconnection forces at joint A need not be considered. u (the extension of the spring) is given by

$$u = 2L - 2L\cos\theta = 2L(1 - \cos\theta)$$

Let θ be the generalized coordinate. Give a virtual angular displacement $\delta\theta$ to the coordinate θ as shown in Figure 2.17b, where the displaced configuration of the system is shown by a dotted line. Now, we need to determine the virtual displacements of the active forces, which are δy_A and δx_B. We can write the coordinates of points A and B, from which the virtual displacements can be found using the rules of normal differentiation.

$$y_A = L\sin\theta \quad ; \quad \delta y_A = L\cos\theta \, \delta\theta$$

Similarly,

$$x_B = 2L\cos\theta \quad ; \quad \delta x_B = -2L\sin\theta \, \delta\theta.$$

δy_A and δx_B represent the virtual displacements in the positive direction of the y and x axes, respectively, and the forces P and Ku also act in those directions, respectively. Hence, the total virtual work of the active forces is

$$\delta W = P.\delta y_A + Ku.\delta x_B$$

Applying the virtual work principle $\delta W = 0$, we get

$$P.\delta y_A + Ku.\delta x_B = 0$$

Substituting the relevant quantities in the above equation, we obtain

$$\left(P.L\cos\theta - K.2L(1 - \cos\theta).2L\sin\theta\right)\delta\theta = 0$$

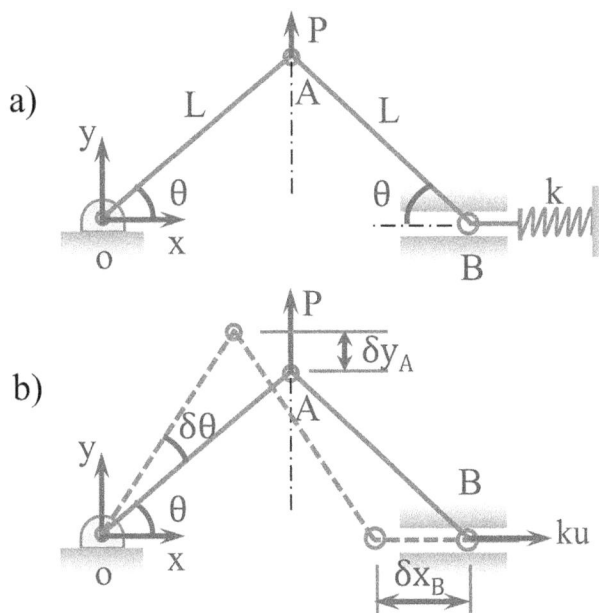

FIGURE 2.17 (a) Interconnected system of two massless rigid links with a spring and (b) active forces and virtual displacements.

Since $\delta\theta$ is non-zero and arbitrary, the expression in the bracket must be zero, which yields the force P necessary to hold the links at the angle θ :

$$P = \frac{4KL(1-\cos\theta)\sin\theta}{\cos\theta}$$

2.8.3 Principle of Virtual Work for Dynamics

In this section, we extend the principle of virtual work to dynamics problems. According to D'Alembert's principle, the inertia forces are included in a dynamic system to consider the system in a state of dynamic equilibrium. The inertia forces also do virtual work due to virtual displacement, and therefore Eq. (2.18) can be modified as follows for the case of dynamics:

$$\delta W = \sum_{i=1}^{m}(F_i - m_i\ddot{u}_i).\delta u_i = 0 \tag{2.20}$$

Damping forces, if any, also do virtual work and, therefore, should also be treated as active forces as part of the forces F_i.

Example 2.7: Principle of Virtual Work for Dynamics Applied to a Pendulum System

A simple pendulum connected to a spring is shown in Figure 2.18a. A force P(t) acts on the mass. Find the mathematical equation governing the pendulum's motion using the principle of virtual work. Friction is negligible.

Solution:

Let θ, which is the pendulum's angle from the y axis, be the generalized coordinate. Consider a displaced configuration of the system at an angle θ as shown in Figure 2.18b and let $\ddot{\theta}$ be the angular acceleration. The external force P(t), the gravity force mg, the force due to spring extension kx_A, and the inertia torque due to the angular acceleration $\ddot{\theta}$ are the active forces/torque, as shown. The system is displaced through a virtual angular displacement $\delta\theta$, indicated by a dotted line. We first work out the virtual displacements for the active forces and torque by writing the coordinates of points A and B at the configuration θ and differentiating them.

$$x_A = a\sin\theta \; ; \qquad \delta x_A = a\cos\theta\,\delta\theta$$

$$y_B = L\cos\theta \; ; \qquad \delta y_B = -L\sin\theta\,\delta\theta$$

$$x_B = L\sin\theta \; ; \qquad \delta x_B = L\cos\theta\,\delta\theta$$

The virtual displacements found above are in the positive directions of the corresponding axes. If an active force is in the direction of the corresponding virtual displacement, then the virtual work is positive else, negative. The total virtual work of the active forces/torque is

$$\delta W = P.\delta x_B - Kx_A.\delta x_A + mg.\delta y_B - I\ddot{\theta}.\delta\theta$$

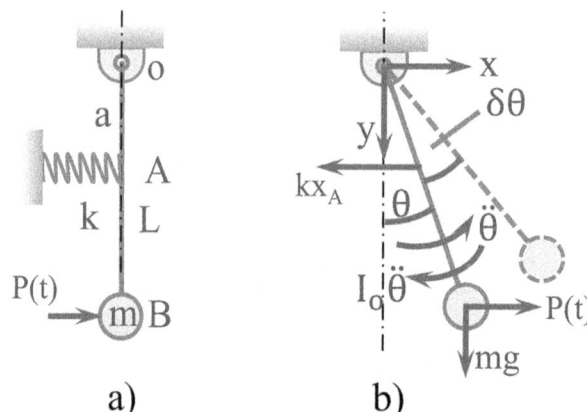

FIGURE 2.18 (a) Simple pendulum and (b) active forces and virtual displacements.

Applying the virtual work principle, i.e. $\delta W = 0$, we get,

$$P.\delta x_B - Kx_A.\theta x_A + mg.\delta y_B - I\ddot{\theta}.\delta\theta = 0$$

Substituting the expressions for the virtual displacements in the above equation and noting that $I = mL^2$, we get

$$\left(P.L\cos\theta - K.a\sin\theta.a\cos\theta - mg.L\sin\theta - mL^2\ddot{\theta}\right).\delta\theta = 0$$

Since $\delta\theta$ is non-zero and arbitrary, the expression in the bracket must be zero for the the virtual work to be zero. It leads to the following equation, which is the governing equation of motion of the system:

$$mL^2\ddot{\theta} + K.a^2 \sin\theta\cos\theta + mg.L\sin\theta = P.L\cos\theta$$

Due to the terms like $\sin\theta$ and $\cos\theta$, the differential equation is nonlinear. If we assume that the angle θ is small during the oscillations, then we can approximate $\sin\theta \approx \theta$ and $\cos\theta \approx 1$, leading to the linearized equation

$$mL^2\ddot{\theta} + K.a^2 \theta + mg.L \theta = P.L$$

2.8.4   PRINCIPLE OF VIRTUAL WORK IN TERMS OF GENERALIZED FORCES

Sections 2.8.2 and 2.8.3 present the principle of virtual work, and Examples 2.6 and 2.7 demonstrate the steps to apply the principle. The idea is to choose the generalized coordinates $\{q_1, q_2,.. q_n\}$ and then find δu_i in terms of the virtual displacement of the generalized coordinates (i.e. δq_k). The equilibrium equations are obtained using the fact that the generalized displacements are arbitrary. We would now see that when this process is expressed mathematically, we get an equivalent statement of the virtual work principle in terms of new quantities called **generalized forces**.

The virtual displacements can be expressed in terms of the virtual changes in the generalized coordinates as

$$\delta u_i = \sum_{k=1}^{n} \frac{\partial u_i}{\partial q_k} \delta q_k \tag{2.21}$$

Substituting Eq. (2.21) into Eq. (2.18) (which can also represent Eq. (2.20) if the inertia forces are included as active forces), we get

$$\delta W = \sum_{i} F_i . \sum_{k=1}^{n} \frac{\partial u_i}{\partial q_k} \delta q_k = 0 \tag{2.22}$$

$$\delta W = \sum_{k=1}^{n} \left(\sum_{i} F_i . \frac{\partial u_i}{\partial q_k} \right) \delta q_k = 0 \tag{2.23}$$

$$\delta W = \sum_{k=1}^{n} Q_k \delta q_k = 0 \tag{2.24}$$

Where,

$$Q_k = \sum_{i} F_i . \frac{\partial u_i}{\partial q_k} \tag{2.25}$$

Q_k are called generalized forces. Since, δq_k are arbitrary, from Eq. (2.24), we must have

$$Q_k = 0 \quad (k = 1, 2, ..., n) \tag{2.26}.$$

Therefore, Eq. (2.26) is equivalent to the statement of the virtual work principle and says that under the equilibrium, whether static or dynamic, the generalized force along each generalized coordinate must be zero.

2.9 LAGRANGE'S EQUATIONS

In Section 2.8, we studied the principle of virtual work and saw how to use it to derive the equation of motion. The principle does depend on the scalar variable, the virtual work of forces, but the computation of virtual work still requires identifying and dealing with the forces and displacements, which are vector quantities. The conservation of energy principle was a fully energy-based approach, but the equation like Eq. (2.8) leads to only one equation which doesn't provide a complete mathematical description of an MDOF system. Lagrange's equation, to be studied in this section, is a fully energy-based approach and provides an equivalent and an alternative approach to developing the governing equations of motion. Lagrange's equations can be derived from the conservation of energy or the principle of virtual work and D'Alembert's principle. The derivation via the former route is presented here.

A system in a state of motion has kinetic energy due to its motion and potential energy due to work done against the conservative forces. Lagrange's equation is based on expressing energies in terms of the generalized coordinates. Therefore, we first see how the kinetic and potential energies are related to the generalized coordinates. We first look at an example to know these relationships and then formally present the relationships.

Example 2.8: Kinetic and Potential Energies in Terms of Generalized Coordinates

Figure 2.19a shows a system of a cart with a pendulum. It is a simple model of an overhead crane moving an object held by a rope. One choice of the coordinates to describe the displaced configuration of the system includes the x-coordinate of the cart (x_1) and the x and y coordinates of the point mass (x_2, y_2) of the pendulum, making a total of three coordinates. These coordinates are dependent.

The kinetic and potential energies of the system are given by

$$T = \frac{1}{2}m_1\dot{x}_1^2 + \frac{1}{2}m_2\dot{x}_2^2 + \frac{1}{2}m_2\dot{y}_2^2$$

$$U = \frac{1}{2}kx_1^2 + m_2g(L - y_2)$$

If we select x_1 and θ as the coordinates, then they are independent and completely describe the displaced configuration of the system and hence qualify as the generalized coordinates. Therefore, the degree of freedom of the system is two.

The coordinates x_2 and y_2 can be expressed in terms of the generalized coordinates as follows:

$$x_2 = x_1 + L\sin\theta \quad \text{and} \quad y_2 = L\cos\theta$$

Differentiating with time, we obtain

$$\dot{x}_2 = \dot{x}_1 + L\cos\theta\,\dot{\theta} \quad \text{and} \quad \dot{y}_2 = -L\sin\theta\,\dot{\theta}$$

After substituting the above expressions and simplifying, we get the kinetic and potential energies in terms of the generalized coordinates,

$$T = \frac{1}{2}(m_1 + m_2)\dot{x}_1^2 + \frac{1}{2}m_2L^2\dot{\theta}^2 + m_2L\cos\theta\,\dot{x}_1\dot{\theta}$$

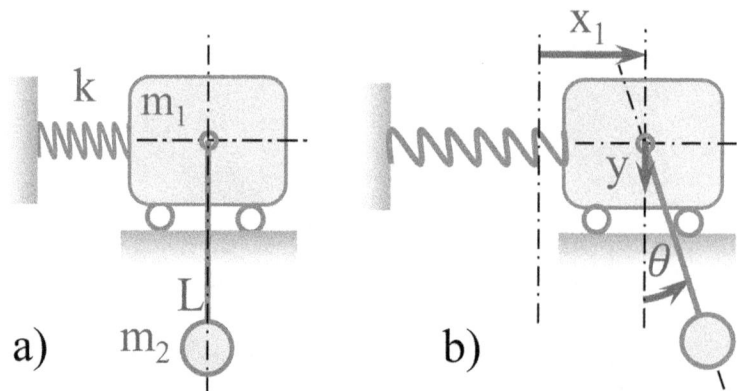

FIGURE 2.19 A cart with a pendulum (a) mean position and (b) displaced position.

$$U = \frac{1}{2}kx_1^2 + m_2g(L - L\cos\theta)$$

It is observed from the above expressions that the kinetic energy is not only a function of the generalized velocities (like $\dot{x}_1$ and $\dot{\theta}$ in this example) but it can also be a function of the generalized coordinates (like θ in this example). The potential energy is a function of only the generalized coordinates. Based on these observations, we can represent, in general, the functional relationship of the kinetic and the potential energies with the generalized variables as

$$T = T(q_1, q_2,.. q_n, \dot{q}_1, \dot{q}_2,.. \dot{q}_n)$$

$$U = U(q_1, q_2,.. q_n)$$

We can rewrite the kinetic energy expression in the following form:

$$T = \frac{1}{2}\begin{Bmatrix} \dot{x}_1 \\ \dot{\theta} \end{Bmatrix}^T \begin{bmatrix} m_1 + m_2 & m_2L\cos\theta \\ m_2L\cos\theta & m_2L^2 \end{bmatrix} \begin{Bmatrix} \dot{x}_1 \\ \dot{\theta} \end{Bmatrix}$$

$$T = \frac{1}{2}\{\dot{q}\}^T [M^q]\{\dot{q}\}$$

Thus, the kinetic energy can be expressed as a quadratic function of the vector of generalized velocities and a matrix $[M^q]$ called the generalized mass matrix.

We can approximate the potential energy expression also by a quadratic function. For motions about the equilibrium position ($\theta = 0$), we can approximate $\cos\theta$ using the power series as

$$\cos\theta \cong 1 - \frac{\theta^2}{2}$$

and with this, we get

$$U \cong \frac{1}{2}kx_1^2 + m_2gL\frac{\theta^2}{2}$$

which in the matrix form is

$$U \cong \frac{1}{2}\begin{Bmatrix} x_1 \\ \theta \end{Bmatrix}^T \begin{bmatrix} k & 0 \\ 0 & m_2gL \end{bmatrix} \begin{Bmatrix} x_1 \\ \theta \end{Bmatrix}$$

$$U \cong \frac{1}{2}\{q\}^T [K^q]\{q\}$$

Thus, we see that the potential energy can be expressed (or approximated) as a quadratic function of the vector of generalized coordinates and a matrix $[K^q]$ called the generalized stiffness matrix.

2.9.1 Kinetic Energy in Terms of Generalized Coordinates

Let m physical coordinates $\{u\} = \{u_1, u_2,.. u_m\}$ are selected to describe the displaced configuration of a system. In general, these coordinates could be dependent.

The kinetic energy of the system in these coordinates is given by

$$T = \frac{1}{2}\sum_{i=1}^{m} M_i .\dot{u}_i^2 \tag{2.27}$$

In Eq. (2.27), M_i and $\dot{u}_i$ may also indicate the moment of inertia and angular velocity, respectively, for a rotational coordinate.

Let the system is described by n number of generalized coordinates $\{q\} = \{q_1, q_2,.. q_n\}$. To express T in terms of the generalized coordinates, we must relate $\{\dot{u}\}$ with $\{\dot{q}\}$. We can write

$$du_i = \sum_{k=1}^{n} \frac{\partial u_i}{\partial q_k} dq_k \quad \text{for} \quad i = 1, 2, ..., m \tag{2.28}$$

Dividing by dt, Eq. (2.28) can be written in terms of the velocities as

$$\dot{u}_i = \sum_{k=1}^{n} \frac{\partial u_i}{\partial q_k} \dot{q}_k \tag{2.29}$$

Substituting Eq. (2.29) into Eq. (2.27), we get

$$T = \frac{1}{2} \sum_{i=1}^{m} M_i \cdot \left(\sum_{k=1}^{n} \frac{\partial u_i}{\partial q_k} \dot{q}_k \right) \cdot \left(\sum_{k=1}^{n} \frac{\partial u_i}{\partial q_k} \dot{q}_k \right) \tag{2.30}$$

$$T = \frac{1}{2} \sum_{i=1}^{m} M_i \cdot \sum_{j=1}^{n} \sum_{k=1}^{n} \frac{\partial u_i}{\partial q_k} \dot{q}_k \cdot \frac{\partial u_i}{\partial q_j} \dot{q}_j \tag{2.31}$$

Moving inside the summation over the index 'i', we get

$$T = \frac{1}{2} \sum_{j=1}^{n} \sum_{k=1}^{n} \left(\sum_{i=1}^{m} M_i \frac{\partial u_i}{\partial q_k} \cdot \frac{\partial u_i}{\partial q_j} \right) \dot{q}_k \dot{q}_j \tag{2.32}$$

Eq. (2.32) can be written as

$$T = \frac{1}{2} \sum_{j=1}^{n} \sum_{k=1}^{n} M_{kj}^q \cdot \dot{q}_k \dot{q}_j \tag{2.33}$$

where,

$$M_{kj}^q = \sum_{i=1}^{m} M_i \frac{\partial u_i}{\partial q_k} \cdot \frac{\partial u_i}{\partial q_j} \quad \text{for} \quad k = 1, 2, ..., n \quad \text{and} \quad j = 1, 2, ..., n \tag{2.34}$$

Eq. (2.33) can be written in matrix form as

$$T = \frac{1}{2} \{\dot{q}\}^T \left[M^q \right] \{\dot{q}\} \tag{2.35}$$

where $\left[M^q \right]$ is the generalized mass matrix whose elements M_{kj}^q are given by Eq. (2.34). Thus, we get the expression for kinetic energy in terms of the generalized velocities. It may also be a function of generalized coordinates, if $\left[M^q \right]$ is dependent on the generalized coordinates.

We can also work out the relationship between the generalized mass matrix and the mass matrix in the coordinates $\{u\}$. We can write Eq. (2.27) in the matrix form as

$$T = \frac{1}{2} \{\dot{u}\}^T [M] \{\dot{u}\} \tag{2.36}$$

Similarly, Eq. (2.29), which represents m equations in n generalized coordinates, can be expressed as

$$\{\dot{u}\} = [G]\{\dot{q}\} \tag{2.37}$$

where [G] is the transformation matrix between the generalized coordinates and the coordinates $\{u\}$. Substituting Eq. (2.37) into Eq. (2.36) gives

$$T = \frac{1}{2} \{\dot{q}\}^T [G]^T [M][G]\{\dot{q}\} \tag{2.38}$$

Comparing Eqs. (2.38) and (2.35) we get

$$\left[M^q \right] = [G]^T [M][G] \tag{2.39}$$

2.9.2 POTENTIAL ENERGY IN TERMS OF GENERALIZED COORDINATES

The stiffness is the elastic force developing due to the unit displacement of a system. The system stores potential energy in the process. Therefore, the stiffness and potential energy are related, and the changes in the potential energy due to the displacement of the system from its static equilibrium position can be related to the stiffness.

We know that

$$U = U(q_1, q_2, .. q_n) \tag{2.40}$$

Let the static equilibrium position is given by $\{q_1, q_2, .. q_n\}^T = \{0, 0, .. 0\}^T$. Using Taylor's series to expand U about this equilibrium position, we can write

$$U(q) = U(0) + \sum_{k=1}^{n} \left(\frac{\partial U}{\partial q_k}\right)_0 (q_k - 0) + \frac{1}{2} \sum_{j=1}^{n} \sum_{k=1}^{n} \left(\frac{\partial^2 U}{\partial q_j \partial q_k}\right)_0 (q_j - 0)(q_k - 0) + .. \tag{2.41}$$

In Eq. (2.41), $U(0)$ is the value of the potential energy at the static equilibrium position. $U(0)$ depends on the reference with which the potential energy is calculated. Since the choice of the reference is arbitrary, the value of $U(0)$ is also arbitrary. We choose $U(0)$ to be zero. The quantity $\left(\frac{\partial U}{\partial q_k}\right)_0$ is the first derivative of the potential energy with the kth generalized coordinate, calculated at the static equilibrium position. This derivative is zero for every k, as this is the first-order necessary condition for the static equilibrium. Neglecting the third and higher-order terms, we get,

$$U(q) \cong \frac{1}{2} \sum_{j=1}^{n} \sum_{k=1}^{n} \left(\frac{\partial^2 U}{\partial q_j \partial q_k}\right)_0 q_j \, q_k \tag{2.42}$$

Define

$$K_{kj}^q = \left(\frac{\partial^2 U}{\partial q_j \partial q_k}\right)_0 \quad \text{for} \quad k = 1, 2, ..., n \quad \text{and} \quad j = 1, 2, ..., n \tag{2.43}$$

With this, Eq. (2.42) can be written in matrix form as,

$$U \cong \frac{1}{2} \{q\}^T \left[K^q\right]\{q\} \tag{2.44}$$

Thus, the variation of the potential energy around the static equilibrium position is approximately a quadratic function of the vector of generalized coordinates and a matrix $\left[K^q\right]$ called the generalized stiffness matrix. Equation (2.43) shows how to determine the stiffness matrix elements. Since [G] is the matrix of transformation between the coordinates $\{u\}$ and the generalized coordinates, it can be shown that the generalized stiffness matrix $\left[K^q\right]$ and the stiffness matrix in the coordinates $\{u\}$ are related by

$$\left[K^q\right] = [G]^T [K][G] \tag{2.45}$$

2.9.3 DERIVATION OF LAGRANGE'S EQUATIONS

A system has kinetic energy due to its motion and potential energy due to conservative forces. Forces due to elasticity and gravity are two examples of conservative forces. External forces (not associated with any potential energy) and the dissipative/ damping forces, either internal or external to the system, are non-conservative or non-potential forces.

From the work-energy theorem, we know that

$$dT = dW_c + dW_d + dW_e \tag{2.46}$$

where dT is the change in kinetic energy of the system while dW_c, dW_d and dW_e are the works done on the system by the conservative, dissipative, and external forces, respectively.

The work done by the conservative forces determines the change in potential energy, and the relationship between the two is given by

$$dU = -dW_c \tag{2.47}$$

If dW_c is positive, then the conservative forces do positive work on the system. Since the energy to carry out this work is derived from the system's potential energy, the system's potential energy reduces, and the change in the potential energy is therefore negative, indicated by the negative sign in Eq. (2.47).

Substituting Eq. (2.47) into Eq. (2.46) gives

$$dT + dU = dW_d + dW_e \tag{2.48}$$

a. Lagrange's equations for conservative systems

We first derive Lagrange's equations, assuming no dissipative and external forces. In this case, the system's total energy is conserved or constant, and such a system is called a **conservative system.** As a result, the terms on the RHS of Eq. (2.48) are zero, leading to

$$dT + dU = 0 \tag{2.49}$$

Eq. (2.49) says that the change in the total energy is zero for a conservative system, though the kinetic and potential energies can change individually.

We discussed the following functional relationship earlier:

$$T = T\left(q_1, q_2, .. q_n, \quad \dot{q}_1, \dot{q}_2, .. \dot{q}_n\right) \tag{2.50}$$

Therefore, dT can be written as

$$dT = \sum_{j=1}^{n} \frac{\partial T}{\partial q_j}.dq_j + \sum_{j=1}^{n} \frac{\partial T}{\partial \dot{q}_j}.d\dot{q}_j \tag{2.51}$$

The second term on the RHS in Eq. (2.51) needs to be expressed in terms of dq_i in place of $d\dot{q}_i$. The following expression for the kinetic energy is derived earlier:

$$T = \frac{1}{2}\{\dot{q}\}^T [M^q]\{\dot{q}\} \tag{2.52}$$

$$T = \frac{1}{2}\begin{Bmatrix} \dot{q}_1 \\ \dot{q}_2 \\ \vdots \\ \dot{q}_n \end{Bmatrix}^T [M^q] \begin{Bmatrix} \dot{q}_1 \\ \dot{q}_2 \\ \vdots \\ \dot{q}_n \end{Bmatrix} \tag{2.53}$$

Differentiating Eq. (2.53) with $\dot{q}_1$, we get

$$\frac{\partial T}{\partial \dot{q}_1} = \frac{1}{2}\left(\frac{\partial}{\partial \dot{q}_1}\begin{Bmatrix} \dot{q}_1 \\ \dot{q}_2 \\ \vdots \\ \dot{q}_n \end{Bmatrix}^T\right)[M^q]\begin{Bmatrix} \dot{q}_1 \\ \dot{q}_2 \\ \vdots \\ \dot{q}_n \end{Bmatrix} + \frac{1}{2}\begin{Bmatrix} \dot{q}_1 \\ \dot{q}_2 \\ \vdots \\ \dot{q}_n \end{Bmatrix}^T [M^q]\left(\frac{\partial}{\partial \dot{q}_1}\begin{Bmatrix} \dot{q}_1 \\ \dot{q}_2 \\ \vdots \\ \dot{q}_n \end{Bmatrix}\right) \tag{2.54}$$

where $[M^q]$ is not a function of the generalized velocities and hence treated as constant.

$$\frac{\partial T}{\partial q_1} = \frac{1}{2} \left\{ \begin{array}{c} 1 \\ 0 \\ \vdots \\ 0 \end{array} \right\}^T \left[M^q \right] \left\{ \begin{array}{c} \dot{q}_1 \\ \dot{q}_2 \\ \vdots \\ \dot{q}_n \end{array} \right\} + \frac{1}{2} \left\{ \begin{array}{c} \dot{q}_1 \\ \dot{q}_2 \\ \vdots \\ \dot{q}_n \end{array} \right\}^T \left[M^q \right] \left\{ \begin{array}{c} 1 \\ 0 \\ \vdots \\ 0 \end{array} \right\} \qquad (2.55)$$

Taking transpose of the first term and noting that $\left[M^q \right]$ is symmetric, we obtain

$$\frac{\partial T}{\partial \dot{q}_1} = \left\{ \begin{array}{c} \dot{q}_1 \\ \dot{q}_2 \\ \vdots \\ \dot{q}_n \end{array} \right\}^T \left[M^q \right] \left\{ \begin{array}{c} 1 \\ 0 \\ \vdots \\ 0 \end{array} \right\} \qquad (2.56)$$

$$\frac{\partial T}{\partial \dot{q}_1} = \sum_{k=1}^{n} M_{1k}^q \cdot \dot{q}_k \qquad (2.57)$$

Multiplying Eq. (2.57) by $\dot{q}_1$ gives

$$\frac{\partial T}{\partial \dot{q}_1} \dot{q}_1 = \sum_{k=1}^{n} M_{1k}^q \cdot \dot{q}_k \dot{q}_1 \qquad (2.58)$$

Equations similar to Eq. (2.58) are obtained by differentiating Eq. (2.53) with $\dot{q}_2$, .., $\dot{q}_n$. Adding all these equations gives

$$\sum_{j=1}^{n} \frac{\partial T}{\partial \dot{q}_j} \dot{q}_j = \sum_{j=1}^{n} \sum_{k=1}^{n} M_{jk}^q \cdot \dot{q}_k \dot{q}_j \qquad (2.59)$$

In light of Eq. (2.33), the RHS of Eq. (2.59) is nothing but 2T, and hence we get

$$\sum_{j=1}^{n} \frac{\partial T}{\partial \dot{q}_j} \dot{q}_j = 2T \qquad (2.60)$$

Taking the first-order change on both sides of Eq. (2.60), we can write

$$\sum_{j=1}^{n} d\left(\frac{\partial T}{\partial \dot{q}_j} \right) \dot{q}_j + \sum_{j=1}^{n} \frac{\partial T}{\partial \dot{q}_j} \cdot d\dot{q}_j = 2 \, dT \qquad (2.61)$$

$$\sum_{j=1}^{n} \frac{\partial T}{\partial \dot{q}_j} \cdot d\dot{q}_j = 2 \, dT - \sum_{j=1}^{n} d\left(\frac{\partial T}{\partial \dot{q}_j} \right) \dot{q}_j \qquad (2.62)$$

The LHS of Eq. (2.62) is nothing but the second term on the RHS in Eq. (2.51), and hence making substitution into that equation, we get

$$dT = \sum_{j=1}^{n} \frac{\partial T}{\partial q_j} \cdot dq_j + 2 \, dT - \sum_{j=1}^{n} d\left(\frac{\partial T}{\partial \dot{q}_j} \right) \dot{q}_j \qquad (2.63)$$

which leads to

$$dT = \sum_{j=1}^{n} d\left(\frac{\partial T}{\partial \dot{q}_j}\right) \dot{q}_j - \sum_{j=1}^{n} \frac{\partial T}{\partial q_j} . dq_j \qquad (2.64)$$

By writing $\dot{q}_j = \dfrac{dq_j}{dt}$ and simplifying Eq. (2.64), we get

$$dT = \sum_{j=1}^{n} \left(\frac{d}{dt}\left(\frac{\partial T}{\partial \dot{q}_j}\right) - \frac{\partial T}{\partial q_j}\right) dq_j \qquad (2.65)$$

We know that

$$U = U\left(q_1,\ q_2, .. \ q_n\right) \qquad (2.66)$$

dU can be written as

$$dU = \sum_{j=1}^{n} \frac{\partial U}{\partial q_j} . dq_j \qquad (2.67)$$

Substituting Eqs. (2.65) and (2.67) into Eq. (2.49), we get

$$dT + dU = \sum_{j=1}^{n} \left(\frac{d}{dt}\left(\frac{\partial T}{\partial \dot{q}_j}\right) - \frac{\partial T}{\partial q_j}\right) dq_j + \sum_{j=1}^{n} \frac{\partial U}{\partial q_j} . dq_j = 0 \qquad (2.68)$$

$$dT + dU = \sum_{j=1}^{n} \left(\frac{d}{dt}\left(\frac{\partial T}{\partial \dot{q}_j}\right) - \frac{\partial T}{\partial q_j} + \frac{\partial U}{\partial q_j}\right) dq_j = 0 \qquad (2.69)$$

Since the generalized coordinates are independent and each can be changed arbitrarily and independently, the only way Eq. (2.69) is satisfied would be when the expression in the bracket is zero for all the values of j. Hence,

$$\frac{d}{dt}\left(\frac{\partial T}{\partial \dot{q}_j}\right) - \frac{\partial T}{\partial q_j} + \frac{\partial U}{\partial q_j} = 0 \qquad j = 1,\ 2,\ ...,\ n \qquad (2.70)$$

These are **Lagrange's equations for the conservative system**. Noting that the potential energy U is not a function of $\dot{q}_j$, Eq. (2.70) can be written as

$$\frac{d}{dt}\left(\frac{\partial T}{\partial \dot{q}_j} - \frac{\partial U}{\partial \dot{q}_j}\right) - \left(\frac{\partial T}{\partial q_j} - \frac{\partial U}{\partial q_j}\right) = 0 \qquad j = 1,\ 2,\ ...,\ n \qquad (2.71)$$

Defining the Lagrangian

$$L = T - U \qquad (2.72)$$

Eq. (2.71) can be written as

$$\frac{d}{dt}\left(\frac{\partial L}{\partial \dot{q}_j}\right) - \frac{\partial L}{\partial q_j} = 0 \qquad j = 1,\ 2,\ ...,\ n \qquad (2.73)$$

Thus, Eq. (2.116) is an alternative form of Lagrange's equation for a conservative system.

b. Lagrange's equation for non-conservative systems

A non-conservative system has dissipative and/or non-potential external forces, so we start from Eq. (2.48). In Section 2.8.4, we saw how work done by physical forces can be expressed in terms of generalized forces. Proceeding similarly, the work done by the dissipative and non-potential external forces due to arbitrary and independent changes dq_1, dq_2,.. dq_n in the generalized coordinates can be expressed in terms of generalized forces (denoted by Q_j^{nc}, with the superscript 'nc' referring to non-conservative forces). Therefore,

$$dW_d + dW_e = \sum_{j=1}^{n} Q_j^{nc}.dq_j \tag{2.74}$$

Substituting Eqs. (2.65), (2.67), and (2.74) into Eq. (2.48), we obtain

$$\sum_{j=1}^{n} \left(\frac{d}{dt}\left(\frac{\partial T}{\partial \dot{q}_j} \right) - \frac{\partial T}{\partial q_j} + \frac{\partial U}{\partial q_j} \right) dq_j = \sum_{j=1}^{n} Q_j^{nc}.dq_j \tag{2.75}$$

Since the generalized coordinates can be changed arbitrarily and independently, we must have

$$\frac{d}{dt}\left(\frac{\partial T}{\partial \dot{q}_j} \right) - \frac{\partial T}{\partial q_j} + \frac{\partial U}{\partial q_j} = Q_j^{nc} \qquad j = 1, 2, ..., n \tag{2.76}$$

which are **Lagrange's equations for a non-conservative system**. In terms of the Lagrangian, Lagrange's equation for non-conservative systems can be written as

$$\frac{d}{dt}\left(\frac{\partial L}{\partial \dot{q}_j} \right) - \frac{\partial L}{\partial q_j} = Q_j^{nc} \qquad j = 1, 2, ..., n \tag{2.77}$$

Therefore, Lagrange's equations for both the conservative and non-conservative systems, described by n generalized coordinates, are a set of n differential equations in terms of the kinetic energy, potential energy, and generalized forces corresponding to the non-potential forces.

2.10   HAMILTON'S PRINCIPLE

Hamilton's principle is an integral equation, while Lagrange's equations, studied in the previous section, are in the form of differential equations. Hamilton's principle prescribes a condition on the path a system would follow while undergoing displacement between two instants of time, t_1 and t_2, while being in the dynamic equilibrium under the action of the external forces. It says that of all the possible paths between the two instants consistent with the constraints, the system follows the path that makes the integral stationery.

From the virtual work principle for dynamics, we have

$$\sum_{i=1}^{m} (F_i - M_i \ddot{u}_i).\delta u_i = 0 \tag{2.78}$$

Let δW_a denote the virtual work done by the active forces F_i^s. That is

$$\delta W_a = \sum_{i=1}^{m} F_i.\delta u_i \tag{2.79}$$

Therefore, Eq. (2.78) becomes

$$\delta W_a = \sum_{i=1}^{m} M_i \ddot{u}_i.\delta u_i \tag{2.80}$$

We also note that

$$\frac{d}{dt}(\dot{u}_i.\delta u_i) = \ddot{u}_i.\delta u_i + \dot{u}_i.\frac{d}{dt}(\delta u_i) \tag{2.81}$$

Multiplying by M_i on both sides and summing over $i = 1$ to m and simplifying, we get

$$\sum_{i=1}^{m} M_i.\ddot{u}_i.\delta u_i = \sum_{i=1}^{m} M_i.\frac{d}{dt}(\dot{u}_i.\delta u_i) - \sum_{i=1}^{m} M_i.\dot{u}_i.\frac{d}{dt}(\delta u_i) \qquad (2.82)$$

The term in the last summation on the RHS can be simplified as

$$M.\dot{u}.\frac{d}{dt}(\delta u) = M.\dot{u}.\delta\left(\frac{d}{dt}u\right) \qquad (2.83)$$

$$= M.\dot{u}.\delta(\dot{u})$$

$$= M.\delta\left(\frac{\dot{u}}{2}\right)$$

Substituting Eq. (2.83) into Eq. (2.82) gives

$$\sum_{i=1}^{m} M_i.\ddot{u}_i.\delta u_i = \sum_{i=1}^{m} M_i.\frac{d}{dt}(\dot{u}_i.\delta u_i) - \sum_{i=1}^{m} \delta(T_i) \qquad (2.84)$$

$$\sum_{i=1}^{m} M_i.\ddot{u}_i.\delta u_i = \sum_{i=1}^{m} M_i.\frac{d}{dt}(\dot{u}_i.\delta u_i) - \delta T \qquad (2.85)$$

Substituting Eq. (2.85) into Eq. (2.80), we get

$$\delta W_a = \sum_{i=1}^{m} M_i.\frac{d}{dt}(\dot{u}_i.\delta u_i) - \delta T \qquad (2.86)$$

The virtual work done by the active forces can be written as the sum of the virtual work done by the conservative and non-conservative forces. Furthermore, using Eq. (2.47), we can write

$$\delta W_a = -\delta U + \delta W_{nc} \qquad (2.87)$$

Substituting Eq. (2.87) into Eq. (2.86) gives

$$\delta T - \delta U + \delta W_{nc} = \sum_{i=1}^{m} M_i.\frac{d}{dt}(\dot{u}_i.\delta u_i) \qquad (2.88)$$

The virtual displacement of the ith coordinate at times between t_1 and t_2 together defines a varied path for that coordinate. If the actual path or equilibrium path followed by the ith coordinate under the dynamic equilibrium is $u_i(t)$ then the varied path due to virtual displacement $\delta u_i(t)$ is $u_i(t) + \delta u_i(t)$. Figure 2.20 shows the equilibrium and varied paths for coordinate u_i.

At t_1 and t_2, the virtual displacements must be zero (for the varied path to be consistent with the constraints). Note that since each coordinate has its path of motion between t_1 and t_2, so each coordinate has a varied path. Hence,

$$\delta u_i(t_1) = 0 \quad \text{and} \quad \delta u_i(t_2) = 0 \quad \text{for } i = 1, 2, .., m \qquad (2.89)$$

Integration of Eq. (2.88) is carried out from t_1 to t_2. It effectively integrates the terms on the two sides of the equation due to virtual displacement $\delta u_i(t)$ for $i = 1, 2, .., m$, i.e., due to the variation of the path from the equilibrium path to the varied path.

$$\int_{t_1}^{t_2} (\delta T - \delta U + \delta W_{nc})\, dt = \int_{t_1}^{t_2} \sum_{i=1}^{m} \left(M_i.\frac{d}{dt}(\dot{u}_i.\delta u_i) \right) dt \qquad (2.90)$$

Interchanging the integration and summation operators and performing the integration, we get

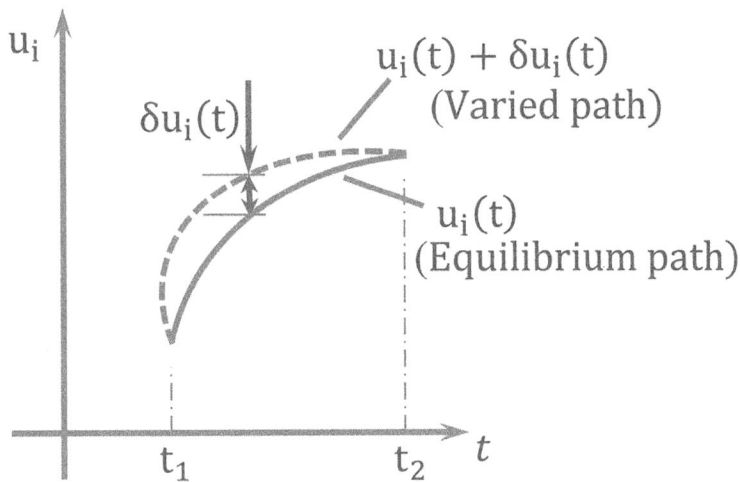

FIGURE 2.20 The equilibrium and varied paths for i^{th} coordinate

$$\int_{t_1}^{t_2} \left(\delta T - \delta U + \delta W_{nc}\right) dt = \sum_{i=1}^{m} M_i \cdot \left(\dot{u}_i . \delta u_i\right)_{t_1}^{t_2} \tag{2.91}$$

In light of the constraints in Eq. (2.89), the RHS of Eq. (2.91) is zero, leading to

$$\int_{t_1}^{t_2} \left(\delta(T - U) + \delta W_{nc}\right) dt = 0 \tag{2.92}$$

Eq. (2.92) is the integral equation of **Hamilton's principle for non-conservative systems**.
   If the system is conservative, $\delta W_{nc} = 0$, and then,

$$\int_{t_1}^{t_2} \delta(T - U) \, dt = 0 \tag{2.93}$$

Since T and U are point functions, we can write,

$$\delta \int_{t_1}^{t_2} (T - U) \, dt = \delta I = 0 \tag{2.94}$$

where,

$$I = \int_{t_1}^{t_2} (T - U) \, dt = \int_{t_1}^{t_2} L \, dt \tag{2.95}$$

Thus, the first variation of I should be zero. Eq. (2.94) is the statement of **Hamilton's principle for conservative systems**. It states that of all the possible paths between the time instants t_1 and t_2, the equilibrium path of motion of the system would be the one for which the integral has a stationary value.

Example 2.9: Application of Newton's Law

A 3-DOF lumped parameter model of the compressor, turbine, and generator in a power plant to analyze torsional vibrations is shown in Figure 2.21a. J_1, J_2 and J_3 are the polar moment of inertia of the rotating components of the compressor, turbine, and generator, respectively. L_j and d_j are the length and diameter of the shaft sections, respectively. $T_1(t)$, $T_2(t)$ and $T_3(t)$ are the torques acting on the three inertias, respectively. Using Newton's law, derive the governing equations of motion.

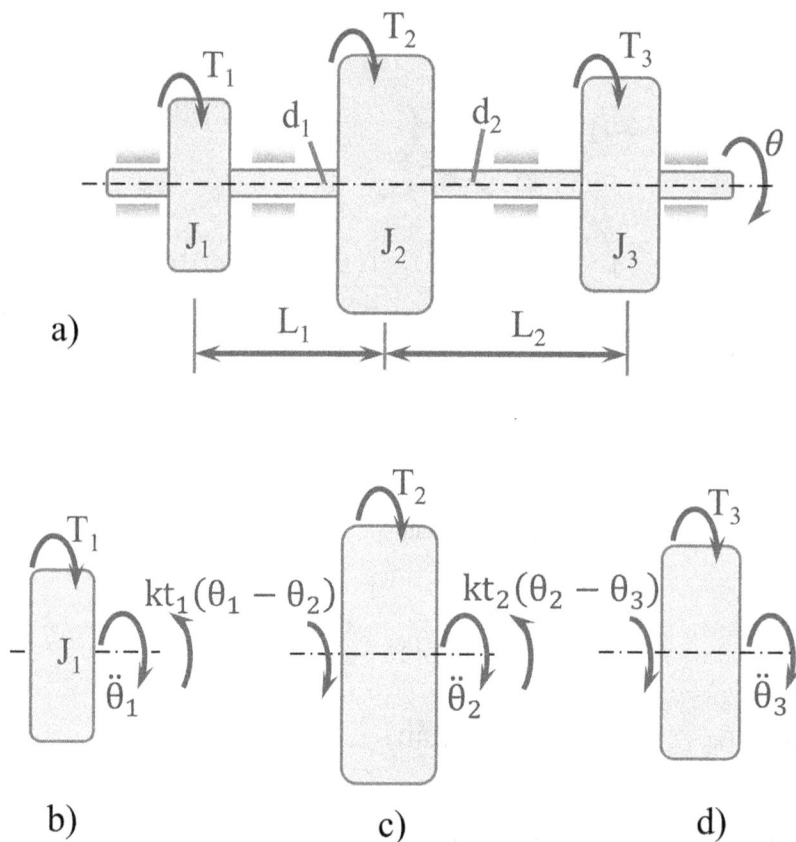

FIGURE 2.21   (a) Three degree of freedom lumped parameter model of the compressor, turbine, and generator in a power plant, (b), (c), and (d): free body diagrams.

Solution

Let θ_1, θ_2 and θ_3 be the angular displacements of J_1, J_2 and J_3, respectively. The shafts provide torsional stiffness, and their inertias are assumed to be either suitably lumped or negligible. The torsional stiffness (defined as torque to produce a unit twist) of the two shaft sections is given by

$$kt_1 = \frac{G_1 I_1}{L_1} \quad \text{and} \quad kt_2 = \frac{G_2 I_2}{L_2}$$

where G_j and I_j are shear modulus for the shaft material and polar second moment of area of the shaft cross-section, respectively. Note $I_j = \frac{\pi d_j^4}{32}$.

Considering the free-body diagram of inertia J_1 (as shown in Figure 2.21b) and applying Newton's law for angular motion, we can write

$$J_1 \ddot{\theta} = \sum \text{Torques}$$

$$J_1 \ddot{\theta}_1 = T_1(t) - kt_1(\theta_1 - \theta_2)$$

The free-body diagrams for J_2 and J_3 are shown in Figure 2.21c and d, respectively. Applying Newton's law gives

$$J_2 \ddot{\theta}_2 = T_2(t) + kt_1(\theta_1 - \theta_2) - kt_2(\theta_2 - \theta_3)$$

$$J_3 \ddot{\theta}_3 = T_3(t) + kt_2(\theta_2 - \theta_3)$$

The above equations can be simplified and combined to obtain the equation of motion in matrix form as

$$\begin{bmatrix} J_1 & 0 & 0 \\ 0 & J_2 & 0 \\ 0 & 0 & J_3 \end{bmatrix} \begin{Bmatrix} \ddot{\theta}_1 \\ \ddot{\theta}_2 \\ \ddot{\theta}_3 \end{Bmatrix} + \begin{bmatrix} kt_1 & -kt_1 & 0 \\ -kt_1 & kt_1+kt_2 & -kt_2 \\ 0 & -kt_2 & kt_2 \end{bmatrix} \begin{Bmatrix} \theta_1 \\ \theta_2 \\ \theta_3 \end{Bmatrix} = \begin{Bmatrix} T_1(t) \\ T_2(t) \\ T_3(t) \end{Bmatrix}$$

Example 2.10: Application of Lagrange's Equation

Figure 2.11, given earlier, shows a lumped parameter model for the longitudinal vibrations of a propulsion system. Derive the governing equations of motion using Lagrange's equation.

Solution

Let the longitudinal displacements of the three masses (denoted by x_1, x_2 and x_3), are the coordinates selected to describe their position in the longitudinal direction. Since these coordinates are independent, they are also generalized coordinates. Thus $q_1 = x_1$, $q_2 = x_2$ and $q_3 = x_3$. Let $\dot{x}_1$, $\dot{x}_2$ and $\dot{x}_3$ are the longitudinal velocities.

Lagrange's equation for a non-conservative system is used since the external and damping forces exist.

$$\frac{d}{dt}\left(\frac{\partial T}{\partial \dot{q}_j}\right) - \frac{\partial T}{\partial q_j} + \frac{\partial U}{\partial q_j} = Q_j^{nc} \quad j = 1, 2, 3$$

We first determine T, U and Q_j^{nc}. The kinetic energy T of the system is

$$T = \frac{1}{2}m_1\dot{x}_1^2 + \frac{1}{2}m_2\dot{x}_2^2 + \frac{1}{2}m_3\dot{x}_3^2$$

The terms in Lagrange's equation related to T are:

$$\frac{d}{dt}\left(\frac{\partial T}{\partial \dot{x}_1}\right) = \frac{d}{dt}(m_1\dot{x}_1) = m_1\ddot{x}_1$$

(Note that $\frac{\partial T}{\partial \dot{x}_1}$ is the partial derivative with $\dot{x}_1$ and therefore, all the other variables are treated constants.)

$$\frac{d}{dt}\left(\frac{\partial T}{\partial \dot{x}_2}\right) = \frac{d}{dt}(m_2\dot{x}_2) = m_2\ddot{x}_2$$

$$\frac{d}{dt}\left(\frac{\partial T}{\partial \dot{x}_3}\right) = \frac{d}{dt}(m_3\dot{x}_3) = m_3\ddot{x}_3$$

Also,

$$\frac{\partial T}{\partial x_j} = 0 \text{ for } j = 1, 2, 3$$

The potential energy U of the system is

$$U = \frac{1}{2}k_1(x_1 - x_2)^2 + \frac{1}{2}k_1(x_2 - x_3)^2 + \frac{1}{2}k_3x_3^2$$

$$\frac{\partial U}{\partial x_1} = k_1(x_1 - x_2)$$

$$\frac{\partial U}{\partial x_2} = -k_1(x_1 - x_2) + k_2(x_2 - x_3)$$

$$\frac{\partial U}{\partial x_3} = -k_2(x_2 - x_3) + k_3x_3$$

The generalized forces are given by

$$Q_j^{nc} = \sum_{i=1}^{3} F_i \cdot \frac{\partial u_i}{\partial q_j} \qquad j = 1, 2, 3$$

F_i in the above equation includes the external forces and damping forces along the ith coordinate. Moreover, the coordinates u_i (represented by x_i in this example) and generalized coordinates are the same. Thus, the above equation can be written as

$$Q_{x_1}^{np} = F_{x_1} \frac{\partial x_1}{\partial x_1} + F_{x_2} \frac{\partial x_2}{\partial x_1} + F_{x_3} \frac{\partial x_3}{\partial x_1}$$

$$Q_{x_1}^{np} = F_{x_1} = P(t) - c_1 (\dot{x}_1 - \dot{x}_2)$$

In the same way, we get

$$Q_{x_2}^{np} = F_{x_2} = c_1 (\dot{x}_1 - \dot{x}_2) - c_2 (\dot{x}_2 - \dot{x}_3)$$

$$Q_{x_3}^{np} = F_{x_3} = c_2 (\dot{x}_2 - \dot{x}_3) - c_3 \dot{x}_3$$

Making substitutions in Lagrange's equation, we get

$$m_1 \ddot{x}_1 + k_1 (x_1 - x_2) = P(t) - c_1 (\dot{x}_1 - \dot{x}_2)$$

$$m_2 \ddot{x}_2 - k_1 (x_1 - x_2) + k_2 (x_2 - x_3) = c_1 (\dot{x}_1 - \dot{x}_2) - c_2 (\dot{x}_2 - \dot{x}_3)$$

$$m_3 \ddot{x}_3 - k_2 (x_2 - x_3) + k_3 x_3 = c_2 (\dot{x}_2 - \dot{x}_3) - c_3 \dot{x}_3$$

The three equations can be combined and written as

$$\begin{bmatrix} m_1 & 0 & 0 \\ 0 & m_2 & 0 \\ 0 & 0 & m_3 \end{bmatrix} \begin{Bmatrix} \ddot{x}_1 \\ \ddot{x}_2 \\ \ddot{x}_3 \end{Bmatrix} + \begin{bmatrix} c_1 & -c_1 & 0 \\ -c_1 & c_1+c_2 & -c_2 \\ 0 & -c_2 & c_2+c_3 \end{bmatrix} \begin{Bmatrix} \dot{x}_1 \\ \dot{x}_2 \\ \dot{x}_3 \end{Bmatrix} + \begin{bmatrix} k_1 & -k_1 & 0 \\ -k_1 & k_1+k_2 & -k_2 \\ 0 & -k_2 & k_2+k_3 \end{bmatrix} \begin{Bmatrix} x_1 \\ x_2 \\ x_3 \end{Bmatrix} = \begin{Bmatrix} P(t) \\ 0 \\ 0 \end{Bmatrix}$$

The above equation is the equation of motion in matrix form. The coefficient matrices are the mass, damping, and stiffness matrices. They are of the order 3×3 since the degree of freedom of the system is 3.

Example 2.11: Application of Lagrange's Equation

Figure 2.22 shows an electric motor driving a compressor, both being mounted on an inertia block isolated from the ground by isolators. A two-degree-of-freedom lumped parameter model of the system to study the bouncing and rocking motion is shown in Figure 2.23. M and J are the mass and polar moment of inertia, respectively, of the motor-compressor-inertia-block system. The stiffness and damping properties of the isolators are modeled by two pairs of spring and dashpot with stiffness and equivalent viscous damping coefficient k_1, c_1 and k_2, c_2, respectively. A disturbing force $P(t)$ acts at a distance b from the CG. The CG of the system and other distances are shown in the figure. Derive the governing mathematical equations using Lagrange's equation.

Solution

Let x and θ be generalized coordinates, as shown in Figure 2.23. x is the vertical displacement of the CG measured from the static equilibrium position, while θ is the rotation of the mass.
The kinetic energy T of the system is

$$T = \frac{1}{2} m \dot{x}^2 + \frac{1}{2} J \dot{\theta}^2$$

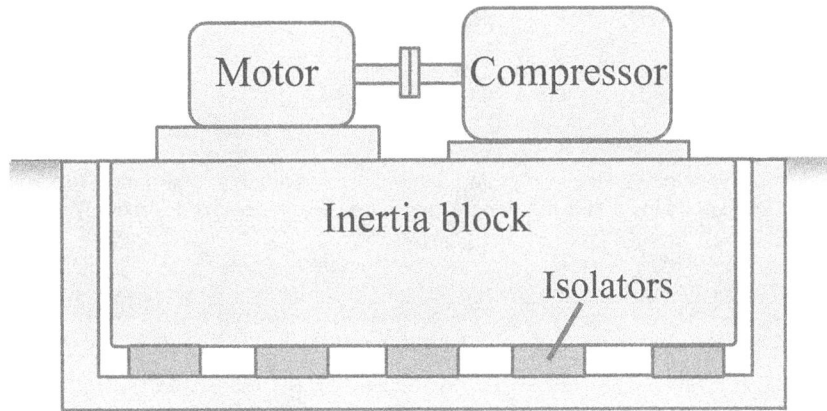

FIGURE 2.22 An electric motor driving a compressor.

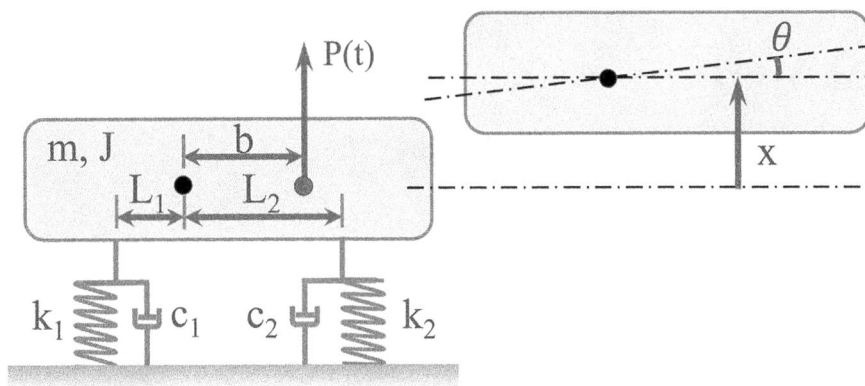

FIGURE 2.23 Two-degree of freedom lumped parameter model.

The terms in Lagrange's equation related to T are:

$$\frac{d}{dt}\left(\frac{\partial T}{\partial \dot{x}}\right) = \frac{d}{dt}(m\dot{x}) = m\ddot{x}$$

$$\frac{d}{dt}\left(\frac{\partial T}{\partial \dot{\theta}}\right) = \frac{d}{dt}\left(J\dot{\theta}\right) = J\ddot{\theta}$$

Also,

$$\frac{\partial T}{\partial x} \text{ and } \frac{\partial T}{\partial \theta} = 0$$

The potential energy U is due to the deflection of the springs and is equal to,

$$U = \frac{1}{2}k_1\left(x - L_1\theta\right)^2 + \frac{1}{2}k_2\left(x + L_2\theta\right)^2$$

The potential energy due to the gravitational force on the mass is not taken above since the force of gravity balances the spring force due to static deflection.

$$\frac{\partial U}{\partial x} = k_1\left(x - L_1\theta\right) + k_2\left(x + L_2\theta\right)$$

$$\frac{\partial U}{\partial \theta} = -k_1 L_1\left(x - L_1\theta\right) + k_2 L_2\left(x + L_2\theta\right)$$

The generalized forces are given by

$$Q_j^{nc} = \sum_{i=1}^{2} F_i \cdot \frac{\partial u_i}{\partial q_j} \qquad j = 1, 2$$

F_i in the above equation includes the external forces and damping forces along i^{th} coordinate. Moreover, the coordinates u_i (representing x and θ) and generalized coordinates are the same. The generalized forces Q_1^{nc} and Q_2^{nc} are nothing but Q_x^{nc} and Q_θ^{nc}, respectively.

Thus, we get

$$Q_x^{nc} = F_x \frac{\partial x}{\partial x} + M_\theta \frac{\partial \theta}{\partial x}$$

$$Q_x^{nc} = F_x$$

F_x is the algebraic sum of all forces along coordinate x. Therefore,

$$Q_x^{nc} = P(t) - c_1\left(\dot{x} - L_1\dot{\theta}\right) - c_2\left(\dot{x} + L_2\dot{\theta}\right)$$

Similarly,

$$Q_\theta^{nc} = F_x \frac{\partial x}{\partial \theta} + M_\theta \frac{\partial \theta}{\partial \theta}$$

$$Q_\theta^{nc} = M_\theta$$

M_θ is the algebraic sum of all moments along the coordinate θ. Therefore,

$$Q_\theta^{nc} = P(t).b + c_1 L_1\left(\dot{x} - L_1\dot{\theta}\right) - c_2 L_2\left(\dot{x} + L_2\dot{\theta}\right)$$

Lagrange's equation is given by,

$$\frac{d}{dt}\left(\frac{\partial T}{\partial \dot{q}_j}\right) - \frac{\partial T}{\partial q_j} + \frac{\partial U}{\partial q_j} = Q_j^{nc} \qquad j = 1, 2$$

For coordinate x

$$\frac{d}{dt}\left(\frac{\partial T}{\partial \dot{\theta}}\right) + \frac{\partial U}{\partial \theta} = Q_x^{nc}$$

and for coordinate θ

$$\frac{d}{dt}\left(\frac{\partial T}{\partial \dot{\theta}}\right) + \frac{\partial U}{\partial \theta} = Q_\theta^{nc}$$

Substituting the relevant quantities in the above two equations, we get

$$m\ddot{x} + k_1\left(x - L_1\theta\right) + k_2\left(x + L_2\theta\right) = P(t) - c_1\left(\dot{x} - L_1\dot{\theta}\right) - c_2\left(\dot{x} + L_2\dot{\theta}\right)$$

$$J\ddot{\theta} - k_1 L_1\left(x - L_1\theta\right) + k_2 L_2\left(x + L_2\theta\right) = P(t).b + c_1 L_1\left(\dot{x} - L_1\dot{\theta}\right) - c_2 L_2\left(\dot{x} + L_2\dot{\theta}\right)$$

The above two equations are the governing equations of motion and can be written in matrix form as

$$\begin{bmatrix} m & 0 \\ 0 & J \end{bmatrix}\begin{Bmatrix} \ddot{x} \\ \ddot{\theta} \end{Bmatrix} + \begin{bmatrix} c_1 + c_2 & -c_1 L_1 + c_2 L_2 \\ -c_1 L_1 + c_2 L_2 & c_1 L_1^2 + c_2 L_2^2 \end{bmatrix}\begin{Bmatrix} \dot{x} \\ \dot{\theta} \end{Bmatrix} + \begin{bmatrix} k_1 + k_2 & -k_1 L_1 + k_2 L_2 \\ -k_1 L_1 + k_2 L_2 & k_1 L_1^2 + k_2 L_2^2 \end{bmatrix}\begin{Bmatrix} x \\ \theta \end{Bmatrix} = \begin{Bmatrix} P(t) \\ P(t)b \end{Bmatrix}$$

Example 2.12: Application of the Principle of Virtual Work

Figure 2.24a shows a cart-compound-pendulum system with force $P(t)$ acting as shown. The cart can move on its rails, while the compound pendulum can swing in the vertical plane about the point it is pinned to the cart. Derive the governing equations of motion of the system using the principle of virtual work. m_1 and m_2 are the cart and pendulum masses, respectively.

Solution

Let x and θ be generalized coordinates, where x is the cart's displacement from the equilibrium position, and θ is the pendulum angle from the vertical. Figure 2.24b shows a displaced configuration of the system from the equilibrium position. Let $\ddot{x}$ be the linear acceleration of mass m_1 and $\ddot{\theta}$ be the angular acceleration of the compound pendulum. Figure 2.24c shows all the forces and moments acting on the system at the displaced configuration. These are spring force kx, gravity force $m_2 g$ and the external force $P(t)$. In addition, following D'Alembert's principle, the inertia forces/moments are shown in the opposite directions of the corresponding accelerations, as explained below.

- Mass m_1 has linear acceleration $\ddot{x}$ and hence an inertia force $m_1\ddot{x}$ is applied on m_1 opposite to acceleration.
- The compound pendulum is hinged on mass m_1. Since m_1 is accelerating, the compound pendulum is in a non-inertial frame (accelerating with $\ddot{x}$) and hence an inertia force equal to $m_2\ddot{x}$ is applied on m_2 opposite to $\ddot{x}$.
- Point C, the CG of the compound pendulum, has a centripetal acceleration $\dot{\theta}^2\dfrac{L}{2}$ towards point O and hence an inertia force $m_2\dot{\theta}^2\dfrac{L}{2}$ is applied on the compound pendulum opposite to the acceleration.
- Point C also has tangential acceleration $\ddot{\theta}\dfrac{L}{2}$ perpendicular to OC and hence an inertia force $m_2\ddot{\theta}\dfrac{L}{2}$ is applied on the compound pendulum opposite to the acceleration.
- Due to the angular acceleration $\ddot{\theta}$ of the compound pendulum, an inertia torque $I_C\ddot{\theta}$ is applied opposite to the angular acceleration. I_C is the moment of inertia of the compound pendulum about an axis through C perpendicular to the paper.

To apply the principle of virtual work, we consider the virtual displacement of the generalized coordinates one at a time. Therefore, let δx be the virtual displacement given to the coordinate x, with no virtual displacement of the coordinate θ as

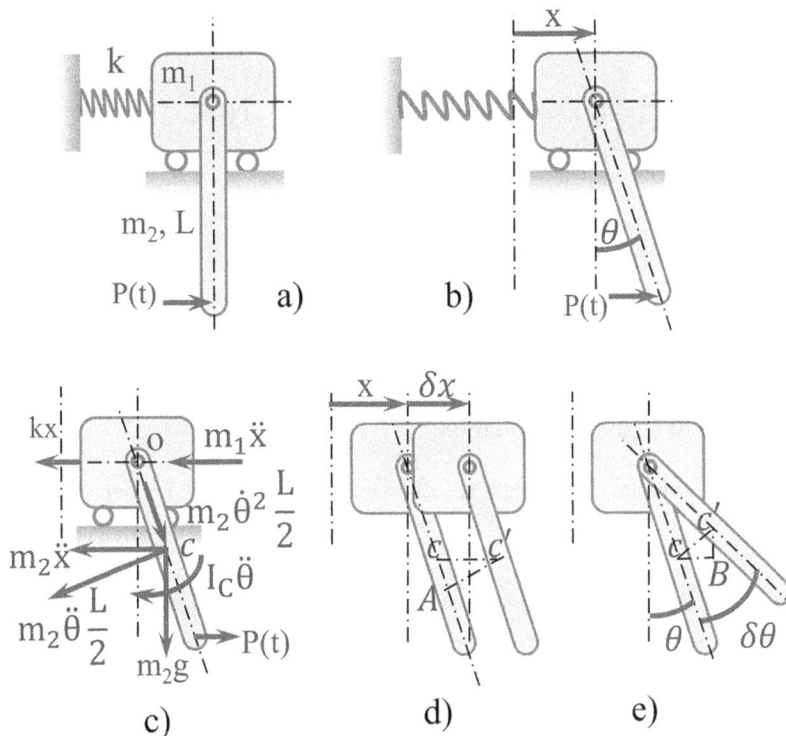

FIGURE 2.24   (a) Cart with a compound pendulum, (b) displaced configuration of the system, (c) forces and moments acting on the system, (d) virtual displacement δx given to the coordinate x, (e) virtual displacement $\delta\theta$ given to the coordinate θ.

shown in Figure 2.24d. The virtual displacements along the axis of each force can be worked out. The virtual work, due to all the forces and moments, is

$$\delta W_1 = -kx.\delta x + m_2g.0 + P(t).\delta x - m_1\ddot{x}.\delta x - m_2\ddot{x}.\delta x + m_2\dot{\theta}^2\frac{L}{2}.\delta x\sin\ddot{\theta} - m_2\ddot{\theta}\frac{L}{2}.\delta x\cos\theta + I_C\ddot{\theta}.0$$

$$\delta W_1 = \left(-kx + P(t) - (m_1 + m_2)\ddot{x} + m_2\dot{\theta}^2\frac{L}{2}\sin\theta - m_2\ddot{\theta}\frac{L}{2}\cos\theta\right)\delta x$$

Similarly, let $\delta\theta$ be the virtual displacement given to coordinate θ, with no virtual displacement of the coordinate x (Figure 2.24e). Again, the virtual displacements along the axis of each force can be worked out. The virtual work, due to all the forces and moments, is

$$\delta W_2 = kx.0 - m_2g.\delta\theta\frac{L}{2}\sin\theta + P(t).\delta\theta\ L\cos\theta + m_1\ddot{x}.0 - m_2\ddot{x}.\delta\theta\frac{L}{2}\cos\theta + m_2\dot{\theta}^2\frac{L}{2}.0 - m_2\ddot{\theta}\frac{L}{2}.\delta\theta\frac{L}{2} - I_C\ddot{\theta}.\delta\theta$$

$$\delta W_2 = \left(-m_2g.\frac{L}{2}\sin\theta + P(t).\ L\cos\theta - m_2\ddot{x}.\frac{L}{2}\cos\theta - m_2\ddot{\theta}\frac{L}{2}.\frac{L}{2} - I_C\ddot{\theta}\right)\delta\theta$$

By the principle of virtual work

$$\delta W = \delta W_1 + \delta W_2 = 0$$

Substituting and noting that the virtual displacements δx and $\delta\theta$ are arbitrary, we get

$$(m_1 + m_2)\ddot{x} + \left(m_2\frac{L}{2}\ \cos\theta\right)\ddot{\theta} - \left(m_2\frac{L}{2}\ \sin\theta\right)\dot{\theta}^2 + kx = P(t)$$

$$\left(I_C + m_2\frac{L^2}{4}\right)\ddot{\theta} + \left(m_2.\frac{L}{2}\cos\theta\right)\ddot{x} + m_2g.\frac{L}{2}\sin\theta = P(t).\ L\cos\theta$$

The equations are nonlinear. If the displacements x and θ are assumed to be small, then we can approximate $\sin\theta \approx \theta$ and $\cos\theta \approx 1.0$. The higher-order term involving $\theta.\dot{\theta}^2$ can also be neglected. These approximations give us a linearized set of governing equations as

$$(m_1 + m_2)\ddot{x} + \left(m_2\frac{L}{2}\right)\ddot{\theta} + kx = P(t)$$

$$\left(I_C + m_2\frac{L^2}{4}\right)\ddot{\theta} + \left(m_2\frac{L}{2}\right)\ddot{x} + m_2g\frac{L}{2}\theta = P(t).\ L$$

The two equations can be written in the following matrix form,

$$\begin{bmatrix} m_1 + m_2 & m_2\frac{L}{2} \\ m_2\frac{L}{2} & I_C + m_2\frac{L^2}{4} \end{bmatrix}\begin{Bmatrix} \ddot{x} \\ \ddot{\theta} \end{Bmatrix} + \begin{bmatrix} k & 0 \\ 0 & m_2g\frac{L}{2} \end{bmatrix}\begin{Bmatrix} x \\ \theta \end{Bmatrix} = \begin{Bmatrix} P(t) \\ P(t)L \end{Bmatrix}$$

Example 2.13: Application of Lagrange's Equation

Figure 2.25a shows a simple pendulum AB of length L, with mass m at its free end, connected to a rotating disc of radius R through a revolute joint. The disc is rotating with a constant angular velocity ω. Derive the equation of motion of the pendulum using Lagrange's equation.

Solution

Figure 2.25b shows a displaced configuration of the system at time t. Since the wheel is rotating with a constant angular velocity, only the angle θ of the pendulum is required to specify the displaced configuration. Let θ be the generalized coordinate. From the figure, we can write

$$x_B = x_A + L\sin\theta = R\sin\phi + L\sin\theta$$

$$y_B = y_A + L\cos\theta = R\cos\phi + L\cos\theta$$

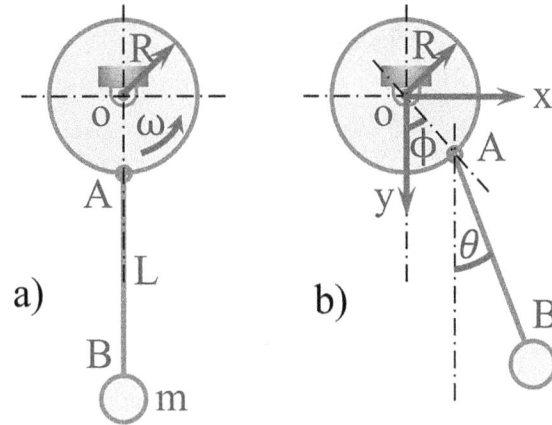

FIGURE 2.25 (a) Rotating simple pendulum and (b) displaced configuration.

Differentiating with time t, we get

$$\dot{x}_B = R\dot{\phi}\cos\phi + L\dot{\theta}\cos\theta$$

$$\dot{y}_B = -\left(R\dot{\phi}\sin\phi + L\dot{\theta}\sin\theta\right)$$

The kinetic energy T of the pendulum is

$$T = \frac{1}{2}m\dot{x}_B^2 + \frac{1}{2}m\dot{y}_B^2$$

$$T = \frac{1}{2}m\left(R^2\dot{\phi}^2 + L^2\dot{\theta}^2 + 2RL\dot{\phi}\,\dot{\theta}\cos(\phi-\theta)\right)$$

The terms in Lagrange's equation related to T are:

$$\frac{\partial T}{\partial\dot{\theta}} = \frac{1}{2}m\left(2L^2\dot{\theta} + 2RL\dot{\phi}\cos(\phi-\theta)\right)$$

$$\frac{d}{dt}\left(\frac{\partial T}{\partial\dot{\theta}}\right) = \frac{1}{2}m\left(2L^2\ddot{\theta} - 2RL\dot{\phi}\left(\dot{\phi}-\dot{\theta}\right)\sin(\phi-\theta)\right)$$

Also,

$$\frac{\partial T}{\partial\theta} = \frac{1}{2}m\left(2RL\dot{\phi}\,\dot{\theta}\sin(\phi-\theta)\right)$$

The potential energy U of the pendulum mass m is

$$U = mg\left((R+L) - \left(R\cos\phi + L\cos\theta\right)\right)$$

$$\frac{\partial U}{\partial\theta} = mgL\sin\theta$$

Lagrange's equation is given by,

$$\frac{d}{dt}\left(\frac{\partial T}{\partial\dot{q}_j}\right) - \frac{\partial T}{\partial q_j} + \frac{\partial U}{\partial q_j} = Q_j^{np} \qquad j = 1$$

Since there is no non-potential external force and/or damping force, $Q_j^{np} = 0$. Note that mg is a potential force and is already accounted for in the potential energy U. Therefore,

$$\frac{d}{dt}\left(\frac{\partial T}{\partial \dot\theta}\right) - \frac{\partial T}{\partial \theta} + \frac{\partial U}{\partial \theta} = 0$$

Substituting the relevant quantities from the previous equations, we get

$$\frac{1}{2}m\left(2L^2\ddot\theta - 2RL\dot\phi(\dot\phi - \dot\theta)\sin(\phi-\theta)\right) - \frac{1}{2}m\left(2RL\dot\phi\dot\theta\sin(\phi-\theta)\right) + mgL\sin\theta = 0$$

On simplification and noting that $\dot\phi = \omega$, and $\phi = \omega t$, we get

$$\ddot\theta + \frac{g}{L}\sin\theta - \frac{R}{L}\omega^2\sin(\omega t - \theta) = 0$$

The solution of the above equation gives the angle θ of the pendulum. If required, we can obtain the coordinates of the pendulum's mass from the previous equations.

2.11   FLEXIBILITY INFLUENCE COEFFICIENTS

Flexibility influence coefficients relate the system's displacement to the static forces acting on it. For dynamics, the net forces incorporating the inertia forces must be taken.

For an SDOF linear vibrating system, the displacement is proportional to the external force,

$$x \propto f \tag{2.96}$$

$$x = a.f \tag{2.97}$$

The constant of proportionality 'a' in Eq. (2.97) is called the flexibility influence coefficient (having units m/N). It is defined as the displacement due to a unit force.

For an MDOF system with n DOFs, flexibility influence coefficients can be defined for different pairs of force and displacement DOFs. The flexibility influence coefficient a_{ir} is defined as the displacement at the ith DOF due to a unit force at the rth DOF, with no force at other DOFs. Thus,

$$a_{ir} = \frac{x_i}{f_r} \quad \text{such that} \quad f_i = 0 \quad \text{for } i \neq r \tag{2.98}$$

Let $\{f\} = \{f_1,\ f_2,\ ...f_n\}^T$ define the forces at n DOFs. The force at each DOF causes displacements at all the DOFs, and therefore the displacement at the ith DOF is given by the sum of the displacements due to the forces at all the DOFs. Hence,

$$x_i = a_{i1}.f_1 + a_{i2}.f_2 + ... + a_{in}.f_n \qquad \text{for } i = 1,\ 2,...n \tag{2.99}$$

Eq. (2.99) represents n equations and can be written in matrix form as

$$\begin{Bmatrix} x_1 \\ x_2 \\ \vdots \\ \vdots \\ x_n \end{Bmatrix} = \begin{bmatrix} a_{11} & a_{12} & \cdots & \cdots & a_{1n} \\ a_{21} & a_{22} & \cdots & \cdots & a_{2n} \\ \vdots & \vdots & \cdots & \cdots & \vdots \\ \vdots & \vdots & \cdots & \cdots & \vdots \\ a_{n1} & a_{n2} & \cdots & \cdots & a_{nn} \end{bmatrix} \begin{Bmatrix} f_1 \\ f_2 \\ \vdots \\ \vdots \\ f_n \end{Bmatrix} \tag{2.100}$$

$$\{x\} = [a]\{f\} \tag{2.101}$$

where $[a]$ is the flexibility matrix of order $n \times n$ and consists of flexibility influence coefficients.

Equations (2.98) and (2.100) indicate how the flexibility influence coefficients can be determined. If we take $f_1 = 1.0$ with no force at other DOFs, the displacement vector $\{x\}$ obtained is nothing but the first column of the flexibility matrix. The rth column of the flexibility matrix is the displacement vector obtained by applying a unit force at the rth DOF with no force at other DOFs.

One important property of linear systems, we prove in Section 2.13, is that the flexibility matrix is symmetric, that is

$$a_{ir} = a_{ri} \tag{2.102}$$

2.12 STIFFNESS INFLUENCE COEFFICIENTS

Stiffness influence coefficients provide an alternative form of the relation between the displacements and external forces on a system.

For an SDOF linear vibrating system, this relationship is

$$f = k\,x \tag{2.103}$$

where k is called the stiffness influence coefficient (or just stiffness) (with units N/m) and is equal to the force required to cause unit displacement of the system.

For an MDOF system with n DOFs, stiffness influence coefficients can be defined for different pairs of force and displacement DOFs. The stiffness influence coefficient k_{ir} is defined as the force required at the ith DOF to cause unit displacement at the rth DOF, with no displacement at the remaining DOFs. Thus,

$$k_{ir} = \frac{f_i}{x_r} \quad \text{such that} \quad x_i = 0 \quad \text{for} \quad i \neq r \tag{2.104}$$

If $\{x\} = \{x_1, \ x_2, \ \dots x_n\}^T$ defines the displacements at n DOFs, what force is required at each DOF to realize this?

Due to displacement x_1 at the first DOF, the force required at the ith DOF is $k_{i1}x_1$. Similarly, due to displacement x_2 at the second DOF, the force required at the ith DOF is $k_{i2}x_2$. Therefore due to the displacements defined by the vector $\{x\}$, the total force required at the ith DOF would be,

$$f_i = k_{i1}.x_1 + k_{i2}.x_2 + \dots + k_{in}.x_n \qquad \text{for} \quad i = 1, \ 2, \dots n \tag{2.105}$$

Eq. (2.105) represents n equations and can be written in matrix form as

$$
\begin{Bmatrix} f_1 \\ f_2 \\ \vdots \\ \vdots \\ f_n \end{Bmatrix}
=
\begin{bmatrix}
k_{11} & k_{12} & \dots & \dots & k_{1n} \\
k_{21} & k_{22} & \dots & \dots & k_{2n} \\
\vdots & \vdots & \dots & \dots & \vdots \\
\vdots & \vdots & \dots & \dots & \vdots \\
k_{n1} & k_{n2} & \dots & \dots & k_{nn}
\end{bmatrix}
\begin{Bmatrix} x_1 \\ x_2 \\ \vdots \\ \vdots \\ x_n \end{Bmatrix}
\tag{2.106}
$$

$$\{f\} = [k]\{x\} \tag{2.107}$$

The matrix $[k]$ above is called the matrix of stiffness influence coefficients (also called the stiffness matrix) of the system. It is an $n \times n$ matrix and contains stiffness influence coefficients. Equations (2.104) and (2.106) indicate how stiffness influence coefficients can be determined.

If $x_1 = 1.0$ with the displacements at other DOFs maintained at zero value, that is, those DOFs are constrained, then the force vector $\{f\}$ required to achieve such a displacement configuration is nothing but the first column of the Stiffness matrix.

In general, the rth column of the stiffness matrix is the force vector to obtain a unit displacement at the rth DOF with all the remaining DOFs constrained to have zero displacements.

One important property of linear systems, we prove in Section 2.14, is that the stiffness matrix is symmetric, that is

$$k_{ir} = k_{ri} \tag{2.108}$$

Substituting Eq. (2.101) into Eq. (2.107), we get

$$\{f\} = [k][a]\{f\} \tag{2.109}$$

$$[k][a] = [I] \quad \text{or} \quad [k] = [a]^{-1} \tag{2.110}$$

Thus, we see that the stiffness matrix is the inverse of the flexibility matrix and vice versa. But, note that the flexibility matrix is defined only for systems constrained in space though the stiffness matrix can be defined even for unconstrained systems. The stiffness matrix is singular for the unconstrained systems, and hence its inverse, i.e., the flexibility matrix, does not exist.

2.13   RECIPROCITY THEOREM

In this section, we look at the principle of reciprocity, an important principle for modal analysis. It is also known as Maxwell's reciprocity theorem. We first consider the reciprocity of the flexibility influence coefficients and then that of the stiffness influence coefficients.

2.13.1   RECIPROCITY OF FLEXIBILITY INFLUENCE COEFFICIENTS

Consider a linear structure and determine the strain energy of the structure when static forces f_1 and f_2 are applied gradually at arbitrary points 1 and 2, respectively. We apply these forces in two alternative sequences.

   A. First, apply f_1 at point 1, followed by f_2 at point 2
   B. First, apply f_2 at point 2, followed by f_1 at point 1

Let us determine the strain energy of the structure in both of these cases.

   A. Figure 2.26a shows the force f_1 being applied at point 1 on the structure. Figure 2.26b shows the resulting displacement x_{11} and x_{21} at points 1 and 2, respectively. Here the symbol x_{ir} is used to represent the displacement at point i due to the force at point r.
      The strain energy is calculated as the area under the force-deflection graph at each point. Since the force is gradually applied, the area is half the product of the force and the displacement. Therefore, the strain energy stored by the structure due to the application of force f_1 is

$$U_{A1} = \frac{1}{2} f_1 x_{11} \tag{2.111}$$

Note that the displacement x_{21} at point 2 does not contribute any energy as there is no external force at that point.
      Figure 2.26c shows that force f_2 is applied at point 2 in the presence of force f_1 at point 1. Figure 2.26d shows the resulting displacements. The dotted line shows the displacement curve due to f_1, while the continuous line shows the displacement curve at the end of the application of force f_2. The incremental displacements due to f_2 are x_{12} and x_{22} at points 1 and 2, respectively, as shown in the figure.
      The strain energy stored by the structure due to the application of force f_2 is

$$U_{A2} = \frac{1}{2} f_2 x_{22} + f_1 x_{12} \tag{2.112}$$

Note that the first term is the strain energy due to incremental displacement at point 2 due to the gradual application of force f_2 at that point. The second term is the strain energy due to the incremental displacement at point 1. Since the force at this point is f_1, present in its whole magnitude, the area under the force-deflection graph at this point is $f_1 x_{12}$.
      Thus, the total strain energy of the structure with both forces acting is $U_{A1} + U_{A2}$ and is given by,

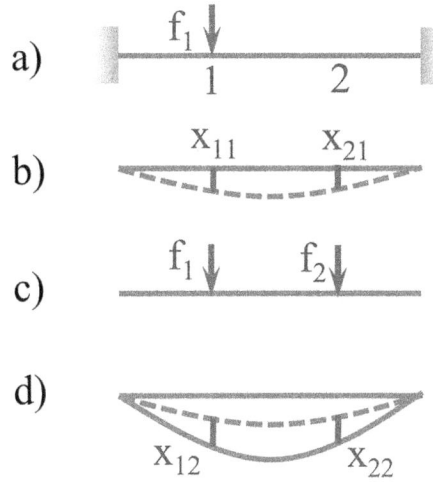

FIGURE 2.26 (a) Force f_1 at point 1, (b) displaced configuration, (c) force f_2 at point 2 in the presence of f_1, (d) the displacement curve due to f_1 (dotted line) and the displacement curve at the end of the application of force f_2 (continuous line).

$$U_A = \frac{1}{2}f_1x_{11} + \frac{1}{2}f_2x_{22} + f_1x_{12} \qquad (2.113)$$

B. Let us now determine the strain energy when f_2 is applied at point 2 first, followed by f_1 at point 1. Figure 2.27a shows the force f_2 being applied at point 2 on the structure. Figure 2.27b shows the resulting displacement x_{12} and x_{22} at points 1 and 2, respectively.

Proceeding as in part A, the strain energy stored due to the application of force f_2 is

$$U_{B1} = \frac{1}{2}f_2x_{22} \qquad (2.114)$$

In this case, the displacement x_{12} at point 1 doesn't contribute any energy as there is no external force at that point.

Figure 2.27c shows that force f_1 is applied at point 1 in the presence of force f_2 at point 2. Figure 2.27d shows the resulting displacements. The dotted line shows the displacement curve due to f_2, while the continuous line shows the displacement curve at the end of the application of force f_1. The incremental displacements due to f_1 are x_{11} and x_{21} at points 1 and 2, respectively, as shown in the figure.

The strain energy stored by the structure due to the application of force f_1 is

$$U_{B2} = \frac{1}{2}f_1x_{11} + f_2x_{21} \qquad (2.115)$$

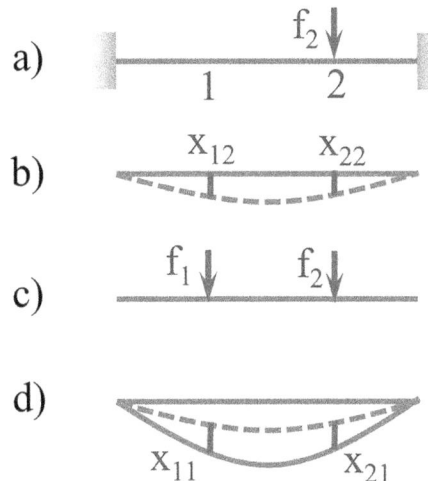

FIGURE 2.27 (a) Force f_2 at point 2, (b) displaced configuration, (c) force f_1 at point 1 in the presence of f_2, (d) the displacement curve due to f_2 (dotted line) and the displacement curve at the end of the application of force f_1 (continuous line).

Note that the first term is the strain energy due to the incremental displacement at point 1 due to the gradual application of force f_1 at that point. The second term is the strain energy due to the incremental displacement at point 2. Since the force at this point is f_2 which is present in its whole magnitude, the area under the force-deflection graph for this point is $f_2 x_{21}$.

Thus, the total strain energy of the structure with both forces acting is $U_{B1} + U_{B2}$ and is given by

$$U_B = \frac{1}{2} f_2 x_{22} + \frac{1}{2} f_1 x_{11} + f_2 x_{21} \tag{2.116}$$

C. In cases A and B, the final loads and displacements are the same, and therefore the strain energies in both cases must be equal. Equating Eqs. (2.113) and (2.116), we get

$$\frac{1}{2} f_1 x_{11} + \frac{1}{2} f_2 x_{22} + f_1 x_{12} = \frac{1}{2} f_2 x_{22} + \frac{1}{2} f_1 x_{11} + f_2 x_{21} \tag{2.117}$$

which gives

$$f_1 x_{12} = f_2 x_{21} \tag{2.118}$$

The two displacements in Eq. (2.118) can be written in terms of the flexibility influence coefficients. Since x_{12} is displacement due to f_2 alone, we can write $x_{12} = a_{12}.f_2$. Similarly, x_{21} is displacement due to f_1 alone, and hence we can write $x_{21} = a_{21}.f_1$. Making these substitutions, we get

$$f_1.(a_{12}.f_2) = f_2.(a_{21}.f_1) \tag{2.119}$$

$$a_{12} = a_{21} \tag{2.120}$$

The above result can be generalized as

$$a_{ir} = a_{ri} \tag{2.121}$$

Eq. (2.121) is the statement of the reciprocity theorem. It states that if a unit force is applied at the rth DOF leading to a certain displacement at the ith DOF, then the same displacement would occur at the rth DOF if a unit force is applied at the ith DOF. It makes the flexibility matrix symmetric. Thus, only the lower or upper triangular flexibility matrix must be determined to know the flexibility matrix fully.

Since the stiffness matrix is the inverse of the flexibility matrix, the stiffness matrix must also be symmetric, as the inverse of a symmetric matrix is also symmetric. However, the following section independently derives the reciprocity property of stiffness influence coefficients.

2.13.2 Reciprocity of Stiffness Influence Coefficients

Consider a linear structure and determine the strain energy of the structure when static forces are applied at arbitrary points 1 and 2 to realize the displacements x_1 and x_2 at those points, respectively. We achieve this displaced configuration of the structure in two alternative sequences.

A. First, obtain displacement x_1 at point 1 with no displacement at point 2. Then obtain displacement x_2 at point 2 with no change in the displacement at point 1.

B. First, obtain displacement x_2 at point 2 with no displacement at point 1. Then obtain displacement x_1 at point 1 with no change in the displacement at point 2.

Let us determine the strain energy of the structure in both of these cases.

A. Figure 2.28a shows the forces f_{11} and f_{21} applied at points 1 and 2, respectively, on the structure to produce displacement x_1 at point 1 with no displacement at point 2. Here the symbol f_{ir} is used to represent the force at point i to produce a given displacement at point r with no displacement at the other points. Thus, force f_{21} provides a constraint at point 2 so that there is no displacement at this point. Figure 2.28b shows the displaced configuration.

The strain energy stored by the structure to achieve the displaced configuration is,

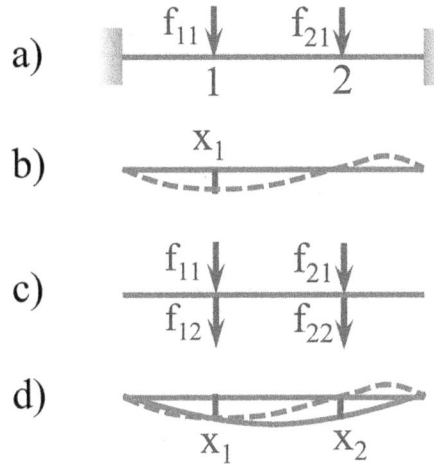

FIGURE 2.28 (a) Forces f_{11} and f_{21} applied at points 1 and 2, respectively, (b) displaced configuration, (c) forces f_{12} and f_{22} applied at points 1 and 2, respectively, (d) displacement x_1 at point 1 and no displacement at 2 (dotted line), and displacements x_1 and x_2 at points 1 and 2 (continuous line).

$$U_{A1} = \frac{1}{2} f_{11} x_1 \qquad (2.122)$$

Note that force f_{21} at point 2 does not contribute any energy as there is no displacement at that point.

Figure 2.28c shows that forces f_{12} and f_{22} are applied at points 1 and 2, respectively, on the structure to produce displacement x_2 at point 2 with no change in the displacement at point 1. Note that the forces f_{11} and f_{21} applied in the first step are also present.

Figure 2.28d shows the displaced configuration. The dotted line shows the displacement curve that was there, while the continuous line shows the final displacement curve showing displacements x_1 and x_2 at points 1 and 2, respectively.

The strain energy stored by the structure in the second step is,

$$U_{A2} = \frac{1}{2} f_{22} x_2 + f_{21} x_2 \qquad (2.123)$$

Note that there is no change in displacement at point 1, and therefore the forces acting there don't contribute any energy. Both the terms in Eq. (2.123) arise due to displacement x_2 at point 2 and forces at that point. The first term is the strain energy due to the force f_{22} applied gradually. The second term is due to the force f_{21} at that point that was applied in step 1 and therefore is present in full magnitude when the displacement x_2 develops.

The total strain energy of the structure, when the desired displacements are realized, is equal to $U_{A1} + U_{A2}$ and is given by,

$$U_A = \frac{1}{2} f_{11} x_1 + \frac{1}{2} f_{22} x_2 + f_{21} x_2 \qquad (2.124)$$

B. In this case, first obtain displacement x_2 at point 2 with no displacement at point 1. Then obtain displacement x_1 at point 1 with no change in displacement at point 2.

Figure 2.29a shows the forces f_{12} and f_{22} applied at points 1 and 2 on the structure to produce displacement x_2 at point 2 with no displacement at point 1. Figure 2.29b shows the displaced configuration.

The strain energy stored by the structure to achieve the displaced configuration is,

$$U_{B1} = \frac{1}{2} f_{22} x_2 \qquad (2.125)$$

Figure 2.29c shows that forces f_{11} and f_{21} are applied at points 1 and 2 on the structure to produce displacement x_1 at point 1 with no change in displacement at point 2. Note that the forces f_{12} and f_{22} applied in the first step are also present. Figure 2.29d shows the final displaced configuration.

The strain energy stored by the structure in the second step is,

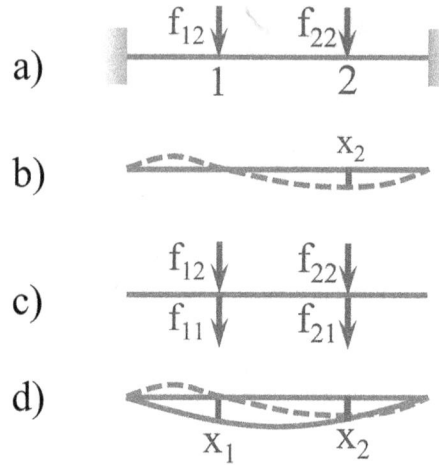

FIGURE 2.29 (a) Forces f_{12} and f_{22} applied at points 1 and 2, respectively, (b) displaced configuration, (c) forces f_{11} and f_{21} applied at points 1 and 2, respectively, and (d) the displacements x_2 at point 2 and no displacement at 1 (dotted line), and the displacements x_1 and x_2 at points 1 and 2 (continuous line).

$$U_{B2} = \frac{1}{2}f_{11}x_1 + f_{12}x_1 \tag{2.126}$$

Thus the total strain energy of the structure, when the desired displacements are realized, is $U_{B1} + U_{B2}$ and is given by,

$$U_B = \frac{1}{2}f_{22}x_2 + \frac{1}{2}f_{11}x_1 + f_{12}x_1 \tag{2.127}$$

C. In cases A and B, the final loads and displacements are the same, and therefore the strain energies in both cases must be equal. Equating Eqs. (2.124) and (2.127), we get

$$\frac{1}{2}f_{11}x_1 + \frac{1}{2}f_{22}x_2 + f_{21}x_2 = \frac{1}{2}f_{22}x_2 + \frac{1}{2}f_{11}x_1 + f_{12}x_1 \tag{2.128}$$

which yields

$$f_{21}x_2 = f_{12}x_1 \tag{2.129}$$

The two displacements in Eq. (2.129) can be written in terms of the stiffness influence coefficients. Since, f_{21} is the force at point 2 to achieve displacement x_1 at point 1, with no displacement at the other points, we can write $f_{21} = k_{21}.x_1$. Similarly, f_{12} is the force at point 1 to achieve displacement x_2 at point 2 with no displacement at the other points, and hence we can write $f_{12} = k_{12}.x_2$. Making these substitutions in Eq. (2.129), we get

$$(k_{21}.x_1).x_2 = (k_{12}.x_2).x_1 \tag{2.130}$$

$$k_{21} = k_{12} \tag{2.131}$$

This can be generalized as

$$k_{ir} = k_{ri} \tag{2.132}$$

Eq. (2.132) is the reciprocity theorem for the stiffness influence coefficients. It states that the force required at the ith DOF to produce a unit displacement at the rth DOF with no displacement at the other DOFs is equal to the force required at the rth DOF to produce a unit displacement at the ith DOF with no displacement at the other DOFs.

Thus, the stiffness matrix is also symmetric, and only the lower or upper triangular stiffness matrix must be determined to know the full stiffness matrix.

Example 2.14: Equations of Motion Using Flexibility Influence Coefficients

Figure 2.30a shows a 3 DOF undamped model for longitudinal vibrations of a propulsion system (Figures 2.11). Assume, $k_1 = 2k$, $k_2 = k$, $k_3 = 6k$ and $m_1 = 2m$, $m_2 = m$, $m_3 = 3m$. Determine the flexibility matrix and derive the equations of motion.

Solution

Let x_1, x_2 and x_3 be the DOFs representing the longitudinal displacements of the three masses, respectively. Using the procedure explained in Section 2.12, the various columns of the flexibility matrix ([a]) can be determined. To determine the first column of [a], apply a unit force at the first DOF with no forces at the other DOFs, and solve for x_1, x_2 and x_3. Figure 2.30b shows the applied and spring forces at the three DOFs. Considering the static equilibrium, we can write

$$k_1(x_1 - x_2) = 1.0$$

$$k_2(x_2 - x_3) = k_1(x_1 - x_2)$$

$$k_3 x_3 = k_2(x_2 - x_3)$$

Substituting the values of stiffnesses as given and solving the above equations, we get the three displacements, which are nothing but the elements of the first column of [a].

$$\{a_{11}, a_{21}, a_{31}\}^T = \left\{\frac{5}{3k}, \frac{7}{6k}, \frac{1}{6k}\right\}^T$$

To determine the second column of [a] apply a unit force at the second DOF and no forces at the other DOFs and solve for x_1, x_2 and x_3. Figure 2.30c shows the applied force and the spring forces. Considering the static equilibrium, we can write

$$x_1 = x_2$$

$$k_2(x_2 - x_3) = 1.0$$

$$k_3 x_3 = k_2(x_2 - x_3)$$

On substituting the values of stiffnesses as given and solving the above equations, we get the second column of [a].

$$\{a_{12}, a_{22}, a_{32}\}^T = \left\{\frac{7}{6k}, \frac{7}{6k}, \frac{1}{6k}\right\}^T$$

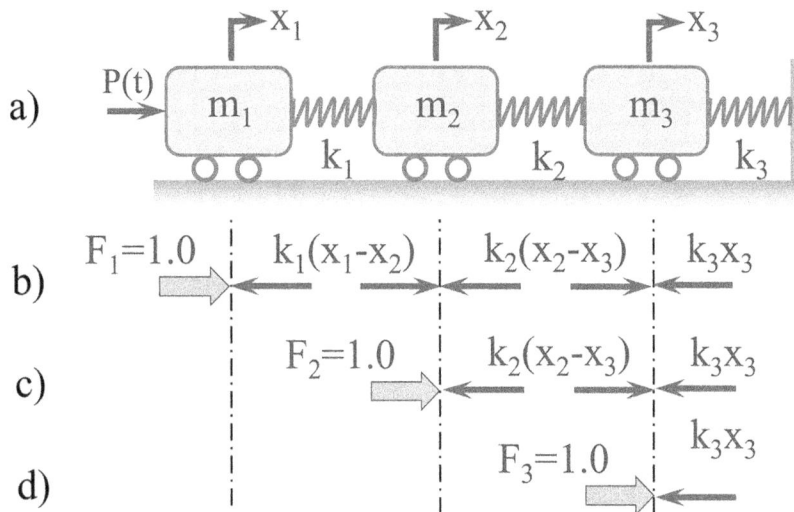

FIGURE 2.30 (a) 3 DOF undamped model for longitudinal vibrations of the propulsion system, (b) unit force at the first DOF along with the spring forces, (c) unit force at the second DOF along with the spring forces, and (d) unit force at the third DOF along with the spring forces.

To determine the third column of $[a]$ apply a unit force at the third DOF and no forces at the other DOFs and solve for x_1, x_2 and x_3. Figure 2.30d shows the applied force and the spring forces. Considering the static equilibrium, we can write

$$x_1 = x_2$$

$$x_2 = x_3$$

$$k_3 x_3 = 1.0$$

Substituting the values of stiffnesses as given and solving the above equations, we get,

$$\{a_{13},\ a_{23},\ a_{33}\}^T = \left\{ \frac{1}{6k},\ \frac{1}{6k},\ \frac{1}{6k} \right\}^T$$

In this way, the flexibility matrix for the model is obtained as

$$[a] = \begin{bmatrix} \dfrac{5}{3k} & \dfrac{7}{6k} & \dfrac{1}{6k} \\[2mm] \dfrac{7}{6k} & \dfrac{7}{6k} & \dfrac{1}{6k} \\[2mm] \dfrac{1}{6k} & \dfrac{1}{6k} & \dfrac{1}{6k} \end{bmatrix}$$

The flexibility matrix relates the displacements and the forces at the three DOFs through the following relationship:

$$\{x\} = [a]\{f\}$$

Applying inertia forces opposite to the direction of the accelerations at the DOFs, the total forces at the three DOFs, which include the external forces and the inertia forces, are

$$\{f\} = \left\{ \ (P(t) - 2m\ddot{x}_1) \quad -m\ddot{x}_2 \quad -3m\ddot{x}_3 \ \right\}^T$$

Substituting $\{f\}$ and $[a]$, we get after simplification, the following equations of motion

$$\begin{bmatrix} \dfrac{10m}{3k} & \dfrac{7m}{6k} & \dfrac{m}{2k} \\[2mm] \dfrac{7m}{3k} & \dfrac{7m}{6k} & \dfrac{m}{2k} \\[2mm] \dfrac{m}{3k} & \dfrac{m}{6k} & \dfrac{m}{2k} \end{bmatrix} \left\{ \begin{array}{c} \ddot{x}_1 \\ \ddot{x}_2 \\ \ddot{x}_3 \end{array} \right\} + \begin{bmatrix} 1 & 0 & 0 \\ 0 & 1 & 0 \\ 0 & 0 & 1 \end{bmatrix} \left\{ \begin{array}{c} x_1 \\ x_2 \\ x_3 \end{array} \right\} = \left\{ \begin{array}{c} \dfrac{5P(t)}{3k} \\[2mm] \dfrac{7P(t)}{6k} \\[2mm] \dfrac{P(t)}{6k} \end{array} \right\}$$

It is observed that the mass matrix is fully populated while the stiffness matrix is diagonal.

Example 2.15: Equations of Motion Using Stiffness Influence Coefficients

A lumped parameter model of a system with three rail bogies each of mass m connected with springs representing the stiffness k of the connections between them is shown in Figure 2.31a. Determine the stiffness matrix using the concept of the stiffness influence coefficients and use that to derive the equations of motion.

Solution

Let x_1, x_2 and x_3 be the DOFs representing the longitudinal displacements of the three bogies, respectively. To determine the first column of $[k]$, the forces at the three DOFs are applied to obtain unit displacement at the first DOF with no displacement at the other DOFs. Figure 2.31b shows the forces and the reaction forces due to constraints necessary to achieve this configuration. These forces are the elements of the first column of the matrix $[k]$. Hence,

$$\{k_{11},\ k_{21},\ k_{31}\}^T = \{k,\ -k,\ 0\}^T$$

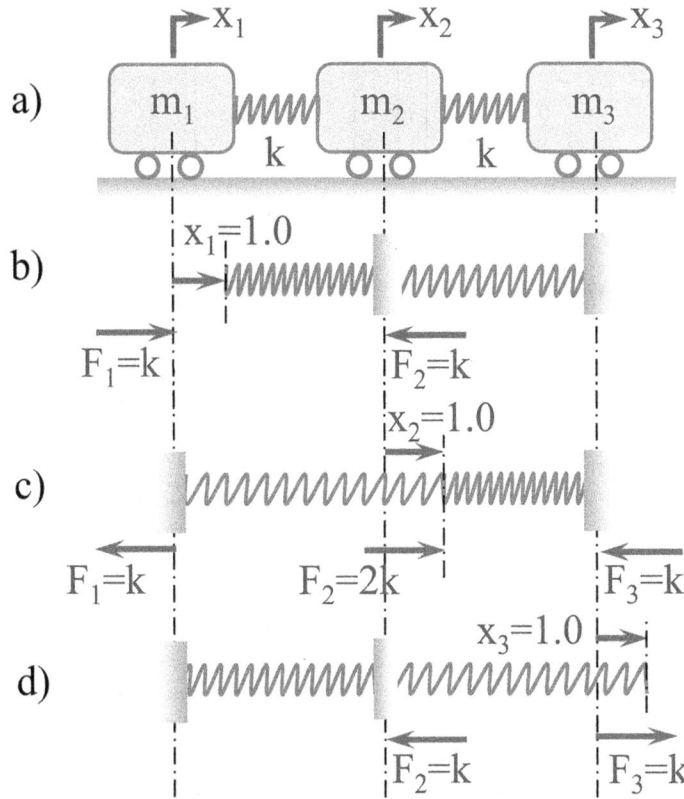

FIGURE 2.31 (a) A lumped parameter model of three rail bogies, (b) forces needed to have a unit displacement at the first DOF with no displacement at the other DOFs, (c) forces needed to have a unit displacement at the second DOF with no displacement at the other DOFs, and (d) forces needed to have a unit displacement at the third DOF with no displacement at the other DOFs.

To determine the second column of $[k]$, the forces at the three DOFs are applied to obtain unit displacement at the second DOF with no displacement at the other DOFs. Figure 2.31c shows the forces and the reaction forces due to constraints necessary to achieve this configuration. These forces are nothing but the elements of the second column of the matrix $[k]$. Hence,

$$\{k_{12}, k_{22}, k_{32}\}^T = \{-k, \quad 2k, \quad -k\}^T$$

To determine the third column of $[k]$, the forces at the three DOFs are applied to obtain unit displacement at the third DOF with no displacement at the other DOFs. Figure 2.31d shows the forces and the reaction forces due to constraints necessary to achieve this configuration. These forces are nothing but the elements of the third column of the matrix $[k]$. Hence,

$$\{k_{13}, k_{23}, k_{33}\}^T = \{0, \quad -k, \quad k\}^T$$

In this way, we get the stiffness matrix for the model as

$$[k] = \begin{bmatrix} k & -k & 0 \\ -k & 2k & -k \\ 0 & -k & k \end{bmatrix}$$

The stiffness matrix relates the displacements and the forces at the three DOFs through the following relationship:

$$[k]\{x\} = \{f\}$$

In this example, since there are no external forces, it is a case of free vibration. Applying inertia forces opposite to the direction of the accelerations at the DOFs, the force vector is

$$\{f\} = \{ \quad -m\ddot{x}_1 \quad -m\ddot{x}_2 \quad -m\ddot{x}_3 \quad \}^T$$

Substituting $\{f\}$ and $[k]$, we get the following equation of motion:

$$
\begin{bmatrix} m & 0 & 0 \\ 0 & m & 0 \\ 0 & 0 & m \end{bmatrix} \begin{Bmatrix} \ddot{x}_1 \\ \ddot{x}_2 \\ \ddot{x}_3 \end{Bmatrix} + \begin{bmatrix} k & -k & 0 \\ -k & 2k & -k \\ 0 & -k & k \end{bmatrix} \begin{Bmatrix} x_1 \\ x_2 \\ x_3 \end{Bmatrix} = \begin{Bmatrix} 0 \\ 0 \\ 0 \end{Bmatrix}
$$

We see that the stiffness matrix is not diagonal while the mass matrix is diagonal.

2.14   LUMPED PARAMETER MODELING OF SYSTEMS WITH BEAM-LIKE MEMBERS

From the many examples presented, we observe that the basic form of the governing equations of motion is the same and involves the mass, stiffness, and damping matrices. It is also observed from the examples that if the stiffness arises from the longitudinal deformation or torsion, the lumped stiffness could be easily determined. If physical springs are parts of the system, then the lumping of the stiffness is straightforward. In other examples, the restoring forces arose due to gravity and could be easily accounted for.

Many structures in practice are made up of beam-like members as structural elements. These members undergo bending due to the loads on the structure. We now consider the lumped parameter modeling of the systems where the system's stiffness originates from the flexure or bending of its members. The determination of the stiffness of such systems is relatively not so straightforward, as the transverse deflection and bending moment are related by a more complex relationship due to the coupling between displacement and slope.

One approach for lumped parameter modeling of such systems has been to use the available formulas for the standard beam types like a cantilever, simply supported, and fixed beam to determine the flexibility and stiffness influence coefficients of such systems. But if the boundary conditions and configuration of the beam members of the problem are non-standard, the standard formulas cannot be used.

Because of this, a general procedure is presented to determine the stiffness influence coefficient, which can also be applied to beams and structural configurations with non-standard beams and non-standard boundary conditions. The procedure is based on the differential equation of the elastic curve. This procedure also yields the stiffness matrix directly and avoids the calculation of the inverse, which is required in the case when the flexibility influence coefficients are computed, yielding the flexibility matrix. The basic procedure of determining the stiffness matrix of systems involving beams and beam-like structural members is demonstrated through the examples below.

Example 2.16: Lumped Parameter Model and Equations of Motion of a Drilling Machine

Figure 2.32a shows a schematic of a drilling machine. Develop an undamped lumped parameter model to predict the displacement response of the drilling head and worktable in the lateral direction due to the transverse excitation of the drilling head due to the driving motor. The drilling head and worktable are located from the bottom at distances L and L/2, respectively. Assume E and I to be the modulus of elasticity of the column material and the second moment of area of the column cross-section, respectively. Obtain the equation of motion using Lagrange's equation.

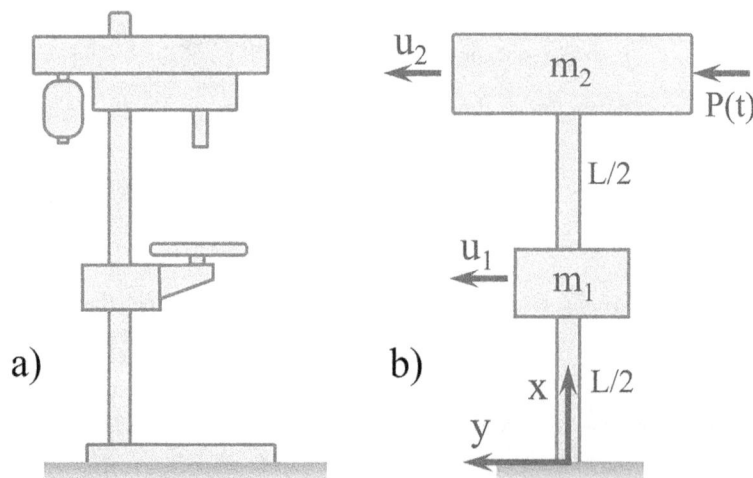

FIGURE 2.32   (a) Schematic of a drilling machine and (b) 2 DOF lumped parameter model.

Solution

Figure 2.32b shows a 2 DOF lumped parameter model of the drilling machine with the lateral displacements of the worktable and drilling head, u_1 and u_2, respectively, as the DOFs. The rotational inertias are neglected.

Let m_w and m_d be the worktable and drilling head masses, respectively and ρ_c be the mass per unit length of the column. The distributed mass of the column is lumped by dividing the mass of each column section equally at its ends. The lumped masses m_1 and m_2 in the model are approximated as

$$m_1 = m_w + \frac{1}{2}\rho_c\frac{L}{2} + \frac{1}{2}\rho_c\frac{L}{2}$$

$$m_2 = m_d + \frac{1}{2}\rho_c\frac{L}{2}$$

The stiffness matrix $[k]$ is of order 2×2. The DOFs u_1 and u_2 are denoted in $[k]$ by indices 1 and 2, respectively.

To determine the first column of $[k]$, we apply the forces and constraints so that we have $u_1 = 1$ and $u_2 = 0$. Therefore, we apply force f along u_1 to cause its unit displacement while DOF u_2 is constrained. The displaced configuration of the beam is shown in Figure 2.33a in which the constraint at $x = L$ prevents displacement u_2 but permits rotation at this point. (The rotation is permitted at the worktable and drilling head locations since the slope of the column is unrestrained.)

The static equilibrium and beam differential equation are used to determine the forces needed to attain the displaced configuration. Figure 2.33b shows the free-body diagram where the constraints on the beam are replaced with the corresponding reactions.

- R_1 and M_1 are the reaction force and moment, respectively, due to the fixed support at $x = 0$.
- R_2 is the reaction force by the constraint at $x = L$. There is no reactionary moment as the beam is free to rotate at this point.

Note that the directions of the reactionary forces and moments are assumed in their positive directions; their correct directions would be revealed by the algebraic signs of their calculated values.

For the static equilibrium of the beam, the summation of forces must be zero. Hence,

$$R_1 + R_2 + f = 0$$

Also, the summation of moments about any point on the beam must be zero. Taking moments of the forces at $x = 0$, we get

$$M_1 + R_2.L + f.\frac{L}{2} = 0$$

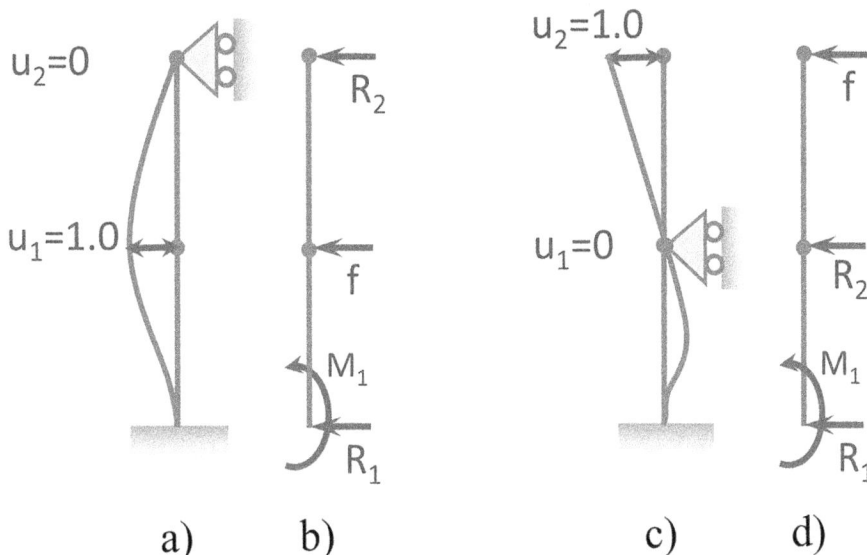

FIGURE 2.33 Displaced configurations.

The above two equations can't be solved to find all the unknown reactions since the beam in Figure 2.33a is statically indeterminate due to the redundant constraint at $x = L$. We consider the transverse deflection of the beam to generate additional equations based on constraints.

The differential equation for the deflection of a beam is given by

$$EI\frac{d^2y}{dx^2} = M_x$$

where y and M_x are the transverse deflection and bending moment at location x on the beam, respectively. Since a concentrated force f is present at $x = \frac{L}{2}$, the equation for M_x depends on whether $x \le \frac{L}{2}$ or $x > \frac{L}{2}$. We use the singularity function to write a single equation for M_x.

The argument of a singularity function is put in angle brackets $\langle\ \rangle$. A singularity function $\langle x-a \rangle^n$ for $n > 0$ is defined as

$$\langle x-a \rangle^n = (x-a)^n \quad \text{for} \quad x > a$$
$$= 0 \quad \text{for} \quad x \le a$$

Thus, a singularity function is to be treated as a normal function for positive values of its argument; else, it is equal to zero.

From Figure 2.33b, we see that to find M_x the moment due to f also needs to be included if $x > \frac{L}{2}$, and hence, it is written as $f\left\langle x - \frac{L}{2} \right\rangle$.

Therefore,

$$EI\frac{d^2y}{dx^2} = R_1.x - M_1 + f\left\langle x - \frac{L}{2} \right\rangle$$

Integrating with x

$$EI\frac{dy}{dx} = \frac{1}{2}R_1.x^2 - M_1.x + f\frac{1}{2}\left\langle x - \frac{L}{2} \right\rangle^2 + c_1$$

where c_1 is a constant of integration. Applying the boundary condition that at $x = 0$, $\frac{dy}{dx} = 0$, we get $c_1 = 0$. Integrating once more with x gives

$$EI.y = \frac{1}{6}R_1.x^3 - \frac{1}{2}M_1.x^2 + f\frac{1}{6}\left\langle x - \frac{L}{2} \right\rangle^3 + c_2$$

where c_2 is a constant of integration. Applying the boundary condition that at $x = 0$, $y = 0$, we get $c_2 = 0$. The boundary condition at $x = L$ is $y = 0$. Imposing this, we get

$$\frac{R_1 L}{6} - \frac{M_1}{2} + \frac{fL}{48} = 0$$

The solution of the above equation and the static equilibrium equations yields the unknown reactions as

$$R_1 = -\frac{11}{16}f \ ; \quad R_2 = -\frac{5}{16}f \ ; \quad M_1 = -\frac{3}{16}fL$$

By substituting the reactionary forces and moments in the equation for y, the deflection u_1 at $x = \frac{L}{2}$ is obtained as

$$u_1 = \frac{7\ fL^3}{768\ EI}$$

Therefore, the force f required to cause unit deflection ($u_1 = 1.0$) is given by

$$f = \frac{768\ EI}{7\ L^3}$$

and is nothing but the stiffness influence coefficient k_{11}.

From f, reaction R_2 is obtained as

$$R_2 = -\frac{240\ EI}{7\ L^3}$$

R_2 is the force required in the direction of u_2 to cause $u_1 = 1.0$. Hence, it is the stiffness influence coefficient k_{21}. In this way, we get the first column of the stiffness matrix as

$$\left\{ \begin{array}{c} k_{11} \\ k_{21} \end{array} \right\} = \left\{ \begin{array}{c} \dfrac{768\ EI}{7\ L^3} \\[2mm] -\dfrac{240\ EI}{7\ L^3} \end{array} \right\}$$

To determine the second column of $[k]$, we apply the forces and constraints so that we have $u_2 = 1$ and $u_1 = 0$. Therefore, we apply force f along u_2 to cause its unit displacement while u_1 is constrained. The resulting displaced configuration of the beam is shown in Figure 2.33c, in which the constraint at $x = \dfrac{L}{2}$ prevents displacement u_1, though the rotation at this point is permitted.

Figure 2.33d shows the free-body diagram of the beam. R_1 and M_1 are the reaction force and moment, respectively, by the fixed support at $x = 0$. R_2 is the reaction force by the constraint at $x = \dfrac{L}{2}$. The beam is free to rotate at this point, and hence there is no reactionary moment.

By following the procedure used to find the first column of the stiffness matrix, the second column of the stiffness matrix is obtained as

$$\left\{ \begin{array}{c} k_{12} \\ k_{22} \end{array} \right\} = \left\{ \begin{array}{c} -\dfrac{240\ EI}{7\ L^3} \\[2mm] \dfrac{96\ EI}{7\ L^3} \end{array} \right\}$$

Thus, the stiffness matrix of the model is

$$[k] = \left[\begin{array}{cc} k_{11} & k_{12} \\ k_{21} & k_{22} \end{array} \right] = \left[\begin{array}{cc} \dfrac{768\ EI}{7\ L^3} & -\dfrac{240\ EI}{7\ L^3} \\[2mm] -\dfrac{240\ EI}{7\ L^3} & \dfrac{96\ EI}{7\ L^3} \end{array} \right]$$

We can write the equations of motion directly, as it is straightforward to write the mass matrix and the force vector in this example. But let us derive the governing equations of motion using Lagrange's equation.

The coordinates u_1 and u_2 are independent and therefore represent a set of generalized coordinates. Thus, $q_1 = u_1$ and $q_2 = u_2$. The potential energy is given by

$$U = \frac{1}{2}\{u\}^T [k]\{u\}$$

Substituting $[k]$, we get

$$U = \frac{1}{2}\left(k_{11}u_1{}^2 + k_{22}u_2{}^2 + 2k_{12}u_1u_2 \right)$$

$$\frac{\partial U}{\partial u_1} = k_{11}u_1 + k_{12}u_2$$

$$\frac{\partial U}{\partial u_2} = k_{22}u_2 + k_{12}u_1$$

The kinetic energy T for the model is

$$T = \frac{1}{2}m_1\dot{u}_1{}^2 + \frac{1}{2}m_2\dot{u}_2{}^2$$

We can work out the terms appearing in Lagrange's equation as follows.

$$\frac{d}{dt}\left(\frac{\partial T}{\partial \dot{u}_1}\right) = \frac{d}{dt}(m_1\dot{u}_1) = m_1\ddot{u}_1$$

$$\frac{d}{dt}\left(\frac{\partial T}{\partial \dot{u}_2}\right) = \frac{d}{dt}(m_2\dot{u}_2) = m_2\ddot{u}_2$$

The generalized forces are given by

$$Q_j^{nc} = \sum_{i=1}^{2} F_i \cdot \frac{\partial u_i}{\partial q_j} \qquad j = 1, 2$$

$$Q_{u_1}^{nc} = F_{u_1}\frac{\partial u_1}{\partial u_1} + F_{u_2}\frac{\partial u_2}{\partial u_1}$$

$$Q_{u_1}^{nc} = F_{u_1} = P(t)$$

In the same way, we get

$$Q_{u_2}^{nc} = F_{u_1}\frac{\partial u_1}{\partial u_2} + F_{u_2}\frac{\partial u_2}{\partial u_2}$$

$$Q_{u_1}^{nc} = F_{u_2} = 0$$

Lagrange's equation for the non-conservative system is

$$\frac{d}{dt}\left(\frac{\partial T}{\partial \dot{q}_j}\right) - \frac{\partial T}{\partial q_j} + \frac{\partial U}{\partial q_j} = Q_j^{nc} \qquad j = 1, 2$$

Substituting the relevant quantities from the previous equations, we get

$$m_1\ddot{u}_1 - 0 + k_{11}u_1 + k_{12}u_2 = P(t)$$

$$m_2\ddot{u}_2 - 0 + k_{22}u_2 + k_{12}u_1 = 0$$

Writing in matrix form and substituting the stiffness coefficients, we obtain the equation of motion as

$$\begin{bmatrix} m_1 & 0 \\ 0 & m_2 \end{bmatrix}\begin{Bmatrix} \ddot{u}_1 \\ \ddot{u}_2 \end{Bmatrix} + \begin{bmatrix} \dfrac{768\,EI}{7\,L^3} & -\dfrac{240\,EI}{7\,L^3} \\ -\dfrac{240\,EI}{7\,L^3} & \dfrac{96\,EI}{7\,L^3} \end{bmatrix}\begin{Bmatrix} u_1 \\ u_2 \end{Bmatrix} = \begin{Bmatrix} P(t) \\ 0 \end{Bmatrix}$$

Example 2.17: Lumped Parameter Model and Equations of Motion of a Rotor System

A shaft carries three rotating elements of masses m_1, m_2 and m_3 as shown in Figure 2.34a. The ends are simply supported. Transverse forces P_1, P_2 and P_3 due to unbalance act on the three masses in the vertical plane. Write equations of motion for the transverse vibration of the shaft using Newton's law.

Solution

Let the transverse displacements, u_1, u_2 and u_3 of the three rotating elements be the degrees of freedom.

The stiffness matrix $[k]$ is of order 3×3. The DOFs u_1, u_2 and u_3 are denoted in $[k]$ by indices 1, 2, and 3, respectively. To determine the first column of $[k]$, we apply the forces and constraints so that we have $u_1 = 1$, $u_2 = 0$ and $u_3 = 0$. Therefore, we apply force f along u_1 to cause its unit displacement while u_2 and u_3 are constrained. Figure 2.34b shows the resulting displaced configuration of the beam. The constraints are applied only on the transverse displacements while the rotations are free.

Figure 2.34c shows the free-body diagram of the beam. R_1 to R_4 are the reaction forces at the supports. For the static equilibrium of the beam, the summation of forces must be zero. Hence,

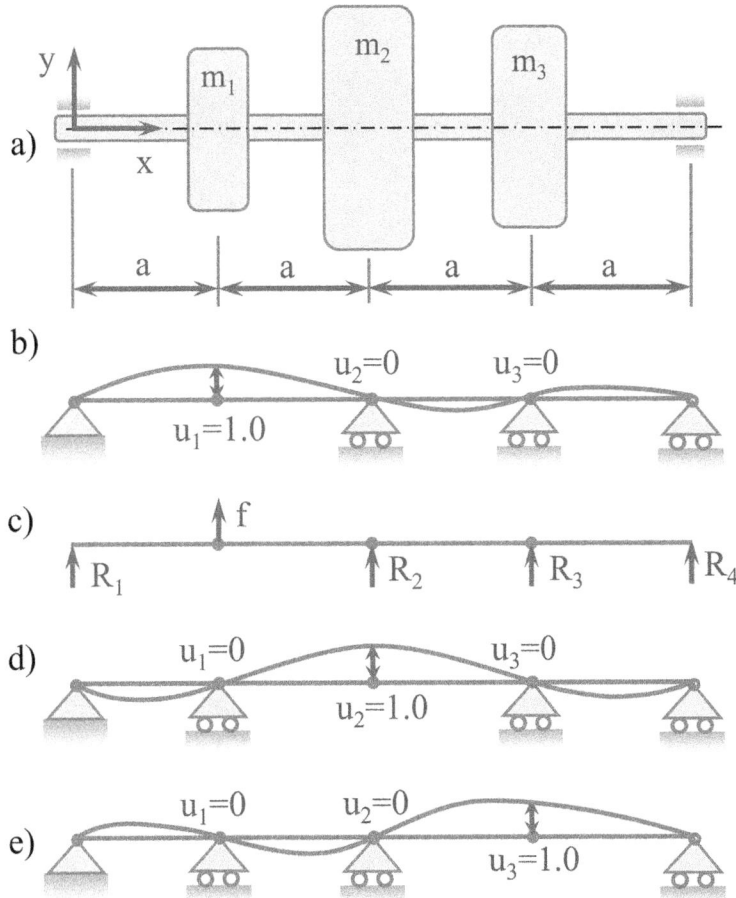

FIGURE 2.34 (a) System with three rotors, (b) constraints to get the displaced configuration $u_1 = 1$, $u_2 = 0$ and $u_3 = 0$, (c) free-body diagram for the displaced configuration in (b), (d) constraints to get the displaced configuration $u_1 = 0$, $u_2 = 1.0$ and $u_3 = 0$, (e) constraints to get the displaced configuration $u_1 = 0$, $u_2 = 0$ and $u_3 = 1.0$.

$$R_1 + R_2 + R_3 + R_4 + f = 0$$

Also, the summation of moments about any point on the beam must be zero. Taking moments of the forces at $x = 0$, we get

$$R_2.2a + R_3.3a + R_4.4a + f.a = 0$$

The above two equations can't be solved to find all the unknown reactions since the beam is statically indeterminate. The differential equation for the deflection of the beam is

$$EI\frac{d^2y}{dx^2} = M_x$$

Making use of the singularity functions, as discussed in the previous example, we can write

$$EI\frac{d^2y}{dx^2} = R_1.x + f.\langle x - a \rangle + R_2.\langle x - 2a \rangle + R_3.\langle x - 3a \rangle$$

Integrating twice with x leads to

$$EI.y = \frac{1}{6}R_1.x^3 + f.\frac{1}{6}\langle x - a \rangle^3 + R_2.\frac{1}{6}\langle x - 2a \rangle^3 + R_3.\frac{1}{6}\langle x - 3a \rangle^3 + c_1 x + c_2$$

Where c_1 and c_2 are constants of integration. Applying the boundary condition that at $x = 0$, $y = 0$, we get $c_2 = 0$. Applying the boundary condition that at $x = 2a$, $y = 0$, we get

60

Analytical and Experimental Modal Analysis

$$c_1 = \frac{1}{2a}\left(-\frac{1}{6}R_1.(2a)^3 - \frac{1}{6}f.(a)^3\right)$$

Applying the boundary condition that at $x = 3a$, $y = 0$, and making the substitution for c_1, we get

$$\frac{5}{2}R_1 + \frac{1}{6}R_2 + \frac{13}{12}f = 0$$

Similarly, applying the boundary condition that at $x = 4a$, $y = 0$, and making the substitution for c_1, we get,

$$8\,R_1 + \frac{4}{3}R_2 + \frac{1}{6}R_3 + \frac{25}{6}f = 0$$

Thus, we have four linear algebraic Eqs., which can be solved for the four unknown reactions in terms of the force f. A more straightforward approach could be to write these equations in matrix form

$$\begin{bmatrix} 1 & 1 & 1 & 1 \\ 0 & 2 & 3 & 4 \\ 5/2 & 1/6 & 0 & 0 \\ 8 & 4/3 & 1/6 & 0 \end{bmatrix} \begin{Bmatrix} R_1 \\ R_2 \\ R_3 \\ R_4 \end{Bmatrix} = \begin{Bmatrix} -1 \\ -1 \\ -13/12 \\ -25/6 \end{Bmatrix} f$$

and solve them in MATLAB® using the command 'inv()', which yields

$$\begin{Bmatrix} R_1 \\ R_2 \\ R_3 \\ R_4 \end{Bmatrix} = \begin{Bmatrix} -0.3696 \\ -0.9565 \\ 0.3913 \\ -0.0652 \end{Bmatrix} f$$

The value of c_1 is calculated as

$$c_1 = 0.163\ fa^2$$

The transverse displacement at $x = a$ can be now calculated from the equation for y and is given by

$$u_1 = 0.1014\ \frac{fa^3}{EI}$$

Therefore, force f to get $u_1 = 1$ is

$$f = 9.85\frac{EI}{a^3}$$

which represents the stiffness influence coefficient k_{11}. Using f, the reactions R_2 and R_3 can be obtained and are nothing but the other two stiffness coefficients in the first column. In this way, we get the first column of the stiffness matrix as

$$\begin{Bmatrix} k_{11} \\ k_{21} \\ k_{31} \end{Bmatrix} = \begin{Bmatrix} 9.85 \\ -9.42 \\ 3.85 \end{Bmatrix}\frac{EI}{a^3}$$

The second and third columns of the stiffness matrix are similarly found by determining the forces needed to achieve the displaced configurations shown in Figures 2.34d and e, respectively.

In this way, we get the stiffness matrix of the model as

$$[k] = \begin{bmatrix} 9.85 & -9.42 & 3.85 \\ -9.42 & 13.71 & -9.42 \\ 3.85 & -9.42 & 9.85 \end{bmatrix} \frac{EI}{a^3}$$

Applying Newton's law to mass m_1

$$m_1 \ddot{u}_1 = \sum_i F_i$$

The forces acting on m_1 are the external and elastic forces due to deflections at the three DOFs. Therefore,

$$m_1 \ddot{u}_1 = P_1(t) - k_{11}u_1 - k_{12}u_2 - k_{13}u_3$$

Similarly, for the other two masses, we can write

$$m_2 \ddot{u}_2 = P_2(t) - k_{21}u_1 - k_{22}u_2 - k_{23}u_3$$

$$m_3 \ddot{u}_3 = P_3(t) - k_{31}u_1 - k_{32}u_2 - k_{33}u_3$$

Writing the above three equations in matrix form and substituting the values of the stiffness coefficients, we get the equation of motion as

$$\begin{bmatrix} m_1 & 0 & 0 \\ 0 & m_2 & 0 \\ 0 & 0 & m_3 \end{bmatrix} \begin{Bmatrix} \ddot{u}_1 \\ \ddot{u}_2 \\ \ddot{u}_3 \end{Bmatrix} + \frac{EI}{a^3} \begin{bmatrix} 9.85 & -9.42 & 3.85 \\ -9.42 & 13.71 & -9.42 \\ 3.85 & -9.42 & 9.85 \end{bmatrix} \begin{Bmatrix} u_1 \\ u_2 \\ u_3 \end{Bmatrix} = \begin{Bmatrix} P_1(t) \\ P_2(t) \\ P_2(t) \end{Bmatrix}$$

Example 2.18: Lumped Parameter Model and Equations of Motion of a Two-Storied Building

A two-storied building is shown in Figure 2.35a. All columns are of length L and flexural rigidity EI. Make a lumped parameter model of the building for studying response due to displacement of the ground in the horizontal direction.

Solution

The floors are assumed to be rigid masses, mass m each, while the columns being slender, are considered to be contributing stiffness to the system in the lateral direction. A two-DOF model with lateral displacements u_2 and u_3 as the DOFs is shown in Figure 2.35b. u_1 represents the displacement excitation of the ground.

The columns are assumed to be identical and symmetrically placed. Since the longitudinal deformation of the columns is negligible, it is assumed that the floors maintain horizontal orientation even while undergoing lateral displacements. It effectively constrains the rotational degrees of freedom of the columns.

An equivalent model by combining the stiffnesses of the columns for a floor is shown in Figure 2.35c. We first obtain the stiffness matrix $[k]$ for the DOFs u_1, u_2 and u_3. Coordinate u_1 is included to model the ground displacement excitation.

To determine the first column of $[k]$, we apply the forces and constraints so that we have $u_1 = 1$, $u_2 = 0$ and $u_3 = 0$. Therefore, we apply force f along u_1 to cause its unit displacement while u_2 and u_3 are constrained. Figure 2.36a shows the

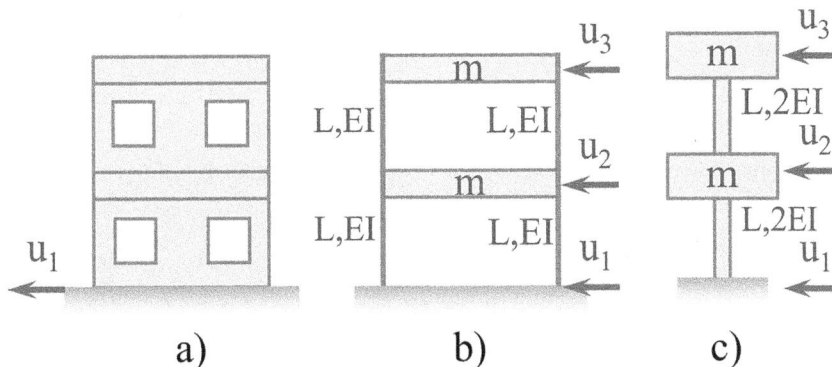

FIGURE 2.35 (a) A two-storied building, (b) two-DOF model, and (c) the equivalent model.

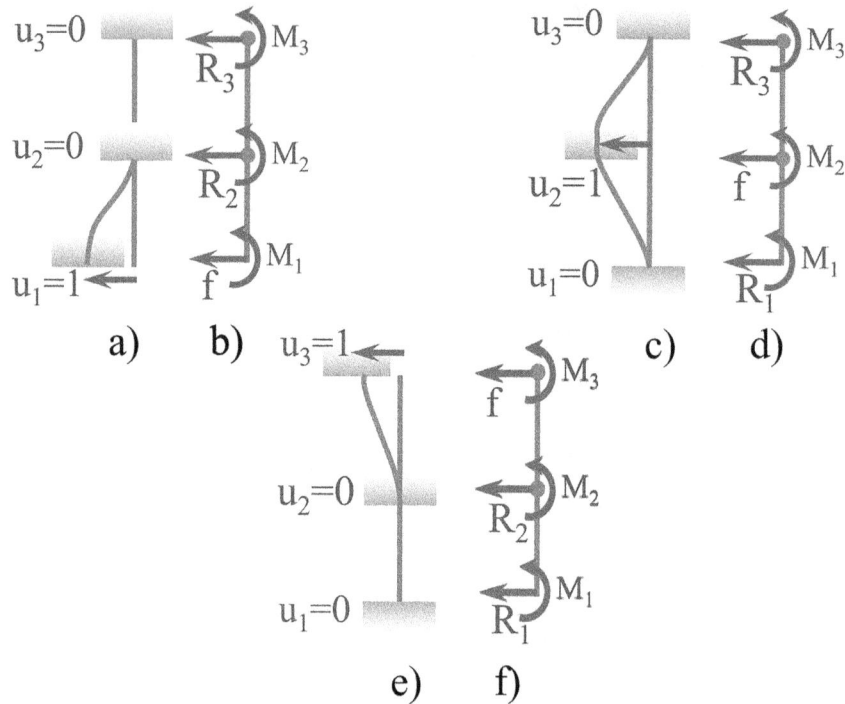

FIGURE 2.36 Displaced configurations and the corresponding free-body diagrams.

displaced configuration of the columns, where transverse displacements u_2 and u_3 are constrained along with the rotations at all DOFs. Figure 2.36b shows the free-body diagram of the columns. R_2 and R_3 are reaction forces, while M_1, M_2 and M_3 are reaction moments. From Figure 2.36a, it is clear that reactions R_3 and M_3 must be zero.

For static equilibrium, we must have

$$R_2 = -f$$

$$M_1 + M_2 + R_2.L = 0$$

From the differential equation of the deflection of the beam,

$$2EI\frac{d^2y}{dx^2} = M_x = f.x - M_1$$

Integrating with x

$$2EI\frac{dy}{dx} = \frac{1}{2}f.x^2 - M_1.x + c_1$$

where c_1 is a constant of integration. Applying the boundary condition that at $x = 0$, $\frac{dy}{dx} = 0$, we get $c_1 = 0$. Integrating once more with x

$$2EI.y = \frac{1}{6}f.x^3 - \frac{1}{2}M_1.x^2 + c_2$$

where c_2 is a constant of integration. Applying the boundary condition that at $x = L$, $y = 0$, we get

$$c_2 = -\frac{1}{6}f.L^3 + \frac{1}{2}M_1.L^2$$

The boundary condition at $x = L$ is $\frac{dy}{dx} = 0$. Imposing this, we get

$$M_1 = \frac{fL}{2}$$

With the reactionary forces and moments known, we obtain the deflection u_1 at $x = 0$ as

$$u_1 = \frac{fL^3}{24\,EI}$$

Therefore, force f to get unit deflection ($u_1 = 1.0$) is

$$f = \frac{24\,EI}{L^3}$$

which is the stiffness influence coefficient k_{11}. Reaction R_2 is given by

$$R_2 = -\frac{24\,EI}{L^3}$$

R_2 is the force required in the direction of u_2 to obtain $u_1 = 1$ and hence it is nothing but the stiffness influence coefficient k_{21}. R_3 is the force required in the direction of u_3 to obtain $u_1 = 1$ and hence it is nothing but the stiffness influence coefficient k_{31}. Since $R_3 = 0$, $k_{31} = 0$.

In this way, we get the first column of the stiffness matrix as

$$\begin{Bmatrix} k_{11} \\ k_{21} \\ k_{31} \end{Bmatrix} = \begin{Bmatrix} \dfrac{24\,EI}{L^3} \\ -\dfrac{24\,EI}{L^3} \\ 0 \end{Bmatrix}$$

To determine the second column of $[k]$, we apply a force f along DOF u_2 to cause its unit displacement while the DOFs u_1 and u_3 are constrained. Figure 2.36c shows the displaced configuration of the columns. Figure 2.36d shows the free-body diagram. The second column of the stiffness matrix is obtained by following the procedure used to find the first column of the stiffness matrix.

To determine the third column of $[k]$, we apply a force f along DOF u_3 to cause its unit displacement while the DOFs u_1 and u_3 are constrained. Figure 2.36e shows the displaced configuration of the columns. Figure 2.36f shows the free-body diagram. The third column of the stiffness matrix is obtained by following the procedure used to find the first column of the stiffness matrix.

The stiffness matrix of the model is obtained as

$$[k] = \frac{EI}{L^3} \begin{bmatrix} 24 & -24 & 0 \\ -24 & 48 & -24 \\ 0 & -24 & 24 \end{bmatrix}$$

Applying Newton's second law to mass m corresponding to DOF u_2, we get

$$m\ddot{u}_2 = -k_{21}u_1 - k_{22}u_2 - k_{23}u_3$$

Similarly, applying Newton's second law to mass m corresponding to DOF u_3, we get

$$m\ddot{u}_3 = -k_{31}u_1 - k_{32}u_2 - k_{33}u_3$$

Displacement u_1 is the input. Writing the above two equations in matrix form and substituting the stiffness influence coefficients, we obtain the equation of motion as

$$\begin{bmatrix} m & 0 \\ 0 & m \end{bmatrix} \begin{Bmatrix} \ddot{u}_2 \\ \ddot{u}_3 \end{Bmatrix} + \frac{EI}{L^3} \begin{bmatrix} 48 & -24 \\ -24 & 24 \end{bmatrix} \begin{Bmatrix} u_2 \\ u_3 \end{Bmatrix} = \frac{EI}{L^3} \begin{Bmatrix} -24 \\ 0 \end{Bmatrix} u_1$$

Example 2.19: Lumped Parameter Model and Equations of Motion of a Machine Frame

Figure 2.37 shows a machine frame with a motor at the end of its horizontal arm. An excitation force P(t) in the vertical direction is generated at the location of the motor. The vertical column and horizontal arm are of length L, area of cross-section A, and flexural rigidity EI each. Develop a lumped parameter model of the system, choosing the degrees of freedom. Also, obtain the governing equations of motion.

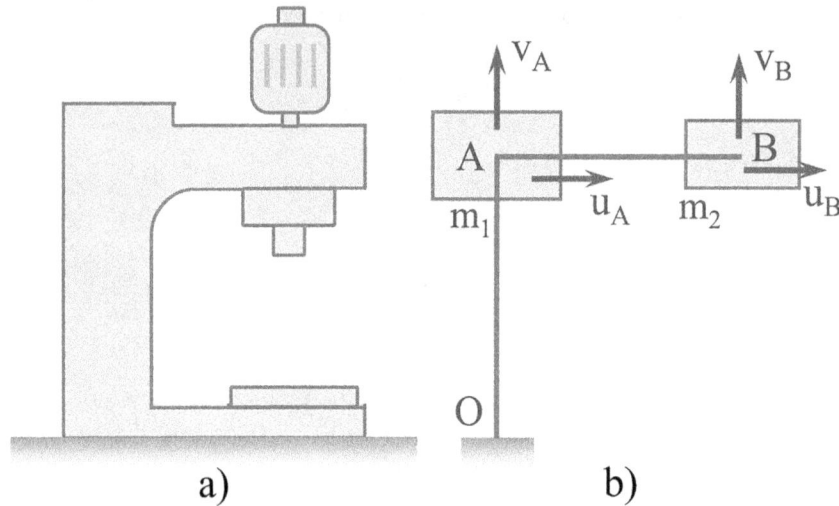

FIGURE 2.37   (a) Machine frame and (b) lumped parameter model of the system.

Solution

The machine structure is in the form of an inverted-L shape frame. A lumped parameter model of the system with two lumped masses is shown in Figure 2.37b.

Mass m_1 is taken equal to the sum of half of the masses of arms AB and OA. Mass m_2 is taken equal to the sum of the masses of the motor and other fixtures at location B and half of the mass of arm AB. Thus, the lumped masses are approximated as

$$m_1 = \frac{m_{AB}}{2} + \frac{m_{OA}}{2}$$

$$m_2 = m_{motor} + m_{fixtures} + \frac{m_{AB}}{2}$$

The displacements u_A, v_A and u_B, v_B, are the coordinates of interest. The rotational inertia is assumed negligible, and since no moment excitations are there, the rotational DOFs are not included.

The axial stiffness of the members OA and AB is much more than their lateral stiffness. Therefore, we assume that the members are rigid in the axial direction and undergo no axial deformation. Hence, we have

$$v_A = 0$$

$$u_B = u_A$$

Due to the above two kinematic constraints, the four coordinates chosen are not independent. Two coordinates can be eliminated, and the remaining two can be chosen as generalized coordinates. Two choices are possible, u_A and v_B or u_B and v_B. We choose u_A and v_B.

The stiffness matrix $[k]$ is of order 2×2. The DOFs u_A and v_B are denoted in $[k]$ by indices 1 and 2, respectively.

To determine the first column of $[k]$, we apply a force f along DOF u_A to cause its unit displacement while the DOF v_B is constrained. The resulting displaced configuration of the frame is shown in Figure 2.38a in which the constraint at point B prevents displacement v_B, but the rotation and displacement parallel to the support are permitted since these DOFs are unconstrained in the machine.

Figure 2.38b shows the free-body diagrams of the beam members OA and AB separately. O_x, O_y and M_o are the reaction forces and moment on OA applied by the clamped end. B_y is the reaction force on AB applied by the simply supported end. There is no reactionary moment since the beam is free to rotate at B. The members are shown separated at the common point A and A_x, A_y and M_A are the interconnection forces and moments.

The coordinate systems for the members OA and AB are shown in Figure 2.38b. y and x are geometric coordinates of points on OA and AB, respectively, while u and v are lateral displacements on them, respectively.

We consider member OA first.

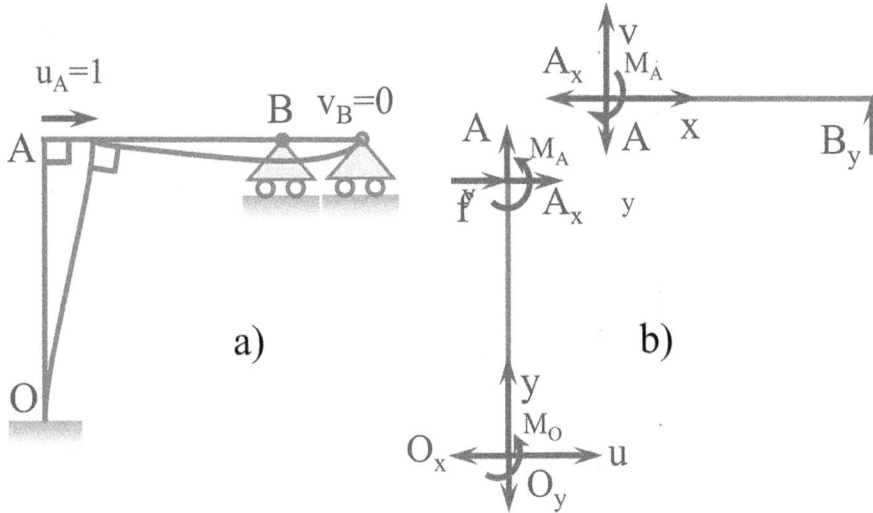

FIGURE 2.38 (a) Displaced configuration and (b) free body diagram.

For static equilibrium, the summation of forces along y and u directions must be zero. The summation of moments at any point on the member should also be zero. The force f at A is assumed to be applied on OA. Hence,

$$A_y = O_y$$

$$A_x - O_x + f = 0$$

$$M_o + M_A - A_x.L - f.L = 0$$

If M_y is the moment at location y on OA, then

$$EI\frac{d^2u}{dy^2} = M_y = M_o - O_x.y$$

Integrating with y,

$$EI\frac{du}{dy} = M_o.y - \frac{1}{2}O_x.y^2 + c_1$$

The boundary condition at $y = 0$ is $\frac{du}{dy} = 0$. It gives $c_1 = 0$. Integrating once more with x,

$$EIu = \frac{1}{2}M_o.y^2 - \frac{1}{6}O_x.y^3 + c_2$$

The boundary condition at $y = 0$ is $u = 0$. It gives $c_2 = 0$.

Now, consider member AB. For static equilibrium, the summation of forces along x and v directions must be zero. The summation of moments at any point on the member should also be zero. Thus,

$$A_x = 0 \; ; \quad A_y = B_y \; ; \quad M_A - B_y.L = 0$$

If M_x is the moment at location x on AB, then

$$EI\frac{d^2v}{dx^2} = M_x = M_A - A_y.x$$

Integrating with x

$$EI\frac{dv}{dx} = M_A.x - \frac{1}{2}A_y.x^2 + d_1$$

The boundary condition at $x = 0$ is $\dfrac{dv}{dx} = \theta_A{}^{AB}$. $\theta_A{}^{AB}$ is the slope at point A when considered lying on AB. It gives,

$$d_1 = EI\,\theta_A{}^{AB}$$

Integrating once more with x,

$$EIv = \frac{1}{2}M_A.x^2 - \frac{1}{6}A_y.x^3 + d_1 x + d_2$$

The boundary condition at $x = 0$ is $v = 0$. It gives $d_2 = 0$.
   Another boundary condition is that at $x = L$, $v = 0$. It leads to

$$\frac{1}{2}M_A.L^2 - \frac{1}{6}B_y.L^3 = -EI\,\theta_A{}^{AB}$$

Solving the two equations involving M_A and B_y, we get

$$B_y = -\frac{3EI}{L^2}\,\theta_A{}^{AB} \quad ; \quad M_A = -\frac{3EI}{L}\,\theta_A{}^{AB}$$

The slope at point A when considered lying on OA (denoted by $\theta_A{}^{OA}$) can be found from the equation for the slope of OA by substituting $y = L$. Solving for M_o and O_x from the equations for OA and substituting into the equation for the slope, we get

$$EI\,\theta_A{}^{OA} = 3EI\,\theta_A{}^{AB} + \frac{1}{2}f.L^2$$

Since the members are assumed rigid axially, the angle between OA and AB is preserved after the deformation. We see from Figures 2.38a and b that the following kinematic condition should be satisfied:

$$\theta_A{}^{OA} = -\,\theta_A{}^{AB}$$

After solving the above equations for $\theta_A{}^{AB}$, B_y and M_A can be found. As all the reactions and moments are known, the displacement u_A at A is obtained as

$$u_A = \frac{7\,fL^3}{48\,EI}$$

Therefore, force f to get $u_A = 1.0$ is

$$f = \frac{48\,EI}{7\,L^3}$$

which is the stiffness influence coefficient k_{11}. From f and $\theta_A{}^{AB}$, the reaction B_y is obtained as

$$B_y = \frac{18\,EI}{7\,L^3}$$

B_y is the force required in the direction of v_B to get $u_A = 1$ and hence is nothing but the stiffness influence coefficient k_{21}.
   In this way, we get the first column of the stiffness matrix as,

$$\left\{ \begin{array}{c} k_{11} \\ k_{21} \end{array} \right\} = \left\{ \begin{array}{c} \dfrac{48\,EI}{7\,L^3} \\ \dfrac{18\,EI}{7\,L^3} \end{array} \right\}$$

To determine the second column of $[k]$, we apply a force f along DOF v_B to cause its unit displacement while the DOF u_A is constrained. The resulting displaced configuration is shown in Figure 2.39a in which the constraint at point A prevents displacement u_A but permits the rotation and displacement parallel to the support.

Figure 2.39b shows the free-body diagrams of the beam members OA and AB separately. O_x, O_y and M_o and A_x, A_y and M_A are the reaction forces and moment as defined earlier. The reaction force R due to the constraint at A is assumed to be acting on OA.

Considering OA first, the equations of static equilibrium give

$$A_y = O_y \quad ; \quad A_x - O_x + R = 0 \quad ; \quad M_o + M_A - A_x.L - R.L = 0$$

We also have

$$EI\frac{d^2u}{dy^2} = M_o - O_x.y$$

Integrating with y,

$$EI\frac{du}{dy} = M_o.y - \frac{1}{2}O_x.y^2 + c_1$$

Applying the boundary condition $\frac{du}{dy} = 0$ at $y = 0$ gives $c_1 = 0$. Integrating once more with x,

$$EIu = \frac{1}{2}M_o.y^2 - \frac{1}{6}O_x.y^3 + c_2$$

Applying the boundary condition $u = 0$ at $y = 0$ gives $c_2 = 0$. Applying the boundary condition $u = 0$ at $y = L$ gives

$$M_o = \frac{1}{3}O_x.L$$

Now, consider AB. The equations of static equilibrium give

$$A_x = 0 \quad ; \quad A_y = f \quad ; \quad M_A - f.L = 0$$

We also have

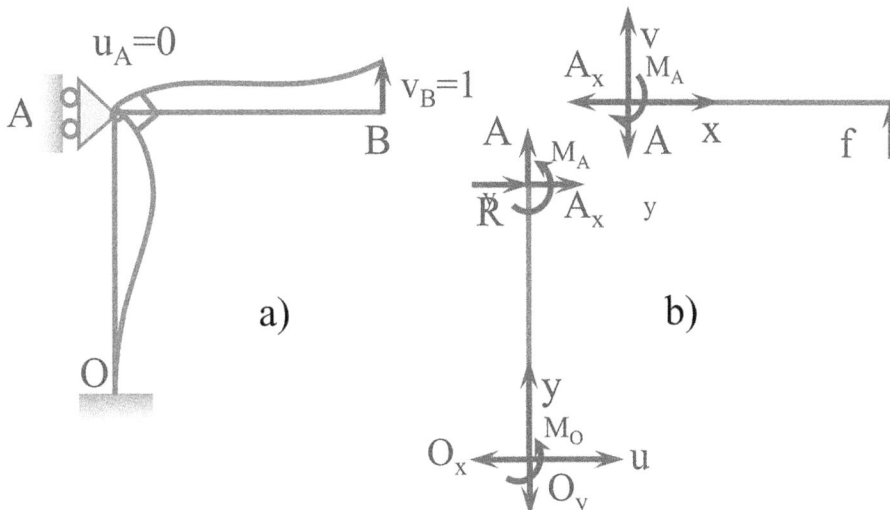

FIGURE 2.39 (a) Displaced configuration and (b) free body diagram.

$$EI\frac{d^2v}{dx^2} = M_A - A_y.x$$

Integrating with x,

$$EI\frac{dv}{dx} = M_A.x - \frac{1}{2}A_y.x^2 + d_1$$

Let $x = 0$ the slope $\frac{dv}{dx} = \theta_A{}^{AB}$. It gives

$$d_1 = EI\,\theta_A{}^{AB}$$

Integrating once more with x,

$$EIv = \frac{1}{2}M_A.x^2 - \frac{1}{6}A_y.x^3 + d_1x + d_2$$

The boundary condition at $x = 0$ is $v = 0$. It gives $d_2 = 0$.
　　Solving the above equations, we get

$$O_x = \frac{3}{2}f$$

The slope at point A when considered lying on OA (denoted by $\theta_A{}^{OA}$) is found from the slope equation by substituting $y = L$. Substituting M_o and O_x, we get

$$EI\,\theta_A{}^{OA} = -\frac{1}{4}f.L^2$$

Using the above equation and the kinematic condition $\theta_A{}^{OA} = -\,\theta_A{}^{AB}$, we get

$$d_1 = \frac{1}{4}f.L^2$$

As all the reactions and moments are now known, the displacement v_B is obtained as

$$v_B = \frac{7\,fL^3}{12\,EI}$$

Therefore, the force f to cause unit deflection ($v_B = 1.0$) is

$$f = \frac{12\,EI}{7\,L^3}$$

which is the stiffness influence coefficient k_{22}. The reaction R is obtained as

$$R = \frac{18\,EI}{7\,L^3}$$

R is the force required in the direction of u_A to get $v_B = 1$ and hence is nothing but the stiffness influence coefficient k_{12}. In this way, we get the second column of the stiffness matrix as

$$\left\{ \begin{array}{c} k_{12} \\ k_{22} \end{array} \right\} = \left\{ \begin{array}{c} \dfrac{18\,EI}{7\,L^3} \\ \dfrac{12\,EI}{7\,L^3} \end{array} \right\}$$

The full stiffness matrix therefore is

$$[k] = \begin{bmatrix} \dfrac{48\,EI}{7\,L^3} & \dfrac{18\,EI}{7\,L^3} \\[2ex] \dfrac{18\,EI}{7\,L^3} & \dfrac{12\,EI}{7\,L^3} \end{bmatrix}$$

Applying Newton's law to mass m_1,

$$m_1 \ddot{u}_A = \sum_i F_i$$

$$m_1 \ddot{u}_A = -k_{11}u_A - k_{12}v_B$$

Similarly for mass m_2, which also has an external force, we can write

$$m_2 \ddot{v}_B = P(t) - k_{21}u_A - k_{22}v_B$$

Writing the above two equations in matrix form and substituting the values of the stiffness coefficients, we obtain the equation of motion as

$$\begin{bmatrix} m_1 & 0 \\ 0 & m_2 \end{bmatrix} \begin{Bmatrix} \ddot{u}_A \\ \ddot{v}_B \end{Bmatrix} + \begin{bmatrix} \dfrac{48\,EI}{7\,L^3} & \dfrac{18\,EI}{7\,L^3} \\[2ex] \dfrac{18\,EI}{7\,L^3} & \dfrac{12\,EI}{7\,L^3} \end{bmatrix} \begin{Bmatrix} u_A \\ v_B \end{Bmatrix} = \begin{Bmatrix} P(t) \\ 0 \end{Bmatrix}$$

REVIEW QUESTIONS

1. What is a lumped parameter model? What is the meaning of lumping of physical properties?
2. Justify with examples that a given system is unique, but a model of the system is not unique.
3. List the different dynamics principles that can be used to derive equations of motion of a vibratory system.
4. Can the energy conservation principle be used to derive governing equations of an MDOF undamped system?
5. Define the flexibility and stiffness influence coefficients for an MDOF system.
6. Show that the flexibility matrix for an MDOF system is symmetric.
7. Show that the stiffness matrix for an MDOF system is symmetric.

PROBLEMS

Problem 2.1: Figure 2.40 shows a spring-mass system in which the mass can slide on a frictionless and massless rod oscillating in the vertical plane. Derive the governing equation of motion using Lagrange's equation.

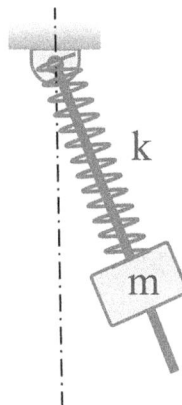

FIGURE 2.40 Spring-mass system on a frictionless and massless oscillating rod.

Problem 2.2: Figure 2.41 shows a simple pendulum whose point of suspension is moved horizontally by a distance x such that $x = x_0 \sin \omega t$. Derive the governing equation of motion using Lagrange's equation.

Problem 2.3: Derive the governing equation of motion for the 4 DOF lumped parameter model of a car shown in Figure 2.4 to predict pitching and bouncing motions if a vertical force P(t) and a moment M(t) acts on the car at its center of mass.

Problem 2.4: Figure 2.42 shows a single-storied building. Its foundation is subjected to a harmonic displacement in the horizontal direction. The foundation stiffness is represented by a linear and a torsional spring of stiffnesses k and kt, respectively. Make a lumped parameter model and derive the equation of motion.

Problem 2.5: The frame of a lathe machine is shown in Figure 2.43. Treat the machine bed as rigid with mass M and moment of inertia I about an axis through its center of mass and normal to the plane of the paper. The distance between the legs is L, while the bed's center of mass is at a distance L/3 from the left end.

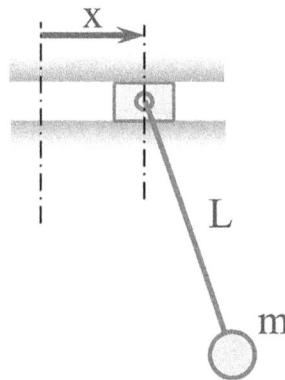

FIGURE 2.41 Pendulum with oscillating point of suspension.

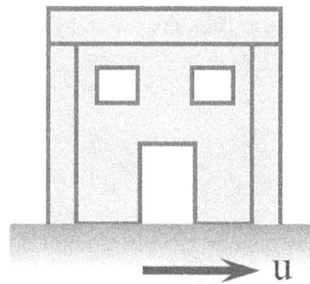

FIGURE 2.42 Single-storied building subjected to ground excitation.

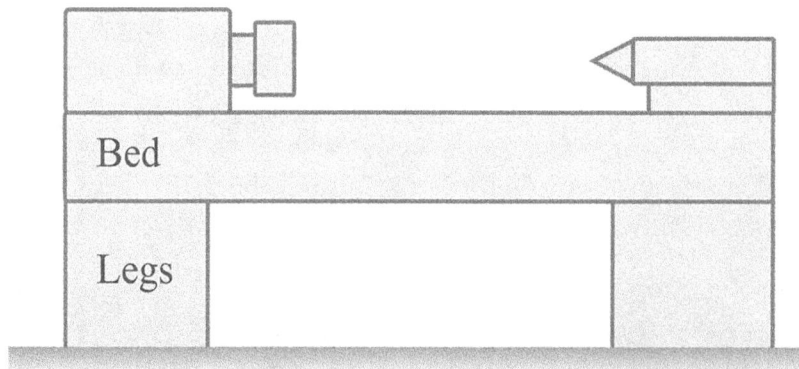

FIGURE 2.43 Lathe machine frame.

Considering the legs to be flexible in the vertical direction with stiffness k each, derive the governing equation of motion.

Problem 2.6: In problem 2.5, along with the flexibility of the legs, treat the bed as flexible with flexural rigidity EI with its mass lumped at two legs such that the position of the center of mass is undisturbed. Neglect rotational inertia. Find the stiffness matrix of the model considering the vertical displacement of the lumped masses as DOFs. Also, obtain the equations of motion.

Problem 2.7: A simplified sketch of a Gantry crane used for material handling is shown in Figure 2.44a. It consists of a rectangular frame with a trolley that can move over the horizontal girder. The frame's cross-section area and flexural rigidity are A and EI, respectively. The legs and horizontal girder lengths are L and 2L, respectively. The mass of the trolley is M_t, while the density of the beam material is P. Figure 2.44b shows a lumped parameter model of the system, with the lumped masses at the two ends of the horizontal girder and the two vertical members clamped at the bottom. All the members are considered rigid axially. If the lateral motion of the two lumped masses, given by u_1 and u_2, due to an excitation force P(t) in the horizontal direction due to the trolley being of interest, find the stiffness matrix of the system. Also, derive governing equations of motion using Newton's law.

Problem 2.8: Figure 2.45a shows a cantilever-type bracket supporting an electric motor. A lumped parameter model of the system is made as shown in Figure 2.45b, where the mass of the motor and beam are lumped as mass m and rotational inertia J at the end. A traverse force due to unbalance in the motor acts on the system. Derive the equation of motion by taking the transverse displacement and slope at the end as DOFs.

Problem 2.9: Find the lumped parameter model and the equation of motion in Problem 2.8 if the rotational inertia is negligible. Comment on the result.

Problem 2.10: Figure 2.46 shows a beam of length L supporting a pump-motor assembly of mass m at a distance of L/3 from its clamped end. Find the stiffness considering the lateral displacement at the pump-motor location as the DOF.

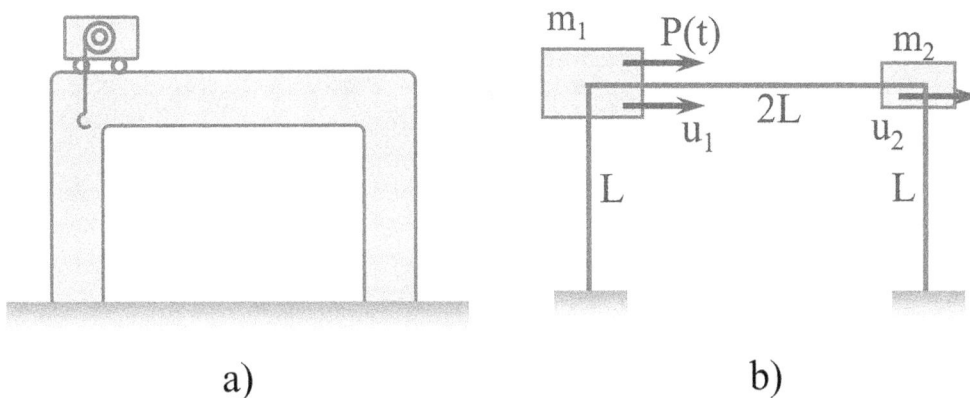

a)   b)

FIGURE 2.44 (a) Gantry crane and (b) lumped parameter model.

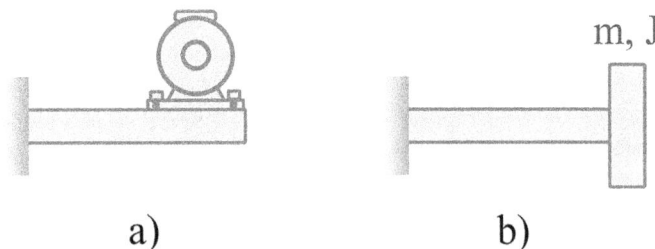

a)   b)

FIGURE 2.45 Cantilever-type bracket with a motor.

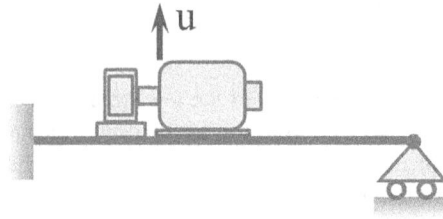

FIGURE 2.46   Beam supporting pump-motor assembly.

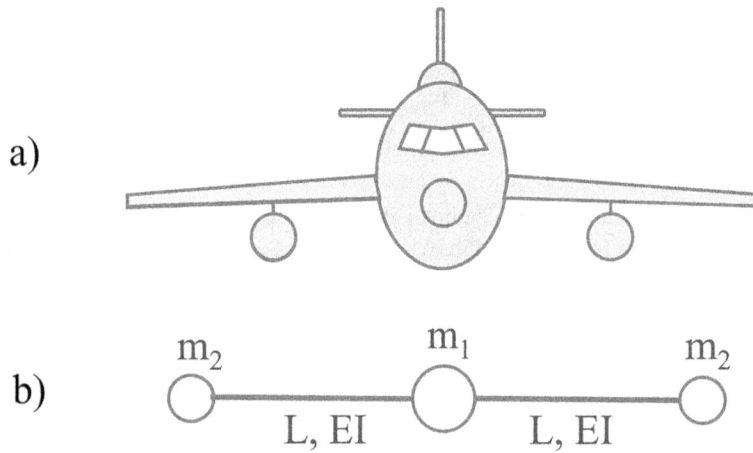

FIGURE 2.47   (a) Airplane and (b) lumped parameter model.

Problem 2.11: Figure 2.47a shows a sketch of an airplane. The mass of the wings per unit length is ρ, while the fuselage mass is m_0. Make a lumped parameter model of the system with three lumped masses, as shown in Figure 2.47b. The flexural rigidity and length of the wings are EI and L, respectively. Find the stiffness matrix and equations of motion of the system.

3 Finite Element Modeling of Vibrating Systems

3.1 INTRODUCTION

In Chapter 2, we studied the analytical modeling of vibrating systems using the lumped parameter approach. The lumped parameter models are generally effective for inherently lumped systems. In other cases, they can be used to obtain initial estimates of the dynamic characteristics before a more refined model is used. In many practical systems, the mass, stiffness, and damping are distributed continuously, and the lumped parameter models may not give accurate results.

The finite element method (FEM) offers a systematic approach to obtain discrete mathematical models of continuous physical systems. FEM is widely used to model systems for analytical modal analysis. The objective of this chapter is to present the finite element formulation of problems in dynamics and develop some of the standard finite elements used to model continuous dynamic systems.

We first present Hamilton's principle, a fundamental principle for deriving governing equations in dynamics. The use of Hamilton's principle to obtain the equations of motion and natural boundary conditions for a continuous system is demonstrated for the longitudinal vibration of a rod. It is shown that for a discrete system, the application of Hamilton's principle to obtain equations of motion can be accomplished by finding the energy functions of the system. We then introduce the FEM by discretizing the continuous domain into finite elements, followed by derivation of standard finite elements such as like the rod, beam, and frame elements. In the end, we discuss the assembly of the elemental matrices and the incorporation of the boundary conditions in the system matrices.

3.2 IMPORTANT ASPECTS OF FEM

This section briefly highlights some crucial aspects of FEM.

3.2.1 NEED FOR NUMERICAL SOLUTIONS

A physical problem in the continuous domain is described by differential equations with relevant boundary conditions. The solution of the differential equations gives the exact solution of the problem within the boundaries of the assumptions and approximations made to derive the differential equations. The exact or analytical solutions of the differential equations can be obtained for problems involving simple geometries and ideal boundary conditions (BCs) and material properties. However, a practical problem often has complex geometry and BCs, and therefore obtaining the exact solution to the problem is either difficult or not possible. This has led to the development of numerical techniques for solving differential equations.

3.2.2 FEM

The FEM is a numerical technique to solve differential equations approximately. The basic idea in FEM is to discretize the physical domain of the problem into small regions, called finite elements. A finite element is a small volume with simple geometry and a finite number of nodes. The adjoining elements are connected at nodes. Thus, the physical domain is represented by an assemblage of finite-size elements, whose equilibrium is sought under the action of external forces.

3.2.3 PIECEWISE APPROXIMATION

How the FEM determines the unknown field (like the displacement field in a solid mechanics problem) over the physical domain? After discretization, the piecewise approximation of the unknown field is another fundamental aspect of FEM. It is called piecewise because it is not a single approximation over the whole domain but an approximation over each finite element separately.

DOI: 10.1201/9780429454783-3

3.2.4   IMPORTANCE OF PIECEWISE APPROXIMATION

How does the piecewise approximation help to simplify the problem? The unknown field variable at an interior point of a finite element is approximated by interpolating the values of the field variables at the finite element nodes. The interpolating functions are in the form of polynomials called **shape functions**. Thus, the equilibrium equations for a finite element are derived using the polynomial approximation over the finite element. When the equilibrium equations for all the finite elements are assembled, the unknowns of the problem are the field variable at the nodes of the finite element mesh. Thus, the original continuous problem with infinite DOFs is effectively converted into a problem with finite DOFs. It has the advantage that the differential equations for the original problem are replaced by a set of linear algebraic equations for the corresponding discrete problem, which can be solved easily.

3.2.5   APPROXIMATE NATURE OF THE FE SOLUTION

What about the correctness of the FE solution? FEM does not give an exact but approximate solution to the problem. It imposes equilibrium requirements on a finite-size element instead of on an infinitesimally small differential element sought by the differential equations. This approach dilutes the equilibrium requirements of the problem and makes the resulting solution approximate.

3.2.6   CONVERGENCE OF THE FE SOLUTION

Because the approximation of the unknown field is over a small region, the approximation error is small even though the approximating polynomial is simple. As the number of finite elements increases, the element size reduces, and the error of approximation reduces. In the limit, the solution converges to the exact solution of the problem.

3.2.7   FORMULATION OF FEM

FEM was developed in the aircraft industry to solve static equilibrium problems in structural mechanics and was then extended to other application domains. The differential equation of the problem is said to be a strong form of the statement of equilibrium as it demands a higher level of continuity requirements on the field variable. However, the solution of a differential equation is equivalent to finding the stationary value of a functional, which is an integral over the problem domain. For the problems in statics, this functional is the potential energy functional, while Hamilton's principle provides such a functional for the problems in dynamics. The advantage of the functional formulation is that it puts less stringent continuity requirements on the field variable. Because of this, the finite element equations are often derived from the variational formulation of the problem. The weighted residual method is another approach to derive finite element equations.

FEM is a versatile method and can easily incorporate complex geometries, BCs, and material properties. Hence, it is widely used in many fields such as solid mechanics, dynamics, heat transfer, fluid flow problems, and electromagnetics.

3.3   HAMILTON'S PRINCIPLE

We studied Hamilton's principle in Chapter 2. The integral equation of Hamilton's principle for the nonconservative systems is given by

$$\int_{t_1}^{t_2} \left(\delta(T - U) + \delta W_{nc} \right) dt = 0 \tag{3.1}$$

where T and U are kinetic and potential energies and δW_{nc} is the work done by the nonconservative forces due to virtual displacement. Hamilton's principle states that of all possible paths between the time instants t_1 and t_2, the equilibrium path of motion of the system would be the one for which the integral has a stationary value. Hamilton's principle can be used to obtain the governing equations of motion for both discrete and continuous systems.

For a conservative system, i.e., a system with no external forces and dissipation forces, $\delta W_{nc} = 0$, and Hamilton's principle becomes

$$\delta \int\limits_{t_1}^{t_2} (T - U)\ dt = \delta I = 0 \tag{3.2}$$

3.4   HAMILTON'S PRINCIPLE APPLIED TO A DISCRETE/MDOF SYSTEM

In this section, we use Hamilton's principle to derive the equations of motion of a discrete/multi-degree-of-freedom (MDOF) system. The conclusions drawn provide the basis for a simple procedure to derive the governing equations for a discrete system.

We consider a linear N-DOF system with viscous damping subjected to external forces. Let the system is described by N generalized coordinates, $\{q\} = \{q_1,\ q_2,..\ q_n\}^T$. To apply Hamilton's principle (Eq. (3.1)), we need to find the quantities T, U, and δW_{nc}. Since we are not considering a specific system, we use general expressions for these quantities to work out each term in Eq. (3.1).

The kinetic energy T can be expressed as

$$T = \frac{1}{2}\{\dot{q}\}^T [M]\{\dot{q}\} \tag{3.3}$$

The virtual change in T due to the virtual changes in the generalized coordinates is

$$\delta T = \sum_{r=1}^{N} \frac{\partial T}{\partial \dot{q}_r} \delta \dot{q}_r \tag{3.4}$$

$$\delta T = \sum_{r=1}^{N} \frac{\partial}{\partial \dot{q}_r} \left(\frac{1}{2}\{\dot{q}\}^T [M]\{\dot{q}\} \right) \delta \dot{q}_r \tag{3.5}$$

$$\delta T = \sum_{r=1}^{N} \frac{\partial}{\partial \dot{q}_r} \left(\frac{1}{2} \sum_{i=1}^{N} \dot{q}_{1i}^T \sum_{j=1}^{N} M_{ij} \dot{q}_{j1} \right) \delta \dot{q}_r \tag{3.6}$$

$$\delta T = \sum_{r=1}^{N} \frac{1}{2} \left(\sum_{j=1}^{N} M_{rj} \dot{q}_{j1} + \sum_{i=1}^{N} \dot{q}_{1i}^T M_{ir} \right) \delta \dot{q}_r = \sum_{r=1}^{N} \sum_{j=1}^{N} M_{rj} \dot{q}_{j1} \delta \dot{q}_r \tag{3.7}$$

Therefore, the term related to the kinetic energy in Hamilton's integral is

$$\int\limits_{t_1}^{t_2} \delta T\ dt = \sum_{r=1}^{N} \left(\int\limits_{t_1}^{t_2} \sum_{j=1}^{N} M_{rj} \dot{q}_{j1} \delta \dot{q}_r\ dt \right) \tag{3.8}$$

Integrating by parts the RHS of Eq. (3.8), we get

$$\int\limits_{t_1}^{t_2} \delta T\ dt = \sum_{r=1}^{N} \left(\left(\sum_{j=1}^{N} M_{rj} \dot{q}_{j1}.\delta q_r \right)_{t_1}^{t_2} - \int\limits_{t_1}^{t_2} \frac{d}{dt} \left(\sum_{j=1}^{N} M_{rj} \dot{q}_{j1} \right).\delta q_r\ dt \right) \tag{3.9}$$

The virtual changes at t_1 and t_2 have to be zero since the varied path must coincide with the actual path at these instants. Therefore,

$$\delta q_r(t_1) = 0 \quad \text{and} \quad \delta q_r(t_2) = 0 \tag{3.10}$$

Because of the conditions in Eq. (3.10), the first term on the RHS of Eq. (3.9) is zero, and we get,

$$\int_{t_1}^{t_2} \delta T \; dt = -\int_{t_1}^{t_2} \sum_{r=1}^{N} \left(\sum_{j=1}^{N} M_{rj} \ddot{q}_{jl} \right) . \delta q_r \; dt \tag{3.11}$$

The potential energy U can be expressed as

$$U = \frac{1}{2} \{q\}^T [K] \{q\} \tag{3.12}$$

The virtual change in U due to the virtual changes in the generalized coordinates is

$$\delta U = \sum_{r=1}^{N} \frac{\partial U}{\partial q_r} \delta q_r \tag{3.13}$$

$$\delta U = \sum_{r=1}^{N} \frac{\partial}{\partial q_r} \left(\frac{1}{2} \{q\}^T [K] \{q\} \right) \delta q_r \tag{3.14}$$

$$\delta U = \sum_{r=1}^{N} \frac{\partial}{\partial q_r} \left(\frac{1}{2} \sum_{i=1}^{N} q_{1i}^T \sum_{j=1}^{N} K_{ij} q_{jl} \right) \delta q_r \tag{3.15}$$

Differentiating like in Eq. (3.7), we get

$$\delta U = \sum_{r=1}^{N} \sum_{j=1}^{N} K_{rj} q_{jl} \delta q_r \tag{3.16}$$

Therefore, the term related to the potential energy in the Hamilton's integral is

$$-\int_{t_1}^{t_2} \delta U \; dt = -\int_{t_1}^{t_2} \sum_{r=1}^{N} \left(\sum_{j=1}^{N} K_{rj} q_{jl} \right) . \delta q_r \; dt \tag{3.17}$$

The work done by the nonconservative forces (the external forces and the energy-dissipating forces) due to the virtual changes in the generalized coordinates can be expressed as

$$\delta W_{nc} = \{F\}^T \{\delta q\} - \left([C] \{\dot{q}\} \right)^T \{\delta q\} = \sum_{r=1}^{N} F_r \delta q_r - \sum_{r=1}^{N} \left(\sum_{j=1}^{N} C_{rj} \dot{q}_{jl} \right) \delta q_r \tag{3.18}$$

Therefore, the term related to the nonconservative forces in Hamilton's integral is

$$\int_{t_1}^{t_2} \delta W_{nc} \; dt = \int_{t_1}^{t_2} \left(\sum_{r=1}^{N} F_r \delta q_r - \sum_{r=1}^{N} \left(\sum_{j=1}^{N} C_{rj} \dot{q}_{jl} \right) \delta q_r \right) dt \tag{3.19}$$

In this manner, we have determined the individual terms in the Hamilton integral. Combining the results in Eqs. (3.11), (3.17), and (3.19), as per the statement of Hamilton's principle, we get

$$-\int_{t_1}^{t_2} \sum_{r=1}^{N} \left(\sum_{j=1}^{N} M_{rj} \ddot{q}_{jl} \right) . \delta q_r \; dt - \int_{t_1}^{t_2} \sum_{r=1}^{N} \left(\sum_{j=1}^{N} K_{rj} q_{jl} \right) . \delta q_r \; dt + \int_{t_1}^{t_2} \left(\sum_{r=1}^{N} F_r \delta q_r - \sum_{r=1}^{N} \left(\sum_{j=1}^{N} C_{rj} \dot{q}_{jl} \right) \delta q_r \right) dt = 0 \tag{3.20}$$

which can be rewritten after taking out the common factor δq_r, as

$$\int_{t_1}^{t_2}\left(\sum_{r=1}^{N}\left(F_r + \sum_{j=1}^{N} -M_{rj}\ddot{q}_{jl} - K_{rj}q_{jl} - C_{rj}\dot{q}_{jl} \right).\delta q_r \right) dt = 0 \tag{3.21}$$

Since the virtual changes δq_r, $r = 1, 2, ..., N$ are independent and arbitrary, the integral equation (3.21) is satisfied only when the coefficient of each δq_r is zero. Thus, we get

$$F_r + \sum_{j=1}^{N} -M_{rj}\ddot{q}_{jl} - K_{rj}q_{jl} - C_{rj}\dot{q}_{jl} = 0 \quad (\text{for } r = 1, 2, ..., N) \tag{3.22}$$

Equation (3.22) represents N equations, which can be written in the following form:

$$[M]\{\ddot{q}\} + [C]\{\dot{q}\} + [K]\{q\} = \{F\} \tag{3.23}$$

Eq. (3.23) is the outcome of applying Hamilton's principle to MDOF systems. It is the equation of motion of an MDOF system involving the mass ([M]), stiffness ([K]), and damping ([C]) matrices and the vector of the external forces ({F}).

The following conclusions can be drawn from the above development:

- The matrices $[M]$, $[K]$ and $[C]$ and the vector $\{F\}$ determine the equation of motion.
- The matrices, $[M]$, $[K]$ and $[C]$ appear in the expressions for the kinetic energy (Eq. 3.3), potential energy (Eq. 3.12), and the work done by the dissipative forces (Eq. 3.18), respectively. Similarly, the vector $\{F\}$ appears in the expression for the work done by the external forces (Eq. 3.18).
- Therefore, the matrices and force vector appearing in the equation of motion obtained by the application of Hamilton's principle can be obtained directly from the expressions of T, U, and δW_{nc}, expressed in matrix form. This observation provides a simple procedure for obtaining the mass, stiffness, damping matrices, and force vector for a discrete system, yielding the governing equations of motion. We use this approach for formulating the mass and stiffness matrices and the force vector using FEM.

3.5 HAMILTON'S PRINCIPLE APPLIED TO A CONTINUOUS DYNAMIC SYSTEM

In this section, the equation of motion for a continuous dynamic system is obtained using Hamilton's principle. A continuous system has a continuous distribution of mass, stiffness, and damping. An infinite number of coordinates need to be specified to describe its displaced configuration; hence, such a system has infinite degrees of freedom. We take the example of the longitudinal vibration of a rod. The rod is of length L, area of cross-section A and density ρ, and is fixed at one end and subjected to a distributed axial body force F_b N/m, as shown in Figure 3.1. The problem is treated as a 1D problem, with u representing the axial displacement at a point x. The rod is assumed to be undamped.

Similar to the approach followed in the previous section for the MDOF system, we determine the expressions for T, U, and δW_{nc} to apply Hamilton's principle. The kinetic energy of a differential element dx is

$$dT = \frac{1}{2}\left(\rho A dx\right)\dot{u}^2 \tag{3.24}$$

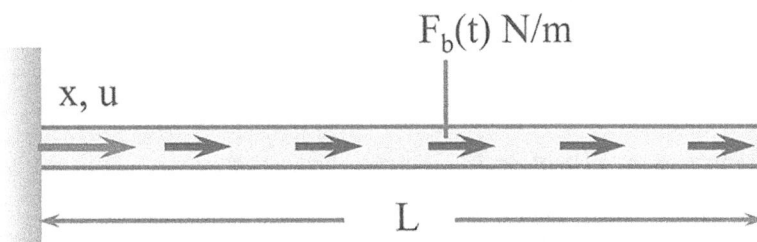

FIGURE 3.1 Longitudinal vibration of a bar.

Therefore, the kinetic energy for the whole rod is

$$T = \int_0^L dT = \int_0^L \frac{1}{2}\rho A\dot{u}^2 \, dx \tag{3.25}$$

Similarly, the strain energy of a differential element dx is

$$dU = \frac{1}{2}\sigma_x\epsilon_x dV = \frac{1}{2}\epsilon_x E.\epsilon_x \ .Adx = \frac{1}{2}EA.\left(\frac{\partial u}{\partial x}\right)^2 .dx \tag{3.26}$$

where σ_x and ϵ_x are the longitudinal stress and strain, respectively, and we have the relation $\epsilon_x = \dfrac{\partial u}{\partial x}$. Note that the partial derivative is used since u depends on more than one independent variable (x and t in this case). The potential energy for the whole rod is

$$U = \int_0^L dU = \int_0^L \frac{1}{2}EA\left(\frac{\partial u}{\partial x}\right)^2 .dx \tag{3.27}$$

The work done by the nonconservative forces due to a virtual displacement can be calculated as

$$\delta W_{nc} = \int_0^L F_b dx \delta u \tag{3.28}$$

Substituting Eqs. (3.25), (3.27), and (3.28) into the statement of Hamilton's principle, we get

$$\int_{t_1}^{t_2} \left(\delta(T-U) + \delta W_{nc}\right) dt = 0 \tag{3.29}$$

$$\int_{t_1}^{t_2} \left(\delta \left(\int_0^L \frac{1}{2}\rho A\dot{u}^2 \, dx - \int_0^L \frac{1}{2}EA\left(\frac{\partial u}{\partial x}\right)^2 .dx \right) + \int_0^L F_b dx \delta u \right) dt = 0 \tag{3.30}$$

$$\int_{t_1}^{t_2} \left(\int_0^L \rho A\dot{u}\delta\dot{u} \, dx - \int_0^L EA\frac{\partial u}{\partial x}\delta\left(\frac{\partial u}{\partial x}\right).dx + \int_0^L F_b dx \delta u \right) dt = 0 \tag{3.31}$$

$$\int_{t_1}^{t_2}\int_0^L \rho A\dot{u}\delta\dot{u} \, dx \, dt - \int_{t_1}^{t_2}\int_0^L EA\frac{\partial u}{\partial x}\delta\left(\frac{\partial u}{\partial x}\right).dx \, dt + \int_{t_1}^{t_2}\int_0^L F_b dx \delta u \, dt = 0 \tag{3.32}$$

The first term in Eq. (3.32) after integration by parts with 't' gives

$$\int_{t_1}^{t_2}\int_0^L \rho A\dot{u}\delta\dot{u} \, dx \, dt = \left(\int_0^L \rho A\dot{u}\delta u \, dx \right)_{t_1}^{t_2} - \int_{t_1}^{t_2}\int_0^L \rho A\ddot{u}\delta u \, dx \, dt \tag{3.33}$$

The virtual changes in u at t_1 and t_2 have to be zero since the varied path must coincide with the true path at these instants. Hence,

$$\delta u(t_1) = 0 \quad \text{and} \quad \delta u(t_2) = 0 \tag{3.34}$$

Because of (3.34), the first term on the RHS of Eq. (3.33) is zero, and we get

$$\int_{t_1}^{t_2}\int_0^L \rho A\dot{u}\delta\dot{u} \, dx \, dt = -\int_{t_1}^{t_2}\int_0^L \rho A\ddot{u}\delta u \, dx \, dt \tag{3.35}$$

The second term in Eq. (3.32) after integration by parts with 'x' gives

$$\int\limits_{t_1}^{t_2}\int\limits_{0}^{L} EA\frac{\partial u}{\partial x}\frac{\partial(\delta u)}{\partial x}.dx\,dt = \int\limits_{t_1}^{t_2}\left(EA\frac{\partial u}{\partial x}.\delta u\right)_{0}^{L}dt - \int\limits_{t_1}^{t_2}\int\limits_{0}^{L} EA\frac{\partial^2 u}{\partial x^2}\delta u.dx\,dt \tag{3.36}$$

Substituting the results in Eqs. (3.35) and (3.36) into Eq. (3.32) lead to

$$-\int\limits_{t_1}^{t_2}\int\limits_{0}^{L}\rho A\ddot{u}\delta u\,dx\,dt - \int\limits_{t_1}^{t_2}\left(EA\frac{\partial u}{\partial x}.\delta u\right)_{0}^{L}dt + \int\limits_{t_1}^{t_2}\int\limits_{0}^{L} EA\frac{\partial^2 u}{\partial x^2}\delta u.dx\,dt + \int\limits_{t_1}^{t_2}\int\limits_{0}^{L} F_b dx\delta u\,dt = 0 \tag{3.37}$$

which can be further written as

$$\int\limits_{t_1}^{t_2}\left(\int\limits_{0}^{L}\left(-\rho A\ddot{u} + EA\frac{\partial^2 u}{\partial x^2} + F_b\right)\delta u\,dx - \left(EA\frac{\partial u}{\partial x}.\delta u\right)_{0}^{L}\right)dt = 0 \tag{3.38}$$

Since the virtual change δu is arbitrary, the integral equation (3.38) is satisfied when the integrand of the second term and the coefficient of δu in the integrand of the first term are individually zero. Thus, we get

$$-\rho A\ddot{u} + EA\frac{\partial^2 u}{\partial x^2} + F_b = 0 \tag{3.39}$$

and

$$\left(EA\frac{\partial u}{\partial x}.\delta u\right)_{0}^{L} = 0 \tag{3.40}$$

Eq. (3.40) can be written as

$$EA\frac{\partial u}{\partial x}(L).\delta u(L) - EA\frac{\partial u}{\partial x}(0).\delta u(0) = 0 \tag{3.41}$$

For the present problem, we have the geometric boundary condition that $u(0) = 0$ and therefore $\delta u(0) = 0$. Since $u(L) \neq 0$, $\delta u(L)$ is non-zero, and hence Eq. (3.41) leads to

$$EA\frac{\partial u}{\partial x}(L) = 0 \tag{3.42}$$

The LHS of Eq. (3.42) physically represents the internal force in the rod at $x = L$. Equation (3.42) essentially says that the internal force at $x = L$ must be zero since no external point force exists on the rod at $x = L$. It represents a natural boundary condition of the problem.

Thus, the application of Hamilton's principle leads to the equation of motion given by Eq. (3.39), a partial differential equation, and the natural boundary condition the solution of the equation must satisfy.

3.6 FEM FOR DYNAMICS

In the last section, we discussed that the equation of motion of a continuous system is in the form of a partial differential equation with natural BCs. Therefore, the solution of vibration of continuous dynamic systems requires the solution of partial differential equations. Except for some simple problems, the solution is difficult for problems with complex geometries, BCs, and material properties.

The FEM seeks to solve the problem approximately through the following steps.

- Represent the continuous system by an assemblage of finite elements connected at nodes.
- Approximate the displacement field over each finite element by a simple polynomial. The approximation essentially interpolates the nodal values of the displacement to define the displacement field over the finite element.

- For each finite element, obtain the equilibrium equations, and assemble them to obtain the equilibrium equations (called the global equations) for the assemblage of the finite elements. The unknowns are the values of the displacement at the nodes of the finite element assembly.
- Impose the BCs on the nodal variables.
- Solve the global equations to get the unknown displacement at the nodes.

3.6.1   Strategy for the Formulation of FE Matrices in Dynamics

One crucial step in FEM is to obtain the elemental matrices. We have shown in Section 3.4 that for the dynamics of discrete/MDOF systems, the structural matrices and the force vector that make up the equilibrium equation can be obtained directly from the expressions of T, U, and δW_{nc} written in matrix form. This approach is also applicable to obtaining the structural matrices and the force vector for a finite element since the finite element representation of a continuous system constitutes a discrete system. In the following sections, we use this approach to formulate the mass and stiffness matrices for finite elements such as the rod, beam, and frame elements. Generally, a proportional damping assumption is used in modeling damping in FE models. In such cases, the damping matrix becomes known once the mass and stiffness matrices are established.

3.6.2   Order of the Polynomial Approximation

In FEM, the unknown displacement field over each finite element is approximated by a simple function, generally a polynomial. But what should be the order of this polynomial? Let q be the order of the highest order derivative appearing in the integral of Hamilton's principle. Then, for the convergence of the FEM solution, the polynomial should be continuous and have derivatives continuous up to order (q-1). This continuity is also required at nodes common to adjacent elements.

3.7   BAR ELEMENT

The bar element is a 1D finite element that can be used to model the axial or longitudinal vibration of a bar or rod. In this section, we formulate the stiffness and mass matrices and the force vector for the bar element.

Figure 3.1 shows a longitudinal bar subjected to an axial/longitudinal body force. The longitudinal displacement over the length of the bar (denoted by $u(x,t)$) is the unknown displacement field. The cross-sectional dimensions are assumed to be smaller than the wavelength of the longitudinal wave motion; hence it is treated as a 1D object, and a 1D finite element with two nodes is used to model the bar.

We need to decide the variables at the nodes of the 1D finite element. The energy expressions for a bar due to the longitudinal vibration (given below) indicate that the order of the highest order derivative is 1. Hence, in light of Section 3.6.2, the derivative of the displacement field up to order 0 must be continuous inside the finite element and at the boundary between the adjacent finite elements. Therefore, we take the longitudinal displacement u as the DOF at the two nodes of the bar element. Figure 3.2a shows the longitudinal bar represented by an assemblage of bar elements. It also shows the DOFs at the nodes of the finite elements.

Figure 3.2b shows one finite element. The element is shown in the global coordinate system, with the coordinates x_i and x_{i+1} being the global coordinates of its nodes, and u_i and u_{i+1} being the global displacement/DOFs at the nodes.

For ease of formulation of the equilibrium equations, the element is mapped into a local coordinate system with coordinate ξ, as shown in Figure 3.2c. The values of ξ for the two nodes are -1 and $+1$, while the DOFs in the local system are denoted by q_1 and q_2 respectively.

3.7.1   Element Stiffness Matrix

We discussed in Chapter 2 that the stiffness matrix of a discrete/MDOF system is related to the potential/strain energy. Therefore, we start from the expression for the strain energy of the element for formulating the element stiffness matrix.

The expression for the strain energy due to longitudinal displacement is derived earlier (Eq. 3.27). Therefore, the strain energy of a bar element is given by

$$U_e = \int_{x_i}^{x_{i+1}} \frac{1}{2} E_e A_e \left(\frac{\partial u}{\partial x} \right)^2 dx \tag{3.43}$$

The subscript 'e' indicates association with a finite element.

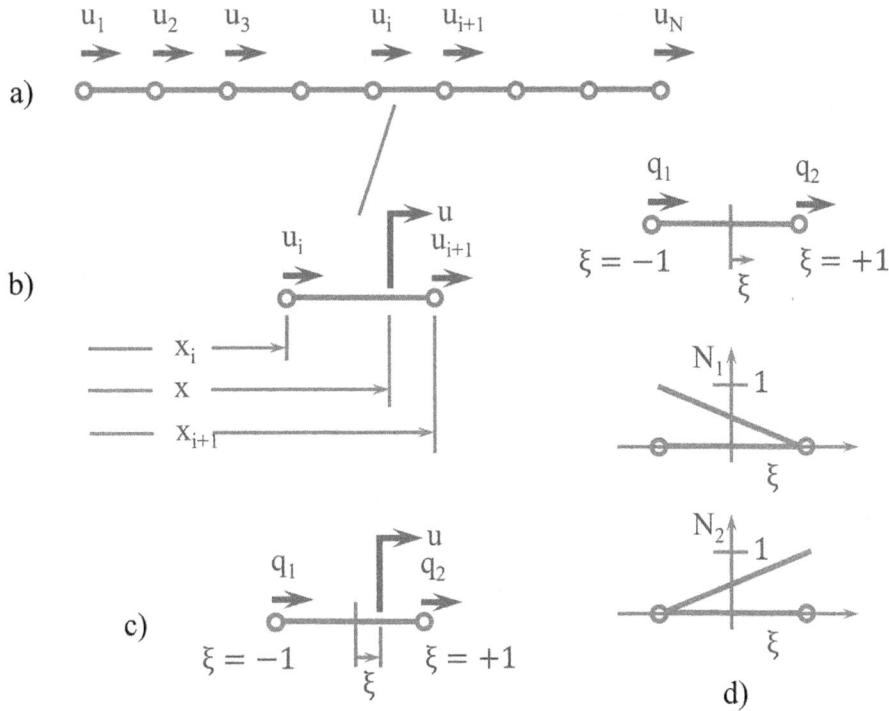

FIGURE 3.2 (a) FE discretization, (b) bar element in the global coordinate system, (c) bar element in the local coordinate system, and (d) shape functions.

Since there is one DOF at every node, two DOFs are associated with a two-node bar element. Hence, we use a linear polynomial to approximate the displacement field over the finite element. In the local coordinate system, the linear polynomial is

$$u = a_0 + a_1\xi \tag{3.44}$$

In light of Figure 3.2c, Eq. (3.44) must satisfy the following conditions:

$$\begin{aligned} \text{at } \xi = -1: \quad & u = q_1 \\ \text{at } \xi = +1: \quad & u = q_2 \end{aligned} \tag{3.45}$$

Applying the conditions in Eq. (3.45) to Eq. (3.44), we get

$$a_0 = \frac{1}{2}(q_2 + q_1)$$

$$a_1 = \frac{1}{2}(q_2 - q_1) \tag{3.46}$$

After substituting Eq. (3.46) into Eq. (3.44), the resulting equation can be written as

$$u = N_1 q_1 + N_2 q_2 \tag{3.47}$$

where N_1 and N_2 are given by

$$N_1 = \frac{1}{2}(1 - \xi)$$

$$N_2 = \frac{1}{2}(1 + \xi) \tag{3.48}$$

Figure 3.2d shows the plots of N_1 and N_2. N_i represents the shape of the displacement field over the finite element when the ith DOF of the element is displaced by unity with the displacements at the remaining DOFs of the element being zero, and that is why the functions N_i are called **shape functions**.

Equation (3.47) can be written as

$$u = \underset{1 \times 2}{[N]} \underset{2 \times 1}{\{q\}} \tag{3.49}$$

where $[N] = \begin{bmatrix} N_1 & N_2 \end{bmatrix}$ is the shape function matrix and $\{q\} = \{q_1 \quad q_2\}^T$ is the vector of elemental DOFs in the local coordinates.

We also need the relationship between the local and global coordinates, that is ξ and x. From Figure 3.2b and c, we can write

$$\xi = \frac{x - \frac{1}{2}(x_i + x_{i+1})}{x_2 - \frac{1}{2}(x_i + x_{i+1})} = \frac{2x - (x_i + x_{i+1})}{(x_i - x_{i+1})} \tag{3.50}$$

Differentiating Eq. (3.50) with x gives

$$\frac{d\xi}{dx} = \frac{2}{(x_i - x_{i+1})} = \frac{2}{L_e} \tag{3.51}$$

where L_e is the length of the bar element.

The derivative $\partial u / \partial x$ is obtained as

$$\frac{\partial u}{\partial x} = \frac{\partial u}{\partial \xi} \cdot \frac{\partial \xi}{\partial x} = \frac{\partial u}{\partial \xi} \cdot \frac{2}{L_e} \tag{3.52}$$

The expression for the strain energy (Eq. 3.43) is converted to the local coordinate system using Eqs. (3.51) and (3.52). The strain energy expression in the local coordinate system is

$$U_e = \frac{1}{2} \int_{-1}^{+1} \frac{2E_e A_e}{L_e} \left(\frac{\partial u}{\partial \xi}\right)^T \left(\frac{\partial u}{\partial \xi}\right) . d\xi \tag{3.53}$$

It follows from Eq. (3.49) that

$$\frac{\partial u}{\partial \xi} = \frac{\partial [N]}{\partial \xi} \{q\} \tag{3.54}$$

Substituting Eq. (3.54) into Eq. (3.53), we get

$$U_e = \frac{1}{2} \{q\}^T \left(\frac{2E_e A_e}{L_e} \int_{-1}^{+1} \frac{\partial [N]^T}{\partial \xi} \frac{\partial [N]}{\partial \xi} d\xi \right) \{q\} \tag{3.55}$$

Comparing the expression in (3.55) with the expression for the strain/potential energy of an MDOF system, $U = \frac{1}{2} \{q\}^T [K] \{q\}$, the expression for the element stiffness matrix is obtained as

$$[K]_e = \frac{2E_e A_e}{L_e} \int_{-1}^{+1} \frac{\partial [N]^T}{\partial \xi} \frac{\partial [N]}{\partial \xi} d\xi \tag{3.56}$$

From Eq. (3.48), the derivative of the shape function matrix is

$$\frac{\partial [N]}{\partial \xi} = \frac{1}{2} \begin{bmatrix} -1 & 1 \end{bmatrix} \tag{3.57}$$

Substituting Eq. (3.57) into Eq. (3.56) and integrating, we get the stiffness matrix for the bar element:

$$[K]_e = \frac{E_e A_e}{L_e} \begin{bmatrix} 1 & -1 \\ -1 & 1 \end{bmatrix} \tag{3.58}$$

3.7.2 Element Mass Matrix

We discussed in Chapter 2 that the mass matrix of a discrete/MDOF system is related to kinetic energy. Therefore, to formulate the element mass matrix, we start with the expression for the kinetic energy of the element.

The expression for the kinetic energy due to longitudinal displacement is derived earlier (Eq. 3.25). Therefore, the kinetic energy of a bar element is given by

$$T_e = \int_{x_i}^{x_{i+1}} \frac{1}{2} \rho_e A_e \dot{u}^2 \, dx = \int_{x_i}^{x_{i+1}} \frac{1}{2} \rho_e A_e \left(\frac{\partial u}{\partial t} \right)^T \left(\frac{\partial u}{\partial t} \right) dx \tag{3.59}$$

Using Eq. (3.51), Eq. (3.59) can be written in the local coordinate system as

$$T_e = \frac{1}{2} \int_{-1}^{+1} \frac{\rho_e A_e L_e}{2} \left(\frac{\partial u}{\partial t} \right)^T \left(\frac{\partial u}{\partial t} \right) d\xi \tag{3.60}$$

The derivative $\partial u / \partial t$ is obtained by differentiating Eq. (3.49) with 't', leading to

$$\frac{\partial u}{\partial t} = [N] \frac{\partial}{\partial t} \{q\} = [N] \{\dot{q}\} \tag{3.61}$$

Substituting Eq. (3.61) into Eq. (3.60) gives

$$T_e = \frac{1}{2} \{\dot{q}\}^T \left(\frac{\rho_e A_e L_e}{2} \int_{-1}^{+1} [N]^T [N] d\xi \right) \{\dot{q}\} \tag{3.62}$$

Comparing the expression in (3.62) with the expression for the kinetic energy of an MDOF system, $T = \frac{1}{2} \{\dot{q}\}^T [M] \{\dot{q}\}$, we get the following expression for the element mass matrix:

$$[M]_e = \frac{\rho_e A_e L_e}{2} \int_{-1}^{+1} [N]^T [N] d\xi \tag{3.63}$$

Substituting the shape functions from Eq. (3.48) into Eq. (3.63), we get

$$[M]_e = \frac{\rho_e A_e L_e}{2} \int_{-1}^{+1} \begin{bmatrix} \dfrac{(1-\xi)^2}{4} & \dfrac{1-\xi^2}{4} \\ \dfrac{1-\xi^2}{4} & \dfrac{(1+\xi)^2}{4} \end{bmatrix} d\xi \tag{3.64}$$

After integration, we get the mass matrix of the bar element:

$$[M]_e = \frac{\rho_e A_e L_e}{6} \begin{bmatrix} 2 & 1 \\ 1 & 2 \end{bmatrix} \tag{3.65}$$

3.7.3 ELEMENT FORCE VECTOR

The work done by the external force, the body force F_b N/m, due to virtual displacement, can be written for an element as

$$\delta W_{F,e} = \int_{x_i}^{x_{i+1}} F_b \, dx \, \delta u \tag{3.66}$$

Making use of Eq. (3.51) and substituting Eq. (3.49), Eq. (3.66) becomes

$$\delta W_{F,e} = \left(\frac{F_b L_e}{2} \int_{-1}^{+1} [N] \, d\xi \right) \{\delta q\} \tag{3.67}$$

Comparing the expression in (3.67) with the expression for the virtual work done by the external forces in an MDOF system, $\delta W_{nc} = \{F\}^T \{\delta q\}$, we get the force vector due to the body force on the bar element as

$$\{F_b\}_e^T = \frac{F_b L_e}{2} \int_{-1}^{+1} [N] \, d\xi \tag{3.68}$$

Substituting the shape functions from Eq. (3.48) into Eq. (3.68) and integrating, we get

$$\{F_b\}_e = \frac{F_b L_e}{2} \begin{Bmatrix} 1 \\ 1 \end{Bmatrix} \tag{3.69}$$

It is seen that the total body force is divided equally among the nodal DOFs. If there are point forces at nodes, they can be incorporated directly into the global force vector.

3.8 BEAM ELEMENT

The beam element is a 1D finite element that can be used to model the transverse vibrations of beams, shafts, or similar members. In this section, we formulate the element stiffness and mass matrices and the force vector for the beam element.

A beam subjected to external forces and moments is shown in Figure 3.3a. The transverse displacement over the length of the beam (denoted by $v(x,t)$) is the unknown displacement field. In the theory of pure bending of beams, the plane sections normal to the centroidal axis of the beam before bending are assumed to remain plane and normal to the centroidal axis after bending. The beam element presented in this section is based on this assumption. The shear deformation and rotational inertia are neglected.

We need to decide the variables at the nodes of the 1D beam element. The energy expressions for the beam (given below) indicate that the order of the highest order derivative is two. Hence, the derivative of the displacement field up to order one must be continuous inside the finite element and at the boundary between the adjacent finite elements. Therefore, we take the transverse displacement and slope as the DOFs at the nodes of the beam element. Figure 3.3b shows the beam represented by an assemblage of beam elements. It also shows the DOFs at the nodes.

Figure 3.3c shows a beam element with nodes at its ends. The element is shown in the global coordinate system, with the coordinates x_i and x_{i+1} being the global coordinates of its nodes, and v_{2i-1}, v_{2i}, v_{2i+1} and v_{2i+2} being the global displacements and slopes at the nodes.

For ease of formulation of the equilibrium equations, the element is mapped into a local coordinate system with coordinate ξ as shown in Figure 3.3d. The value of ξ for the two nodes are -1 and $+1$, and the displacements and slopes in the local system are denoted by d_1, θ_1, d_2 and θ_2.

3.8.1 ELEMENT STIFFNESS MATRIX

The stiffness matrix of a discrete/MDOF system is related to the potential/strain energy. Therefore, let us first obtain the expression for the strain energy due to transverse displacement. We know that

$$U = \int_V \frac{1}{2} \sigma_x \epsilon_x dV \tag{3.70}$$

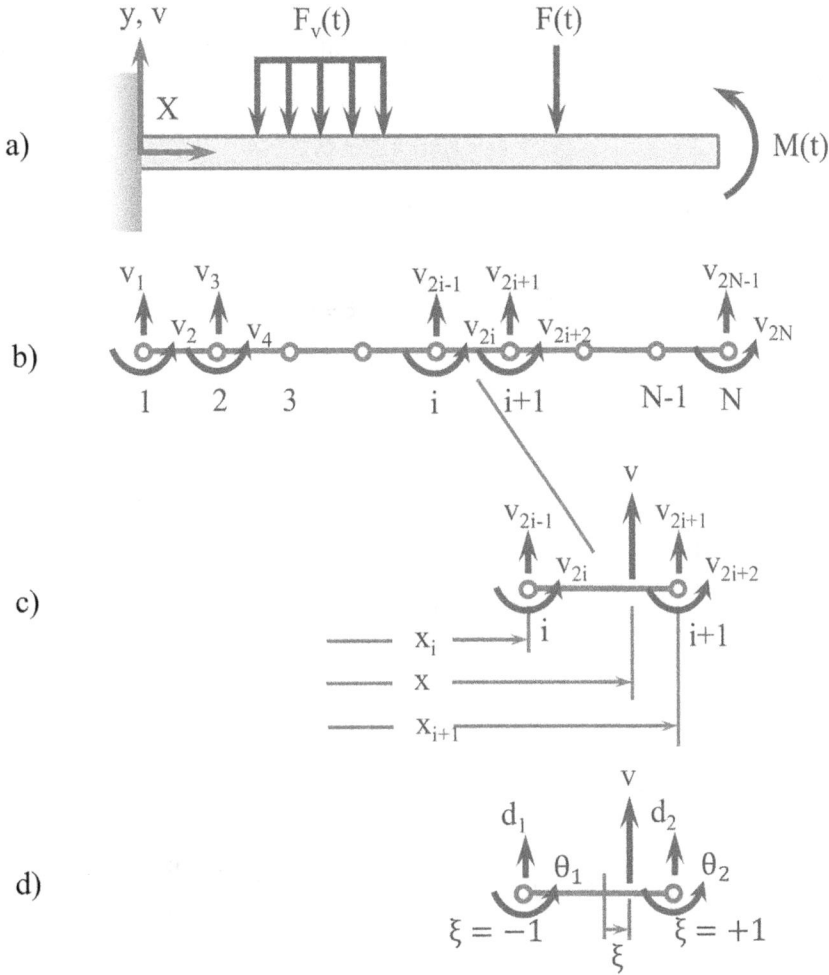

FIGURE 3.3 (a) Transverse vibration of a beam, (b) FE discretization, (c) beam element in the global coordinate system, and (d) beam element in the local coordinate system.

We also know from the bending of beams that

$$\frac{\sigma_x}{y} = \frac{E}{R} \quad \text{and} \quad \frac{1}{R} = \frac{\partial^2 v}{\partial x^2} \tag{3.71}$$

where R is the radius of curvature of the beam section. Using Hook's law ($\epsilon_x = \sigma_x /E$) and substituting σ_x from Eq. (3.71) into Eq. (3.70), we get

$$U = \iint_{L\,A} \frac{1}{2E} \sigma_x^2 dA dx = \iint_{L\,A} \frac{1}{2E} \left(yE \frac{\partial^2 v}{\partial x^2} \right)^2 dA dx \tag{3.72}$$

$$U = \int_L \frac{E}{2} \left(\frac{\partial^2 v}{\partial x^2} \right)^2 \left(\int_A y^2 dA \right) dx \tag{3.73}$$

The second integral in Eq. (3.73) is the second moment of area (I) about the neutral axis. Therefore, the strain energy of a finite element can be written (after adding subscript 'e' denoting association with a finite element) as

$$U_e = \int_{x_i}^{x_{i+1}} \frac{1}{2} E_e I_e \left(\frac{\partial^2 v}{\partial x^2} \right)^2 dx \tag{3.74}$$

Since there are two DOFs at a node, there are four DOFs associated with the two-node beam element, and hence we use a cubic polynomial to approximate the displacement field over the finite element. The cubic polynomial in the local coordinate system is written as

$$v = a_0 + a_1\xi + a_2\xi^2 + a_3\xi^3 \tag{3.75}$$

In light of Figure 3.3d, Eq. (3.75) must satisfy the following conditions:

$$\text{at } \xi = -1: \quad \begin{array}{l} v = d_1 \\ \partial v/\partial \xi = \theta_1 \end{array}$$

$$\text{at } \xi = +1: \quad \begin{array}{l} v = d_2 \\ \partial v/\partial \xi = \theta_2 \end{array} \tag{3.76}$$

Applying the conditions in Eq. (3.76) to Eq. (3.75), the resulting equations can be written as

$$\begin{bmatrix} 1 & -1 & 1 & -1 \\ 0 & 1 & -2 & 3 \\ 1 & 1 & 1 & 1 \\ 0 & 1 & 2 & 3 \end{bmatrix} \begin{Bmatrix} a_0 \\ a_1 \\ a_2 \\ a_3 \end{Bmatrix} = \begin{Bmatrix} d_1 \\ \theta_1 \\ d_2 \\ \theta_2 \end{Bmatrix} \tag{3.77}$$

which gives

$$\begin{Bmatrix} a_0 \\ a_1 \\ a_2 \\ a_3 \end{Bmatrix} = \frac{1}{4} \begin{bmatrix} 2 & 1 & 2 & -1 \\ -3 & -1 & 3 & -1 \\ 0 & -1 & 0 & 1 \\ 1 & 1 & -1 & 1 \end{bmatrix} \begin{Bmatrix} d_1 \\ \theta_1 \\ d_2 \\ \theta_2 \end{Bmatrix} \tag{3.78}$$

Substituting the polynomial coefficients from Eq. (3.78) into Eq. (3.75), the polynomial can be expressed as

$$v = N_1 d_1 + N_2 \theta_1 + N_3 d_2 + N_4 \theta_2 \tag{3.79}$$

where N_1, N_2, N_3 and N_4 are given by

$$N_1 = \frac{1}{4}\left(2 - 3\xi + \xi^3\right)$$

$$N_2 = \frac{1}{4}\left(1 - \xi - \xi^2 + \xi^3\right)$$

$$N_3 = \frac{1}{4}\left(2 + 3\xi - \xi^3\right)$$

$$N_4 = \frac{1}{4}\left(-1 - \xi + \xi^2 + \xi^3\right) \tag{3.80}$$

N_i represents the shape of the displacement field over the element when the ith DOF of the element is unity with the remaining three DOFs of the element being zero, and hence the functions N_i are called **shape functions**.

Equation (3.79) can be written as

$$v = \underset{1\times4}{[N]} \underset{4\times1}{\{d\}} \tag{3.81}$$

where $[N] = \begin{bmatrix} N_1 & N_2 & N_3 & N_4 \end{bmatrix}$ is the shape function matrix and $\{d\} = \begin{Bmatrix} d_1 & \theta_1 & d_2 & \theta_2 \end{Bmatrix}^T$ is the vector of elemental DOFs in the local coordinates.

The relationship between the local and global coordinates that is ξ and x is given by Eqs. (3.50) and (3.51).

The second derivative $\partial^2 v/\partial x^2$ appearing in the strain energy expression can be obtained as

$$\frac{\partial v}{\partial x} = \frac{\partial v}{\partial \xi} \cdot \frac{\partial \xi}{\partial x} = \frac{\partial v}{\partial \xi} \cdot \frac{2}{L_e} \tag{3.82}$$

$$\frac{\partial^2 v}{\partial x^2} = \frac{\partial}{\partial \xi}\left(\frac{\partial v}{\partial x}\right) \cdot \frac{\partial \xi}{\partial x} = \frac{\partial^2 v}{\partial \xi^2} \cdot \frac{4}{L_e^2} \tag{3.83}$$

Using Eqs. (3.51) and (3.83), the strain energy expression (Eq. 3.74) is converted to the local coordinate system. The strain energy in the local coordinate system is

$$U_e = \frac{1}{2}\int_{-1}^{+1} \frac{8E_e I_e}{L_e^3}\left(\frac{\partial^2 v}{\partial \xi^2}\right)^T \left(\frac{\partial^2 v}{\partial \xi^2}\right).d\xi \tag{3.84}$$

The derivative $\partial^2 v/\partial \xi^2$ is obtained by differentiating Eq. (3.81) twice giving

$$\frac{\partial^2 v}{\partial \xi^2} = \frac{\partial^2 [N]}{\partial \xi^2}\{d\} \tag{3.85}$$

Substituting Eq. (3.85) into Eq. (3.84), we get

$$U_e = \frac{1}{2}\{d\}^T\left(\frac{8E_e I_e}{L_e^3}\int_{-1}^{+1} \frac{\partial^2 [N]}{\partial \xi^2}^T \frac{\partial^2 [N]}{\partial \xi^2}d\xi\right)\{d\} \tag{3.86}$$

Comparing the expression in Eq. (3.86) with the expression for the potential energy of an MDOF system, $U = \frac{1}{2}\{d\}^T[K]\{d\}$, we get the following integral expression for the element stiffness matrix:

$$[K]_e = \frac{8E_e I_e}{L_e^3}\int_{-1}^{+1} \frac{\partial^2 [N]}{\partial \xi^2}^T \frac{\partial^2 [N]}{\partial \xi^2}d\xi \tag{3.87}$$

Substituting the shape functions from Eq. (3.80) and integrating Eq. (3.87), we get the stiffness matrix for the beam element as

$$[K]_e = \frac{E_e I_e}{L_e^3}\begin{bmatrix} 12 & 6L_e & -12 & 6L_e \\ 6L_e & 4L_e^2 & -6L_e & 2L_e^2 \\ -12 & -6L_e & 12 & -6L_e \\ 6L_e & 2L_e^2 & -6L_e & 4L_e^2 \end{bmatrix} \tag{3.88}$$

3.8.2 ELEMENT MASS MATRIX

The mass matrix of a discrete/MDOF system is related to kinetic energy. The kinetic energy for a finite element due to transverse velocity can be written as

$$T_e = \int_{x_i}^{x_{i+1}} \frac{1}{2}\rho_e A_e \dot{v}^2\, dx = \int_{x_i}^{x_{i+1}} \frac{1}{2}\rho_e A_e \left(\frac{\partial v}{\partial t}\right)^T\left(\frac{\partial v}{\partial t}\right) dx \tag{3.89}$$

Using Eq. (3.51), Eq. (3.89) is expressed in the local coordinate system given by

$$T_e = \frac{1}{2}\int_{-1}^{+1} \frac{\rho_e A_e L_e}{2}\left(\frac{\partial v}{\partial t}\right)^T\left(\frac{\partial v}{\partial t}\right) d\xi \tag{3.90}$$

The derivative $\partial v/\partial t$ is obtained by differentiating Eq. (3.81) with 't'.

$$\frac{\partial v}{\partial t} = [N]\frac{\partial}{\partial t}\{d\} = [N]\{\dot{d}\} \tag{3.91}$$

Substituting Eq. (3.91) into Eq. (3.90), we get

$$T_e = \frac{1}{2}\{\dot{d}\}^T\left(\frac{\rho_e A_e L_e}{2}\int_{-1}^{+1}[N]^T[N]d\xi\right)\{\dot{d}\} \tag{3.92}$$

Comparing the expression in Eq. (3.92) with the expression for the kinetic energy of an MDOF system, $T = \frac{1}{2}\{\dot{d}\}^T[M]\{\dot{d}\}$, we get an integral expression for the mass matrix given by

$$[M]_e = \frac{\rho_e A_e L_e}{2}\int_{-1}^{+1}[N]^T[N]d\xi \tag{3.93}$$

Substituting the shape functions from Eq. (3.80) into Eq. (3.93), and carrying out the integration, we get the mass matrix for the beam element as

$$[M]_e = \frac{\rho_e A_e L_e}{420}\begin{bmatrix} 156 & 22L_e & 54 & -13L_e \\ 22L_e & 4L_e^2 & 13L_e & -3L_e^2 \\ 54 & 13L_e & 156 & -22L_e \\ -13L_e & -3L_e^2 & -22L_e & 4L_e^2 \end{bmatrix} \tag{3.94}$$

3.8.3 ELEMENT FORCE VECTOR

The work done by the uniformly distributed force, F_v N/m, due to virtual displacement is given by

$$\delta W_{F,e} = \int_{x_i}^{x_{i+1}} F_v \, dx \, \delta v \tag{3.95}$$

Using Eq. (3.51), and substituting Eq. (3.81) into Eq. (3.95), we obtain the equation in the local coordinate system.

$$\delta W_{F,e} = \left(\frac{F_v L_e}{2}\int_{-1}^{+1}[N] \, d\xi\right)\{\delta d\} \tag{3.96}$$

Comparing the expression in Eq. (3.96) with the expression for the virtual work done by the external forces in an MDOF system, $\delta W_{nc} = \{F\}^T\{\delta d\}$, we get the force vector due to the uniformly distributed force on the finite element as

$$\{F_v\}_e^T = \frac{F_u L_e}{2}\int_{-1}^{+1}[N] \, d\xi \tag{3.97}$$

Substituting the shape function matrix from Eq. (3.80) into Eq. (3.97) and carrying out the integration, we get

$$\{F_v\}_e = \left\{ \begin{array}{cccc} \dfrac{F_v L_e}{2} & \dfrac{F_v L_e^2}{12} & \dfrac{F_v L_e}{2} & -\dfrac{F_v L_e^2}{12} \end{array} \right\}^T \tag{3.98}$$

Thus, the finite element approximation converts a uniformly distributed force into discrete forces and moments at the DOFs of the element. As discussed in the assembly procedure later, the point forces and moments at the nodes can be incorporated directly into the global force vector.

3.9 FRAME ELEMENT

A frame or frame-like structure consists of slender members rigidly connected. The individual members of the frame may undergo both bending and axial deformation. A plane frame subjected to external forces and moments is shown in Figure 3.4a. Each point on the frame can undergo displacements in the global coordinate directions X and Y. In addition, each point is also associated with the local slope/rotation of the member. Hence, the displacement fields denoted by $U(X,Y,t)$ and $V(X,Y,t)$ and the slope $\theta(X,Y,t)$ over the frame are the unknowns of the problem. The displacements and slope at a point on the frame can also be defined in terms of the transverse and longitudinal displacements and slope in the local coordinate system. Because of this, the analysis of frames can be conducted by combining the 1D bar and beam elements presented in the previous two sections. The element resulting from this combination is referred to as a 1D frame element. In this section, we present the formulation of the element stiffness and mass matrices and the force vector for the planar 1D frame element.

Figure 3.4b shows the frame represented by an assemblage of frame elements. It also shows the global DOFs U, V, and θ at the nodes.

Figure 3.4c shows a frame element in the global coordinate system, with U_i, V_i, θ_i and U_{i+1}, V_{i+1}, θ_{i+1} being the global DOFs at the two nodes. Figure 3.4d shows the frame element in the local coordinate system of the element, with q_1, d_1, θ_1 and q_2, d_2, θ_2 being the local DOFs at the two nodes. q_1, and q_2 are the longitudinal displacements, d_1, and d_2 are the transverse displacements, and θ_1 and θ_2 are the rotations. For ease of formulation of the equilibrium equations, the element's geometry is mapped into a local coordinate system with coordinate ξ, as shown in the figure.

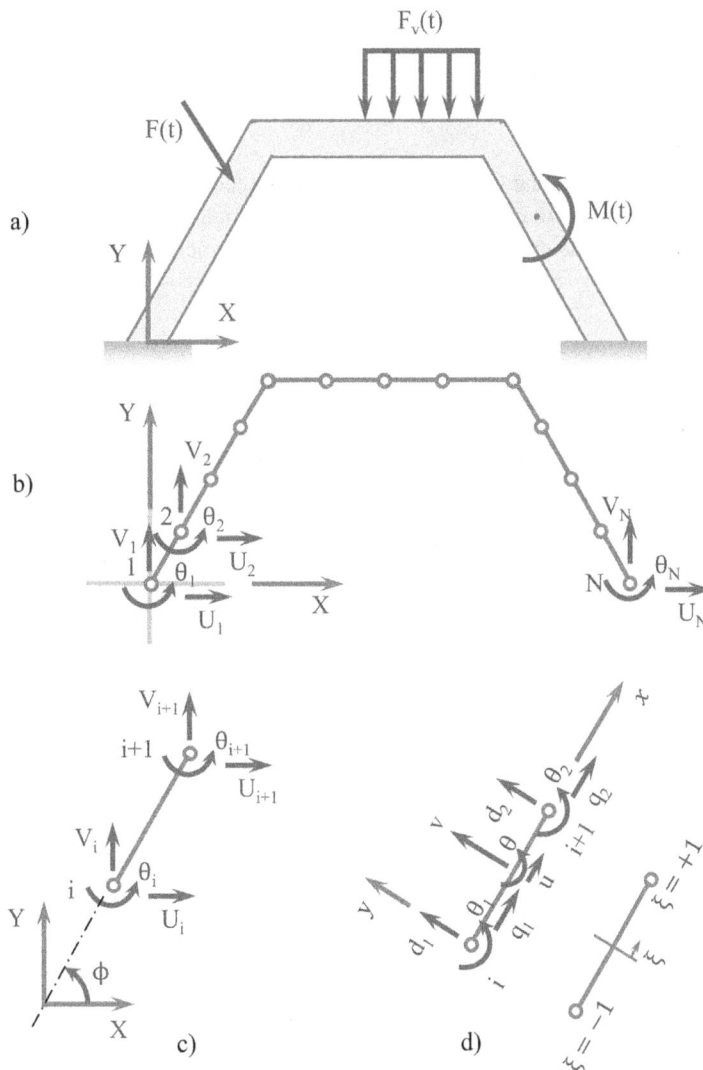

FIGURE 3.4 (a) Vibration of a frame, (b) FE discretization, (c) frame element in the global coordinate system, and (d) frame element in the local coordinate system.

TABLE 3.1

Summary of Notations Used for Frame Element

X, Y	Global coordinates
x, y	Local coordinates
U, V	Global displacements
u, v	Local displacements
U_i, V_i, θ_i and U_{i+1}, V_{i+1}, θ_{i+1}	Global DOFs
q_1, d_1, θ_1 and q_2, d_2, θ_2	Local DOFs

For clarity, the notations used are summarized in Table 3.1. Note that the notations for the local displacement and DOFs are the same as they were for the bar and beam elements.

3.9.1 Element Stiffness Matrix

The energy functions can be written in terms of the global or local DOFs. In the present case, it would be convenient to write the expressions using local DOFs, since they consist of the DOFs of the bar and beam elements. Since the DOFs of the bar and beam elements are uncoupled, the strain energy of a frame element is equal to the sum of the strain energies of the bar and beam elements. Hence, making use of the strain energy expressions from the previous two sections, we get

$$U_e = \int_{x_i}^{x_{i+1}} \frac{1}{2} E_e A_e \left(\frac{\partial u}{\partial x} \right)^2 . dx + \int_{x_i}^{x_{i+1}} \frac{1}{2} E_e I_e \left(\frac{\partial^2 v}{\partial x^2} \right)^2 dx \tag{3.99}$$

where the coordinate x is the local coordinate aligned with the element length, as shown in Figure 3.4d. We approximate the longitudinal displacement u using the shape functions for the bar element (represented below by $[N_u]$). The transverse displacement v is approximated using the shape functions for the beam element (represented below by $[N_v]$).

Thus, we have

$$u = N_{u1} q_1 + N_{u2} q_2 = \underset{1 \times 2}{[N_u]} \underset{2 \times 1}{\{q\}} \tag{3.100}$$

$$v = N_{v1} d_1 + N_{v2} \theta_1 + N_{v3} d_2 + N_{v4} \theta_2 = \underset{1 \times 4}{[N_v]} \underset{4 \times 1}{\{d\}} \tag{3.101}$$

We know the following relationships from the previous two sections,

$$\frac{d\xi}{dx} = \frac{2}{(x_i - x_{i+1})} = \frac{2}{L_e} \tag{3.102}$$

$$\frac{\partial u}{\partial x} = \frac{\partial u}{\partial \xi} \cdot \frac{2}{L_e} \tag{3.103}$$

$$\frac{\partial^2 v}{\partial x^2} = \frac{\partial^2 v}{\partial \xi^2} \cdot \frac{4}{L_e^2} \tag{3.104}$$

Making use of Eqs. (3.102)–(3.104), the strain energy of the frame element (Eq. 3.99) can be expressed in the ξ coordinate system as

$$U_e = \frac{1}{2} \int_{-1}^{+1} \frac{2 E_e A_e}{L_e} \left(\frac{\partial u}{\partial \xi} \right)^T \left(\frac{\partial u}{\partial \xi} \right) . d\xi + \frac{1}{2} \int_{-1}^{+1} \frac{8 E_e I_e}{L_e^3} \left(\frac{\partial^2 v}{\partial \xi^2} \right)^T \left(\frac{\partial^2 v}{\partial \xi^2} \right) . d\xi \tag{3.105}$$

Substituting Eqs. (3.100) and (3.101) into Eq. (3.105), we get

$$U_e = \frac{1}{2}\{q\}^T\left(\frac{2E_eA_e}{L_e}\int_{-1}^{+1}\frac{\partial[N_u]^T}{\partial\xi}\frac{\partial[N_u]}{\partial\xi}d\xi\right)\{q\} + \frac{1}{2}\{d\}^T\left(\frac{8E_eI_e}{L_e^3}\int_{-1}^{+1}\frac{\partial^2[N_v]^T}{\partial\xi^2}\frac{\partial^2[N_v]}{\partial\xi^2}d\xi\right)\{d\} \quad (3.106)$$

We now define the vectors of local DOFs ($\{g'\}$) and global DOFs ($\{g\}$) of the frame element as

$$\{g'\} = \left\{\begin{matrix} q_1 & d_1 & \theta_1 & q_2 & d_2 & \theta_2 \end{matrix}\right\}^T$$
$$\{g\} = \left\{\begin{matrix} U_i & V_i & \theta_i & U_{i+1} & V_{i+1} & \theta_{i+1} \end{matrix}\right\}^T \quad (3.107)$$

The shape function matrices $[N_u]$ (of the bar element) and $[N_v]$ (of the beam element) given in the previous two sections are substituted in Eq. (3.106), and the integrations are carried out. The two terms of the strain energy are combined into a single term using the vector $\{g'\}$. The resulting equation for the strain energy can be expressed in the following form:

$$U_e = \frac{1}{2}\{g'\}^T[K']_e\{g'\} \quad (3.108)$$

where

$$[K']_e = \begin{bmatrix} \dfrac{E_eA_e}{L_e} & 0 & 0 & -\dfrac{E_eA_e}{L_e} & 0 & 0 \\[2mm] 0 & \dfrac{12E_eI_e}{L_e^3} & \dfrac{6E_eI_e}{L_e^2} & 0 & -\dfrac{12E_eI_e}{L_e^3} & \dfrac{6E_eI_e}{L_e^2} \\[2mm] 0 & \dfrac{6E_eI_e}{L_e^2} & \dfrac{4E_eI_e}{L_e} & 0 & -\dfrac{6E_eI_e}{L_e^2} & \dfrac{2E_eI_e}{L_e} \\[2mm] -\dfrac{E_eA_e}{L_e} & 0 & 0 & \dfrac{E_eA_e}{L_e} & 0 & 0 \\[2mm] 0 & -\dfrac{12E_eI_e}{L_e^3} & -\dfrac{6E_eI_e}{L_e^2} & 0 & \dfrac{12E_eI_e}{L_e^3} & -\dfrac{6E_eI_e}{L_e^2} \\[2mm] 0 & \dfrac{6E_eI_e}{L_e^2} & \dfrac{2E_eI_e}{L_e} & 0 & -\dfrac{6E_eI_e}{L_e^2} & \dfrac{4E_eI_e}{L_e} \end{bmatrix} \quad (3.109)$$

The element stiffness matrix $[K']_e$ relates to the local DOF vector. By coordinate transformation, it can be shown that

$$\{g'\} = [L] \ \{g\} \quad (3.110)$$
$$\phantom{\{g'\} = }{}_{6\times1} \quad {}_{6\times6} \quad {}_{6\times1}$$

where $[L]$ is the transformation matrix between the local and global DOF vectors. If $l = \cos\phi$ and $m = \sin\phi$, with ϕ being the angle made by the element from the +X axis, then the matrix $[L]$ is given by

$$[L] = \begin{bmatrix} l & m & 0 & 0 & 0 & 0 \\ -m & l & 0 & 0 & 0 & 0 \\ 0 & 0 & 1 & 0 & 0 & 0 \\ 0 & 0 & 0 & l & m & 0 \\ 0 & 0 & 0 & -m & l & 0 \\ 0 & 0 & 0 & 0 & 0 & 1 \end{bmatrix} \quad (3.111)$$

Substituting Eq. (3.110) into Eq. (3.108), we get

$$U_e = \frac{1}{2}\{g\}^T [L]^T [K']_e [L]\{g\} \tag{3.112}$$

Comparing the expression in (3.112) with the expression for the potential energy of an MDOF system, $U = \frac{1}{2}\{q\}^T [K]\{q\}$, we get the stiffness matrix for the frame element relating to the global DOF vector as

$$[K]_e = [L]^T [K']_e [L] \tag{3.113}$$

3.9.2 ELEMENT MASS MATRIX

The kinetic energy of a frame element is due to longitudinal and transverse velocities and is given by

$$T_e = \int\limits_{x_i}^{x_{i+1}} \frac{1}{2} \rho_e A_e \dot{u}^2 \, dx + \int\limits_{x_i}^{x_{i+1}} \frac{1}{2} \rho_e A_e \dot{v}^2 \, dx \tag{3.114}$$

Using Eq. (3.102), Eq. (3.114) can be written in the local coordinate system as

$$T_e = \frac{1}{2} \int\limits_{-1}^{+1} \frac{\rho_e A_e L_e}{2} \left(\frac{\partial u}{\partial t}\right)^T \left(\frac{\partial u}{\partial t}\right) d\xi + \frac{1}{2} \int\limits_{-1}^{+1} \frac{\rho_e A_e L_e}{2} \left(\frac{\partial v}{\partial t}\right)^T \left(\frac{\partial v}{\partial t}\right) d\xi \tag{3.115}$$

Differentiating Eqs. (3.100) and (3.101) with time and substituting them in Eq. (3.115), we get

$$T_e = \frac{1}{2}\{\dot{q}\}^T \left(\frac{\rho_e A_e L_e}{2} \int\limits_{-1}^{+1} [N_u]^T [N_u] d\xi\right)\{\dot{q}\} + \frac{1}{2}\{\dot{d}\}^T \left(\frac{\rho_e A_e L_e}{2} \int\limits_{-1}^{+1} [N_v]^T [N_v] d\xi\right)\{\dot{d}\} \tag{3.116}$$

The shape function matrices $[N_u]$ and $[N_v]$ are substituted in Eq. (3.116), and the integrations are carried out. The two terms of the kinetic energy are combined into a single term using the vector $\{g'\}$. The resulting equation for the kinetic energy can be expressed in the following form:

$$T_e = \frac{1}{2}\{g'\}^T [M']_e \{g'\} \tag{3.117}$$

where

$$[M']_e = \begin{bmatrix} 2a & 0 & 0 & a & 0 & 0 \\ 0 & 156b & 22L_e b & 0 & 54b & -13L_e b \\ 0 & 22L_e b & 4L_e^2 b & 0 & 13L_e b & -3L_e^2 b \\ a & 0 & 0 & 2a & 0 & 0 \\ 0 & 54b & 13L_e b & 0 & 156b & -22L_e b \\ 0 & -13L_e b & -3L_e^2 b & 0 & -22L_e b & 4L_e^2 b \end{bmatrix} \tag{3.118}$$

and $a = \dfrac{\rho_e A_e L_e}{6}$ and $b = \dfrac{\rho_e A_e L_e}{420}$. The element mass matrix $[M']_e$ is related to the local DOF vector. Substituting Eq.(3.110) into Eq. (3.117), we get

$$T_e = \frac{1}{2}\{g\}^T [L]^T [M']_e [L]\{g\} \tag{3.119}$$

Comparing the expression in Eq. (3.119) with the expression for the kinetic energy of an MDOF system, $T = \frac{1}{2}\{q\}^T[M]\{q\}$, we get the mass matrix for the frame element relating to the global DOF vector as

$$[M]_e = [L]^T[M']_e[L] \tag{3.120}$$

The load vector also can be similarly obtained by summing the virtual work of the external forces.

3.10   OBTAINING EQUATIONS OF MOTION OF THE SYSTEM

In Sections 3.7–3.9, we looked at the formulation of the bar, beam, and frame elements. The element stiffness and mass matrices were obtained from the strain and kinetic energy functions. However, our objective is to obtain the dynamic equilibrium equations for the whole system represented by the FE model, and hence we need to obtain the stiffness and mass matrices and the force vector for the FE model. We follow the same basic approach used for deriving the elemental matrices from the energy functions to obtain the FE model matrices.

We determine the energy functions for the whole system by summing the energy functions for the individual finite elements. We then express the total energy in terms of the vector of the global DOFs of the system, which yields the matrices for the whole system. This process is called the 'assembly' of the elemental matrices.

3.10.1   ASSEMBLY OF THE ELEMENTAL MATRICES

Let us consider the assembly of the elemental stiffness matrices. The procedure can be extended to the assembly of the elemental mass matrices.

The strain energy of a finite element is given by (Eq. 3.55)

$$U_e = \frac{1}{2}\{q\}^T[K]_e\{q\} \tag{3.121}$$

where $\{q\}$ is the vector of the local DOFs of the finite element. Different finite elements are associated with a different set of global DOFs. Therefore, when Eq. (3.121) corresponding to all the finite elements is added to obtain the total energy, each elemental energy term is associated with a different set of DOFs, and therefore all the terms cannot be directly combined into a single expression. To combine the elemental energy functions, we first relate the vector of elemental DOFs to the global DOF vector. We then use this relationship to express the elemental energy functions in terms of the global DOF vector.

To illustrate the whole process, let us take an example of a longitudinal bar modeled with two bar elements, as shown in Figure 3.5.

The element stiffness matrices of the two elements are

$$[K]_1 = \frac{E_1 A_1}{L_1}\begin{bmatrix} 1 & -1 \\ -1 & 1 \end{bmatrix} \tag{3.122}$$

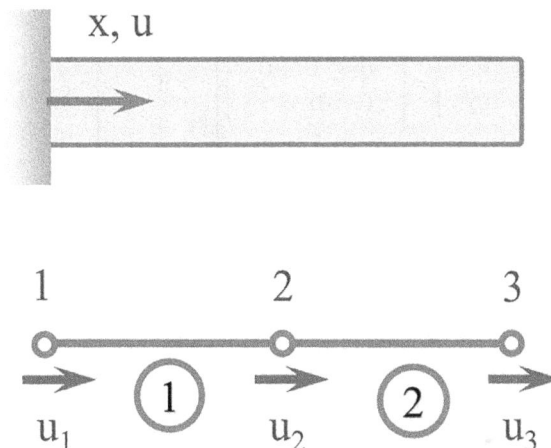

FIGURE 3.5   Longitudinal bar modeled with two bar elements.

$$[K]_2 = \frac{E_2 A_2}{L_2}\begin{bmatrix} 1 & -1 \\ -1 & 1 \end{bmatrix} \tag{3.123}$$

The global DOF vector for the FE model (shown in Figure 3.5) is

$$\{u\} = \{u_1 \quad u_2 \quad u_3\}^T \tag{3.124}$$

The local DOF vector for element 1 is

$$\{q\} = \{q_1 \quad q_2\}^T \tag{3.125}$$

Now, we relate $\{u\}$ and $\{q\}$ as follows.

$$\{q\} = \left\{ \begin{array}{c} q_1 \\ q_2 \end{array} \right\} = \begin{bmatrix} 1 & 0 & 0 \\ 0 & 1 & 0 \end{bmatrix} \left\{ \begin{array}{c} u_1 \\ u_2 \\ u_3 \end{array} \right\} = [A]_1\{u\} \tag{3.126}$$

Substituting Eq. (3.126) into Eq. (3.121), the strain energy for element 1 in terms of the global DOF vector is obtained as

$$U_1 = \frac{1}{2}\{u\}^T [A]_1^T [K]_1 [A]_1 \{u\} \tag{3.127}$$

The matrix $[A]_1^T [K]_1 [A]_1$ is the stiffness matrix of the first element expanded to the global size (denoted by $[K^G]_1$) and is given by

$$[K^G]_1 = [A]_1^T [K]_1 [A]_1 = \begin{bmatrix} \dfrac{E_1 A_1}{L_1} & -\dfrac{E_1 A_1}{L_1} & 0 \\ -\dfrac{E_1 A_1}{L_1} & \dfrac{E_1 A_1}{L_1} & 0 \\ 0 & 0 & 0 \end{bmatrix} \tag{3.128}$$

Similarly, the local DOF vector for element 2 is

$$\{q\} = \{q_1 \quad q_2\}^T \tag{3.129}$$

We can relate $\{u\}$ and $\{q\}$ as

$$\{q\} = \left\{ \begin{array}{c} q_1 \\ q_2 \end{array} \right\} = \begin{bmatrix} 0 & 1 & 0 \\ 0 & 0 & 1 \end{bmatrix} \left\{ \begin{array}{c} u_1 \\ u_2 \\ u_3 \end{array} \right\} = [A]_2\{u\} \tag{3.130}$$

Substituting Eq. (3.130) into Eq. (3.121), the strain energy for element 2 in terms of the global DOF vector is obtained as

$$U_2 = \frac{1}{2}\{u\}^T [A]_2^T [K]_2 [A]_2 \{u\} \tag{3.131}$$

The matrix $[A]_2^T [K]_2 [A]_2$ is the stiffness matrix of the second element expanded to the global size (denoted by $[K^G]_2$) and is given by

$$\left[K^G\right]_2 = [A]_2^T [K]_2 [A]_2 = \begin{bmatrix} 0 & 0 & 0 \\ 0 & \dfrac{E_2A_2}{L_2} & -\dfrac{E_2A_2}{L_2} \\ 0 & -\dfrac{E_2A_2}{L_2} & \dfrac{E_2A_2}{L_2} \end{bmatrix} \tag{3.132}$$

The total strain energy is

$$U = U_1 + U_2 = \frac{1}{2}\{u\}^T \left[K^G\right]_1 \{u\} + \frac{1}{2}\{u\}^T \left[K^G\right]_2 \{u\} = \frac{1}{2}\{u\}^T \left(\left[K^G\right]_1 + \left[K^G\right]_2\right)\{u\} \tag{3.133}$$

Substituting $\left[K^G\right]_1$ and $\left[K^G\right]_2$ into Eq. (3.133), we get

$$U = \frac{1}{2}\left\{\begin{matrix} u_1 \\ u_2 \\ u_3 \end{matrix}\right\}^T \left(\begin{bmatrix} \dfrac{E_1A_1}{L_1} & -\dfrac{E_1A_1}{L_1} & 0 \\ -\dfrac{E_1A_1}{L_1} & \dfrac{E_1A_1}{L_1} & 0 \\ 0 & 0 & 0 \end{bmatrix} + \begin{bmatrix} 0 & 0 & 0 \\ 0 & \dfrac{E_2A_2}{L_2} & -\dfrac{E_2A_2}{L_2} \\ 0 & -\dfrac{E_2A_2}{L_2} & \dfrac{E_2A_2}{L_2} \end{bmatrix}\right)\left\{\begin{matrix} u_1 \\ u_2 \\ u_3 \end{matrix}\right\} \tag{3.134}$$

which gives

$$U = \frac{1}{2}\left\{\begin{matrix} u_1 \\ u_2 \\ u_3 \end{matrix}\right\}^T \begin{bmatrix} \dfrac{E_1A_1}{L_1} & -\dfrac{E_1A_1}{L_1} & 0 \\ -\dfrac{E_1A_1}{L_1} & \dfrac{E_1A_1}{L_1}+\dfrac{E_2A_2}{L_2} & -\dfrac{E_2A_2}{L_2} \\ 0 & -\dfrac{E_2A_2}{L_2} & \dfrac{E_2A_2}{L_2} \end{bmatrix}\left\{\begin{matrix} u_1 \\ u_2 \\ u_3 \end{matrix}\right\} \tag{3.135}$$

Thus, the stiffness matrix of the FE model is

$$\left[\bar{K}\right] = \begin{bmatrix} \dfrac{E_1A_1}{L_1} & -\dfrac{E_1A_1}{L_1} & 0 \\ -\dfrac{E_1A_1}{L_1} & \dfrac{E_1A_1}{L_1}+\dfrac{E_2A_2}{L_2} & -\dfrac{E_2A_2}{L_2} \\ 0 & -\dfrac{E_2A_2}{L_2} & \dfrac{E_2A_2}{L_2} \end{bmatrix} \tag{3.136}$$

We draw from the above the following conclusions for assembling the element matrices.

- An element stiffness matrix ($[K]_e$) can be expanded to the global size ($\left[K^G\right]_e$) based on the correspondence between the local and global DOFs of the element.
- The assembled stiffness matrix ($\left[\bar{K}\right]$) is the sum of the element stiffness matrices expanded to the global size (i.e. $\left[K^G\right]_1$ and $\left[K^G\right]_2$). This observation forms the basis of the assembly of the element stiffness matrices to obtain the global stiffness matrix for the FE model.
- A more efficient procedure that doesn't require the element matrices to be explicitly expanded to the global size can be adopted. The rows and columns of an element matrix correspond to local DOFs. Each local DOF has a corresponding index in the global DOF vector. Therefore, for assembling an element matrix to the global matrix, each entry of the element matrix is added to the corresponding entry in the global matrix based on the correspondence between the local and global DOFs.
- The above procedure can also be used to assemble the element force vectors corresponding to the body and distributed forces. The points forces, if any, at the nodes of the FE model are directly added to the relevant DOFs in the assembled force vector.

3.10.2 INCORPORATING BCs

Once the global mass and stiffness matrices and the force vector are obtained by assembling the elemental matrices and vectors, the next step is to incorporate the BCs of the problem. These are often in the form of constraints on displacements and slopes at specific nodes.

For example, in the case of the longitudinal bar in Figure 3.5, the left end is fixed. Therefore, the corresponding node, i.e., node 1, is fixed, and thus, $u_1 = 0$ is the boundary condition. The assembled matrices and vectors need to be modified to satisfy this condition. To incorporate this condition for the stiffness matrix, we substitute $u_1 = 0$ in the total strain energy expression (Eq. (3.135). Simplification of the equation results in the following expression for the strain energy of the system:

$$U = \frac{1}{2} \left\{ \begin{array}{c} u_2 \\ u_3 \end{array} \right\}^{T} \times \left[\begin{array}{cc} \dfrac{E_1 A_1}{L_1} + \dfrac{E_2 A_2}{L_2} & -\dfrac{E_2 A_2}{L_2} \\ -\dfrac{E_2 A_2}{L_2} & -\dfrac{E_2 A_2}{L_2} \end{array} \right] \left\{ \begin{array}{c} u_2 \\ u_3 \end{array} \right\} \tag{3.137}$$

Therefore, the stiffness matrix, after incorporating the boundary condition, is

$$[K] = \left[\begin{array}{cc} \dfrac{E_1 A_1}{L_1} + \dfrac{E_2 A_2}{L_2} & -\dfrac{E_2 A_2}{L_2} \\ -\dfrac{E_2 A_2}{L_2} & -\dfrac{E_2 A_2}{L_2} \end{array} \right] \tag{3.138}$$

If we compare the matrix $[K]$ obtained after applying the boundary condition with the one before it, i.e. $[\bar{K}]$, then we observe that the first row and column of $[\bar{K}]$ are eliminated. This result can be generalized, and we say that if any DOF is fixed, then the corresponding row and column from the stiffness matrix should be eliminated. The BCs should also be applied to the mass matrix to obtain the matrix $[M]$ from $[\bar{M}]$. For the force vector, only the corresponding rows in the vector $\{\bar{F}\}$ need to be eliminated to obtain $\{F\}$.

Once all the BCs are imposed, the equation of motion based on the finite element model, assuming proportional damping, is obtained as

$$[M]\{\ddot{u}\} + [C]\{\dot{u}\} + [K]\{u\} = \{F\} \tag{3.139}$$

3.10.3 DECIDING THE NUMBER OF FINITE ELEMENTS

The number of finite elements to be used in an FE model should be enough so that there is negligible discretization error in the model for the output to be predicted. The number of finite elements can be decided using a convergence study. The FE models with an increasing number of finite elements can be solved for the output. The number of finite elements, when the change in the output of interest becomes negligible, can be considered adequate.

Example 3.1: FE Model for the Longitudinal Vibrations Using Bar Element

Obtain the equation of motion for the longitudinal vibrations of the propulsion system shown in Figure 2.10 (in Chapter 2). Use three bar elements to model the system.

Solution

The propulsion system modeled using three bar elements is shown in Figure 3.6. The global DOFs are also shown. Each section of the propeller shaft is modeled using only one element for illustration purposes. The masses of the propeller blades (m_b) and the two couplings (m_{c1} and m_{c2}) are lumped at nodes 1, 2, and 3, respectively. The propeller is assumed to be fixed at the thrust bearing location.

The cross-sectional area, modulus of elasticity, second moment of area, and density are assumed to be the same for the three elements, and only the lengths are different.

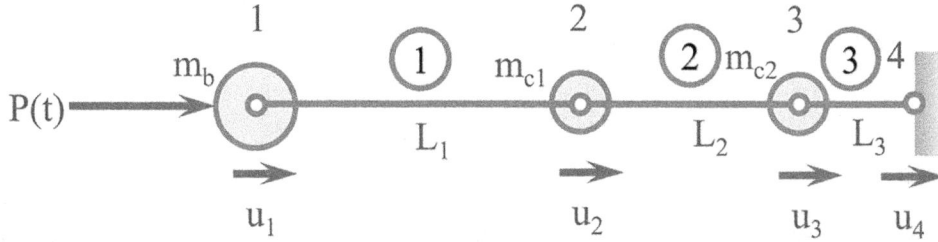

FIGURE 3.6   FE model for the longitudinal vibrations of the propulsion system.

a. Element stiffness matrices

The element stiffness matrices are obtained using the expression given in Eq. (3.58). Let, $k_1 = EA/L_1, k_2 = EA/L_2$ and $k_3 = EA/L_3$. The elemental DOF vector in the global coordinates for each element is also given below, which is later needed to assemble the element matrix.

$$[K]_1 = k_1 \begin{bmatrix} 1 & -1 \\ -1 & 1 \end{bmatrix}, \quad \{q\}_1 = \{u_1 \quad u_2\}^T$$

$$[K]_2 = k_2 \begin{bmatrix} 1 & -1 \\ -1 & 1 \end{bmatrix}, \quad \{q\}_2 = \{u_2 \quad u_3\}^T$$

$$[K]_3 = k_3 \begin{bmatrix} 1 & -1 \\ -1 & 1 \end{bmatrix}, \quad \{q\}_3 = \{u_3 \quad u_4\}^T$$

b. Element mass matrices

The element stiffness matrices are obtained using the expression given in Eq. (3.65). Let, $m_1 = \rho A L_1$, $m_2 = \rho A L_2$ and $m_3 = \rho A L_3$.

$$[M]_1 = \frac{m_1}{6} \begin{bmatrix} 2 & 1 \\ 1 & 2 \end{bmatrix}, \quad [M]_2 = \frac{m_2}{6} \begin{bmatrix} 2 & 1 \\ 1 & 2 \end{bmatrix} \quad \text{and} \quad [M]_3 = \frac{m_3}{6} \begin{bmatrix} 2 & 1 \\ 1 & 2 \end{bmatrix}$$

c. Global stiffness matrix

The global DOF vector is

$$\{\bar{u}\} = \{u_1 \quad u_2 \quad u_3 \quad u_4\}^T$$

Element stiffness matrices are assembled using the procedure explained in Section 3.10.1 to obtain $[\bar{K}]$.

$$[\bar{K}] = \begin{bmatrix} k_1 & -k_1 & 0 & 0 \\ -k_1 & k_1 + k_2 & -k_2 & 0 \\ 0 & -k_2 & k_2 + k_3 & -k_3 \\ 0 & 0 & -k_3 & k_3 \end{bmatrix}$$

d. Global mass matrix

Element mass matrices are assembled to obtain $[\bar{M}]$. In addition, the lumped masses m_b, m_{c1} and m_{c2} are assembled at DOFs 1, 2, and 3, respectively. It gives

$$[\bar{M}] = \begin{bmatrix} \dfrac{2m_1}{6} + m_b & \dfrac{m_1}{6} & 0 & 0 \\[2ex] \dfrac{m_1}{6} & \dfrac{2m_1 + 2m_2}{6} + m_{c1} & \dfrac{m_2}{6} & 0 \\[2ex] 0 & \dfrac{m_2}{6} & \dfrac{2m_2 + 2m_3}{6} + m_{c2} & \dfrac{m_3}{6} \\[2ex] 0 & 0 & \dfrac{m_3}{6} & \dfrac{2m_3}{6} \end{bmatrix}$$

e. Force vector

There is no distributed/body force, but one point force is at the 1st DOF. Therefore,

$$\{\bar{F}\} = \{P(t) \quad 0 \quad 0 \quad 0\}^T$$

f. Impose BCs

There is one boundary condition, $u_4 = 0$. The fourth row and column of the stiffness and mass matrices are eliminated to incorporate this boundary condition. The fourth row of the force vector is also eliminated. The global matrices and force vector after incorporating the boundary condition are

$$[K] = \begin{bmatrix} k_1 & -k_1 & 0 \\ -k_1 & k_1 + k_2 & -k_2 \\ 0 & -k_2 & k_2 + k_3 \end{bmatrix}$$

$$[M] = \begin{bmatrix} \dfrac{2m_1}{6} + m_b & \dfrac{m_1}{6} & 0 \\ \dfrac{m_1}{6} & \dfrac{2m_1 + 2m_2}{6} + m_{c1} & \dfrac{m_2}{6} \\ 0 & \dfrac{m_2}{6} & \dfrac{2m_2 + 2m_3}{6} + m_{c2} \end{bmatrix}$$

$$\{F\} = \{P(t) \quad 0 \quad 0\}^T$$

The global DOF vector is

$$\{u\} = \{u_1 \quad u_2 \quad u_3\}^T$$

The equation of motion is given by

$$[M]\{\ddot{u}\} + [K]\{u\} = \{F\}$$

Example 3.2: FE Model for Transverse Vibrations Using Beam Elements

Using beam elements, obtain the equations of motion for the transverse vibrations of the drilling machine shown in Figure 2.32 (in Chapter 2).

Solution

For illustration purposes, the drilling machine in this example is modeled using two beam elements, as shown in Figure 3.7. The global DOFs are also shown. The mass of the drilling head (m_d) and table (m_t) are lumped at nodes 1 and 2, respectively. The machine column is assumed to be fixed at the bottom.

The cross-sectional area, modulus of elasticity, second moment of area of the column cross-section, density, and length are assumed to be the same for the two elements.

a. Element stiffness matrices

The element stiffness matrices are obtained using the expression given in Eq. (3.88). Let $k_0 = EA/L^3$. The elemental DOF vector in the global coordinates for each element is also given below, which is later needed to assemble the element matrix.

$$[K]_1 = k_0 \begin{bmatrix} 12 & 6L & -12 & 6L \\ 6L & 4L^2 & -6L & 2L^2 \\ -12 & -6L & 12 & -6L_e \\ 6L & 2L^2 & -6L & 4L^2 \end{bmatrix},$$

$$\{d\}_1 = \{v_1 \quad v_2 \quad v_3 \quad v_4\}^T$$

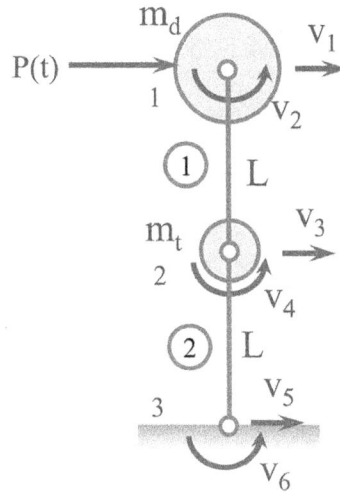

FIGURE 3.7 FE model for the transverse vibrations of the drilling machine.

$$[K]_2 = [K]_1, \quad \{d\}_2 = \{v_3 \quad v_4 \quad v_5 \quad v_6\}^T$$

b. Element mass matrices

The element mass matrices are obtained using the expression given in Eq. (3.94). Let $a = \rho AL/420$.

$$[M]_1 = \begin{bmatrix} 156a & 22La & 54a & -13La \\ 22La & 4L^2a & 13La & -3L^2a \\ 54a & 13La & 156a & -22La \\ -13La & -3L^2a & -22La & 4L^2a \end{bmatrix}$$

$$[M]_2 = [M]_1$$

c. Global stiffness matrix

The global DOF vector is

$$\{\bar{u}\} = \{v_1 \quad v_2 \quad v_3 \quad v_4 \quad v_5 \quad v_6\}^T$$

Element stiffness matrices are assembled using the procedure explained in Section 3.10.1 to obtain $[\bar{K}]$.

$$[\bar{K}] = k_0 \begin{bmatrix} 12 & 6L & -12 & 6L & 0 & 0 \\ 6L & 4L^2 & -6L & 2L^2 & 0 & 0 \\ -12 & -6L & 24 & 0 & -12 & 6L \\ 6L & 2L^2 & 0 & 8L^2 & -6L & 2L^2 \\ 0 & 0 & -12 & -6L & 12 & -6L \\ 0 & 0 & 6L & 2L^2 & -6L & 4L^2 \end{bmatrix}$$

d. Global mass matrix

Element mass matrices are assembled to obtain the global mass matrix $[\bar{M}]$. In addition, the lumped masses m_d and m_t are assembled at DOFs 1 and 3, respectively. The rotational inertia of these masses is neglected. We get

$$[\bar{M}] = \begin{bmatrix} 156a + m_d & 22La & 54a & -13La & 0 & 0 \\ 22La & 4L^2a & 13La & -3L^2a & 0 & 0 \\ 54a & 13La & 312La + m_t & 0 & 54a & -13La \\ -13La & -3L^2a & 0 & 8L^2a & 13La & -3L^2a \\ 0 & 0 & 54a & 13La & 156a & -22La \\ 0 & 0 & -13La & -3L^2a & -22La & 4L^2a \end{bmatrix}$$

e. Force vector

There is no distributed/body force, but there is one point force at the first DOF. Therefore,

$$\{\bar{F}\}=\{P(t) \quad 0 \quad 0 \quad 0 \quad 0 \quad 0\}^T$$

f. Impose BCs

Since the drilling machine base is fixed, the DOFs at node 3 are also fixed. Hence, the BCs are, $v_5 = 0$ and $v_6 = 0$. Therefore, the row and column numbers 5 and 6 of the stiffness and mass matrices are eliminated to incorporate the BCs. The corresponding rows of the force vector are also eliminated. After incorporating the boundary condition, the global matrices and force vector are

$$[K]=k_0 \begin{bmatrix} 12 & 6L & -12 & 6L \\ 6L & 4L^2 & -6L & 2L^2 \\ -12 & -6L & 24 & 0 \\ 6L & 2L^2 & 0 & 8L^2 \end{bmatrix}$$

$$[M]= \begin{bmatrix} 156a+m_d & 22La & 54a & -13La \\ 22La & 4L^2a & 13La & -3L^2a \\ 54a & 13La & 312a+m_t & 0 \\ -13La & -3L^2a & 0 & 8L^2a \end{bmatrix}$$

$$\{F\}=\{P(t) \quad 0 \quad 0 \quad 0\}^T$$

The global DOF vector is

$$\{u\}=\{v_1 \quad v_2 \quad v_3 \quad v_4\}^T$$

The equation of motion is

$$[M]\{\ddot{u}\}+[K]\{u\}=\{F\}$$

Example 3.3: FE Model for the Transverse Vibrations Using Frame Elements

Using frame elements, obtain the governing equations of motion for the transverse vibrations of the machine frame shown in Figure 2.39 (in Chapter 2).

Solution

For illustration purposes, the machine frame in this example is modeled using two frame elements, as shown in Figure 3.8. The global DOFs are also shown. The mass of the electric motor (m_m) is lumped at node 3. The base of the machine is assumed rigid, and the machine column is assumed fixed at the bottom. The cross-sectional area, modulus of elasticity, second moment of area of cross-section, density, and length are assumed to be the same for the two elements.

a. Element stiffness matrices

The element stiffness matrices are obtained using Eqs. (3.109) and (3.113). Let, $k_1 = EA/L$ and $k_2 = EI/L^3$. The elemental DOF vector in the global coordinates for each element is also given below, which is later needed to assemble the element matrix.

Element 1

The stiffness matrix for element 1 in the local coordinates is

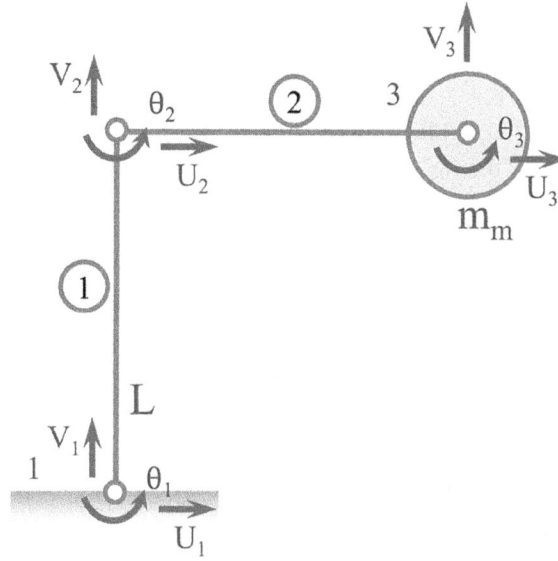

FIGURE 3.8 FE model for the transverse vibrations of the machine frame.

$$[K']_1 = \begin{bmatrix} k_1 & 0 & 0 & -k_1 & 0 & 0 \\ 0 & 12k_2 & 6k_2L & 0 & -12k_2 & 6k_2L \\ 0 & 6k_2L & 4k_2L^2 & 0 & -6k_2L & 2k_2L^2 \\ -k_1 & 0 & 0 & k_1 & 0 & 0 \\ 0 & -12k_2 & -6k_2L & 0 & 12k_2 & -6k_2L \\ 0 & 6k_2L & 2k_2L^2 & 0 & -6k_2L & 4k_2L^2 \end{bmatrix}$$

The transformation matrix between the local and global DOF vectors is given by Eq. (3.111). Since $\phi = 90°$, the direction cosines are, $l = \cos\phi = 0$ and $m = \sin\phi = 1$, the transformation matrix is

$$[L] = \begin{bmatrix} 0 & 1 & 0 & 0 & 0 & 0 \\ -1 & 0 & 0 & 0 & 0 & 0 \\ 0 & 0 & 1 & 0 & 0 & 0 \\ 0 & 0 & 0 & 0 & 1 & 0 \\ 0 & 0 & 0 & -1 & 0 & 0 \\ 0 & 0 & 0 & 0 & 0 & 1 \end{bmatrix}$$

The element stiffness matrix in the global coordinates is computed from Eq. (3.113) as

$$[K]_1 = [L]^T[K']_1[L] = \begin{bmatrix} 12k_2 & 0 & -6k_2L & -12k_2 & 0 & -6k_2L \\ 0 & k_1 & 0 & 0 & -k_1 & 0 \\ -6k_2L & 0 & 4k_2L^2 & 6k_2L & 0 & 2k_2L^2 \\ -12k_2 & 0 & 6k_2L & 12k_2 & 0 & 6k_2L \\ 0 & -k_1 & 0 & 0 & k_1 & 0 \\ -6k_2L & 0 & 2k_2L^2 & 6k_2L & 0 & 4k_2L^2 \end{bmatrix}$$

The elemental DOF vector in the global coordinates is

$$\{g\} = \{U_1 \quad V_1 \quad \theta_1 \quad U_2 \quad V_2 \quad \theta_2\}^T$$

Element 2

The stiffness matrix for element 2 in the local coordinates is

$$
\left[K'\right]_2 =
\begin{bmatrix}
k_1 & 0 & 0 & -k_1 & 0 & 0 \\
0 & 12k_2 & 6k_2L & 0 & -12k_2 & 6k_2L \\
0 & 6k_2L & 4k_2L^2 & 0 & -6k_2L & 2k_2L^2 \\
-k_1 & 0 & 0 & k_1 & 0 & 0 \\
0 & -12k_2 & -6k_2L & 0 & 12k_2 & -6k_2L \\
0 & 6k_2L & 2k_2L^2 & 0 & -6k_2L & 4k_2L^2
\end{bmatrix}
$$

Since $\phi = 0°$, the stiffness matrices for element 2 in the local and global coordinates are the same. Hence,

$$\left[K\right]_2 = \left[K'\right]_2$$

The elemental DOF vector in the global coordinates is

$$\{g\} = \left\{U_2 \quad V_2 \quad \theta_2 \quad U_3 \quad V_3 \quad \theta_3\right\}^T$$

b. Element mass matrices

The element mass matrices are obtained using Eqs. (3.118) and (3.120).

Element 1

The mass matrix for element 1 in the local coordinates is

$$
\left[M'\right]_1 =
\begin{bmatrix}
2a & 0 & 0 & a & 0 & 0 \\
0 & 156b & 22L_e b & 0 & 54b & -13L_e b \\
0 & 22L_e b & 4L_e^2 b & 0 & 13L_e b & -3L_e^2 b \\
a & 0 & 0 & 2a & 0 & 0 \\
0 & 54b & 13L_e b & 0 & 156b & -22L_e b \\
0 & -13L_e b & -3L_e^2 b & 0 & -22L_e b & 4L_e^2 b
\end{bmatrix}
$$

where $a = \dfrac{\rho_e A_e L_e}{6}$ and $b = \dfrac{\rho_e A_e L_e}{420}$.

The transformation matrix between the local and global DOF vectors is the same as it was for finding the element stiffness matrix for this element.

The element mass matrix in the global coordinates is computed from Eq. (3.120) as

$$
\left[M\right]_1 = \left[L\right]^T\left[M'\right]_1\left[L\right] =
\begin{bmatrix}
156b & 0 & -22L_e b & 54b & 0 & 13L_e b \\
0 & 2a & 0 & 0 & a & 0 \\
-22L_e b & 0 & 4L_e^2 b & -13L_e b & 0 & -3L_e^2 b \\
54b & 0 & -13L_e b & 156b & 0 & 22L_e b \\
0 & a & 0 & 0 & 2a & 0 \\
13L_e b & 0 & -3L_e^2 b & 22L_e b & 0 & 4L_e^2 b
\end{bmatrix}
$$

Element 2

The mass matrix for element 2 in the local coordinates is

$$
\left[M'\right]_2 =
\begin{bmatrix}
2a & 0 & 0 & a & 0 & 0 \\
0 & 156b & 22L_e b & 0 & 54b & -13L_e b \\
0 & 22L_e b & 4L_e^2 b & 0 & 13L_e b & -3L_e^2 b \\
a & 0 & 0 & 2a & 0 & 0 \\
0 & 54b & 13L_e b & 0 & 156b & -22L_e b \\
0 & -13L_e b & -3L_e^2 b & 0 & -22L_e b & 4L_e^2 b
\end{bmatrix}
$$

Since $\phi = 0^\circ$, the mass matrices for element 2 in the local and global coordinates are the same. Hence,

$$[M]_2 = [M']_2$$

c. Global stiffness matrix
The global DOF vector is

$$\{\bar{u}\} = \{U_1 \quad V_1 \quad \theta_1 \quad U_2 \quad V_2 \quad \theta_2 \quad U_3 \quad V_3 \quad \theta_3\}^T$$

Element stiffness matrices are assembled using the procedure explained in Section 3.10.1 to obtain $[\bar{K}]$.

$$[\bar{K}] = \begin{bmatrix}
12k_2 & 0 & -6k_2L & -12k_2 & 0 & -6k_2L & 0 & 0 & 0 \\
0 & k_1 & 0 & 0 & -k_1 & 0 & 0 & 0 & 0 \\
-6k_2L & 0 & 4k_2L^2 & 6k_2L & 0 & 2k_2L^2 & 0 & 0 & 0 \\
-12k_2 & 0 & 6k_2L & 12k_2+k_1 & 0 & 6k_2L & -k_1 & 0 & 0 \\
0 & -k_1 & 0 & 0 & k_1+12k_2 & 6k_2L & 0 & -12k_2 & 6k_2L \\
-6k_2L & 0 & 2k_2L^2 & 6k_2L & 6k_2L & 8k_2L^2 & 0 & -6k_2L & 2k_2L^2 \\
0 & 0 & 0 & -k_1 & 0 & 0 & k_1 & 0 & 0 \\
0 & 0 & 0 & 0 & -12k_2 & -6k_2L & 0 & 12k_2 & -6k_2L \\
0 & 0 & 0 & 0 & 6k_2L & 2k_2L^2 & 0 & -6k_2L & 4k_2L^2
\end{bmatrix}$$

d. Global mass matrix
Element mass matrices are assembled to obtain $[\bar{M}]$. In addition, the lumped mass m_m is assembled at the DOFs U_3 and V_3. The rotational inertia of the mass is neglected. We get

$$[\bar{M}] = \begin{bmatrix}
156b & 0 & -22L_eb & 54b & 0 & 13L_eb & 0 & 0 & 0 \\
0 & 2a & 0 & 0 & a & 0 & 0 & 0 & 0 \\
-22L_eb & 0 & 4L_e^2b & -13L_eb & 0 & -3L_e^2b & 0 & 0 & 0 \\
54b & 0 & -13L_eb & 156b+2a & 0 & 22L_eb & a & 0 & 0 \\
0 & a & 0 & 0 & 2a+156b & 22L_eb & 0 & 54b & -13L_eb \\
13L_eb & 0 & -3L_e^2b & 22L_eb & 22L_eb & 8L_e^2b & 0 & 13L_eb & -3L_e^2b \\
0 & 0 & 0 & a & 0 & 0 & 2a+m_m & 0 & 0 \\
0 & 0 & 0 & 0 & 54b & 13L_eb & 0 & 156b+m_m & -22L_eb \\
0 & 0 & 0 & 0 & -13L_eb & -3L_e^2b & 0 & -22L_eb & 4L_e^2b
\end{bmatrix}_{9\times 9}$$

e. Force vector
There is no distributed/body force, but one point force acts along V_3. Therefore,

$$\{\bar{F}\} = \{0 \quad 0 \quad 0 \quad 0 \quad 0 \quad 0 \quad 0 \quad P(t) \quad 0\}^T$$

f. Impose BCs
Since the machine is fixed at the bottom, the DOFs at node 1 are fixed. Hence, the BCs are, $U_1 = 0$, $V_1 = 0$ and $\theta_1 = 0$. Therefore, the row and column numbers 1, 2, and 3 of the stiffness and mass matrices are eliminated to incorporate the BCs. The corresponding rows of the force vector are also eliminated. After incorporating the boundary condition, the global matrices and force vector are

$$[K] = \begin{bmatrix}
12k_2+k_1 & 0 & 6k_2L & -k_1 & 0 & 0 \\
0 & k_1+12k_2 & 6k_2L & 0 & -12k_2 & 6k_2L \\
6k_2L & 6k_2L & 8k_2L^2 & 0 & -6k_2L & 2k_2L^2 \\
-k_1 & 0 & 0 & k_1 & 0 & 0 \\
0 & -12k_2 & -6k_2L & 0 & 12k_2 & -6k_2L \\
0 & 6k_2L & 2k_2L^2 & 0 & -6k_2L & 4k_2L^2
\end{bmatrix}_{6\times 6}$$

$$[M] = \begin{bmatrix} 156b+2a & 0 & 22L_eb & a & 0 & 0 \\ 0 & 2a+156b & 22L_eb & 0 & 54b & -13L_eb \\ 22L_eb & 22L_eb & 8L_e^2b & 0 & 13L_eb & -3L_e^2b \\ a & 0 & 0 & 2a+m_m & 0 & 0 \\ 0 & 54b & 13L_eb & 0 & 156b+m_m & -22L_eb \\ 0 & -13L_eb & -3L_e^2b & 0 & -22L_eb & 4L_e^2b \end{bmatrix}_{6\times 6}$$

$$\{F\} = \begin{Bmatrix} 0 & 0 & 0 & 0 & P(t) & 0 \end{Bmatrix}^T_{6\times 1}$$

The global DOF vector is

$$\{u\} = \begin{Bmatrix} U_2 & V_2 & \theta_2 & U_3 & V_3 & \theta_3 \end{Bmatrix}^T$$

The equation of motion is given by

$$[M]\{\ddot{u}\} + [K]\{u\} = \{F\}$$

REVIEW QUESTIONS

1. What is the need for a numerical modeling technique, like FEM, for vibration analysis?
2. What do you mean by the piecewise approximation?
3. What is the advantage of the piecewise approximation compared to a single approximation function over the whole domain?
4. What are shape functions?
5. How can the adequacy of the number of finite elements used in a model be decided?
6. Why is the slope also a DOF at each node in a beam element?

PROBLEMS

Problem 3.1: Using four beam elements, obtain the governing equations of motion for the transverse vibrations of the shaft with three rotors shown in Figure 2.34 (in Chapter 2). Assume simply supported BCs. Treat the three rotors as lumped masses. The vertical components of the unbalance forces in the rotors are the excitation forces.

Problem 3.2: Using three frame elements, obtain the governing equations of motion for the in-plane transverse vibrations of the overhead crane shown in Figure 2.44 a) (in Chapter 2). Treat the trolley with the motor as a lumped mass.

Problem 3.3: Using two beam elements, obtain the governing equations of motion for the free transverse vibrations of the system shown in Figure 2.46 (in Chapter 2). Treat the motor as a lumped mass and lumped rotational inertia.

4 Analytical Modal Analysis of SDOF Systems

4.1 INTRODUCTION

The analytical modal analysis is the modal analysis of a system using its mathematical model. In Chapters 2 and 3, we looked at how a mathematical model of a dynamic system can be obtained by the lumped parameter approach and finite element method, respectively. The mathematical model obtained is in the form of simultaneous ordinary differential equations. These equations need to be solved to study the system's dynamic behavior and predict the system's response to the dynamic forces acting on the system. We study the solution of these differential equations through modal analysis in this and Chapters 5 and 6. This chapter deals with the analytical modal analysis of the single degree of freedom (SDOF) systems.

An SDOF system has only one mode of vibration; that way, the modal analysis approach has not much relevance since the system response is always due to the only mode the system has. However, the modal analysis is also concerned with determining the dynamic characteristics, which describe its natural vibration and how these properties affect the free and forced response. These issues are also relevant to SDOF systems and are dealt with in this chapter. We consider the free and forced vibration of undamped, and viscously and hysteretically damped systems. In this process, we come across the concepts of natural frequency, damping factor, frequency response function, and Impulse response function, which are the system descriptors relevant to the modal analysis. We also introduce the Laplace transform approach to characterize the system in terms of the transfer function and determine its response to given inputs.

4.2 FREE VIBRATION WITHOUT DAMPING

The vibration of a system due to an external dynamic force continuously acting on it is called the **forced vibration** of the system. The system can also vibrate without such a force if the system is just initially disturbed. The initial disturbance imparts energy to the system causing the system to vibrate, which is referred to as **free vibration**. It is also called **natural vibration**, as the system is vibrating naturally in a state it can, without any constraint dictated by external forces.

While every system in practice always has some damping mechanisms by which the system's energy is dissipated, it may be helpful to study the vibration if these energy-dissipating mechanisms are neglected. One motivation is that for systems with low damping, the accuracy of the results may not be much compromised. The second motivation is that the dynamic properties of the system, in the absence of damping mechanisms, play a vital role in describing its dynamics. The system that has no damping or in which the damping is neglected for analysis is called the **undamped system**. In this section, we study the free vibration of an undamped SDOF system.

Figure 4.1 shows an undamped SDOF system with no external force acting. The governing equation of motion can be derived, as studied in Chapter 2, as

$$m\ddot{x} + kx = 0 \tag{4.1}$$

The equation is a linear, second-order, homogeneous, ordinary differential equation with x and t as the dependent and independent variables, respectively. The motion represented by the equation is simple harmonic, as rearranging the terms in the equation indicates that the acceleration is proportional to the displacement and opposite to it. Since the

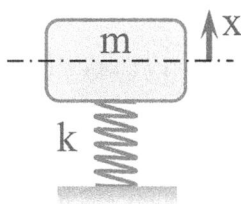

FIGURE 4.1 An undamped SDOF system.

DOI: 10.1201/9780429454783-4

motion is simple harmonic, two possible forms of the solutions are $A\cos\omega t$ and $B\sin\omega t$. Substituting any of these into Eq. (4.1) yields

$$\omega = \sqrt{\frac{k}{m}} \quad \text{rad/sec} \tag{4.2}$$

Thus, the functions $A\cos\omega t$ or $B\sin\omega t$, with ω given by Eq. (4.2), are the solutions of Eq. (4.1). We draw the following conclusions from the above analysis:

- The motion represents vibration as the two functions are oscillatory; it is free vibration since no external force is acting on the system.
- The coefficient of variable t in the functions $\cos\omega t$ and $\sin\omega t$ represents the frequency of the motion in rad/sec; therefore ω is the frequency of the motion.
- Since the vibration occurs naturally, the frequency of motion is called **natural frequency**. It is the **undamped natural frequency** since the system is undamped. It is denoted by ω_n. Thus,

$$\omega_n = \sqrt{\frac{k}{m}} \quad \text{rad/sec} \tag{4.3}$$

- The system vibrates under free vibration at a particular frequency, called the natural frequency, since the dynamic equilibrium of the elastic and inertia forces is possible only at that frequency.
- The natural frequency is a system property since it depends only on the system parameters. The undamped natural frequency depends on the stiffness and mass characteristics of the system.

A linear combination of the two solutions also satisfies Eq. (4.1) and is called the general solution, representing all possible vibratory motions of the system under the free state. The general solution is

$$x(t) = A\cos\omega_n t + B\sin\omega_n t \tag{4.4}$$

where A and B are arbitrary constants whose values depend upon the initial state of the system, referred to as **initial conditions** on the system. A system described by a second-order differential equation requires two variables to describe its state at any instant. It is also reflected by the fact that there are two arbitrary constants in its general solution. The displacements and velocities are commonly used as state variables. For the SDOF system, we choose the initial displacement $(x(t=0)=x_0)$ and initial velocity $(\dot{x}(t=0)=v_0)$ as the state variables. Solving Eq. (4.4) for A and B using the initial conditions and substituting them back in that equation gives

$$x(t) = x_0\cos\omega_n t + \frac{v_0}{\omega_n}\sin\omega_n t \tag{4.5}$$

Figure 4.2 shows the free vibration response for the given values of (x_0, v_0). It shows that the system oscillates with a constant amplitude at its natural frequency. However, we observe in practice that the free vibration response of the system decays with time due to damping. The undamped model of the system could not predict this behavior, so damping modeling is necessary to predict the decaying nature of the free vibration.

4.3 FREE VIBRATION WITH VISCOUS DAMPING

Figure 4.3 shows an SDOF system with viscous damping but no external force. The governing equation of motion is

$$m\ddot{x} + c\dot{x} + kx = 0 \tag{4.6}$$

4.3.1 DERIVATION OF THE FREE RESPONSE

Equation (4.6) does not represent a simple harmonic motion, and hence the functions $A\cos\omega t$ or $B\sin\omega t$ considered in the previous section are no longer the solutions. Due to damping, the amplitude of the free vibration response reduces with

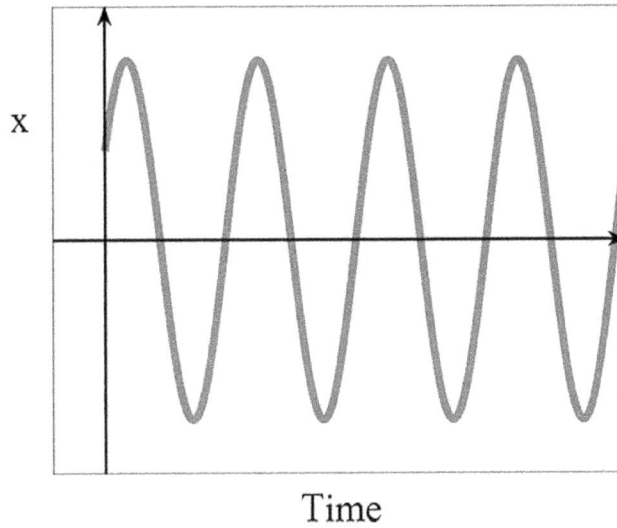

FIGURE 4.2 The free vibration response of the undamped system.

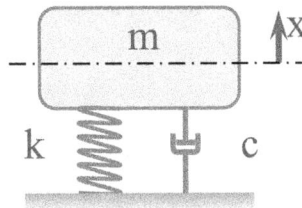

FIGURE 4.3 An SDOF system with viscous damping.

time as the initial energy given to the system is dissipated by the damper. Because of this, a trial solution in the form of an exponential function is assumed,

$$x(t) = Ae^{st} \tag{4.7}$$

Substituting Eq. (4.7) into Eq. (4.6), we get

$$Ae^{st}(ms^2 + cs + k) = 0 \tag{4.8}$$

The possibility $Ae^{st} = 0$ gives a trivial solution and does not represent the vibratory motion; therefore, the term in the bracket must be zero,

$$ms^2 + cs + k = 0 \tag{4.9}$$

The solution of this quadratic equation gives the following roots:

$$s_1, \ s_2 = -\frac{c}{2m} \pm \sqrt{\left(\frac{c}{2m}\right)^2 - \frac{k}{m}} \tag{4.10}$$

The general solution of the differential equation (4.6) is a linear combination of the solutions due to these two roots and can be written as

$$x(t) = A_1 e^{s_1 t} + A_2 e^{s_2 t} \tag{4.11}$$

where A_1 and A_2 are arbitrary constants that depend on the initial conditions on the state of the system. The nature of the solution x(t) depends on the nature of the roots s_1 and s_2, which in turn depends on the sign of the term in the square root in Eq. (4.10). The cases possible are

- If $\left(\dfrac{c}{2m}\right)^2 > \dfrac{k}{m}$, then the sign of the term is positive, giving real and unequal roots.

- If $\left(\dfrac{c}{2m}\right)^2 < \dfrac{k}{m}$, then the sign of the term is negative, giving a pair of complex conjugate roots.

- If $\left(\dfrac{c}{2m}\right)^2 = \dfrac{k}{m}$, then the roots are real but equal. This equality is the boundary between the first two cases. This equality gives the value of the damping coefficient (c) for which the two roots would be real and equal and is called the critical damping coefficient (c_c) and is obtained as

$$c_c = 2\sqrt{km} \quad \text{N-sec/m} \tag{4.12}$$

A non-dimensional measure of damping is defined by taking the ratio of the actual damping coefficient in the system and its critical value. This ratio is called the damping factor ξ and is given by

$$\xi = \frac{c}{c_c} \tag{4.13}$$

Making use of Eqs. (4.3), (4.12), and (4.13), the roots given by Eq. (4.10) can be expressed in terms of ω_n and ξ as

$$s_{1,2} = -\xi\omega_n \pm \omega_n\sqrt{\xi^2 - 1} \tag{4.14}$$

The nature of the free vibration, therefore, depends upon the value of ξ, as discussed below.

a. $\xi > 1.0$ (Overdamped system)

If the actual damping coefficient is more than the critical value, the system is said to be an overdamped system. As the damping factor is more than 1.0, the roots s_1 and s_2 are real and unequal, and the solution x(t) (from Eq. 4.11) is given by

$$x(t) = e^{-\xi\omega_n t}\left(A_1 e^{+\omega_n\sqrt{\xi^2-1}t} + A_2 e^{-\omega_n\sqrt{\xi^2-1}t}\right) \tag{4.15}$$

To understand the nature of the motion, we take a specific choice of the initial conditions. Let the initial displacement is nonzero ($x(0) = x_0$), but the initial velocity is zero ($\dot{x}(0) = 0$). Solving for A_1 and A_2, we get the following free vibration response:

$$x(t) = \frac{x_0}{2\sqrt{\xi^2-1}}\left[\left(\xi + \sqrt{\xi^2-1}\right)e^{\left(-\xi+\sqrt{\xi^2-1}\right)\omega_n t} + \left(-\xi + \sqrt{\xi^2-1}\right)e^{\left(-\xi-\sqrt{\xi^2-1}\right)\omega_n t}\right] \tag{4.16}$$

The response $x(t)$ consists of two exponential terms in which the coefficients of $\omega_n t$ in the exponents are always negative real numbers since $\xi > 1.0$. Hence, each of these terms and, therefore, the response $x(t)$ decrease exponentially with time t. However, the response is not oscillatory.

The equation for $x(t)$ can similarly be obtained for the general case when the initial state of the system is described by initial displacement x_0 and initial velocity v_0. Figure 4.4 shows the effect of initial conditions on the free vibration response of overdamped systems. The plots correspond to different values of v_0 but a common value of x_0. For some initial conditions, the response may cross the mean position once before it gradually decays to zero. Figure 4.5 shows the effect of the damping factor ξ on the free vibration response of overdamped systems. As ξ increases, the rate of decay of response reduces.

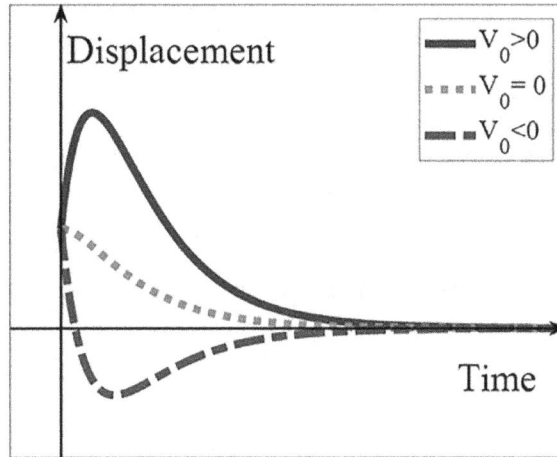

FIGURE 4.4   Effect of initial conditions on the free vibration response of overdamped systems.

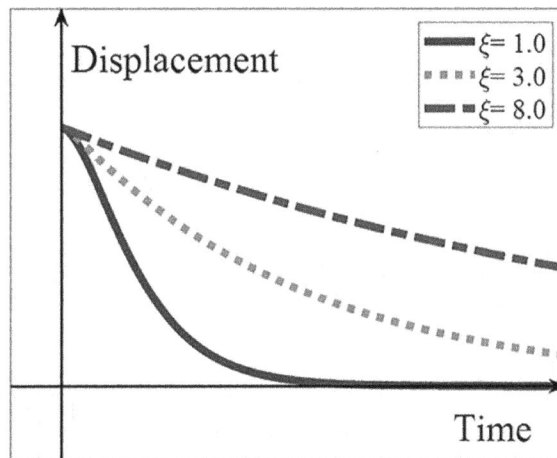

FIGURE 4.5   Effect of the damping factor on the free vibration response of overdamped systems.

b. $\xi = 1.0$ (Critically damped system)

If the actual damping coefficient equals its critical value, the system is said to be critically damped. The roots s_1 and s_2 are real and equal, and when they are substituted into Eq. (4.11), the resulting equation effectively has only one arbitrary constant, which doesn't represent the general solution. For the critically damped case, the general solution is given by

$$x(t) = e^{-\xi\omega_n t}(A_1 + A_2 t) \tag{4.17}$$

For the initial conditions $x(0) = x_0$ and $\dot{x}(0) = 0$, the response is obtained as

$$x(t) = x_0(1 + \omega_n t)e^{-\xi\omega_n t} \tag{4.18}$$

It is seen that the slope $\left(\dfrac{dx(t)}{dt}\right)$ is always negative for any t, and therefore the response continuously decreases exponentially. Equation (4.18) also does not indicate oscillations. We observe from Figure 4.5 that the critically damped system has the fastest decay.

c. $\xi < 1.0$ (Underdamped system)

If the actual damping coefficient is less than the critical value, the system is said to be an underdamped system. As the damping factor is less than 1.0, the roots s_1 and s_2 are complex conjugates given by

$$s_{1,2} = -\xi\omega_n \pm i\omega_n\sqrt{1-\xi^2} \tag{4.19}$$

The solution x(t) is

$$x(t) = e^{-\xi\omega_n t}\left(A_1 e^{+i\omega_n\sqrt{1-\xi^2}t} + A_2 e^{-i\omega_n\sqrt{1-\xi^2}t}\right) \tag{4.20}$$

For initial conditions x_0 and v_0, the constants A_1 and A_2 are obtained as

$$A_1 = \frac{x_0}{2} - i\frac{v_0 + x_0\xi\omega_n}{2\omega_n\sqrt{1-\xi^2}} \quad \text{and}$$

$$A_2 = \frac{x_0}{2} + i\frac{v_0 + x_0\xi\omega_n}{2\omega_n\sqrt{1-\xi^2}} \tag{4.21}$$

The initial conditions for the vibration of a physical system are described by the real values of x_0 and v_0 and therefore, the arbitrary constants in Eq. (4.21) are complex conjugates. Substituting the arbitrary constants into (4.20) and making use of Euler's identities, $e^{\pm i\omega_n t} = \cos\omega_n t \pm i\sin\omega_n t$, we get $x(t)$ as

$$x(t) = e^{-\xi\omega_n t} \times \left(x_0\cos\omega_n\sqrt{1-\xi^2}t + \frac{v_0 + x_0\xi\omega_n}{\omega_n\sqrt{1-\xi^2}}\sin\omega_n\sqrt{1-\xi^2}t\right) \tag{4.22}$$

Due to the presence of 'cos' and 'sin' terms $x(t)$ is oscillatory, and the coefficient of time t, i.e., $\omega_n\sqrt{1-\xi^2}$ represents the frequency of oscillations in rad/sec. Since $x(t)$ represents the free vibration with damping, the frequency of oscillations is called the damped natural frequency (often denoted by ω_d). The coefficient in the expression for $x(t)$, i.e., $e^{-\xi\omega_n t}$, indicates that the amplitude of the response exponentially decays with time.

Figure 4.6 shows the effect of the damping factor on the free vibration response of underdamped systems. All the graphs correspond to the same initial conditions. For the underdamped systems, the higher the damping factor, the higher the rate of decay of response amplitude.

d. Rate of decay of energy

Figure 4.7 shows the total energy (i.e., the sum of the potential and kinetic energies) of the system as a function of time for some selected values of the damping factor. The mass, stiffness, and initial conditions are identical for all these plots, with only the damping coefficient being different. Based on this figure and the analysis in the previous sections, the following conclusions are drawn:

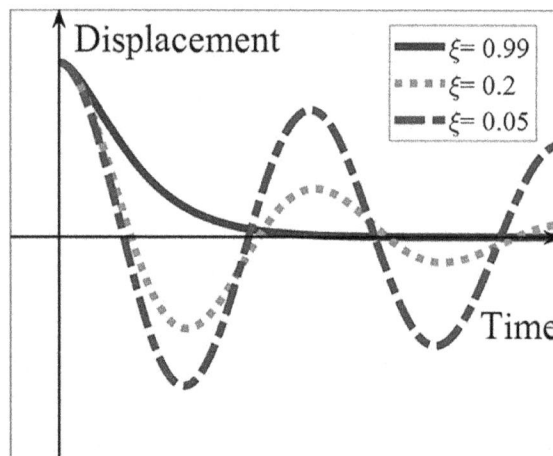

FIGURE 4.6   Effect of the damping factor on the free vibration response of underdamped systems.

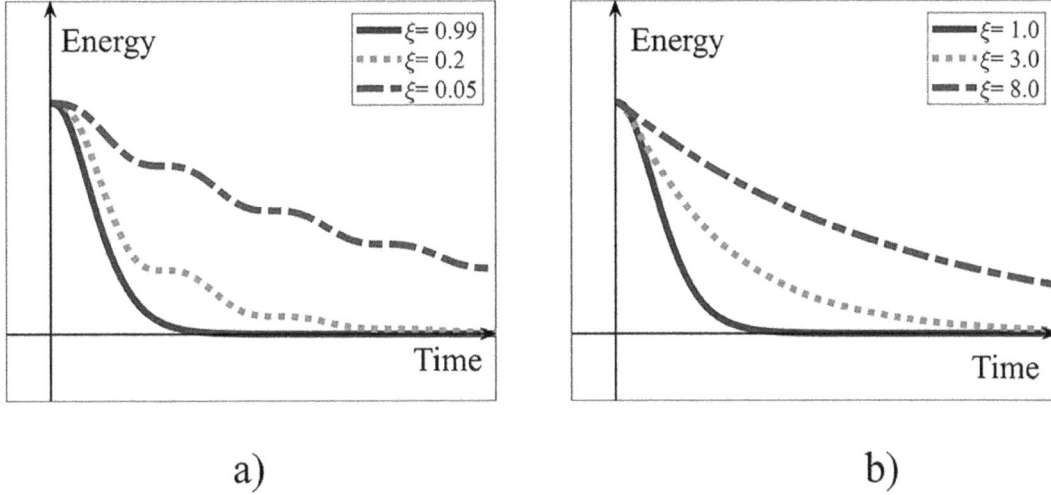

a)   b)

FIGURE 4.7 Effect of damping factor on the decay of the total energy of the system (a) underdamped systems, (b) overdamped systems.

- For the underdamped systems, the decay rate increases with an increase in the damping factor.
- For the overdamped systems, the decay rate decreases with an increase in the damping factor.
- The decay rate is fastest when the damping factor is equal to one; that is, the system is critically damped.

It should be noted that the rate of decay of system energy at an instant is the product of damping force (which is the product of the damping factor and the velocity) and the velocity. Thus, the magnitude of the damping factor and the magnitude of the instantaneous velocity both affect the decay rate.

4.3.2 LOGARITHMIC DECREMENT

The logarithmic decrement is the relationship between the ratio of amplitudes in the free vibration response and the damping factor of an underdamped system.

After combining the 'cos' and 'sin' terms in Eq. (4.22), the free vibration response of an SDOF underdamped system takes the form of

$$x(t) = Ae^{-\xi\omega_n t}\cos(\omega_d t - \phi) \tag{4.23}$$

We find the ratio of displacements at t and $t + T$, where T is the time period of motion. That ratio is

$$\frac{x(t)}{x(t+T)} = \frac{x_1}{x_2} = \frac{Ae^{-\xi\omega_n t}\cos(\omega_d t - \phi)}{Ae^{-\xi\omega_n(t+T)}\cos(\omega_d(t+T) - \phi)} \tag{4.24}$$

The 'cos' function repeats after an integer number of cycles, and noting that $T = \dfrac{2\pi}{\omega_d}$, we get

$$\frac{x_1}{x_2} = \frac{Ae^{-\xi\omega_n t}}{Ae^{-\xi\omega_n(t+T)}} = e^{\frac{2\pi\xi}{\sqrt{1-\xi^2}}} \tag{4.25}$$

Taking the natural logarithm of both sides, we obtain

$$\log_e \frac{x_1}{x_2} = \delta = \frac{2\pi\xi}{\sqrt{1-\xi^2}} \tag{4.26}$$

The parameter δ is called the logarithmic decrement and depends only on the damping factor. The above analysis can be extended to show that the ratio of amplitudes at time t and after an elapse of n periods are related by

$$\delta = \frac{1}{n} \log_e \frac{x_0}{x_n} \qquad (4.27)$$

The logarithmic decrement can be used to identify the damping factor of the system from the free vibration response measurement.

4.4 FORCED VIBRATION OF AN UNDAMPED SYSTEM UNDER HARMONIC EXCITATION

In the previous two sections, we studied the free vibration of the system, where there was no external force on the system, and the vibration resulted from an initial disturbance to the system. In this section, we look at the forced vibration of an undamped SDOF system when a harmonic force $F(t) = F_o \cos \omega t$ acts on the system. Therefore, in the system shown in Figure 4.1, we also have a harmonic force acting on the mass. Applying Newton's second law, we get

$$m\ddot{x} + kx = F(t) \qquad (4.28)$$

4.4.1 GENERAL SOLUTION

Equation (4.28) is a linear, second-order nonhomogeneous differential equation, and the general solution or complete solution is given by

$$x(t) = x_c(t) + x_p(t) \qquad (4.29)$$

where $x_c(t)$ is the complementary function (or homogeneous solution) and is nothing but the general solution of the corresponding homogenous equation (the equation obtained by making F(t) zero in Eq. (4.28) and $x_p(t)$ is the particular solution of Eq. (4.28).

The solution $x_c(t)$ is already obtained in Section 4.2 (Eq. 4.4) and is given by

$$x_c(t) = A \cos \omega_n t + B \sin \omega_n t \qquad (4.30)$$

$x_p(t)$ is the solution that satisfies Eq. (4.28). Considering a trial solution $x_p(t) = x_{po} \cos \omega t$ and substituting into (4.28) and solving for x_{po} gives

$$x_{po} = \frac{F_o}{k - m\omega^2} \qquad (4.31)$$

Making use of the fact that the undamped natural frequency $\omega_n = \sqrt{\dfrac{k}{m}}$ and noting that $\dfrac{F_o}{k}$ represents the deflection under a static force F_o, denoted by x_{st}, Eq. (4.31) becomes

$$x_{po} = \frac{x_{st}}{1 - \left(\dfrac{\omega}{\omega_n}\right)^2} \qquad (4.32)$$

Thus, the general or complete solution is obtained as

$$x(t) = A \cos \omega_n t + B \sin \omega_n t + \frac{x_{st}}{1 - \left(\dfrac{\omega}{\omega_n}\right)^2} \cos \omega t \qquad (4.33)$$

where A and B are arbitrary constants and are obtained from the system's initial conditions.

4.4.2 PARTICULAR SOLUTION BY THE COMPLEX EXPONENTIAL METHOD

The particular solution to the given harmonic force can be alternatively found by the complex exponential method. In this method, the idea is first to find a particular solution ($\overline{x}_p(t)$) to the complex exponential force, whose real part is the harmonic force. The particular solution ($x_p(t)$) to the harmonic force is then obtained as the real part of $\overline{x}_p(t)$.

Consider a forcing function in the form

$$\overline{F}(t) = F_0\, e^{i\omega t} \tag{4.34}$$

$\overline{F}(t)$ is a vector in the complex plane rotating in the counter-clockwise direction with an angular velocity ω. The real part of this vector, as shown in Figure 4.8, is nothing but the harmonic force $F(t)$ acting on the mass. Thus,

$$F(t) = \mathrm{Re}\left(\overline{F}(t)\right) \tag{4.35}$$

Let us first find the response to $\overline{F}(t)$. Let

$$\overline{x}_p(t) = \overline{x}_{p0} e^{i\omega t} \tag{4.36}$$

where $\overline{x}_{p0}$ is the amplitude of the complex displacement response. Substitute $\overline{F}(t)$ and $\overline{x}_p(t)$ in Eq. (4.28) in places of $F(t)$ and $x(t)$, respectively. Here we see the advantage of the complex exponential method as the complex exponential can be differentiated (or integrated) easily. For example,

$$\dot{\overline{x}}_p(t) = \frac{d\left(\overline{x}_p(t)\right)}{dt} = i\omega \overline{x}_p(t) \tag{4.37}$$

and

$$\ddot{\overline{x}}_p(t) = \frac{d\left(\dot{\overline{x}}_p(t)\right)}{dt} = (i\omega)^2\, \overline{x}_p(t) \tag{4.38}$$

Thus, n-times differentiation (integration) of a complex exponential vector with time is equal to multiplication (division) of the vector by $(i\omega)^n$. The multiplication and division operations of two complex exponential vectors are also straightforward. Substituting Eqs. (4.36) and (4.38) into Eq. (4.28) and making use of the expressions for the undamped natural frequency and static deflection, we get

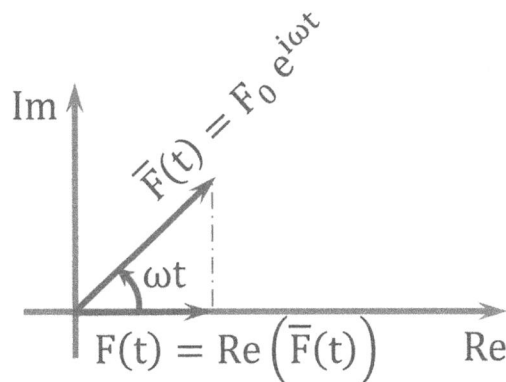

FIGURE 4.8 Complex exponential force.

$$\overline{x}_{p0} = \frac{F_o}{k - m\omega^2} = \frac{x_{st}}{1 - \left(\dfrac{\omega}{\omega_n}\right)^2} \tag{4.39}$$

$x_p(t)$ is obtained as the real part of $\overline{x}_{p(t)}$

$$x_p(t) = \mathrm{Re}\left(\overline{x}_p(t)\right) = \mathrm{Re}\left(\overline{x}_{p0}e^{i\omega t}\right) = \frac{x_{st}}{1 - \left(\dfrac{\omega}{\omega_n}\right)^2}\cos\omega t \tag{4.40}$$

which is the same as obtained by the direct method in Section 4.4.1. This result shows that the complex exponential method can be used to determine the vibration response to the harmonic forces. In this method, the algebraic simplifications are easier to carry out; hence, this method is used in this and the following chapters to find the response to harmonic forces.

Thus, the general or total solution is

$$x(t) = A\cos\omega_n t + B\sin\omega_n t + \frac{x_{st}}{1 - \left(\dfrac{\omega}{\omega_n}\right)^2}\cos\omega t \tag{4.41}$$

4.4.3 Nature of the Response

We observe the following from the complete response (Eq. 4.41) of the undamped SDOF system to a harmonic force:

- There are two parts to the response.
- The complementary part, for the undamped case, does not decay. When the damping is included, this part decays with time and eventually becomes zero. Therefore, it is called the transient part.
- The complementary part is at the natural frequency (ω_n) of the system. This part arises due to the initial state of the system.
- The particular solution is called the steady-state part and is at the frequency (ω) of the external harmonic force.

4.4.4 Nature of the Particular Solution

As seen from Eq. (4.40), the amplitude of the particular part of the solution varies with the frequency ratio $\dfrac{\omega}{\omega_n}$. Figure 4.9 shows the amplitude and phase lag of the response $x_p(t)$ as a function of the frequency ratio $\dfrac{\omega}{\omega_n}$.

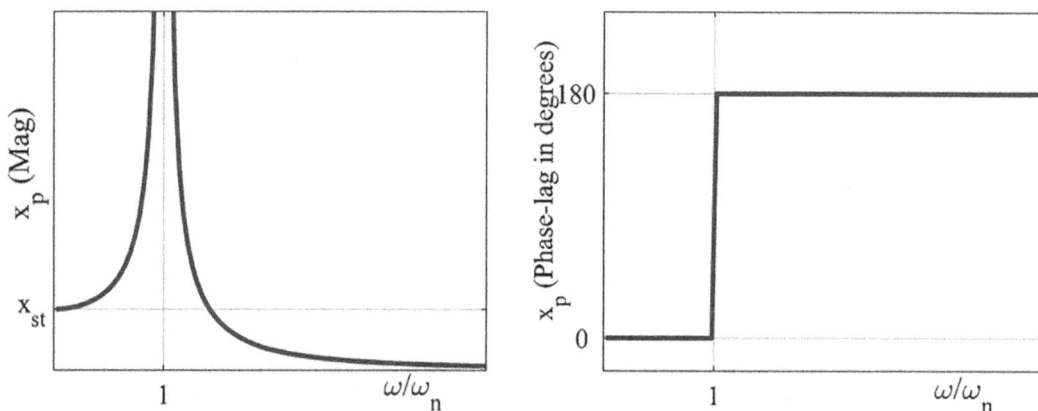

FIGURE 4.9 Magnitude and phase lag of the steady-state response.

We observe the following:

- If $\dfrac{\omega}{\omega_n} \ll 1$, then $x_{p0} \approx x_{st}$. Thus, if the forcing frequency ω is much smaller than ω_n, then the external force behaves like a static force and produces a displacement equal to the static deflection of the system. The phase angle is $0°$ which means that the response $x_p(t)$ is in phase with force $F(t)$.

- If $\dfrac{\omega}{\omega_n} \gg 1$, then the response $x_{p0} \approx 0$. Thus, if the forcing frequency ω is much higher than the natural frequency, the displacement tends to zero due to the very high vibration frequency. The phase lag of the displacement is $180°$, which means that the displacement is lagging the force by $180°$.

- If $\dfrac{\omega}{\omega_n} = 1$, i.e., when the frequency of the excitation force is equal to the system's natural frequency, then Eq. (4.40) says that the amplitude is infinite at the steady state. We see that there is a component of the response arising out from the homogeneous solution at this frequency. Therefore, to capture the total effect of the force Eq. (4.41) must be solved. If the initial displacement and velocity of the mass are x_o and v_o respectively, then solving Eq. (4.41) for A and B and substituting back their values in this equation gives the following response:

$$x(t) = x_o \cos\omega_n t + \frac{v_o}{\omega_n}\cos\omega_n t + \frac{x_{st}}{1 - \left(\dfrac{\omega}{\omega_n}\right)^2}(\cos\omega t - \cos\omega_n t) \tag{4.42}$$

The last term in the RHS is due to the external force and its limit when $\omega \to \omega_n$ would be the particular solution at $\omega = \omega_n$. When $\omega \to \omega_n$, both the numerator and denominator individually tend to zero, and hence obtaining the limit using L' Hospital's rule, we get

$$x_p(t) = \lim_{\omega \to \omega_n} x_{st} \frac{\dfrac{d}{d\omega}(\cos\omega t - \cos\omega_n t)}{\dfrac{d}{d\omega}\left(1 - \left(\dfrac{\omega}{\omega_n}\right)^2\right)} \tag{4.43}$$

$$x_p(t) = \frac{1}{2} t\,\omega_n\, x_{st}\, \sin\omega_n t \tag{4.44}$$

The response $x_p(t)$ in Eq. (4.44) is plotted in Figure 4.10. As t increases, $x_p(t)$ also increases and approaches infinity gradually and not instantaneously.

Thus, in an undamped system, the response theoretically reaches infinity when the external force frequency coincides with the natural frequency. The coincidence of the external force frequency with the natural frequency

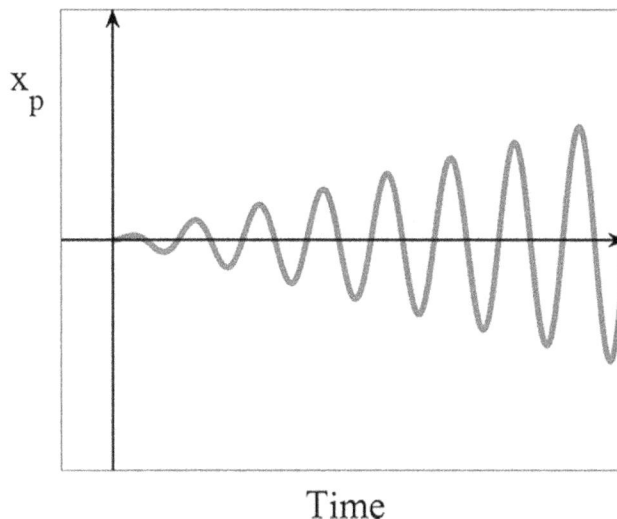

FIGURE 4.10   The particular part of the response when $\omega = \omega_n$.

is called **resonance**. The occurrence of the resonance leads to excessive vibration amplitude and is undesirable in most applications.

The infinite amplitude predicted by the undamped model is only a theoretical possibility for several reasons. In practice, some damping is always present and limits the response at resonance, as we see in the next section. Second, as the response builds beyond the range of small amplitude motion, the linearized model of the system may no longer be valid, and hence the predictions made by such a model may not be correct. Also, if the amplitude were to rise indefinitely, the system would break down before the amplitude could reach substantial values.

Equation (4.44) shows that the displacement is described by the function $\sin\omega_n t$ while the external force is $\cos\omega_n t$. Thus, at $\omega = \omega_n$, the displacement lags the external force by a phase of $90°$.

4.4.5 Vector Representation of the Forces Under the Dynamic Equilibrium

An alternative way to understand the system's response to different excitation force frequencies can be by considering the dynamic equilibrium between the forces acting on the mass. We focus on the particular part of the solution. For brevity, the suffix 'p' in $\overline{x}_p$ and $\ddot{\overline{x}}_p$ is dropped from the discussion below. The equilibrium of forces given by Eq. (4.28) can be written as

$$F_o\cos\omega t + (-m\ddot{x}) + (-kx) = 0 \tag{4.45}$$

$$\text{External force} + \text{Inertia force} + \text{Elastic force} = 0 \tag{4.46}$$

Expressing these forces as the real parts of their corresponding complex exponential vectors, we get

$$\text{Re}\left(F_o\,e^{i\omega t}\right) + \text{Re}\left(-m\ddot{\overline{x}}\right) + \text{Re}(-k\overline{x}) = 0 \tag{4.47}$$

$$\text{Re}\left(F_o\,e^{i\omega t} + \left(-m\ddot{\overline{x}}\right) + (-k\overline{x})\right) = 0 \tag{4.48}$$

The sum in the bracket must be zero for Eq. (4.48) to be true for any arbitrary value of t, giving

$$F_o\,e^{i\omega t} + \left(-m\ddot{\overline{x}}\right) + (-k\overline{x}) = 0 \tag{4.49}$$

$$\text{Complex External force} + \text{Complex Inertia force} + \text{Complex Elastic force} = 0 \tag{4.50}$$

Thus, the equilibrium of the actual forces represented by Eq. (4.45) is equivalent to the equilibrium of the complex forces represented by Eq. (4.49). Therefore, the equilibrium of the forces can be studied by studying the equilibrium of the corresponding complex forces. The complex forces are complex exponential vectors and rotate in the complex plane with angular velocity ω as time t progresses. Their projections on the real axis give the corresponding actual forces at that instant in time.

By using Eqs. (4.36), (4.38), and (4.39), the complex inertia and elastic forces are obtained as

$$\text{Complex Inretia force} = -m\ddot{\overline{x}} = \frac{F_o\left(\dfrac{\omega}{\omega_n}\right)^2}{1-\left(\dfrac{\omega}{\omega_n}\right)^2}\,e^{i\omega t} \tag{4.51}$$

$$\text{Complex Elastic force} = -k\overline{x} = -\frac{F_o}{1-\left(\dfrac{\omega}{\omega_n}\right)^2}\,e^{i\omega t} \tag{4.52}$$

By representing the complex forces as vectors in the complex plane, we analyze the effects of forcing frequency on the response.

a. When $\dfrac{\omega}{\omega_n} < 1$:

- Figure 4.11a shows the displacement, velocity, and acceleration vectors. The displacement is in phase with the external force while the acceleration leads the external force by 180°.
- Figure 4.11b shows the complex force vectors. Note that the elastic and inertia forces act opposite to the displacement and acceleration.
- Figure 4.11c shows the vector polygon of the complex forces. The elastic force is more than the inertia force and balances the sum of the external and inertia forces.
- When $\dfrac{\omega}{\omega_n} \ll 1$, it is seen from Eqs. (4.51–4.52) that the elastic force is almost equal to the external force, while the inertia force is negligibly small. Therefore, the mass has a nonzero displacement (equal to the static deflection due to the external force) but a negligible acceleration.

b. The case when $\dfrac{\omega}{\omega_n} < 1$:

- Figure 4.12a shows the displacement, velocity, and acceleration vectors and their phases. The displacement lags the external force by 180° while the acceleration is in phase with the external force.
- Figure 4.12b shows the vectors of complex forces, and as before, the elastic and inertia forces act opposite to the displacement and acceleration.
- Figure 4.12c shows the vector polygon of the complex forces. The inertia force is more than the elastic force and balances the sum of the external and elastic forces.

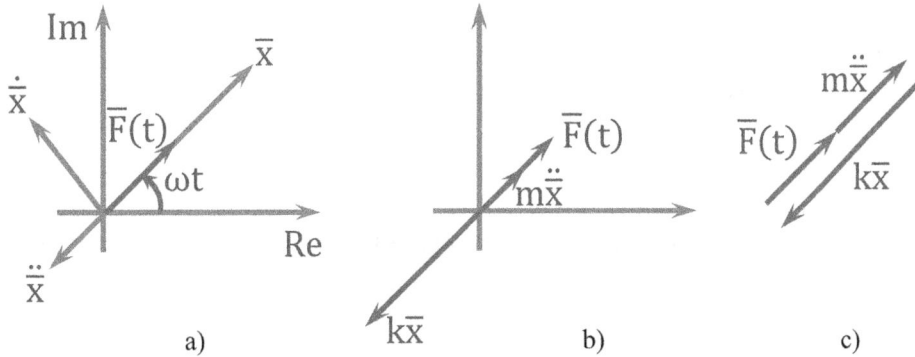

FIGURE 4.11 For $\dfrac{\omega}{\omega_n} < 1$ (a) displacement, velocity, and acceleration vectors, (b) complex force vectors, and (c) vector polygon of the complex forces.

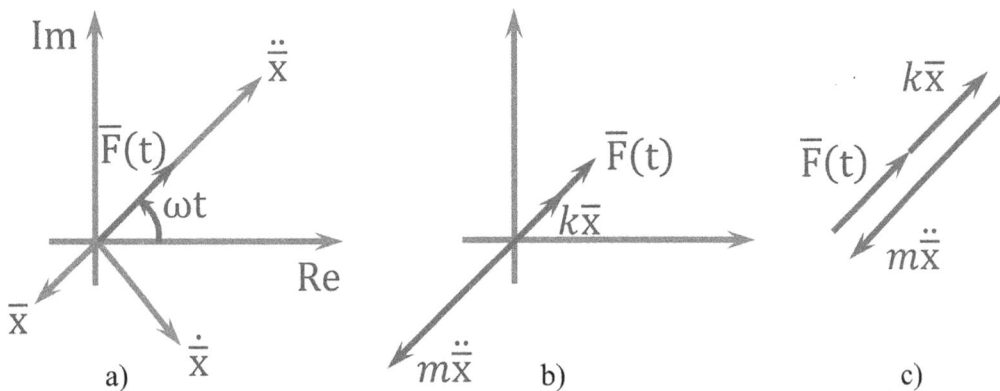

FIGURE 4.12 For $\dfrac{\omega}{\omega_n} > 1$ (a) displacement, velocity, and acceleration vectors, (b) complex force vectors, and (c) vector polygon of the complex forces.

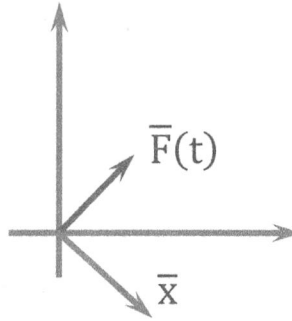

FIGURE 4.13   For $\dfrac{\omega}{\omega_n} = 1$, the phase between displacement and the external force.

- When $\dfrac{\omega}{\omega_n} \gg 1$, it is seen from Eqs. (4.51–4.52) that it is now the inertia force that is nearly equal to the external force, while the elastic force is negligible. Therefore, the mass has a nonzero acceleration but a negligible
  displacement.

c. We also note that when $\dfrac{\omega}{\omega_n} > 1$ or $\dfrac{\omega}{\omega_n} < 1$ the external force and velocity have a phase difference of 90°. As
   a result, the work done by the external force on the system is zero over a cycle. Therefore, the amplitude of the
   vibration remains constant with time.

d. When $\dfrac{\omega}{\omega_n} = 1$, i.e., the case of resonance:

- We discussed in Section 4.4.4 that in this case, the displacement lags the force by 90° and the same is shown
  in Figure 4.13.
- In this case, the phase between the external force and velocity is no longer 90°. Hence, the external force
  does positive work on the system every cycle. Therefore, the energy and response of the system continuously
  increase with time (as seen in Figure 4.10).

4.5   FORCED VIBRATION OF A VISCOUSLY DAMPED SYSTEM UNDER HARMONIC EXCITATION

The equation of motion of an SDOF system with viscous damping is given by

$$m\ddot{x} + c\dot{x} + kx = F(t) \tag{4.53}$$

4.5.1   General Solution

The general solution is again equal to the sum of the complementary function and particular solution,

$$x(t) = x_c(t) + x_p(t) \tag{4.54}$$

$x_c(t)$ is the solution of the corresponding homogeneous equation and, assuming the system to be underdamped, is given
by Eq. (4.20) or (4.22), which can also be written in the following form:

$$x_c(t) = Ae^{-\xi\omega_n t} \cos\left(\omega_n \sqrt{1 - \xi^2}\, t - \phi\right) \tag{4.55}$$

where A and ϕ are arbitrary constants.

   We use the complex exponential method discussed in the previous section to determine the particular solution. Assume
that, as before, the forcing function is $\bar{F}(t) = F_o\, e^{i\omega t}$ and the particular solution is $\bar{x}_p(t) = \bar{x}_{po} e^{i\omega t}$. Substituting $\bar{F}(t)$ and
$\bar{x}_p(t)$ in places of $F(t)$ and $x(t)$, respectively, in Eq. (4.53), we get

$$\overline{x}_{p0} = \frac{F_o}{k - m\omega^2 + i\omega c} \qquad (4.56)$$

Thus, the amplitude of the particular solution is complex and can be written as

$$\overline{x}_{p0} = \frac{F_o}{\sqrt{\left(k - m\omega^2\right)^2 + (\omega c)^2}}\, e^{-i\psi} \qquad (4.57)$$

with phase lag ψ given by

$$\psi = \tan^{-1} \frac{\omega c}{k - m\omega^2} \qquad (4.58)$$

The complex amplitude essentially indicates a phase difference (ψ) between the response and external force. Often, the terms phase or phase lag and their signs create confusion in their interpretation. In Eq. (4.57), the positive value of ψ is phase lag since there is a negative sign in the power of the exponential $e^{-j\psi}$. In this case, the response lags the force by angle ψ. If ψ is negative, then the response leads the force by angle $|\psi|$.

Making use of the expressions for the undamped natural frequency and damping factor, Eqs. (4.57–4.58) can be further written as

$$\overline{x}_{p0} = \frac{x_{st}}{\sqrt{\left(1 - \left(\frac{\omega}{\omega_n}\right)^2\right)^2 + \left(2\xi\frac{\omega}{\omega_n}\right)^2}}\, e^{-i\psi} \qquad (4.59)$$

and

$$\psi = \tan^{-1} \frac{2\xi\dfrac{\omega}{\omega_n}}{1 - \left(\dfrac{\omega}{\omega_n}\right)^2} \qquad (4.60)$$

The particular solution to the external harmonic force is given by

$$x_p(t) = \mathrm{Re}\left(\overline{x}_p(t)\right) = \mathrm{Re}\left(\overline{x}_{p0}e^{i\omega t}\right) = \frac{x_{st}}{\sqrt{\left(1 - \left(\frac{\omega}{\omega_n}\right)^2\right)^2 + \left(2\xi\frac{\omega}{\omega_n}\right)^2}}\cos(\omega t - \psi) \qquad (4.61)$$

Substituting Eqs. (4.55) and (4.61) into Eq. (4.54), we get the complete solution of the system,

$$x(t) = Ae^{-\xi\omega_n t}\cos\left(\omega_n\sqrt{1 - \xi^2}\,t - \phi\right) + \frac{x_{st}}{\sqrt{\left(1 - \left(\frac{\omega}{\omega_n}\right)^2\right)^2 + \left(2\xi\frac{\omega}{\omega_n}\right)^2}}\cos(\omega t - \psi) \qquad (4.62)$$

The arbitrary constants A and ϕ can be found from the above equation using the initial conditions on the system.

4.5.2 Nature of the Response

Figure 4.14 shows the complementary function, particular solution, and total response for certain initial conditions. For an underdamped system, the complementary function occurs at the damped natural frequency of the system and decays to zero exponentially. It is referred to as the transient response. The particular solution has a constant amplitude and occurs at the frequency of the excitation force. It stays as long as the forcing function is there. The total response consists of only the particular part when the transient part has decayed and hence is referred to as the steady-state response.

4.5.3 Nature of the Particular Solution

From Eq. (4.61), we see that the amplitude of the steady-state part varies with frequency ω and damping factor ξ. Figure 4.15a shows plots of the amplitude (x_{po}) of the response as a function ω for some selected values of the damping factor ξ. The following observations are made from the plots.

- Similar to the case of the undamped system, the magnitude of the response is higher at excitation frequencies closer to the natural frequency and decreases as the excitation frequency moves away from it. Thus, the resonance occurs in a damped system as well.
- The peak response, however, doesn't occur at either $\omega = \omega_n$ or $\omega = \omega_d$, but at a frequency slightly lower. This frequency corresponding to the maximum magnitude of the steady-state amplitude can be found by differentiating Eq. (4.50) with ω and solving for ω. The frequency corresponding to the peak response amplitude is: $\omega_{peak} = \omega_n\sqrt{1-2\xi^2}$. Thus, the damped resonant frequency, i.e., the frequency corresponding to the peak response, is less than the undamped resonant frequency in a viscously damped system.
- As the damping factor increases, the amplitude of the response reduces. The damping influences the response more in the frequency range around the natural frequency.
- Irrespective of the damping level, $x_{po} \approx x_{st}$ for forcing frequencies $\omega \ll \omega_n$.

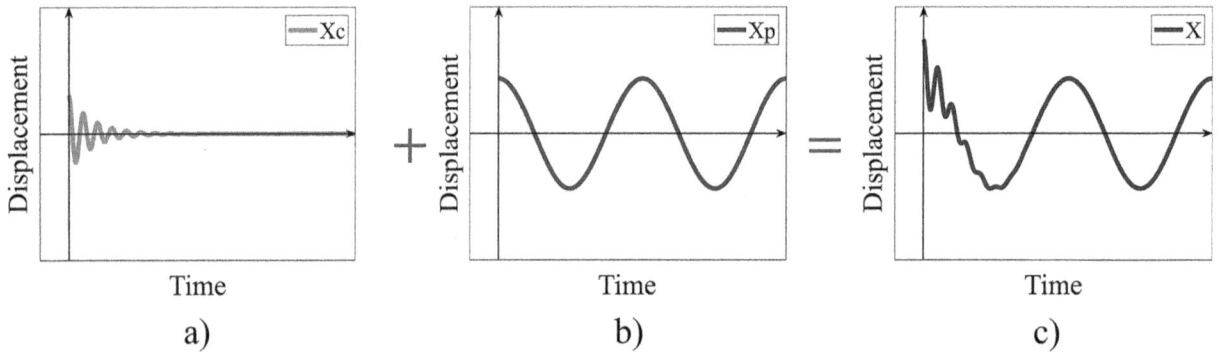

FIGURE 4.14 (a) Complimentary function, (b) particular solution, and (c) total response.

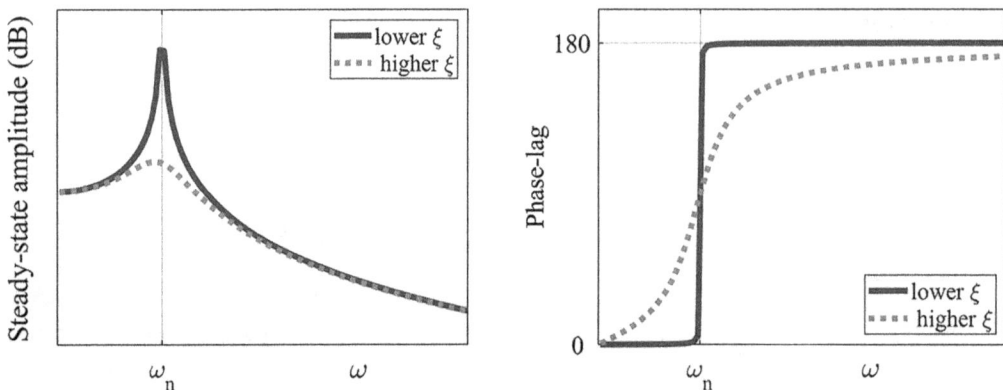

FIGURE 4.15 Variation of steady-state amplitude and phase with frequency and damping factor

- The ratio of the dynamic displacement to the static displacement (i.e. $\frac{x_{po}}{x_{st}}$) is called the magnification factor. It indicates the magnification of the response due to the dynamic/oscillatory nature of the force. In the resonance region, the magnification factor is high.

Figure 4.15b shows plots of the phase lag (ψ) of the steady-state response ($x_p(t)$) as a function of the frequency ω for some selected values of the damping factor ξ. The following observations are made from the plots.

- For forcing frequencies $\omega < \omega_n$ the phase lag is $< 90°$.
- At $\omega = \omega_n$ the phase lag is $90°$ irrespective of the value of the damping factor.
- For frequencies $\omega > \omega_n$ the phase lag is $> 90°$.
- As the damping factor increases, the phase lag increases for a given excitation frequency (except at $\omega = \omega_n$).

4.5.4   Vector Representation of the Dynamic Equilibrium of the Forces

In this section, we consider the dynamic equilibrium of the forces in a viscously damped system to understand the system's response to different excitation force frequencies. The equilibrium of forces given by Eq. (4.53) is

$$F_o \cos\omega t + (-m\ddot{x}) + (-c\dot{x}) + (-kx) = 0 \qquad (4.63)$$

$$\text{External force} + \text{Inertia force} + \text{Damping force} + \text{Elastic force} = 0 \qquad (4.64)$$

These forces are written as the real parts of their corresponding complex exponential forces.

$$\text{Re}\left(F_o\, e^{i\omega t}\right) + \text{Re}\left(-m\ddot{\bar{x}}\right) + \text{Re}\left(-c\dot{\bar{x}}\right) + \text{Re}(-k\bar{x}) = 0 \qquad (4.65)$$

$$\text{Re}\left(F_o\, e^{i\omega t} + \left(-m\ddot{\bar{x}}\right) + \left(-c\dot{\bar{x}}\right) + (-k\bar{x})\right) = 0 \qquad (4.66)$$

The sum in the bracket must be zero for Eq. (4.66) to be true for any arbitrary value of t, giving

$$F_o\, e^{i\omega t} + \left(-m\ddot{\bar{x}}\right) + \left(-c\dot{\bar{x}}\right) + (-k\bar{x}) = 0 \qquad (4.67)$$

$$\text{Complex extrenal force} + \text{Complex inertia force} + \text{Complex damping force} + \text{Complex elastic force} = 0 \qquad (4.68)$$

Thus, the equilibrium of the actual forces represented by Eq. (4.63) is equivalent to the equilibrium of the complex forces represented by Eq. (4.67).

The complex inertia, damping, and elastic forces are obtained as

$$\text{Complex inertia force} = -m\ddot{\bar{x}} = \frac{F_o\left(\dfrac{\omega}{\omega_n}\right)^2}{\sqrt{\left(1 - \left(\dfrac{\omega}{\omega_n}\right)^2\right)^2 + \left(2\xi\dfrac{\omega}{\omega_n}\right)^2}}\, e^{i(\omega t - \psi)} \qquad (4.69)$$

$$\text{Complex damping force} = -c\dot{\bar{x}} = \frac{-i2\xi\dfrac{\omega}{\omega_n}F_o}{\sqrt{\left(1 - \left(\dfrac{\omega}{\omega_n}\right)^2\right)^2 + \left(2\xi\dfrac{\omega}{\omega_n}\right)^2}}\, e^{i(\omega t - \psi)} \qquad (4.70)$$

$$\text{Complex elastic force} = -k\overline{x} = \dfrac{-F_o}{\sqrt{\left(1-\left(\dfrac{\omega}{\omega_n}\right)^2\right)^2 + \left(2\xi\dfrac{\omega}{\omega_n}\right)^2}}\, e^{i(\omega t-\psi)} \tag{4.71}$$

By representing the complex forces as vectors in the complex plane, we analyze the effects of the forcing frequency on the response.

a. When $\dfrac{\omega}{\omega_n} < 1$:

- Figure 4.16a shows the displacement, velocity, and acceleration vectors. The displacement lags the external force.
- Figure 4.16b shows the complex force vectors. The elastic and inertia forces are opposite to displacement and acceleration vectors, while the damping force is opposite to the velocity vector.
- Figure 4.16c shows the vector polygon of the complex forces. The force component along the velocity vector balances the damping force, while the perpendicular component combines with the inertia force to balance the elastic force.

b. When $\dfrac{\omega}{\omega_n} = 1$:

- Figure 4.17 shows that the velocity is in phase with the external force vector.
- The inertia and elastic forces balance each other, while the external force is balanced only by the damping force, leading to a dominant response at this frequency.

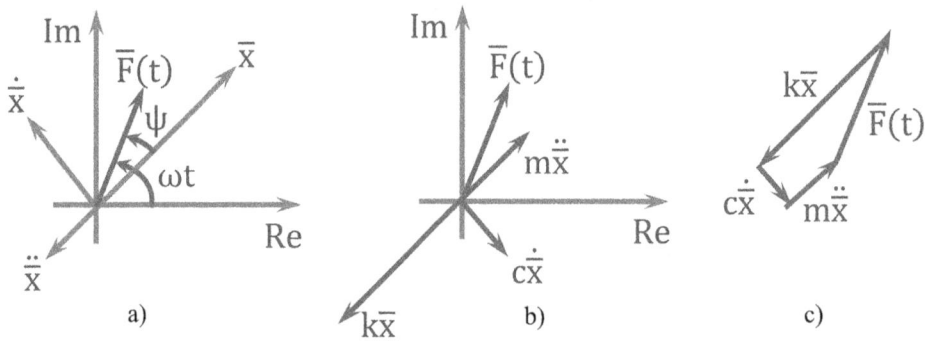

FIGURE 4.16 For $\dfrac{\omega}{\omega_n} < 1$ (a) displacement, velocity, and acceleration vectors, (b) complex force vectors, and (c) vector polygon of the complex forces.

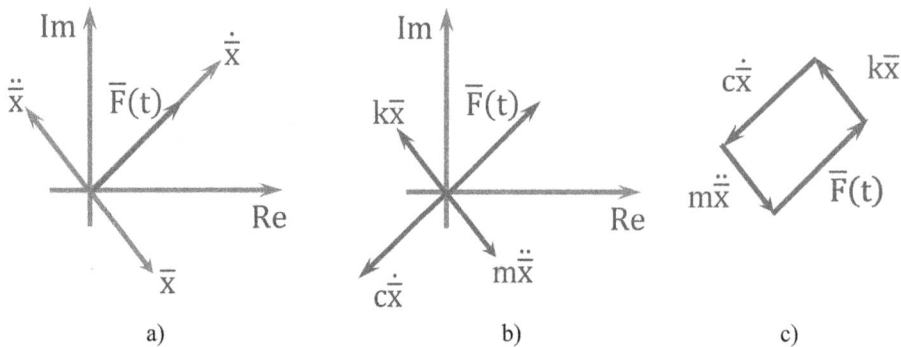

FIGURE 4.17 For $\dfrac{\omega}{\omega_n} = 1$ (a) displacement, velocity, and acceleration vectors, (b) complex force vectors, and (c) vector polygon of the complex forces.

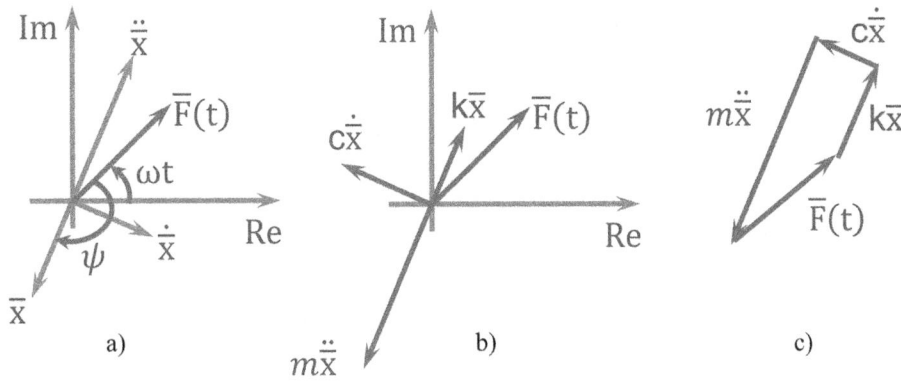

FIGURE 4.18 For $\dfrac{\omega}{\omega_n} > 1$ (a) displacement, velocity, and acceleration vectors, (b) complex force vectors, and (c) vector polygon of the complex forces.

- At resonance, since the damping force is there to counter the external force, the response is finite and doesn't increase indefinitely as in an undamped system.

c. When $\dfrac{\omega}{\omega_n} > 1$:

- Figure 4.18 shows that the force component along the velocity vector balances the damping force, while the perpendicular component combines with the elastic force to balance the inertia force.

4.6 VIBRATION WITH STRUCTURAL DAMPING

4.6.1 STRUCTURAL DAMPING

Every structural member undergoing vibration loses some energy in every cycle of vibration. Decay of fee vibration amplitude with time when damping mechanisms like viscous or Coulomb or other mechanisms external to the structure are not there points to the presence of damping mechanisms internal to the structure. It is seen that the energy loss also occurs inside a structural member due to molecular friction and other complex processes.

In the cyclic loading of a structure, the plot of stress versus strain produces a closed loop called the hysteresis loop, as shown in Figure 4.19. The hysteresis loop results from the phase lag between the stress and strain due to energy-dissipating forces or mechanisms internal to the material. The area enclosed by the loop represents the energy dissipated in the material per unit volume during one loading cycle. Therefore, a structural member undergoing cyclic changes in stress and strain during vibration dissipates energy due to hysteresis and is referred to as **structural, hysteretic, or internal damping**.

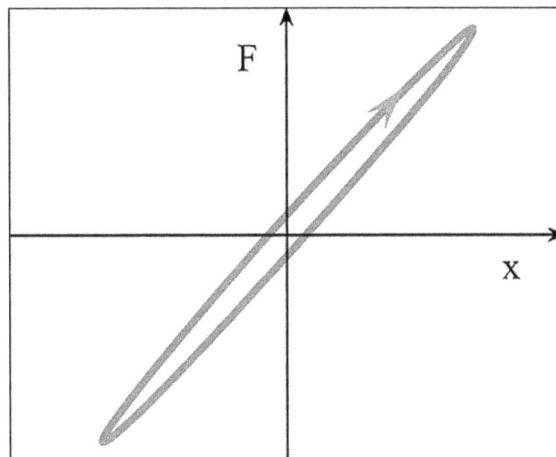

FIGURE 4.19 Hysteresis loop with structural damping.

How to quantify structural damping? The area estimat offers a basis to quantify the structural damping. **Loss factor** (η) is such a measure and is defined as the energy dissipated in a cycle (E_d), per unit radian, per unit energy at the beginning of the cycle (U).

$$\eta = \frac{E_d}{2\pi U} \tag{4.72}$$

Typical values of the loss factor for structures made of some common engineering materials are: aluminum (0.00002–0.002), steel (0.002–0.02), concrete (0.02–0.06), and rubber (0.1–0.9). Thus, structures made up of metals have low internal damping while structures made up of materials like concrete and rubber have higher levels of internal damping.

4.6.2   SDOF SYSTEM WITH STRUCTURAL DAMPING

In this section, we develop the governing equation of motion of an SDOF system with structural damping. Writing the dynamic equilibrium equation for a system with structural damping requires quantifying the damping force. However, the loss factor defined above is an energy-based index and doesn't quantify the damping force. In the absence of a suitable model for the damping force, the strategy is to define an equivalent viscous damping coefficient as the corresponding damping force is known. The equivalence is established by equating the energy dissipated via hysteresis (E_{dh}) with the energy dissipated by a viscous damper (E_{dv}). In experiments with the cyclic loading tests on specimens, the energy dissipated via hysteresis is found to be proportional to the square of the displacement amplitude (A) and independent of the frequency of the cyclic loading. Thus, E_{dh} can be written as

$$E_{dh} = \alpha A^2 \tag{4.73}$$

where α is the constant of proportionality.

The energy dissipated by a viscous damper can be determined as follows:

$$E_{dv} = \int_0^T F.\dot{x}.dt = \int_0^T c\dot{x}.\dot{x}.dt \tag{4.74}$$

If the steady-state response is given by $x = A\cos(\omega t - \psi)$, we get after simplification

$$E_{dv} = \pi c \omega A^2 \tag{4.75}$$

Thus, the energy dissipated by a viscous damper is proportional to the frequency of oscillations and the square of the displacement amplitude. Equating E_{dh} and E_{dv}, we get an **equivalent value of the viscous damping coefficient (c_{eq})**,

$$c_{eq} = c = \frac{\alpha}{\pi\omega} = \frac{h}{\omega} \tag{4.76}$$

where h is the **structural or hysteretic damping coefficient**. The equivalent viscous damping coefficient is frequency-dependent. Therefore, the equivalent value, c_{eq}, can be used for vibration analysis in the frequency domain but not in the time domain to analyze the system's response to an arbitrary forcing function.

We can find the loss factor for a viscous damper by substituting E_{dv} from Eq. (4.75) in place of E_d in Eq. (4.72). It gives

$$\eta_v = \eta = \frac{\pi c \omega A^2}{2\pi \frac{1}{2}kA^2} = \frac{c\omega}{k} \tag{4.77}$$

Similarly, the loss factor for a structural damper is obtained by substituting E_{dh} from Eq. (4.73) in place of E_d in Eq. (4.72). It gives

$$\eta_h = \eta = \frac{\alpha A^2}{2\pi \frac{1}{2}kA^2} = \frac{\alpha}{\pi k} = \frac{h}{k} \tag{4.78}$$

To analyze an SDOF system with structural damping, we approximate the structural damping force as an equivalent viscous damping force based on the equivalent viscous damping coefficient (Eq. 4.76). The equation of motion for an SDOF system with viscous damping subjected to an external force F(t) is

$$m\ddot{x} + c\dot{x} + kx = F(t) \tag{4.79}$$

We can't replace c in Eq. (4.79) by c_{eq} because the equation is in the time domain with a general excitation forcing function. Taking the Fourier transform (covered in Chapter 8) of Eq. (4.79), we get

$$-m\omega^2 x(\omega) + i\omega c x(\omega) + kx(\omega) = F(\omega) \tag{4.80}$$

Replacing c by c_{eq}, and using Eq. (4.76),

$$-m\omega^2 x(\omega) + ihx(\omega) + kx(\omega) = F(\omega) \tag{4.81}$$

$-ihx(\omega)$ can be considered the structural damping force in the frequency domain. It is proportional to the displacement but acts opposite to the velocity (indicated by the presence of 'i' and the negative sign).

$$-m\omega^2 x(\omega) + (k + ih)x(\omega) = F(\omega) \tag{4.82}$$

Using Eq. (4.78), we get

$$-m\omega^2 x(\omega) + k(1 + i\eta_h)x(\omega) = F(\omega) \tag{4.83}$$

$$x(\omega) = \frac{F(\omega)}{k(1 + i\eta_h) - m\omega^2} \tag{4.84}$$

The quantity $(k + ih)$ or $k(1 + i\eta_h)$ is called **complex stiffness**. The complex stiffness represents a combined resistance of elasticity and structural damping of the system to its displacement. If the input is a harmonic force $F(t) = Re(F_o e^{i\omega t})$ then from Eq. (4.84), the steady-state response can be obtained as

$$x(t) = Re(x(\omega)\, e^{i\omega t}) = Re\left(\frac{F_o}{k(1 + i\eta_h) - m\omega^2}\, e^{i\omega t}\right) = \frac{x_{st}}{\sqrt{\left(1 - \left(\frac{\omega}{\omega_n}\right)^2\right)^2 + \eta_h^2}}\cos(\omega t - \beta) \tag{4.85}$$

with the phase lag β given by

$$\beta = \tan^{-1}\frac{\eta_h}{1 - \left(\frac{\omega}{\omega_n}\right)^2} \tag{4.86}$$

Figure 4.20 shows the amplitude and phase lag of the steady-state response as a function of the harmonic excitation frequency for two values of the loss factor. The following observations are made from the graphs:

- As the structural loss factor increases, the peak response decreases.
- The peak response occurs at $\omega = \omega_n$ irrespective of the loss factor.
- When the excitation frequency $\omega \ll \omega_n$ the response of the system is $\approx \dfrac{x_{st}}{\sqrt{1 + \eta_h^2}}$.
- The phase lag of the response increases with an increase in the loss factor.

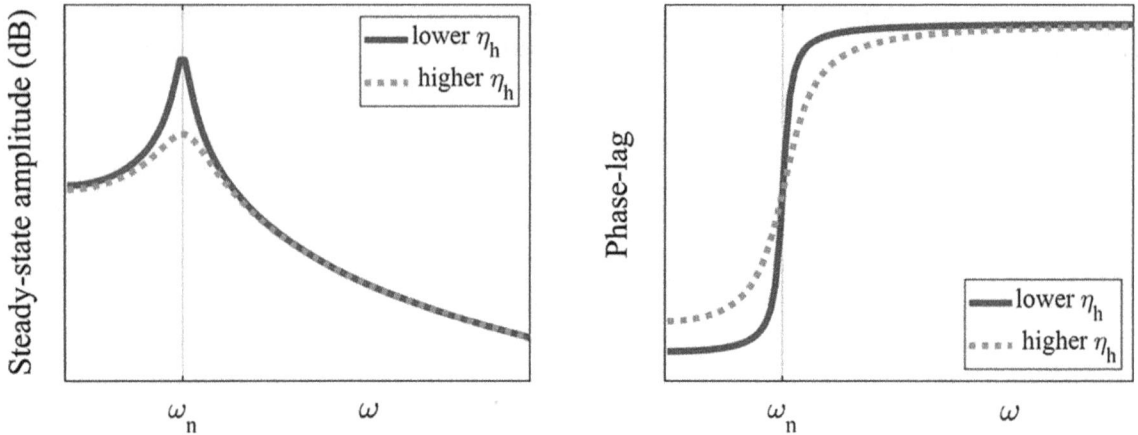

FIGURE 4.20   Amplitude and phase lag of the steady-state response with structural damping.

4.6.3   Response Analysis with Structural Damping

In principle, the differential equation of motion of an SDOF system with structural damping can be obtained by the inverse Fourier transform of the equation in the frequency domain (like Eq. 4.81). But for the loss factor model, presented in Section 4.6.2, the inverse Fourier transform doesn't represent a physically realistic damping mechanism (Crandall [31]). Thus, the time-domain representation of structural damping poses a challenge, and the response analysis of non-harmonic inputs in the time domain is difficult. For the same reason, the complementary part of the response, which arises due to initial conditions and requires the solution of the homogeneous form of the differential equation, is again difficult. But the steady-state part of the response to harmonic inputs can be determined using the frequency domain equation (Eq. 4.81).

4.6.4   Approximation by an SDOF System with Viscous Damping

Can we approximate a hysteretically damped system by an equivalent viscously damped system for easier analysis? Let us see what can be done.

From Eq. (4.76), it is clear that the equivalent viscous damping coefficient (c_{eq}) for a given structural damping coefficient (h) is frequency-dependent. Let us express that equation in terms of the viscous and structural damping factors. Substitute Eq. (4.78) into Eq. (4.76) to obtain

$$c_{eq} = \frac{h}{\omega} = \frac{\eta_h \, k}{\omega} \tag{4.87}$$

Dividing by the critical viscous damping coefficient (c_c) on both sides, and noting that $c_c = 2\sqrt{km}$ and $\xi_{eq} = \frac{c_{eq}}{c_c}$, leads to

$$\xi_{eq} = \frac{\eta_h \, \omega_n}{2\omega} \tag{4.88}$$

Eq. (4.88) indicates that the damping factor (ξ_{eq}) of the viscously damped system equivalent to a given hysteretically damped system (with a loss factor η_h) is frequency-dependent. In other words, for each excitation frequency (ω), a distinct value of the viscous damping factor needs to be used in the equivalent model. Thus, a single equivalent viscous damping system for the frequency range doesn't exist.

One approximation could be to use the equivalent viscous damping factor corresponding to a single chosen frequency. Since the influence of damping is more pronounced at and around the natural frequency, $\omega = \omega_n$ could be a suitable choice of frequency. Therefore, from Eq. (4.88), we get

$$\xi_{eq} \cong \frac{\eta_h}{2} \tag{4.89}$$

Thus, a hysteretically damped system can be approximated by an equivalent viscously damped system with the equivalent viscous damping factor (ξ_{eq}) given by Eq. (4.89). It also allows an approximate analysis of the hysteretically damped systems in the time domain.

4.7 FREQUENCY RESPONSE FUNCTION

In Sections 4.4–4.6, we looked at the forced response analysis of the undamped, viscously damped, and hysteretically damped SDOF systems. We studied how the system's response is affected by the frequency and amplitude of the harmonic force and by the natural frequency and damping/loss factor of the system. Therefore, the output response is affected by both the input and system characteristics.

Can we have a descriptor that depends only on the system parameters but not on the magnitude of the input? The frequency response function (FRF) is one such descriptor. Since the experimental modal analysis involves the identification of the dynamic characteristics of a system, the knowledge of a system descriptor, like an FRF, offers the basis to identify these characteristics.

The FRF of an SDOF system is the system's steady-state response to a unit harmonic input force. For a linear system, the frequency of the steady-state response is the same as the harmonic force frequency. Note that the FRF is defined using the steady-state response of the system, which requires that the transient response has decayed completely. For brevity, we may omit the reference 'steady-state' in the discussion about the FRF.

Assume a complex exponential harmonic force

$$F(t) = F_o \, e^{i\omega t} \tag{4.90}$$

Let the resulting steady-state displacement response is

$$x(t) = X e^{i\omega t} \tag{4.91}$$

Then the FRF ($\alpha(\omega)$) is the ratio of the steady-state response and force amplitudes. Hence,

$$\alpha(\omega) = \frac{X}{F_o} \tag{4.92}$$

Note that the response amplitude may be complex, making the FRF complex. Though an FRF is defined in the frequency domain and thus for harmonic inputs, it can be used to determine the system's response to non-harmonic inputs like periodic, transient, and non-deterministic using the Fourier transform.

4.7.1 FRF OF AN UNDAMPED SDOF SYSTEM

The steady-state response amplitude of an undamped SDOF system to a complex harmonic force is given by Eq. (4.39). Hence, the FRF is

$$\alpha(\omega) = \frac{\dfrac{F_o}{k - m\omega^2}}{F_o} \tag{4.93}$$

$$\alpha(\omega) = \frac{1}{k - m\omega^2} = \frac{1/m}{\omega_n^2 - \omega^2} \tag{4.94}$$

4.7.2 FRF OF AN SDOF SYSTEM WITH VISCOUS DAMPING

Using the steady-state response amplitude obtained in Eq. (4.56), we get the FRF as

$$\alpha(\omega) = \frac{\dfrac{F_o}{k - m\omega^2 + i\omega c}}{F_o} \tag{4.95}$$

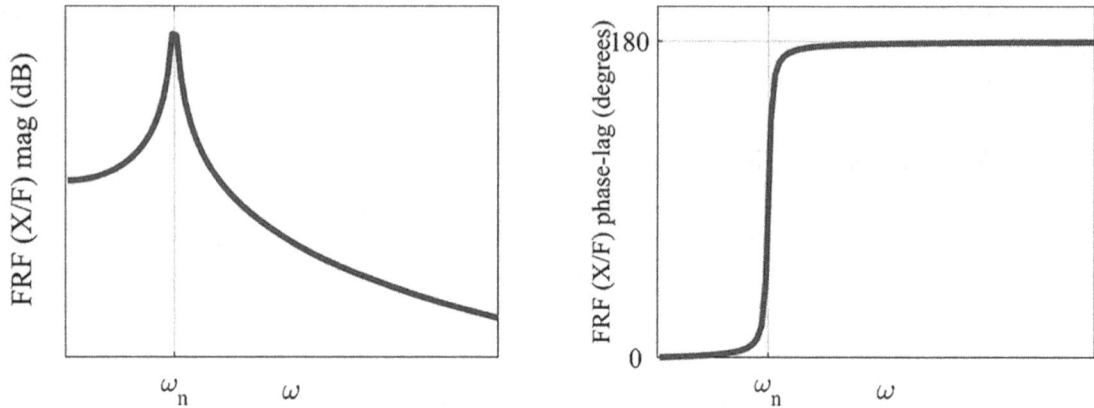

FIGURE 4.21 Amplitude and phase lag of the FRF of an SDOF system with viscous damping.

$$\alpha(\omega) = \frac{1}{k - m\omega^2 + i\omega c} \tag{4.96}$$

$$\alpha(\omega) = \frac{1/m}{\omega_n^2 - \omega^2 + i2\xi\omega\omega_n} \tag{4.97}$$

Thus, the FRF of a damped system is complex, indicating that it has both magnitude and phase. Figure 4.21 shows the magnitude and phase lag of the FRF. The magnitude is shown on a dB scale calculated as $20\text{Log}_{10}|\alpha(\omega)|$.

4.7.3 FRF OF AN SDOF SYSTEM WITH STRUCTURAL DAMPING

Using the steady-state response amplitude obtained in Eq. (4.84), we get

$$\alpha(\omega) = \frac{\dfrac{F_0}{k(1+i\eta_h) - m\omega^2}}{F_0} = \frac{1}{k - m\omega^2 + ik\eta_h} \tag{4.98}$$

$$\alpha(\omega) = \frac{1}{k - m\omega^2 + ih} \tag{4.99}$$

$$\alpha(\omega) = \frac{1/m}{\omega_n^2 - \omega^2 + i\eta_h\omega_n^2} \tag{4.100}$$

4.8 RESPONSE TO HARMONIC EXCITATION USING FRF

If the FRF of a system is given, then the response to a given harmonic force can be determined. Note that the FRF magnitude at the harmonic force frequency gives the steady-state response to a unit amplitude harmonic force. Hence, the FRF magnitude multiplied by the amplitude of the harmonic force yields the amplitude of the response to the given force.

Let $F(t) = F_0\cos\omega_0 t$ is the harmonic force acting on the system. If the FRF magnitude is in dB, it should be converted to the absolute value. Let it be $|\alpha(\omega_0)|$. Then the amplitude of the steady-state response is equal to $F_0|\alpha(\omega_0)|$. Similarly, the phase of the steady-state response would be the sum of the phase lag of the FRF and the harmonic force.

4.9 RESPONSE TO PERIODIC EXCITATION USING FRF

Using the Fourier series expansion (Chapter 8), a periodic force can be represented as a sum of the harmonic forces. Thus, the system acted upon by a periodic force can be viewed as being acted upon by several harmonic forces of different amplitudes, frequencies, and phases as determined by the Fourier series expansion. A linear system follows the superposition principle, which means that the response to the periodic force is the same as the sum of the responses due to the constituent harmonic forces. The steady-state response to each harmonic forcing function can be determined either by directly solving the equation of motion for the steady-state response or by using the FRF of the system.

Let F(t) be the periodic force. The equation of motion, assuming viscous damping, is

$$m\ddot{x} + c\dot{x} + kx = F(t) \tag{4.101}$$

Using the Fourier series, we can write

$$F(t) = f_0 + \sum_{r=1}^{\infty} f_r \cos(r\omega_o t - \phi_r) \tag{4.102}$$

where ω_o is the fundamental frequency given by $\dfrac{2\pi}{T_o}$, and T_o is the period of the periodic force. f_0 is a constant and represents the average value of the periodic force. f_r and ϕ_r are the amplitude and phase, respectively, of the rth harmonic force in the Fourier series expansion. The frequencies of the harmonic forces in the Fourier series expansion are related to the fundamental frequency by an integer relationship.

From the FRF of the system, Eq. (4.96), the magnitude and phase of the FRF are obtained as

$$\left|\alpha(r\omega_o)\right| = \left|\frac{1}{k - m(r\omega_o)^2 + ir\omega_o c}\right| \tag{4.103}$$

$$\text{Phase lag of } \alpha(r\omega_o) = \psi_r = \tan^{-1}\frac{r\omega_o c}{k - m(r\omega_o)^2} \tag{4.104}$$

The phase lag of the steady-state response is equal to the phase lag of the FRF plus the phase lag of the force. Therefore the steady-state response to the rth harmonic force is

$$x_r(t) = \left|\alpha(r\omega_o)\right| f_r \cos(r\omega_o t - \phi_r - \psi_r) \tag{4.105}$$

Therefore, the steady-state response to the periodic force is

$$x_r(t) = x_o + \sum_{r=1}^{\infty} x_r(t) \tag{4.106}$$

$$x_r(t) = \frac{f_0}{k} + \sum_{r=1}^{\infty} \left|\alpha(r\omega_o)\right| f_r \cos(r\omega_o t - \phi_r - \psi_r) \tag{4.107}$$

4.10 IMPULSE RESPONSE FUNCTION

The FRF characterizes a system's dynamic behavior in the frequency domain. The **impulse response function (IRF)** (also called **unit impulse response**) characterizes a system's dynamic behavior in the time domain. The IRF is the response of the system to a unit impulse. Let us first define unit impulse.

4.10.1 UNIT IMPULSE

An ideal unit impulse can be defined using the **Dirac delta function** ($\delta(t)$). $\delta(t)$ is defined as

$$\delta(t) \quad \begin{aligned} &= 0 \quad \text{for } t \neq 0 \\ &\neq 0 \quad \text{for } t = 0 \end{aligned} \tag{4.108}$$

and

$$\int_{-\infty}^{+\infty} \delta(t)\, dt = 1.0 \tag{4.109}$$

Eq. (4.108) implies that the function is zero everywhere except at t=0, while Eq. (4.109) implies that the area under the function $\delta(t)$ is unity. Thus, $\delta(t)$ can be imagined as a rectangular pulse located at the origin such that the width of the pulse is infinitesimally small, its height is infinitely large, and the area of the pulse is unity. The function $\delta(t)$ can be used to model the forces applied over a short duration.

Function $\delta(t)$ delayed by time τ is represented by $\delta(t-\tau)$. Figure 4.22 shows the functions $\delta(t)$ and $\delta(t-\tau)$, respectively. Thus, $\delta(t-\tau)$ is nothing but $\delta(t)$ delayed by time τ.

To create an impulse of 1 N-sec on the mass, a force defined by $\delta(t)$ is applied to the mass, which corresponds to a unit impulse input as a result of Eq. (4.109).

4.10.2 Unit Impulse Response

The equation of motion of an SDOF system with viscous damping with a unit impulse input can be written as

$$m\ddot{x} + c\dot{x} + kx = \delta(t) \tag{4.110}$$

We are interested to know the resulting response under zero initial conditions. The impulse applied is equal to the change of momentum. Since the velocity is zero before the application of the impulse, the velocity after the application of the unit impulse would be

$$\dot{x} = \frac{1}{m} \tag{4.111}$$

Since the displacement remains zero at t=0, the new set of initial conditions is: $x(0) = 0$ and $\dot{x}(0) = 1/m$. The response to these initial conditions for an underdamped system with viscous damping can be obtained using Eq. (4.22). This response is the unit impulse response or IRF of the system (represented by $h(t)$) and is given by

$$h(t) = \frac{e^{-\xi\omega_n t}}{m\omega_n\sqrt{1-\omega^2}} \sin\omega_n\sqrt{1-\xi^2}\, t \qquad (\text{for } t > 0) \tag{4.112}$$

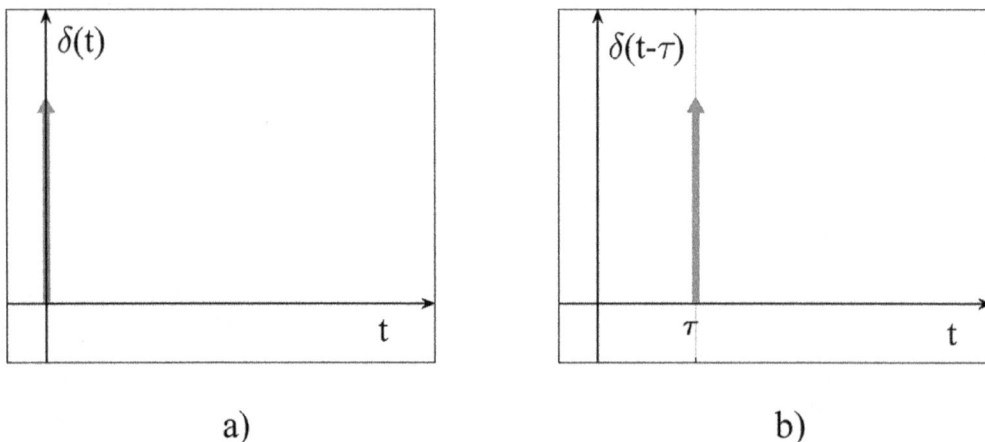

FIGURE 4.22   (a) $\delta(t)$, (b) $\delta(t-\tau)$.

or

$$h(t) = \frac{1}{m\omega_d} e^{-\xi\omega_n t} \sin\omega_d t \quad (\text{for } t > 0) \tag{4.113}$$

Figure 4.23 shows the unit impulse response given by Eq. (4.113). Note that it starts from x=0 but has a nonzero slope equal to the initial velocity of the mass imparted by the unit impulse.

If the system is undamped, then the unit impulse response or IRF is

$$h(t) = \frac{1}{m\omega_n} \sin\omega_n t \quad (\text{for } t > 0) \tag{4.114}$$

Figure 4.24 shows the unit impulse response of an undamped system. Thus, once initiated, the response continues with a constant amplitude.

The Fourier transform relates the IRF and FRF. The Fourier transform of IRF is FRF, and the inverse Fourier transform of FRF is IRF.

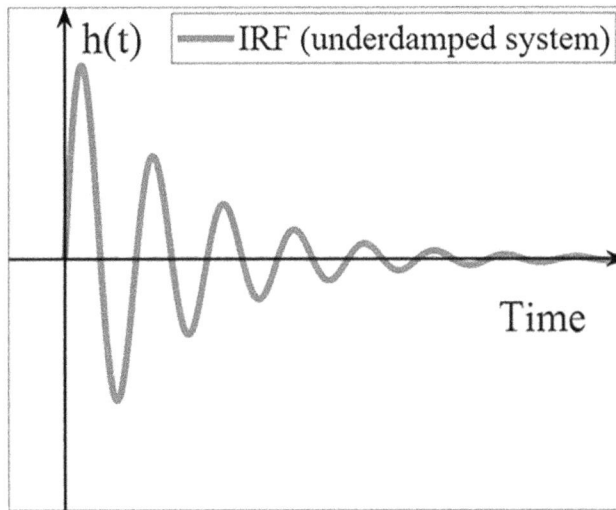

FIGURE 4.23 The unit impulse response of an underdamped SDOF system with viscous damping.

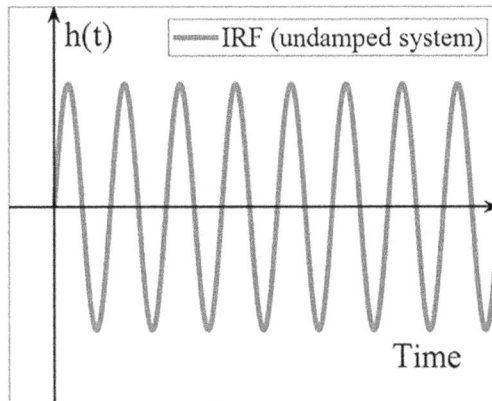

FIGURE 4.24 The unit impulse response of an undamped SDOF system.

4.11 RESPONSE TO TRANSIENT EXCITATION (CONVOLUTION INTEGRAL)

The forces that act on the system for short/finite duration are called transient forces. Forces due to impact and collusion generate transient excitation. It is possible to find the response to a transient force in the frequency domain, but in this section, we find the response in the time domain using the IRF presented in the previous section.

Figure 4.25 shows a transient force $F(\tau)$ acting on the system. The idea is to approximate this force as a series of constant magnitude forces acting over small time steps (of width $\Delta\tau$) as shown in Figure 4.26. Each force step has an impulse strength, and the response due to each can be found using the IRF. The response to the transient force can be obtained by superposing the responses due to individual force steps since the system is assumed to be linear.

Figure 4.27 shows one such force step at a time τ. The strength of the impulse due to the force step is $F(\tau)\Delta\tau$.

Now, the response due to the impulse function is simply the product of the unit impulse response $h(t-\tau)$, and $F(\tau)\Delta\tau$. If this response is denoted by Δx then, we have

$$\Delta x(t) = F(\tau)\Delta\tau\, h(t-\tau) \tag{4.115}$$

Figure 4.28 shows the unit impulse response $(h(t-\tau))$. Figure 4.29 shows the response Δx. Superposing the responses due to all the force steps gives

$$x(t) = \sum h(t-\tau)F(\tau)\,\Delta\tau \tag{4.116}$$

If the time step $\Delta\tau \to 0$ then Eq. (4.116) is converted into an integral equation given by

$$x(t) = \int_0^t h(t-\tau)F(\tau)\,d\tau \tag{4.117}$$

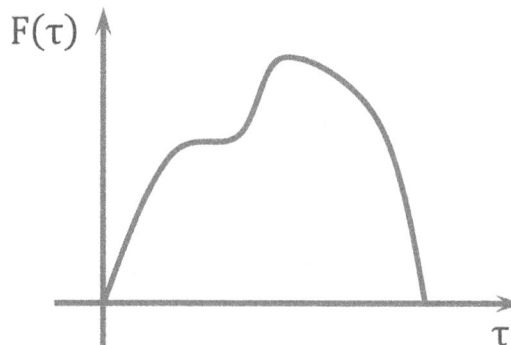

FIGURE 4.25   A transient force.

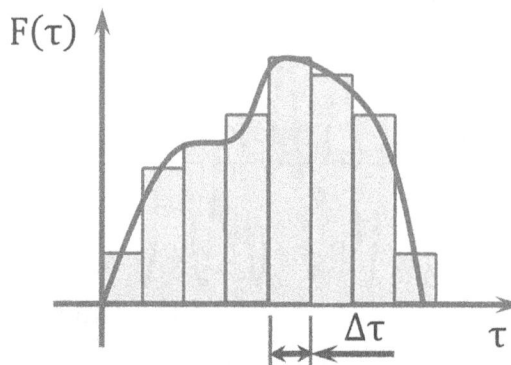

FIGURE 4.26   The transient force as a series of constant magnitude force steps.

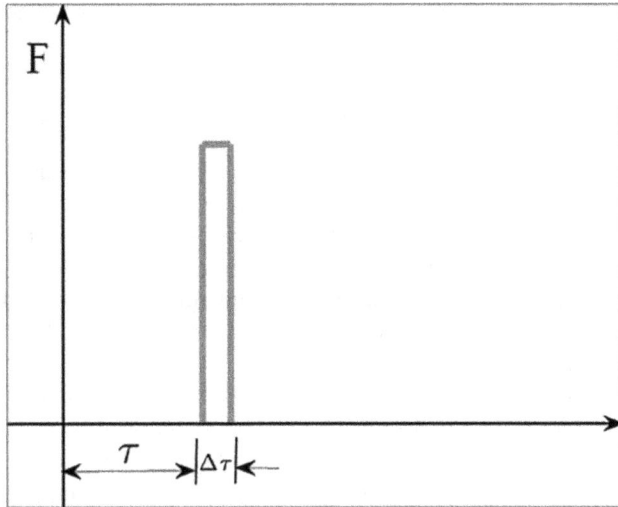

FIGURE 4.27 A force step at a time τ.

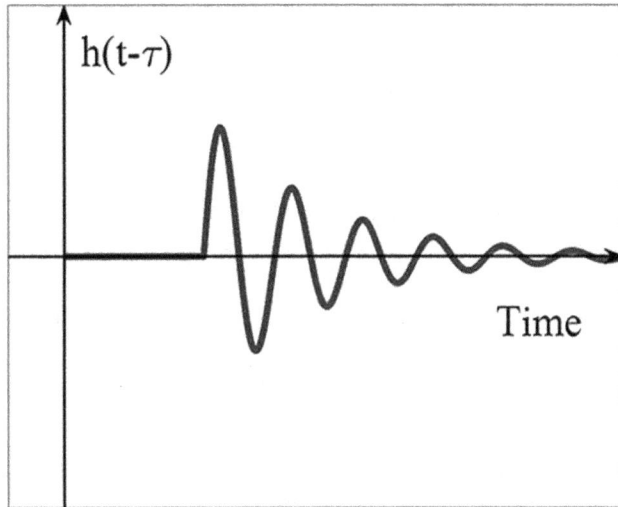

FIGURE 4.28 Unit impulse response $h(t-\tau)$.

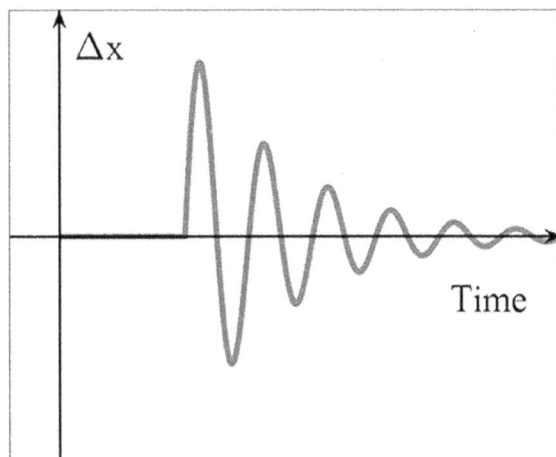

FIGURE 4.29 Response Δx.

which is called the **convolution integral**. The name convolution indicates the mathematical process of the superposition of the scaled impulse responses whose starting points are delayed from one another.

4.12   RESPONSE ANALYSIS BY THE LAPLACE TRANSFORM APPROACH

The previous sections looked at the response to transient excitation in the time domain using the IRF and convolution integral. However, evaluating the convolution integral may not be easy for many forcing functions. In this section, we present the Laplace transform approach, which may be easier to follow to determine the response.

Laplace transform converts a time-domain function or signal to the complex frequency domain. When applied to a differential equation, it results in an algebraic equation that can be solved more easily for the desired response than the direct solution of the differential equation. The response in the time domain is then obtained by the inverse Laplace transform of the response in the frequency domain.

We first define the Laplace transform and then look at the Laplace transform of some standard signals and relevant Laplace transform theorems. The transforms and theorems are derived from the fundamental definition. The objective is to cover the essential elements of the Laplace transform approach to analyze the response of vibratory systems. Another objective is to provide the background necessary to understand many concepts in modal analysis and modal parameter estimation methods based on the Laplace transform.

4.12.1   LAPLACE TRANSFORM

Laplace transform of a function $x(t)$ is defined as

$$L\big(x(t)\big) = x(s) = \int_{-\infty}^{\infty} x(t)e^{-st}\, dt \tag{4.118}$$

where 's' is a complex variable given by $s = \sigma + j\omega$, in which σ is a real constant and ω is the frequency in rad/sec. Equation (4.118) is referred to as the **two-sided Laplace transform**. In practice, excitation and the resulting response begin at a specific time, and hence for all such signals, by a proper choice of time origin, it can be said that $x(t) = 0$ for $t < 0$. Because of this, the so-called **single-sided Laplace transform** is defined as

$$L\big(x(t)\big) = x(s) = \int_{0}^{\infty} x(t)e^{-st}\, dt \tag{4.119}$$

To be precise, the lower limit of the integral represents 0^-, which is an infinitesimal time just before $t = 0$. The Fourier transform is a particular case of the Laplace transform defined by $\sigma = 0$.

4.12.2   LAPLACE TRANSFORM OF SOME STANDARD FUNCTIONS

a. Laplace transform of the Dirac delta function

$$x(t) = \delta(t) \tag{4.120}$$

Starting from the definition of the Laplace transform, and in light of the definition of the Dirac delta function (see Section 4.10.1), we get

$$x(s) = \int_{0}^{\infty} x(t)e^{-st}\, dt = \int_{0}^{\infty} \delta(t)e^{-st}\, dt = 1 \tag{4.121}$$

b. Laplace transform of e^{-at}

$$\begin{aligned} x(t) &= e^{-at} \qquad t \geq 0 \\ &= 0 \qquad\quad t < 0 \end{aligned} \tag{4.122}$$

From the definition of the Laplace transform,

$$x(s) = \int_0^\infty e^{-at} e^{-st} \, dt = \left[\frac{e^{-(s+a)t}}{-(s+a)} \right]_0^\infty = -\frac{1}{s+a}[0-1] = \frac{1}{s+a} \qquad (4.123)$$

c. Laplace transform of the unit step function
   This function is defined as

$$\begin{aligned} x(t) &= 1 & t \geq 0 \\ &= 0 & t < 0 \end{aligned} \qquad (4.124)$$

Taking a = 0, the exponential function (Eq. 4.122) represents the unit step function, and hence the Laplace transform can be obtained from Eq. (4.123) by substituting a = 0, yielding

$$x(s) = \frac{1}{s} \qquad (4.125)$$

d. Laplace transform of the ramp function
   This function is defined as,

$$\begin{aligned} x(t) &= A\,t & t \geq 0 \\ &= 0 & t < 0 \end{aligned} \qquad (4.126)$$

Starting from the definition of the Laplace transform,

$$x(s) = \int_0^\infty x(t) e^{-st} \, dt = \int_0^\infty At e^{-st} \, dt \qquad (4.127)$$

Performing the integration by parts gives,

$$x(s) = A\left[\left(t.\frac{e^{-st}}{-s} \right)_0^\infty - \int_0^\infty 1.\frac{e^{-st}}{-s} \, dt \right] = \frac{A}{s^2} \qquad (4.128)$$

e. Laplace transform of $\sin\omega_0 t$

$$\begin{aligned} x(t) &= A \sin\omega_0 t & t \geq 0 \\ &= 0 & t < 0 \end{aligned} \qquad (4.129)$$

Starting from the definition of the Laplace transform,

$$x(s) = \int_0^\infty A \sin\omega_0 t \, e^{-st} \, dt = A \int_0^\infty \frac{e^{i\omega_0 t} - e^{-i\omega_0 t}}{2i} e^{-st} \, dt \qquad (4.130)$$

$$x(s) = \frac{A}{2j} \int_0^\infty \left(e^{-(s-i\omega_o)t} - e^{-(s+i\omega_o)t} \right) dt = \frac{A\,\omega_o}{s^2 + \omega_o^2} \tag{4.131}$$

f. Laplace transform of $\cos\omega_o t$

$$\begin{aligned} x(t) &= A\,\cos\omega_o t && t \geq 0 \\ &= 0 && t < 0 \end{aligned} \tag{4.132}$$

Proceeding similarly to case e), we obtain

$$x(s) = \frac{A\,s}{s^2 + \omega_o^2} \tag{4.133}$$

g. Laplace transform of $e^{-at}\sin\omega_o t$

$$\begin{aligned} x(t) &= e^{-at}\sin\omega_o t && t \geq 0 \\ &= 0 && t < 0 \end{aligned} \tag{4.134}$$

Expressing $\sin\omega_o t$ using Euler's identity and then using the result in Eq. (4.123), we get

$$x(s) = \frac{\omega_o}{(s+a)^2 + \omega_o^2} \tag{4.135}$$

h. Laplace transform of $e^{-at}\cos\omega_o t$

$$\begin{aligned} x(t) &= e^{-at}\cos\omega_o t && t \geq 0 \\ &= 0 && t < 0 \end{aligned} \tag{4.136}$$

Expressing $\cos\omega_o t$ using Euler's identity and then using the result in Eq. (4.123), we get

$$x(s) = \frac{s+a}{(s+a)^2 + \omega_o^2} \tag{4.137}$$

i. Laplace transform of any function delayed by a time t_o
   Let there is a function $x(t)$ defined as

$$\begin{aligned} x(t) &= f(t) && t \geq 0 \\ &= 0 && t < 0 \end{aligned} \tag{4.138}$$

Let the function is delayed by a time t_o, such that

$$\begin{aligned} x(t-t_o) &= f(t) && t \geq t_o \\ &= 0 && t < t_o \end{aligned} \tag{4.139}$$

Then what is the Laplace transform of $x(t - t_0)$?

Starting from the definition of Laplace transform,

$$L(x(t - t_0)) = \int_0^\infty x(t - t_0) \, e^{-st} \, dt \tag{4.140}$$

$$L(x(t - t_0)) = \int_0^{t_0} x(t - t_0) \, e^{-st} \, dt + \int_{t_0}^\infty x(t - t_0) \, e^{-st} \, dt \tag{4.141}$$

Based on the definition of $x(t - t_0)$ the first integral is zero. Making a change of variable, $t - t_0 = \tau$ and noting that the new limits change from (t_0, ∞) to $(0, \infty)$, we get

$$L(x(t - t_0)) = e^{-st_0} \int_0^\infty x(\tau) \, e^{-s\tau} \, d\tau = e^{-st_0} x(s) \tag{4.142}$$

Thus, the Laplace transform of any function delayed by a time t_0 is equal to the product of the Laplace transform of the function without delay and e^{-st_0}.

j. Real differentiation theorem

This theorem is about the Laplace transform of differentiation of a function. Starting from the definition of the Laplace transform and performing the integration by parts gives

$$x(s) = \int_0^\infty x(t) e^{-st} \, dt = \left(x(t) \cdot \frac{e^{-st}}{-s} \right)_0^\infty - \int_0^\infty \frac{dx(t)}{dt} \cdot \frac{e^{-st}}{-s} \, dt \tag{4.143}$$

$$x(s) = \left(x(t) \cdot \frac{e^{-st}}{-s} \right)_0^\infty - \int_0^\infty \frac{dx(t)}{dt} \cdot \frac{e^{-st}}{-s} \, dt \tag{4.144}$$

$$-sx(s) = (0 - x(0)) - L\left(\frac{dx(t)}{dt} \right) \tag{4.145}$$

$$L\left(\frac{dx(t)}{dt} \right) = sx(s) - x(0) \tag{4.146}$$

where $x(0)$ is the initial value of $x(t)$.

By proceeding similarly, it can be shown that

$$L\left(\frac{d^2 x(t)}{dt^2} \right) = s^2 x(s) - sx(0) - \dot{x}(0) \tag{4.147}$$

where $\dot{x}(0)$ is the initial value of $\dot{x}(t)$.

4.12.3 INVERSE LAPLACE TRANSFORM

The inverse Laplace transform of $x(s)$ yields the time domain function $x(t)$ and is given by

$$x(t) = \frac{1}{2\pi i} \int_{\sigma - i\infty}^{\sigma + i\infty} x(s) e^{st} \, ds \tag{4.148}$$

The integral in Eq. (4.148) is along a line parallel to the imaginary axis in the complex plane. However, the integral need not be evaluated in practice in most cases as the inverse Laplace transform can be found from the knowledge of the Laplace transform of the commonly encountered functions available in the form of Laplace transform tables. Some of these results we got in the previous section. If we want to know the Laplace inverse of $x(s)$ then we must look for that $x(t)$ whose Laplace transform is $x(s)$. This $x(t)$ would be nothing but the Laplace inverse of $x(s)$.

4.12.4 DETERMINATION OF RESPONSE USING LAPLACE TRANSFORM

Using the Laplace transform, let us find the response of an SDOF system with viscous damping subjected to a force $F(t)$. The equation of motion is given by

$$m\ddot{x} + c\dot{x} + kx = F(t) \tag{4.149}$$

Taking the Laplace transform of both sides and using the real differentiation theorem, we get

$$mL(\ddot{x}) + cL(\dot{x}) + kL(x) = L\big(F(t)\big) \tag{4.150}$$

$$m\big(s^2 x(s) - sx(0) - \dot{x}(0)\big) + c\big(sx(s) - x(0)\big) + kx(s) = F(s) \tag{4.151}$$

$$x(s)\big(ms^2 + cs + k\big) = F(s) + (ms + c)x(0) + m\dot{x}(0) \tag{4.152}$$

$$x(s) = \frac{F(s)}{\big(ms^2 + cs + k\big)} + \frac{(ms + c)}{\big(ms^2 + cs + k\big)}x(0) + \frac{m}{\big(ms^2 + cs + k\big)}\dot{x}(0) \tag{4.153}$$

Eq. (4.153) is the total response in the Laplace domain. It includes the response due to the external force and the initial conditions. The response in the time domain can be found by taking the inverse Laplace transform. Eq. (4.153) is valid for any general excitation function $F(t)$. It requires finding out $F(s)$ for the given $F(t)$. A higher-order fraction or function for which the Laplace transform is unknown can be broken down into a sum of standard functions, for which the Laplace transforms are already known, and since the Laplace transform is a linear transform, the Laplace transform of the given function is equal to the sum of the Laplace transform of the standard functions.

4.12.5 UNIT IMPULSE RESPONSE OF AN SDOF SYSTEM WITH VISCOUS DAMPING

The unit impulse response of an SDOF system with viscous damping was determined in Section 4.10.2. Let us determine the same using the Laplace transform. To find the unit impulse response, we have $F(t) = 1.\delta(t)$ and the initial conditions are zero. Therefore,

$$F(s) = 1, \quad x(0) = 0 \text{ and } \dot{x}(0) = 0 \tag{4.154}$$

Substituting (4.154) into Eq. (4.153), we get

$$x(s) = \frac{1}{\big(ms^2 + cs + k\big)} \tag{4.155}$$

$$x(s) = \frac{1/m}{\left(s^2 + \dfrac{c}{m}s + \dfrac{k}{m}\right)} \tag{4.156}$$

Noting that $\dfrac{c}{m} = 2\xi\omega_n$ and $\dfrac{k}{m} = \omega_n^2$, we get

$$x(s) = \dfrac{1/m}{\left(s^2 + 2\xi\omega_n s + \omega_n^2\right)} \qquad (4.157)$$

Comparing the RHS of Eq. (4.157) with the RHS of Eq. (4.135), we get, $a = \xi\omega_n$ and $\omega_o = \omega_n\sqrt{1-\xi^2} = \omega_d$. Therefore Eq. (4.157) can be written as

$$x(s) = \dfrac{1}{m\omega_d} \cdot \dfrac{\omega_d}{\left(s + \xi\omega_n\right)^2 + \omega_d^2} \qquad (4.158)$$

Taking the inverse Laplace transform of Eq. (4.158) using the results in Eqs. (4.134) and (4.135), we get

$$h(t) = x(t) = \dfrac{1}{m\omega_d}\ e^{-\xi\omega_n t}\ \sin\omega_d t \qquad (\text{for } t > 0) \qquad (4.159)$$

The result in Eq. (4.159) is identical to the one found by direct solution (Eq. 4.113). The specification of the range of t over which the equation is valid can be done away with by multiplying the RHS of Eq. (4.159) with the unit step function $u(t)$ as follows:

$$h(t) = \dfrac{1}{m\omega_d}\ e^{-\xi\omega_n t}\ \sin\omega_d t\ u(t) \qquad (4.160)$$

The presence of $u(t)$ automatically ensures that $x(t) = 0$ for $t < 0$.

4.12.6   TRANSFER FUNCTION

The transfer function is the ratio of the Laplace transforms of the output and input with zero initial conditions. This ratio obtained from Eq. (4.153) gives the transfer function of an SDOF system with viscous damping and is given by

$$\dfrac{x(s)}{F(s)} = \dfrac{1}{\left(ms^2 + cs + k\right)} \qquad (4.161)$$

The transfer function is defined in the complex frequency domain and forms another system descriptor. If the transfer function of a system is known, then the system's response with zero initial conditions can be determined.

The FRF is a particular case of the transfer function and can be obtained by substituting $s = i\omega$ in the transfer function. The FRF obtained from Eq. (4.161) matches the result given in Eq. (4.96).

Example 4.1

A vehicle represented by an SDOF undamped model moves with a speed of V m/sec over a speed breaker in the form of a sine pulse, as shown in Figure 4.30. The base of the speed breaker extends over a length L while its height is Y. Find the response using the Laplace transform approach, assuming zero initial conditions.

Solution

The upward displacement of the wheel due to the speed breaker can be expressed as

$$y(t) = Y \sin\omega_o t \qquad \text{for } t \leq T_o$$
$$= 0 \qquad\qquad \text{for } t > T_o$$

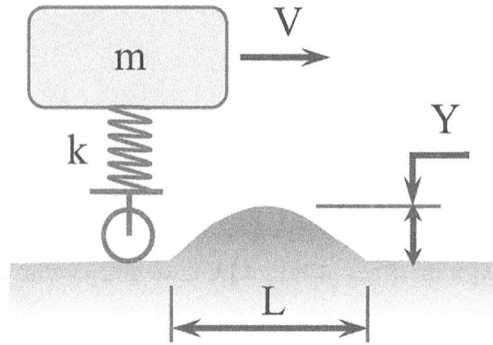

FIGURE 4.30   SDOF model of a vehicle moving over a speed breaker.

where

$$\omega_o = \frac{2\pi}{2T_o} \quad \text{and} \quad T_o = \frac{L}{V}$$

If x is the vertical displacement of the mass, the equation of motion can be written as

$$m\ddot{x} = -k(x-y) \quad \text{or} \quad m\ddot{x} + kx = ky$$

Taking the Laplace transform of both sides and noting that the initial conditions on x are zero, we get

$$x(s) = \frac{k\,y(s)}{(ms^2 + k)} = \frac{\omega_n^2\,y(s)}{(s^2 + \omega_n^2)}$$

Let us first find y(s).

$$y(s) = \int_0^\infty y(t)e^{-st}\,dt = \int_0^{T_o} Y\,\sin\omega_o t\,e^{-st}\,dt$$

$$y(s) = \int_0^{T_o} Y\frac{e^{i\omega_o t} - e^{-i\omega_o t}}{2j}e^{-st}\,dt = \frac{Y}{2j}\int_0^{T_o}\left(e^{-(s-i\omega_o)t} - e^{-(s+i\omega_o)t}\right)dt$$

After integration and simplification, we get

$$y(s) = \frac{Y\omega_o\left(1 + e^{-sT_o}\right)}{\left(s^2 + \omega_o^2\right)}$$

Substituting y(s) into the expression for x(s),

$$x(s) = \frac{\omega_n^2\,Y\omega_o\left(1 + e^{-sT_o}\right)}{\left(s^2 + \omega_n^2\right)\left(s^2 + \omega_o^2\right)}$$

which can be written as

$$x(s) = x_1(s) + x_2(s)$$

where

$$x_1(s) = \frac{\omega_n^2 \, Y\omega_o}{\left(s^2 + \omega_n^2\right)\left(s^2 + \omega_o^2\right)} \quad \text{and} \quad x_2(s) = e^{-sT_o} x_1(s)$$

Let us first find the inverse Laplace transform of $x_1(s)$. Splitting it into partial fractions, we get

$$x_1(s) = \frac{\omega_n^2 \, H\omega_o}{\left(\omega_o^2 - \omega_n^2\right)} \left(\frac{1}{s^2 + \omega_n^2} - \frac{1}{s^2 + \omega_o^2} \right)$$

Taking the inverse Laplace transform and using the result derived earlier, we get

$$x_1(t) = \frac{\omega_n^2 \, H\omega_o}{\left(\omega_o^2 - \omega_n^2\right)} \left(\frac{1}{\omega_n} \sin \omega_n t - \frac{1}{\omega_o} \sin \omega_o t \right) u(t)$$

Note that $u(t)$ is also there in the above equation to ensure that $x_1(t) = 0$ for $t < 0$.

In the expression of $x_2(s)$, multiplication of $x_1(s)$ by e^{-sT_o} indicates that $x_2(t)$ is $x_1(t)$ delayed by time T_o. Therefore, replacing t with $(t - T_o)$ in the equation of $x_1(t)$ gives $x_2(t)$.

$$x_2(t) = \frac{\omega_n^2 \, H\omega_o}{\left(\omega_o^2 - \omega_n^2\right)} \times \left(\frac{1}{\omega_n} \sin \omega_n (t - T_o) - \frac{1}{\omega_o} \sin \omega_o (t - T_o) \right) u(t - T_o)$$

Therefore, the response $x(t)$ would be

$$x(t) = x_1(t) + x_2(t)$$

$$x(t) = \frac{\omega_n^2 \, H\omega_o}{\left(\omega_o^2 - \omega_n^2\right)} \left[\left(\frac{1}{\omega_n} \sin \omega_n t - \frac{1}{\omega_o} \sin \omega_o t \right) u(t) + \left(\frac{1}{\omega_n} \sin \omega_n (t - T_o) - \frac{1}{\omega_o} \sin \omega_o (t - T_o) \right) u(t - T_o) \right]$$

This example shows that the Laplace transform approach can be used easily to find response to transient excitation.

Example 4.2

A machine is subjected to a step force $F(t)$ given below. Determine using the convolution integral the displacement response of the machine modeled as an SDOF system with viscous damping.

$$F(t) = F_o \quad t \geq 0$$
$$= 0 \quad t < 0$$

Solution

The convolution integral is

$$x(t) = \int_0^t h(t - \tau) F(\tau) \, d\tau$$

It would be convenient to evaluate the integral by changing the variable of integration. Let $t-\tau=\eta$. We get after simplification

$$x(t)=\int_0^t h(\eta)F(t-\eta)\,d\eta$$

Substituting the IRF $h(\eta)$ from Eq. (4.113),

$$x(t)=\int_0^t \frac{1}{m\omega_d}\,e^{-\xi\omega_n\eta}\,\sin\omega_d\eta\;F_o\,d\eta$$

$$x(t)=\int_0^t \frac{1}{m\omega_d}\,e^{-\xi\omega_n\eta}\,\frac{e^{i\omega_d\eta}-e^{-i\omega_d\eta}}{2i}F_o\,d\eta$$

$$x(t)=\frac{F_o}{2im\omega_d}\int_0^t\left(e^{-(\xi\omega_n-i\omega_d)\eta}-e^{-(\xi\omega_n+i\omega_d)\eta}\right)d\eta$$

$$x(t)=\frac{F_o}{2im\omega_d}\left[\left(\frac{e^{-(\xi\omega_n-i\omega_d)t}}{-(\xi\omega_n-i\omega_d)}-\frac{e^{-(\xi\omega_n+i\omega_d)t}}{-(\xi\omega_n+i\omega_d)}\right)-\left(\frac{1}{-(\xi\omega_n-i\omega_d)}-\frac{1}{-(\xi\omega_n+i\omega_d)}\right)\right]$$

On simplification, we get the step response as

$$x(t)=\frac{F_o}{k}\left[1-e^{-\xi\omega_n t}\cos\omega_d t-\frac{\xi}{\sqrt{1-\xi^2}}e^{-\xi\omega_n t}\sin\omega_d t\right]$$

Example 4.3

An electric motor is mounted on a heavy inertia block isolated from the surroundings via isolators to reduce the force transmitted to the ground, as shown in Figure 2.7 (in Chapter 2). Figure 2.8 shows a lumped parameter model of the system. Let the mass of the motor is 50 kg, and the mass of the inertia block is 1,000 kg. The speed of the motor is 287 RPM, and it has an unbalance of 0.1333 kg-m. The viscous damping factor of the system is 0.09, while the stiffness of the isolators is 2.95e+05 N/m.

A. Find the displacement amplitude of the steady-state response of the motor and the force transmitted to the ground.
B. Find the quantities in (A) if the motor is mounted only with isolators but without the inertia block.

Solution

A.

The displacement amplitude of the steady-state response is given by

$$x_o=\frac{x_{st}}{\sqrt{\left(1-\left(\frac{\omega}{\omega_n}\right)^2\right)^2+\left(2\xi\frac{\omega}{\omega_n}\right)^2}}$$

For the present problem,

$$\omega = \frac{2\pi N}{60} = \frac{2\pi \times 287}{60} = 30 \text{ rad/sec}$$

$$x_{st} = \frac{F_o}{k} = \frac{me\omega^2}{k} = \frac{0.1333 \times 30^2}{295,000} = 0.00041 \text{ m}$$

$$\omega_n = \sqrt{\frac{k}{m}} = \sqrt{\frac{295,000}{50+1,000}} = 16.75 \text{ rad/sec}$$

$$\xi = 0.09$$

Substituting the values we get

$$x_o = 0.000183 \text{ m}$$

The force transmitted is the sum of the force due to the stiffness and damping of the isolator

$$F_t(t) = kx(t) + c\dot{x}(t)$$

Representing the unbalance force on the motor and the resulting displacement in complex forms as

$$\bar{F}_t(t) = F_o e^{i\omega t} \quad ; \qquad \bar{x}(t) = x_o e^{i(\omega t - \psi)}$$

Then,

$$\dot{\bar{x}}(t) = i\omega x_o e^{i(\omega t - \psi)}$$

The actual force transmitted $F_t(t)$ is the real part of the complex force $\bar{F}_t(t)$. Since $\bar{F}_t(t)$ is a rotating vector, the maximum value of $F_t(t)$ is nothing but the amplitude of the force $\bar{F}_t(t)$. Therefore,

$$\bar{F}_t(t) = k\bar{x}(t) + c\dot{\bar{x}}(t) = kx_o e^{i(\omega t - \psi)} + ic\omega x_o e^{i(\omega t - \psi)}$$

Therefore, the amplitude of the force $\bar{F}_t(t)$ is

$$F_{to} = \sqrt{(kx_o)^2 + (c\omega x_o)^2}$$

Damping coefficient c is,

$$c = \xi \times c_c = 0.09 \times 2\sqrt{km} = 3165 \text{ N} - \text{sec/m}$$

Therefore, we get the maximum transmitted force as $F_{to} = 56.5 \text{ N}$.

B.

Now the inertia block is not used, and therefore, only the motor contributes to the mass of the system,

$$m = 50 \text{ kg}$$

Since the same isolator is used, the stiffness and damping coefficient remains the same, and the damping factor becomes

$$\xi = \frac{c}{c_c} = \frac{3,165}{2\sqrt{km}} = 0.4124$$

Since the unbalance and speed are the same, x_{st} also remains the same.
The natural frequency of the system is

$$\omega_n = \sqrt{\frac{k}{m}} = \sqrt{\frac{295,000}{50}} = 76.75 \text{ rad/sec}$$

Substituting all the above values, we get, $x_o = 0.00045$ m.
From the equation of F_{to}, we get, $F_{to} = 139.1$ N.
Thus, we see that the transmitted force and the displacement amplitude of the steady-state response are higher without the inertia block. One of the critical variables affecting the amplitude of the transmitted force is the frequency ratio $\frac{\omega}{\omega_n}$. For an SDOF system, it can be shown that if $\frac{\omega}{\omega_n} > \sqrt{2}$, then the transmitted force is smaller than the force applied; else, it is higher. The condition $\frac{\omega}{\omega_n} > \sqrt{2}$ is satisfied with the inertia block in part A while not in part B. Due to this, the amount of the transmitted force is higher than the applied force in part B.

Example 4.4

Figure 2.45 (in Chapter 2) shows a steel bracket in the form of a cantilever supporting an electric motor of mass 2 kg. The bracket's length, width, and thickness are 0.4, 0.05, and 0.005 m, respectively. The modulus of elasticity of the material of the bracket is 2×10^{11} N/m². Due to unbalance, the electric motor operating at 1,416 RPM applies a dynamic force of 20 N amplitude on the bracket. Neglect the mass of the bracket and moment of inertia of the motor and assume a structural damping loss factor of 0.02.

A. Find the displacement amplitude of the steady-state response in the lateral direction.
B. Evaluate the amplitude if an equivalent SDOF model with viscous damping, with a constant viscous damping factor, is used. How much is the error in the prediction?

Solution

A.
The system is represented by an SDOF system with the following properties:

$$k = \frac{3EI}{L^3} = \frac{3 \times 2 \times 10^{11} \times 0.05 \times 0.005^3}{(0.4)^3 \times 12} = 4,882.8 \text{ N/m}$$

$$m = 2 \text{ kg}$$

$$\eta_h = 0.02$$

$$\omega = \frac{2\pi N}{60} = \frac{2\pi \times 1,416}{60} = 148.2 \text{ rad/sec}$$

$$\omega_n = \sqrt{\frac{k}{m}} = \sqrt{\frac{4,882.8}{2}} = 49.4 \text{ rad/sec}$$

$$x_{st} = \frac{F_o}{k} = \frac{20}{4,882.8} = 0.004096 \text{ m}$$

The steady-state amplitude of the displacement response is

$$x_{os} = \frac{x_{st}}{\sqrt{\left(1-\left(\dfrac{\omega}{\omega_n}\right)^2\right)^2 + \eta_h^2}}$$

where the suffix 's' represents association with structural damping case. Substituting the relevant values, we get, $x_{os} = 0.00051199$ m.

B.

Using Eq. (4.89), the structural damping is approximated with a constant viscous damping factor,

$$\xi_{eq} \cong \frac{\eta_h}{2} = \frac{0.02}{2} = 0.01$$

The steady-state amplitude of the displacement response is

$$x_{ov} = \frac{x_{st}}{\sqrt{\left(1-\left(\dfrac{\omega}{\omega_n}\right)^2\right)^2 + \left(2\xi\dfrac{\omega}{\omega_n}\right)^2}}$$

where the suffix 'v' represents association with the viscous damping case. Substituting the relevant values, we get, $x_{ov} = 0.00051198$ m.

Thus % error in predicted amplitude is,

$$\% \text{ error} = \frac{x_{ov} - x_{os}}{x_{os}} \times 100 = -0.0024 \%$$ (l)

The error is negligibly small.

Example 4.5

A vehicle is modeled as an SDOF spring-mass-damper system. If the vehicle is moving over a wavy road, find the transfer function and FRF between the displacement x of the mass treated as the output and the road waviness y as the input.

Solution

The equation of motion can be written as

$$m\ddot{x} + c(\dot{x} - \dot{y}) + k(x - y) = 0$$

$$m\ddot{x} + c\dot{x} + kx = ky + c\dot{y}$$

Taking the Laplace transform of both sides,

$$m\left(s^2 x(s) - sx(0) - \dot{x}(0)\right) + c\left(sx(s) - x(0)\right) + kx(s) = ky(s) + c\left(sy(s) - y(0)\right)$$

To find the transfer function, we take zero initial conditions. Hence,

$$x(s)\left(ms^2 + cs + k\right) = (cs + k)y(s)$$

The desired transfer function is

$$\frac{x(s)}{y(s)} = \frac{cs + k}{ms^2 + cs + k}$$

The FRF is obtained by substituting $s = i\omega$ in the transfer function

$$\frac{x(\omega)}{y(\omega)} = \frac{i\omega c + k}{-m\omega^2 + i\omega c + k}$$

It can be written in a more familiar form by dividing the numerator and denominator by k, leading to

$$\frac{x(\omega)}{y(\omega)} = \frac{i2\xi\dfrac{\omega}{\omega_n} + 1}{-\left(\dfrac{\omega}{\omega_n}\right)^2 + i2\xi\dfrac{\omega}{\omega_n} + 1}$$

Writing the complex quantities in phasor form, we get

$$\frac{x(\omega)}{y(\omega)} = \frac{\left(\sqrt{\left(2\xi\dfrac{\omega}{\omega_n}\right)^2 + 1}\right)e^{i\theta}}{\left(\sqrt{\left(1-\left(\dfrac{\omega}{\omega_n}\right)^2\right)^2 + \left(2\xi\dfrac{\omega}{\omega_n}\right)^2} + \right)e^{i\phi}}$$

Therefore, the magnitude and the phase of the FRF (representing the motion transmissibility) are given by

$$\left|\frac{x(\omega)}{y(\omega)}\right| = \frac{\left(\sqrt{\left(2\xi\dfrac{\omega}{\omega_n}\right)^2 + 1}\right)}{\left(\sqrt{\left(1-\left(\dfrac{\omega}{\omega_n}\right)^2\right)^2 + \left(2\xi\dfrac{\omega}{\omega_n}\right)^2} + \right)}$$

$$\text{Phase} = \theta - \phi = \tan^{-1} 2\xi\frac{\omega}{\omega_n} - \tan^{-1}\frac{2\xi\dfrac{\omega}{\omega_n}}{1-\left(\dfrac{\omega}{\omega_n}\right)^2}$$

Example 4.6

The logarithmic decrement was derived in section 4.3.2 using the for free vibration displacement response. However, often an accelerometer is used to measure the vibration in practice giving a record of the acceleration. Can the concept of the logarithmic decrement be applied to the acceleration history to find the damping factor?

Solution

The free vibration displacement response of an SDOF system is

$$x(t) = Ae^{-\xi\omega_n t}\cos(\omega_d t - \phi)$$

Differentiating the above equation twice with time and using trigonometry, the acceleration can be written in the following form with B as its amplitude and phase lag $(\phi + \theta)$,

$$\ddot{x}(t) = Be^{-\xi\omega_n t}\cos(\omega_d t - \phi - \theta)$$

The ratio of the accelerations at times t and $t + T$, if T is the time period, is

$$\frac{\ddot{x}(t)}{\ddot{x}(t+T)} = \frac{\ddot{x}_1}{\ddot{x}_2} = \frac{Be^{-\xi\omega_n t}\cos(\omega_d t - \phi - \theta)}{Be^{-\xi\omega_n(t+T)}\cos(\omega_d(t+T) - \phi - \theta)}$$

Since the 'cos' function repeats after T, the 'cos' terms cancel. Taking the 'log' of both sides and noting that $T = \dfrac{2\pi}{\omega_n\sqrt{1-\xi^2}}$, we obtain

$$\log_e \frac{\ddot{x}_1}{\ddot{x}_2} = \frac{2\pi\xi}{\sqrt{1-\xi^2}}$$

The RHS of the above equation is nothing but the logarithmic decrement ξ. The result shows that the logarithmic decrement can be applied to the acceleration history to find the damping factor.

Example 4.7

Figure 4.31 shows a record of the acceleration time history of free vibration measured at a point on a structure. Estimate the damping factor assuming the structure is an SDOF system with viscous damping.

Solution

From the figure, we note that the accelerations at the beginning and end of the first cycle are,

$$\ddot{x}_1 = 3.0 \text{ m/sec}^2 \text{ and } \ddot{x}_2 = 1.6 \text{ m/sec}^2$$

Substituting these values in the expression for the logarithmic decrement derived in Example 4.6 and solving for ξ gives

$$\log_e \frac{3.0}{1.6} = \frac{2\pi\xi}{\sqrt{1-\xi^2}}$$

$$\xi = 0.1$$

Example 4.8

Figure 4.32 shows a periodic force acting on an SDOF system with viscous damping. The system parameters are $m = 2$ kg, $k = 492.98$ N/m, and $c = 5.652$ N $-$ sec/m. Find the steady-state displacement response.

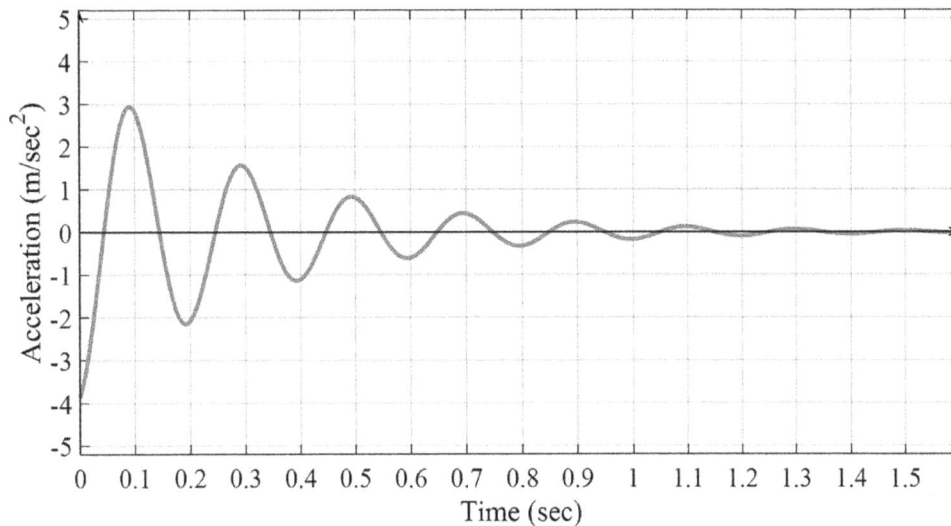

FIGURE 4.31   Free vibration acceleration response.

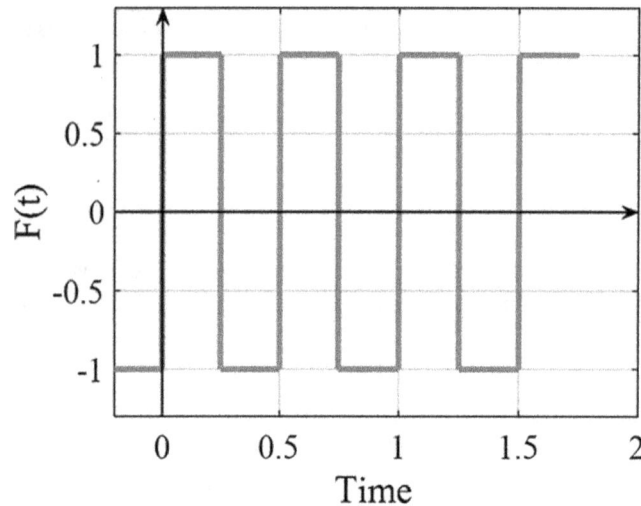

FIGURE 4.32 Periodic force.

Solution

The equation of motion is

$$m\ddot{x} + c\dot{x} + kx = F(t)$$

where $F(t)$ is a periodic force. We use Fourier series expansion (covered in Chapter 8) to express it as a sum of the harmonic components. From the figure, the period (T) of $F(t)$ is 0.5 second and can be expressed over one period as

$$F(t) = 1.0 \quad \text{for} \quad 0 \le t \le 0.25 \text{ sec}$$
$$= -1.0 \quad \text{for} \quad 0.25 \text{ sec} \le t \le 0.5 \text{ sec}$$

By Fourier series expansion

$$F(t) = f_0 + \sum_{r=1}^{\infty}(a_r \cos r\omega_0 t + b_r \sin r\omega_0 t)$$

The fundamental frequency ω_0 is

$$\omega_0 = \frac{2\pi}{T} = \frac{2\pi}{0.5} = 12.56 \text{ rad/sec}$$

The coefficient f_0 is obtained as follows.

$$f_0 = \frac{1}{T}\int_0^T F(t)\, dt$$

$$f_0 = \frac{1}{T}\left(\int_0^{T/2}F(t)\, dt + \int_{T/2}^T F(t)\, dt\right) = \frac{1}{T}\left(\int_0^{T/2}1\, dt + \int_{T/2}^T -1\, dt\right) = 0$$

The coefficient a_r is obtained as follows.

$$a_r = \frac{2}{T}\int_0^T F(t).\cos r\omega_0 t\, dt$$

$$a_r = \frac{2}{T}\left(\int_0^{T/2}F(t).\cos r\omega_0 t\, dt + \int_{T/2}^T F(t).\cos r\omega_0 t\, dt\right)$$

$$a_r = \frac{2}{T}\left(\left[\frac{\sin r\omega_0 t}{r\omega_0}\right]_0^{T/2} - \left[\frac{\sin r\omega_0 t}{r\omega_0}\right]_{T/2}^T\right) = \frac{2}{r\omega_0 T}\left(2\sin\frac{r\omega_0 T}{2} - \sin r\omega_0 T\right)$$

$$a_r = \frac{1}{\pi r}(2\sin\pi r - \sin 2\pi r) = 0 \qquad (\text{since } r \text{ is an integer})$$

Similarly, the coefficient b_r is obtained as follows.

$$b_r = \frac{2}{T}\int_0^T F(t).\sin rt \; dt$$

$$b_r = \frac{2}{T}\left(\int_0^{T/2} F(t).\sin r\omega_0 t \; dt + \int_{T/2}^T F(t).\sin r\omega_0 t \; dt\right)$$

$$b_r = \frac{2}{T}\left(-\left[\frac{\cos r\omega_0 t}{r\omega_0}\right]_0^{T/2} + \left[\frac{\cos r\omega_0 t}{r\omega_0}\right]_{T/2}^T\right) = \frac{2}{r\omega_0 T}\left(1 - 2\cos\frac{r\omega_0 T}{2} + \cos r\omega_0 T\right)$$

$$b_r = \frac{1}{\pi r}(1 - 2\cos\pi r + \cos 2\pi r)$$

The coefficient b_r is nonzero only for the odd values of the integer r.
Therefore, the Fourier series expansion of $F(t)$ is

$$F(t) = \sum_{r=1}^{\infty} b_r \sin r\omega_0 t = \frac{4}{\pi}\sin 1\omega_0 t + \frac{4}{3\pi}\sin 3\omega_0 t + \frac{4}{5\pi}\sin 3\omega_0 t + \ldots$$

The steady-state displacement response to each harmonic force in the expansion is obtained using the following expression.

$$x_r(t) = \frac{1/k}{\sqrt{\left(1 - \left(\frac{\omega}{\omega_n}\right)^2\right)^2 + \left(2\xi\frac{\omega}{\omega_n}\right)^2}} b_r \sin r\omega_0 t$$

The results for the first three harmonic components are as follows:

- Frequencies (rad/sec): 12.65, 37.95, 63.25
- The corresponding force amplitudes (N): 1.27, 0.42, 0.25
- The corresponding steady-state displacement response amplitudes (m): 6.8e-03, 1.77e-04, 3.387e-05

The harmonic responses are algebraically added to obtain the steady-state displacement response to $F(t)$.

- The maximum value of the resultant steady-state displacement response (m): 6.7e-03

REVIEW QUESTIONS

1. Can an overdamped system cross the mean position under free vibration? If yes, then under what conditions?
2. If the damping in a critically damped system is increased or decreased, why is the decay rate of the response reduced in both cases?
3. Why is it difficult to do a rigorous time domain analysis of systems with structural damping?
4. Define logarithmic decrement and explain its application.
5. Why the steady state response of an SDOF undamped system subjected to harmonic excitation reaches infinity at resonance? Explain this result on physical grounds.

6. Why is the phase of the steady-state response of an SDOF undamped system subjected to harmonic excitation is 180° when $\dfrac{\omega}{\omega_n} > 1$. Explain this result on physical grounds.

7. Why is the steady state response of an SDOF system with viscous damping subjected to harmonic excitation finite at resonance (as opposed to infinite response if damping is not there)?

8. Why can't a system vibrate naturally at a frequency other than its natural frequency? Explain based on an SDOF undamped system.

9. What are transient and steady-state responses, and what are the differences between the two?

10. What is complex stiffness, and what does the complex nature of the stiffness represent?

11. Define transfer function and FRF. How are they related?

PROBLEMS

Problem 4.1: For the bracket-motor problem with structural damping in Example 4.4, find the steady-state response amplitude if there are two harmonics of the unbalance force: one at 1,416 RPM of 20 N amplitude and the second at 2,832 RPM of 15 N amplitude.

Problem 4.2: A spring-mass system is subjected to a rectangular pulse force, as shown in Figure 4.33. Find the response of the system using the Laplace transform approach.

Problem 4.3: Find the response in Problem P4.2 using the system's impulse response and convolution integral.

Problem 4.4: Figure 4.34 shows the free vibration displacement response of an SDOF system with viscous damping. Estimate the undamped natural frequency and damping factor.

Problem 4.5: Show by taking an SDOF system with viscous damping that the IRF of the system is the inverse Laplace transform of its transfer function.

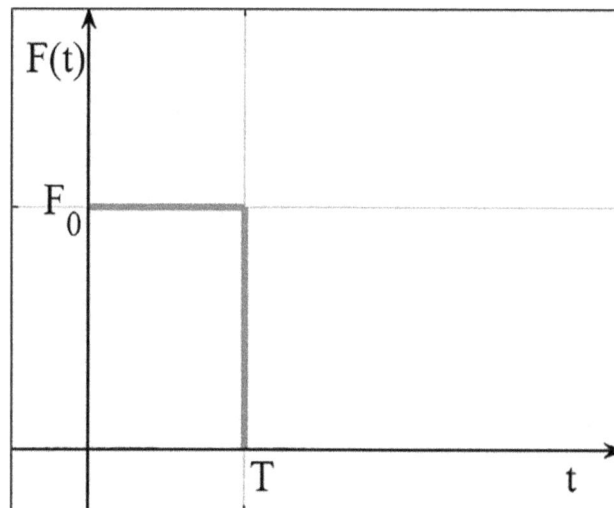

FIGURE 4.33 Rectangular pulse force.

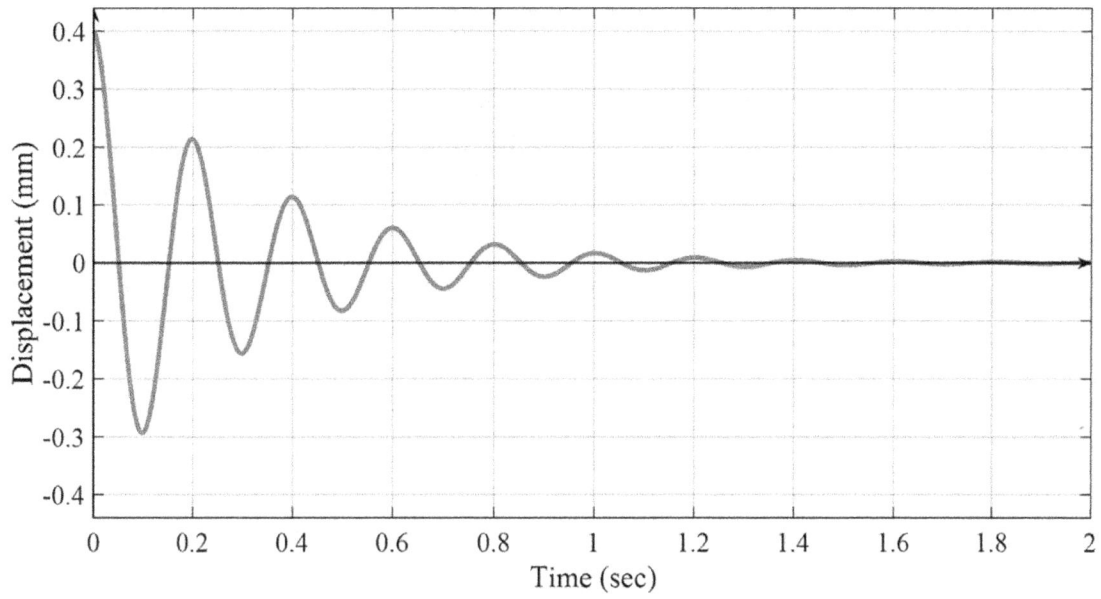

FIGURE 4.34 Free vibration displacement response.

5 Analytical Modal Analysis of Undamped MDOF Systems

5.1 INTRODUCTION

A multi degree of freedom (MDOF) system has as many natural vibration modes as it has degrees of freedom. The existence of multiple vibration modes of a system has led to the concept of modal analysis. This chapter addresses the analytical modal analysis of MDOF systems, focusing on undamped systems. The focus on undamped systems is motivated by two reasons. In many systems, the extent of damping is small, and in such cases, the results based on an undamped MDOF model may give reasonably accurate estimates of the quantities of interest such as the natural frequencies, mode shapes, and dynamic response. The second reason is that the concept of modal analysis can be introduced and understood more easily using the undamped MDOF models. The ideas developed can then be extended to MDOF damped systems.

We first introduce the basic concepts such as the eigenvalue problem, natural modes of vibration, orthogonality of modes, and mode shape normalization. Then we look at the expansion theorem, which provides the basis for modal analysis. The concepts of modal space and modal and spatial models of an MDOF system are then discussed. We then extend the concepts of frequency response function (FRF) and impulse response function (IRF), introduced in Chapter 4 for SDOF systems, to MDOF systems. They make up the response models in frequency and time domains, respectively. Then the response analysis to harmonic and transient excitations through modal analysis is considered.

5.2 FREE VIBRATION OF UNDAMPED MDOF SYSTEMS

In this section, we consider the free vibration of undamped MDOF systems. We use a two-DOF system (shown in Figure 5.1a) to introduce the concepts since they can be generalized for systems with more DOFs.

x_1 and x_2 are the displacements of the two masses measured from their static equilibrium positions and make up the two DOFs. The equations of motion can be written using any of the dynamics principles studied in Chapter 2. Newton's law is used in the present case. Figure 5.1b shows the free-body diagrams of the two masses displaced from their equilibrium positions. Applying Newton's law to the two masses individually, we get

$$m_1\ddot{x}_1 = -k_1 x_1 + k_2 (x_2 - x_1) \tag{5.1}$$

$$m_2\ddot{x}_2 = -k_2 (x_2 - x_1) - k_3 x_2 \tag{5.2}$$

which can be combined into a matrix equation

$$\begin{matrix} m_1 & 0 \\ 0 & m_2 \end{matrix} \begin{matrix} \ddot{x}_1 \\ \ddot{x}_2 \end{matrix} + \begin{matrix} k_1 + k_2 & -k_2 \\ -k_2 & k_2 + k_3 \end{matrix} \begin{matrix} x_1 \\ x_2 \end{matrix} = \begin{matrix} 0 \\ 0 \end{matrix} \tag{5.3}$$

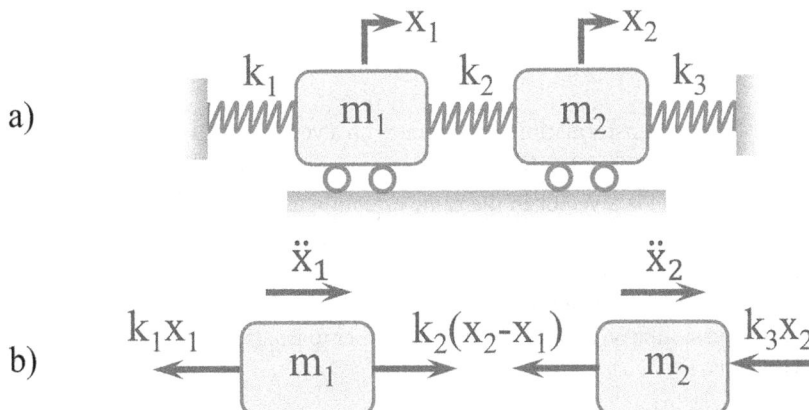

FIGURE 5.1 Two-DOF system.

DOI: 10.1201/9780429454783-5

and represented symbolically as

$$[M]\{\ddot{x}\} + [K]\{x\} = \{0\} \tag{5.4}$$

$[M]$ and $[K]$ are the mass and stiffness matrices and $\{x\}$ and $\{\ddot{x}\}$ are the displacement and acceleration vectors. They are given by

$$[M] = \begin{bmatrix} m_1 & 0 \\ 0 & m_2 \end{bmatrix}, [K] = \begin{bmatrix} k_1 + k_2 & -k_2 \\ -k_2 & k_2 + k_3 \end{bmatrix}, \{x\} = \begin{Bmatrix} x_1 \\ x_2 \end{Bmatrix} \text{ and } \{\ddot{x}\} = \begin{Bmatrix} \ddot{x}_1 \\ \ddot{x}_2 \end{Bmatrix} \tag{5.5}$$

Thus, the system matrices for a two-DOF system are of order 2×2. Equation (5.3) or (5.4) is a set of two simultaneous or coupled ordinary second-order differential equations.

Note that the mass and stiffness matrices are symmetric due to the reciprocity principle (discussed in Chapter 2). The mass matrix is positive-definite since any system with a nonzero velocity always has nonzero kinetic energy. However, the stiffness matrix can be positive-definite or positive semi-definite. If a system has rigid body modes, then the strain energy of the system would be zero even for a nonzero displacement in these modes, which is reflected in the positive semi-definite nature of the stiffness matrix. If the rigid body modes are not there, then the stiffness matrix is positive-definite.

5.2.1 EIGENVALUE PROBLEM

Let the solution to Eq. (5.4) be given by

$$\{x\} = \{X\}\cos\omega t \tag{5.6}$$

The question now is what values the frequency ω and the amplitude vector $\{X\}$ should have so that Eq. (5.6) represents the free vibration of the system. Substituting Eq. (5.6) into Eq. (5.4) gives

$$\left([K]\{X\} - \omega^2[M]\{X\}\right)\cos\omega t = \{0\} \tag{5.7}$$

$\cos\omega t$ being nonzero for an arbitrary value of t, the vector in the bracket must be zero. Using $\lambda = \omega^2$, we get

$$[K]\{X\} = \lambda[M]\{X\} \tag{5.8}$$

Thus, for Eq. (5.6) to be a free vibration solution, frequency ω, and amplitude vector $\{X\}$ should satisfy Eq. (5.8). Equation (5.8) is referred to as a **Generalized eigenvalue problem** in linear algebra. It is a generalization of the **Standard eigenvalue problem** given by

$$[A]\{X\} = \lambda\{X\} \tag{5.9}$$

In Eq. (5.9), $[A]$ is a matrix of linear transformation that operates on a vector $\{X\}$. In general, the linear transformation of a vector rotates the vector and changes its length, thus yielding a new vector. However, Eq. (5.9) seeks to determine a vector $\{X\}$ that, when operated by $[A]$, yields a vector in the same direction as that of $\{X\}$ but allows a change in its length by some factor λ. Such a vector is called an **eigenvector** of the transformation matrix $[A]$ while λ is said to be the corresponding **eigenvalue**. There exist as many eigenvector-eigenvalue pairs as is the order of the matrix.

The generalized eigenvalue problem has, additionally, a non-identify matrix $[M]$ on the RHS, and it is not in the standard form but can be interpreted similarly. Henceforth, we shall refer to Eq. (5.8) simply as an **eigenvalue problem (EVP)**. Equation (5.8) can be written as

$$\left([K] - \lambda[M]\right)\{X\} = \{0\} \tag{5.10}$$

which represents a set of homogeneous algebraic equations in $\{X\}$. For a nontrivial solution, the determinant of the coefficient matrix must be zero,

$$\left|[K] - \lambda[M]\right| = 0 \qquad (5.11)$$

Eq. (5.11) represents a polynomial in λ and is called the characteristic equation. For a two-DOF system, it is a quadratic equation in λ, which, when solved, yields the two eigenvalues (λ_1, λ_2) of the system.

Equation (5.10) is solved for $\{X\}$ to obtain the eigenvector by substituting $\lambda = \lambda_1$. If we denote an eigenvector by the symbol $\{\psi\}$, then the eigenvector corresponding to λ_1 is denoted by $\{\psi\}_1$. Similarly, the eigenvector $\{\psi\}_2$ corresponding to λ_2 is found. It should be noted that Eq. (5.10) is a set of homogeneous equations that cannot be solved uniquely for $\{X\}$. But we can assume one of the displacements arbitrarily and solve for the remaining elements of $\{X\}$ in terms of the arbi-

trarily chosen displacement. For the two-DOF system, we have $\{X\} = \begin{Bmatrix} x_1 \\ x_2 \end{Bmatrix}$, and if we choose x_2 arbitrarily and find x_1,

then we essentially obtain the ratio x_1/x_2 and not the absolute value of x_1. Thus, the elements of an eigenvector are not the absolute values but the ratios of the displacements.

5.2.2 Natural Modes of Vibration

In Section 5.2.1, we discussed that the solution of the equations of motion for the free vibration led to the eigenvalue problem, whose solution yielded pairs of eigenvalues and eigenvectors. But how do these eigenvalues and eigenvectors relate to the system?

The eigenvalue problem was arrived at by starting from an assumed solution (Eq. 5.6) for the free vibration problem. Therefore, if an eigenvalue/eigenvector pair is substituted into Eq. (5.6), it must represent the free vibration of the system. Substituting $\{X\} = \{\psi\}_1$ and $\omega = \sqrt{\lambda_1}$ into Eq. (5.6), we get

$$\{x\} = \{\psi\}_1 \cos\sqrt{\lambda_1}\,t \qquad (5.12)$$

It is noted that the eigenvalues of a vibratory system are non-negative since the mass matrix is positive-definite, and the stiffness matrix is either positive-semi-definite or positive-definite.

Eq. (5.12) shows that the free vibration of an undamped system is a harmonic motion at a frequency $\sqrt{\lambda_1}$ rad/sec. Since the free vibration is the natural vibration of the system, with no external excitation, this frequency is called the **natural frequency** of the system. It is equal to the square root of the eigenvalue. Denoting it by ω_1, Eq. (5.12) can be written as

$$\{x\} = \{\psi\}_1 \cos\omega_1 t \qquad (5.13)$$

The eigenvector $\{\psi\}_1$, consisting of the elements representing the ratio of displacements, represents the shape of the displacement profile of the system when it vibrates at the natural frequency ω_1. Therefore, it is also called the **mode shape** or **modal vector**. The displacement $\{x\} = \{\psi\}_2 \cos\omega_2 t$ describes another distinct way the system can have harmonic motion.

Thus, the vibration of a system at one of its natural frequencies, with its displacement governed by the corresponding mode shape, represents a principal 'mode' in which a system can vibrate naturally with harmonic motion and is called a **principal or natural mode of vibration**. Thus, a principal mode of vibration of an undamped system is characterized by its natural frequency and mode shape. The two-DOF system has two principal/natural modes, which are described by the pairs ω_1, $\{\psi\}_1$ and ω_2, $\{\psi\}_2$.

5.2.3 Free Vibration Response

It is noted that in addition to Eq. (5.13), the equation $\{x\} = \{\psi\}_1 \sin\omega_1 t$ is another valid solution representing the same natural mode. Hence, a general equation describing the response in the first mode can be written as a linear combination of these two solutions,

$$\{x\} = A_1\{\psi\}_1 \cos\omega_1 t + B_1\{\psi\}_1 \sin\omega_1 t \qquad (5.14)$$

where A_1, B_1 are arbitrary constants.

Similarly, if the system vibrates naturally in its second mode, then its free vibration response can be written as

$$\{x\} = A_2\{\psi\}_2 \cos\omega_2 t + B_2\{\psi\}_2 \sin\omega_2 t \qquad (5.15)$$

In general, the displacement of a system, vibrating naturally, consists of a combination of its motion in its various natural modes. For the two-DOF system, the response can be written as the sum of the responses given by Eqs. (5.14) and (5.15). Therefore, the general response is

$$\{x\} = A_1\{\psi\}_1 \cos\omega_1 t + B_1\{\psi\}_1 \sin\omega_1 t + A_2\{\psi\}_2 \cos\omega_2 t + B_2\{\psi\}_2 \sin\omega_2 t \qquad (5.16)$$

or,

$$\{x\} = \sum_{r=1}^{2} A_r\{\psi\}_r \cos\omega_r t + B_r\{\psi\}_r \sin\omega_r t \qquad (5.17)$$

The arbitrary constants A_r and B_r can be obtained from Eq. (5.17) from the initial state of the system.

5.2.4 GENERALIZATION TO MDOF SYSTEMS WITH N DOFS

There would be N number of simultaneous ordinary differential equations for an N-DOF system, and the order of the mass and stiffness matrices would be $N \times N$. The eigenvalue problem leads to a characteristic equation of order N, whose solution yields N eigenvalues and N eigenvectors. Thus, there are N natural modes of vibrations. The general free vibration response of the system can be written as a linear combination of the responses in N modes,

$$\{x\} = \sum_{r=1}^{N} A_r\{\psi\}_r \cos\omega_r t + B_r\{\psi\}_r \sin\omega_r t \qquad (5.18)$$

We take an example to consolidate the idea of the eigenvalues and eigenvectors.

Example 5.1

For the two-DOF system shown in Figure 5.1, taking $m_1 = m_2 = m$ and $k_1 = k_2 = k_3 = k$, find the natural frequencies and mode shapes.

Solution

From Eq. (5.5), the mass and stiffness matrices are

$$[M] = \begin{bmatrix} m & 0 \\ 0 & m \end{bmatrix}, \ [K] = \begin{bmatrix} 2k & -k \\ -k & 2k \end{bmatrix}$$

The eigenvalue problem is given by Eq. (5.10). As indicated in Eq. (5.11), the determinant of the coefficient matrix must be zero, which gives

$$\det\left(\begin{bmatrix} 2k & -k \\ -k & 2k \end{bmatrix} - \lambda \begin{bmatrix} m & 0 \\ 0 & m \end{bmatrix} \right) = 0$$

$$2k - \lambda m = \pm k$$

The eigenvalues are

$$\lambda_1 = \frac{k}{m} \ \text{and} \ \lambda_2 = \frac{3k}{m}$$

The corresponding natural frequencies in rad/sec are

$$\omega_1 = \sqrt{\frac{k}{m}} \ \text{and} \ \omega_2 = \sqrt{\frac{3k}{m}}$$

To find the first mode shape or eigenvector, substitute $\lambda = \lambda_1$ into Eq. (5.10), we get

$$\left(\begin{bmatrix} 2k & -k \\ -k & 2k \end{bmatrix} - \lambda_1 \begin{bmatrix} m & 0 \\ 0 & m \end{bmatrix}\right)\{X\} = \{0\}$$

$$\begin{bmatrix} k & -k \\ -k & k \end{bmatrix}\begin{Bmatrix} X_1 \\ X_2 \end{Bmatrix} = \{0\}$$

Using any of the aforementioned two equations, we get

$$X_2 = X_1 \quad \text{or} \quad \frac{X_2}{X_1} = 1$$

Let us treat X_1 as arbitrary and take $X_1 = 1.0$. Then the first mode shape is obtained as

$$\{\psi\}_1 = \begin{Bmatrix} \psi_{11} \\ \psi_{21} \end{Bmatrix} = \begin{Bmatrix} X_1 \\ X_2 \end{Bmatrix} = \begin{Bmatrix} 1 \\ 1 \end{Bmatrix}$$

To find the second mode shape, substitute $\lambda = \lambda_2$ into Eq. (5.10), we get

$$\begin{bmatrix} 2k - \lambda_2 m & -k \\ -k & 2k - \lambda_2 m \end{bmatrix}\begin{Bmatrix} X_1 \\ X_2 \end{Bmatrix} = \{0\}$$

$$\begin{bmatrix} -k & -k \\ -k & -k \end{bmatrix}\begin{Bmatrix} X_1 \\ X_2 \end{Bmatrix} = \{0\}$$

Using any of the above two equations, we get

$$X_2 = -X_1 \quad \text{or} \quad \frac{X_2}{X_1} = -1$$

Again treating X_1 as arbitrary and taking $X_1 = 1.0$, the second mode shape is obtained as

$$\{\psi\}_2 = \begin{Bmatrix} \psi_{12} \\ \psi_{22} \end{Bmatrix} = \begin{Bmatrix} X_1 \\ X_2 \end{Bmatrix} = \begin{Bmatrix} 1 \\ -1 \end{Bmatrix}$$

Note that if a different value is assigned to the arbitrarily chosen variable, then the two eigenvectors get multiplied by a new scalar. However, the shape of the displacement profile they represent remains unchanged.

The two modes are shown in Figures 5.2 and 5.3, respectively. The (a) part of the figures shows the displacement of the masses, (b) shows the mode shape, and (c) shows the displacement profile at discrete times when vibrating in that mode. In both modes, the masses attain extreme and mean positions simultaneously.

5.2.5 ORTHOGONALITY PROPERTIES

The eigenvectors have the properties that they are orthogonal with respect to the mass and stiffness matrices. Let us derive these properties considering an N-DOF vibratory system. There are N pairs of eigenvalues and eigenvectors. Each eigenvalue-eigenvector pair satisfies the eigenvalue problem given by Eq. (5.8). Writing this equation for the rth and s^{th} pairs, we get

$$[K]\{\psi\}_r = \lambda_r [M]\{\psi\}_r \tag{5.19}$$

$$[K]\{\psi\}_s = \lambda_s [M]\{\psi\}_s \tag{5.20}$$

Pre-multiplying Eq. (5.19) by $\{\psi\}_s^T$ and Eq. (5.20) by $\{\psi\}_r^T$ leads to

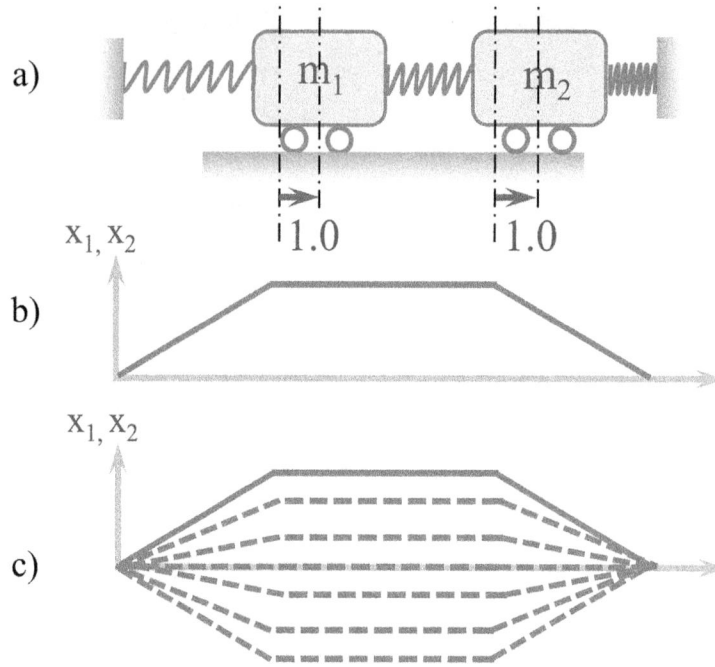

FIGURE 5.2 First mode (a) displacement of the masses, (b) mode shape, and (c) displacement profile at discrete times when vibrating in the first mode.

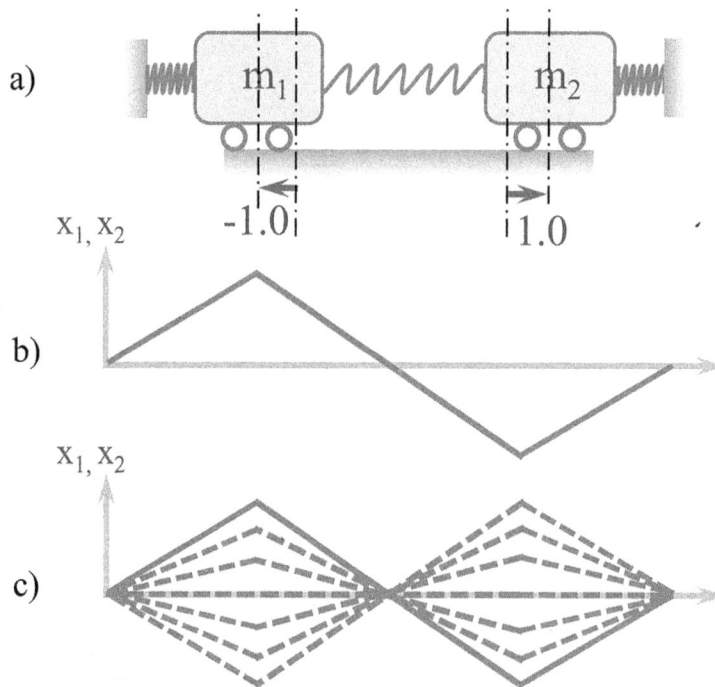

FIGURE 5.3 Second mode (a) displacement of the masses, (b) mode shape, and (c) displacement profile at discrete times when vibrating in the second mode.

$$\{\psi\}_s^T [K] \{\psi\}_r = \lambda_r \{\psi\}_s^T [M] \{\psi\}_r \tag{5.21}$$

$$\{\psi\}_r^T [K] \{\psi\}_s = \lambda_s \{\psi\}_r^T [M] \{\psi\}_s \tag{5.22}$$

Taking transpose of Eq. (5.21) and noting that the matrices $[K]$ and $[M]$ are symmetric, we get

$$\{\psi\}_r^T [K]\{\psi\}_s = \lambda_r \{\psi\}_r^T [M]\{\psi\}_s \tag{5.23}$$

where we have used the result from the matrix algebra that $([A][B])^T = [B]^T[A]^T$. Subtracting Eq. (5.22) from Eq. (5.23) gives

$$(\lambda_r - \lambda_s)\left(\{\psi\}_r^T [M]\{\psi\}_s\right) = 0 \tag{5.24}$$

If the eigenvalues are distinct, then for $r \neq s$, $\lambda_r \neq \lambda_s$. Hence, we must have

$$\{\psi\}_r^T [M]\{\psi\}_s = 0 \qquad \text{for } r \neq s \tag{5.25}$$

and

$$\{\psi\}_r^T [M]\{\psi\}_s \neq 0 \qquad \text{for } r = s \tag{5.26}$$

Substituting Eq. (5.25) into Eq. (5.22), we get

$$\{\psi\}_r^T [K]\{\psi\}_s = 0 \qquad \text{for } r \neq s \tag{5.27}$$

and substituting Eq. (5.26) into Eq. (5.22), we get

$$\{\psi\}_r^T [K]\{\psi\}_s \neq 0 \qquad \text{for } r = s \tag{5.28}$$

Thus, Eqs. (5.25) and (5.27) covey that the two eigenvectors are orthogonal with respect to the mass and stiffness matrices. Since the mass and stiffness matrices are symmetric, the eigenvectors corresponding to even the non-distinct eigenvalues satisfy the orthogonality properties. The orthogonality of eigenvectors enables the decoupling of the differential equations of motion. This aspect forms the basis for modal analysis.

Eigenvector matrix or modal matrix, denoted by $[\psi]$, is formed with the eigenvectors as its columns. The mass-orthogonality properties (Eqs. 5.25–5.26) can be combined and expressed as

$$[\psi]^T [M][\psi] = [\ulcorner m_r \lrcorner] \tag{5.29}$$

where $[\ulcorner m_r \lrcorner]$ is a diagonal matrix given by

$$[\ulcorner m_r \lrcorner] = \begin{bmatrix} m_1 & 0 & 0 & 0 \\ 0 & m_2 & 0 & 0 \\ 0 & 0 & \ddots & 0 \\ 0 & 0 & 0 & m_N \end{bmatrix} \tag{5.30}$$

The N diagonal values are called modal masses. m_r is the modal mass of the rth mode.

Similarly, the stiffness-orthogonality properties (Eqs. 5.27–5.28) can be combined and expressed as

$$[\psi]^T [K][\psi] = [\ulcorner k_r \lrcorner] \tag{5.31}$$

where $[\ulcorner k_r \lrcorner]$ is a diagonal matrix given by

$$[\ulcorner k_r \lrcorner] = \begin{bmatrix} k_1 & 0 & 0 & 0 \\ 0 & k_2 & 0 & 0 \\ 0 & 0 & \ddots & 0 \\ 0 & 0 & 0 & k_N \end{bmatrix} \tag{5.32}$$

The N diagonal values are called modal stiffnesses. k_r is the modal stiffness of the rth mode. The subscript 'r' in m_r and k_r represent the mode number.

If we substitute the pair $\left(\lambda_r, \{\psi\}_r\right)$ in Eq. (5.8) (the eigenvalue problem) and pre-multiply by $\{\psi\}_r^T$, then we obtain

$$\lambda_r = \frac{k_r}{m_r} \tag{5.33}$$

and, since $\lambda = \omega^2$, we get an important result relating the natural frequency of a mode with the modal mass and modal stiffness of that mode,

$$\omega_r = \sqrt{\frac{k_r}{m_r}} \tag{5.34}$$

5.2.6 Normalization of the Mode Shapes

As mentioned in Section 5.2.1, the mode shapes multiplied or scaled by scalars don't change the displacement shape described by them. The following methods can be used to normalize or scale the eigenvectors.

a. Unit maximum element
In this method, the eigenvector is divided by the maximum element of the eigenvector. It forces the maximum element to be unity, while the rest of the elements become less than unity. This method is useful in the graphical display of the mode shape.

b. Unit length
In this method, the eigenvector is divided by its length or Euclidean norm, giving the normalized vector a unit length.

c. Unit modal mass
In this method, the eigenvector is divided by the square root of the corresponding modal mass. This step makes the modal mass of the normalized vector unity, and the eigenvector is said to be a mass-normalized eigenvector. This method of normalization is used in modal analysis. From Eq. (5.29), the modal mass of the rth mode is given by

$$\{\psi\}_r^T [M]\{\psi\}_r = m_r \tag{5.35}$$

The mass-normalized eigenvector is obtained as

$$\{\phi\}_r = \frac{\{\psi\}_r}{\sqrt{m_r}} \tag{5.36}$$

It gives

$$\{\phi\}_r^T [M]\{\phi\}_r = 1 \tag{5.37}$$

The mass-normalization also leads to an interesting result for the orthogonality expression involving the stiffness matrix. Substituting the pair $\left(\lambda_r, \{\phi\}_r\right)$ in Eq. (5.8) and pre-multiplying by $\{\phi\}_r^T$ gives

$$\{\phi\}_r^T [K]\{\phi\}_r = \lambda_r \{\phi\}_r^T [M]\{\phi\}_r \tag{5.38}$$

Making use of Eq. (5.37), we get

$$\{\phi\}_r^T [K]\{\phi\}_r = \lambda_r \tag{5.39}$$

We can build a mass-normalized eigenvector matrix, $[\phi]$, using the mass-normalized eigenvectors. Eq. (5.29) written in terms of $[\phi]$ takes the form

$$[\phi]^T [M][\phi] = [I] \tag{5.40}$$

where $[I]$ is an identity matrix, indicating that all the modal masses are unity.

Similarly, due to Eq. (5.39), the stiffness-orthogonality relation (Eq. 5.31) can be expressed in terms of $[\phi]$ as

$$[\phi]^T[K][\phi] = [\,^\backprime\lambda_{r_\backprime}]$$ (5.41)

where $[\,^\backprime\lambda_{r_\backprime}]$ is the eigenvalue matrix whose diagonal elements are the eigenvalues. Thus, when the eigenvector matrix is mass-normalized, the modal masses are unity, and the modal stiffnesses are equal to the eigenvalues.

5.3   EXPANSION THEOREM

Before discussing the expansion theorem for the modal analysis of a vibratory system, let us look at the representation of a vector in an N-dimensional space.

5.3.1   EXPANSION THEOREM FOR A VECTOR IN AN N-DIMENSIONAL SPACE

A vector in an N-dimensional space can be expressed as a linear combination of N linearly independent vectors. A set of linearly independent vectors is said to form a basis that spans the vector space. The vectors are linearly independent if none can be expressed as a linear combination of the other vectors in the set.

Let us take an example of a vector in a 3D space, as it can also be visualized geometrically. The position vector of a point P in the space, with coordinates (x, y, and z), is the vector $\overrightarrow{OP}$ from the origin O to the point P. $\overrightarrow{OP}$ can be expressed as

$$\overrightarrow{OP} = x\,\hat{i} + y\,\hat{j} + z\,\hat{k}$$ (5.42)

where $\left(\hat{i},\ \hat{j},\ \hat{k}\right)$ are unit vectors in the three orthogonal directions spanning the three-dimensional space. In Eq. (5.42), (x, y, z) are the components of the vector $\overrightarrow{OP}$ along the three unit vectors, respectively. The unit vectors are given by

$$\{\hat{i}\} = \begin{Bmatrix} 1 \\ 0 \\ 0 \end{Bmatrix}, \quad \{\hat{j}\} = \begin{Bmatrix} 0 \\ 1 \\ 0 \end{Bmatrix}, \quad \{\hat{k}\} = \begin{Bmatrix} 0 \\ 0 \\ 1 \end{Bmatrix}$$ (5.43)

Now, the vector $\overrightarrow{OP}$ can also be represented as a column vector $\{P\}$, with its elements being the components of $\overrightarrow{OP}$ along the three unit vectors,

$$\{P\} = \begin{Bmatrix} x \\ y \\ z \end{Bmatrix}$$ (5.44)

Therefore, Eq. (5.42) can be written as

$$\begin{Bmatrix} x \\ y \\ z \end{Bmatrix} = x\begin{Bmatrix} 1 \\ 0 \\ 0 \end{Bmatrix} + y\begin{Bmatrix} 0 \\ 1 \\ 0 \end{Bmatrix} + z\begin{Bmatrix} 0 \\ 0 \\ 1 \end{Bmatrix}$$ (5.45)

Thus, the vector $\{x\ \ y\ \ z\}^T$ is represented as a linear combination of the vectors $\{1\ \ 0\ \ 0\}^T$ $\{0\ \ 1\ \ 0\}^T$ and $\{0\ \ 0\ \ 1\}^T$. The three vectors, $\{1\ \ 0\ \ 0\}^T$, $\{0\ \ 1\ \ 0\}^T$, and $\{0\ \ 0\ \ 1\}^T$, are the basis vectors whose linear combination can be used to express the position vector of any point in the 3D space. Eq. (5.45) represents the expansion theorem in 3D space.

The vectors, $\{1\ \ 0\ \ 0\}^T$, $\{0\ \ 1\ \ 0\}^T$, and $\{0\ \ 0\ \ 1\}^T$, are linearly independent, which makes them qualify as the basis vectors. In the present case, the basis vectors are also orthogonal, i.e., the dot product between any two of them is zero. Since they are of unit lengths, they are also said to be orthonormal.

For example, if we take the three vectors as $\{u\}_1 = \{2 \quad -1 \quad 0\}^T, \{u\}_2 = \{-1 \quad 2 \quad -1\}^T$ and $\{u\}_3 = \{0 \quad -1 \quad 1\}^T$, it also forms a basis as they are linearly independent. The vector $\{x \quad y \quad z\}^T$ can be represented as

$$\begin{Bmatrix} x \\ y \\ z \end{Bmatrix} = q_1 \begin{Bmatrix} 2 \\ -1 \\ 0 \end{Bmatrix} + q_2 \begin{Bmatrix} -1 \\ 2 \\ -1 \end{Bmatrix} + q_3 \begin{Bmatrix} 0 \\ -1 \\ 1 \end{Bmatrix} \qquad (5.46)$$

or,

$$\{P\} = q_1\{u\}_1 + q_2\{u\}_2 + q_3\{u\}_3 \qquad (5.47)$$

In Eq. (5.47), q_1, q_2 and q_3 represent the participation factors of the three basis vectors to construct the vector $\{P\}$. But what are the values of the participation factors? Eq. (5.47) can be written as

$$\{P\} = \begin{bmatrix} \{u\}_1 & \{u\}_2 & \{u\}_3 \end{bmatrix} \begin{Bmatrix} q_1 \\ q_2 \\ q_3 \end{Bmatrix} . \qquad (5.48)$$

$$\{P\} = [u]\{q\} \qquad (5.49)$$

Thus, for a given $\{P\}$, the vector of participation factors $\{q\}$ is

$$\{q\} = [u]^{-1}\{P\} \qquad (5.50)$$

Note that the three basis vectors, $\{u\}_1$, $\{u\}_2$ and $\{u\}_3$, are not orthogonal, unlike the basis vectors in Eq. (5.45). Thus, the choice of basis vectors for expansion is not unique.

5.3.2 EXPANSION THEOREM FOR AN MDOF SYSTEM

The basis vectors, like the orthogonal vectors in Eq. (5.45) or any arbitrary linearly independent vectors like the vectors in Eq. (5.46), do not provide any helpful interpretation or insight after the expansion.

The eigenvectors of the system are linearly independent and can be chosen as the basis vectors. Using the mass-normalized eigenvectors, the response vector $\{x\}$ can be expanded as

$$\{x\} = y_1\{\phi\}_1 + y_2\{\phi\}_2 + \dots + y_N\{\phi\}_N \qquad (5.51)$$

In Eq. (5.51), y_1, y_2, …,y_N represent the participation of N eigenvectors making up the response. They can also be viewed as the components of $\{x\}$ along the N eigenvectors, respectively. Equation (5.51) is the **expansion theorem** for an MDOF system. It can be written as

$$\{x\} = \begin{bmatrix} \{\phi\}_1 & \{\phi\}_2 & \cdots & \{\phi\}_N \end{bmatrix} \begin{Bmatrix} y_1 \\ y_2 \\ \vdots \\ y_N \end{Bmatrix} \qquad (5.52)$$

$$\{x\} = [\phi]\{y\} \qquad (5.53)$$

Eq. (5.53) is a compact representation of the expansion theorem. It represents a linear transformation in which the eigenvector matrix $[\phi]$ operates on the vector of participation factors $\{y\}$ to produce the vector of responses $\{x\}$. The participation factors represent the contribution of various eigenvectors to the response. It provides insight into the role of each eigenvector in the response, which is why expanding the system response in terms of its eigenvectors is beneficial.

The participation factors can be directly obtained from Eq. (5.53) as

$$\{y\} = [\phi]^{-1}\{x\}$$ (5.54)

But a better approach is to use the orthogonality properties of the eigenvectors. To determine y_r, pre-multiply Eq. (5.51) by $\{\phi\}_r^T[M]$,

$$\{\phi\}_r^T[M]\{x\} = \{\phi\}_r^T[M]\left(y_1\{\phi\}_1 + y_2\{\phi\}_2 + \dots + y_N\{\phi\}_N\right)$$ (5.55)

Due to the orthogonality properties (Eq. 5.25), all the terms on the RHS except the rth term are zero, and making use of the result in Eq. (5.37), we get

$$y_r = \{\phi\}_r^T[M]\{x\} \qquad r = 1,\ 2,\ \dots,\ N$$ (5.56)

which can be written in matrix form as

$$\{y\} = [\phi]^T[M]\{x\}$$ (5.57)

The use of Eq. (5.57) doesn't require the computation of the inverse of the eigenvector matrix as needed in the direct approach (Eq. 5.54). It is another advantage of expanding the response in terms of the eigenvectors of the system.

5.4 PHYSICAL SPACE AND MODAL SPACE

This section introduces the concepts of physical space and modal space using the free vibration of an MDOF undamped system.

5.4.1 FREE VIBRATION RESPONSE BY MODAL ANALYSIS

Let the initial disturbance on the system is specified in terms of its initial displacement and velocity,

$$\{x(0)\} = \{x_0\}$$ (5.58)

$$\{\dot{x}(0)\} = \{v_0\}$$ (5.59)

The equation of motion of the system is

$$[M]\{\ddot{x}\} + [K]\{x\} = \{0\}$$ (5.60)

The objective is to solve Eq. (5.60) for the response $\{x\}$ due to the initial disturbance defined by Eqs. (5.58) and (5.59). Using the expansion theorem, the response $\{x\}$ is expressed as

$$\{x(t)\} = [\psi]\{y(t)\}$$ (5.61)

Note that, in Eq. (5.61), we have taken a general case where the eigenvectors $[\psi]$ are used instead of the mass-normalized eigenvectors $[\phi]$. $\{x(t)\}$ being a vector of physical or actual displacements, its elements $x_i(t)$, $i = 1,\ 2,\dots,\ N$ are referred to as physical coordinates, and the description of the system response using $\{x(t)\}$ is said to be a description in **physical space**.

But Eq. (5.61) shows that $\{x(t)\}$ can be expressed in terms of the vector $\{y(t)\}$. What is $\{y(t)\}$ and what does it represent? Substituting Eq. (5.61) into Eq. (5.60) gives

$$[M][\psi]\{\ddot{y}\} + [K][\psi]\{y\} = \{0\}$$ (5.62)

Pre-multiplying by $[\psi]^T$ gives

$$[\psi]^T[M][\psi]\{\ddot{y}\}+[\psi]^T[K][\psi]\{y\}=\{0\} \tag{5.63}$$

Using the orthogonality properties (Eqs. 5.29 and 5.31), we get

$$[\,`m_r\,]\{\ddot{y}\}+[\,`k_r\,]\{y\}=\{0\} \tag{5.64}$$

Since the matrices $[\,`m_r\,]$ and $[\,`k_r\,]$ are diagonal, Eq. (5.64) represents N decoupled or independent second-order differential equations. The rth differential equation can be written as

$$m_r y_r + k_r y_r = 0 \tag{5.65}$$

The solution of Eq. (5.65) gives the participation of the rth mode, represented by y_r, in the physical response, and therefore this equation describes the dynamic equilibrium of the system in its rth natural mode.

We know that the equation of motion for free vibration of an SDOF spring-mass system, with displacement q, can be written as

$$m\ddot{q}+kq=0 \tag{5.66}$$

Eqs. (5.65) and (5.66) are of the same form. Thus, the motion described by Eq. (5.65) can be interpreted as the free vibration of an SDOF spring-mass system.

Therefore, the contribution of the rth natural mode to the response can be obtained from the analysis of an equivalent hypothetical SDOF spring-mass system with the mass and stiffness equal to m_r and k_r, respectively. For this reason, the quantities m_r and k_r are called the 'modal' mass and 'modal' stiffness, respectively, for the rth mode. The rth mode modal properties determine the motion of the system in its rth mode of vibration. Therefore, each equation in Eq. (5.64) represents the dynamic motion in a particular vibration mode.

The system response can also be represented by the responses in the system's natural modes, represented by the participation factors y_1, y_2, ...,y_N and these are referred to as **modal responses**. Thus, for an MDOF system, the N eigenvectors constitute its **modal space or modal domain,** and the response in its modal space is referred to as modal response. Equation (5.64) represents the equations of motion for free vibration in modal space. We should know the initial excitation in modal space to solve these equations. We can determine these as follows.

From Eq. (5.61), at $t = 0$,

$$\{x(0)\}=[\psi]\{y(0)\} \tag{5.67}$$

$$\{x(0)\}=\sum_{j=1}^{N}\{\psi\}_j\, y_j(0) \tag{5.68}$$

Pre-multiplying by $\{\psi\}_r^T[M]$

$$\{\psi\}_r^T[M]\{x(0)\}=\{\psi\}_r^T[M]\left(\sum_{j=1}^{N}\{\psi\}_j\, y_j(0)\right) \tag{5.69}$$

Using the orthogonality properties, all the terms except the rth term on the RHS is zero, and we get

$$\{\psi\}_r^T[M]\{x(0)\}=m_r y_r(0) \tag{5.70}$$

Making the substitution for $\{x(0)\}$ from Eq. (5.58), we get

$$y_r(0)=\frac{1}{m_r}\{\psi\}_r^T[M]\{x_0\} \tag{5.71}$$

which is the initial condition on $y_r(t)$.

Differentiating Eq. (5.61) once with time and writing it for $t = 0$, we get

$$\{\dot{x}(0)\} = [\psi]\{\dot{y}(0)\} \tag{5.72}$$

$$\{\dot{x}(0)\} = \sum_{j=1}^{N} \{\psi\}_j \dot{y}_j(0) \tag{5.73}$$

Pre-multiplying by $\{\psi\}_r^T [M]$, using the orthogonality properties and substituting for $\{\dot{x}(0)\}$, we get

$$\dot{y}_r(0) = \frac{1}{m_r} \{\psi\}_r^T [M]\{v_0\} \tag{5.74}$$

which is the initial condition on $\dot{y}_r(t)$. Thus, Eqs. (5.71) and (5.74) represent the initial conditions in modal space.

The solution of the rth differential equation in the modal domain, Eq. (5.65), subject to the initial modal conditions $y_r(0)$ and $\dot{y}_r(0)$ can be written (similar to Eq. 4.5 in Chapter 4) as

$$y_r(t) = y_r(0)\cos\omega_r t + \frac{\dot{y}_r(0)}{\omega_r}\sin\omega_r t \quad \text{for } r = 1, 2, ..., N \tag{5.75}$$

where $\omega_r = \sqrt{\dfrac{k_r}{m_r}}$. Substituting initial modal conditions, we get the modal response

$$y_r(t) = \frac{1}{m_r}\{\psi\}_r^T [M]\{x_0\}\cos\omega_r t + \frac{1}{\omega_r m_r}\{\psi\}_r^T [M]\{v_0\}\sin\omega_r t \quad \text{for } r = 1, 2, ..., N \tag{5.76}$$

Substituting Eq. (5.76) into Eq. (5.61), we get the response in physical space,

$$\{x(t)\} = \sum_{r=1}^{N} \{\psi\}_r y_r(t) \tag{5.77}$$

$$\{x(t)\} = \sum_{r=1}^{N} \{\psi\}_r \left(\frac{1}{m_r}\{\psi\}_r^T [M]\{x_0\}\cos\omega_r t + \frac{1}{\omega_r m_r}\{\psi\}_r^T [M]\{v_0\}\sin\omega_r t \right) \tag{5.78}$$

$$\{x(t)\} = \sum_{r=1}^{N} \frac{1}{m_r}\{\psi\}_r \{\psi\}_r^T [M] \left(\{x_0\}\cos\psi_r t + \frac{1}{\omega_r}\{v_0\}\sin\omega_r t \right) \tag{5.79}$$

If the mass-normalized eigenvectors are used, then the modal masses are unity leading to

$$\{x(t)\} = \sum_{r=1}^{N} \{\phi\}_r \{\phi\}_r^T [M] \left(\{x_0\}\cos\omega_r t + \frac{1}{\omega_r}\{v_0\}\sin\omega_r t \right) \tag{5.80}$$

We see that the initial disturbance, in general, excites simultaneously multiple modes of vibration.

5.4.2 Comparison of Physical Space and Modal Space

We have two ways of describing the system's dynamic equilibrium and initial excitation: in physical space or modal space. All the relevant terms and equations of these two descriptions are summarized in Table 5.1.

Figure 5.4a shows the description of the system in physical space. In physical space, there are distinct physical degrees of freedom. But the response at each DOF consists of the sum of the responses in all the modes. Equation (5.60) is the governing equation of motion in physical space. Each equation represents the dynamic equilibrium of forces at a particular DOF. But the equations are coupled since either the mass matrix, stiffness matrix, or both are non-diagonal, and hence,

TABLE 5.1

System Description in Physical Space and Modal Space

	Physical Space	**Modal Space**
Basis vectors	Coordinate vectors $$\left\{\begin{matrix}1\\0\\\vdots\\0\end{matrix}\right\}_1 ,\left\{\begin{matrix}0\\1\\\vdots\\0\end{matrix}\right\}_2 ,\ldots,\left\{\begin{matrix}0\\0\\\vdots\\1\end{matrix}\right\}_N$$	Eigenvectors $\{\psi\}_1 ,\{\psi\}_2 ,\ldots, \{\psi\}_N$
Response coordinates	Physical coordinates $\{x\} = \{x_1, x_2, \ldots, x_N\}^T$	Modal coordinates $\{y\} = \{y_1, y_2, \ldots, y_N\}^T$
Model parameters	Lumped masses and lumped stiffnesses	Modal masses m_r and modal stiffnesses k_r, with $r = 1, 2, \ldots, N$
Equation of motion for free vibration	$[M]\{\ddot{x}\}+[K]\{x\} = \{0\}$ (The matrices $[M]$ and $[K]$ are, in general, non-diagonal)	$[\,\check{\,}m_{r_\backslash}]\{\ddot{y}\}+[\,\check{\,}k_{r_\backslash}]\{y\} = \{0\}$ ($[\,\check{\,}m_{r_\backslash}]$ and $[\,\check{\,}k_{r_\backslash}]$ are diagonal matrices)
Initial conditions or excitation	Vector of physical displacements $\{x(0)\} = \{x_0\}$ Vector of physical velocities $\{\dot{x}(0)\} = \{v_0\}$	Vector of modal displacements $\{y(0)\} = [\,\check{\,}m_{r_\backslash}]^{-1}[\psi]^T[M]\{x_0\}$ Vector of modal Velocities $\{\dot{y}(0)\} = [\,\check{\,}m_{r_\backslash}]^{-1}[\psi]^T[M]\{v_0\}$

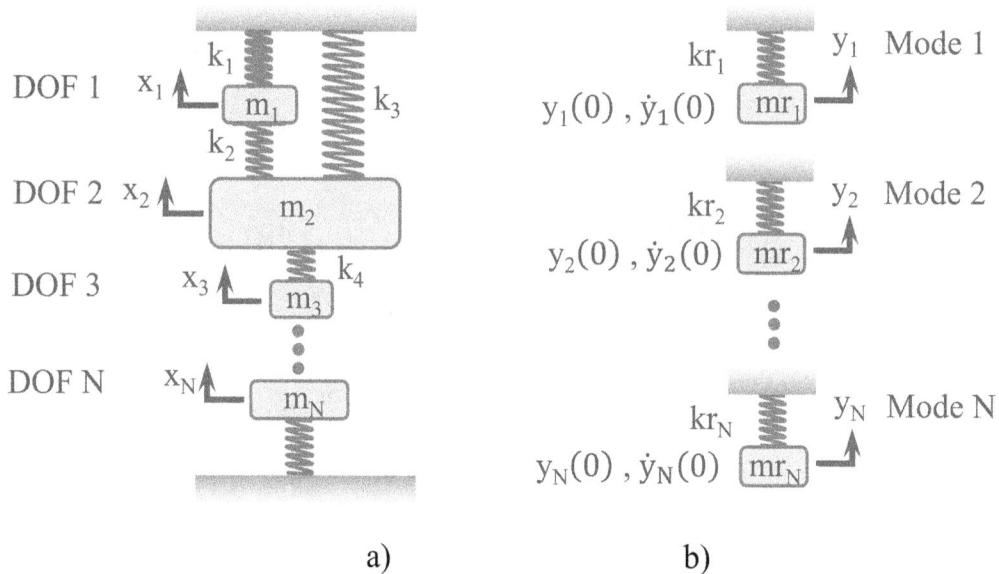

FIGURE 5.4   System in (a) physical space and (b) modal space in the free vibration.

they need to be solved simultaneously. The initial conditions are on the physical coordinates. Due to coupling, the initial conditions on each DOF cause the response at all the DOFs.tt

Figure 5.4b shows the system in modal space. It consists of N independent SDOF systems. Each SDOF system corresponds to a particular mode of vibration but gives information about the corresponding response at all the physical coordinates. Equation (5.64) is the equation of motion in modal space. Each equation represents dynamic equilibrium corresponding to a particular mode of vibration. The equations for all the modes are uncoupled since the modal mass, and modal stiffness matrices are diagonal. Thus, the equations in the modal domain have the advantage that each equation can be solved independently. The initial conditions are on the modal coordinates, and due to decoupling, the initial conditions on a mode don't cause the response in other modes.

5.5 FORCED VIBRATION RESPONSE TO HARMONIC FORCES

We now consider the forced vibration response of an undamped MDOF system subjected to harmonic forces. The equation of motion is

$$[M]\{\ddot{x}\}+[K]\{x\}=\{F\} \tag{5.81}$$

where $\{F\}$ is the vector of the external harmonic forces. The general solution of Eq. (5.81) consists of complementary and particular parts. We first determine the particular part of the solution in the following two subsections, using direct and modal analysis approaches, and then determine the total solution.

We use the complex exponential method (discussed in Section 4.4.2, Chapter 4) to obtain the particular part of the solution (or the steady-state response). Hence,

$$\{F\}=\begin{Bmatrix} F_{01} \\ F_{02} \\ \vdots \\ F_{0N} \end{Bmatrix}e^{i\omega t}=\{F_0\}e^{i\omega t} \tag{5.82}$$

where F_{0j} is the amplitude of the force at the jth DOF.

5.5.1 DIRECT SOLUTION IN PHYSICAL SPACE

Let the particular solution of Eq. (5.81) is

$$\{x_p\}=\begin{Bmatrix} x_{01} \\ x_{02} \\ \vdots \\ x_{0N} \end{Bmatrix}e^{i\omega t}=\{x_0\}e^{i\omega t} \tag{5.83}$$

Substituting Eqs. (5.82) and (5.83) into Eq (5.81), we get

$$-\omega^2[M]\{x_0\}+[K]\{x_0\}=\{F_0\} \tag{5.84}$$

$$\left([K]-\omega^2[M]\right)\{x_0\}=\{F_0\} \tag{5.85}$$

where the matrix $\left([K]-\omega^2[M]\right)$ is called the **dynamic stiffness matrix**. The dynamic stiffness matrix of an undamped system is the effective stiffness matrix that considers both the elastic and inertia forces. It depends on the frequency ω.

$$\{x_0\}=\left([K]-\omega^2[M]\right)^{-1}\{F_0\} \tag{5.86}$$

$$\{x_0\}=[\alpha]\{F_0\} \tag{5.87}$$

where

$$[\alpha]=\left([K]-\omega^2[M]\right)^{-1} \tag{5.88}$$

$[\alpha]$ is called the **FRF matrix**. FRF matrix is another description of an MDOF system and is further discussed later in the chapter. In this way, Eq. (5.86) gives the amplitude of the particular part of the response to the harmonic excitation forces.

5.5.2 SOLUTION BY MODAL ANALYSIS

This section presents an alternative procedure based on the modal analysis. In this procedure, we first obtain the particular part of the response in modal space, from which the response in physical space is obtained.

We use the expansion theorem to express the particular part of the response in modal space,

$$\{x_p(t)\}=[\psi]\{y(t)\} \tag{5.89}$$

Substituting Eq. (5.89) into Eq. (5.81) (after taking $\{x(t)\} = \{x_p(t)\}$), and pre-multiplying by $[\psi]^T$ gives

$$[\psi]^T[M][\psi]\{\ddot{y}\} + [\psi]^T[K][\psi]\{y\} = [\psi]^T\{F\}$$

(5.90)

Using the orthogonality properties, we get

$$[\text{`}m_{r\smallsetminus}]\{\ddot{y}\} + [\text{`}k_{r\smallsetminus}]\{y\} = \{f_r\}$$

(5.91)

where

$$\{f_r\} = [\psi]^T\{F\}$$

(5.92)

$\{f_r\}$ is called the vector of **modal forces**. The rth element of this vector represents the modal force for the rth mode and is given by

$$f_r = \{\psi\}_r^T\{F\} \qquad r = 1, 2,..., N$$

(5.93)

Thus, the modal force for the rth mode is obtained by the dot product between the rth eigenvector and the physical force vector and can be viewed as the effective force exciting the rth mode.

Since the modal matrices, $[\text{`}m_{r\smallsetminus}]$ and $[\text{`}k_{r\smallsetminus}]$, are diagonal, Eq. (5.91) represents N decoupled or independent second-order differential equations. The rth differential equation can be written as

$$m_r\ddot{y}_r + k_ry_r = f_r$$

(5.94)

This equation represents the forced vibration of the rth mode and can be viewed as the forced vibration of an SDOF spring-mass system with the mass and stiffness equal to the modal mass m_r and modal stiffness k_r, respectively, subjected to a force equal to the modal force f_r. Equation (5.91) represents the dynamics of the undamped system in modal space. Figure 5.5 shows the forced vibration of the system in physical and modal spaces.

Since the external force is harmonic, the modal forces are also harmonic.

$$f_r = \{\psi\}_r^T\{F\} = \{\psi\}_r^T\{F_0\}e^{j\omega t} = f_{r0}e^{j\omega t}$$

(5.95)

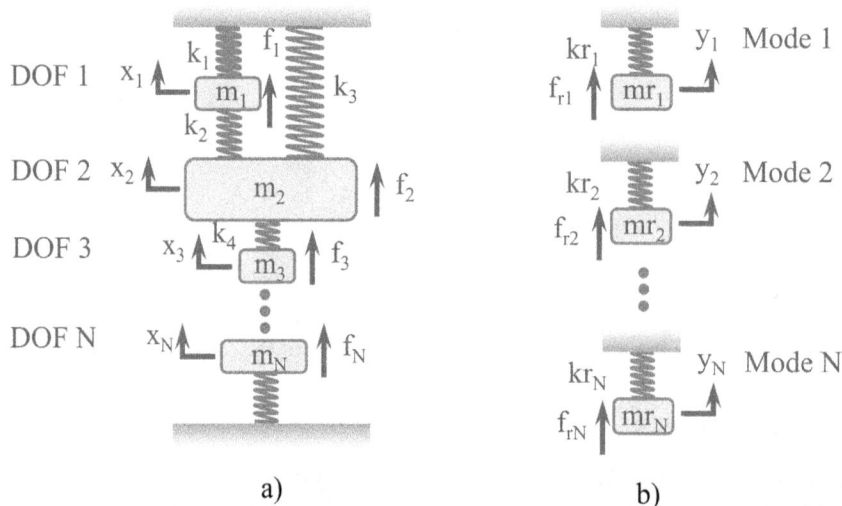

a)   b)

FIGURE 5.5   The forced vibration of a system in (a) physical space and (b) modal space.

Assuming

$$y_r(t) = y_{r0}e^{j\omega t} \tag{5.96}$$

and substituting Eqs. (5.95) and (5.96) into Eq. (5.94), we get

$$y_{r0} = \frac{f_{r0}}{k_r - \omega^2 m_r} \tag{5.97}$$

Dividing the numerator and denominator in Eq. (5.97) by m_r, substituting y_{r0} into Eq. (5.96) and making use of Eq. (5.95) gives

$$y_r(t) = \frac{\dfrac{1}{m_r} f_r(t)}{\omega_r^2 - \omega^2} \tag{5.98}$$

Substituting Eq. (5.93) into Eq. (5.98) gives the modal response of rth mode

$$y_r(t) = \frac{\dfrac{1}{m_r} \{\psi\}_r^T \{F\}}{\omega_r^2 - \omega^2} \qquad r = 1, 2,..., N \tag{5.99}$$

It is seen from Eq. (5.99) that the modal response of a mode depends only on the properties of that mode apart from the external force vector and its frequency. The physical response is obtained by substituting Eq. (5.99) into Eq. (5.89). Equation (5.89) is written as

$$\{x_p(t)\} = \sum_{r=1}^{N} \{\psi\}_r \, y_r(t) \tag{5.100}$$

Substituting Eq. (5.99)

$$\{x_p(t)\} = \sum_{r=1}^{N} \frac{\{\psi\}_r \dfrac{1}{m_r} \{\psi\}_r^T \{F\}}{\omega_r^2 - \omega^2} \tag{5.101}$$

Since the mass-normalized eigenvectors are given by $\{\phi\}_r = \dfrac{\{\psi\}_r}{\sqrt{m_r}}$, we get

$$\{x_p(t)\} = \sum_{r=1}^{N} \frac{\{\phi\}_r \{\phi\}_r^T \{F\}}{\omega_r^2 - \omega^2} \tag{5.102}$$

Substituting Eq. (5.82)

$$\{x_p(t)\} = \left(\sum_{r=1}^{N} \frac{\{\phi\}_r \{\phi\}_r^T \{F_0\}}{\omega_r^2 - \omega^2} \right) e^{j\omega t} \tag{5.103}$$

The term in the bracket is the amplitude of the particular part or steady-state part of the response,

$$\{x_0\} = \sum_{r=1}^{N} \frac{\{\phi\}_r \{\phi\}_r^T \{F_0\}}{\omega_r^2 - \omega^2} = \left(\sum_{r=1}^{N} \frac{\{\phi\}_r \{\phi\}_r^T}{\omega_r^2 - \omega^2} \right) \{F_0\} \tag{5.104}$$

We can use direct or modal analysis methods to find the response. However, the advantage of the modal analysis method is that it also gives the contributions of various modes making up the response. This insight may be valuable in diagnosing the cause of any excessive vibration and undertaking vibration control measures.

5.5.3 COMPLETE RESPONSE

The general solution is

$$\{x(t)\} = \{x_c(t)\} + \{x_p(t)\} \tag{5.105}$$

The complementary part, $\{x_c(t)\}$, is the solution of the homogeneous equation and is given by Eq. (5.18). We use that equation but with the mass-normalized eigenvectors. Let us assume that the actual force is $\{F_0\}\cos\omega t$, which is nothing but the real part of the force in Eq. (5.82). Therefore, $\{x_p(t)\}$ to be used in Eq. (5.105) should be the real part of Eq. (5.103).

Making these substitutions, we get

$$\{x(t)\} = \sum_{j=1}^{N} \left(\{\phi\}_j A_j \cos\omega_j t + \{\phi\}_j B_j \sin\omega_j t\right) + \sum_{k=1}^{N} \frac{\{\phi\}_k \{\phi\}_k^T}{\omega_k^2 - \omega^2} \{F_0\}\cos\omega t \tag{5.106}$$

where A_j and B_j are arbitrary constants and depend on the initial conditions of the system. One approach to finding them could be applying the initial conditions directly to Eq. (5.106). This approach results in a system of 2N linear equations in 2N unknowns (A_j and B_j), which can be solved to find the unknowns. It requires the inversion of a $2N \times 2N$ matrix or the solution of the 2N linear equations by some alternative method.

Alternatively, the arbitrary constants can be found using the orthogonality properties of the eigenvectors. This approach, given below, is computationally more efficient.

Imposing the initial condition $\{x(0)\} = \{x_0\}$ on Eq. (5.106), pre-multiplying by $\{\phi\}_r^T[M]$, and using the orthogonality properties, we get

$$A_r = \{\phi\}_r^T[M]\{x_0\} - \{\phi\}_r^T[M]\left(\sum_{k=1}^{N} \frac{\{\phi\}_k \{\phi\}_k^T}{\omega_k^2 - \omega^2} \{F_0\}\right) \quad r = 1, 2,..., N \tag{5.107}$$

Differentiating Eq. (5.106) once with time and imposing the initial condition $\{\dot{x}(0)\} = \{v_0\}$, we get

$$\{v_0\} = \left(\sum_{j=1}^{N} B_j \omega_j \{\phi\}_j\right) \tag{5.108}$$

Again, pre-multiplying by $\{\phi\}_r^T[M]$, and using the orthogonality properties, we get

$$B_r = \frac{1}{\omega_r}\{\phi\}_r^T[M]\{v_0\} \qquad r = 1, 2,..., N \tag{5.109}$$

Substituting Eqs. (5.107) and (5.109) into Eq. (5.106), we obtain the total response as

$$\{x(t)\} = \sum_{j=1}^{N} \left(\{\phi\}_j \{\phi\}_j^T[M]\left(\{x_0\} - \sum_{k=1}^{N} \frac{\{\phi\}_k \{\phi\}_k^T}{\omega_k^2 - \omega^2} \{F_0\}\right)\cos\omega_j t + \frac{1}{\omega_j}\{\phi\}_j \{\phi\}_j^T[M]\{v_0\}\sin\omega_j t\right)$$

$$+ \sum_{k=1}^{N} \frac{\{\phi\}_k \{\phi\}_k^T}{\omega_k^2 - \omega^2} \{F_0\}\cos\omega t \tag{5.110}$$

which can be simplified to

$$\{x(t)\} = \sum_{j=1}^{N} \{\phi\}_j \{\phi\}_j^T [M] \left(\{x_0\} \cos \omega_j t + \frac{1}{\omega_j} \{v_0\} \sin \omega_j t \right) + \left(\cos \omega t - \sum_{j=1}^{N} \{\phi\}_j \{\phi\}_j^T [M] \cos \omega_j t \right) \left(\sum_{k=1}^{N} \frac{\{\phi\}_k \{\phi\}_k^T}{\omega_k^2 - \omega^2} \right) \{F_0\}$$

(5.111)

5.6 SPATIAL MODEL AND MODAL MODEL

a. We note from the previous sections that one way a system's dynamic response to an initial disturbance or external dynamic forces can be found analytically is by a direct solution of the differential equations of motion of the system in the physical coordinates. For an undamped MDOF system, the direct solution in physical space needs the mass and stiffness matrices of the system. In Chapters 2 and 3, we studied how these matrices can be obtained using lumped parameter and finite element approaches. The description of an undamped system using the mass and stiffness matrices is called the **spatial model** of the system. Figure 5.6a depicts the dynamic response analysis of a system using its spatial model.

b. We also saw in the previous sections that the response of a system could also be obtained through modal analysis, using the knowledge of the eigenvalues and eigenvectors of the system. Hence, the eigenvalues and eigenvectors together form an alternative description of the system. This description is referred to as the **modal model** of the system. The eigenvalue matrix $[\lambda]$ and mass-normalized eigenvector matrix $[\phi]$ (a real matrix for an undamped system) constituting the modal model are:

$$[\lambda] = \begin{bmatrix} \omega_{n1}^2 & 0 & 0 & 0 \\ 0 & \omega_{n2}^2 & 0 & 0 \\ 0 & 0 & \ddots & 0 \\ 0 & 0 & 0 & \omega_{nN}^2 \end{bmatrix}$$

(5.112)

$$[\phi] = \begin{bmatrix} \{\phi\}_1 & \{\phi\}_2 & \cdots & \{\phi\}_N \end{bmatrix} = \begin{bmatrix} \phi_{11} & \phi_{12} & \cdots & \phi_{1N} \\ \phi_{21} & \phi_{22} & \cdots & \phi_{2N} \\ \vdots & \vdots & \vdots & \vdots \\ \phi_{N1} & \phi_{N2} & \cdots & \phi_{NN} \end{bmatrix}$$

(5.113)

Figure 5.6b depicts the dynamic response analysis of a system using its modal model.

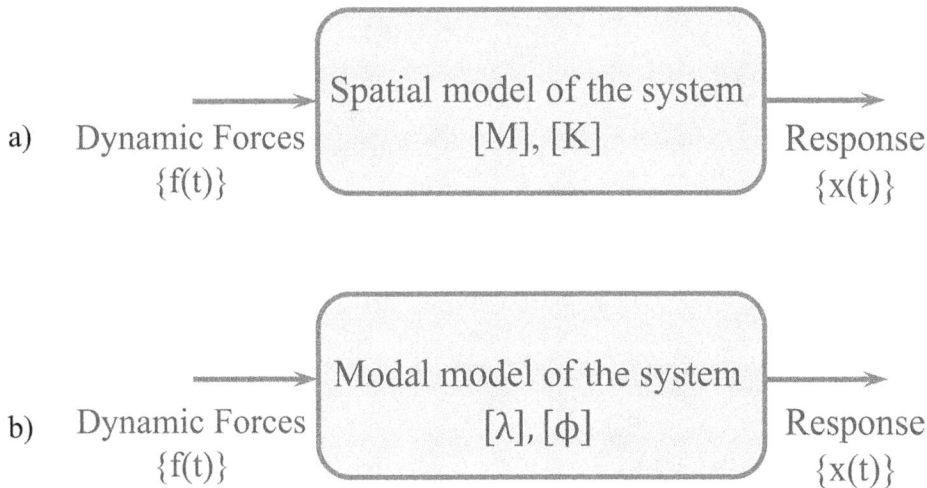

a) Dynamic Forces $\{f(t)\}$ → Spatial model of the system $[M], [K]$ → Response $\{x(t)\}$

b) Dynamic Forces $\{f(t)\}$ → Modal model of the system $[\lambda], [\phi]$ → Response $\{x(t)\}$

FIGURE 5.6 Dynamic response analysis using (a) spatial model and (b) modal model of the system.

5.7 FRF MATRIX

We defined the FRF of an SDOF system in Chapter 4. For an MDOF system with N DOFs, there are N choices each for the force and response DOFs, and hence N^2 number of FRFs can be defined. These N^2 FRFs together make up the FRF matrix of order $N \times N$.

The FRF matrix is of the form

$$[\alpha] = \begin{bmatrix} \alpha_{11} & \alpha_{12} & \cdots & \alpha_{1N} \\ \alpha_{21} & \alpha_{22} & \cdots & \alpha_{2N} \\ \vdots & \vdots & \vdots & \vdots \\ \alpha_{N1} & \alpha_{N2} & \cdots & \alpha_{NN} \end{bmatrix} \tag{5.114}$$

What does the FRF α_{jk} represent? Extending the definition of the FRF for SDOF systems, the FRF α_{jk} is the ratio of the steady-state response amplitude at the jth DOF and the harmonic force amplitude applied at the kth DOF with no force at the rest of the DOFs. Mathematically,

$$\alpha_{jk}(\omega) = \frac{x_j(\omega)}{F_k(\omega)} \qquad \text{with } F_q(\omega) = 0 \text{ for all q except q = k} \tag{5.115}$$

where $F_k(\omega)$ represents the amplitude of the harmonic force with frequency ω applied at the kth DOF while $x_j(\omega)$ is the resulting steady-state response at the jth DOF.

For a linear system, the principle of reciprocity, discussed in Chapter 2, is valid and therefore $\alpha_{jk}(\omega) = \alpha_{kj}(\omega)$ and the FRF matrix is symmetric.

5.8 RELATIONSHIP BETWEEN THE FRF AND MODAL MODEL

For an undamped system, Eq. (5.88) gives the relationship between the FRF matrix and the spatial model (i.e., the system matrices $[K]$ and $[M]$). Is there any relationship between the FRF matrix and the modal model?

If we compare the steady-state response amplitude given by Eqs. (5.87) and (5.104), then the coefficients of the vector $\{F_0\}$ in the two equations must be equal. Hence, we obtain

$$[\alpha] = \sum_{r=1}^{N} \frac{\{\phi\}_r \{\phi\}_r^T}{\omega_r^2 - \omega^2} \tag{5.116}$$

Any particular FRF, α_{jk}, can be extracted from Eq. (5.116) as

$$\alpha_{jk}(\omega) = \sum_{r=1}^{N} \frac{\phi_{jr} \phi_{kr}}{\omega_r^2 - \omega^2} \tag{5.117}$$

Eqs. (5.116) and (5.117) form the relationships between the FRFs and the modal model. In Chapter 7, this relationship is derived from an alternative approach.

5.9 RESPONSE MODEL (FREQUENCY DOMAIN)

If the FRF matrix for a system is known, then the system response to the harmonic forces can be determined. Let us say, $F_{01}(\omega)$, $F_{02}(\omega)$, ..., $F_{0N}(\omega)$ be the amplitude of the harmonic forces at frequency ω, acting at N DOFs of an MDOF system. For a linear system, the response due to individual forces can be added to find the total response. Thus, the response at the jth DOF can be obtained as

$$x_{0j}(\omega) = \alpha_{j1}(\omega) F_{01}(\omega) + \alpha_{j2}(\omega) F_{02}(\omega) + \ldots + \alpha_{jN}(\omega) F_{0N}(\omega) \qquad \text{for } j = 1, 2, \ldots, N \tag{5.118}$$

Thus, by the knowledge of the FRF matrix $[\alpha]$ of the system, the system response can be determined. The FRF matrix $[\alpha]$, therefore, constitutes an alternative model of the system dynamics and is called the **response model**. Since it is defined for the harmonic forces, characterized by frequency ω, it is a **frequency domain response model**.

5.10 RESPONSE TO PERIODIC FORCES BY MODAL ANALYSIS

When the force acting on the system is periodic, then using the Fourier series expansion (covered in Chapter 8), the periodic force can be represented as a sum of harmonic forces. The response of the system to the periodic force is obtained by the superposition of the responses to the individual harmonic forces. This approach is illustrated for an SDOF system in Example 4.8 (Chapter 4), and the same can be extended to MDOF systems.

The response of an MDOF system to a periodic force can also be found through modal analysis. For this, the response to each harmonic is first found by modal analysis as described in Section 5.5.2, and then the responses are algebraically added to find the system response.

5.11 RESPONSE TO TRANSIENT EXCITATION BY MODAL ANALYSIS

The solution consists of complementary and particular parts. We obtain below the particular part of the response. The complementary part can be obtained by proceeding as given in Section 5.5.3.

The particular part of the response to the transient excitation can be found through the frequency domain approach using the Fourier transform. In the time domain, it can be found through two alternative approaches:

- The first approach is based on modal analysis and uses the definition of the IRF for an SDOF system.
- The second approach uses the IRF matrix of the system.

We present the first approach in this section, while the second approach is presented in Section 5.13.

The equation of motion is given by Eq. (5.81), with the vector $\{F\}$ representing the vector of the transient forces acting on the system. With this change, we can transform the equation from physical to modal domain as given by Eqs. (5.89–5.94) in Section 5.5.2.

The modal force for the rth mode, i.e., f_r, is now a transient force. Since the equations in the modal domain are decoupled, the equation for any particular mode, represented by Eq. (5.94), can be solved independently. We solve this equation using the convolution integral (as given in Section 4.10 in Chapter 4).

The modal equation for the rth mode represents the dynamics of a hypothetical SDOF spring-mass system. Using Eq. (4.114), the modal IRF for this system can be written as

$$h_r(t) = \frac{1}{m_r \omega_r} \sin \omega_r t \quad (\text{for } t > 0) \tag{5.119}$$

The modal response in the rth mode is

$$y_r(t) = \int_0^t h_r(t-\tau) f_r(\tau) \, d\tau \tag{5.120}$$

The integral in Eq. (5.120) needs to be evaluated to find $y_r(t)$. The response in physical coordinates is

$$\{x(t)\} = [\psi]\{y(t)\} \tag{5.121}$$

$$\{x(t)\} = \sum_{r=1}^N \{\psi\}_r \, y_r(t) \tag{5.122}$$

Making substitutions for $y_r(t)$ and $h_r(t)$, we get

$$\{x(t)\} = \sum_{r=1}^N \{\psi\}_r \left(\int_0^t \frac{1}{m_r \omega_r} \sin \omega_r(t-\tau) f_r(\tau) \, d\tau \right) \tag{5.123}$$

Further, by writing the modal force as $f_r(\tau) = \{\psi\}_r^T \{F(\tau)\}$, we get a consolidated expression for the particular part of the response to the transient excitation,

$$\{x(t)\} = \sum_{r=1}^N \frac{1}{m_r \omega_r} \{\psi\}_r \{\psi\}_r^T \left(\int_0^t \sin \omega_r(t-\tau) \{F(\tau)\} \, d\tau \right) \tag{5.124}$$

If the mass-normalized eigenvectors $[\phi]$ are used in place of $[\psi]$, then the modal mass (m_r) for all the modes is unity, and Eq. (5.124) takes the following form:

$$\{x(t)\} = \sum_{r=1}^{N} \frac{1}{\omega_r} \{\phi\}_r \{\phi\}_r^T \left(\int_0^t \sin\omega_r (t-\tau) \{F(\tau)\} \, d\tau \right) \qquad (5.125)$$

5.12   IRF MATRIX

In Chapter 4, we defined the IRF for an SDOF system. It is the response of the system to a unit impulse. For an MDOF system with N DOFs, there are N choices each for the force and response DOFs, and hence N^2 number of IRFs can be defined. These N^2 IRFs together make up the IRF matrix of order $N \times N$.

The IRF $h_{jk}(t)$ is the response at the jth DOF due to a unit impulse at the kth DOF with no impulses applied at the rest of the DOFs. If a force $F_k(\tau)$ acts at k^{th} DOF with no force acting at other DOFs, then the response at the jth DOF is given by the following convolution integral:

$$x_j(t) = \int_0^t h_{jk}(t-\tau) F_k(\tau) \, d\tau \quad \text{with } F_q(\tau) = 0 \text{ for all q except q} = k \qquad (5.126)$$

Eq. (5.126) is symbolically written using the symbol '*' to denote the convolution operation as

$$x_j(\tau) = h_{jk} * F_k \qquad \text{with } F_q(\tau) = 0 \text{ for all q except q} = k \qquad (5.127)$$

If $j = 1, 2, \ldots, N$ and $k = 1, 2, \ldots, N$, Eq. (5.127) defines IRFs for all the combinations of the input and output DOFs. These IRFs together constitute the IRF matrix $[h(t)]$ given below.

$$[h(t)] = \begin{bmatrix} h_{11} & h_{12} & \cdots & h_{1N} \\ h_{21} & h_{22} & \cdots & h_{2N} \\ \vdots & \vdots & \vdots & \vdots \\ h_{N1} & h_{N2} & \cdots & h_{NN} \end{bmatrix} \qquad (5.128)$$

5.13   RELATIONSHIP BETWEEN THE IRFS AND THE MODAL MODEL

Can the IRFs be found from the modal model? We can relate the two using Eq. (5.125), which allows finding the response of a system to a transient or arbitrary excitation.

Let us apply a unit impulse at the kth DOF. The impulse is realized by applying a force at the kth DOF in the form of a Dirac delta function with no force at the rest of the DOFs. Since, $\int_{-\infty}^{\infty} \delta(\tau)d\tau = 1.0$, we get a unit impulse at k^{th} DOF. Thus, the force vector given by

$$\{F(\tau)\} = \{F_1(\tau), \, F_2(\tau), \, \ldots, \, F_k(\tau), \, \ldots, F_N(\tau)\}^T \qquad (5.129)$$

is selected as

$$\{F(\tau)\} = \{0, \, 0, \, \ldots, \, \delta(\tau), \, \ldots, 0\}^T \qquad (5.130)$$

Eq. (5.125) can be rewritten as

$$\{x(t)\} = \sum_{r=1}^{N} \frac{\{\phi\}_r}{\omega_r} \left(\int_0^t \sin\omega_r (t-\tau) \{\phi\}_r^T \{F(\tau)\} \, d\tau \right) \qquad (5.131)$$

Substituting Eq. (5.130) into Eq. (5.131) and noting that this makes $\{\phi\}_r^T \{F(\tau)\} = \phi_{kr}\delta(\tau)$, we get

$$\{x(t)\} = \sum_{r=1}^{N} \frac{\{\phi\}_r \phi_{kr}}{\omega_r} \left(\int_0^t \sin\omega_r (t-\tau)\delta(\tau) \, d\tau \right) \qquad (5.132)$$

Given the definition of $\delta(\tau)$ (which says that $\delta(\tau) = 0$ for $\tau \neq 0$) the integral in the bracket is equal to the value of $\sin \omega_r (t - \tau)$ at $\tau = 0$. Also, since the response $\{x(t)\}$ is due to a unit impulse at the kth DOF, $\{x(t)\}$ represents the kth column of the IRF matrix (denoted by $\{h(t)\}^k$). Given these observations, Eq. (5.132) reduces to

$$\{h(t)\}^k = \sum_{r=1}^{N} \frac{\{\phi\}_r \phi_{kr}}{\omega_r} \sin \omega_r t \tag{5.133}$$

For $k = 1, 2, \ldots, N$, Eq. (5.133) yields all the columns of the IRF matrix, and thus this equation gives the full IRF matrix. The desired IRF $h_{jk}(t)$ is nothing but the jth element of $\{h(t)\}^k$ and is given by

$$h_{jk}(t) = \sum_{r=1}^{N} \frac{\phi_{jr} \phi_{kr}}{\omega_r} \sin \omega_r t \tag{5.134}$$

Eq. (5.134) is the relationship between the IRFs and the modal model. Therefore, if the modal model is available, the IRFs can be determined. For a linear system, due to the principle of reciprocity, discussed in Chapter 2, $h_{jk}(t) = h_{kj}(t)$ and the IRF matrix is symmetric.

5.14   RESPONSE MODEL (TIME DOMAIN)

If the IRF matrix for a system is given, then the response to any set of arbitrary forces can be determined using the convolution integral. Let $F_1(\tau)$, $F_2(\tau)$, ..., $F_N(\tau)$ be the forces in the time domain acting at the N DOFs of the system. The responses due to individual forces can be algebraically added for a linear system to find the system response. The response at the jth DOF can be written as

$$x_j(t) = h_{j1} * F_1 + h_{j2} * F_2 + \ldots + h_{jN} * F_N \qquad \text{for } j = 1, 2, \ldots, N \tag{5.135}$$

Thus, if the IRF matrix $[h(t)]$ of the system is known, the system response can be determined using Eq. (5.135). Therefore, the IRF matrix constitutes an alternative system model and is the **Response model in the time domain**.

The response models FRF and IRF matrices form a Fourier transform pair. Representing the Fourier transform by the symbol $\mathcal{F}$, we have

$$\alpha_{jk}(\omega) = \mathcal{F}\left(h_{jk}(t)\right) \tag{5.136}$$

and

$$[\alpha(\omega)] = \mathcal{F}\left([h(t)]\right) \tag{5.137}$$

Representing the inverse Fourier transform by the symbol $\mathcal{F}^{-1}$, we have

$$h_{jk}(t) = \mathcal{F}^{-1}\left(\alpha_{jk}(\omega)\right) \tag{5.138) and}$$

$$[h(t)] = \mathcal{F}^{-1}\left([\alpha(\omega)]\right) \tag{5.139}$$

We also note that in the frequency domain, the output equals the product of the input with the FRF. However, in the time domain, the output is equal to the convolution of the input with the IRF.

Example 5.2

The two-DOF system shown in Figure 5.1 has the following properties: m = 2 kg and k = 1,000 N/m. If the initial displacement and velocity are: $\{x(0)\} = \{0.01 \quad 0\}^T$, and $\{\dot{x}(0)\} = \{0 \quad 0\}^T$, determine the modal and physical responses.

Solution

From Example 5.1, the mass and stiffness matrices, natural frequencies, and eigenvectors are:

$$[M] = \begin{bmatrix} m & 0 \\ 0 & m \end{bmatrix} = \begin{bmatrix} 2 & 0 \\ 0 & 2 \end{bmatrix}, [K] = \begin{bmatrix} 2k & -k \\ -k & 2k \end{bmatrix} = \begin{bmatrix} 2,000 & -1,000 \\ -1,000 & 2,000 \end{bmatrix},$$

$$\omega_1 = \sqrt{\frac{k}{m}} = 22.3 \text{ rad/sec and } \omega_2 = \sqrt{\frac{3k}{m}} = 38.7 \text{ rad/sec}$$

$$\{\psi\}_1 = \begin{Bmatrix} 1 \\ 1 \end{Bmatrix} \text{ and } \{\psi\}_2 = \begin{Bmatrix} 1 \\ -1 \end{Bmatrix}$$

Calculate modal mass matrix:

$$[\acute{}m_{r\backslash}] = \begin{bmatrix} m_1 & 0 \\ 0 & m_2 \end{bmatrix} = [\psi]^T [M][\psi] = \begin{bmatrix} 1 & 1 \\ 1 & -1 \end{bmatrix}^T \begin{bmatrix} 2 & 0 \\ 0 & 2 \end{bmatrix} \begin{bmatrix} 1 & 1 \\ 1 & -1 \end{bmatrix} = \begin{bmatrix} 4 & 0 \\ 0 & 4 \end{bmatrix}$$

Calculate the initial condition on modal displacement (using Eq. 5.71):

$$y_r(0) = \frac{1}{m_r} \{\psi\}_r^T [M]\{x_0\}$$

$$\text{for } r = 1: \quad y_1(0) = \frac{1}{m_1} \{\psi\}_1^T [M]\{x_0\} = 0.005$$

$$\text{for } r = 2: \quad y_2(0) = \frac{1}{m_2} \{\psi\}_2^T [M]\{x_0\} = 0.005$$

Calculate the initial condition on modal velocities (using Eq. 5.74):

$$\dot{y}_r(0) = \frac{1}{m_r} \{\Psi\}_r^T [M]\{v_0\}$$

Since, $\{v_0\} = \{0\}$, $\dot{y}_1(0) = \dot{y}_2(0) = 0$

Calculate modal displacement response (using Eq. 5.75)

$$y_r(t) = y_r(0)\cos\omega_r t + \frac{1}{\omega_r}\dot{y}_r(0)\sin\omega_r t$$

$$\text{for } r = 1: \quad y_1(t) = 0.005\cos\omega_1 t$$

$$\text{for } r = 2: \quad y_2(t) = 0.005\cos\omega_2 t$$

The modal response amplitudes are equal; hence, both modes contribute equally to the response.
Calculate the physical response (using Eq. 5.77):

$$\{x(t)\} = \sum_{r=1}^{2} \{\psi\}_r y_r(t) = \begin{Bmatrix} 1 \\ 1 \end{Bmatrix} 0.005\cos 22.3\, t + \begin{Bmatrix} 1 \\ -1 \end{Bmatrix} 0.005\cos 38.7\, t$$

$$x_1(t) = 0.005\cos 22.3\, t + 0.005\cos 38.7\, t$$

$$x_2(t) = 0.005\cos 22.3\, t - 0.005\cos 38.7\, t$$

Example 5.3

In Example 5.2, if harmonic forces of amplitudes −5N and 5N of frequency 24 rad/sec act on the two masses, determine the modal forces and steady-state modal and physical responses. Justify the results.

Solution

The force vector amplitude is: $\{f_0\} = \{-5 \quad 5\}^T$

Calculate the amplitude of the modal force vector:

$$\{f_{r0}\} = [\psi]^T \{f_0\} = \begin{bmatrix} 1 & 1 \\ 1 & -1 \end{bmatrix}^T \begin{Bmatrix} -5 \\ 5 \end{Bmatrix} = \begin{Bmatrix} 0 \\ -10 \end{Bmatrix}$$

Calculate modal stiffness matrix:

$$[`k_r] = \begin{bmatrix} k_1 & 0 \\ 0 & k_2 \end{bmatrix} = [\psi]^T [K][\psi] = \begin{bmatrix} 1 & 1 \\ 1 & -1 \end{bmatrix}^T \begin{bmatrix} 2,000 & -1,000 \\ -1,000 & 2,000 \end{bmatrix} \begin{bmatrix} 1 & 1 \\ 1 & -1 \end{bmatrix} = \begin{bmatrix} 2,000 & 0 \\ 0 & 6,000 \end{bmatrix}$$

Calculate the amplitude of the modal response vector (using Eq. 5.97)

$$y_{r0} = \frac{f_{r0}}{k_r - \omega^2 m_r}$$

$$\text{for } r = 1: \quad y_{10} = \frac{0}{2,000 - 24^2 \times 4} = 0$$

$$\text{for } r = 2: \quad y_{10} = \frac{-10}{6,000 - 24^2 \times 4} = -0.0027$$

Calculate the amplitude of the physical response vector (using Eq. 5.100)

$$\{x_0\} = \sum_{r=1}^{2} \{\psi\}_r y_r(t) = \begin{Bmatrix} 1 \\ 1 \end{Bmatrix} \times 0 - \begin{Bmatrix} 1 \\ -1 \end{Bmatrix} 0.0027 = \begin{Bmatrix} -0.0027 \\ 0.0027 \end{Bmatrix}$$

The modal force for mode 1 is zero, which means that mode 1 is not excited by the external force. The zero modal force leads to a zero modal response of this mode. Thus, the physical response is contributed by only mode 2, and as a result, the physical response is identical to the second mode shape within a scaling factor.

Example 5.4

The equation of motion for the longitudinal vibrations of a propulsion system using a lumped parameter model is obtained in Example 2.10 in Chapter 2. The mass and stiffness matrices for the system are:

$$[M] = \begin{bmatrix} m_1 & 0 & 0 \\ 0 & m_2 & 0 \\ 0 & 0 & m_3 \end{bmatrix}, [K] = \begin{bmatrix} k_1 & -k_1 & 0 \\ -k_1 & k_1+k_2 & -k_2 \\ 0 & -k_2 & k_2+k_3 \end{bmatrix}$$

m_1, m_2 and m_3 are 100, 70, and 90 kg, respectively. k_1, k_2 and k_3 are 2E+8, 6E+7, and 5E+9 N/m, respectively.

Using a MATLAB® program, determine the natural frequencies and mode shapes. Also, find the modal mass and modal stiffness. Are the mode shapes mass-normalized?

Solution

MATLAB program 5.1 is used to solve the problem. The results obtained are given. The modal masses are unity because the MATLAB function 'eig(k, m)' normalizes the eigenvectors in m-norm for a symmetric k and symmetric positive-definite m, as in the present case. Therefore, the eigenvectors obtained are already mass-normalized.

```
%#######################
%   MATLAB PROGRAM 5.1
%#######################
clear all;

%DEFINE MASS MATRIX
m1=100; m2=70; m3=90;
m=[m1  0  0;
   0  m2 0;
   0       0  m3];

%DEFINE STIFFNESS MATRIX
k1=2e+8;  k2=6e+7; k3=5e+9;
k=[ k1  −k1    0;
   −k1  k1+k2  −k2;
    0   −k2    k2+k3];

%SOLVE EIGEN-VALUE PROBLEM
[Evec,Eval]=eig(k,m);

%FIND NATURAL FREQUENCY IN %rad/sec
NF_rad_sec=sqrt(diag(Eval))

%FIND NATURAL FREQUENCY IN Hz
NF_Hz=NF_rad_sec/(2*pi)

%FIND MODAL MASS AND MODAL %STIFFNESS
mr=diag(Evec'*m*Evec)
kr=diag(Evec'*k*Evec)
%#######################
```

Results:

NATURAL FREQUENCIES IN rad/sec :
560.5   2321.4   7498.8

NATURAL FREQUENCIES IN Hz :
89.2   369.4   1193.4

EIGEN VECTORS :
0.081721   0.057634   0.000064
0.068880   -0.097664   −0.001724
0.000821   −0.001281   0.105398

MODAL MASSES:
1.0   1.0 1.0

MODAL STIFFNESSES:
0.0314 e+7   0.5389 e+7   5.6233 e+7

Example 5.5

For the propulsion system in Example 5.4, if a longitudinal excitation force of amplitude 75 N at frequency 110 Hz acts on mass m1, determine using a MATLAB program the steady-state modal responses and longitudinal displacement of mass m1.

Solution

MATLAB program 5.2 is used to solve the problem. The results obtained are also given.

```
%#######################
%   MATLAB PROGRAM 5.2
%#######################

%PROGRAM CONTINUING FROM %MATLAB PROGRAM 5.1

%DEFINE EXCITATION FREQUENCY
fexc=2*pi*110;

%DEFINE AMPLITUDE OF %EXCITATION VECTOR
f0=[75 0 0]';

%FIND AMPLITUDE OF MODAL %FORCE VECTOR
fr0=Evec'*f0
```

Results:

MODAL RESPONSES AMPLITUDES
−3.75e-05   8.8e-7   8.55e-11

PHYSICAL RESPONSE AMPLITUDE (MASS 1)
−3.0142e-6

(Continued)

```
%FIND MODAL RESPONSES %AMPLITUDES
for j=1:3
    yr0(j)=fr0(j)/(kr(j)−mr(j)*(fexc^2))
end
yr0=yr0';

%FIND PHYSICAL RESPONSES %AMPLITUDES
x0=Evec*yr0
%AMPLITUDE OF MASS 1
x0(1)
%#######################
```

Example 5.6

A two-DOF lumped parameter model of a car is shown in Figure 5.7. Find the response of the car body (mass m_1) if the car moves over a speed breaker with velocity V_0 m/sec. The length and height of the speed breaker are L and H m, respectively, and its shape can be approximated as a half sine-pulse. Use the convolution-integral approach.

Solution

Write the equilibrium equations:

$$m_1\ddot{x}_1 = -k_1(x_1 - x_2)$$

$$m_2\ddot{x}_2 = k_1(x_1 - x_2) - k_2(x_2 - x_3)$$

$$T = \frac{L}{V_0}$$

$$x_3 = H\sin\frac{2\pi}{2T}t = H\sin\frac{\pi}{L}V_0t = H\sin\omega_0t$$

$$\begin{bmatrix} m_1 & 0 \\ 0 & m_2 \end{bmatrix}\begin{Bmatrix} \ddot{x}_1 \\ \ddot{x}_2 \end{Bmatrix} + \begin{bmatrix} k_1 & -k_1 \\ -k_1 & k_1+k_2 \end{bmatrix}\begin{Bmatrix} x_1 \\ x_2 \end{Bmatrix} = \begin{Bmatrix} 0 \\ k_2x_3 \end{Bmatrix} = \begin{Bmatrix} 0 \\ F(t) \end{Bmatrix}$$

where $F(t) = k_2x_3 = k_2H\sin\omega_0t = F_0\sin\omega_0t$
Eigenvalues and eigenvectors are found by solving the eigenvalue problem.
Response x_1 using the IRFs:

$$x_1(t) = h_{11}*F_1 + h_{12}*F_2 = h_{11}*0 + h_{12}*F(t) = h_{12}*F(t)$$

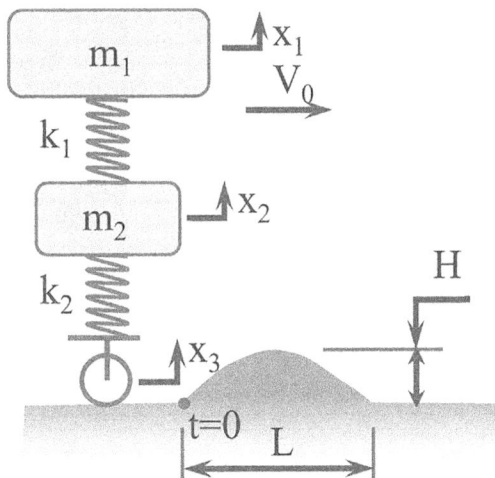

FIGURE 5.7   Two-DOF model of a car.

IRF h_{12}:

$$h_{12}(t) = \sum_{r=1}^{2} \frac{\phi_{1r}\phi_{2r}}{\omega_r} \sin\omega_r t = A_1 \sin\omega_1 t + A_2 \sin\omega_2 t$$

where $A_1 = \dfrac{\phi_{11}\phi_{21}}{\omega_1}$ and $A_2 = \dfrac{\phi_{12}\phi_{22}}{\omega_2}$

By convolution integral:

$$x_1(t) = \int_0^T h_{12}(t-\tau)F(\tau)\,d\tau = \int_0^T A_1\sin\omega_1(t-\tau)k_2 H\sin\omega_0\tau\,d\tau$$

$$x_1(t) = \int_0^T A_1\sin\omega_1(t-\tau)k_2 H\sin\omega_0\tau\,d\tau + \int_0^T A_2\sin\omega_2(t-\tau)k_2 H\sin\omega_0\tau\,d\tau$$

$$x_1(t) = X_1(t) + X_2(t)$$

Find the first integral (referred as $X_1(t)$):

$$X_1(t) = \int_0^T A_1\sin\omega_1(t-\tau)k_2 H\sin\omega_0\tau\,d\tau$$

$$X_1(t) = \frac{1}{2}A_1 k_2 H\int_0^T \Big(\cos\big(\omega_1(t-\tau)-\omega_0\tau\big) - \cos\big(\omega_1(t-\tau)+\omega_0\tau\big)\Big)d\tau$$

$$X_1(t) = \frac{1}{2}A_1 k_2 H\left(\frac{\sin\omega_1 t - \sin\big(\omega_1 t-(\omega_1+\omega_0)T\big)}{\omega_1+\omega_0} - \frac{\sin\omega_1 t - \sin\big(\omega_1 t-(\omega_1-\omega_0)T\big)}{\omega_1-\omega_0}\right)$$

Similarly, the second integral gives

$$X_2(t) = \frac{1}{2}A_2 k_2 H\left(\frac{\sin\omega_2 t - \sin\big(\omega_2 t-(\omega_2+\omega_0)T\big)}{\omega_2+\omega_0} - \frac{\sin\omega_2 t - \sin\big(\omega_2 t-(\omega_2-\omega_0)T\big)}{\omega_2-\omega_0}\right)$$

Response of the car body (mass m_1):

$$x_1(t) = \frac{k_2 H}{2}\left(\left(A_1\left(\frac{\sin\omega_1 t - \sin\big(\omega_1 t-(\omega_1+\omega_0)T\big)}{\omega_1+\omega_0} - \frac{\sin\omega_1 t - \sin\big(\omega_1 t-(\omega_1-\omega_0)T\big)}{\omega_1-\omega_0}\right)\right.\right.$$

$$\left.\left. + A_2\left(\frac{\sin\omega_2 t - \sin\big(\omega_2 t-(\omega_2+\omega_0)T\big)}{\omega_2+\omega_0} - \frac{\sin\omega_2 t - \sin\big(\omega_2 t-(\omega_2-\omega_0)T\big)}{\omega_2-\omega_0}\right)\right)\right)$$

Example 5.7

The modal model of an undamped two-DOF system is given below. Find the displacement amplitude at the first DOF if a harmonic force of amplitude 20 N at 16 Hz acts at the second DOF.

NATURAL FREQUENCIES IN Hz :	MASS-NORMALIZED EIGENVECTORS :	
9.0770	−0.4538	−0.2098
18.2688	−0.1586	0.3431

Solution

We find the FRF α_{12} to find the displacement amplitude at the first DOF due to force at the second DOF.
Find FRF α_{12}:

$$\alpha_{12}(\omega) = \sum_{r=1}^{2} \frac{\phi_{1r}\,\phi_{2r}}{\omega_r^2 - \omega^2}$$

$$\alpha_{12}(\omega) = \frac{-0.4538 \times -0.1586}{(2\pi 9.07)^2 - (2\pi 16)^2} + \frac{-0.2098 \times 0.3431}{(2\pi 18.26)^2 - (2\pi 16)^2} = -2.3451e - 05 \text{ m/N}$$

Displacement amplitude:

$$X_1 = \sigma_{12}(\omega)F_2 = -2.3451e - 05 \times 20 = -4.6903e - 04 \text{ m}$$

REVIEW QUESTIONS

1. What are the standard and generalized eigenvalue problems?
2. Why the mass matrix is always positive-definite while the stiffness matrix can be positive-definite or semi-definite?
3. Define the terms 'physical space' and 'modal space'? What are the differences between them?
4. What are spatial, modal, and response models? State relationships between them.
5. What do you mean by modal force? Can the modal force be zero? Answer with reasons.
6. What do you mean by modal mass and modal stiffness?
7. Can the modal mass or modal stiffness be zero? Answer with reasons.
8. What is the FRF matrix for an MDOF system?
9. Define the FRF $\alpha_{jk}(\omega)$.
10. Define the IRF $h_{jk}(t)$.
11. Why is the FRF matrix of a system said to form a model of the system?
12. Why is the IRF matrix of a system said to form a model of the system?
13. Why an MDOF system disturbed arbitrarily vibrates with a combination of its principle modes? Justify on physical grounds, assuming the system to be undamped.

PROBLEMS

Problem 5.1: The mass and stiffness matrix of a two-DOF system are:

MASS	MATRIX :	STIFFNESS	MATRIX :
1	0	1,500	−1,000
0	2	−1,000	2,000

If the initial displacement and velocity of the masses are: $\{x(0)\} = \{0.01 \quad 0.02\}^T$, and $\{\dot{x}(0)\} = \{2 \quad 0\}^T$, determine the modal response amplitudes. Is there any dominant mode in the response?

Problem 5.2: Example 3.2 in Chapter 3 gives the FE model mass and stiffness matrices for the transverse vibrations of a drilling machine. The diameter of the machine column is 10 cm. Take the following parameters: L = 0.9 m ρ = 7,800 kg/m3, E = 2.0e +11 N/m2, m_d = 40 kg, m_t = 30 kg. Using a MATLAB program, determine the natural frequencies and mode shapes of the drilling machine.

Problem 5.3: A 3-DOF lumped parameter model of a system has the mass and stiffness matrices given below. The corresponding mode shapes are also given. Find the natural frequencies of the system without solving the eigenvalue problem.

MASS MATRIX :			STIFFNESS MATRIX :			MODE SHAPES :		
2	0	0	310,000	−140,000	0	0.9923	0.4063	−0.1264
0	8	0	−140,000	150,000	−80,000	−0.1216	0.8598	−0.1611
0	0	4	0	−80,000	250,000	0.0241	0.3092	0.9788

Problem 5.4: The modal model of an undamped system is given below. Find the acceleration amplitude of the response at the third DOF if a sinusoidal force of amplitude 50 N at 60 Hz is acting at the first DOF.

NATURAL FREQUENCIES IN Hz :	MASS-NORMALIZED EIGEN VECTORS :		
18.2657	−0.2292	−0.1272	−0.1769
57.7753	−0.2141	−0.0434	0.3085
117.8930	−0.1309	0.3028	−0.0482

6 Analytical Modal Analysis of Damped MDOF Systems

6.1 INTRODUCTION

We studied the analytical modal analysis of undamped MDOF systems in Chapter 5. The undamped representation of a system doesn't consider the damping present, and the predicted results may not reveal the system's dynamic behavior accurately, especially when the damping is not negligible. In this chapter, we study the analytical modal analysis of damped MDOF systems.

Eigenvalue problems for the systems with structural and viscous damping are presented, showing how the damping affects the dynamic characteristics like the eigenvalues and eigenvectors. The concept of modal space is extended to the damped systems. The free and forced vibration analysis of MDOF systems with structural and viscous damping using the direct and modal analysis methods is carried out. Systems with both proportional and non-proportional damping are considered. Modal models and response models in the frequency and time domains are presented. The relationship between the FRFs and the modal model presented in this chapter forms the fundamental basis for modal parameter estimation in experimental modal analysis.

6.2 EIGENVALUE ANALYSIS OF SYSTEMS WITH STRUCTURAL DAMPING

In Chapter 4, we discussed the analysis of SDOF systems with structural damping. It was noted that the time-domain representation of the structural damping poses a challenge, and hence the free vibration analysis, like finding the response to initial excitation, is difficult. This is also true for the MDOF systems with structural damping, and therefore, our goal is to see how steady-state response analysis to harmonic forces can be carried out. It is also possible to define the eigenvalue problem for an MDOF system with structural damping using the inexact model, as presented in this section. The knowledge of the eigenvalues and eigenvectors of the system enables steady-state response analysis to harmonic forces through modal analysis.

In Chapter 5, we learned that an undamped MDOF system with N DOFs is described by N simultaneous ordinary differential equations involving mass and stiffness matrices of order N × N. We also discussed in Chapter 4 that for an SDOF system with structural damping, the combined resistance of elasticity and structural damping for harmonic vibrations could be represented by the complex stiffness given by $(k + ih)$ where 'h' is the structural damping coefficient. By extending the idea to MDOF systems, the combined resistance of elasticity and structural damping can be represented by a complex stiffness matrix given by $[K] + i[D]$, where $[D]$ is the structural damping matrix.

Therefore, similar to Eq. (4.82) in Chapter 4 for an SDOF system, the equation of motion for an MDOF system with structural damping subjected to harmonic forces can be written in the frequency domain as

$$-\omega^2[M]\{x(\omega)\} + ([K] + i[D])\{x(\omega)\} = \{F(\omega)\} \tag{6.1}$$

To define the eigenvalue problem, we need to write the equation of motion for the free vibration of the system in the time domain. But the free vibration with damping is non-harmonic, so the time-domain equations cannot be strictly written using complex stiffness, which is discussed in Chapter 4. However, with the limited objective of defining the eigenvalue problem and obtaining estimates of the characteristics of the natural modes of vibration of the system, the time-domain equation of motion is written in the following way by assuming that the structural damping force can be incorporated through the complex stiffness:

$$[M]\{\ddot{x}\} + ([K] + i[D])\{x\} = \{0\} \tag{6.2}$$

If the damping is distributed independently of the system's mass and stiffness distributions, it is called **nonproportional damping**. However, if the damping in the system is proportional to the mass or stiffness matrices or the combination of the two, then it is called **proportional damping** (also called Rayleigh damping). We consider both these cases.

DOI: 10.1201/9780429454783-6

6.2.1 Proportional Structural Damping

In this case, the damping matrix $[D]$ can be written as a linear combination of the mass and stiffness matrices,

$$[D] = \beta_k[K] + \beta_m[M] \tag{6.3}$$

where β_k and β_m are the constants of proportionality corresponding to the stiffness and mass matrices, respectively.

Let us assume the solution to Eq. (6.2) as

$$\{x\} = \{X\}e^{i\omega t} \tag{6.4}$$

The frequency ω in Eq. (6.4) is complex to account for vibration decay with time due to damping.

Substituting Eq. (6.4) into Eq. (6.2), we get

$$\left(([K] + i[D])\{X\} - \omega^2[M]\{X\}\right)e^{i\omega t} = \{0\} \tag{6.5}$$

$e^{i\omega t}$ is non-zero for an arbitrary t, and hence the vector in the bracket must be zero. Using, $\lambda = \omega^2$, we get

$$([K] + i[D])\{X\} = \lambda[M]\{X\} \tag{6.6}$$

Eq. (6.6) is an eigenvalue problem (EVP) in the generalized form involving a complex stiffness matrix and a real mass matrix. It is similar to the EVP for an undamped MDOF system. The eigenvalues corresponding to Eq. (6.6) are complex and can be expressed in the form

$$\lambda_r = p_r(1 + iq_r) \tag{6.7}$$

and the corresponding eigenvectors are denoted by $\{\psi\}_r$, with r = 1, 2, ..., N.

What do p_r and q_r represent, and are the eigenvectors $\{\psi\}_r$ related to their undamped system counterparts? To answer these questions, we first check for orthogonality between the eigenvectors.

Writing Eq. (6.6) for the rth and s^{th} eigenvalue-eigenvector pairs and proceeding like in Eqs. (5.19)–(5.28) in Chapter 5, we get the following orthogonality relationships/properties between the eigenvectors:

$$\{\psi\}_r^T[M]\{\psi\}_s = 0 \quad \text{for } r \neq s \tag{6.8}$$

$$\{\psi\}_r^T[M]\{\psi\}_s \neq 0 \quad \text{for } r = s \tag{6.9}$$

and,

$$\{\psi\}_r^T([K] + i[D])\{\psi\}_s = 0 \quad \text{for } r \neq s \tag{6.10}$$

$$\{\psi\}_r^T([K] + j[D])\{\psi\}_s \neq 0 \quad \text{for } r = s \tag{6.11}$$

If a matrix $[\psi]$, called the eigenvector matrix, is formed with its columns as eigenvectors, the mass-orthogonality properties (Eqs. 6.8 and 6.9) can be combined and expressed in terms of $[\psi]$ as

$$[\psi]^T[M][\psi] = [\,\ddot{}\,m_r\,\ddot{}\,] \tag{6.12}$$

where $[\,\ddot{}\,m_r\,\ddot{}\,]$ is the modal mass matrix, a diagonal matrix, with m_r being the modal mass of the rth mode.

Similarly, the stiffness-orthogonality properties (Eqs. 6.10 and 6.11) can be combined and expressed in terms of $[\psi]$ as

$$[\psi]^T([K] + i[D])[\psi] = [\,\ddot{}\,\bar{k}_r\,\ddot{}\,] \tag{6.13}$$

where $[\,\ddot{}\,\bar{k}_r\,\ddot{}\,]$ is the complex modal stiffness matrix, a diagonal matrix, with $\bar{k}_r$ being the complex modal stiffness of the rth mode.

It is found that the eigenvectors $[\psi]$ are real, i.e., the relative phase between any two elements of an eigenvector is either $0°$ or $180°$. Also, the eigenvector matrix $[\psi]$ not only diagonalizes the mass matrix and the complex stiffness matrix, as seen from Eqs. (6.12) and (6.13), but also the stiffness matrix $[K]$. These observations point to the fact that the eigenvectors $[\psi]$ are the same as the eigenvectors of the corresponding undamped system. Given this, the matrix $[\psi]^T[K][\psi]$ is also a diagonal matrix and is nothing but the modal stiffness matrix of the undamped model,

$$[\psi]^T[K][\psi] = [\ulcorner k_{r \urcorner}] \tag{6.14}$$

Since the damping matrix $[D]$ is a linear combination of the mass and stiffness matrices, $[\psi]^T[D][\psi]$ is a diagonal matrix and represents a matrix of modal structural damping coefficients,

$$[\psi]^T[D][\psi] = [\ulcorner h_{r \urcorner}] \tag{6.15}$$

Because of Eqs. (6.14) and (6.15), Eq. (6.13) can be written as

$$[\ulcorner \bar{k}_{r \urcorner}] = [\psi]^T ([K] + i[D])[\psi]$$

$$= [\psi]^T[K][\psi] + i[\psi]^T[D][\psi] = [\ulcorner k_{r \urcorner}] + i[\ulcorner h_{r \urcorner}] \tag{6.16}$$

If the mass-normalized eigenvector matrix $[\phi]$ is used, then the modal matrices take the following form:

$$[\ulcorner m_{r \urcorner}] = [I] \tag{6.17}$$

$$[\ulcorner k_{r \urcorner}] = [\ulcorner \omega_{r \urcorner}^2] \tag{6.18}$$

$$[\ulcorner \bar{k}_{r \urcorner}] = [\ulcorner \lambda_{r \urcorner}] \tag{6.19}$$

where $[\ulcorner \omega_{r \urcorner}^2]$ is a diagonal matrix of the eigenvalues of the corresponding undamped system and $[\ulcorner \lambda_{r \urcorner}]$ is a diagonal matrix of the complex eigenvalues of the damped system.

From Chapter 4 (Eq. 4.78), we know that for an SDOF system, the structural damping coefficient h can be written as $h = k.\eta$, and if this relationship is used at the modal level, then the matrix $[\ulcorner h_{r \urcorner}]$ can be written as

$$[\ulcorner h_{r \urcorner}] = [\ulcorner k_r.\eta_{r \urcorner}] = [\ulcorner k_{r \urcorner}][\ulcorner \eta_{r \urcorner}] = [\ulcorner \omega_{r \urcorner}^2][\ulcorner \eta_{r \urcorner}] \tag{6.20}$$

where $[\ulcorner \eta_{r \urcorner}]$ is a diagonal matrix of the structural modal damping factors (or modal loss factors), with η_r being the structural modal damping factor for the rth mode. In light of Eqs. (6.17)–(6.20), Eq. (6.16) becomes

$$[\ulcorner \lambda_{r \urcorner}] = [\ulcorner \omega_{r \urcorner}^2] + i[\ulcorner \omega_{r \urcorner}^2][\ulcorner \eta_{r \urcorner}] = [\ulcorner \omega_{r \urcorner}^2]([I] + i[\ulcorner \eta_{r \urcorner}]) \tag{6.21}$$

Thus, the rth complex eigenvalue can be written as

$$\lambda_r = \omega_r^2 (1 + i\eta_r) \quad r = 1, 2, \dots, N \tag{6.22}$$

where ω_r is nothing but the undamped natural frequency of the rth mode.

Modal Model

Thus, the dynamic characteristics of the vibration modes are natural frequencies (ω_r, $r = 1, 2, \dots, N$), structural modal damping factors (η_r, $r = 1, 2, \dots, N$) and mass-normalized eigenvectors ($\{\phi\}_r$, $r = 1, 2, \dots, N$) and collectively describe the system in the modal domain. This description is the modal model of the system. For the proportional damping, the natural frequencies and eigenvectors are the same as the corresponding quantities for the undamped version of the system.

6.2.2 Nonproportional Structural Damping

If the damping is nonproportional, then the damping matrix can no longer be expressed as a linear combination of the mass and stiffness matrices. Equation (6.2) still defines the EVP with the difference that the matrix $[D]$ is nonproportional.

The solution of the EVP yields the eigenvalues $\left(\left[\,^\diagdown\lambda_{r\diagdown}\right]\right)$ and eigenvectors $\left(\left[\psi\right]\right)$. The eigenvectors satisfy the orthogonality with respect to the mass matrix and complex stiffness matrix. It is seen that with the nonproportional damping, the eigenvectors $\left[\psi\right]$ are complex, i.e., the relative phase between any two elements of an eigenvector is not necessarily 0° or 180°. No longer, the eigenvector matrix $\left[\psi\right]$ diagonalizes the stiffness matrix $[K]$. Therefore, $\left[\psi\right]^{T}[K][\psi]$ and $\left[\psi\right]^{T}[D][\psi]$ are not diagonal matrices. Thus, the eigenvectors for the nonproportional and corresponding undamped systems are different.

If the eigenvectors are mass-normalized, then the modal matrices are

$$\left[\,^\diagdown m_{r\diagdown}\right]=\left[I\right] \tag{6.23}$$

$$\left[\,^\diagdown \overline{k}_{r\diagdown}\right]=\left[\,^\diagdown \lambda_{r\diagdown}\right] \tag{6.24}$$

where $\left[\,^\diagdown\lambda_{r\diagdown}\right]$ is a diagonal matrix of complex eigenvalues. This matrix can be written as a sum of the real and imaginary parts. We write the real part as $\left[\,^\diagdown\omega'^{2}_{r\diagdown}\right]$. The imaginary part is written as a product of two diagonal matrices, $\left[\,^\diagdown\omega'^{2}_{r\diagdown}\right]$ and $\left[\,^\diagdown\eta_{r\diagdown}\right]$. Thus,

$$\left[\,^\diagdown\lambda_{r\diagdown}\right]=\left[\,^\diagdown\omega'^{2}_{r\diagdown}\right]+i\left[\,^\diagdown\omega'^{2}_{r\diagdown}\right]\left[\,^\diagdown\eta_{r\diagdown}\right]=\left[\,^\diagdown\omega'^{2}_{r\diagdown}\right]\left(\left[I\right]+i\left[\,^\diagdown\eta_{r\diagdown}\right]\right) \tag{6.25}$$

where $\left[\,^\diagdown\eta_{r\diagdown}\right]$ is the diagonal matrix of structural modal damping factors for the nonproportional case.

Therefore, the rth complex eigenvalue is

$$\lambda_r = \omega'^2_r\left(1+i\eta_r\right) \quad r = 1, 2, ..., N \tag{6.26}$$

Frequency ω'_r is, in general, different than the undamped natural frequency ω_r (and hence denoted by a different symbol ω'_r).

Modal Model

The modal model of the system consists of the natural frequencies (ω'_r, $r = 1, 2, ..., N$), structural modal damping factors (η_r, $r = 1, 2, ..., N$) and mass-normalized complex eigenvectors ($\{\phi\}_r$, $r = 1, 2, ..., N$). Note that the natural frequencies and eigenvectors are different from the corresponding quantities for the undamped version of the system.

6.3   FORCED VIBRATION RESPONSE OF SYSTEMS WITH STRUCTURAL DAMPING

In this section, we consider the forced vibration response of a system with structural damping subjected to harmonic forces of frequency ω. From Section 6.2, the equation of motion for the system in the frequency domain is

$$-\omega^2[M]\{x(\omega)\}+\left([K]+i[D]\right)\{x(\omega)\}=\{F(\omega)\} \tag{6.27}$$

where $\{F(\omega)\}$ and $\{x(\omega)\}$ represent the amplitude vectors of the harmonic forces and steady-state displacements, respectively. The general solution consists of complementary and particular parts. As discussed, determining the complementary function requiring the free vibration solution in the time domain is difficult for systems with structural damping. The complementary part dies out after some time, and only the steady-state part is left. We find out the steady-state part of the response.

6.3.1   DIRECT SOLUTION IN PHYSICAL SPACE

Equation (6.27) can be written as

$$\left([K]+i[D]-\omega^2[M]\right)\{x(\omega)\}=\{F(\omega)\} \tag{6.28}$$

The steady-state response amplitude is obtained as

$$\{x(\omega)\}=\left[\left([K]+i[D]-\omega^2[M]\right)\right]^{-1}\{F(\omega)\} \tag{6.29}$$

while the steady-state response in the time domain is

$$\{x(t)\} = \{x(\omega)\} e^{i\omega t} \tag{6.30}$$

Writing Eq. (6.29) as

$$\{x(\omega)\} = [\alpha]\{F(\omega)\} \tag{6.31}$$

where $[\alpha]$ is the frequency response function (FRF) matrix of the structurally damped system. It is given by

$$[\alpha] = \left([K] + i[D] - \omega^2[M]\right)^{-1} \tag{6.32}$$

Note that the FRFs are defined in the frequency domain (for the harmonic forces); therefore, they could be defined even for systems with structural damping.

All the above equations, including the solution (Eq. 6.29), are valid for proportional and nonproportional structural damping systems.

6.3.2 SOLUTION BY MODAL ANALYSIS

In this section, we obtain the forced vibration response via modal analysis.

a. Proportional damping case

Consider the modal transformation in the frequency domain using the mass-normalized eigenvectors

$$\{x(\omega)\} = [\phi]\{y(\omega)\} \tag{6.33}$$

Substituting Eq. (6.33) into Eq. (6.27) and pre-multiplying by $[\phi]^T$ gives

$$-\omega^2[\phi]^T[M][\phi]\{y(\omega)\} + [\phi]^T([K] + i[D])[\phi]\{y(\omega)\} = [\phi]^T\{F(\omega)\} \tag{6.34}$$

Using the orthogonality properties, Eq. (6.34) becomes

$$-\omega^2[I]\{y(\omega)\} + [\lambda_r \diagdown]\{y(\omega)\} = \{f_r(\omega)\} \tag{6.35}$$

where $\{f_r(\omega)\}$ is the vector of modal forces, given by

$$\{f_r(\omega)\} = [\phi]^T\{F(\omega)\} \tag{6.36}$$

Eq. (6.35), using Eq. (6.25), can be written as

$$\left([\diagdown \omega_r'^2 \diagdown]([I] + i[\diagdown \eta_r \diagdown]) - \omega^2[I]\right)\{y(\omega)\} = \{f_r(\omega)\} \tag{6.37}$$

$$\left([\diagdown(\omega_r^2 - \omega^2)\diagdown] + i[\diagdown \eta_r\omega_r^2 \diagdown]\right)\{y(\omega)\} = \{f_r(\omega)\} \tag{6.38}$$

$$\left[\diagdown(\omega_r^2 - \omega^2 + i\eta_r\omega_r^2)\diagdown\right]\{y(\omega)\} = \{f_r(\omega)\} \tag{6.39}$$

$$\{y(\omega)\} = \left[\diagdown(\omega_r^2 - \omega^2 + i\eta_r\omega_r^2)\diagdown\right]^{-1}\{f_r(\omega)\} \tag{6.40}$$

which gives the frequency domain response in modal space as,

$$\{y(\omega)\} = \left[\diagdown\left(\frac{1}{\omega_r^2 - \omega^2 + i\eta_r\omega_r^2}\right)\diagdown\right]\{f_r(\omega)\} \tag{6.41}$$

From Eq. (6.41), the modal response in the rth mode is

$$y_r(\omega) = \frac{f_r(\omega)}{\omega_r^2 - \omega^2 + i\eta_r\omega_r^2} \qquad r = 1,\, 2,\ldots,\, N \tag{6.42}$$

The physical responses are obtained by substituting Eq. (6.41) into Eq. (6.33), followed by substituting Eq. (6.36)

$$\{x(\omega)\} = [\phi]\left[\cdot\left(\frac{1}{\omega_r^2 - \omega^2 + i\eta_r\omega_r^2} \right) \cdot \right][\phi]^T\{F(\omega)\} \tag{6.43}$$

Alternatively, Eq. (6.33) can be first rewritten as

$$\{x(\omega)\} = \sum_{r=1}^{N}\{\phi\}_r\, y_r(\omega) \tag{6.44}$$

Substituting Eq. (6.42) into (6.44) and noting that the rth modal force is $f_r(\omega) = \{\phi\}_r^T\{F(\omega)\}$, we get the amplitude of the steady-state response

$$\{x(\omega)\} = \left(\sum_{r=1}^{N} \frac{\{\phi\}_r\{\phi\}_r^T}{\omega_r^2 - \omega^2 + i\eta_r\omega_r^2} \right)\{F(\omega)\} \tag{6.45}$$

The steady-state response in the time domain to the complex exponential force is

$$\{x(t)\} = \{x(\omega)\}e^{i\omega t} \tag{6.46}$$

b. Nonproportional damping case

For the nonproportional damping, as noted in Section 6.2.2, the eigenvalues $(\lceil \lambda_r \rfloor)$ and the eigenvectors $([\phi])$ are different from those for the corresponding undamped system. By following the steps as done in Eqs. (6.33)–(6.45), we get the following expressions (corresponding to Eqs. 6.43 and 6.45):

$$\{x(\omega)\} = [\phi]\left[\cdot\left(\frac{1}{\omega_r'^2 - \omega^2 + i\eta_r\omega_r'^2} \right) \cdot \right][\phi]^T\{F(\omega)\} \tag{6.47}$$

and,

$$\{x(\omega)\} = \left(\sum_{r=1}^{N} \frac{\{\phi\}_r\{\phi\}_r^T}{\omega_r'^2 - \omega^2 + i\eta_r\omega_r'^2} \right)\{F(\omega)\} \tag{6.48}$$

The steady-state response to harmonic forces is given by Eq. (6.46).

6.4   FREE VIBRATION RESPONSE OF MDOF SYSTEMS WITH PROPORTIONAL VISCOUS DAMPING

In this section, we consider the free vibration of an MDOF system with viscous damping, assuming the damping to be proportional. We first define the eigenvalue Problem and then determine the free vibration response to the given initial conditions by modal analysis.

6.4.1   EIGENVALUE PROBLEM

The equation of motion for a two-DOF system with viscous damping was derived in Chapter 2 (Eq. 2.8). For an N DOF system, the matrix equation is of the same form, but the system matrices are of order $N \times N$. For the free vibration, the force vector is zero, and the equation of motion in matrix form is

$$[M]\{\ddot{x}\} + [C]\{\dot{x}\} + [K]\{x\} = \{0\} \tag{6.49}$$

The study of free vibration enables studying the natural modes of vibration of the system. Let the free vibration response is

$$\{x\} = \{X\}e^{st} \tag{6.50}$$

Substituting into Eq. (6.49) gives

$$\left(s^2[M] + s[C] + [K]\right)\{X\}e^{st} = \{0\} \tag{6.51}$$

Since, $e^{st} \neq 0$ for an arbitrary t, we have

$$\left(s^2[M] + s[C] + [K]\right)\{X\} = \{0\} \tag{6.52}$$

A comparison with the EVP for the undamped system indicates that we now have an additional term $s[C]$ due to viscous damping. Equation (6.52) is a quadratic eigenvalue problem in the eigenvalue 's' and is more challenging to solve than the generalized linear EVP.

However, for the proportional damping case, we can more easily determine the system's eigenvalues and the eigenvectors by following a modal approach. Let s_r be the unknown r^{th} eigenvalue of the system. Let $\{\phi\}_r$ be the rth mass-normalized eigenvector of the corresponding undamped system, obtained by solving its EVP. We assume the free vibration response in the following form and check whether it forms a solution:

$$\{x\} = \{\phi\}_r e^{s_r t} \tag{6.53}$$

Substituting Eq. (6.53) into Eq. (6.51) and pre-multiplying by $\{\phi\}_r^T$ gives

$$\left(\{\phi\}_r^T[M]\{\phi\}_r \, s_r^2 + \{\phi\}_r^T[C]\{\phi\}_r \, s_r + \{\phi\}_r^T[K]\{\phi\}_r\right)e^{s_r t} = \{0\} \tag{6.54}$$

From the orthogonality properties of the eigenvectors, we note that

$$\{\phi\}_r^T[M]\{\phi\}_r = 1.0 \tag{6.55}$$

and

$$\{\phi\}_r^T[K]\{\phi\}_r = \omega_r^2 \tag{6.56}$$

where ω_r is the rth natural frequency of the undamped system.

It can be shown that any viscous damping matrix $[C]$ that satisfies the condition $[K][M]^{-1}[C] = [C][M]^{-1}[K]$ qualifies as a proportional viscous damping matrix. However, a more commonly used model of the proportional viscous damping matrix, known as the Rayleigh damping model, is based on expressing the damping matrix as a linear combination of the mass and stiffness matrices,

$$[C] = v_m[M] + v_k[K] \tag{6.57}$$

with v_m and v_k being the constants of proportionality for the two matrices, respectively. Given Eqs. (6.55)–(6.57), we obtain

$$\{\phi\}_r^T[C]\{\phi\}_r = v_m + v_k\omega_r^2 \tag{6.58}$$

Substituting Eqs. (6.55), (6.56), and (6.58) into Eq. (6.54), we get

$$\left(s_r^2 + \left(v_m + v_k\omega_r^2\right)s_r + \omega_r^2\right)e^{s_r t} = \{0\} \tag{6.59}$$

Since $e^{s_r t} \neq 0$ for an arbitrary t, we have

$$s_r^2 + \left(v_m + v_k\omega_r^2\right)s_r + \omega_r^2 = 0 \tag{6.60}$$

Eq. (6.60) is similar to the characteristic equation for an SDOF viscously damped system. For example, Eq. (4.9) in Chapter 4 divided by mass m transforms to $s^2 + 2\xi\omega_n s + \omega_n^2 = 0$. With this analogy, Eq. (6.60) can be written as

$$s_r^2 + 2\xi_r\omega_r s_r + \omega_r^2 = 0 \tag{6.61}$$

with

$$\xi_r = \frac{1}{2}\left(\frac{v_m}{\omega_r} + v_k\omega_r\right) \tag{6.62}$$

Assuming that $\xi_r < 1.0$ for all r, which is often the case in practice, the roots of Eq. (6.61) are

$$s_{r\,1,2} = -\xi_r\omega_r \pm i\omega_r\sqrt{1-\xi_r^2} \tag{6.63}$$

We can summarize the outcome of the above steps as follows. With the eigenvalues given by Eq. (6.63), along with the eigenvectors of the corresponding undamped model, Eq. (6.53) is indeed a solution to the free vibration problem (Eq. 6.49) and, hence, also of the corresponding EVP (Eq. 6.52). Therefore, the two values of s_r together with the eigenvector $\{\phi\}_r$ are nothing but the eigen-properties of the rth mode of vibration of the MDOF system with proportional viscous damping.

With the eigenvalue $s_r = s_{r1}$ or $s_r = s_{r2}$, the response given by Eq. (6.53) would be complex. The real solution for $\{x\}$ can be represented by a combination of the responses due to both the eigenvalues (s_{r1} and s_{r2}). Since the eigenvalues are complex conjugate, we denote them by s_r and s_r^*. (An overhead '*' is used to denote the complex conjugate operation). Thus,

$$\{x\} = A_r\{\phi\}_r e^{s_r t} + B_r\{\phi\}_r e^{s_r^* t} \tag{6.64}$$

where we also have $B_r = A_r^*$ for a real $\{x\}$.

$$\{x\} = A_r\{\phi\}_r e^{s_r t} + A_r^*\{\phi\}_r e^{s_r^* t} \tag{6.65}$$

Substituting Eq. (6.63) into Eq. (6.65) and simplifying, we get the following free vibration response in the rth mode of vibration of the damped system, with D_r and E_r being the arbitrary constants.

$$\{x\} = e^{-\xi_r\omega_r t}\{\phi\}_r\left(D_r\cos\omega_r\sqrt{1-\xi_r^2}t + E_r\sin\omega_r\sqrt{1-\xi_r^2}t\right) \tag{6.66}$$

6.4.2 Modal Model

It is seen from Eq. (6.66) that the free vibration response in the rth mode of vibration of a proportional viscously damped system is determined by the rth mass-normalized eigenvector of the corresponding undamped system and a pair of complex conjugate eigenvalues. This result is true for r = 1, 2,..., N, thus, defining the eigenvalues and eigenvectors of all the modes of the system. The modal model of the system can be presented in two alternative forms.

One method to define the modal model of a proportional viscously damped system consists of prescribing N pairs of complex conjugate eigenvalues and the corresponding eigenvectors of the undamped system. The complex conjugate eigenvalues for a mode are related to the undamped natural frequency and modal viscous damping factor of that mode. The modal viscous damping factor for a mode is related to the undamped natural frequency of the mode and the damping matrix proportionality constants v_m and v_k.

Equation (6.66) points to an alternative method to define the modal model where each mode of vibration is described by the natural frequency, corresponding undamped eigenvector, and modal viscous damping factor. These two methods are shown in Table 6.1.

6.4.3 Response to Initial Conditions by Modal Analysis

Let the initial conditions on the system be prescribed in terms of initial displacements and velocities.

$$\{x(0)\} = \{x_0\} \tag{6.67}$$

$$\{\dot{x}(0)\} = \{v_0\} \tag{6.68}$$

TABLE 6.1
Modal Model Descriptions of a Proportional Viscously Damped System

First Method	Second Method
N pairs of Complex conjugate eigenvalues:	**N natural frequencies:**
$s_r = -\xi_r\omega_r + i\omega_r\sqrt{1-\xi_r^2}$	ω_r
$s_r^* = -\xi_r\omega_r - i\omega_r\sqrt{1-\xi_r^2}$	where
with r = 1, 2,..., N	ω_r : r^{th} undamped natural frequency
where	with r = 1, 2,..., N
ω_r : r^{th} undamped natural frequency	**N modal viscous damping factors:**
ξ_r : r^{th} modal viscous damping factor	$\xi_r = \dfrac{1}{2}\left(\dfrac{v_m}{\omega_r} + v_k\omega_r\right)$
N eigenvectors of the undamped system:	where
$\{\phi\}_1, \{\phi\}_2, \ldots, \{\phi\}_N$	ξ_r : r^{th} modal viscous damping factor
	v_m and v_k are damping matrix proportionality constants
	N eigenvectors of the undamped system:
	$\{\phi\}_1, \{\phi\}_2, \ldots, \{\phi\}_N$

Consider transformation to modal coordinates using the mass-normalized eigenvectors

$$\{x(t)\} = [\phi]\{y(t)\} \tag{6.69}$$

We first find the modal response $\{y(t)\}$. It requires initial conditions on the modal response, obtained by proceeding as we did for the undamped MDOF system in Chapter 5.

From Eq. (6.69), at t = 0,

$$\{x(0)\} = [\phi]\{y(0)\} \tag{6.70}$$

$$\{x(0)\} = \sum_{j=1}^{N}\{\phi\}_j\, y_j(0) \tag{6.71}$$

Pre-multiplying by $\{\phi\}_r^T[M]$, using orthogonality properties and substituting for $\{x(0)\}$, we get

$$y_r(0) = \{\phi\}_r^T[M]\{x_0\} \quad \text{for} \quad r = 1, 2, \ldots, N \tag{6.72}$$

Differentiating Eq. (6.69) once with time and writing it for t = 0, we get

$$\{\dot{x}(0)\} = [\phi]\{\dot{y}(0)\} \tag{6.73}$$

$$\{\dot{x}(0)\} = \sum_{j=1}^{N}\{\phi\}_j\, \dot{y}_j(0) \tag{6.74}$$

Again pre-multiplying by $\{\phi\}_r^T[M]$, using orthogonality properties and substituting for $\{\dot{x}(0)\}$, we get

$$\dot{y}_r(0) = \{\phi\}_r^T[M]\{v_0\} \quad \text{for} \quad r = 1, 2, \ldots, N \tag{6.75}$$

Thus, Eqs. (6.72) and (6.75) represent the initial conditions in modal space or modal coordinates.

We now transform the equation of motion to modal space by substituting Eq. (6.69) into (6.49). Pre-multiplication by $[\phi]^T$ gives

$$[\phi]^T[M][\phi]\{\ddot{y}\} + [\phi]^T[C][\phi]\{\dot{y}\} + [\phi]^T[K][\phi]\{y\} = \{0\} \tag{6.76}$$

Since the damping matrix is proportional, the matrix $[\phi]^T[C][\phi]$ is diagonal. In light of Eqs. (6.57) and (6.62), we can write

$$[\phi]^T[C][\phi] = \lceil 2\xi_r\omega_r \rfloor \tag{6.77}$$

Given Eq. (6.77) and making use of the orthogonality properties, Eq. (6.76) becomes

$$[I]\{\ddot{y}\} + \lceil 2\xi_r\omega_r \rfloor\{\dot{y}\} + \lceil \omega_r^2 \rfloor\{y\} = \{0\} \tag{6.78}$$

Eq. (6.78) represents N uncoupled differential equations in the modal domain corresponding to N modes; hence, each can be solved independently. The equation corresponding to the rth mode is

$$\ddot{y}_r + 2\delta_r\xi_r\dot{y}_r + \omega_r^2 y_r = 0 \tag{6.79}$$

which is similar to the governing equation of an SDOF system with viscous damping. Its roots are given by Eq. (6.63), which are complex, as we have assumed $\xi_r < 1.0$ for all r. Since Eq. (6.79) is a homogeneous equation, its general solution is given by

$$y_r(t) = A_{1r}e^{\left(-\xi_r\omega_r + i\omega_r\sqrt{1-\xi_r^2}\right)t} + A_{2r}e^{\left(-\xi_r\omega_r - i\omega_r\sqrt{1-\xi_r^2}\right)t} \tag{6.80}$$

Arbitrary constants A_{1r} and A_{2r} are obtained from the knowledge of the initial conditions $y_r(0)$ and $\dot{y}_r(0)$ (given by Eqs. 6.72 and 6.75). Using the result we obtained earlier in Chapter 4 for the SDOF system (Eq. 4.22), we can write

$$y_r(t) = e^{-\xi_r\omega_r t}\left(y_r(0)\cos\omega_r\sqrt{1-\xi_r^2}\,t + \frac{\dot{y}_r(0) + y_r(0)\xi_r\omega_r}{\omega_r\sqrt{1-\xi_r^2}}\sin\omega_r\sqrt{1-\xi_r^2}\,t\right) \tag{6.81}$$

Substituting Eqs. (6.72) and (6.75) into Eq. (6.81), we get the modal response in the rth mode

$$y_r(t) = e^{-\xi_r\omega_r t}\left(\{\phi\}_r^T[M]\{x_0\}\cos\omega_r\sqrt{1-\xi_r^2}\,t + \frac{\{\phi\}_r^T[M](\{v_0\} + \{x_0\}\xi_r\omega_r)}{\omega_r\sqrt{1-\xi_r^2}}\sin\omega_r\sqrt{1-\xi_r^2}\,t\right) \quad \text{for } r = 1, 2, ..., N \tag{6.82}$$

Substituting Eq. (6.82) into Eq. (6.69), we get the response in physical space

$$\{x(t)\} = \sum_{r=1}^{N}\{\phi\}_r\, y_r(t) \tag{6.83}$$

$$\{x(t)\} = \sum_{r=1}^{N}e^{-\xi_r\omega_r t}\{\phi\}_r\{\phi\}_r^T[M]\left(\{x_0\}\cos\omega_r\sqrt{1-\xi_r^2}\,t + \frac{\{v_0\} + \{x_0\}\xi_r\omega_r}{\omega_r\sqrt{1-\xi_r^2}}\sin\omega_r\sqrt{1-\xi_r^2}\,t\right) \tag{6.84}$$

Thus, the free vibration response of the system consists of a sum of the free responses in various modes. The free vibration response in each mode is at the damped natural frequency ($\omega_r\sqrt{1-\xi_r^2}$) of that mode, with the amplitude of the response decaying exponentially, with the decay rate governed by the modal damping factor (ξ_r) of that mode. As we observed for an undamped system, Eq. (6.84) indicates that the initial disturbance may simultaneously excite multiple modes of vibration of the damped system.

6.5 FORCED VIBRATION RESPONSE OF MDOF SYSTEMS WITH PROPORTIONAL VISCOUS DAMPING

Let the system is subjected to harmonic forces at frequency ω. The equation of motion of the system is

$$[M]\{\ddot{x}\} + [C]\{\dot{x}\} + [K]\{x\} = \{F\} \tag{6.85}$$

We take the harmonic force vector $\{F\}$ in complex form as

$$\{F\} = \{F_0\} e^{i\omega t} \tag{6.86}$$

The total response consists of complementary and particular parts. The particular part of the response is determined first in the following two subsections using direct and modal analysis approaches. The total solution is presented in the subsection after that.

6.5.1 DIRECT METHOD

Let the particular solution of Eq. (6.85) is

$$\{x_p\} = \{x_0\} e^{i\omega t} \tag{6.87}$$

Substituting it into Eq. (6.85), we get

$$-\omega^2 [M]\{x_0\} + i\omega [C]\{x_0\} + [K]\{x_0\} = \{F_0\} \tag{6.88}$$

$$\left([K] - \omega^2 [M] + i\omega [C]\right)\{x_0\} = \{F_0\} \tag{6.89}$$

where the matrix $\left([K] - \omega^2 [M] + i\omega [C]\right)$ is called the dynamic stiffness matrix, and for the damped case, it also accounts for the damping forces along with the elastic and inertia forces.

$$\{x_0\} = \left([K] - \omega^2 [M] + i\omega [C]\right)^{-1} \{F_0\} \tag{6.90}$$

$$\{x_0\} = [\alpha]\{F_0\} \tag{6.91}$$

where

$$[\alpha] = \left([K] - \omega^2 [M] + j\omega [C]\right)^{-1} \tag{6.92}$$

$[\alpha]$ is the FRF matrix of the damped system.

Thus, Eq. (6.90) gives the amplitude of the particular part of the response to the harmonic excitation forces, which, when substituted in Eq. (6.87), gives the particular solution in the time domain.

6.5.2 MODAL ANALYSIS METHOD

In this section, we determine the particular part of the response using modal analysis. Consider the transformation to modal coordinates using the mass-normalized eigenvectors

$$\{x_p(t)\} = [\phi]\{y(t)\} \tag{6.93}$$

Substituting into Eq. (6.85) and pre-multiplying by $[\phi]^T$ gives

$$[\phi]^T [M][\phi]\{\ddot{y}\} + [\phi]^T [C][\phi]\{\dot{y}\} + [\phi]^T [K][\phi]\{y\} = [\phi]^T \{F\} \tag{6.94}$$

Using the orthogonality properties, we get

$$[I]\{\ddot{y}\} + [`2\xi_r\omega_r`]\{\dot{y}\} + [`\omega_r^2`]\{y\} = \{f_r\} \tag{6.95}$$

where

$$\{f_r\} = [\phi]^T\{F\} \tag{6.96}$$

is a vector of modal forces, with the rth modal force given by

$$f_r = \{\phi\}_r^T\{F\} \qquad\qquad r = 1, 2,..., N \tag{6.97}$$

Eq. (6.94) represents N uncoupled differential equations in the modal domain corresponding to the N modes of the proportional viscously damped system. Let us consider the equation corresponding to the rth mode,

$$\ddot{y}_r + 2\xi_r\omega_r\dot{y}_r + \omega_r^2 y_r = f_r \tag{6.98}$$

Equation (6.98) is similar to the governing equation of the forced vibration of an SDOF system with viscous damping. As the external force is harmonic, the modal forces are also harmonic. Hence, from Eq. (6.97), we can write

$$f_r(t) = \{\phi\}_r^T\{F_0\}e^{i\omega t} = f_{r0}e^{i\omega t} \tag{6.99}$$

The corresponding modal response is also harmonic and can be written as

$$y_r(t) = y_{r0}e^{i\omega t} \tag{6.100}$$

Substituting Eqs. (6.99) and (6.100) into Eq. (6.98) gives

$$y_{r0} = \frac{f_{r0}}{\omega_r^2 - \omega^2 + i2\xi_r\omega_r\omega} \tag{6.101}$$

$$y_r(t) = \frac{f_r(t)}{\omega_r^2 - \omega^2 + i2\xi_r\omega_r\omega} \tag{6.102}$$

Substituting Eq. (6.97) into Eq. (6.102) gives the modal response in the rth mode

$$y_r(t) = \frac{\{\phi\}_r^T\{F\}}{\omega_r^2 - \omega^2 + i2\xi_r\omega_r\omega} \tag{6.103}$$

and with $r = 1, 2,..., N$, we get modal responses for the N modes.

Eq. (6.103) is now substituted into Eq. (6.93) to obtain the physical response.

$$\{x_p(t)\} = \sum_{r=1}^{N}\{\phi\}_r \, y_r(t) \tag{6.104}$$

$$\{x_p(t)\} = \sum_{r=1}^{N}\frac{\{\phi\}_r\{\phi\}_r^T\{F\}}{\omega_r^2 - \omega^2 + i2\xi_r\omega_r\omega} \tag{6.105}$$

$$\{x_p(t)\} = \left(\sum_{r=1}^{N}\frac{\{\phi\}_r\{\phi\}_r^T\{F_0\}}{\omega_r^2 - \omega^2 + i2\xi_r\omega_r\omega}\right)e^{i\omega t} \tag{6.106}$$

The term in the bracket is the amplitude ($\{x_0\}$) of the particular part or steady-state part of the response. Hence,

$$\{x_0\} = \left(\sum_{r=1}^{N} \frac{\{\phi\}_r \{\phi\}_r^T}{\omega_r^2 - \omega^2 + i2\xi_r\omega_r\omega} \right)\{F_0\} \tag{6.107}$$

Unlike the case of an undamped system, the amplitude is now complex. The complex amplitude represents that the phase of the response with force is no longer either $0°$ or $180°$.

6.5.3 Total Response

The general solution of the MDOF system is given by

$$\{x(t)\} = \{x_c(t)\} + \{x_p(t)\} \tag{6.108}$$

The complementary part, $\{x_c(t)\}$, is the solution of the homogeneous equation and consists of a linear combination of the free vibration response in all the modes. Equation (6.66) is the free vibration response in a mode of vibration and, therefore, $\{x_c(t)\}$ is given by

$$\{x_c(t)\} = \sum_{j=1}^{N} e^{-\xi_j\omega_j t} \{\phi\}_j \left(D_j \cos\omega_j\sqrt{1-\xi_j^2}\,t + E_j \sin\omega_j\sqrt{1-\xi_j^2}\,t \right) \tag{6.109}$$

We obtained the particular solution to the complex exponential force vector. Assuming the actual force to be $\{F_0\}\cos\omega t$, which is the real part of the complex force, $\{x_p(t)\}$ to be used in Eq. (6.108) is also the real part of Eq. (6.106). Since the amplitude is complex, we first write Eq. (6.107) as

$$\{x_p(t)\} = \left(\sum_{r=1}^{N} \frac{\{\phi\}_r \{\phi\}_r^T \{F_0\}}{\sqrt{\left(\omega_r^2 - \omega^2\right)^2 + \left(2\xi_r\omega_r\omega\right)^2} \ e^{i\theta_r}} \right) e^{i\omega t} \tag{6.110}$$

where

$$\theta_r = \tan^{-1}\frac{2\xi_r\omega_r\omega}{\omega_r^2 - \omega^2} \tag{6.111}$$

Substituting Eq. (6.109) and the real part of $\{x_p(t)\}$ in Eq. (6.110) into Eq. (6.108), we get

$$\{x(t)\} = \sum_{j=1}^{N} e^{-\xi_r\omega_r t} \{\phi\}_j \left(D_j \cos\omega_r\sqrt{1-\xi_j^2}\,t + E_j \sin\omega_j\sqrt{1-\xi_j^2}\,t \right) + \sum_{r=1}^{N} \frac{\{\phi\}_r \{\phi\}_r^T \{F_0\}}{\sqrt{\left(\omega_r^2 - \omega^2\right)^2 + \left(2\xi_r\omega_r\omega\right)^2}} \cos(\omega t - \theta_r) \tag{6.112}$$

D_j and E_j are arbitrary constants and are to be found from the given initial conditions. They can be found either by directly applying the initial conditions on Eq. (6.112) or by using the orthogonality properties, as done for the undamped system (in Section 5.5.3 in Chapter 5).

We note the following from Eq. (6.112):

- The first part is a sum of the free-damped responses in N modes. It is transient as the responses decay with time. This part contributes negligibly after some initial time.
- The second part is a sum of the steady-state responses in N modes. The frequency of this part is the same as the frequency of the external harmonic forces. After some initial time, the steady-state part dominates the total response.

6.6 FREE VIBRATION OF SYSTEMS WITH NONPROPORTIONAL VISCOUS DAMPING

For many systems in practice, the assumption of proportional damping may not be valid, and it may not accurately represent the damping distribution in the system. For nonproportional damping also, the Quadratic EVP (Eq. 6.52) is applicable and can be solved. But the modal analysis theory of systems with general/nonproportional viscous damping is based on state-space equations, and we present this approach in the following sections. The state-space approach transforms the second-order differential equations of motion into a set of first-order differential equations based on which the EVP is defined. The EVP can be formulated in two alternative ways, as presented in the following sections. Based on the EVPs, the modal models are defined, and then the free vibration response of the system to given initial conditions by modal analysis is presented.

6.6.1 EIGENVALUE PROBLEM (FORMULATION I)

The equation of motion for an MDOF system with viscous damping under the free vibration is

$$[M]\{\ddot{x}\}+[C]\{\dot{x}\}+[K]\{x\}=\{0\} \tag{6.113}$$

where the damping matrix $[C]$ is nonproportional.

Equation (6.113) is a set of N second-order differential equations; hence, 2N variables are required to specify the system's state at any time. We choose the displacements and velocities at the N DOFs (denoted by the vectors $\{z_d\}$ and $\{z_v\}$, respectively) as the state variables. Thus,

$$\{z_d\}=\{x\} \tag{6.114}$$

$$\{z_v\}=\{\dot{x}\} \tag{6.115}$$

Differentiating with time gives

$$\{\dot{z}_d\}=\{\dot{x}\} \tag{6.116}$$

$$\{\dot{z}_v\}=\{\ddot{x}\} \tag{6.117}$$

Given Eqs. (6.114)–(6.117), Eq. (6.113) can be written as

$$[M]\{\dot{z}_v\}+[C]\{\dot{z}_d\}+[K]\{z_d\}=\{0\} \tag{6.118}$$

Substituting $\{\dot{x}\}$ from Eq. (6.115) into Eq. (6.116) and pre-multiplying by $[M]$ gives

$$[M]\{\dot{z}_d\}-[M]\{z_v\}=\{0\} \tag{6.119}$$

Defining a state vector $\{z\}$ consisting of all the state variables, we have

$$\{z\}=\begin{Bmatrix} z_d \\ z_v \end{Bmatrix} \tag{6.120}$$

Now, Eqs. (6.118) and (6.119) can be combined into a single matrix equation in terms of the state vector as follows:

$$\begin{bmatrix} C & M \\ M & 0 \end{bmatrix}\begin{Bmatrix} \dot{z}_d \\ \dot{z}_v \end{Bmatrix}+\begin{bmatrix} K & 0 \\ 0 & -M \end{bmatrix}\begin{Bmatrix} z_d \\ z_v \end{Bmatrix}=\begin{Bmatrix} 0 \\ 0 \end{Bmatrix} \tag{6.121}$$

Eq. (6.121) can be symbolically expressed as

$$[A]\{\dot{z}\}+[B]\{z\}=\{0\} \tag{6.122}$$

where

$$[A] = \begin{bmatrix} C & M \\ M & 0 \end{bmatrix} \quad \text{and} \quad [B] = \begin{bmatrix} K & 0 \\ 0 & -M \end{bmatrix} \tag{6.123}$$

Eq. (6.122) is a set of 2N first-order differential equations, called the **State equation,** and is equivalent to Eq. (6.113). It represents the free vibration as there is no forcing term. Let its solution is

$$\{z\} = \{\bar{\psi}\}e^{st} \tag{6.124}$$

Substituting Eq. (6.124) into Eq. (6.122) gives

$$\left([B]\{\bar{\psi}\} + s[A]\{\bar{\psi}\}\right)e^{st} = \{0\} \tag{6.125}$$

e^{st} is non-zero for an arbitrary t, and hence the vector in the bracket must be zero. Therefore, we get

$$[B]\{\bar{\psi}\} = s\left(-[A]\right)\{\bar{\psi}\} \tag{6.126}$$

Eq. (6.126) represents an EVP in state-space in terms of the matrices $-[A]$ and $[B]$, and $\{\bar{\psi}\}$ represents the eigenvector in state-space with s being the eigenvalue. These matrices are of order $2N \times 2N$, and therefore the EVP solution yields 2N eigenvalues and 2N eigenvectors.

a. Nature of the eigenvalues and eigenvectors

Let the 2N eigenvalues and the corresponding eigenvectors be denoted as

- 2N eigenvalues : $s_1, s_2, \ldots, s_N, s_{N+1}, s_{N+2} \ldots, s_{2N}$ (6.127)

- 2N eigenvectors : $\{\bar{\psi}\}_1, \{\bar{\psi}\}_2, \ldots, \{\bar{\psi}\}_N, \{\bar{\psi}\}_{N+1}, \{\bar{\psi}\}_{N+2}, \ldots, \{\bar{\psi}\}_{2N}$ (6.128)

The free vibration response due to the eigenvalue s_r and the corresponding eigenvector $\{\bar{\psi}\}_r$, obtained from Eq. (6.124), is

$$\{z\} = \{\bar{\psi}\}_r e^{s_r t} \tag{6.129}$$

Using Eq. (6.120) and partitioning $\{\bar{\psi}\}_r$ into displacement and velocity parts, we get

$$\left\{ \begin{array}{c} z_d \\ z_v \end{array} \right\} = \left\{ \begin{array}{c} \bar{\psi}_d \\ \bar{\psi}_v \end{array} \right\}_r e^{s_r t} \tag{6.130}$$

which gives

$$\{z_d\} = \{\bar{\psi}_d\}_r e^{s_r t} \tag{6.131}$$

and,

$$\{z_v\} = \{\bar{\psi}_v\}_r e^{s_r t} \tag{6.132}$$

Differentiating Eq. (6.131) with t leads to

$$\{\dot{z}_d\} = s_r \{\bar{\psi}_d\}_r e^{s_r t} \tag{6.133}$$

Vector $\{\dot{z}_d\}$ is nothing but $\{z_v\}$ and hence comparing Eqs. (6.132) and (6.133), we get the result that

$$\{\bar{\psi}_v\}_r = s_r \{\bar{\psi}_d\}_r \tag{6.134}$$

With the above result, we can simplify the notations. Let us denote $\{\overline{\psi}_d\}_r$ as $\{\psi\}_r$, then $\{\overline{\psi}_v\}_r$ is simply $s_r\{\psi\}_r$. Hence, the state-space eigenvector $\{\overline{\psi}\}_r$ is of the form,

$$\{\overline{\psi}\}_r = \left\{ \begin{array}{c} \{\overline{\psi}_d\}_r \\ \{\overline{\psi}_v\}_r \end{array} \right\} = \left\{ \begin{array}{c} \{\psi\}_r \\ s_r\{\psi\}_r \end{array} \right\} \qquad r = 1,\ 2,\ ...,N,\ N+1,\ N+2,...,\ 2N \tag{6.135}$$

Note that $\{\psi\}_r$ is part of the eigenvector related to the displacement coordinates.

Since the eigenvalues and eigenvectors are complex, the nature of the response due to an eigenvalue-eigenvector pair, given by Eq. (6.129), is also complex. However, since the system matrices $[M]$, $[C]$, and $[K]$ are real, the free vibration response should also be real. While one complex eigenvalue and the corresponding complex eigenvector cannot generate a real response, two complex conjugate eigenvalues and their corresponding complex conjugate eigenvectors can make up a real response. It is found that the solution of the EVP (Eq. 6.126) yields complex conjugate eigenvalues and complex conjugate eigenvectors.

To summarize, an N DOF system has N natural modes of vibrations. With viscous damping, each natural mode is represented by two complex conjugate eigenvalues and the corresponding complex conjugate eigenvectors. This outcome is similar to an SDOF system with viscous damping (covered in Chapter 4), where two complex conjugate eigenvalues describe the free/natural response.

Thus, the 2N eigenvalues and the corresponding eigenvectors in Eqs. (6.127) and (6.128) are

- 2N eigenvalues : $\{s_1,\ s_2,...,\ s_N,\ s_1^*,\ s_2^*,...,\ s_N^*\}$
- 2N eigenvectors : $\left(\{\overline{\psi}\}_1,\ \{\overline{\psi}\}_2,\ ...,\ \{\overline{\psi}\}_N,\ \{\overline{\psi}\}_1^*,\ \{\overline{\psi}\}_2^*,...,\ \{\overline{\psi}\}_N^*\right)$

In light of the observations, the free vibration response in the rth mode can be written as

$$\{z\} = c_r\{\overline{\psi}\}_r e^{s_r t} + c_r^*\{\overline{\psi}\}_r^* e^{s_r^* t} \tag{6.136}$$

Because of Eq. (6.135) and the conclusions drawn above regarding the complex conjugate nature of the eigenvalues and the eigenvectors, the state-space eigenvector matrix $[\overline{\psi}]$, formed by the eigenvectors as its columns, has the following structure:

$$[\overline{\psi}] = \begin{bmatrix} [\psi] & [\psi]^* \\ [\psi][`s_r`] & [\psi]^*[`s_r`]^* \end{bmatrix} \tag{6.137}$$

Note that all the sub-matrices in Eq. (6.137) are of order $N \times N$. $[`s_r`]$ is a diagonal matrix of N eigenvalues (s_1 to s_N), while $[`s_r`]^*$ is the matrix of corresponding complex conjugate eigenvalues (s_1^* to s_N^*). Thus,

$$[`s_r`] = \begin{bmatrix} s_1 & 0 & 0 & 0 \\ 0 & s_2 & 0 & 0 \\ 0 & 0 & \ddots & 0 \\ 0 & 0 & 0 & s_N \end{bmatrix} \tag{6.138}$$

$$[`s_r`]^* = \begin{bmatrix} s_1^* & 0 & 0 & 0 \\ 0 & s_2^* & 0 & 0 \\ 0 & 0 & \ddots & 0 \\ 0 & 0 & 0 & s_N^* \end{bmatrix} \tag{6.139}$$

b. Orthogonality properties

Every eigenvalue-eigenvector pair satisfies the EVP given by Eq. (6.126). Writing this equation for the pth and q^{th} eigenvalue-eigenvector pairs, we get

$$[B]\{\overline{\psi}\}_p = -s_p[A]\{\overline{\psi}\}_p \tag{6.140}$$

$$[B]\{\overline{\psi}\}_q = -s_q[A]\{\overline{\psi}\}_q \tag{6.141}$$

Pre-multiplying Eq. (6.140) by $\{\overline{\psi}\}_q^T$ and Eq. (6.141) by $\{\overline{\psi}\}_p^T$, we obtain

$$\{\overline{\psi}\}_q^T[B]\{\overline{\psi}\}_p = -s_p\{\overline{\psi}\}_q^T[A]\{\overline{\psi}\}_p \tag{6.142}$$

$$\{\overline{\psi}\}_p^T[B]\{\overline{\psi}\}_q = -s_q\{\overline{\psi}\}_p^T[A]\{\overline{\psi}\}_q \tag{6.143}$$

Taking transpose of Eq. (6.142) and noting that the matrices $[A]$ and $[B]$ are symmetric, we get

$$\{\overline{\psi}\}_p^T[B]\{\overline{\psi}\}_q = -s_p\{\overline{\psi}\}_p^T[A]\{\overline{\psi}\}_q \tag{6.144}$$

Subtracting Eq. (6.144) from Eq. (6.143) leads to

$$(s_p - s_q)\{\overline{\psi}\}_p^T[A]\{\overline{\psi}\}_q = 0 \tag{6.145}$$

If the eigenvalues are distinct, then for $p \neq q$, $s_p \neq s_q$, then we must have

$$\{\overline{\psi}\}_p^T[A]\{\overline{\psi}\}_q = 0 \quad \text{for } p \neq q \tag{6.146}$$

and

$$\{\overline{\psi}\}_p^T[A]\{\overline{\psi}\}_q \neq 0 \quad \text{for } p = q, \tag{6.147}$$

Substituting the results in Eqs. (6.146) and (6.147) into Eq. (6.144), we get

$$\{\overline{\psi}\}_p^T[B]\{\overline{\psi}\}_q = 0 \quad \text{for } p \neq q \tag{6.148}$$

and

$$\{\overline{\psi}\}_p^T[B]\{\overline{\psi}\}_q \neq 0 \quad \text{for } p = q, \tag{6.149}$$

Eqs. (6.146)–(6.149) form the orthogonality relations in state-space for the state-space model given by Eq. (6.122). Thus, the eigenvectors corresponding to any two eigenvalues (chosen from 2N eigenvalues) are orthogonal to the matrices $[A]$ and $[B]$. The orthogonality relationships for all the eigenvectors can be combined and written in terms of the eigenvector matrix $[\overline{\psi}]$ as

$$[\overline{\psi}]^T[A][\overline{\psi}] = \lceil a_r \rfloor \tag{6.150}$$

$$[\overline{\psi}]^T[B][\overline{\psi}] = \lceil b_r \rfloor \tag{6.151}$$

If we substitute the pair $(s_r, \{\overline{\psi}\}_r)$ in the EVP (Eq. 6.126) and pre-multiply by $\{\overline{\psi}\}_r^T$, then we obtain the following result:

$$\frac{b_r}{a_r} = -s_r \tag{6.152}$$

c. Normalization of eigenvectors

We now normalize the eigenvector $\{\bar{\psi}\}_r$ to obtain the normalized eigenvector $\{\bar{\phi}\}_r$ such that $\{\bar{\phi}\}_r^T[A]\{\bar{\phi}\}_r = 1$ for all r. The normalized eigenvector is obtained by

$$\{\bar{\phi}\}_r = \frac{\{\bar{\psi}\}_r}{\sqrt{a_r}} \tag{6.153}$$

The orthogonality properties in terms of the normalized eigenvector matrix $[\bar{\phi}]$ can be stated as

$$[\bar{\phi}]^T[A][\bar{\phi}] = [I] \tag{6.154}$$

$$[\bar{\phi}]^T[B][\bar{\phi}] = -[\,\grave{s}_{r\diagdown}] \tag{6.155}$$

Note that the matrix $[\,\grave{s}_{r\diagdown}]$ in Eq. (6.155) is a diagonal matrix of 2N eigenvalues (s_1 to s_N and then s_1^* to s_N^*). Thus,

$$[\,\grave{s}_{r\diagdown}] = \begin{bmatrix} [\,\grave{s}_{r\diagdown}] & [0] \\ [0] & [\,\grave{s}_{r\diagdown}]^* \end{bmatrix} \tag{6.156}$$

where the matrices $[\,\grave{s}_{r\diagdown}]$ and $[\,\grave{s}_{r\diagdown}]^*$ are defined in Eqs. (6.138) and (6.139), respectively.

We also note from Eq. (6.153) that the following relationship holds for the displacement part of the eigenvector:

$$\{\phi\}_r = \frac{\{\psi\}_r}{\sqrt{a_r}} \tag{6.157}$$

d. Orthogonality properties with the system matrices

In the previous section, we derived the orthogonality properties with the state-space matrices $[A]$ and $[B]$. The eigenvector matrix $[\bar{\psi}]$, given by Eq. (6.137), has a unique structure due to the complex conjugate nature of the eigenvalues and eigenvectors. Therefore, the matrix $[\bar{\psi}]$ (and the 2N eigenvalues) are entirely defined by the sub-matrices $[\psi]$ and $[\,\grave{s}_{r\diagdown}]$. Because of this, we expect some form of orthogonality of the displacement part of the eigenvectors, i.e. $[\psi]$, with the system matrices $[M]$, $[C]$ and $[K]$, as we see below.

The solution given by Eq. (6.129) satisfies the equation of motion for free vibration. The displacement part of this solution is

$$\{x\} = \{z_d\} = \{\psi\}_r e^{s_r t} \tag{6.158}$$

Eq. (6.158) must satisfy the equation of motion in x-coordinates (i.e., Eq. 6.113). Let us write $\{x\}$ corresponding to the pth and qth eigenvalue-eigenvector pairs and substitute the vector $\{x\}$ in Eq. (6.113). We obtain

$$s_p^2[M]\{\psi\}_p + s_p[C]\{\psi\}_p + [K]\{\psi\}_p = 0 \tag{6.159}$$

$$s_q^2[M]\{\psi\}_q + s_q[C]\{\psi\}_q + [K]\{\psi\}_q = 0 \tag{6.160}$$

Pre-multiplying Eq. (6.159) by $\{\psi\}_q^T$ and Eq. (6.160) by $\{\psi\}_p^T$, we get

$$s_p^2\{\psi\}_q^T[M]\{\psi\}_p + s_p\{\psi\}_q^T[C]\{\psi\}_p + \{\psi\}_q^T[K]\{\psi\}_p = 0 \tag{6.161}$$

$$s_q^2\{\psi\}_p^T[M]\{\psi\}_q + s_q\{\psi\}_p^T[C]\{\psi\}_q + \{\psi\}_p^T[K]\{\psi\}_q = 0 \tag{6.162}$$

Taking the transpose of Eq. (6.161) and subtracting Eq. (6.162) from it, we get

$$(s_p^2 - s_q^2)\{\psi\}_p^T[M]\{\psi\}_q + (s_p - s_q)\{\psi\}_p^T[C]\{\psi\}_q = 0 \tag{6.163}$$

$$(s_p + s_q)\{\psi\}_p^T[M]\{\psi\}_q + \{\psi\}_p^T[C]\{\psi\}_q = 0 \tag{6.164}$$

Eq. (6.164) is the orthogonality relation involving mass and damping matrices. Another relation is obtained by multiplying Eq. (6.161) by s_q, taking its transpose and subtracting it from Eq. (6.162) multiplied by s_p. The damping matrix term is canceled out. The equation obtained is the second orthogonality relation involving the mass and stiffness matrices and is given by

$$s_p s_q \{\psi\}_p^T [M] \{\psi\}_q + \{\psi\}_p^T [K] \{\psi\}_q = 0 \qquad (6.165)$$

Note that for an undamped system or system with proportional viscous damping, the terms like $\{\psi\}_p^T [M] \{\psi\}_q$, $\{\psi\}_p^T [K] \{\psi\}_q$ and $\{\psi\}_p^T [C] \{\psi\}_q$ are zero for $p \neq q$, but in the present case, it is not so. As a result, the transformation to modal coordinates using the matrix $\lceil \psi \rfloor$ cannot diagonalize $[M]$, $[K]$, and $[C]$ matrices, and hence the equations of motion in displacement coordinates (Eq. 6.113) cannot be decoupled.

Let s_q and s_p form a complex conjugate eigenvalue pair, i.e., $s_q = s_p^*$, then $\{\psi\}_q = \{\psi\}_p^*$. Eq. (6.164) leads to

$$\left(s_p + s_p^*\right) \{\psi\}_p^T [M] \{\psi\}_p^* + \{\psi\}_p^T [C] \{\psi\}_p^* = 0 \qquad (6.166)$$

For the underdamped case, the eigenvalues can be written as, s_p and $s_p^* = -\psi_p \psi_p \pm i \omega_p \sqrt{1 - \xi_p^{\,2}}$ and hence,

$$\frac{\{\psi\}_p^T [C] \{\psi\}_p^*}{\{\psi\}_p^T [M] \{\psi\}_p^*} = 2 \xi_p \omega_p \qquad (6.167)$$

From Eq. (6.165), we get

$$s_p s_p^* \{\psi\}_p^T [M] \{\psi\}_p^* + \{\psi\}_p^T [K] \{\psi\}_p^* = 0 \qquad (6.168)$$

This gives

$$\frac{\{\psi\}_p^T [K] \{\psi\}_p^*}{\{\psi\}_p^T [M] \{\psi\}_p^*} = \omega_p^{\,2} \qquad (6.169)$$

e. Meaning of complex eigenvectors

What do the complex eigenvectors represent, or how are they different from the real eigenvectors? Let a system is vibrating in a particular mode, which means that the corresponding eigenvector describes the displacement profile. If the eigenvector is complex, each eigenvector element is associated with a different phase angle and the relative phase between the elements can be any, not only 0° or 180°. As a result, the displacements at various DOFs of the system pass through their mean positions at different instants of time. Similarly, they attain their extreme positions at different instants of time.

In contrast, if the eigenvector is real, all the DOFs have a synchronous motion when the system vibrates in that mode, i.e., the displacements at all the DOFs pass through their mean and extreme positions simultaneously.

Therefore, in a real mode, the location of a node stays constant during vibration, while it changes in a complex mode creating the impression of a traveling wave.

6.6.2 Eigenvalue Problem (Formulation II)

In this formulation, the EVP is established from an alternative form of the state-space model. For the equation of motion given by Eq. (6.113), we continue with the choice of the state variables as given by Eqs. (6.114) and (6.115), which, when differentiated, leads to Eqs. (6.116) and (6.117). Substitution of $\{\dot{x}\}$ from Eq. (6.115) into Eq. (6.116) gives,

$$\{\dot{z}_d\} = \{z_v\} \qquad (6.170)$$

Moreover, substituting $\{\ddot{x}\}$ from Eq. (6.113) into Eq. (6.117) gives

$$\{\dot{z}_v\} = -[M]^{-1}[K]\{x\} - [M]^{-1}[C]\{\dot{x}\} \qquad (6.171)$$

$$\{\dot{z}_v\} = -[M]^{-1}[K]\{z_d\} - [M]^{-1}[C]\{z_v\} \qquad (6.172)$$

Eqs. (6.170) and (6.172) can be combined into a single matrix equation in terms of the state vector as

$$\left\{ \begin{array}{c} \dot{z}_d \\ \dot{z}_v \end{array} \right\} = \left[\begin{array}{cc} 0 & I \\ -M^{-1}K & -M^{-1}C \end{array} \right] \left\{ \begin{array}{c} z_d \\ z_v \end{array} \right\} \tag{6.173}$$

Eq. (6.173) can be symbolically written as

$$\{\dot{z}\} = [A_o]\{z\} \tag{6.174}$$

where

$$[A_o] = \left[\begin{array}{cc} 0 & I \\ -M^{-1}K & -M^{-1}C \end{array} \right] \tag{6.175}$$

is called the state matrix. Here it is represented by the symbol $[A_o]$ to differentiate it from the state matrix $[A]$ in formulation I. One difference regarding the nature of the matrices in the two state-space representations is that the matrix $[A_o]$ is unsymmetrical, while the matrices $[A]$ and $[B]$ were symmetric. Let us first consider the EVP associated with Eq. (6.174). Let the solution to this equation is

$$\{z\} = \{\overline{\psi}_R\} e^{st} \tag{6.176}$$

Substitution into Eq. (6.174) gives

$$[A_o]\{\overline{\psi}_R\} = s\{\overline{\psi}_R\} \tag{6.177}$$

Eq. (6.177) is in the form of a standard EVP. Its solution gives 2N complex eigenvalues and 2N complex eigenvectors. Since $[A_o]$ is real, the eigenvalues appear in complex conjugate pairs, with the corresponding eigenvectors also being complex conjugate. The eigenvectors obtained from the solution of Eq. (6.177) are associated with $[A_o]$ and are called right eigenvectors. A suffix 'R' has been used to indicate the same.

An EVP can also be defined in terms of the matrix $[A_o]^T$ as

$$[A_o]^T\{\overline{\psi}_L\} = s\{\overline{\psi}_L\} \tag{6.178}$$

The solution to Eq. (6.178) yields the same set of eigenvalues as obtained from the EVP in Eq. (6.177) since the characteristic polynomials in both problems are identical. However, the eigenvectors are different. The eigenvectors associated with $[A_o]^T$ are called left eigenvectors. A suffix 'L' is used to indicate the same. Eq. (6.178) after taking transpose can also be written as

$$\{\overline{\psi}_L\}^T[A_o] = s\{\overline{\psi}_L\}^T \tag{6.179}$$

We see that $\{\overline{\psi}_L\}$ appears on the left side of the matrix $[A_o]$ and is the reason why the eigenvectors $\{\overline{\psi}_L\}$ are called left eigenvectors. The right eigenvectors of $[A_o]^T$ are the same as the left eigenvectors of $[A_o]$. Since the matrix $[A_o]$ is unsymmetrical, it is characterized by two sets of eigenvectors, the right, and the left.

a. Orthogonality properties

Writing Eq. (6.177) in terms of the pth eigenvalue and the corresponding right eigenvector, we get

$$[A_o]\{\overline{\psi}_R\}_p = s_p\{\overline{\psi}_R\}_p \tag{6.180}$$

Similarly, writing Eq. (6.178) for the qth eigenvalue and the corresponding left eigenvector, we get

$$\{\overline{\psi}_L\}_q^T[A_o] = s_q\{\overline{\psi}_L\}_q^T \tag{6.181}$$

Pre-multiplying Eq. (6.169) by $\{\bar{\psi}_L\}_q^T$ and post-multiplying Eq. (6.170) by $\{\bar{\psi}_R\}_p$, we obtain

$$\{\bar{\psi}_L\}_q^T [A_o] \{\bar{\psi}_R\}_p = s_p \{\bar{\psi}_L\}_q^T \{\bar{\psi}_R\}_p \tag{6.182}$$

$$\{\bar{\psi}_L\}_q^T [A_o] \{\bar{\psi}_R\}_p = s_q \{\bar{\psi}_L\}_q^T \{\bar{\psi}_R\}_p \tag{6.183}$$

Subtracting Eq. (6.183) from Eq. (6.182) gives

$$(s_p - s_q)\{\bar{\psi}_L\}_q^T \{\bar{\psi}_R\}_p = 0 \tag{6.184}$$

If the eigenvalues are distinct, then for $p \neq q$, $s_p \neq s_q$, then we must have

$$\{\bar{\psi}_L\}_q^T \{\bar{\psi}_R\}_p = 0 \quad \text{for } p \neq q \tag{6.185}$$

and for $p = q$,

$$\{\bar{\psi}_L\}_q^T \{\bar{\psi}_R\}_p \neq 0 \quad \text{for } p = q, \tag{6.186}$$

By substituting Eqs. (6.185) and (6.186) into Eq. (6.182), we can write

$$\{\bar{\psi}_L\}_q^T [A_o] \{\bar{\psi}_R\}_p = 0 \quad \text{for } p \neq q \tag{6.187}$$

and

$$\{\bar{\psi}_L\}_q^T [A_o] \{\bar{\psi}_R\}_p \neq 0 \quad \text{for } p = q \tag{6.188}$$

Eqs. (6.187) and (6.188) form the orthogonality relations in state-space for the state-space model given by Eq. (6.174). Thus, the left eigenvector corresponding to any eigenvalue is directly orthogonal to the right eigenvector corresponding to some other eigenvalue. The orthogonality also exists between them with respect to the state matrix $[A_o]$.

Let us form a matrix $[\bar{\psi}_R]$, called the right eigenvector matrix, with columns as the right eigenvectors, and a matrix $[\bar{\psi}_L]$, called the left eigenvector matrix, with columns as the left eigenvectors. The orthogonality relationships for all the eigenvectors can be written as

$$[\bar{\psi}_L]^T [\bar{\psi}_R] = [`b_{or}\backslash] \tag{6.189}$$

$$[\bar{\psi}_L]^T [A_o][\bar{\psi}_R] = [`a_{or}\backslash] \tag{6.190}$$

If we substitute the pair $(s_r, \{\bar{\psi}_R\}_r)$ in the EVP (Eq. 6.177) and pre-multiply by $\{\bar{\psi}_L\}_r^T$, then we get the following result:

$$\frac{a_{or}}{b_{or}} = s_r \tag{6.191}$$

b. Normalization of eigenvectors

Let us normalize the eigenvectors such that $\{\bar{\psi}_L\}_r^T \{\bar{\psi}_R\}_r = 1$ for all r. The normalized eigenvectors are

$$\{\bar{\phi}_R\}_r = \frac{\{\bar{\psi}_R\}_r}{\sqrt{b_{or}}} \tag{6.192}$$

$$\{\bar{\phi}_L\}_r = \frac{\{\bar{\psi}_L\}_r}{\sqrt{b_{or}}} \tag{6.193}$$

The orthogonality properties in terms of the normalized right and left eigenvector matrices, $\left[\bar{\phi}_R\right]$ and $\left[\bar{\phi}_L\right]$ respectively, are

$$\left[\bar{\phi}_L\right]^T \left[\bar{\phi}_R\right] = [I] \tag{6.194}$$

$$\left[\bar{\phi}_L\right]^T [A_0] \left[\bar{\phi}_R\right] = [\,\ddot{s}_r\,] \tag{6.195}$$

where the matrix $[\,\ddot{s}_r\,]$ is a diagonal matrix of 2N eigenvalues (s_1 to s_N and then s_1^* to s_N^*).

6.6.3 MODAL MODEL

We looked at two formulations of the EVP for a system with nonproportional viscous damping. The modal model of the structure can be described using either of the two formulations.

a. Modal model I

It is based on the EVP formulation I. One method is to prescribe 2N complex eigenvalues and the corresponding complex eigenvectors, i.e., prescribe the matrix of complex eigenvalues $[\,\ddot{s}_r\,]$ and the matrix of complex normalized eigenvectors $\left[\bar{\phi}\right]$.

This modal model can also be alternatively specified as follows. The matrix $\left[\bar{\phi}\right]$ can be written as

$$\left[\bar{\phi}\right] = \begin{bmatrix} [\phi] & [\phi]^* \\ [\phi][\,\dot{s}_r\,] & [\phi]^*[\,\dot{s}_r\,]^* \end{bmatrix} \tag{6.196}$$

Thus, we note that

- the matrix $[\,\ddot{s}_r\,]$ is determined entirely by the submatrix $[\,\dot{s}_r\,]$
- the matrix $\left[\bar{\phi}\right]$ is determined entirely by the sub-matrices $[\phi]$ and $[\,\dot{s}_r\,]$.

Since we have assumed an underdamped system, the complex eigenvalue is of the form $s_r = -\xi_r \omega_r + i\omega_r \sqrt{1-\xi_r^2}$. Thus, each s_r is defined by its natural frequency ξ_r and modal damping factor ξ_r. Therefore, the modal model can also be defined in terms of the N natural frequencies, N modal damping factors, and N eigenvectors corresponding to the displacement coordinates.

b. Modal model II

It is based on the EVP formulation II. It consists of prescribing 2N complex eigenvalues (i.e., the matrix $[\,\ddot{s}_r\,]$) and the corresponding 2N right and 2N left eigenvectors (i.e., the matrices $\left[\bar{\phi}_R\right]$ and $\left[\bar{\phi}_L\right]$).

Table 6.2 gives the modal model descriptions of an MDOF system with nonproportional viscous damping.

6.6.4 CHOICE OF THE EVP FORMULATION

We looked at two formulations of the EVPs and the resulting modal model descriptions in the previous two subsections. One advantage of the first formulation is that it requires only one set of eigenvectors to describe the modal model instead of two sets, the right, and the left, as required by the second formulation. We use the first formulation and the associated modal model for the modal analysis of nonproportional viscously damped systems.

6.6.5 RESPONSE TO INITIAL CONDITIONS

Once the modal model is defined, we can determine through modal analysis the free vibration response of the system with nonproportional viscous damping subjected to an initial disturbance. Let Eqs. (6.67) and (6.68) define the initial conditions. The initial condition on the state vector is

$$\{z(0)\} = \begin{Bmatrix} x(0) \\ \dot{x}(0) \end{Bmatrix} = \begin{Bmatrix} x_0 \\ v_0 \end{Bmatrix} = \{z_0\} \tag{6.197}$$

TABLE 6.2

Modal Model Descriptions of an MDOF System with Nonproportional Viscous Damping

Based on the Eigenvalue Problem Formulation I

Method 1

Matrix of 2N complex eigenvalues:

$$[`\bar{s}_{\backprime}] = \begin{bmatrix} [`s_{r\backprime}] & [0] \\ [0] & [`s_{r\backprime}]^* \end{bmatrix} = \begin{bmatrix} s_1 & 0 & 0 & 0 & 0 & 0 \\ 0 & \ddots & 0 & 0 & 0 & 0 \\ 0 & 0 & s_N & 0 & 0 & 0 \\ 0 & 0 & 0 & s_1^* & 0 & 0 \\ 0 & 0 & 0 & 0 & \ddots & 0 \\ 0 & 0 & 0 & 0 & 0 & s_N^* \end{bmatrix}$$

where

$$s_r = -\xi_r\omega_r + i\omega_r\sqrt{1-\omega_r^2}$$

$$s_r^* = -\xi_r\omega_r - i\omega_r\sqrt{1-\xi_r^2}$$

with r = 1, 2,..., N

and

Matrix of normalized 2N complex eigenvectors:

$$[\bar{\phi}] = \begin{bmatrix} [\phi] & [\phi]^* \\ [\phi][`s_{r\backprime}] & [\phi]^*[`s_{r\backprime}]^* \end{bmatrix}$$

Method 2

N natural frequencies:

ξ_r

where

ω_r: r[th] natural frequency

ξ_r: r[th] modal viscous damping factor

N modal viscous damping factors: ξ_r

and

N eigenvectors corresponding to displacement coordinates:

$\{\phi\}_1, \{\phi\}_2, \ldots, \{\phi\}_N$

Based on the Eigenvalue Problem Formulation II

Matrix of 2N complex eigenvalues

$$[`\bar{s}_{\backprime}] = \begin{bmatrix} [`s_{r\backprime}] & [0] \\ [0] & [`s_{r\backprime}]^* \end{bmatrix} = \begin{bmatrix} s_1 & 0 & 0 & 0 & 0 & 0 \\ 0 & \ddots & 0 & 0 & 0 & 0 \\ 0 & 0 & s_N & 0 & 0 & 0 \\ 0 & 0 & 0 & s_1^* & 0 & 0 \\ 0 & 0 & 0 & 0 & \ddots & 0 \\ 0 & 0 & 0 & 0 & 0 & s_N^* \end{bmatrix}$$

and the right and the left normalized eigenvector matrices

$$[\bar{\phi}_R], [\bar{\phi}_L]$$

The free vibration response consists of a linear combination of the free vibration response due to 2N complex modes. Equation (6.129) is the free vibration response in a mode, and therefore a linear combination of the free vibration responses due to all the modes (using the normalized eigenvectors) is

$$\{z(t)\} = \sum_{j=1}^{2N} \{\bar{\phi}\}_j a_j e^{s_j t} \tag{6.198}$$

where s_j and $\{\bar{\phi}\}_j$ represent the jth eigenvalue and the corresponding eigenvector, respectively, based on the eigenvalue problem (formulation I), while a_j represents the participation of the jth mode in the free vibration response. a_j (j = 1,2,...2N) are the arbitrary constants to be determined from the initial conditions.

One approach could be to impose initial conditions on Eq. (6.198) and solve the resulting equations for the unknowns. However, we follow an approach using the orthogonality properties of the eigenvectors.

Applying the initial condition given by Eq. (6.197), we get from Eq. (6.198)

$$\{z_0\} = \sum_{j=1}^{2N} a_j \{\bar{\phi}\}_j \tag{6.199}$$

Pre-multiplying by $\{\bar{\phi}\}_r^T[A]$ and using the orthogonality properties (Eq. 6.154), we can solve for the unknowns as

$$a_r = \{\bar{\phi}\}_r^T[A]\{z_0\} \quad \text{for} \quad r = 1, 2, ..., 2N \tag{6.200}$$

Substituting Eq. (6.200) into Eq. (6.198), we get

$$\{z(t)\} = \sum_{j=1}^{2N} \{\bar{\phi}\}_j \{\bar{\phi}\}_r^T[A]\{z_0\}e^{s_j t} \tag{6.201}$$

which is the free vibration response. $\{z(t)\}$ consists of the displacement and velocity parts; hence, both these responses become known.

6.7 FORCED VIBRATION RESPONSE OF SYSTEMS WITH NONPROPORTIONAL VISCOUS DAMPING

The equation of motion of an MDOF system with viscous damping under the forced vibration is

$$[M]\{\ddot{x}\} + [C]\{\dot{x}\} + [K]\{x\} = \{F\} \tag{6.202}$$

where $[C]$ is a nonproportional damping matrix. The external forces are harmonic with frequency ω. The forced response consists of complementary and particular parts. We determine the particular part in the following two sections, while the complementary part and the total response are determined afterward.

6.7.1 BY DIRECT METHOD

Equations (6.87)–(6.92) are used to find the particular part of the response using the direct method. The only difference is that the nonproportional viscous damping matrix is used in place of the damping matrix in those equations.

6.7.2 BY MODAL ANALYSIS

We now look at how the particular part of the response can be determined using modal analysis. The first step in the analytical modal analysis is the coordinate transformation from the physical to the modal space. Note that, as mentioned in Section 6.6.1 d), the decoupling of the equations of motion for a system with nonproportional viscous damping cannot be done in displacement coordinates, and hence we start from the equations of motion in state space. The state-space equation for the free vibration is presented in Section 6.6.1 (Eq. 6.122). For the forced vibration, the steps in Eqs. (6.112)–(6.122) can be followed with one modification that now there is also a force vector $\{F\}$ to be considered in Eq. (6.113). With this change, the state-space equation for the forced vibration is obtained as

$$[A]\{\dot{z}\} + [B]\{z\} = \{\bar{F}\} \tag{6.203}$$

where $\{\bar{F}\}$ is the force vector in state space and is given by

$$\{\bar{F}\} = \left\{ \begin{array}{c} F \\ 0 \end{array} \right\} \tag{6.204}$$

while the matrices $[A]$ and $[B]$ remain the same as given in Eq. (6.123).

Let $\{z_p(t)\}$ is the particular part of the response. Consider transformation to modal coordinates using the normalized state-space eigenvectors

$$\{z_p(t)\} = [\bar{\phi}]\{\bar{y}(t)\} \tag{6.205}$$

where $\{\bar{y}(t)\}$ is a vector of modal responses. Substituting Eq. (6.205) into Eq. (6.203) and pre-multiplying by $[\bar{\phi}]^T$ gives

$$[\bar{\phi}]^T[A][\bar{\phi}]\{\dot{\bar{y}}\} + [\bar{\phi}]^T[B][\bar{\phi}]\{\bar{y}\} = [\bar{\phi}]^T\{\bar{F}\} \tag{6.206}$$

Using the orthogonality properties (Eqs. 6.150 and 6.151), we get

$$[I]\{\dot{\bar{y}}\} - [\,\diagdown\bar{s}_r\diagdown]\{\bar{y}\} = \{\bar{f}_r\}$$

(6.207)

where

$$\{\bar{f}_r\} = [\bar{\phi}]^T \{\bar{F}\}$$

(6.208)

is a vector of modal forces in state space, with the rth modal force given by

$$\bar{f}_r = \{\bar{\phi}\}_r^T \{\bar{F}\} \qquad\qquad r = 1, \ 2,..., \ 2N$$

(6.209)

Eq. (6.207) is a set of 2N decoupled first-order differential equations in modal state-space. It corresponds to the 2N complex conjugate modes of the system. Since each of these can be solved independently, we consider the rth equation

$$\dot{\bar{y}}_r - \bar{s}_r \bar{y}_r = \bar{f}_r$$

(6.210)

As the external force is harmonic, the modal forces are also harmonic, and hence from Eq. (6.209), we can write

$$\bar{f}_r(t) = \{\bar{\phi}\}_r^T \{\bar{F}_0\} e^{i\omega t} = \bar{f}_{r0} e^{i\omega t}$$

(6.211)

The corresponding modal response is also harmonic and is given by

$$\bar{y}_r(t) = \bar{y}_{r0} e^{i\omega t}$$

(6.212)

Substituting Eqs. (6.211) and (6.212) into Eq. (6.210), we get

$$\bar{y}_{r0} = \frac{\bar{f}_{r0}}{i\omega - \bar{s}_r}$$

(6.213)

$$\bar{y}_r(t) = \frac{\bar{f}_r(t)}{i\omega - \bar{s}_r}$$

(6.214)

Substituting Eq. (6.209) into Eq. (6.214) gives the modal response of the rth mode in state space

$$\bar{y}_r(t) = \frac{\{\bar{\phi}\}_r^T \{\bar{F}\}}{i\omega - \bar{s}_r}$$

(6.215)

and with $r = 1, \ 2,...,2N$, we get modal responses for the 2N modes in state space. Eq. (6.215) is substituted into Eq. (6.205) to obtain the physical response. From Eq. (6.205)

$$\{z_p(t)\} = \sum_{r=1}^{2N} \{\bar{\phi}\}_r \ \bar{y}_r(t)$$

(6.216)

$$\{z_p(t)\} = \sum_{r=1}^{2N} \frac{\{\bar{\phi}\}_r \{\bar{\phi}\}_r^T \{\bar{F}\}}{i\omega - \bar{s}_r}$$

(6.217)

Substituting $\{\bar{F}\} = \{\bar{F}_0\} e^{i\omega t}$ into Eq. (6.217), we get

$$\{z_p(t)\} = \left(\sum_{r=1}^{2N} \frac{\{\bar{\phi}\}_r \{\bar{\phi}\}_r^T \{\bar{F}_0\}}{i\omega - \bar{s}_r} \right) e^{i\omega t} \tag{6.218}$$

The term in the bracket represents the amplitude of the particular or steady-state part of the state-space response,

$$\{z_0\} = \sum_{r=1}^{2N} \frac{\{\bar{\phi}\}_r \{\bar{\phi}\}_r^T \{\bar{F}_0\}}{i\omega - \bar{s}_r} \tag{6.219}$$

$$\{z_0\} = \left(\sum_{r=1}^{N} \frac{\{\bar{\phi}\}_r \{\bar{\phi}\}_r^T}{i\omega - \bar{s}_r} + \sum_{r=N+1}^{2N} \frac{\{\bar{\phi}\}_r \{\bar{\phi}\}_r^T}{i\omega - \bar{s}_r} \right) \{\bar{F}_0\} \tag{6.220}$$

Because of the complex conjugate nature of the eigenvalues and eigenvectors, Eq. (6.220) can be further written as

$$\{z_0\} = \left(\sum_{r=1}^{N} \left(\frac{\{\bar{\phi}\}_r \{\bar{\phi}\}_r^T}{i\omega - s_r} + \frac{\{\bar{\phi}\}_r^* \{\bar{\phi}\}_r^{*T}}{i\omega - s_r^*} \right) \right) \{\bar{F}_0\} \tag{6.221}$$

Making substitution for $\{\bar{\phi}\}_r$ and the force vector $\{\bar{F}_0\}$ (Eq. 6.204), we get

$$\{z_0\} = \left(\sum_{r=1}^{N} \left(\frac{ \left\{ \begin{array}{c} \{\phi\}_r \\ s_r\{\phi\}_r \end{array} \right\} \left[\begin{array}{c} \{\phi\}_r \\ s_r\{\phi\}_r \end{array} \right]^T }{i\omega - s_r} + \frac{ \left\{ \begin{array}{c} \{\phi\}_r^* \\ s_r^*\{\phi\}_r^* \end{array} \right\} \left[\begin{array}{c} \{\phi\}_r^* \\ s_r^*\{\phi\}_r^* \end{array} \right]^T }{i\omega - s_r^*} \right) \right) \left\{ \begin{array}{c} F_0 \\ 0 \end{array} \right\} \tag{6.222}$$

$$\{z_0\} = \left(\sum_{r=1}^{N} \left(\frac{ \left[\begin{array}{cc} \{\phi\}_r \{\phi\}_r^T & s_r\{\phi\}_r\{\phi\}_r^T \\ s_r\{\phi\}_r\{\phi\}_r^T & s_r s_r\{\phi\}_r\{\phi\}_r^T \end{array} \right] }{i\omega - s_r} + \frac{ \left[\begin{array}{cc} \{\phi\}_r^* \{\phi\}_r^{*T} & s_r^*\{\phi\}_r^*\{\phi\}_r^{*T} \\ s_r^*\{\phi\}_r^*\{\phi\}_r^{*T} & s_r^* s_r^*\{\phi\}_r^*\{\phi\}_r^{*T} \end{array} \right] }{i\omega - s_r^*} \right) \right) \left\{ \begin{array}{c} F_0 \\ 0 \end{array} \right\} \tag{6.223}$$

$$\{z_0\} = \sum_{r=1}^{N} \left(\frac{ \left[\begin{array}{c} \{\phi\}_r \{\phi\}_r^T \{F_0\} \\ s_r\{\phi\}_r\{\phi\}_r^T \{F_0\} \end{array} \right] }{i\omega - s_r} + \frac{ \left[\begin{array}{c} \{\phi\}_r^* \{\phi\}_r^{*T} \{F_0\} \\ s_r^*\{\phi\}_r^*\{\phi\}_r^{*T} \{F_0\} \end{array} \right] }{i\omega - s_r^*} \right) \tag{6.224}$$

The first half of the vectors on the RHS corresponds to the displacement response, while the second half corresponds to the velocity response. Therefore, the displacement response is given by

$$\{x_0\} = \sum_{r=1}^{N} \left(\frac{\{\phi\}_r \{\phi\}_r^T \{F_0\}}{i\omega - s_r} + \frac{\{\phi\}_r^* \{\phi\}_r^{*T} \{F_0\}}{i\omega - s_r^*} \right) \tag{6.225}$$

$$\{x_0\} = \left(\sum_{r=1}^{N} \left(\frac{\{\phi\}_r \{\phi\}_r^T}{i\omega - s_r} + \frac{\{\phi\}_r^* \{\phi\}_r^{*T}}{i\omega - s_r^*} \right) \right) \{F_0\} \tag{6.226}$$

6.7.3 TOTAL RESPONSE

The general solution of the state equation is

$$\{z(t)\} = \{z_c(t)\} + \{z_p(t)\} \tag{6.227}$$

The complementary part $\{z_c(t)\}$ is a linear combination of the free vibration response due to 2N modes and is given by Eq. (6.198).

Assuming the actual force to be $\{F_0\}\cos\omega t$, which is nothing but the real part of the force F, $\{z_p(t)\}$ to be used in Eq. (6.227) has to be the real part of Eq. (6.218). Note that since the eigenvectors are complex, the phase angle for the particular part of the response, in general, is different at different DOFs. The total response is

$$\{z(t)\} = \sum_{j=1}^{2N} \{\overline{\phi}\}_j \, a_j \, e^{s_j t} + \mathrm{Re}\left(\left(\sum_{r=1}^{2N} \frac{\{\overline{\phi}\}_r \{\overline{\phi}\}_r^T \{\overline{F}_0\}}{i\omega - \overline{s}_r}\right) e^{i\omega t}\right) \tag{6.228}$$

a_j^s are arbitrary constants and can be found from the given initial conditions.

6.8 FRF MATRIX

The FRF matrix for an undamped MDOF system was introduced in Chapter 5, Section 5.7, and an FRF α_{jk} was defined. We also saw how the FRF matrix is related to the system matrices and the modal model. For the damped MDOF systems, we looked at the relationships between the FRF matrix and the system matrices for systems with structural and viscous damping in the previous sections. In this section, we present the relationships between the FRF matrix and the modal models of the damped systems.

6.8.1 SYSTEM WITH PROPORTIONAL STRUCTURAL DAMPING

If we compare the steady-state response amplitude given by Eqs. (6.31) and (6.45), then the coefficient matrices of the force vector in the two equations must be equal. The equality gives the FRF matrix

$$[\alpha] = \sum_{r=1}^{N} \frac{\{\phi\}_r \{\phi\}_r^T}{\omega_r^2 - \omega^2 + i\eta_r\omega_r^2} = [\phi]\left[\,^\backprime\omega_r^2 - \omega^2 + i\eta_r\omega_{r\backprime}^2\,\right]^{-1}[\phi]^T \tag{6.229}$$

A particular FRF, α_{jk}, can be extracted from Eq. (6.229) as

$$\alpha_{jk}(\omega) = \sum_{r=1}^{N} \frac{\phi_{jr}\,\phi_{kr}}{\omega_r^2 - \omega^2 + i\eta_r\omega_r^2} \tag{6.230}$$

6.8.2 SYSTEM WITH NONPROPORTIONAL STRUCTURAL DAMPING

If we compare the steady-state response amplitude given by Eqs. (6.31) and (6.48), then the coefficient matrices of the force vector in the two equations must be equal. The equality gives the FRF matrix

$$[\alpha] = \sum_{r=1}^{N} \frac{\{\phi\}_r \{\phi\}_r^T}{\omega_r'^2 - \omega^2 + i\eta_r\omega_r'^2} = [\phi]\left[\,^\backprime\omega_r'^2 - \omega^2 + i\eta_r\omega_{r\backprime}'^2\,\right]^{-1}[\phi]^T \tag{6.231}$$

A particular FRF, α_{jk}, can be extracted from Eq. (6.231) as

$$\alpha_{jk}(\omega) = \sum_{r=1}^{N} \frac{\phi_{jr}\,\phi_{kr}}{\omega_r'^2 - \omega^2 + i\eta_r\omega_r'^2} \tag{6.232}$$

6.8.3   System with Proportional Viscous Damping

If we compare the steady-state response amplitude given by Eqs. (6.91) and (6.107), then the coefficient matrices of the force vector in the two equations must be equal. The equality gives the FRF matrix

$$[\alpha] = \sum_{r=1}^{N} \frac{\{\phi\}_r \{\phi\}_r^T}{\omega_r^2 - \omega^2 + i2\xi_r\omega_r\omega} = [\phi] \left[\,`\omega_r^2 - \omega^2 + i2\xi_r\omega_r\omega \, \right]^{-1} [\phi]^T \qquad (6.233)$$

A particular FRF, α_{jk}, can be extracted from Eq. (6.233) as,

$$\alpha_{jk}(\omega) = \sum_{r=1}^{N} \frac{\phi_{jr}\ \phi_{kr}}{\omega_r^2 - \omega^2 + i2\xi_r\omega_r\omega} \qquad (6.234)$$

6.8.4   System with Nonproportional Viscous Damping

If we compare the steady-state response amplitude given by Eqs. (6.91) and (6.226), then the coefficient matrices of the force vector in the two equations must be equal. The equality gives the FRF matrix

$$[\alpha] = \sum_{r=1}^{N} \left(\frac{\{\phi\}_r \{\phi\}_r^T}{i\omega - s_r} + \frac{\{\phi\}_r^* \{\phi\}_r^{*T}}{i\omega - s_r^*} \right) = [\phi] \left[\,`i\omega - s_r \, \right]^{-1} [\phi]^T + [\phi]^* \left[ \,`i\omega - s_r^* \, \right]^{-1} [\phi]^{*T} \qquad (6.235)$$

A particular FRF, α_{jk} can be extracted from Eq. (6.235) as

$$\alpha_{jk}(\omega) = \sum_{r=1}^{N} \left(\frac{\phi_{jr}\ \phi_{kr}}{i\omega - s_r} + \frac{\phi_{jr}^*\ \phi_{kr}^*}{i\omega - s_r^*} \right) \qquad (6.236)$$

6.9   RESPONSE MODEL (FREQUENCY DOMAIN)

The response to harmonic forces can be determined once the FRF matrix for a system is known, as discussed in Section 5.9 (Chapter 5). Thus, the FRF matrices in Eqs. (6.229) and (6.231) represent the frequency domain response models for the systems with proportional and nonproportional structural damping, respectively. Similarly, the FRF matrices in Eqs. (6.233) and (6.235) represent the frequency domain response models for the systems with proportional and nonproportional viscous damping, respectively. The response of the damped systems to harmonic forces can be found at any DOF using Eq. (5.118) (Chapter 5). The response to periodic and aperiodic forces can be determined using the Fourier series, and Fourier transform presented in Chapter 8.

6.10   RESPONSE TO TRANSIENT EXCITATION BY MODAL ANALYSIS

In this section, we analyze the response of viscously damped systems subjected to transient excitation. The solution consists of complementary and particular parts. We obtain below the particular part of the response using modal analysis. The total response, including the complementary part, can be obtained similarly to Section 6.5.3.

6.10.1   System with Proportional Viscous Damping

The equation of motion is given by Eq. (6.85), with the vector $\{F\}$ representing the vector of transient forces acting at N DOFs of the system. We transform the equations to the modal domain as given by Eqs. (6.93)–(6.98), with Eq. (6.98) representing the equilibrium equation for the rth mode. The modal force for the rth mode, f_r, is now a transient force. Since the equations in the modal domain are decoupled, the equation for any particular mode, like Eq. (6.98), can be solved independently (like the solution of an SDOF system).

   The response of an SDOF system with viscous damping subjected to transient excitation can be found using its impulse response function (IRF) and the convolution integral. Assuming that the system is underdamped, the IRF for the vibration in r^{th} mode can be written using the result in Eq. (4.112) (Chapter 4) as

$$h_r(t) = \frac{e^{-\xi_r\omega_r t}}{\omega_r\sqrt{1-\xi_r^2}} \sin \omega_r\sqrt{1-\xi_r^2}\,t \qquad (\text{for } t > 0) \qquad (6.237)$$

Using convolution integral, Eq. (4.117), modal response in the rth mode is given by

$$y_r(t) = \int_0^t h_r(t-\tau)f_r(\tau)\,d\tau \tag{6.238}$$

The response in physical coordinates can be obtained by

$$\{x_p(t)\} = [\phi]\{y(t)\} \tag{6.239}$$

$$\{x_p(t)\} = \sum_{r=1}^N \{\phi\}_r\, y_r(t) \tag{6.240}$$

Making substitutions for $y_r(t)$ and $h_r(t)$ from Eqs. (6.237) and (6.238), respectively, we get

$$\{x_p(t)\} = \sum_{r=1}^N \{\phi\}_r \left(\int_0^t \frac{e^{-\xi_r\omega_r(t-\tau)}}{\omega_r\sqrt{1-\xi_r^2}} \sin\omega_r\sqrt{1-\xi_r^2}\,(t-\tau)f_r(\tau)\,d\tau \right) \tag{6.241}$$

Further, by writing the modal force as, $f_r(\tau) = \{\phi\}_r^T\{F(\tau)\}$, we get a consolidated expression for the particular part of the response to the transient excitation,

$$\{x_p(t)\} = \sum_{r=1}^N \{\phi\}_r \left(\int_0^t \frac{e^{-\xi_r\omega_r(t-\tau)}}{\omega_r\sqrt{1-\xi_r^2}} \sin\omega_r\sqrt{1-\xi_r^2}\,(t-\tau)\{\phi\}_r^T\{F(\tau)\}\,d\tau \right) \tag{6.242}$$

$$\{x_p(t)\} = \sum_{r=1}^N \frac{1}{\omega_r\sqrt{1-\xi_r^2}} \{\phi\}_r\{\phi\}_r^T \left(\int_0^t e^{-\xi_r\omega_r(t-\tau)} \sin\omega_r\sqrt{1-\xi_r^2}\,(t-\tau)\{F(\tau)\}\,d\tau \right) \tag{6.243}$$

6.10.2　System with Nonproportional Viscous Damping

We use the state-space equation (Eq. 6.203) for the forced response analysis of systems with nonproportional viscous damping. The force vector $\{F\}$ in Eq. (6.204) is now a vector of transient forces acting at N DOFs of the system.

We transform the state-space equation to the modal domain as indicated by Eqs. (6.205)–(6.210). Equation (6.210) represents the rth modal equilibrium equation with $\bar{f}_r$ being the corresponding modal force, which now is a transient force. The solution of this equation can be done using the convolution integral, but that requires the knowledge of the unit impulse response. Therefore, we determine the unit impulse response corresponding to Eq. (6.210). To obtain this, replace $\bar{f}_r$ by the Dirac-delta forcing function $\delta(t)$, leading to the following equation:

$$\dot{\bar{y}}_r - \bar{s}_r\bar{y}_r = \delta(t) \tag{6.244}$$

Note that the over-bar over the variables indicates that they are related to the state-space model of the system. We use the Laplace transform method to solve Eq. (6.224). Using the results of the Laplace transform given in Chapter 2, the Laplace transform of this equation, with zero initial conditions, gives

$$s\bar{y}_r(s) - \bar{s}_r\bar{y}_r(s) = 1 \tag{6.245}$$

$$\bar{y}_r(s) = \frac{1}{s - \bar{s}_r} \tag{6.246}$$

Taking the inverse Laplace transform gives the time-domain response. The Laplace inverse of $\bar{y}_r(s)$ is denoted by $\bar{h}_r(t)$ since it is the unit impulse response.

$$\bar{h}_r(t) = e^{\bar{s}_r t} \tag{6.247}$$

Now we can write the solution to Eq. (6.210) using the convolution integral as

$$\bar{y}_r(t) = \int_0^t \bar{h}_r(t-\tau)\bar{f}_r(\tau)\,d\tau \tag{6.248}$$

Eq. (6.248) is substituted into Eq. (6.205) to obtain the response in state space. Equation (6.205) is

$$\{z_p(t)\} = \sum_{r=1}^{2N}\{\bar{\phi}\}_r\,\bar{y}_r(t) \tag{6.249}$$

Making substitutions for $\bar{y}_r(t)$ and $\bar{h}_r(t)$ from Eqs. (6.248) and (6.247), respectively, we get

$$\{z_p(t)\} = \sum_{r=1}^{2N}\{\bar{\phi}\}_r\left(\int_0^t e^{\bar{s}_r(t-\tau)}\,\bar{f}_r(\tau)\,d\tau\right) \tag{6.250}$$

Further, from Eq. (6.209), the modal force is written as $\bar{f}_r = \{\bar{\phi}\}_r^T\{\bar{F}\}$, leading to

$$\{z_p(t)\} = \sum_{r=1}^{2N}\{\bar{\phi}\}_r\{\bar{\phi}\}_r^T\left(\int_0^t e^{\bar{s}_r(t-\tau)}\{\bar{F}(\tau)\}\,d\tau\right) \tag{6.251}$$

Making substitution for $\{\bar{\phi}\}_r$ and the force vector $\{\bar{F}(\tau)\}$, we get

$$\{z_p(t)\} = \sum_{r=1}^{2N}\left\{\begin{array}{c}\{\phi\}_r \\ s_r\{\phi\}_r\end{array}\right\}\left\{\begin{array}{c}\{\phi\}_r \\ s_r\{\phi\}_r\end{array}\right\}^T\left(\int_0^t e^{\bar{s}_r(t-\tau)}\left\{\begin{array}{c}F(\tau) \\ 0\end{array}\right\}\,d\tau\right) \tag{6.252}$$

$$\{z_p(t)\} = \sum_{r=1}^{2N}\left[\begin{array}{cc}\{\phi\}_r\{\phi\}_r^T & s_r\{\phi\}_r\{\phi\}_r^T \\ s_r\{\phi\}_r\{\phi\}_r^T & s_r s_r\{\phi\}_r\{\phi\}_r^T\end{array}\right]\left\{\begin{array}{c}\int_0^t e^{\bar{s}_r(t-\tau)}F(\tau)\,d\tau \\ 0\end{array}\right\} \tag{6.253}$$

$$\{z_p(t)\} = \sum_{r=1}^{2N}\left\{\begin{array}{c}\{\phi\}_r\{\phi\}_r^T\int_0^t e^{\bar{s}_r(t-\tau)}\{F(\tau)\}\,d\tau \\ \\ s_r\{\phi\}_r\{\phi\}_r^T\int_0^t e^{\bar{s}_r(t-\tau)}\{F(\tau)\}\,d\tau\end{array}\right\} \tag{6.254}$$

Since $\{z_p(t)\}$ consists of the displacement and velocity vectors, the displacement response to the transient excitation is obtained as

$$\{x_p(t)\} = \sum_{r=1}^{2N}\{\phi\}_r\{\phi\}_r^T\int_0^t e^{\bar{s}_r(t-\tau)}\{F(\tau)\}\,d\tau \tag{6.255}$$

6.11   IMPULSE RESPONSE FUNCTION

We considered the impulse response of undamped MDOF systems in Chapter 5. There we defined the IRF $h_{jk}(t)$ and also saw that for an MDOF system, there would be a matrix of impulse responses, the IRF matrix, $[h(t)]$. In this section, we determine the expression for the impulse response $h_{jk}(t)$ for an MDOF system with viscous damping. We apply a unit impulse at the kth DOF and find the response at the jth DOF. As explained in Section 5.12, to realize unit impulse, we apply a force at the kth DOF in the form of the Dirac-delta function with no force at the remaining DOFs. Thus, we chose the force vector as

$$\{F(\tau)\} = \{0, \ 0, \ ..., \ F_k(\tau), \ ...,0\}^T \tag{6.256}$$

with

$$F_k(\tau) = \delta(\tau) \tag{6.257}$$

6.11.1 SYSTEM WITH PROPORTIONAL VISCOUS DAMPING

For this case, we derived a general expression for response to transient excitation (Eq. 6.243). Rewriting that equation

$$\{x_p(t)\} = \sum_{r=1}^{N} \frac{1}{\omega_r\sqrt{1-\xi_r^2}}\{\phi\}_r\left(\int_0^t e^{-\xi_r\omega_r(t-\tau)}\sin\omega_r\sqrt{1-\xi_r^2}(t-\tau)\{\phi\}_r^T\{F(\tau)\}\ d\tau\right) \tag{6.258}$$

Substituting Eqs. (6.256) and (6.257), and noting that this makes $\{\phi\}_r^T\{F(\tau)\} = \phi_{kr}\delta(\tau)$, we get

$$\{x_p(t)\} = \sum_{r=1}^{N} \frac{1}{\omega_r\sqrt{1-\xi_r^2}}\{\phi\}_r\left(\int_0^t e^{-\xi_r\omega_r(t-\tau)}\sin\omega_r\sqrt{1-\xi_r^2}(t-\tau)\phi_{kr}\delta(\tau)\ d\tau\right) \tag{6.259}$$

The integral in the bracket is equal to the integrand value at $\tau = 0$. Also, the vector $\{x_p(t)\}$ being the response to a unit impulse at the kth DOF represents the kth column of the IRF matrix (denoted by $\{h(t)\}^k$). Given these observations, Eq. (6.259) reduces to

$$\{h(t)\}^k = \sum_{r=1}^{N} \frac{e^{-\xi_r\omega_r t}}{\omega_r\sqrt{1-\xi_r^2}}\{\phi\}_r\ \phi_{kr}\sin\omega_r\sqrt{1-\xi_r^2}t \tag{6.260}$$

With $k = 1, 2,..., N$, Eq. (6.260) yields all the columns of the IRF matrix. The desired IRF ($h_{jk}(t)$) is the jth element of $\{h(t)\}^k$ and is given by

$$h_{jk}(t) = \sum_{r=1}^{N} \frac{e^{-\xi_r\omega_r t}}{\omega_r\sqrt{1-\xi_r^2}}\phi_{jr}\phi_{kr}\sin\omega_r\sqrt{1-\xi_r^2}t \tag{6.261}$$

6.11.2 SYSTEM WITH NONPROPORTIONAL VISCOUS DAMPING

For this case, we derived a general expression for response to transient excitation (Eq. 6.255). Rewriting that equation

$$\{x_p(t)\} = \sum_{r=1}^{2N}\{\phi\}_r\int_0^t e^{\bar{s}_r(t-\tau)}\{\phi\}_r^T\{F(\tau)\}d\tau \tag{6.262}$$

Substituting Eqs. (6.256) and (6.257), and noting that this makes $\{\phi\}_r^T\{F(\tau)\} = \phi_{kr}\tau(\tau)$, we get

$$\{x_p(t)\} = \sum_{r=1}^{2N}\{\phi\}_r\int_0^t e^{\bar{s}_r(t-\tau)}\phi_{kr}\delta(\tau)d\tau \tag{6.263}$$

As previously noted, the integral in the bracket is equal to the integrand value at $\tau = 0$. Moreover, due to a unit impulse at the kth DOF, the response $\{x_p(t)\}$ represents the kth column of the IRF matrix. Thus, we have

$$\{h(t)\}^k = \sum_{r=1}^{2N}\{\phi\}_r\ \phi_{kr}e^{\bar{s}_r t} \tag{6.264}$$

With k = 1, 2,..., N, Eq. (6.244) yields all the columns of the IRF matrix. The IRF $h_{jk}(t)$ is obtained as

$$h_{jk}(t) = \sum_{r=1}^{2N} \phi_{jr}\phi_{kr}e^{\bar{s}_r t} \qquad (6.265)$$

Because of the complex conjugate nature of the eigenvalues and the corresponding eigenvectors, Eq. (2.65) can also be written in the following form:

$$h_{jk}(t) = \sum_{r=1}^{N}\left(\phi_{jr}\phi_{kr}e^{s_r t} + \phi_{jr}^{*}\phi_{kr}^{*}e^{s_r^{*}t}\right) \qquad (6.266)$$

From the IRF expressions, it is clear that, like the FRFs, the IRFs are also related to the corresponding modal models.

6.12   RESPONSE MODEL (TIME DOMAIN)

The system response to the given input forces can be determined if the IRF matrix $[h(t)]$ is available. The response at any system DOF due to the force at various DOFs can be determined using the convolution integral and the relevant IRFs. All such responses can be added to find the total response of the system (using Eq. 5.135, Chapter 5). Therefore, the IRF matrix $[h(t)]$ constitutes the time-domain response model of the MDOF systems.

Example 6.1

The mass and stiffness matrices of a lumped parameter model of a structure with proportional hysteretic damping are given below. The mass and stiffness proportionality constants for the damping matrix are $\beta_k = 0.01$ and $\beta_m = 1,500$, respectively. Find the natural frequencies, modal loss factors, and mass-normalized mode shapes using a MATLAB® program.

MASS	MATRIX:	STIFFNESS	MATRIX :
6	0	600,000	−400,000
0	7	−400,000	1,100,000

Solution

MATLAB program 6.1 is used to solve the problem. The EVP is solved using the complex stiffness matrix. Note that the eigenvectors obtained are real due to the proportional nature of the damping.

```
%###############################
%   MATLAB PROGRAM 6.1
%###############################
clear all;

%DEFINE MASS MATRIX
m = [6  0
     0  7];

%DEFINE STIFFNESS MATRIX
k = [600,000    −400,000
     −400,000   1,100,000];

%FIND PROPORTIONAL DAMPING MATRIX
d = 0.01*k + 1,500*m;

%FIND COMPLEX STIFFNESS MATRIX
k_cmplx = k + i*d;
```

Results:

NF_Hz:
39.16   70.56

modal_loss_factors:

0.0348   0.0176

Mass-normalized mode shapes:

0.3440 − 0.0000i −0.2198 − 0.0000i
0.2035 + 0.0000i   0.3185 − 0.0000i

(*Continued*)

```
%SOLVE EVP
[Evec,Eval]=eig(k_cmplx,m);

%FIND NATURAL FREQUENCY IN rad/sec
NF_rad_sec=sqrt(real(diag(Eval)))

%FIND NATURAL FREQUENCY IN Hz
NF_Hz=NF_rad_sec/(2*pi)

%FIND MODAL LOSS FACTORS
modal_loss_factor=imag(diag(Eval))./real(diag(Eval))

%FIND MODAL MASS MATRIX
mr=Evec.'*m*Evec;

%FIND MASS-NORMALIZED EIGENVECTORS
for p=1:2
    Evec_Nor(:,p)=Evec(:,p)/sqrt(mr(p,p));
end
Evec_Nor
%############################
```

Example 6.2

In Example 6.1, if the damping in the system is described by the matrix given below, then using a MATLAB program, find the eigenvalues and eigenvectors. What do you infer from the nature of the eigenvectors?

DAMPING	MATRIX :
15,000	−4,000
−4,000	172,000

Solution

MATLAB program 6.1 is used to solve the problem by replacing the proportional damping matrix with the damping matrix of the present example. In addition, we also find the phase of the eigenvector elements using the 'angle()' command in MATLAB. The results obtained are given.

The relative phase angles are neither 0° nor 180°, indicating that the eigenvectors of the system are complex. It is a reflection of the nonproportional damping in the system.

```
Results:

Eigenvalues
1.0e+05 *
   0.6126 + 0.0829i   0.0000 + 0.0000i
   0.0000 + 0.0000i   1.9589 + 0.1879i

Mass-normalized eigenvectors
   0.3461 + 0.0157i   −0.2185 + 0.0248i
   0.2023 − 0.0230i    0.3204 + 0.0145i

Phase of the eigenvectors (in degrees)
   2.5899  173.5234
  −6.4766    2.5899
```

Example 6.3

The mass, stiffness, and damping matrices of a 3-DOF system with viscous damping are given below. Using a MATLAB program, verify whether the damping is proportional or nonproportional.

MASS MATRIX:			STIFFNESS MATRIX :			VISCOUS DAMPING MATRIX:		
2	0	0	260,000	−140,000	0	31.2000	−16.8000	0
0	1	0	−140,000	150,000	−80,000	−16.8000	18.0000	−9.6000
0	0	4	0	−80,000	180,000	0	−9.6000	21.6000

Solution

MATLAB program 6.2 is used to solve the problem. A proportional damping matrix is diagonalized by the eigenvectors of the corresponding undamped model. We use this fact to determine the nature of the damping matrix. We solve the EVP of the corresponding undamped system. We see from the results that the damping matrix in the modal domain is diagonal, which indicates that the given damping matrix is proportional.

```
%###########################          Results:
%   MATLAB PROGRAM 6.2
%###########################          Modal damping matrix
clear all;                            cr =

%DEFINE MASS MATRIX                       1.6766       0  −0.0000
m = [2  0  0                             0.0000   8.0430   0.0000
0 1 0                                   −0.0000   0.0000  29.2803
0 0 4 ];

%DEFINE STIFFNESS MATRIX
k = [260,000  −140,000  0
  −140,000  150,000  −80,000
      0  −80,000  180,000 ];

%DEFINE DAMPING MATRIX
c = [31.2000  −16.8000      0
  −16.8000  18.0000  −9.6000
      0  −9.6000  21.6000];

%SOLVE EVP OF THE UNDAMPED SYSTEM
[Evec,Eval]=eig(k,m);

%FIND MODAL DAMPING MATRIX
cr=Evec'*c*Evec
%###########################
```

Example 6.4

The following matrices describe a two-DOF system with nonproportional viscous damping. Find the eigenvalues and the normalized eigenvectors. Also, find the natural frequencies, modal damping factors, and mass-normalized mode shapes corresponding to the displacement coordinates. Use a MATLAB program.

MASS MATRIX:		STIFFNESS	MATRIX :	VISCOUS DAMPING MATRIX:	
1	0	80,000	−20,000	38.4	−2.4
0	3	−20,000	50,000	−2.4	6.0

Solution

MATLAB program 6.3 is used to solve the problem. Since the damping matrix is nonproportional, we solve the state-space-based EVP of the system. Since the DOF of the system is 2, there are four complex eigenvalues and eigenvectors (in complex conjugate pairs), which form the modal model of the system.

```
%##############################
%    MATLAB PROGRAM 6.3
%##############################
clear all;

%DEFINE MASS MATRIX
m = [1  0
     0  3];

%DEFINE STIFFNESS MATRIX
k=[80,000    −20,000
   −20,000    50,000];

%DEFINE DAMPING MATRIX
c = [38.4000  −2.4000
     −2.4000   6.0000];

%FIND SYSTEM DOF
N = size(m,1);

%FIND STATE SPACE MATRIX A
a = [c m;
     m zeros(size(m))];

%FIND STATE SPACE MATRIX B
b = [k zeros(size(m));
     zeros(size(m)) -m];

%SOLVE EIGENVALUE PROBLEM
[Evec,Eval]=eig(b,-a);
Eval=diag(Eval)

%FIND MODAL A AND B MATRICES
ar=Evec.'*a*Evec;
br=Evec.'*b*Evec;

%NORMALIZE THE EIGENVECTORS
for p=1:2*N
    Evec_nor(:,p)=Evec(:,p)/sqrt(max(ar(:,p)));
end
Evec_nor

%FIND NATURAL FREQUENCY IN rad/sec
for p=1:2*N
 NF_rad_sec(p)=sqrt(real(Eval(p))^2 + imag(Eval(p))^2);
end
%FIND MODAL VISCOUS DAMPING FACTOR
for p=1:2*N
 damping_factor(p)=abs(real(Eval(p)))/NF_rad_sec(p);
end

%EXTRACT N FREQUENCIES AND DAMPING FACTORS
NF_rad_sec=NF_rad_sec(1:2:2*N)
NF_Hz=NF_rad_sec/(2*pi)
damping_factor=damping_factor(1:2:2*N)

%EXTRACT DISPLACEMENT PART OF THE EIGENVECTORS
X_Evec=Evec_nor(1:N,1:2:2*N)
%##############################=
```

Results:

Complex eigenvalues
Eval =
  1.0e+02 *
 −0.1889 + 2.8575i   −0.1889 − 2.8575i   −0.0131 + 1.2096i   −0.0131 − 1.2096i

Normalized complex eigenvectors
Evec_nor =

  0.0292 − 0.0290i   0.0292 + 0.0290i   0.0075 − 0.0083i   0.0075 + 0.0083i
 −0.0026 + 0.0033i −0.0026 − 0.0033i   0.0259 − 0.0258i   0.0259 + 0.0258i
  7.7432 + 8.9050i   7.7432 − 8.9050i   0.9922 + 0.9172i   0.9922 − 0.9172i
 −0.8918 − 0.8023i −0.8918 + 0.8023i   3.0886 + 3.1661i   3.0886 − 3.1661i

NF_rad_sec
 286.3740 120.9643

NF_Hz
 45.5778   19.2521

damping_factor
 0.0660   0.0108

Displacement part of the eigenvectors
 0.0292 − 0.0290i   0.0075 − 0.0083i
−0.0026 + 0.0033i   0.0259 − 0.0258i

Example 6.5

The modal model of a 3-DOF system with structural damping is given below. Using a MATLAB program, find and plot the FRF α_{32}. Also, find the steady-state response amplitude if a harmonic force of amplitude 15 N at 7 Hz acts at the second DOF.

NATURAL FREQUENCIES (Hz)	MODAL LOSS FACTORS	MASS-NORMALIZED MODE SHAPES		
78.6172	0.0071	0.4584	0.4272	0.3277
41.2040	0.0234	−0.7465	0.3844	0.5431
18.8125	0.1084	0.0750	−0.3490	0.3501

Solution

MATLAB program 6.4 is used to solve the problem. The FRF is computed using Eq. (6.232). The results show the plot of the FRF (Figure 6.1). The steady-state response amplitude is complex since the FRF is also complex.

```
%###############################
%   MATLAB PROGRAM 6.4
%###############################
clear all;

%DEFINE NATURAL FREQUENCIES IN Hz
NF_Hz = [ 78.6172  41.2040  18.8125];

%DEFINE MODAL LOSS FACTORS
modal_loss_factor = [ 0.0071  0.0234  0.1084];
```

(*Continued*)

```
%DEFINE MASS-NORMALIZED EIGENVECTORS
Evec_nor=[
  0.4584   0.4272   0.3277
 -0.7465   0.3844   0.5431
  0.0750  -0.3490   0.3501];

%FIND NUMBER OF MODES
N=length(NF_Hz);

%SELECT FREQUENCY RANGE AND RESOLUTION (IN HZ) FOR FRF
fstart=0;
fend=150;
df=0.5;

%GENERTAE FREQUENCY VECTOR
fvec=fstart:df:fend;

%SELECT RESPONSE AND EXCITATION DOFs FOR THE DESIRED FRF
j = 3;
q = 2;

%COMPUTE FRF FROM THE MODAL MODEL
for p=1:length(fvec)
   sum=0;
   for r=1:N
     Num=Evec_nor(j,r)*Evec_nor(q,r);
Den=((2*pi*NF_Hz(r))^2)*(1+i*modal_loss_factor(r))-(2*pi*fvec(p))^2;
     sum= sum + Num/Den;
   end
   FRF(p)=sum;
end

%FIND FRF MAGNITUDE IN dB
FRF_Mag_dB=20*log10(abs(FRF));

%PLOT THE FRF
figure;
plot(fvec,FRF_Mag_dB,'LineWidth',4,'Color','b');
grid on;
xlabel('Frequency (Hz)');
ylabel('FRF Mag dB (Re:1m/N)');
title('FRF(3,2)');
set(gca,'FontSize',16);
set(gca,'FontWeight','bold');
set(gca,'FontName','calibri');

%EXTRACT FRF AT 7 Hz
FRF_Value=FRF(fvec==7)

%DEFINE FORCE AMPLITUDE
Force_amp=15;

%FIND STAEDY_STATE RESPONSE AMPLITUDE
X=FRF_Value*Force_amp
%#############################
```

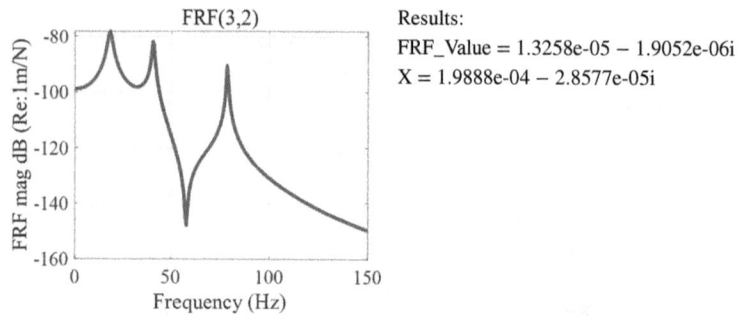

Results:
FRF_Value = 1.3258e-05 − 1.9052e-06i
X = 1.9888e-04 − 2.8577e-05i

FIGURE 6.1   FRF.

Example 6.6

For the two-DOF system with nonproportional viscous damping in Example 6.4, find and plot the IRF h_{21} computed from the modal model of the system. Use a MATLAB program.

Solution

MATLAB program 6.5 is used to solve the problem. The modal model obtained in Example 6.4 is treated as input. The IRF is computed using Eq. (6.265). The results show the plot of the IRF (Figure 6.2).

```
%###############################
%   MATLAB PROGRAM 6.5
%###############################
clear all;

%DEFINE COMPLEX EIGENVALUES
Eval=1.0e+02 *[
 −0.1889 + 2.8575i  −0.1889 − 2.8575i  −0.0131 + 1.2096i
 −0.0131 − 1.2096i];

%DEFINE NORMALIZED COMPLEX EIGENVECTOR MATRIX
Evec_nor=[
  0.0292 − 0.0290i   0.0292 + 0.0290i   0.0075 − 0.0083i   0.0075 + 0.0083i
 −0.0026 + 0.0033i  -0.0026 − 0.0033i   0.0259 − 0.0258i   0.0259 + 0.0258i
  7.7432 + 8.9050i   7.7432 − 8.9050i   0.9922 + 0.9172i   0.9922 − 0.9172i
 −0.8918 − 0.8023i  −0.8918 + 0.8023i   3.0886 + 3.1661i   3.0886 − 3.1661i];

%FIND NUMBER OF COMPLEX MODES
N=length(Eval);

%SELECT LENGTH OF TIME AND RESOLUTION (IN Sec) FOR IRF
tstart=0;
tend=4;
dt=0.001;

%GENERTAE TIME VECTOR
tvec=tstart:dt:tend;

%SELECT RESPONSE AND EXCITATION DOFs FOR THE DESIRED IRF
j=2;
q=1;

%COMPUTE IRF FROM THE MODAL MODEL
for p=1:length(tvec)
   sum=0;
   for r=1:N
```

(Continued)

```
        sum= sum + Evec_nor(j,r)*Evec_nor(q,r)*exp(Eval(r)*tvec(p));
    end
    IRF(p)=sum;
end

%PLOT THE IRF
figure;
plot(tvec,IRF,'LineWidth',1.5,'Color','r');
grid on;
xlabel('Time (Sec)');
ylabel('IRF (m)');
title('IRF(2,1)');

set(gca,'FontSize',16);
set(gca,'FontWeight','bold');
set(gca,'FontName','calibri');
%###############################
```

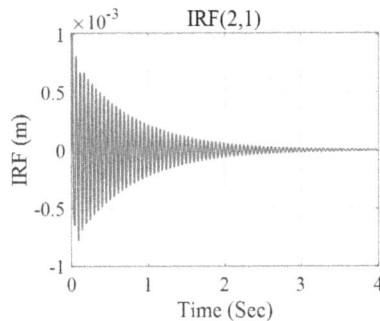

FIGURE 6.2   IRF.

REVIEW QUESTIONS

1. State the eigenvalue problem of an MDOF system with structural damping.
2. What do you mean by modal damping coefficient?
3. What are proportional and nonproportional damping?
4. How does the proportional or nonproportional nature of damping in a system affect the eigenvectors of the system?
5. What is the dynamic stiffness matrix? How is it related to the FRF matrix?
6. Show that the eigenvectors for an MDOF system with proportional viscous damping are the same as those of the corresponding undamped system.
7. State the modal model of an MDOF system with structural damping.
8. Obtain the EVP of an MDOF system with nonproportional viscous damping in state-space.
9. State the modal model of an MDOF system with nonproportional viscous damping.
10. What is a complex mode? How can it be established whether a mode is real or complex?
11. State the IRF $h_{jk}(t)$ for an MDOF system with nonproportional viscous damping.

PROBLEMS

Problem 6.1: The mass and stiffness matrices of a two-DOF system with viscous damping are given below. The damping matrix is proportional to the mass matrix with the constant of proportionality equal to 12. Find the natural frequencies, mass-normalized mode shapes, and modal damping factors.

MASS	MATRIX :		STIFFNESS	MATRIX :
1	0		1,500	−1,000
0	2		−1,000	2,000

Problem 6.2: The eigenvalues of a system with viscous damping are given below. Determine the natural frequencies and modal damping factors.

Eigenvalues
  1.0e+02 *
[−0.1464 + 4.9375i   −0.1464 − 4.9375i   −0.0402 + 2.5886i   −0.0402 − 2.5886i
  −0.0084 + 1.1820i   −0.0084 − 1.1820i]

Problem 6.3: The mass and stiffness matrices of a 2-DOF system with viscous damping are given below. The damping matrix is proportional to the stiffness matrix, with the proportionality constant equal to 0.0002. The mode shapes of the system are also given. Find the system's natural frequencies and modal damping factors (without solving the EVP).

MASS MATRIX :		STIFFNESS MATRIX :		MODE SHAPES:	
5	0	96,000	−24,000	−0.3975	−0.2050
0	2	−24,000	60,000	−0.3241	0.6284

Problem 6.4: For the two-DOF system in Problem 6.1, find and plot the FRF α_{12} using a MATLAB program.

Problem 6.5: For the modal model of a hysteretically damped system given below, find the maximum value of the steady-state response at the first DOF when the harmonic forces of amplitudes 20N and 30N at frequency 39 Hz act at the first and second DOFs, respectively.

NATURAL FREQUENCIES (Hz)	MODAL LOSS FACTORS	MASS-NORMALIZED MODE SHAPES	
	0.0151	0.5861	0.2376
86.3336	0.0391	−0.2008	0.4954
36.1047			

Problem 6.6: For the system in Problem 6.5, find the modal responses corresponding to the excitation frequencies 35, 60, and 83 Hz. Justify the results.

7 Characteristics of Frequency Response Functions

7.1 INTRODUCTION

The analytical modal analyses of SDOF and MDOF undamped and damped systems are presented in Chapters 4–6. The analytical modal analysis approach utilizes the spatial model of the system. On the other hand, as we study in Chapters 9 and 10, the experimental modal analysis (EMA) approach is based on the knowledge of the response model, i.e., the frequency response functions (FRFs). This chapter presents the characteristics of FRFs. First, we look at the types and graphical representations of FRF. Then we study the characteristics of FRFs of SDOF undamped and damped systems with structural and viscous damping. We look at the asymptotic behavior of the FRFs of these systems characterized by stiffness and mass lines and the nature of the FRF plots in the complex plane. The characteristics of the FRFs of MDOF undamped and damped systems with structural and viscous damping are then discussed. It is followed by a discussion of the antiresonance frequencies and their computation. The relationship between the FRFs and the modal model is derived in Chapters 5 and 6 using the response to harmonic forces. In this chapter, we derive this relationship directly from the spatial model. The understanding developed in this chapter provides necessary inputs for understanding the theory of EMA covered in the following chapters.

7.2 FRF TYPES

a. For an SDOF system

An FRF can be defined as Receptance, Mobility, or Inertance. The **receptance** is based on the displacement output of the system and is the ratio of displacement and input force in the frequency domain. For an SDOF system, it is given by

$$\alpha(\omega) = \frac{x(\omega)}{F(\omega)} \tag{7.1}$$

The **mobility** is the ratio of output velocity and input force in the frequency domain and is given by

$$Y(\omega) = \frac{v(\omega)}{F(\omega)} \tag{7.2}$$

Since $v(\omega) = i\omega x(\omega)$, we have the following relationship:

$$Y(\omega) = i\omega\alpha(\omega) \tag{7.3}$$

The **inertance/accelerance** is the ratio of output acceleration and input force in the frequency domain and is given by

$$A(\omega) = \frac{a(\omega)}{F(\omega)} \tag{7.4}$$

Since $a(\omega) = i\omega\, v(\omega) = -\omega^2\, x(\omega)$, we have the following relationships:

$$A(\omega) = i\omega y(\omega) = -\omega^2\alpha(\omega) \tag{7.5}$$

Thus, the three types of FRF are interrelated, and if anyone is available, the other two can be obtained.

The input divided by the output in the frequency domain forms an inverse FRF of the system. The dynamic stiffness $(Z(\omega))$, impedance $(I(\omega))$, and apparent mass $(m^a(\omega))$ are inverse FRFs and are defined as

DOI: 10.1201/9780429454783-7

$$Z(\omega) = \frac{F(\omega)}{x(\omega)} \tag{7.6}$$

$$I(\omega) = \frac{F(\omega)}{v(\omega)} \tag{7.7}$$

$$m^a(\omega) = \frac{F(\omega)}{a(\omega)} \tag{7.8}$$

b. For an MDOF system

For an MDOF system, the receptance $\alpha_{jk}(\omega)$ is the ratio of the displacement amplitude at the jth DOF and the force amplitude at the kth DOF, both in the frequency domain, with no force acting at the other DOFs. Mathematically,

$$\alpha_{jk}(\omega) = \frac{x_j(\omega)}{F_k(\omega)} \quad \text{with } F_q(\omega) = 0 \text{ for all } q \text{ except } q = k \tag{7.9}$$

The FRFs $\alpha_{jk}(\omega)$ for $j = k$ are called **point or direct FRFs** since the response and excitation DOFs are the same. The FRFs $\alpha_{jk}(\omega)$ for $j \neq k$ are called **transfer or cross FRFs** since the response and excitation DOFs are different.

Mobility $\left(Y_{jk}(\omega)\right)$ and Inertance $\left(I_{jk}(\omega)\right)$ are defined using the velocity and acceleration amplitudes, respectively, in the frequency domain. As for an SDOF system, they are related to the receptance by

$$Y_{jk}(\omega) = i\omega\alpha_{jk}(\omega) \tag{7.10}$$

$$A_{jk}(\omega) = i\omega Y_{jk}(\omega) = -\omega^2\alpha_{jk}(\omega) \tag{7.11}$$

The inverse descriptor, the dynamic stiffness, is defined as

$$Z_{jk}(\omega) = \frac{F_j(\omega)}{x_k(\omega)} \quad \text{with } x_q(\omega) = 0 \text{ for all } q \text{ except } q = k \tag{7.12}$$

The impedance and apparent mass for an MDOF system can also be similarly defined.

An important point to note is that for an SDOF system, the receptance FRF (Eq. 7.1) and the dynamic stiffness (Eq. 7.6) are reciprocal of each other, but the same doesn't hold for an MDOF system. It is clear from Eqs. (7.9) and (7.12) that if the receptance $\alpha_{jk}(\omega)$ is known then the $Z_{jk}(\omega)$ can't be obtained simply as $1/\alpha_{jk}(\omega)$ since the conditions to be met at the other DOFs are different for the two quantities.

c. Why are the FRFs preferred quantities for measurement than their inverse counterparts?

We note from Eq. (7.9) that to measure the FRF α_{jk} we need to apply the force at the jth DOF with no force applied at the other DOFs. These conditions are easy to meet in practice, so the FRF can be measured easily. On the other hand, Eq. (7.12) indicates that to measure an inverse descriptor like the dynamic stiffness Z_{jk} a nonzero displacement at the jth DOF should be accompanied by no displacement at the other DOFs. To recognize the implications of this, take the example of a beam. To measure Z_{jk}, all the DOFs, other than the kth DOF, need to be constrained to have no displacement, and it includes both the translational and rotational DOFs. Realizing these constraints is very difficult in practice. It explains why the FRFs are preferred quantities for measurement in EMA than their inverse counterparts.

7.3 GRAPHICAL REPRESENTATION OF FRFs

7.3.1 BODE PLOT

The Bode plot presents an FRF as plots of magnitude versus frequency and phase versus frequency. The x-axis is the frequency axis in linear or log scale, though a linear axis is commonly used in modal analysis. The y-axis for the Bode magnitude plot shows the FRF magnitude in dB and is calculated as $20\log_{10}|\alpha_{jk}(\omega)|$.

The y-axis for the phase plot shows the phase of the FRF in degrees or radians on a linear scale.

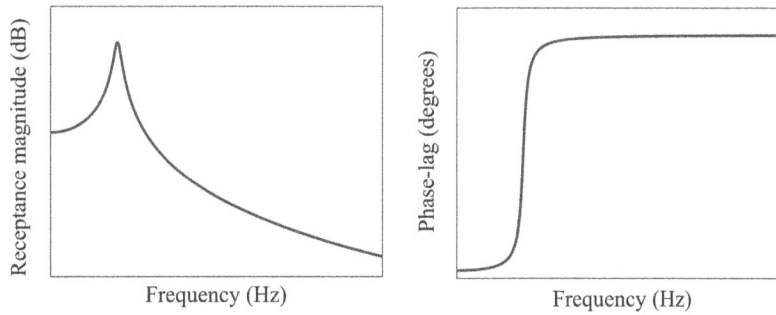

FIGURE 7.1   Bode magnitude and phase plots of receptance.

Figure 7.1 shows the Bode magnitude and phase plots of the receptance of an SDOF system. The advantage of this plot is that it directly gives the magnitude of the FRF at any frequency, which is not available from the Nyquist plot (presented later). It gives a quick idea of how the system responds to harmonic forces at various frequencies. The dB scale compresses the FRF magnitude range and shows the variation of FRF magnitude in the whole frequency range. Similarly, the phase plot directly gives information about the phase of the response with the input force. Due to these advantages, the Bode plots are widely used for presenting the FRFs. A linear frequency axis can be used for shorter frequency ranges, as in Figure 7.1, but a logarithmic axis can be used for larger frequency ranges.

7.3.2   Absolute Magnitude Plot

In this plot (Figure 7.2), the absolute FRF magnitude is plotted as a function of frequency on a linear axis. The peak magnitude dominates the plot, and so this plot helps assess the peaks in the FRF and the corresponding frequencies. However, the variation of the FRF magnitude at other frequencies is eclipsed by the peaks.

7.3.3   Nyquist Plot

The Nyquist plot is a plot between the real and imaginary parts of the FRF, marked on the x and y axes, respectively. Figure 7.3 shows the Nyquist plot of the receptance of an SDOF system. Each point on the Nyquist plot corresponds to a specific excitation frequency; thus, the frequency variable is not explicitly accessible on the plot.

7.3.4   Real and Imaginary FRF Plots

Another alternative is to plot the real and imaginary parts of the FRF separately as a function of the frequency. Figure 7.4 shows these plots for the receptance of an SDOF system. The advantage of these plots is that the real and imaginary parts of the FRF at each frequency can be obtained directly. It is also possible to plot both the real and imaginary parts as a function of frequency in a single 3D plot.

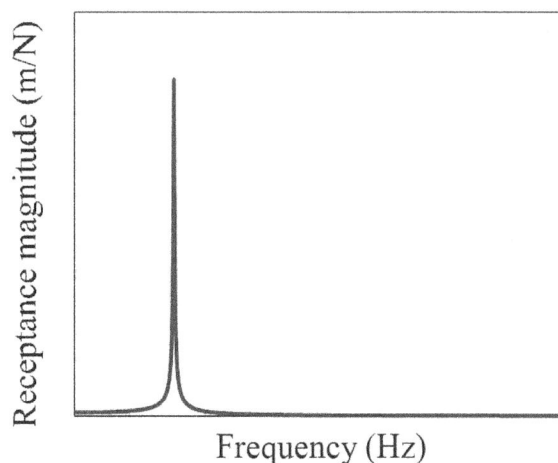

FIGURE 7.2   Absolute magnitude plot of receptance.

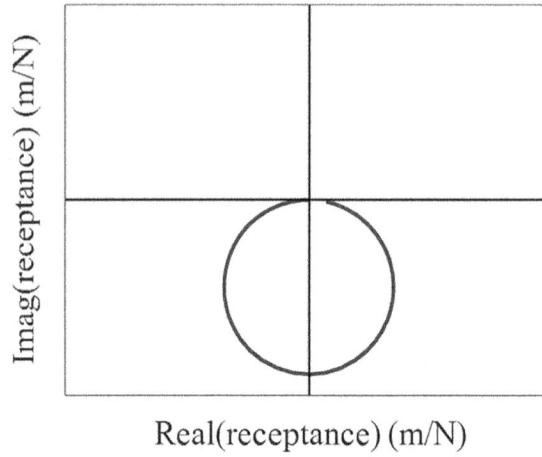

FIGURE 7.3   Nyquist plot of receptance.

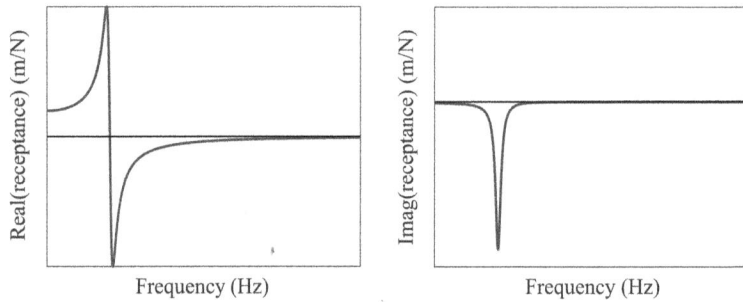

FIGURE 7.4   Real and imaginary receptance plots.

7.4   CHARACTERISTICS OF SDOF SYSTEM FRFs

In this section, we look at the characteristics of the SDOF undamped and damped system FRFs.

7.4.1   UNDAMPED SDOF SYSTEM

a. FRF magnitude and phase

In Chapter 4, the receptance of an SDOF undamped system was derived and is given by

$$\alpha(\omega) = \frac{1}{k - m\omega^2} = \frac{1/m}{\omega_n^2 - \omega^2} = \frac{1/k}{1 - \dfrac{\omega^2}{\omega_n^2}} \tag{7.13}$$

It can be written in the polar form as $\alpha(\omega) = |\alpha(\omega)|e^{-i\phi}$. The magnitude and phase lag of the FRF are

$$|\alpha(\omega)| = \left|\frac{1}{k - m\omega^2}\right| = \left|\frac{1/m}{\omega_n^2 - \omega^2}\right| = \left|\frac{1/k}{1 - \dfrac{\omega^2}{\omega_n^2}}\right| \tag{7.14}$$

$$\phi = \tan^{-1} \frac{0}{1 - \dfrac{\omega^2}{\omega_n^2}} \qquad (7.15)$$

Figure 7.5 shows the magnitude and phase of $\alpha(\omega)$. The following observations are made from these plots:

- The FRF magnitude tends to infinity when the excitation frequency (ω) becomes equal to the natural frequency of the system (ω_n).
- For the forcing frequencies $\omega < \omega_n$ the phase lag is zero, which means that the displacement is in phase with the excitation force.
- At $\omega = \omega_n$ the phase lag is 90°.
- For the frequencies $\omega > \omega_n$ the phase lag is 180°. Thus the displacement lags the excitation force by 180°.

b. Stiffness and mass lines of the receptance

Let us study the behavior of the magnitude of the receptance in two extreme cases.

When $\omega \ll \omega_n$:

In this case, Eq. (7.14) can be approximated as

$$|\alpha(\omega)| = \left| \frac{1/k}{1 - \dfrac{\omega^2}{\omega_n^2}} \right| \approx \frac{1}{k} \qquad (7.16)$$

The FRF magnitude in dB is

$$\text{magnitude in dB} = 20 \log_{10} |\alpha(\omega)| = -20 \log_{10} k \qquad (7.17)$$

Since the RHS in Eq. (7.17) is independent of ω the FRF magnitude in the frequency range $\omega \ll \omega_n$ can be represented by a line parallel to the frequency axis at an intercept of $-20 \log_{10} k$ dB on the y-axis. This line is an asymptote of the FRF magnitude plot. Since it depends on the stiffness of the system, it is called the **stiffness line**.

When $\omega \gg \omega_n$:

In this case, Eq. (7.14) can be approximated as

$$|\alpha(\omega)| = \left| \frac{1/k}{1 - \dfrac{\omega^2}{\omega_n^2}} \right| \approx \left| \frac{1/k}{-\dfrac{\omega^2}{\omega_n^2}} \right| \approx \frac{1}{m\omega^2} \qquad (7.18)$$

The FRF magnitude in dB is

$$\text{magnitude in dB} = 20 \log_{10} |\alpha(\omega)| = -20 \log_{10} m - 40 \log_{10} \omega \qquad (7.19)$$

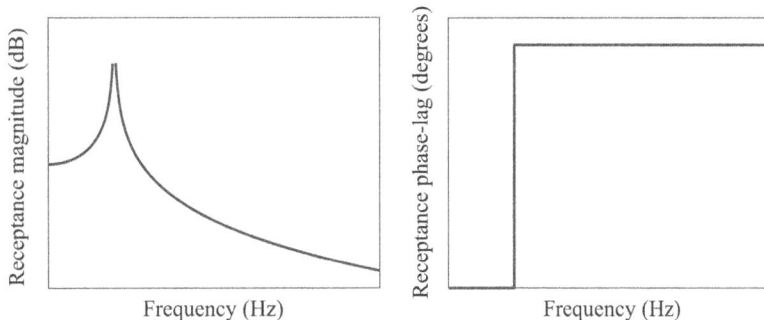

FIGURE 7.5 Magnitude and phase of the receptance of an undamped SDOF system.

The RHS in Eq. (7.19) is dependent on $\log_{10}\omega$. The equation indicates that, on a logarithmic frequency axis, the FRF magnitude in the range $\omega \gg \omega_n$ can be represented by a straight line with a slope of -40dB/decade, making an intercept of $-20\log_{10} m$ dB on the ordinate at $\omega = 1$ rad/sec. It is another asymptote of the FRF magnitude plot. Since it depends on the mass of the system, it is called the **mass line.**

Figure 7.6 shows the magnitude plot of $\alpha(\omega)$ on a logarithmic frequency axis. Also shown are the stiffness and mass lines. In the low-frequency region $(\omega \ll \omega_n)$, the stiffness line or the stiffness of the system largely determines its response; hence, this region is called the **stiffness-controlled region**. Similarly, in the high-frequency region $(\omega \gg \omega_n)$, the mass line or the mass of the system largely determines its response; hence, this region is called the **mass-controlled region**. In the frequency range around the natural frequency $(\omega \approx \omega_n)$, the damping factor of the system largely determines its response; hence, this region is called the **damping-controlled region**.

The concept of stiffness and mass lines can be used to identify the stiffness and mass of an SDOF system from its receptance using Eqs. (7.17) and (7.19).

c. Stiffness and mass lines of mobility

From Eq. (7.3), the mobility is given by $y(\omega) = i\omega\alpha(\omega)$, Substituting $\alpha(\omega)$ from Eq. (7.13), we get the expression for $y(\omega)$. It can be shown that in the range $\omega \ll \omega_n$, the magnitude of $y(\omega)$ in dB can be approximated as

$$|y(\omega)| \text{ in dB} = 20\log_{10}|y(\omega)| = 20\log_{10}\omega - 20\log_{10} k \qquad (7.20)$$

The equation indicates that, on a logarithmic frequency axis, the FRF magnitude in the range $\omega \ll \omega_n$ can be represented by a straight line with a slope of 20dB/decade, making an intercept of $-20\log_{10} k$ dB on the ordinate at $\omega = 1$ rad/sec. It is the stiffness line of the mobility of the system.

Similarly, it can be shown that in the range $\omega \gg \omega_n$, the magnitude of $y(\omega)$ in dB can be approximated as

$$|y(\omega)| \text{ in dB} = 20\log_{10}|y(\omega)| = -20\log_{10} m - 20\log_{10}\omega \qquad (7.21)$$

The equation indicates that, on a logarithmic frequency axis, the FRF magnitude in the range $\omega \gg \omega_n$ can be represented by a straight line with a slope of -20dB/decade, making an intercept of $-20\log_{10} m$ dB on the ordinate at $\omega = 1$ rad/sec. It is the mass line of the mobility of the system. Figure 7.7 shows the stiffness and mass lines of the mobility of an SDOF undamped system overlaid on the mobility magnitude.

d. Stiffness and mass lines of inertance

From Eq. (7.5), the Inertance is given by $A(\omega) = -\omega^2\alpha(\omega)$. Substituting $\alpha(\omega)$ from Eq. (7.13), we get the expression for $A(\omega)$. It can be shown that in the range $\omega \ll \omega_n$, the magnitude of $A(\omega)$ in dB can be approximated as

$$|A(\omega)| \text{ in dB} = 20\log_{10}|A(\omega)| = 40\log_{10}\omega - 20\log_{10} k \qquad (7.22)$$

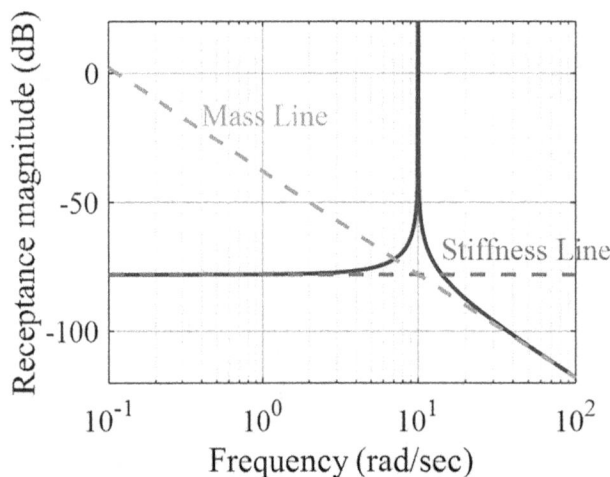

FIGURE 7.6   Stiffness and mass lines of the receptance of an SDOF undamped system.

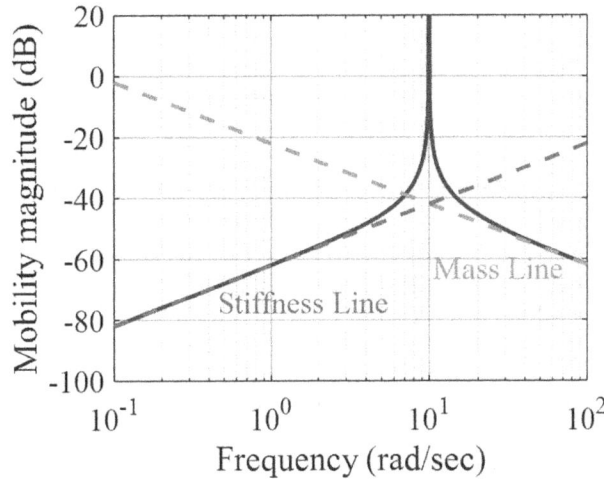

FIGURE 7.7 Stiffness and mass lines of the mobility of an SDOF undamped system.

The equation indicates that, on a logarithmic frequency axis, the FRF magnitude in the range $\omega \ll \omega_n$ can be represented by a straight line with a slope of 40 dB/decade, making an intercept of $-20\log_{10} k$ dB on the ordinate at $\omega = 1$ rad/sec. It is the stiffness line of the inertance of the system.

It can be shown that in the range $\omega \gg \omega_n$, the magnitude of $A(\omega)$ in dB can be approximated as

$$|A(\omega)| \text{ in } dB = 20\log_{10}|A(\omega)| = -20\log_{10} m \tag{7.23}$$

Thus, in the frequency range $\omega \gg \omega_n$, the magnitude of FRF can be represented by a line parallel to the frequency axis, making an intercept of $-20\log_{10} m$ dB on the y-axis. It is the mass line for the Inertance FRF. Figure 7.8 shows the stiffness and mass lines of the inertance of an SDOF undamped system overlaid on the inertance magnitude.

e. Real and imaginary parts of the FRF

The imaginary part of the receptance of an undamped SDOF system is zero. Figure 7.9 shows the real part of the receptance. We see that when the excitation frequency is equal to the natural frequency, the real part approaches infinity. The real part is negative at the frequencies above the natural frequency, indicating that the displacement has a phase lag of 180°.

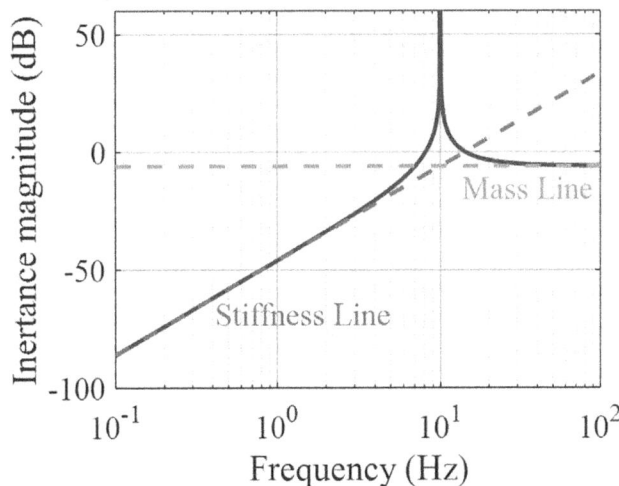

FIGURE 7.8 Stiffness and mass lines of the inertance of an SDOF undamped system.

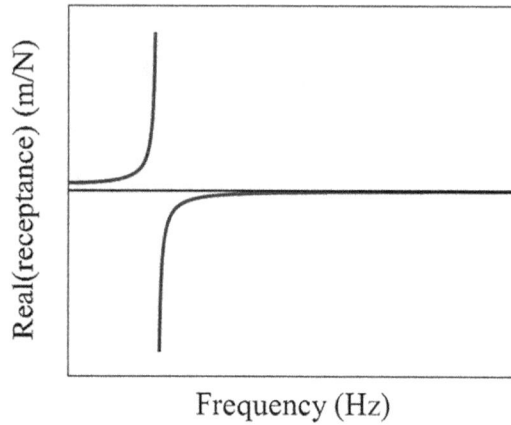

FIGURE 7.9   Real part of the receptance of an SDOF undamped system.

7.4.2   SDOF System with Viscous Damping

a. FRF magnitude and phase

The receptance of an SDOF system with viscous damping is given by

$$\alpha(\omega) = \frac{1}{k - m\omega^2 + i\omega c} = \frac{1/m}{\omega_n^2 - \omega^2 + i2\xi\omega\omega_n} = \frac{1/k}{1 - \dfrac{\omega^2}{\omega_n^2} + i2\xi\dfrac{\omega}{\omega_n}} \tag{7.24}$$

The magnitude and phase lag of the FRF are obtained as

$$|\alpha(\omega)| = \frac{1}{\sqrt{\left(k - m\omega^2\right)^2 + (\omega c)^2}} = \frac{1/m}{\sqrt{\left(\omega_n^2 - \omega^2\right)^2 + (2\xi\omega\omega_n)^2}} = \frac{1/k}{\sqrt{\left(1 - \dfrac{\omega^2}{\omega_n^2}\right)^2 + \left(2\xi\dfrac{\omega}{\omega_n}\right)^2}} \tag{7.25}$$

$$\phi = \tan^{-1}\frac{2\xi\dfrac{\omega}{\omega_n}}{1 - \dfrac{\omega^2}{\omega_n^2}} \tag{7.26}$$

Figure 7.10 shows the magnitude and phase of $\alpha(\omega)$. The plots for the corresponding undamped system are overlaid so that the deviation due to damping can be observed. The following observations can be made from these plots:

- The damping mainly affects the FRF magnitude in a relatively small frequency range around the undamped natural frequency (ω_n) of the system.
- The peak FRF magnitude reduces from the theoretically infinite value for an undamped system to a finite value.

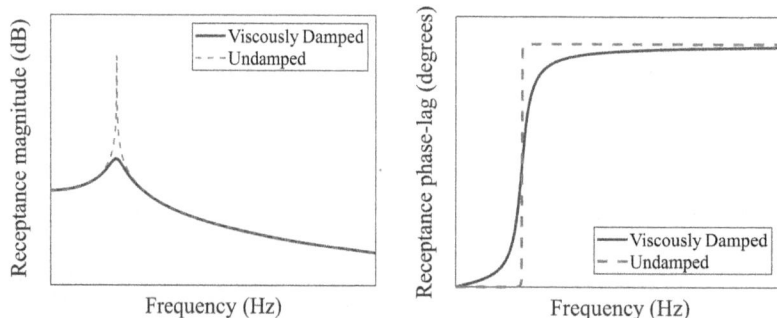

FIGURE 7.10   Magnitude and phase of the receptance of an SDOF system with viscous damping.

- The peak FRF magnitude no longer occurs at ω_n but slightly to its left.
- For forcing frequencies $\omega < \omega_n$, the phase lag is in the range $0° - 90°$, while for $\omega > \omega_n$, it is in the range $90° - 180°$.
- At $\omega = \omega_n$ the phase lag is $90°$.

b. Stiffness and mass lines

Let us study the behavior of the magnitude of the receptance in two extreme cases.

When $\omega \ll \omega_n$:

In this case, the magnitude of the FRF given by Eq. (7.25) can be approximated as

$$|\alpha(\omega)| = \frac{1/k}{\sqrt{\left(1 - \frac{\omega^2}{\omega_n^2}\right)^2 + \left(2\xi\frac{\omega}{\omega_n}\right)^2}} \approx \frac{1}{k} \tag{7.27}$$

$$\text{magnitude in dB} = 20\log_{10}|\alpha(\omega)| = -20\log_{10} k \tag{7.28}$$

which matches the receptance magnitude of the corresponding undamped system. Thus, we conclude that the stiffness line for the system with viscous damping is the same as that of the corresponding undamped system.

When $\omega \gg \omega_n$:

In Eq. (7.25), the term $\left(\frac{\omega^2}{\omega_n^2}\right)^2$ in the denominator dominates, and therefore, in the limit, the magnitude can be approximated as

$$|\alpha(\omega)| = \frac{1/k}{\sqrt{\left(1 - \frac{\omega^2}{\omega_n^2}\right)^2 + \left(2\xi\frac{\omega}{\omega_n}\right)^2}} \approx \frac{1/k}{-\frac{\omega^2}{\omega_n^2}} \approx \frac{1}{m\omega^2} \tag{7.29}$$

$$\text{magnitude in dB} = 20\log_{10}|\alpha(\omega)| = -20\log_{10} m - 40\log_{10}\omega \tag{7.30}$$

which matches the receptance magnitude of the corresponding undamped system. Thus, we conclude that the mass line for the system with viscous damping is the same as that of the corresponding undamped system.

Figure 7.11 shows the stiffness and mass lines overlaid on the plot of the receptance magnitude.

c. Estimation of viscous damping factor

Let us determine the absolute magnitudes of the stiffness and mass lines and the FRF at $\omega = \omega_n$. Denoting these quantities as $\alpha_{sl}(\omega_n)$, $\alpha_{ml}(\omega_n)$ and $\alpha(\omega_n)$ respectively, we obtain the following results:

$$\alpha_{sl}(\omega_n) = \frac{1}{k} \tag{7.31}$$

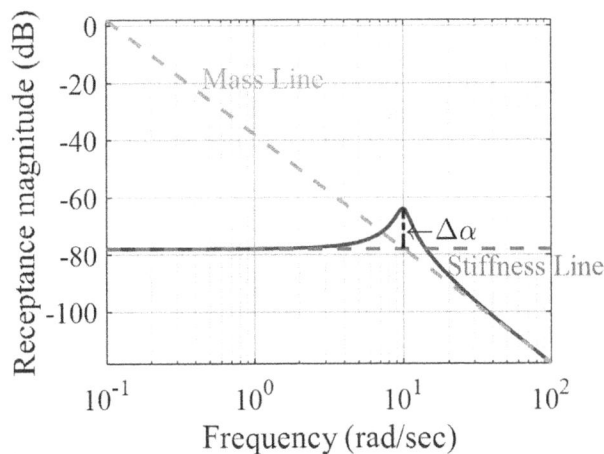

FIGURE 7.11 Stiffness and mass lines of receptance of an SDOF system with viscous damping.

$$\alpha_{ml}(\omega_n) = \frac{1}{k} \tag{7.32}$$

$$\alpha(\omega_n) = \frac{1}{2\xi k} \tag{7.33}$$

Eqs. (7.31) and (7.32) indicate that the stiffness and mass lines intersect at $\omega = \omega_n$. Thus, the intersection of stiffness and mass lines locates the undamped natural frequency of the system. Dividing Eq. (7.33) by Eq. (7.31), we get

$$\frac{\alpha(\omega_n)}{\alpha_{sl}(\omega_n)} = \frac{1}{2\xi} \tag{7.34}$$

$$20\log_{10}\alpha(\omega_n) - 20\log_{10}\alpha_{sl}(\omega_n) = 20\log_{10}\frac{1}{2\xi} \tag{7.35}$$

$$\Delta\alpha = 20\log_{10}\frac{1}{2\xi} \tag{7.36}$$

which finally gives

$$\xi = \frac{1}{2}10^{\frac{-\Delta\alpha}{20}} \tag{7.37}$$

$$c = \sqrt{km}\ 10^{\frac{-\Delta\alpha}{20}} \tag{7.38}$$

where $\Delta\alpha$ represents the difference in dB-magnitudes of the FRF and the stiffness line at $\omega = \omega_n$ (as shown in Figure 7.11). The value of $\Delta\alpha$ can be determined from the FRF, and the viscous damping factor can be estimated from Eq. (7.37). Similarly, the mass and stiffness of the system can be obtained from the mass and stiffness lines, respectively, and the viscous damping coefficient can be obtained from Eq. (7.38). The figure shows that the stiffness and mass lines represent the skeleton of the FRF.

d. Nature of the Nyquist plot of the mobility
  The mobility of an SDOF system with viscous damping has the property that it traces a circular arc in the Nyquist plane. The mobility of FRF is given by

$$y(\omega) = i\omega\alpha(\omega) = \frac{i\omega}{k - m\omega^2 + i\omega c} \tag{7.39}$$

The real and imaginary parts are

$$Re(y(\omega)) = \frac{\omega^2 c}{(k-m\omega^2)^2 + (\omega c)^2} \tag{7.40}$$

$$Im(y(\omega)) = \frac{(k-m\omega^2)\omega}{(k-m\omega^2)^2 + (\omega c)^2} \tag{7.41}$$

It is seen that $Re(y(\omega))$ and $Im(y(\omega))$ satisfy the following equation:

$$\left(Re(y(\omega)) - \frac{1}{2c}\right)^2 + Im(y(\omega))^2 = \left(\frac{1}{2c}\right)^2 \tag{7.42}$$

which represents the equation of a circle since the $\text{Re}(y(\omega))$ and $\text{Im}(y(\omega))$ are nothing but the x and y coordinates of the Nyquist plane, respectively. Thus, the Nyquist plot of the mobility of the system traces a circular trajectory of radius $\dfrac{1}{2c}$ with its center at $\left(\dfrac{1}{2c},0\right)$ as shown in Figure 7.12. This property forms the basis for the circle fit method of modal parameter extraction to be studied in Chapter 11.

e. Real and imaginary parts of the FRF

Figure 7.13 shows the real and imaginary parts of the receptance of an SDOF system with viscous damping. When the excitation frequency is equal to the undamped natural frequency, the real part is zero, while the imaginary part attains a maximum value. These properties are useful in locating the resonant frequency in the FRF of an SDOF system.

7.4.3 SDOF System with Structural Damping

a. FRF magnitude and phase

The receptance of an SDOF system with structural damping is given by

$$\alpha(\omega) = \frac{1}{k - m\omega^2 + ih} = \frac{1/m}{\omega_n{}^2 - \omega^2 + i\eta\omega_n{}^2} = \frac{1/k}{1 - \dfrac{\omega^2}{\omega_n{}^2} + i\eta} \tag{7.43}$$

where η denotes the structural damping loss factor.

The magnitude and phase lag of the FRF are obtained as

$$|\alpha(\omega)| = \frac{1}{\sqrt{\left(k - m\omega^2\right)^2 + h^2}} = \frac{1/m}{\sqrt{\left(\omega_n{}^2 - \omega^2\right)^2 + \left(\eta\omega_n{}^2\right)^2}} = \frac{1/k}{\sqrt{\left(1 - \dfrac{\omega^2}{\omega_n{}^2}\right)^2 + \eta^2}} \tag{7.44}$$

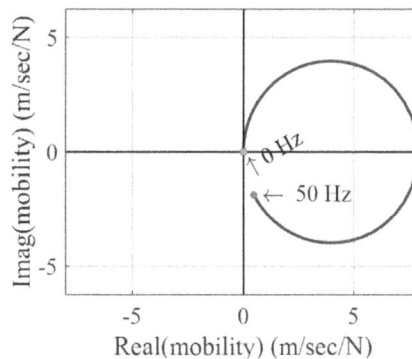

FIGURE 7.12 Nyquist plot of mobility of an SDOF system with viscous damping.

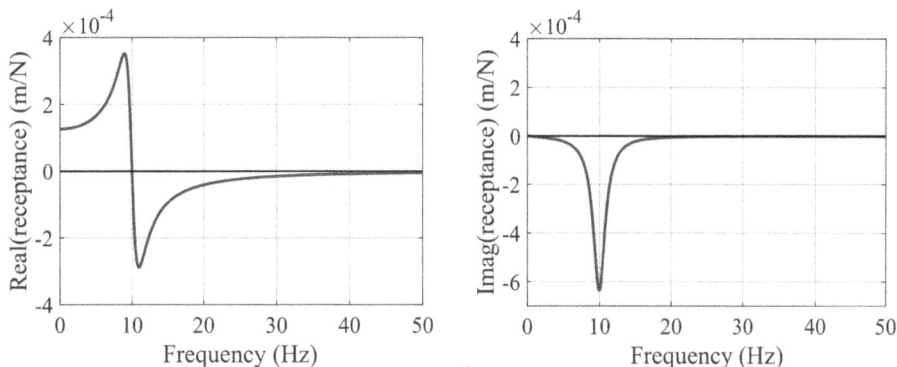

FIGURE 7.13 Real and imaginary parts of the receptance of an SDOF system with viscous damping.

$$\phi = \tan^{-1} \frac{\eta}{1 - \dfrac{\omega^2}{\omega_n^2}} \tag{7.45}$$

Figure 7.14 shows the magnitude and phase of $\alpha(\omega)$ overlaid on the corresponding plots of the undamped system. The following observations are made from these plots:

- The damping affects the FRF magnitude over a relatively narrow frequency range around the undamped natural frequency (ω_n) of the system.
- The peak FRF magnitude is finite and occurs at $\omega = \omega_n$, irrespective of damping.
- For forcing frequencies $\omega < \omega_n$, the phase lag is in the range $0° - 90°$, while for $\omega > \omega_n$, it is in the range of $90° - 180°$.
- At $\omega = \omega_n$ the phase lag is $90°$.

b. Stiffness and mass lines

It is seen that in the range $\omega \ll \omega_n$, the receptance magnitude can be approximated as $|\alpha(\omega)| \approx 1/k\sqrt{1 + \eta^2}$. For systems with small damping (i.e., $\eta \ll 1$) the receptance magnitude in dB can be approximated as the stiffness line of the receptance of the corresponding undamped system. For the range $\omega \gg \omega_n$, the receptance magnitude in dB can be approximated as the mass line of the receptance of the corresponding undamped system.

By following the steps for estimating the viscous damping factor (Section 7.4.2 c), we get

$$\eta = 10^{\frac{-\Delta\alpha}{20}} \tag{7.46}$$

where $\Delta\alpha$ is the difference in the dB-magnitudes of the FRF and the stiffness line at $\omega = \omega_n$, as shown in Figure 7.15. By measuring $\Delta\alpha$ from the FRF, the structural damping loss factor can be obtained from the above equation. Similarly, the mass and stiffness of the system can be obtained from the mass and stiffness lines, respectively, and the structural damping coefficient is obtained as $h = k\eta$.

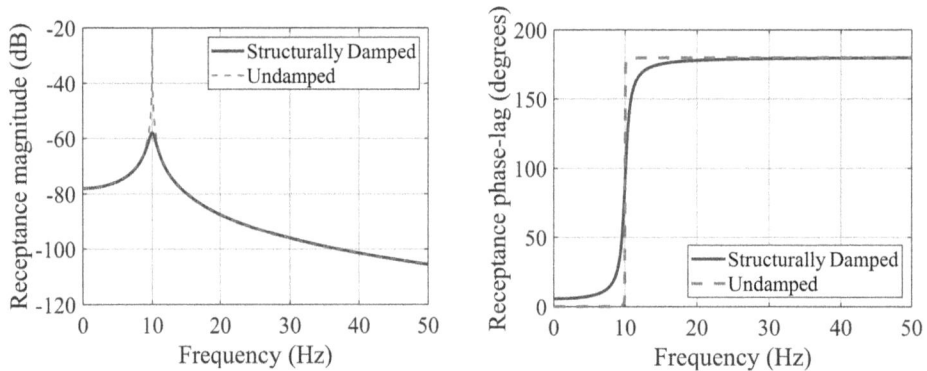

FIGURE 7.14 Magnitude and phase of the receptance of an SDOF system with structural damping.

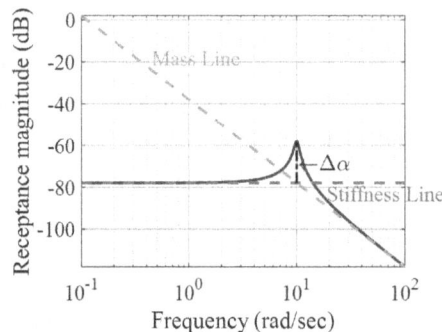

FIGURE 7.15 Stiffness and mass lines of receptance FRF of an SDOF system with structural damping.

c. **Nature of the Nyquist plot of the receptance**

The receptance of an SDOF system with structural damping has the property that it traces a circular trajectory in the Nyquist plane. The real and imaginary parts of the FRF are given by

$$\text{Re}(\alpha(\omega)) = \frac{k - m\omega^2}{\left(k - m\omega^2\right)^2 + h^2} \tag{7.47}$$

$$\text{Im}(\alpha(\omega)) = \frac{-h}{\left(k - m\omega^2\right)^2 + h^2} \tag{7.48}$$

It is seen that $\text{Re}(\alpha(\omega))$ and $\text{Im}(\alpha(\omega))$ satisfy the following equation:

$$\left(\text{Re}(\alpha(\omega))\right)^2 + \left(\text{Im}(\alpha(\omega)) + \frac{1}{2h}\right)^2 = \left(\frac{1}{2h}\right)^2 \tag{7.49}$$

which represents the equation of a circle since the $\text{Re}(\alpha(\omega))$ and $\text{Im}(\alpha(\omega))$ are the x and y coordinates of the Nyquist plane, respectively. Thus, the Nyquist plot of the receptance traces a circular trajectory of radius $\frac{1}{2h}$ with its center at $\left(0, -\frac{1}{2h}\right)$ as shown in Figure 7.16.

d. **Real and imaginary parts of the FRF**

Figure 7.17 shows the real and imaginary parts of the receptance of an SDOF system with structural damping. Like the system with viscous damping, the real part is zero at the undamped natural frequency while the imaginary part has a maximum value.

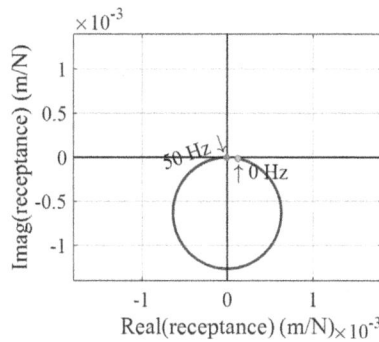

FIGURE 7.16   Nyquist plot of the receptance of an SDOF system with structural damping.

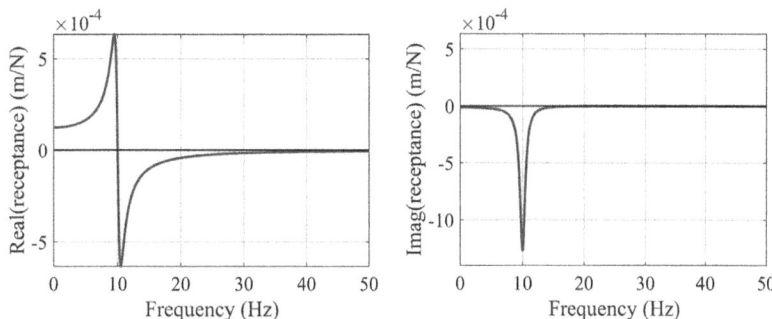

FIGURE 7.17   Real and imaginary parts of the receptance of an SDOF system with structural damping.

7.5   CHARACTERISTICS OF UNDAMPED MDOF SYSTEM FRFs

An MDOF system has multiple vibration modes; therefore, the FRF of an MDOF system has many other characteristics different from the FRF of an SDOF system. This section presents the FRF characteristics of undamped MDOF systems. We illustrate the graphical representation of the FRFs and their characteristics using an undamped 3-DOF system shown in Figure 7.18.

7.5.1   RELATIONSHIP OF FRFs WITH THE MODAL MODEL

The relationship of the FRF matrix with the modal model is derived in Chapter 5 using the response to harmonic forces. In this section, we present the derivation of this relationship by an alternative route by starting from the relationship of the FRF matrix with the spatial model. We know that

$$[\alpha] = \left([K] - \omega^2 [M] \right)^{-1} \tag{7.50}$$

Taking the inverse of Eq. (7.50), pre- and post-multiplying the resulting equation by $[\phi]^T$ and $[\phi]$ (the mass-normalized eigenvector matrix), respectively, and using the orthogonality properties, we get

$$[\phi]^T [\alpha]^{-1} [\phi] = [\,\ddot{}\lambda_r\,] - \omega^2 [I] \tag{7.51}$$

Writing $[\,\ddot{}\lambda_r\,] = [\,\ddot{}\omega_r^2\,]$, with $\lambda_r = \omega_r^2$ being the rth eigenvalue of the undamped system, and taking inverse, leads to

$$[\phi]^{-1} [\alpha] [\phi]^{-T} = \left[\ddot{}\left(\omega_r^2 - \omega^2 \right) \right]^{-1} \tag{7.52}$$

where the relationship from the matrix algebra $\left(([A][B][C])^{-1} = [C]^{-1}[B]^{-1}[A]^{-1} \right)$ is used. Pre and post-multiplying the above equation by $[\phi]$ and $[\phi]^T$, respectively, gives

$$[\alpha] = [\phi] \left[\ddot{}\left(\frac{1}{\omega_r^2 - \omega^2} \right) \right] [\phi]^T \tag{7.53}$$

The above relationship can also be written after expanding the RHS as

$$[\alpha] = \sum_{r=1}^{N} \frac{\{\phi\}_r \{\phi\}_r^T}{\omega_r^2 - \omega^2} \tag{7.54}$$

which is the same as derived in Chapter 5. The FRF $\alpha_{jk}(\omega)$ is given by

$$\alpha_{jk}(\omega) = \sum_{r=1}^{N} \frac{\phi_{jr}\, \phi_{kr}}{\omega_r^2 - \omega^2} \tag{7.55}$$

7.5.2   ANTIRESONANCES

Figure 7.19 shows the magnitude and phase lag of a point receptance $(\alpha_{11}(\omega))$ of the 3 DOF undamped system (Figure 7.18). We see multiple resonance peaks against a single such peak in the FRF of an SDOF system. It is also seen that between every two consecutive peaks, the shape of the FRF plot looks like an 'inverted' resonance peak. The amplitude attains a very large value at a resonance frequency. In contrast, the amplitude attains a very low value at an 'inverted' peak

FIGURE 7.18   An undamped 3-DOF system.

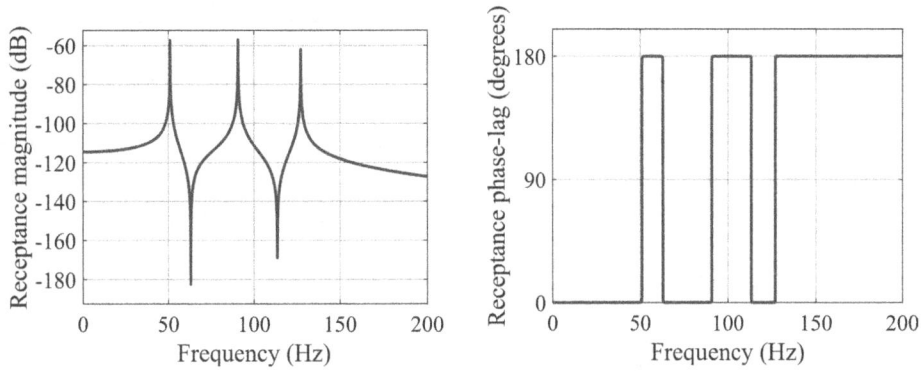

FIGURE 7.19   Magnitude and phase lag of a point receptance of an undamped MDOF system.

frequency. Hence, an 'inverted' peak is called antiresonance, and the corresponding frequency is called antiresonance frequency.

For an undamped system, the magnitude of the FRF at an antiresonance frequency is zero. Thus, with regard to $\alpha_{11}(\omega)$, the response at the first DOF due to a harmonic force applied at the first DOF of frequency coinciding with an antiresonance frequency would be zero. The phase plot for the undamped MDOF system indicates that at every resonance and antiresonance frequency, there is a phase change by $180°$.

Figure 7.20 shows the magnitude and phase lag of a transfer receptance $\left(\alpha_{12}(\omega)\right)$. We observe no antiresonance between the second and third resonance peaks; instead, the FRF magnitude reaches a minimum value. Thus, in a transfer FRF, an antiresonance may not exist between every two consecutive resonance peaks. The antiresonances and the corresponding frequencies vary with the FRF and depend on the response and excitation DOFs. Hence, antiresonance is not a global property of the system like a natural frequency.

Antiresonances can be exploited in the dynamic design of systems. For example, by suitable modifications in the system, an antiresonance frequency for a given pair of excitation and response DOFs can be assigned or located at the desired frequency to ensure lower vibration at the response DOF.

Why do antiresonances occur in MDOF systems? And how can the antiresonance frequencies be determined? We answer these questions in the following sections.

7.5.3   MODAL CONTRIBUTIONS TO AN FRF

The antiresonances in the FRFs of an MDOF system can be explained based on the modal contributions of the various modes of the system. From Eq. (7.55), we see that an FRF of an N-DOF system can be expressed as a modal series in which each term of the series is related to the modal parameters of a mode. Rewriting Eq. (7.55) in the expanded form

$$\alpha_{jk}(\omega) = \frac{\phi_{j1}\ \phi_{k1}}{\omega_1^2 - \omega^2} + \frac{\phi_{j2}\ \phi_{k2}}{\omega_2^2 - \omega^2} + \ldots + \frac{\phi_{jr}\ \phi_{kr}}{\omega_r^2 - \omega^2} + \ldots + \frac{\phi_{jN}\ \phi_{kN}}{\omega_N^2 - \omega^2} \tag{7.56}$$

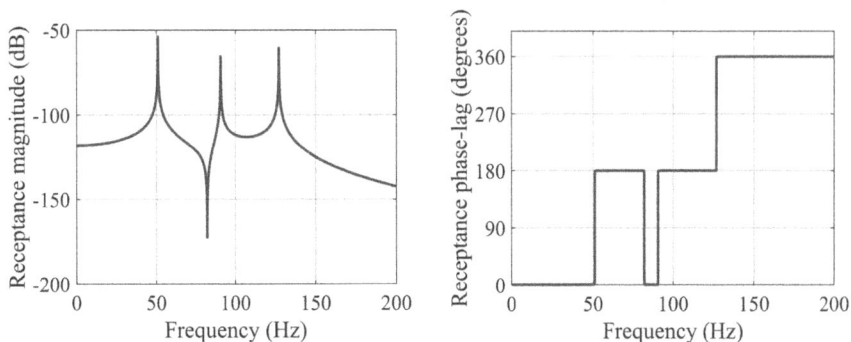

FIGURE 7.20   Magnitude and phase lag of a transfer receptance FRF of an undamped MDOF system.

The first term depends on ϕ_{j1} and ϕ_{k1}, which are the jth and k^{th} elements of the first mode shape, and ω_1, which is the natural frequency of the first mode. Thus, the first term represents the contribution of the first mode to the FRF α_{jk}. We can extend this argument to conclude that the rth term represents the modal contribution of the rth mode to the FRF α_{jk}.

For the 3 DOF system (Figure 7.18), the magnitude of the FRF α_{11}, along with the modal contributions of the three modes of the system, are shown in Figure 7.21. The modal contributions are like the FRFs of SDOF systems. The modal contribution of a particular mode is dominant around the corresponding resonance peak in the FRF, and the contribution reduces as the excitation frequency moves away from the natural frequency of the mode.

7.5.4 WHY DO ANTIRESONANCES EXIST IN MDOF SYSTEM FRFS?

We note from Section 7.5.3 that an FRF at every frequency consists of the sum of modal contributions. How these modal contributions combine, adding or subtracting, determines the FRF amplitude. For an undamped model, the modes are real, and there is no damping; hence the modal contributions are also real. Therefore, the algebraic signs of the modal contributions determine whether they are added or subtracted. The sign of a modal contribution depends on the signs of its numerator and the denominator. The numerator is a product of the eigenvector elements, like $\phi_{ir}\,\phi_{jr}$ for the rth term, referred to as the modal constant.

Let us study the antiresonances of the point FRF α_{11} of the 3 DOF system. To get a clearer picture, we plot the sign of the modal contributions with the FRF magnitude plotted in the background for reference, as shown in Figure 7.22. Let f_{n1}, f_{n2} and f_{n3} be the natural frequencies of the three modes, respectively, and f be the excitation frequency.

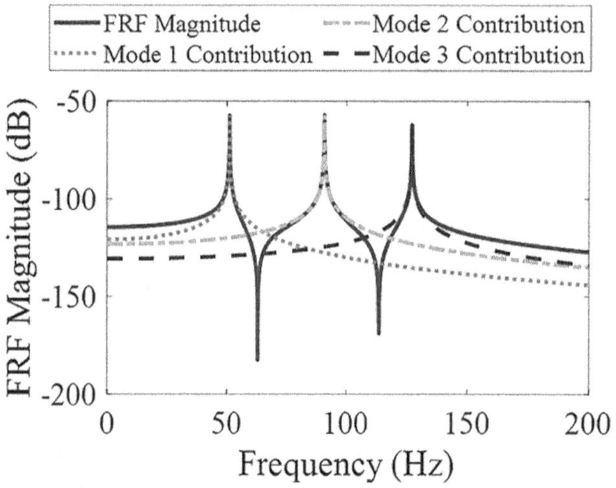

FIGURE 7.21   FRF magnitude and modal contributions.

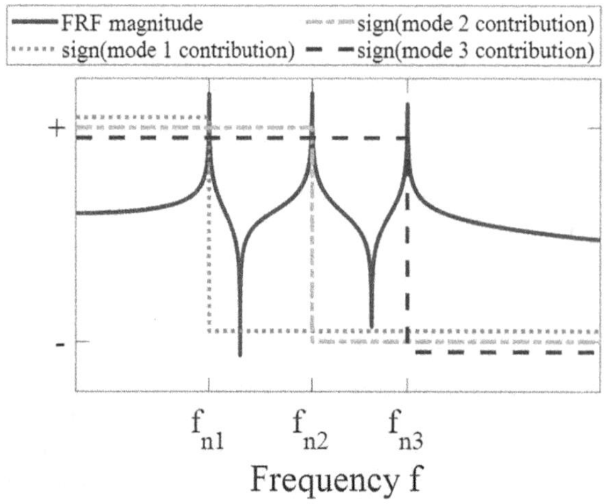

FIGURE 7.22   Signs of the modal contributions of a point FRF.

- For the frequency range $f < f_{n1}$ the three modal contributions have the same sign (positive sign); hence, they add up.
- But for $f_{n1} < f < f_{n2}$, the signs of mode 1 modal contribution and modes 2 and 3 modal contributions are opposite. From Figure 7.21, we see that the contribution of mode 1 decreases with the frequency while that of mode 2 increases. The contribution of mode 3 is relatively less. Due to these reasons, at some frequency in the range $f_{n1} < f < f_{n2}$, the sum of the contributions of the three modes is zero due to mutual cancellation, leading to an antiresonance at that frequency.
- In the range $f_{n2} < f < f_{n3}$, the signs of mode 3 modal contribution and modes 1 and 2 modal contributions are opposite. Thus, the complete mutual cancellation of the contributions occurs at some frequency leading to an antiresonance.

For an undamped system, the point FRF using Eq. (7.56) can be written as

$$\alpha_{ii}(\omega) = \frac{(\phi_{i1})^2}{\omega_1^2 - \omega^2} + \frac{(\phi_{i2})^2}{\omega_2^2 - \omega^2} + \ldots + \frac{(\phi_{ir})^2}{\omega_r^2 - \omega^2} + \ldots + \frac{(\phi_{iN})^2}{\omega_N^2 - \omega^2} \tag{7.57}$$

We see that the numerators or modal constants of all the terms on the RHS are positive, and therefore, the sign of each term (or modal contribution) is determined solely by the sign of its denominator. The sign of the modal contribution of a mode is positive to the left of the peak and negative to its right. Therefore, at a frequency between two adjacent peaks, the sign of the contributions of the modes to the left of the frequency is negative, while that to the right is positive. As a result, a complete mutual cancellation of the modal contributions always occurs at some frequency, leading to an antiresonance between every two consecutive resonance peaks.

We now consider a transfer FRF. The signs of the modal contributions of a transfer FRF, α_{12}, are shown in Figure 7.23.

- In the range $f_{n1} < f < f_{n2}$, the signs of mode 2 modal contribution and modes 1 and 3 contributions are opposite, leading to an antiresonance at some frequency in the range.
- However, in the range $f_{n2} < f < f_{n3}$, the signs of modal contributions of all three modes are identical. As a result, this range has no mutual cancellation and antiresonance. The modal contributions add up and attain a minimum value at some frequency in the range.
- We should note that, for a transfer FRF, the signs of the modal constants would not, in general, be identical for all the modes, as was the case for a point FRF, and their signs would also be a factor in determining the signs of the modal contributions. Therefore, for a transfer FRF, an antiresonance may or may not exist between two adjacent resonance peaks. Generally, the closer the FRF response and excitation DOFs more is the number of antiresonances.

7.5.5 ESTIMATION OF ANTIRESONANCE FREQUENCIES

We can analytically find the antiresonance frequencies in an FRF from the modal or spatial model of the system.

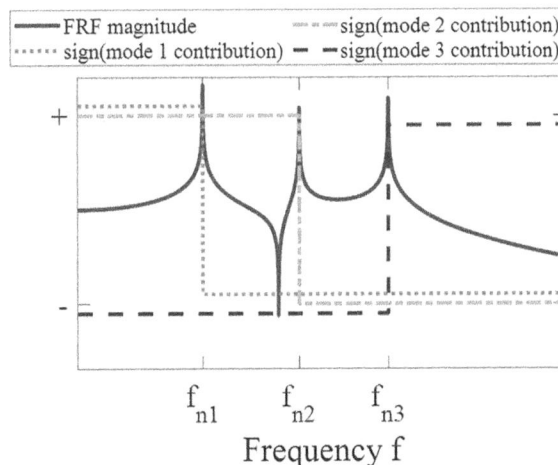

FIGURE 7.23 Sign of the modal contributions in a transfer FRF.

a. From the modal model

Equation (7.55) defines the relationship between the FRF α_{jk} and the modal model of the system. If the modal model is known, then the quantities ω_r^2 and $\{\phi\}_r$ for $r = 1, 2, \ldots N$ appearing on the RHS of this equation are known. To find the antiresonance frequencies of α_{jk}, it is expressed as a ratio of polynomials in ω^2. Thus, Eq. (7.55) is written as

$$\alpha_{jk}(\omega) = \frac{P(\omega^2)}{Q(\omega^2)} \qquad (7.58)$$

The numerator and denominator polynomials orders are $N - 1$ and N, respectively. The FRF magnitude is zero at antiresonance frequencies for an undamped system, and hence these frequencies correspond to the roots of the numerator polynomial $P(\omega^2)$. The roots may be real or complex, but only the positive real roots represent the antiresonance frequencies. Note that the roots of the polynomial $Q(\omega^2)$ are nothing but the natural/resonance frequencies of the system.

b. From the spatial model

The FRF matrix is related to the spatial model by the following relationship:

$$[\alpha] = [Z]^{-1} \qquad (7.59)$$

where $[Z]$ is the dynamic stiffness matrix of the system, and for an undamped system, it is given by

$$[Z] = [K] - \omega^2[M] \qquad (7.60)$$

Eq. (7.59), using the definition of the inverse of a matrix, can be written as

$$[\alpha] = \frac{\text{adjoint of } [Z]}{\det\ [Z]} \qquad (7.61)$$

$$[\alpha] = \frac{\text{Matrix of cofactors of } [Z]}{\det\ [Z]} \qquad (7.62)$$

The cofactor (j, k) of $[Z]$ is given by

$$\text{cofactor}(j, k) = (-1)^{j+k} \det\ [Z]^{jk} \qquad (7.63)$$

where $[Z]^{jk}$ is a matrix of order $N - 1$ obtained by deleting the jth row and the kth column of $[Z]$. We note from Eq. (7.62) that the $(j, k)^{th}$ element of $[\alpha]$ is governed by only the $(j, k)^{th}$ element of the matrix of cofactors of $[Z]$. Hence, in light of Eqs. (7.62) and (7.63), α_{jk} is

$$\alpha_{jk} = \frac{(-1)^{j+k} \det\ [Z]^{jk}}{\det\ [Z]} \qquad (7.64)$$

For an undamped system, the FRF magnitude is zero at the antiresonance frequencies of the FRF. Hence, these frequencies correspond to the positive real roots of the numerator of Eq. (7.64), which can be obtained by solving the following equation:

$$\det\ [Z]^{jk} = 0 \qquad (7.65)$$

Given Eq. (7.60), the above equation can be written as

$$\det\left([K]^{jk} - \omega^2[M]^{jk}\right) = 0 \qquad (7.66)$$

Eq. (7.66) is like a characteristic equation, and its solution is equivalent to the solution of the following eigenvalue problem:

$$[K]^{jk}\{y\} = \omega^2[M]^{jk}\{y\} \qquad (7.67)$$

We can imagine a hypothetical system corresponding to the eigenvalue problem in Eq. (7.67) (see Example 7.2). But the antiresonance frequencies correspond to only the positive eigenvalues of this system. The matrices $[M]^{jk}$ and $[K]^{jk}$, of the hypothetical system are symmetrical for j = k and unsymmetrical for j ≠ k. In the latter case, the eigenvalues may be positive, negative, or complex.

Thus, we can determine the antiresonance frequencies of an FRF by solving an eigenvalue problem in terms of the matrices obtained by deleting the row and column of the mass and stiffness matrices corresponding to the FRF's response and excitation DOFs.

7.5.6 STIFFNESS AND MASS LINES FOR MDOF SYSTEMS

For an SDOF system, we saw that the magnitude of the FRF can be approximated by stiffness and mass lines for frequencies away from the natural frequency. Similar lines could be drawn to approximate the magnitude of an FRF of an MDOF system. Such lines drawn for every resonance peak lead to a cluster of lines forming a skeleton of the FRF. However, due to multiple peaks, formulating the relationships between these lines and the system parameters is complicated for a general system. Therefore, the concept of stiffness and mass lines for the MDOF systems is not of much practical value.

7.6 CHARACTERISTICS OF FRFs OF MDOF SYSTEMS WITH STRUCTURAL DAMPING

We now study the characteristics of FRFs of MDOF systems with structural damping.

7.6.1 RELATIONSHIP OF THE FRFs WITH THE MODAL MODELS

The relationship between the FRF matrix and the modal model was derived in Chapter 6 based on the response to harmonic forces. In this section, we derive the relationship by an alternative route.

For an N-DOF system with proportional structural damping, the FRF matrix is related to the spatial model by the following equation:

$$[\alpha] = \left([K] + i[D] - \omega^2 [M] \right)^{-1} \tag{7.68}$$

Taking the inverse of Eq. (7.68), pre- and post-multiplying the resulting equation by $[\phi]^T$ and $[\phi]$ (the mass-normalized eigenvector matrix), respectively, and using the orthogonality properties, Eq. (7.68) becomes

$$[\phi]^T [\alpha]^{-1} [\phi] = \left[\,^{\backprime}\omega_r^2 (1 + i\eta_r) \,_{\backprime} \right] - \omega^2 [I] \tag{7.69}$$

Taking inverse

$$[\phi]^{-1} [\alpha] [\phi]^{-T} = \left[\,^{\backprime} \left(\omega_r^2 (1 + i\eta_r) - \omega^2 \right) \,_{\backprime} \right]^{-1} \tag{7.70}$$

Pre and post-multiplying by $[\phi]$ and $[\phi]^T$, respectively, gives

$$[\alpha] = [\phi] \left[\,^{\backprime} \left(\frac{1}{\omega_r^2 (1 + i\eta_r) - \omega^2} \right) \,_{\backprime} \right] [\phi]^T \tag{7.71}$$

Expanding the RHS, the above relationship can also be written as

$$[\alpha] = \sum_{r=1}^{N} \frac{\{\phi\}_r \{\phi\}_r^T}{\omega_r^2 - \omega^2 + i\eta_r \omega_r^2} \tag{7.72}$$

which is the same as derived in Chapter 6. The FRF $\alpha_{jk}(\omega)$ can be extracted from the FRF matrix as

$$\alpha_{jk}(\omega) = \sum_{r=1}^{N} \frac{\phi_{jr} \, \phi_{kr}}{\omega_r^2 - \omega^2 + i\eta_r \omega_r^2} \tag{7.73}$$

If the damping is nonproportional, then the expression can be derived following the above steps using the corresponding mass-normalized eigenvector matrix.

7.6.2 Antiresonances

We use the 3-DOF system (Figure 7.18) to study the antiresonances in a damped system by adding proportional structural damping to the system.

Figure 7.24 shows the magnitude and phase lag of the point receptance $\alpha_{11}(\omega)$ of the system. Compared to the corresponding FRF for the undamped system, where the resonance peaks were very sharp, for the damped system, the peaks are relatively less sharp, indicating the presence of damping. We also see 'inverted' peaks in the FRF, representing the antiresonances. Like resonances becoming less sharp in a damped system, the antiresonances are also less sharp than those in an undamped system.

For an undamped system, the magnitude of the FRF at an antiresonance frequency is zero. However, no such frequency exists for a damped system with zero FRF magnitude. The reason for this is as follows. An FRF for a damped system is complex. Therefore, the magnitude of the FRF is given by

$$\left|\alpha_{jk}(\omega)\right| = \sqrt{\left(\mathrm{Re}\left(\alpha_{jk}(\omega)\right)\right)^2 + \left(\mathrm{Im}\left(\alpha_{jk}(\omega)\right)\right)^2} \tag{7.74}$$

Thus, the magnitude of the FRF depends on both the real and imaginary parts. To get a better insight, we plot the magnitude and the real and imaginary parts of the FRF.

- Figure 7.25a shows an overlay of these graphs for $\alpha_{11}(\omega)$. We observe that the real part magnitude does become zero at some frequency between every two consecutive resonance peaks. But, since the imaginary part magnitude is not zero at any of these frequencies, the FRF magnitude is not zero.
- Similarly, for a transfer FRF ($\alpha_{12}(\omega)$) (shown in Figure 7.25b), at no frequency, both the real and imaginary parts of the FRF are zero.

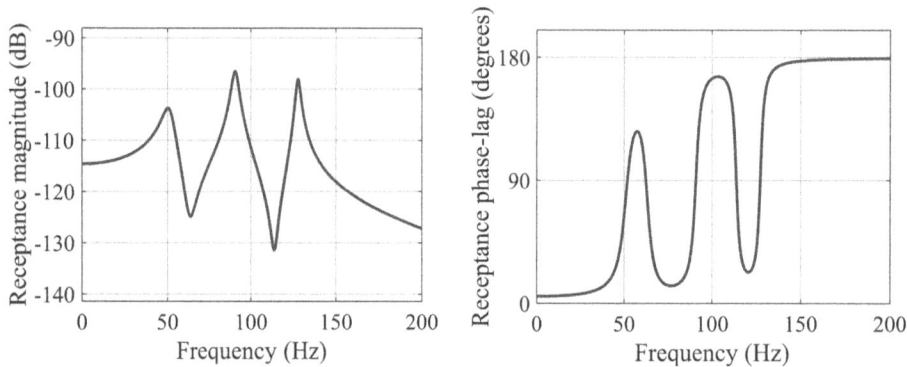

FIGURE 7.24   Magnitude and phase lag of the point receptance $\alpha_{11}(\omega)$ of the 3 DOF system with structural damping.

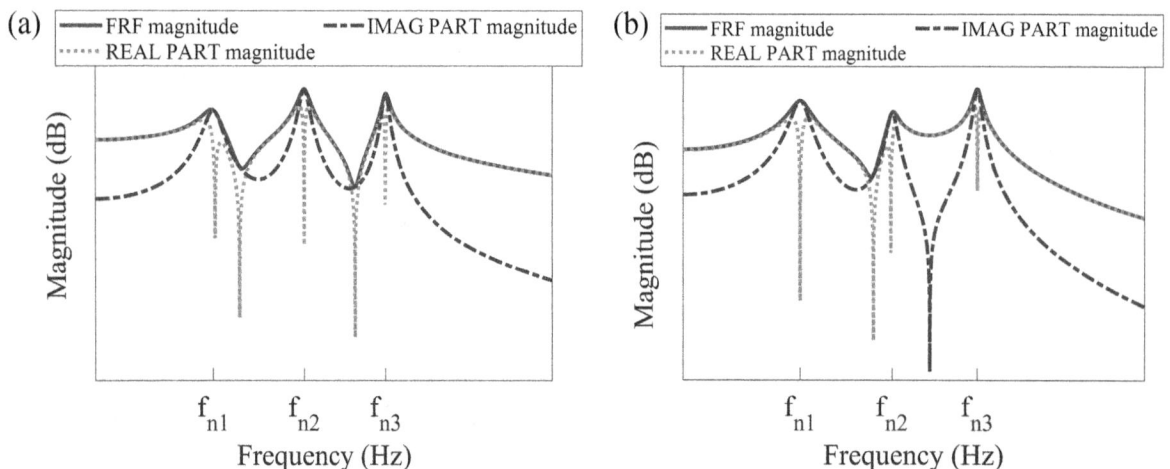

FIGURE 7.25   Overlay of the magnitudes of FRF and its real and imaginary parts (a) $\alpha_{11}(\omega)$, (b) $\alpha_{21}(\omega)$.

- Thus, the FRF magnitude is not zero at an antiresonance in an FRF of a damped system. Note that in an undamped system, the imaginary part of the FRF is zero; hence, the FRF magnitude becomes zero whenever the real part of the FRF becomes zero.

The antiresonance frequencies of an undamped system are determined based on the fact that the FRF magnitude is zero at these frequencies. However, an FRF of a damped system doesn't satisfy this property, and hence the methods used for determining the antiresonance frequencies of an undamped system can't be used in this case to find the antiresonance frequencies. But, the antiresonance frequencies of a damped system can be estimated approximately by treating the system as undamped.

We note from the phase plot for the damped system that no longer the phase changes abruptly at resonances and anti-resonances, but the change is gradual.

7.6.3 REAL, IMAGINARY, AND NYQUIST PLOTS

The real and imaginary plots and the Nyquist plot are based on the absolute (and not dB) values of an FRF. The absolute values are too high around resonance frequencies for an undamped system. Therefore, these plots for undamped systems don't reveal much information about the FRF. However, the amplitudes in the resonance regions reduce due to damping for damped systems, making these plots meaningful.

Figure 7.26 shows the real and imaginary parts of the point receptance $\alpha_{11}(\omega)$ of the 3 DOF system with proportional structural damping. Figure 7.27 shows these plots for the transfer receptance $\alpha_{21}(\omega)$. Like in damped SDOF systems, the real part crosses the zero line in the resonance regions while the imaginary part peaks up in these regions. But there are some deviations compared to damped SDOF systems. The real part no longer crosses the zero line precisely at the undamped natural frequencies of the system. Also, the frequencies at which the imaginary part attains a maximum value are not coincident with the system's undamped natural frequencies. These characteristics are also valid for systems with nonproportional damping.

Figure 7.28 shows the Nyquist plot over the frequency range of 0–200 Hz of the point receptance $\alpha_{11}(\omega)$. The undamped natural frequencies of the three modes are also marked. The plot has three loops corresponding to the three resonance peaks in the FRF. However, these loops are not parts of the exact circles. We can understand this as follows. For the 3 DOF system, $\alpha_{11}(\omega)$ is

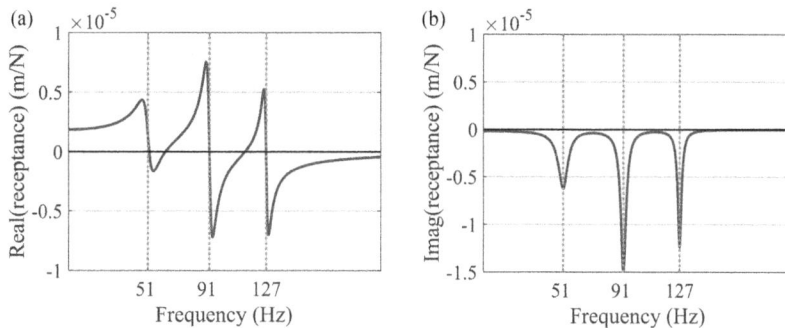

FIGURE 7.26   (a) Real and (b) imaginary parts of the point receptance $\alpha_{11}(\omega)$ of the 3 DOF system with structural damping.

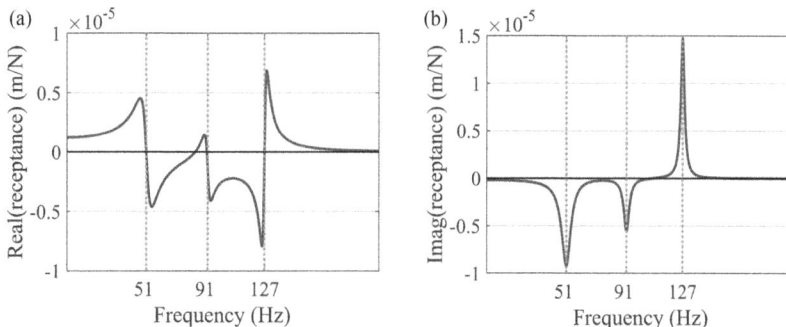

FIGURE 7.27   (a) Real and (b) imaginary parts of the transfer receptance $\alpha_{21}(\omega)$ of the 3 DOF system with structural damping.

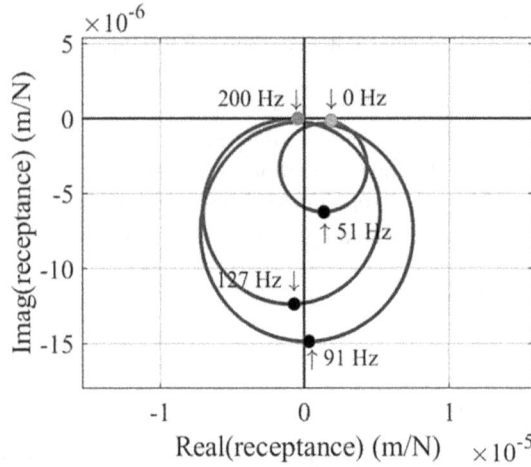

FIGURE 7.28 Nyquist plot of the point receptance $\alpha_{11}(\omega)$ of the 3 DOF system with proportional structural damping.

$$\alpha_{11}(\omega) = \frac{\phi_{11}\,\phi_{11}}{\omega_1^2 - \omega^2 + i\eta_1\omega_1^2} + \frac{\phi_{12}\,\phi_{12}}{\omega_2^2 - \omega^2 + i\eta_2\omega_2^2} + \frac{\phi_{13}\,\phi_{13}}{\omega_3^2 - \omega^2 + i\eta_3\omega_3^2} \tag{7.75}$$

In a narrow band of the frequency range around, say, the first resonance peak, the modal contribution of the first mode, represented by the first term in Eq. (7.75), is dominant and traces a circular arc when plotted on the Nyquist plane. However, the contribution of modes 2 and 3, represented by the two remaining terms, distorts the circular nature of the Nyquist plot of the first term. The extent of the distortion depends on the closeness of the resonance peaks and the damping in the system. The closer the peaks and higher the damping, the higher the deviation of a loop from a circle-like character. The contribution from other modes also displaces a loop from being centered on the imaginary axis.

Figure 7.29 shows the Nyquist plot of the transfer receptance $\alpha_{21}(\omega)$. The loop corresponding to the second mode is shifted opposite to the loops corresponding to modes 1 and 3. We should note that a transfer FRF is associated with different response and excitation DOFs, and hence the sign of the modal constants for different modes may be different, affecting the locations of the various loops of the Nyquist plot.

We now add some nonproportional damping to the 3 DOF system. The Nyquist plot of the transfer receptance $\alpha_{21}(\omega)$ of this system is shown in Figure 7.30. We see that the portions of the Nyquist plot corresponding to different modes rotate significantly in the complex plane. The rotation occurs due to the complex nature of the eigenvectors, which makes the modal constants also complex. However, in general, one can't say much about the nature of the damping, i.e., proportional or nonproportional, just based on the nature of the Nyquist plot.

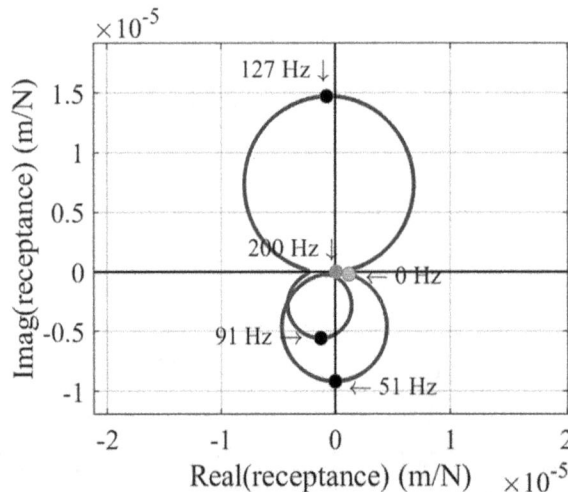

FIGURE 7.29 Nyquist plot of the transfer receptance $\alpha_{21}(\omega)$ of the 3 DOF system with proportional structural damping.

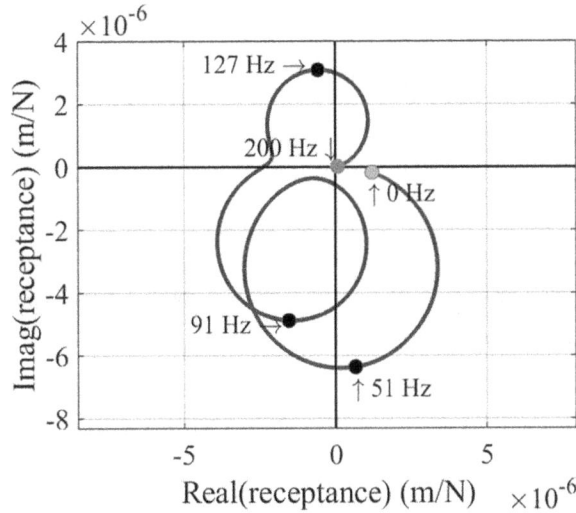

FIGURE 7.30 Nyquist plot of transfer receptance $\alpha_{21}(\omega)$ of the 3 DOF system with nonproportional structural damping.

7.7 CHARACTERISTICS OF FRFs OF MDOF SYSTEMS WITH VISCOUS DAMPING

We now study the characteristics of FRFs of MDOF systems with viscous damping.

7.7.1 RELATIONSHIP OF THE FRFs WITH THE MODAL MODEL

a. Proportional viscous damping

We proceed as in the previous sections to derive the relationship of the FRF matrix with the modal model by an alternative route.

For an N-DOF system with proportional viscous damping, the FRF matrix is related to the spatial model by

$$[\alpha] = \left([K] - \omega^2[M] + i\omega[C]\right)^{-1} \tag{7.76}$$

By following the steps in Section 7.6.1, we get the following relationship between the FRF matrix and the modal model:

$$[\alpha] = [\phi]\left[\left(\frac{1}{\omega_r^2 - \omega^2 + j2\xi_r\omega_r\omega}\right)\right][\phi]^T \tag{7.77}$$

Expanding the RHS, the relationship can be written as

$$[\alpha] = \sum_{r=1}^{N} \frac{\{\phi\}_r\{\phi\}_r^T}{\omega_r^2 - \omega^2 + j2\xi_r\omega_r\omega} \tag{7.78}$$

which is the same as derived in Chapter 6. A particular FRF $\alpha_{jk}(\omega)$ can be extracted from the FRF matrix as

$$\alpha_{jk}(\omega) = \sum_{r=1}^{N} \frac{\phi_{jr}\,\phi_{kr}}{\omega_r^2 - \omega^2 + i2\xi_r\omega_r\omega} \tag{7.79}$$

b. Nonproportional viscous damping

For an N-DOF system with nonproportional viscous damping, Eq. (7.76) is valid with only the difference that the matrix $[C]$ represents the nonproportional viscous damping matrix. However, we can't derive the relationship of the FRF matrix with the modal model by starting from this equation. It is because the mass, stiffness, and damping matrices cannot be diagonalized in physical/displacement coordinates for a system with nonproportional viscous damping. However, we have seen in Chapter 6 that the diagonalization of the matrices in state space can be done. Hence, we start from a similar equation in state space.

For a harmonic excitation force, from Eq. (6.203) (Chapter 6), we get

$$\{z(\omega)\} = \left(i\omega[A]+[B]\right)^{-1}\{\bar{F}(\omega)\} \tag{7.80}$$

Hence, we can write the relationship between the FRF matrix in state space (denoted by $[\bar{\alpha}]$) and the state space matrices as

$$[\bar{\alpha}]^{-1} = i\omega[A]+[B] \tag{7.81}$$

Pre and post-multiplying by $[\bar{\phi}]^T$ and $[\bar{\phi}]$ (the normalized eigenvector matrix in state space), respectively, and using the orthogonality properties, we get

$$[\bar{\phi}]^T[\bar{\alpha}]^{-1}[\bar{\phi}] = i\omega[I]-[\bar{s}_r\backslash] \tag{7.82}$$

Taking the inverse of both the sides and pre and post-multiplying the resulting equation by $[\bar{\phi}]$ and $[\bar{\phi}]^T$, respectively, gives

$$[\bar{\alpha}] = [\bar{\phi}]\left[\left(\frac{1}{i\omega-\bar{s}_r}\right)\right][\bar{\phi}]^T \tag{7.83}$$

Expanding the RHS, the relationship can be written as a sum over 2N complex modes

$$[\bar{\alpha}] = \sum_{r=1}^{2N} \frac{\{\bar{\phi}\}_r\{\bar{\phi}\}_r^T}{i\omega-\bar{s}_r} \tag{7.84}$$

Given the complex conjugate nature of the state space eigenvalues and the eigenvectors, the equation can be written as a sum over N pairs of complex conjugate modes

$$[\bar{\alpha}] = \sum_{r=1}^{N}\left(\frac{\{\bar{\phi}\}_r\{\bar{\phi}\}_r^T}{i\omega-s_r}+\frac{\{\bar{\phi}\}_r^*\{\bar{\phi}\}_r^{*T}}{i\omega-s_r^*}\right) \tag{7.85}$$

The first half of the eigenvectors corresponds to the displacements, while the second half corresponds to the velocities. The FRF matrix corresponding to the displacements (denoted by $[\alpha]$) is given by

$$[\alpha] = \sum_{r=1}^{N}\left(\frac{\{\phi\}_r\{\phi\}_r^T}{i\omega-s_r}+\frac{\{\phi\}_r^*\{\phi\}_r^{*T}}{i\omega-s_r^*}\right) \tag{7.86}$$

which is the same as derived in Chapter 6. A particular FRF $\alpha_{jk}(\omega)$ can be extracted from the FRF matrix as

$$\alpha_{jk}(\omega) = \sum_{r=1}^{N}\left(\frac{\phi_{jr}\,\phi_{kr}}{i\omega-s_r}+\frac{\phi_{jr}^*\,\phi_{kr}^*}{i\omega-s_r^*}\right) \tag{7.87}$$

7.7.2   FRF CHARACTERISTICS

For an SDOF system with viscous damping, we saw that the mobility plot is a circle in the Nyquist plane. But this property is not true for MDOF systems. For the 3 DOF system with proportional viscous damping added, the Nyquist plots for the mobilities $Y_{11}(\omega)$ and $Y_{21}(\omega)$ over the frequency range 0–200 Hz are shown in Figure 7.31a and b, respectively. The undamped natural frequencies of the three modes are also marked. There are three loops in the Nyquist plot corresponding to the three resonance peaks in the FRFs. But each loop is no longer a circular arc.

The observations made in Section 7.6 about antiresonances and real and imaginary parts of the FRFs for an MDOF system with structural damping also apply to systems with viscous damping.

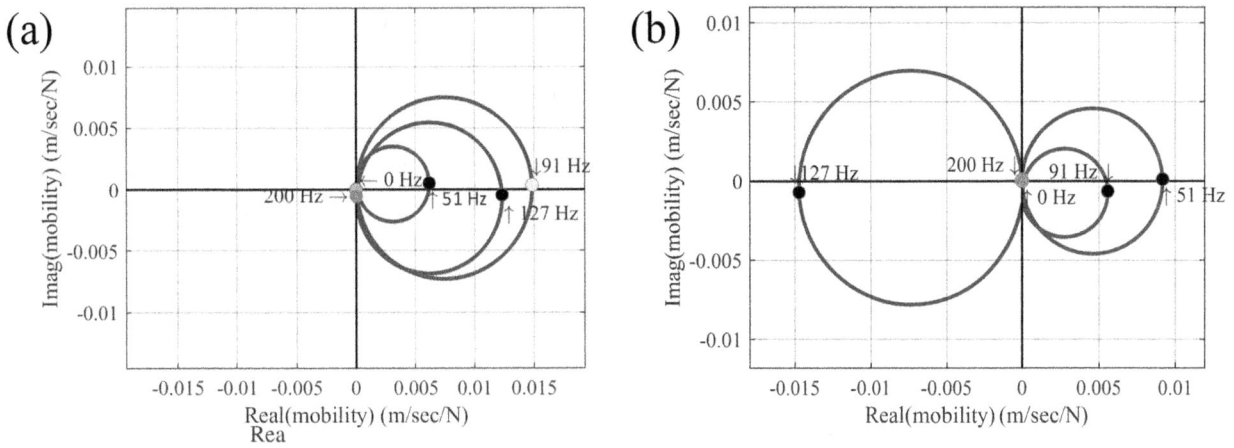

FIGURE 7.31 Nyquist plot of mobilities (a) $Y_{11}(\omega)$, (b) $Y_{21}(\omega)$ of the 3 DOF system with proportional viscous damping.

Example 7.1

Figure 7.32 shows the receptance of an SDOF system with structural damping. Find the mass, stiffness, and damping coefficient of the system using the concepts of stiffness and mass lines.

Solution

- We draw the asymptote to the FRF in the low-frequency range with a slope of 0 dB/decade, giving the stiffness line as shown in Figure 7.33. The intercept on the y-axis is −67.8 dB.

$$\text{magnitude in dB} = -20\log_{10} k$$

$$-67.8 = -20\log_{10} k \qquad \Rightarrow \qquad k = 2454.7 \text{ N/m}$$

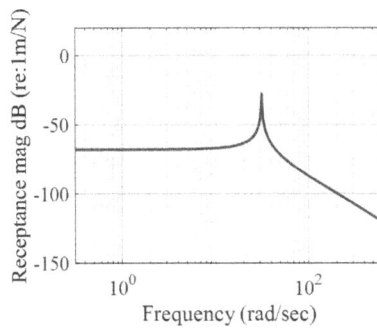

FIGURE 7.32 Receptance of an SDOF system with structural damping.

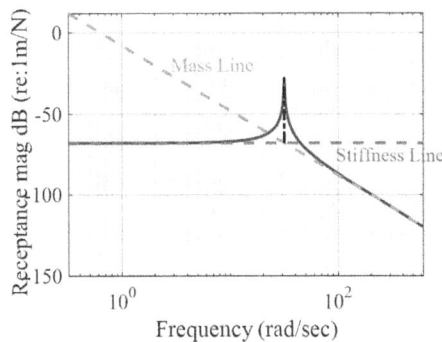

FIGURE 7.33 Stiffness and mass lines.

- We draw the asymptote to the FRF in the high-frequency range with a slope of −40dB/decade, giving the mass line as shown in Figure 7.33. The intercept on the ordinate at $\omega = 1$ rad/sec is −7.9 dB.

$$\text{magnitude in dB} = -20\log_{10} m$$

$$-7.9 = -20\log_{10} m \qquad \Rightarrow \qquad m = 2.4831 \text{ kg}$$

- From the dB plot, determine $\Delta\alpha$, which is the height of the FRF (in dB) above the stiffness line at the frequency corresponding to the intersection of stiffness and mass lines (as shown in the figure).

$$\Delta\alpha = -27.8 - (-67.8) = 40 \text{ dB}$$

$$\eta = 10^{\frac{-\Delta\alpha}{20}} = 10^{\frac{-40}{20}} = 0.01$$

$$h = k\eta = 24.5471 \text{ N/m}$$

Example 7.2

Determine the antiresonance frequencies in the FRFs α_{11}, α_{12} and α_{13} of the 3 DOF undamped system (Figure 7.18) using the spatial model of the system.

Solution

The mass and stiffness matrices and the displacement vector of the system can be written as

$$[M] = \begin{bmatrix} m_1 & 0 & 0 \\ 0 & m_2 & 0 \\ 0 & 0 & m_3 \end{bmatrix}; \quad [K] = \begin{bmatrix} k_1 + k_2 & -k_2 & 0 \\ -k_2 & k_2 + k_3 & -k_3 \\ 0 & -k_3 & k_3 + k_4 \end{bmatrix}; \quad \{x\} = \begin{Bmatrix} x_1 \\ x_2 \\ x_3 \end{Bmatrix}$$

Antiresonance frequencies for the FRF α_{11}: We obtain the matrix $[M]^{11}$ by deleting the 1st row and the 1st column of $[M]$. Similarly, $[K]^{11}$ is obtained from $[K]$. We get

$$[M]^{11} = \begin{bmatrix} m_2 & 0 \\ 0 & m_3 \end{bmatrix} \quad [K]^{11} = \begin{bmatrix} k_2 + k_3 & -k_3 \\ -k_3 & k_3 + k_4 \end{bmatrix}$$

Eq. (7.87) requires that,

$$\begin{vmatrix} k_2 + k_3 - \omega^2 m_2 & -k_3 \\ -k_3 & k_3 + k_4 - \omega^2 m_3 \end{vmatrix} = 0$$

It results in a quadratic equation and yields the following two positive real values of ω:

$$\omega_{1,2} = \sqrt{\frac{k_2 + k_3}{2m_2} + \frac{k_3 + k_4}{2m_3} \pm \sqrt{\left(\frac{k_2 + k_3}{2m_2} + \frac{k_3 + k_4}{2m_3}\right)^2 - \left(\frac{k_2 k_3 + k_2 k_4 + k_3 k_4}{m_2 m_3}\right)}}$$

These are nothing but the antiresonance frequencies of α_{11}. It matches the fact that a point FRF of a 3-DOF system must have two antiresonance frequencies.

These antiresonance frequencies are also the natural frequencies of a system obtained by fixing the DOF x_1 of the system. The resulting system is shown in Figure 7.34, in which only DOFs x_2 and x_3 are active.

Antiresonance frequencies for the FRF α_{12}: We obtain the matrix $[M]^{12}$ by deleting the first row and the second column of $[M]$. Similarly, $[K]^{12}$ is obtained from $[K]$. We get

$$[M]^{12} = \begin{bmatrix} 0 & 0 \\ 0 & m_3 \end{bmatrix} \quad [K]^{12} = \begin{bmatrix} -k_2 & -k_3 \\ 0 & k_3 + k_4 \end{bmatrix}$$

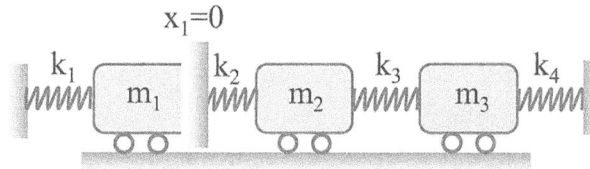

FIGURE 7.34 System obtained by fixing the DOF x_1.

Because of Eq. (7.87)

$$\begin{vmatrix} -k_2 & -k_3 \\ 0 & k_3 + k_4 - \omega^2 m_3 \end{vmatrix} = 0$$

The solution gives only one positive real value of ω, and hence the FRF α_{12} has only one antiresonance frequency, which is

$$\omega = \sqrt{\frac{k_3 + k_4}{m_3}}$$

Antiresonance frequencies for the FRF α_{13}: We obtain the matrix $[M]^{13}$ by deleting the first row and the third column of $[M]$. Similarly, $[K]^{13}$ is obtained from $[K]$. We get

$$[M]^{13} = \begin{bmatrix} 0 & m_2 \\ 0 & 0 \end{bmatrix} \quad [K]^{13} = \begin{bmatrix} -k_2 & k_2 + k_3 \\ 0 & -k_3 \end{bmatrix}$$

Because of Eq. (7.87)

$$\begin{vmatrix} -k_2 & k_2 + k_3 - \omega^2 m_2 \\ 0 & -k_3 \end{vmatrix} = 0$$

There is no root, and hence the FRF α_{13} has no antiresonance.

REVIEW QUESTIONS

1. What are the stiffness and mass lines for an SDOF system?
2. Show that for an SDOF system with viscous damping, the FRF can be approximated by stiffness and mass lines in the low and high-frequency ranges, respectively.
3. Why is it difficult to measure the dynamic stiffness of an MDOF system?
4. What is an antiresonance frequency? Is it a global property of the system, like a resonance frequency? Give reasons for your answer.
5. Why does a point FRF has antiresonance frequencies between every two consecutive resonance peaks?
6. Show that the antiresonance frequencies in a point FRF correspond to the eigenvalues of the system obtained by fixing the response/excitation DOF.
7. For an MDOF system with structural damping, derive the FRF matrix and modal model relationship, starting from the FRF matrix and spatial model relationship.
8. For an MDOF system with nonproportional viscous damping, derive the FRF matrix and modal model relationship, starting from the FRF matrix and spatial model relationship.

PROBLEMS

Problem 7.1: Figure 7.35 shows the mobility of an SDOF system. Find the mass and stiffness of the system using the concepts of stiffness and mass lines.

Problem 7.2: Figure 7.36 shows the receptance of an SDOF system with viscous damping. Find the mass, stiffness, and damping coefficient of the system using the concepts of stiffness and mass lines.

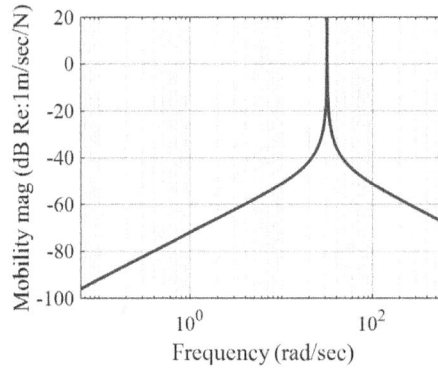

FIGURE 7.35   Mobility of an SDOF system.

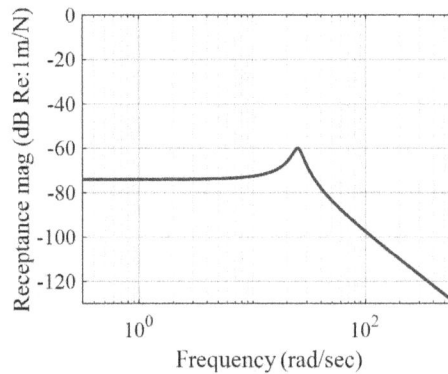

FIGURE 7.36   Receptance of an SDOF system with viscous damping.

Problem 7.3: Determine the antiresonance frequencies in the FRF α_{22} of the undamped 2-DOF system using its modal model given below.

NATURAL FREQUENCIES IN Hz :	MASS-NORMALIZED EIGENVECTORS :
9.0770	−0.4538 −0.2098
18.2688	−0.1586 0.3431

Problem 7.4: Determine the antiresonance frequencies in the FRFs α_{33} and α_{32} of the 3-DOF system using its spatial model given below.

MASS MATRIX :	STIFFNESS MATRIX :
2 0 0	310,000 −140,000 0
0 8 0	−140,000 150,000 −80,000
0 0 4	0 −80,000 250,000

8 Signal Processing for Experimental Modal Analysis

8.1 INTRODUCTION

In Chapter 7, we studied the characteristics of FRFs. They form the fundamental inputs for estimating the modal parameters by experimental modal analysis (EMA). We also studied in the previous chapters that an FRF is the ratio of the response and force in the frequency domain. The frequency-domain representation of the response and force signals can be obtained by transforming them from time to frequency domain. Thus, the transformation from time to frequency domain is crucial for FRF estimation. The signals encountered in EMA may be periodic, transient, or random; hence, understanding the transformation of these signals from time to frequency domain is one of the objectives of this chapter. In practice, the signal while recording is discretized, and it is the discrete signal that is transformed into the frequency domain. Therefore, another objective of this chapter is to introduce the basic theory of digital signal processing (DSP) to obtain the frequency-domain representation of a discrete signal.

This chapter starts with the theory of Fourier analysis of the periodic and aperiodic continuous-time signals. It forms the basis for understanding the Fourier analysis of discrete-time signals. A sound understanding of the Fourier analysis of continuous and discrete-time signals is necessary to interpret the frequency-domain representation correctly and avoid measurement errors.

This chapter then considers the analysis of discrete signals and develops the concepts of discrete Fourier series (DFS), discrete-time Fourier transform (DTFT), and discrete Fourier transform (DFT). The relationships between the parameters of the discrete-time signal and its DFT are presented. The pitfalls of using DSP (such as aliasing, leakage, and quantization noise) and the considerations to minimize the associated errors are discussed. The concept of windowing is described, and its effect on continuous and discrete signals is analyzed.

In the end, we consider the frequency-domain representation of random signals. The concepts of auto/cross-correlation functions and auto/cross-power spectral densities are presented. The estimation of these functions based on discrete signals is described. The concepts of white noise and band-limited white noise signals are also discussed.

The topics covered in this chapter provide a foundation and the necessary inputs for carrying out the signal processing for performing EMA to be covered in the following chapters.

8.2 CONTINUOUS-TIME SIGNALS

A signal conveys information as a function of an independent variable. A continuous-time signal is a signal in which the independent variable is time, and the signal is defined for every instant. Examples are the acceleration of a vibrating structure or the force transmitted by a machine to its support. A continuous-time signal is henceforth simply referred to as a continuous signal.

Continuous signals can be classified, as shown in Figure 8.1. A continuous signal, also called an analog signal, can be periodic or aperiodic (i.e., not periodic). A periodic signal repeats after some time (T) and satisfies the property that $x(t + nT) = x(t)$ for any t, and an integer n. It can be harmonic or non-harmonic.

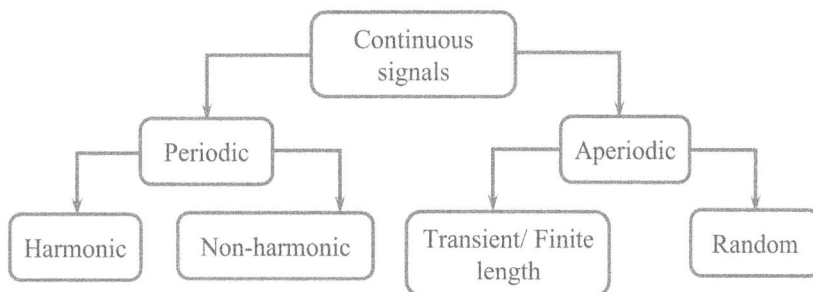

FIGURE 8.1 Classification of continuous signals.

DOI: 10.1201/9780429454783-8

A periodic time signal is harmonic if the second derivative of the signal with time is proportional to but the negative of the signal itself. That is $\ddot{x}(t) = -\omega^2 x(t)$. There are signals that don't satisfy this property and hence they are non-harmonic but are still periodic. For example, a sum of two harmonic signals of different frequencies is not harmonic though periodic. Transient (and any finite length) and random signals are aperiodic signals because they don't repeat.

8.3   FOURIER ANALYSIS OF PERIODIC CONTINUOUS SIGNALS

The Fourier analysis is a crucial tool for signal analysis and can be applied to signals of any physical nature. Hence, it is widely used in many areas of science and engineering. We are generally concerned with continuous signals such as displacement, velocity, acceleration, and force (or the proportionate voltage signals) in EMA. In this section, we consider the Fourier analysis of continuous periodic signals.

8.3.1   FOURIER SERIES EXPANSION (TRIGONOMETRIC FORM)

The Fourier series expresses any periodic signal x(t), with period T seconds, as a sum of sine and cosine signals with frequencies integer multiples of the signal's fundamental frequency. The sine and cosine signals are harmonic signals, and the Fourier analysis expands a given signal in terms of these harmonic signals expressed by the following equation:

$$x(t) = a_o + \sum_{n=1}^{\infty} a_n \cos n\omega_o t + \sum_{n=1}^{\infty} b_n \sin n\omega_o t \tag{8.1}$$

where ω_o is the fundamental frequency of the signal given by

$$\omega_o = \frac{2\pi}{T} \text{ rad/sec} \tag{8.2}$$

In Eq. (8.1), a_o is the constant component in the signal x(t) and is often referred to as the 'dc' component (like the DC voltage, which is a constant voltage). From the RHS of Eq. (8.1), it is clear that the frequencies of the sin and cosine signals/components are ω_o, $2\omega_o$, $3\omega_o$,..... In general, an infinite number of frequency components may be required to expand a given signal using the Fourier series. The harmonic signal of frequency ω_o is called the fundamental harmonic, and those with the frequencies $2\omega_o$, $3\omega_o$,.... are called higher harmonics.

Now, the question is, how do we find out the amplitudes a_o, a_n and b_n? These are found using the orthogonality properties of the sine and cosine functions. If p and q are nonzero positive integers, then we obtain the following results, which together make up the orthogonality properties of the sine and cosine functions:

$$\int_0^T \cos p\omega_o t \cos q\omega_o t.dt = \frac{1}{2}\int_0^T \left(\cos(p+q)\omega_o t + \cos(p-q)\omega_o t \right).dt = 0 \quad \text{for } p \neq q \tag{8.3}$$

$$\int_0^T \sin p\omega_o t \sin q\omega_o t.dt = \frac{1}{2}\int_0^T \left(\cos(p-q)\omega_o t - \cos(p+q)\omega_o t \right).dt = 0 \quad \text{for } p \neq q \tag{8.4}$$

$$\int_0^T \sin p\omega_o t \cos q\omega_o t.dt = 0 = \frac{1}{2}\int_0^T \left(\sin(p+q)\omega_o t + \sin(p-q)\omega_o t \right).dt = 0 \quad \text{for any p and q} \tag{8.5}$$

It is also seen that

$$\int_0^T \sin p\omega_o t.dt = 0 \tag{8.6}$$

$$\int_0^T \cos p\omega_o t.dt = 0 \tag{8.7}$$

To find the coefficient a_r, which is nothing but the amplitude of the harmonic signal $\cos r\omega_o t$, we multiply both sides of Eq. (8.1) by $\cos r\omega_o t$ and integrate over time 0 to T. Making use of the properties in Eqs. (8.3), (8.5), and (8.7), all the terms on the RHS, except the term involving a_r, are integrated to zero, leading to

$$a_r = \frac{2}{T} \int_0^T x(t) \cos r\omega_o t.dt \qquad (8.8)$$

Similarly, to find the coefficient b_r, which is the amplitude of the harmonic signal $\sin r\omega_o t$, we multiply both sides of Eq. (8.1) by $\sin r\omega_o t$ and integrate over time 0 to T. Using the results in Eqs. (8.4)–(8.6), all the terms on the RHS, except the term involving b_r, is integrated to zero, and we get

$$b_r = \frac{2}{T} \int_0^T x(t) \sin r\omega_o t.dt \qquad (8.9)$$

Eqs. (8.8) and (8.9) can be used to find the coefficients corresponding to $r = 1, 2, 3, \ldots$. To find the coefficient a_o, we integrate Eq. (8.1) over time 0 to T. Because of Eqs. (8.6) and (8.7), we obtain

$$a_o = \frac{1}{T} \int_0^T x(t).dt \qquad (8.10)$$

Eq. (8.10) indicates that a_o is the average value of the signal.

The Fourier series expansion (Eq. 8.1) can also be expressed in an alternative trigonometric form by combining the sine and cosine signals corresponding to the same frequency. It leads to

$$x(t) = a_o + \sum_{n=1}^{\infty} d_n \cos(n\omega_o t - \theta_n) \qquad (8.11)$$

where

$$d_n = \sqrt{a_n^2 + b_n^2} \qquad (8.12)$$

and

$$\theta_n = \cos^{-1} \frac{a_n}{\sqrt{a_n^2 + b_n^2}} \qquad (8.13)$$

a. Time and frequency-domain representations, Fourier spectrums

As per Eq. (8.11), each frequency component on the RHS is characterized by its frequency $n\omega_o$, amplitude d_n, and phase lag θ_n. The dc component a_o can be considered the amplitude of a cosine signal with zero frequency and phase. These characteristics of the frequency components making up the original signal are often presented as **Fourier spectrums**, amplitude/magnitude, and phase spectrums. These are referred to as line spectrums since the frequency components exist only at discrete frequencies 0, ω_o, $2\omega_o$, $3\omega_o$,…..

For example, Figure 8.2 shows a square wave periodic signal, and Figure 8.3 shows the Fourier spectrums (magnitude and phase spectrums).

The representation of a signal as a function of time (like in Figure 8.2) is its **time-domain representation**. On the other hand, the representation of the signal in which the frequency, amplitude, and phase of the harmonic components making up the signal are specified is called the **frequency-domain representation**. The details of the frequency components can be in the form of a table or the Fourier magnitude and phase spectrums (as shown in Figure 8.3). In a nutshell, the LHS of Eq. (8.11) is the time-domain representation of the signal, while its RHS is the frequency-domain representation. The two representations are equivalent.

It is generally difficult to infer what the signal consists of from its time-domain representation. But the frequency-domain representation reveals the frequency components making up the signal. This information may give vital clues regarding the origin of these frequency components and the probable causes of excessive vibration.

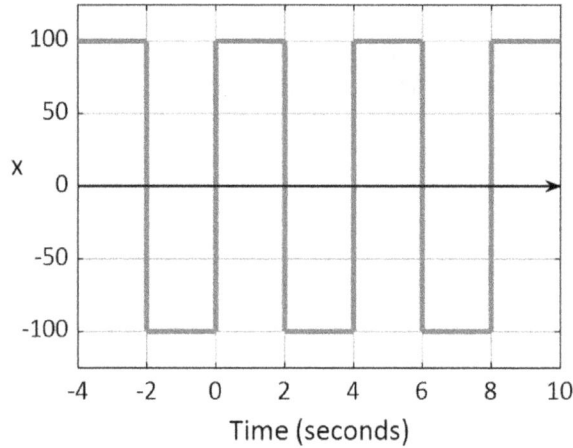

FIGURE 8.2 Square wave periodic signal.

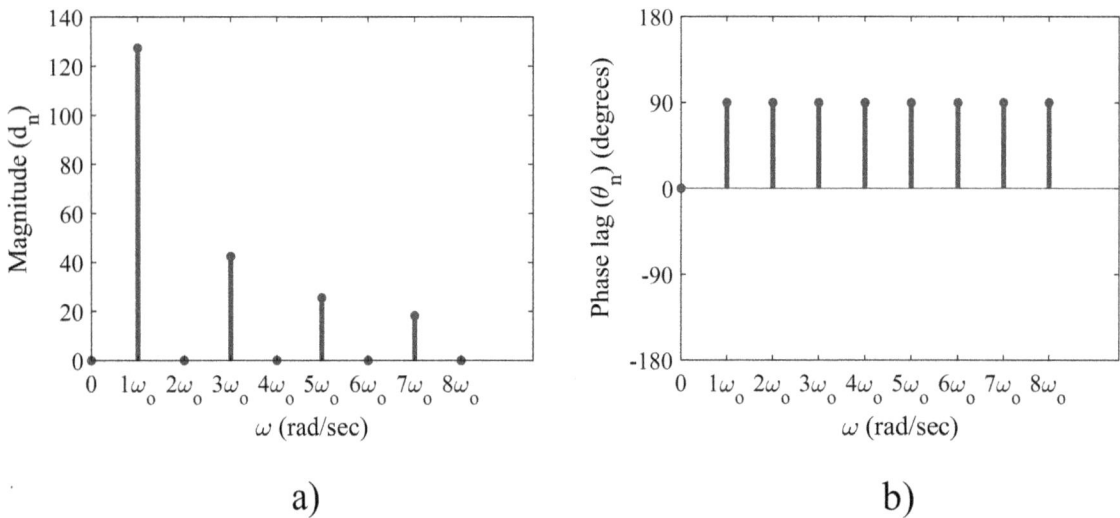

a)   b)

FIGURE 8.3 Fourier spectrums (a) magnitude and (b) phase.

Just as a signal can be expanded using the Fourier series and represented in the frequency domain, it can be reconstructed using its frequency-domain representation through an inverse process. The signal reconstruction is done by substituting the parameters of the frequency-domain representation into the RHS of Eq. (8.11).

In general, an infinite number of frequency components may be required in the Fourier series expansion for the exact signal representation. But, in practice, the series is truncated, and a finite number of components is used to obtain a reasonably accurate representation. The adequacy of the number of frequency components can be judged by comparing the reconstructed and original signals. Figure 8.4 compares the square wave signal with its reconstructed versions when the 2, 4, 8, and 16 nonzero frequency (or Fourier) components are used for the signal reconstruction.

It is observed from the figure that as the number of the frequency components increase, the error of approximation reduces, and the reconstructed signal gets closer to the original signal. We also note that in the present example, the reconstructed signal has an overshoot, and the overshoot persists even when the number of frequency components increases. This behavior is called the **Gibbs phenomenon**, which explains the difficulty in approximating a signal with discontinuities (like the square wave signal which has a jump discontinuity) using a finite number of continuous harmonic signals. However, it is found that as the number of frequency components increases, the overshoot's location moves closer to the jump discontinuity, and the error of approximation reduces.

The sin and cosine functions used in the Fourier series expansion are orthogonal functions, providing the Fourier series a very useful property. The use of orthogonal functions has the advantage that if the number of frequency components in the series is increased, then the amplitude and phase of the current frequency components remain unchanged. In essence, the amplitude and phase of a frequency component are independent of the other frequency components.

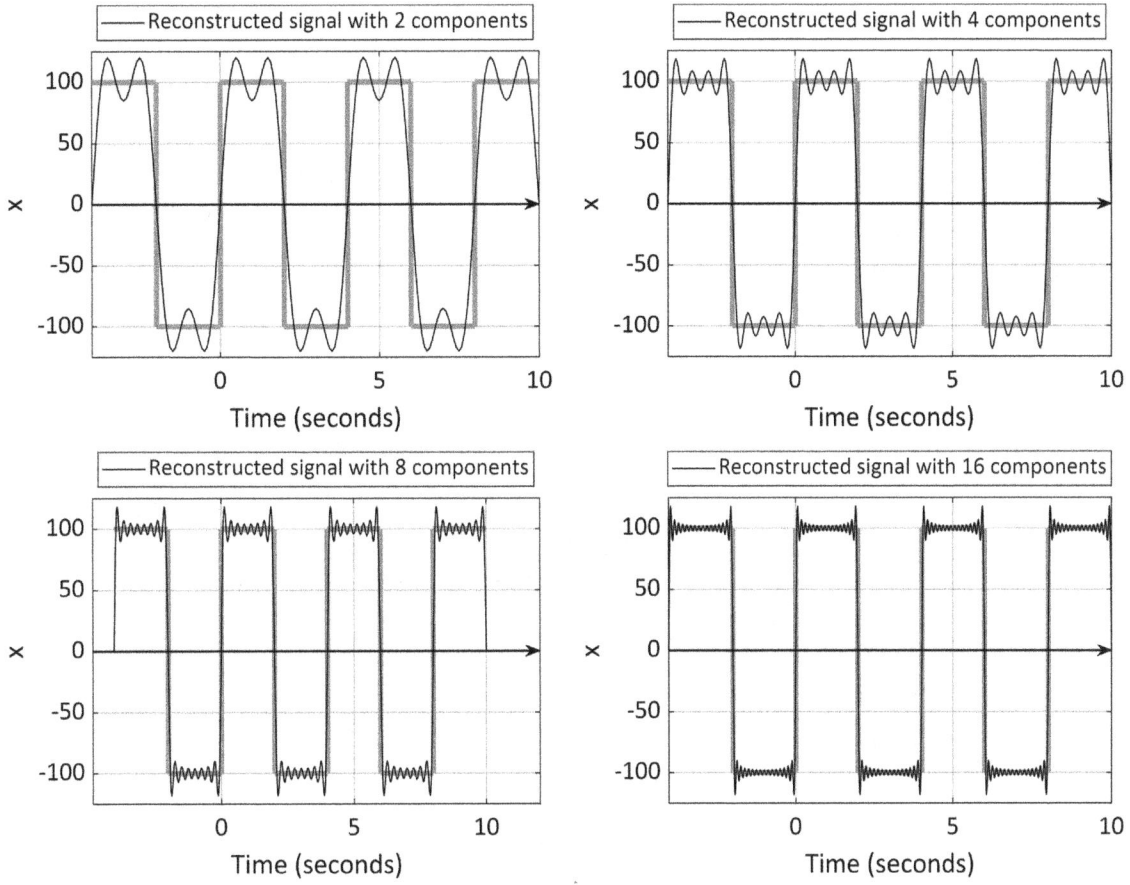

FIGURE 8.4 Comparison of the square wave signal with the reconstructed signals.

8.3.2 FOURIER SERIES EXPANSION (COMPLEX FORM)

The Fourier series expansion can also be written as an infinite sum of complex exponentials. This form of Fourier series is essential, as later, we use this form for the frequency-domain representations of continuous aperiodic and discrete signals.

Using Euler's formula, $e^{\pm in\omega_o t} = \cos n\omega_o t \pm i \sin n\omega_o t$, we can write

$$\cos n\omega_o t = \frac{e^{in\omega_o t} + e^{-in\omega_o t}}{2} \tag{8.14}$$

$$\sin n\omega_o t = \frac{e^{in\omega_o t} - e^{-in\omega_o t}}{2i} \tag{8.15}$$

Substituting the relationships (8.14)–(8.15) into Eq. (8.1) and simplifying, we get

$$x(t) = a_o + \sum_{n=1}^{\infty} \frac{1}{2}(a_n - ib_n)e^{in\omega_o t} + \sum_{n=1}^{\infty} \frac{1}{2}(a_n + ib_n)e^{-in\omega_o t} \tag{8.16}$$

Let $c_n = \frac{1}{2}(a_n - ib_n)$, then $c_n^* = \frac{1}{2}(a_n + ib_n)$, where the superscript '*' represents the complex conjugate operation. Hence,

$$x(t) = a_o e^{i0\omega_o t} + \sum_{n=1}^{\infty} c_n e^{in\omega_o t} + \sum_{n=1}^{\infty} c_n^* e^{-in\omega_o t} \tag{8.17}$$

which can be further written as

$$x(t) = c_0 e^{i0\omega_0 t} + \sum_{n=1}^{\infty} c_n e^{in\omega_0 t} + \sum_{n=-1}^{-\infty} c_n e^{in\omega_0 t} \tag{8.18}$$

where we have used $c_0 = a_o$ and $c_{-n} = c_n^*$. The three terms in Eq. (8.18) can be combined into a single term with an infinite sum with n varying from $-\infty$ to ∞. Eq. (8.18) becomes

$$x(t) = \sum_{n=-\infty}^{\infty} c_n e^{in\omega_0 t} \tag{8.19}$$

Eq. (8.19) is the complex form of the Fourier series.

a. How do the complex exponentials make up a real signal?

The complex form of the Fourier series expansion is in terms of the complex exponential signals instead of sine and cosine signals. Therefore, the question arises: How do the complex exponentials (on the RHS of Eq. 8.19) make up a real signal x(t)? We can understand this by representing the complex exponential signals as vectors in the complex plane. As shown in Figure 8.5a, a complex exponential signal $e^{in\omega_0 t}$ is a vector in the complex plane rotating anticlockwise with angular velocity $n\omega_0$. Since c_n is complex, the vector $c_n e^{in\omega_0 t}$ can be written as $|c_n| e^{i(n\omega_0 t - \theta_n)}$ and represents a vector of length $|c_n|$, having phase lag θ_n (with $e^{in\omega_0 t}$) and rotating anticlockwise with an angular velocity of $n\omega_0$.

Similarly, Figure 8.5b shows the complex exponential signal $c_{-n} e^{-in\omega_0 t}$ written as $|c_{-n}| e^{-i(n\omega_0 t - \theta_n)}$ and represents a vector of length $|c_{-n}| = |c_n|$, having phase lag θ_n (with $e^{-in\omega_0 t}$) and rotating clockwise with an angular velocity of $n\omega_0$.

The addition of vectors $|c_n| e^{i(n\omega_0 t - \theta_n)}$ and $|c_{-n}| e^{-i(n\omega_0 t - \theta_n)}$ results into a real cosine signal of frequency $n\omega_0$, amplitude $2|c_n|$ (which is equal to d_n), and phase lag θ_n, as shown in Figure 8.5c.

Therefore, every pair of complex exponentials in the RHS of Eq. (8.19) corresponding to +n and −n make up a real signal. In other words, for every frequency component in the trigonometric form of the Fourier series, there are two frequency components in the complex Fourier series, one at frequency $+n\omega_0$ and the other at $-n\omega_0$. Thus,

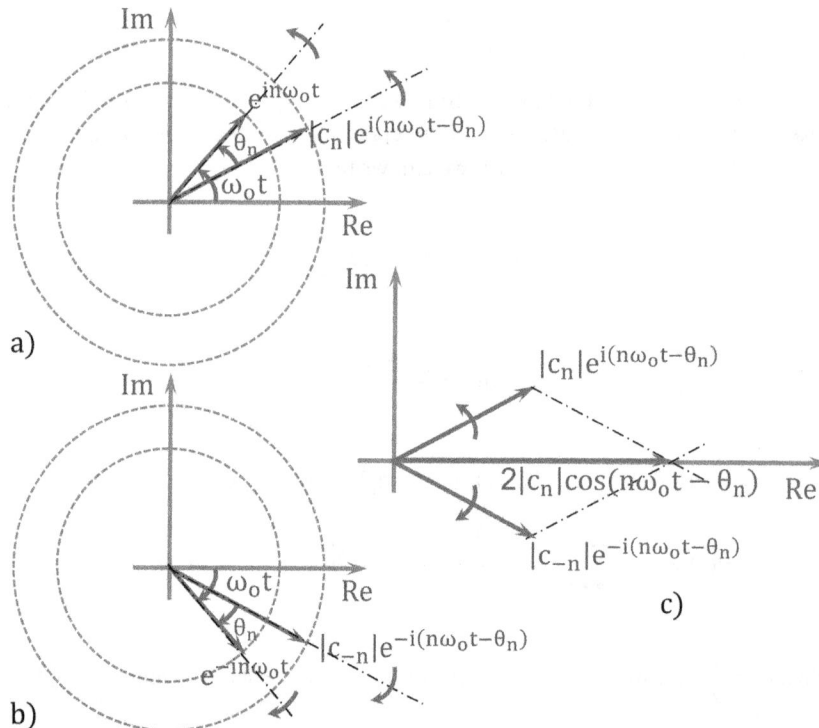

FIGURE 8.5   (a) Complex exponential signal $|c_n| e^{i(n\omega_0 t - \theta_n)}$, (b) complex exponential signal $|c_{-n}| e^{-i(n\omega_0 t - \theta_n)}$, and (c) generation of a real signal from complex exponential signals.

the frequency axis extends over positive and negative frequencies and is often called the complex frequency axis. The frequency sign represents the direction of rotation of the corresponding complex exponential signal; hence, the negative frequency in the complex Fourier series means the corresponding vector is rotating in the clockwise direction.

b. Complex Fourier coefficients

For a given $x(t)$, how to find c_n for the various values of n? To find the coefficient c_r, we multiply both sides of Eq. (8.19) by $e^{-ir\omega_0 t}$ and integrate both sides with time over 0 to T. On the RHS, all the terms except the term with coefficient c_r are integrated to zero leading to

$$c_r = \frac{1}{T} \int_0^T x(t)\ e^{-ir\omega_0 t}.dt \tag{8.20}$$

c. Complex Fourier spectrum

The frequency-domain representation of the periodic signal x(t) using complex Fourier series requires specification of the magnitude and phase of the complex amplitude c_r, for r = ...,−3, − 2, − 1, 0, 1, 2, 3,.... Thus, we have two-sided magnitude and phase spectrums showing the magnitude and phase, respectively of the positive and negative frequency components. The magnitude spectrum is symmetric about the amplitude axis since the magnitudes of the positive and negative frequency components are equal. The phase spectrum is asymmetric as the phases of the positive and negative frequency components differ by 180°. Figure 8.6 shows the complex Fourier magnitude and phase spectrums for the square wave signal considered in the previous section. Alternatively, as shown in Figure 8.7, we can have the spectrums showing the real and imaginary parts of the complex Fourier

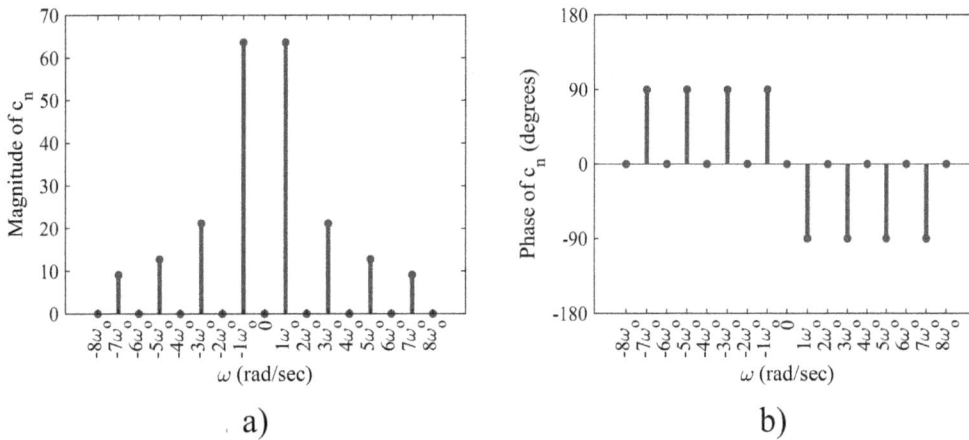

a)   b)

FIGURE 8.6 Complex Fourier spectrums (a) magnitude and (b) phase.

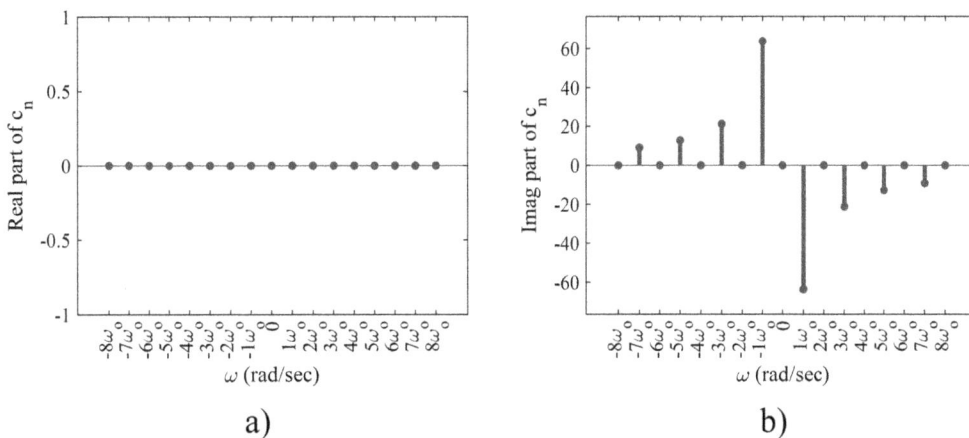

a)   b)

FIGURE 8.7 Complex Fourier spectrums (a) real part and (b) imaginary part.

coefficients as a function of frequency for both positive and negative frequencies. The complex Fourier spectrums are also line spectrums, like the spectrums based on the trigonometric form of the Fourier series.

8.4   FOURIER ANALYSIS OF APERIODIC CONTINUOUS SIGNALS

In this section, we consider the Fourier analysis of aperiodic continuous signals of finite length, like an impact force applied on a structure or the free vibration response of a structure. If the signal is aperiodic, then the Fourier series expansion presented in the previous section cannot be used. The Fourier series involves the sum of harmonic signals of discrete frequencies; hence, the sum yields only a periodic signal and cannot represent the aperiodic, finite-length signal.

We can view an aperiodic signal of finite length as a periodic signal, which repeats after an infinite time. To understand this, let us consider the aperiodic signal shown in Figure 8.8a, a rectangular pulse signal. Figure 8.8b shows a periodic signal, which is nothing but a repetition of the copies of the pulse signal in Figure 8.8a after every time T.

If T is increased, in the limit $T \to \infty$, the periodic signal in Figure 8.8b becomes the aperiodic signal in Figure 8.8a. This observation shows that the frequency-domain representation of an aperiodic signal of finite length can be obtained by the Fourier series expansion of the corresponding periodic signal when the period of the signal approaches infinity.

8.4.1   FOURIER TRANSFORM

We start from the Fourier series expansion of the periodic signal obtained by repetition of the copies of the aperiodic signal after every time T. The complex Fourier coefficients are given by Eq. (8.20). The limits of integration 0 to T can be written as $-T/2$ to $T/2$, giving us

$$c_r = \frac{1}{T} \int_{-T/2}^{T/2} x(t)\, e^{-ir\omega_o t} dt \tag{8.21}$$

The frequency resolution or the gap ($\Delta\omega$) between two consecutive frequency components of the Fourier spectrum is

$$\Delta\omega = \omega_o = \frac{2\pi}{T} \tag{8.22}$$

where ω_o is the fundamental frequency of the periodic signal. Substituting Eq. (8.22) into Eq. (8.21), we get

$$c_r = \frac{\Delta\omega}{2\pi} \int_{-T/2}^{T/2} x(t)\, e^{-ir\Delta\omega t}.dt \tag{8.23}$$

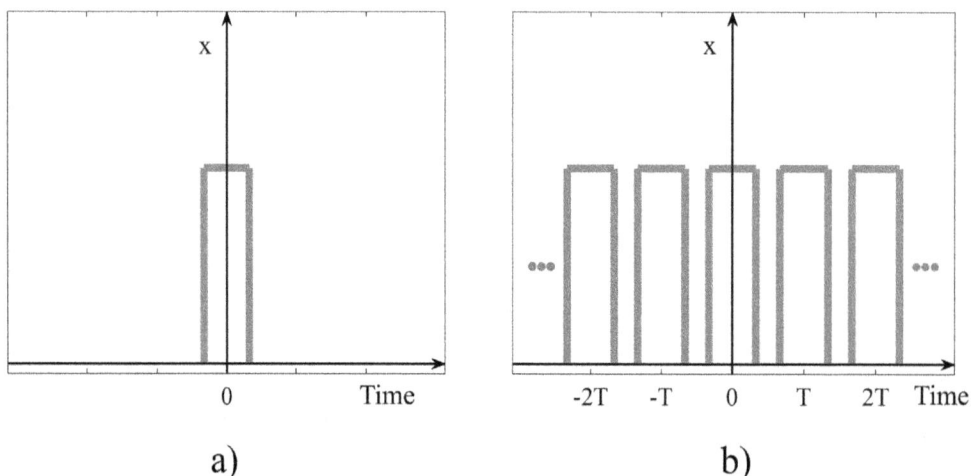

a)   b)

FIGURE 8.8   (a) An aperiodic signal, (b) a periodic signal of period T.

If $T \rightarrow \infty$, the frequency gap tends to an infinitesimally small quantity $d\omega$. As a result, the discrete frequency $r\Delta\omega$ in the exponent also becomes a continuous frequency variable ω. The Fourier coefficient corresponding to a frequency also becomes infinitesimally small, and hence we seek the limit of the quantity $\dfrac{c_r}{\Delta\omega/2\pi}$ rather than c_r.

Taking the limit of both sides of Eq. (8.23) as $T \rightarrow \infty$, we obtain

$$\lim_{T \to \infty} \frac{c_r}{\Delta\omega/2\pi} = \lim_{T \to \infty} \int_{-T/2}^{T/2} x(t)\, e^{-ir\Delta\omega t}.dt \tag{8.24}$$

$$\lim_{T \to \infty} \frac{c_r}{\Delta\omega/2\pi} = \int_{-\infty}^{\infty} x(t)\, e^{-i\omega t}.dt \tag{8.25}$$

The integral on the RHS of Eq. (8.25) is a function of the continuous frequency variable ω; hence the limit on the LHS is denoted by a symbol $X(\omega)$.

$$X(\omega) = \int_{-\infty}^{\infty} x(t)\, e^{-i\omega t}.dt \tag{8.26}$$

Thus, we get an integral expression for $X(\omega)$ and is called the **Fourier transform or Fourier integral**. It is the frequency-domain representation of the continuous aperiodic signal x(t). It is also called the **Forward Fourier transform** as it converts a signal from the time to the frequency domain. Note that the Fourier spectrum $X(\omega)$ is continuous and not discrete. If x(t) is in meters, then the unit of $X(\omega)$ is meter/Hz. Thus, $X(\omega)$ doesn't represent the amplitude of a discrete frequency component but is the value of the amplitude per unit frequency in Hz.

The time-domain signal can be reconstructed from the Fourier transform of the signal. Let us start with the complex Fourier series (Eq. 8.19) to obtain an expression for this.

$$x(t) = \sum_{r=-\infty}^{\infty} c_r e^{ir\omega_0 t} \tag{8.27}$$

$$x(t) = \sum_{r=-\infty}^{\infty} \frac{c_r}{\Delta\omega/2\pi} \cdot \frac{\Delta\omega}{2\pi} e^{ir\Delta\omega t} \tag{8.28}$$

As noted earlier, when $T \rightarrow \infty$, the discrete frequency $r\Delta\omega$ becomes a continuous frequency variable ω. Therefore, the summation operation on the RHS needs to be replaced by an integral operation. Also, as $T \rightarrow \infty$, $\dfrac{c_r}{\Delta\omega/2\pi} \rightarrow X(\omega)$ and $\Delta\omega \rightarrow d\omega$. Hence, we obtain

$$x(t) = \frac{1}{2\pi} \int_{-\infty}^{\infty} X(\omega)\, e^{i\omega t} d\omega \tag{8.29}$$

Eq. (8.29) allows reconstructing or synthesizing the time signal from its Fourier transform. It allows us to go back from the frequency-domain representation to the time-domain representation, and hence is called the **inverse Fourier transform**. Equations (8.26) and (8.29) together make up the Fourier transform pair.

Let us consider the Fourier transform of the rectangular pulse signal shown in Figure 8.8a. Substituting x(t) into Eq. (8.26), we can obtain the Fourier transform (the details of which we can see in an example later in this chapter), whose magnitude and phase are shown in Figure 8.9. The spectrum is now continuous as against a line/discrete spectrum in the case of a periodic signal. Note that both the positive and negative frequencies exist in the spectrum.

8.4.2 Relationship between the Spectrums of an Aperiodic and the Corresponding Periodic Signal

In the previous section, the Fourier transform of an aperiodic signal (like the signal in Figure 8.8a) is obtained as the limit of the Fourier series expansion of the corresponding periodic signal (shown in Figure 8.8b). The question arises, is there any relationship between the frequency-domain representation of these two signals? In other words, is there any

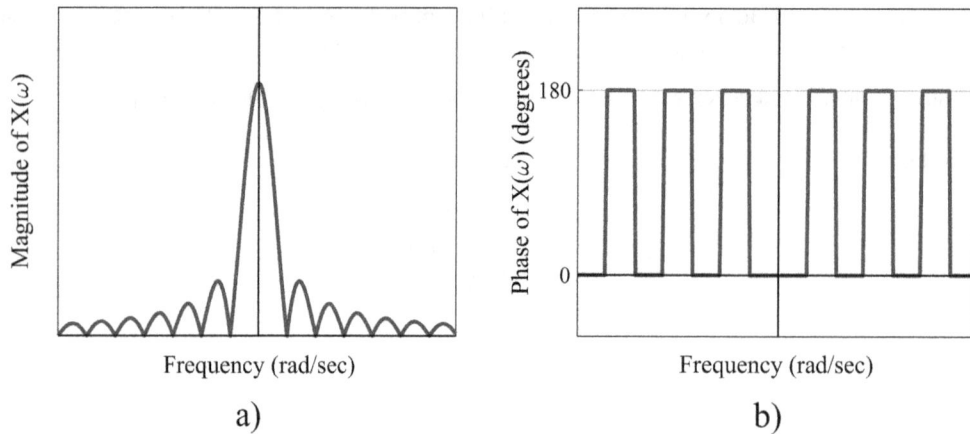

FIGURE 8.9   (a) Magnitude and (b) phase of the Fourier transform of the rectangular pulse.

relationship between the Fourier transform $X(\omega)$ of the signal in Figure 8.8a and the Fourier series coefficients c_r of the signal in Figure 8.8b?

The answer is that the two are related. The discrete/line spectrum of the periodic signal is a sampled version of the continuous spectrum $X(\omega)$. The discrete frequencies in the line spectrum represent the frequency samples drawn from the continuous spectrum. The amplitude coefficients (c_r values) corresponding to the discrete frequencies when divided by the frequency resolution of the discrete spectrum (in Hz) gives the amplitude of the continuous spectrum at the discrete frequencies.

8.5   DISCRETE-TIME SIGNALS

Fourier analyses of the continuous periodic and aperiodic signals are presented in the previous sections. Now we consider discrete-time signals. A discrete-time signal is a signal in which the independent variable, time, is discrete. Thus, the signal is available or defined at only discrete times. The signal amplitude, however, is continuous in a discrete-time signal. If the signal amplitude is also discretized, it becomes a digital signal. The discrete-time signal henceforth is referred to as the discrete signal.

Modern signal analyzers are based on digital signal processing (DSP) technology in which the analog signal is discretized and digitized, and then the digitized version of the continuous signal is processed. Analog signal processing, which processes analog signals, requires analog circuits and hardware and has limitations regarding the scope of tasks and functions by which the signal can be processed. On the other hand, DSP can be implemented using high-speed microprocessors and general-purpose computers, which are programmable, and can be used to implement signal-processing tasks that are difficult to achieve with analog signal processing.

The method used in DSP processors for obtaining a discrete representation of a continuous signal is periodic sampling. In this, the sample of the continuous signal $x(t)$ is recorded after every time interval Δt. Δt is related to the sampling frequency (f_s), which is the number of samples of the signal recorded over a one-second duration and is given by

$$f_s = \frac{1}{\Delta t} \text{ Hz} \tag{8.30}$$

Thus, at the end of the periodic sampling, we have the discrete signal, which consists of the values of the continuous signal at N discrete instants separated by time interval Δt,

$$\{x(n)\} = \{x(0\Delta t) \ \ x(1\Delta t) \ x(2\Delta t) \ \cdots \ x(n\Delta t) \ \cdots \ x((N-1)\Delta t)\} \tag{8.31}$$

Figure 8.10a and b show a continuous signal $(x(t))$ and a discrete version of the signal $(x(n))$, respectively.

Figure 8.11 shows a schematic of the process of measurement and discretization of a continuous signal. A transducer is used to measure the signal (say, $x(t)$). After the necessary signal conditioning, the output is typically in the form of voltage $V(t)$ proportional to the signal being measured. $V(t)$, a continuous/analog signal, is fed to an analog-to-digital (A/D) converter. An A/D converter is a hardware that performs the discretization and digitization of the incoming signal. For explanation, the two processes are shown in the figure as two separate blocks.

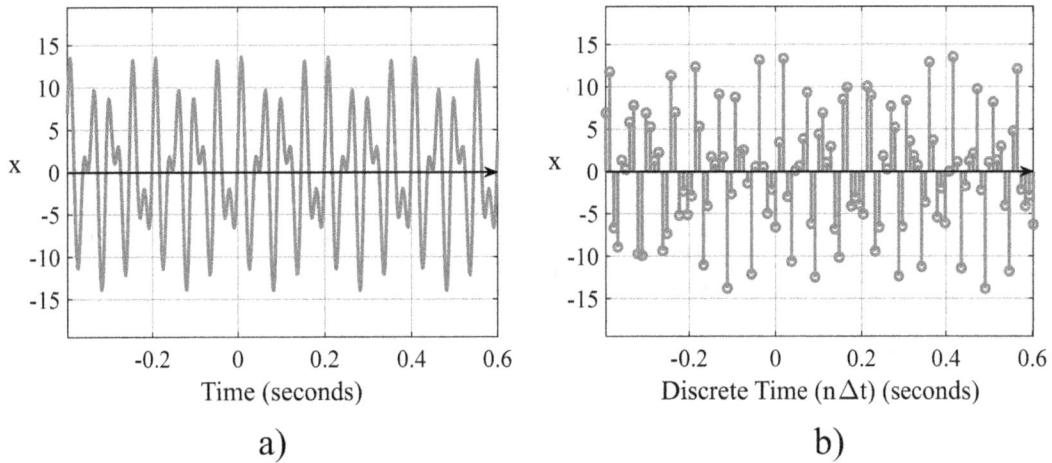

FIGURE 8.10   (a) Continuous signal and (b) a discrete version of the signal.

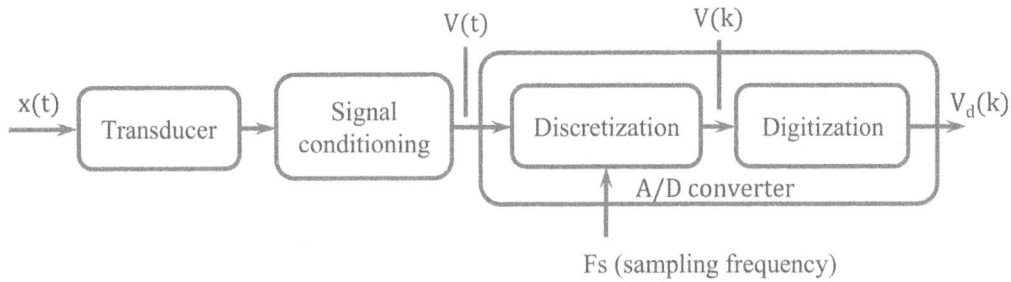

FIGURE 8.11   Measurement and discretization of a continuous signal.

The discretization depends on the sampling frequency, and yields a discrete signal $V(n)$. The amplitude of the discrete signal is quantized based on the number of bits of the A/D converter and the voltage range, giving a digital signal $V_d(n)$. The discrete voltage signal $V(n)$ or $V_d(n)$ can be interpreted in terms of the units of the signal $x(t)$ using the sensitivity of the transducer used to measure $x(t)$.

In the following sections, we present the Fourier analysis of discrete signals.

8.6   FOURIER ANALYSIS OF PERIODIC DISCRETE SIGNALS

The Fourier series representation of a discrete periodic signal can be developed by considering it as a discrete sequence and representing it as a sum of complex exponential discrete sequences. An alternative way is to develop it from the expression for the Fourier series of a continuous signal presented in the previous section by approximating the integral as a discrete sum. This latter approach is followed in the sections below.

8.6.1   DISCRETE FOURIER SERIES (DFS)

From the complex form of the Fourier series for a continuous signal, the amplitude of the kth frequency component can be written using Eq. (8.20) as

$$c_k = \frac{1}{T} \int_0^T x(t)\, e^{-ik\omega_0 t} dt \tag{8.32}$$

We now modify Eq. (8.32) for a discrete signal. Figure 8.12 shows a discrete periodic signal. Let Δt be the sampling time or the time gap between two consecutive samples, and N be the number of samples in one period. The N samples numbered from 0 to N-1 are shown in the figure. N also represents the number of discrete steps in one period.

Due to the discrete nature of the signal, the following changes are made:

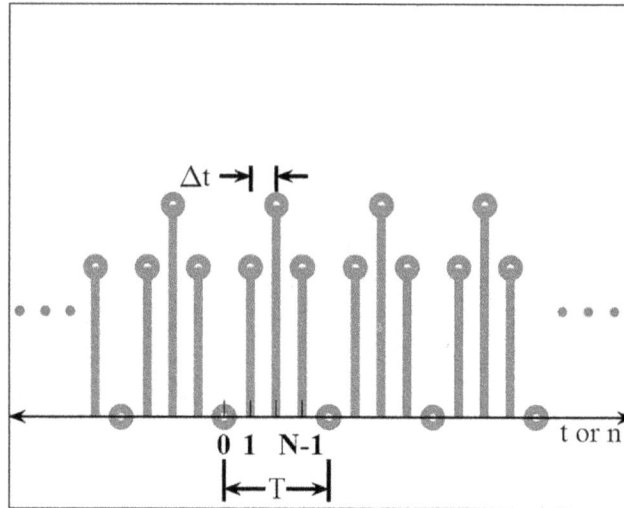

FIGURE 8.12   A discrete periodic signal.

- The independent variable t is no longer a continuous variable but a discrete one given by $t = n\Delta t$.
- As n represents the sample number, $x(t)$ is replaced by $x(n\Delta t)$ (henceforth written as $x(n)$).
- Since there are N time steps in one period, record length $T = N\Delta t$, and the fundamental frequency ω_o is written as $\dfrac{2\pi}{N\Delta t}$.
- The integral operator is replaced by the summation over the N time steps, with n varying from 0 to $N-1$.
- dt, the infinitesimal time step, is replaced by the finite time step Δt.
- c_k is denoted by a new symbol $X(k)$. The lower case is used to represent the time-domain discrete signal $(x(n))$ while the upper case is used to represent the frequency-domain representation $(X(k))$ of the discrete signal.

Incorporating the above changes in Eq. (8.32), we get

$$X(k) = \frac{1}{N\Delta t} \sum_{n=0}^{N-1} x(n)\, e^{-ik\frac{2\pi}{N\Delta t}n\Delta t} \Delta t \tag{8.33}$$

which simplifies to

$$X(k) = \frac{1}{N} \sum_{n=0}^{N-1} x(n)\, e^{-ik\frac{2\pi}{N}n} \tag{8.34}$$

Eq. (8.34) is the frequency-domain representation of the discrete periodic signal $x(n)$ and is called **Discrete Fourier Series (DFS).** This equation gives the complex amplitude $X(k)$ of the kth frequency component having frequency $k\dfrac{2\pi}{N\Delta t}$ rad/sec. Substituting $k = ..., -2, -1, 0, 1, 2, ...$, gives the amplitude of the corresponding frequency components.

8.6.2   Periodicity of DFS

Let us determine the amplitude of the rth and $(r+N)^{th}$ frequency components in the discrete Fourier series. From Eq. (8.34), for the rth component, we get

$$X(r) = \frac{1}{N} \sum_{n=0}^{N-1} x(n)\, e^{-ir\frac{2\pi}{N}n} \tag{8.35}$$

Similarly, for the $(r+N)^{th}$ component, we get

$$X(r+N) = \frac{1}{N}\sum_{n=0}^{N-1} x(n)\, e^{-i(r+N)\frac{2\pi}{N}n} \tag{8.36}$$

$$X(r+N) = \frac{1}{N}\sum_{n=0}^{N-1} x(n)\, e^{-ir\frac{2\pi}{N}n}.e^{-i2\pi n} \tag{8.37}$$

Since n is an integer, $e^{-i2\pi n}=1$ and therefore, comparing Eqs. (8.35) and (8.37), we see that

$$X(r+N) = X(r) \quad \text{(for any r)} \tag{8.38}$$

Eq. (8.38) shows that the frequency component indices separated by N have identical amplitudes. The complex amplitudes of the components k = N, N+1, N+2, ..., 2N−1 are the same as those of k = 0, 1, 2, ..., N−1. Therefore, the DFS is periodic in the frequency domain with a period $N\omega_o$ rad/sec (which is equal to the sampling frequency ω_s). Thus, there are only N distinct frequency components, and they form the frequency-domain representation of the discrete signal.

Note that one period of DFS can be defined either with k = 0, 1, 2...,N−1 or with k = $-\frac{N}{2}$, ...,0,..., $\frac{N}{2}-1$. In the first case, the first half is the positive frequency part, and the next half is the negative frequency part. In the second case, the left part of the spectrum (i.e., corresponding to the negative values of k) is the negative frequency part, and the right part of the spectrum (i.e., corresponding to the positive values of k) is the positive frequency part.

8.6.3 Relationship among Fourier Coefficients

From Eq. (8.38),

$$X(k+N) = X(k) \tag{8.39}$$

For −k

$$X(N-k) = X(-k) \tag{8.40}$$

Also, from Eq. (8.34), we note that

$$X(-k) = X^*(k) \tag{8.41}$$

Therefore, from Eqs. (8.40) and (8.41), we get

$$X(N-k) = X^*(k) \tag{8.42}$$

which leads to

$$|X(N-k)| = |X(k)| \tag{8.43}$$

and

$$\text{Phase of } X(N-k) = -\text{Phase of } X(k) \tag{8.44}$$

Thus, we see that the complex amplitudes of the N distinct frequency components, i.e., X(0), X(1), X(2), ..., X(N−1), are not independent but are related by Eq. (8.42).

For example, if N = 4, then we have four components with amplitudes X(0), X(1), X(2) and X(3).

- For k = 0, Eqs. (8.39) and (8.42) show that X(0) is real.
- For k = 1, Eq. (8.42) shows that $X(3) = X^*(1)$.
- For k = 2, Eq. (8.42) gives $X(2) = X^*(2)$, that is, X(2) is real.

Due to the periodicity of DFS, $X(4) = X(0)$. Now, if we look at the magnitude plot of the amplitudes $X(0)$, $X(1)$, $X(2)$, $X(3)$ and $X(4)$, then it would be symmetric about $X(2)$, while the phase plot would be antisymmetric. This conclusion is also clear from Eqs. (8.43) and (8.44).

8.6.4 INVERSE DFS

An expression for synthesizing or reconstructing the discrete signal from its frequency-domain representation can be obtained by starting from the corresponding expression for the continuous signal (Eq. 8.19),

$$x(t) = \sum_{k=-\infty}^{\infty} c_k e^{ik\omega_0 t} \tag{8.45}$$

In Eq. (8.45), we incorporate the modifications due to the discrete nature of the signal stated in Section 8.6.1. We also know from Section 8.6.2 that only the first N frequency components are needed for the frequency-domain representation of the discrete periodic signal, and hence the summation should extend from $k = 0$ to $N - 1$. Therefore, Eq. (8.45) can be written as

$$x(n) = \sum_{k=0}^{N-1} X(k) e^{ik\frac{2\pi}{N}n} \tag{8.46}$$

Eq. (8.46) is the expression for the **inverse DFS** as it allows synthesizing $x(n)$ from the knowledge of $X(k)$.

Example 8.1

Let us take an example to understand further the characteristics of the DFS. We take the following continuous signal, a sum of two sinusoids, and a dc component.

$$x_c(t) = 5 + 20\cos(2\pi 2t - 30°) + 12\cos(2\pi 6t - 60°)$$

The frequency, amplitude, and phase lag of the two sinusoids in the signal are (2 Hz, 20, 30°) and (6 Hz, 12, 60°) respectively. The dc component magnitude is 5.

Let $x_c(t)$ is sampled at a frequency of 20 Hz, and the discrete signal of length one second is used to obtain its DFS. The DFS is computed using Eq. (8.34). Figure 8.13a and b show the magnitude and phase of the frequency components in the DFS, respectively.

Observations from the figures:

- The magnitude and phase spectrums are periodic with a period of 20 Hz (since the sampling frequency is 20 Hz). One period (from −10 to 10 Hz) is marked with star markers for clarity.
- The spectrums have both positive and negative frequencies. (This is because the DFS is derived from the complex Fourier series).

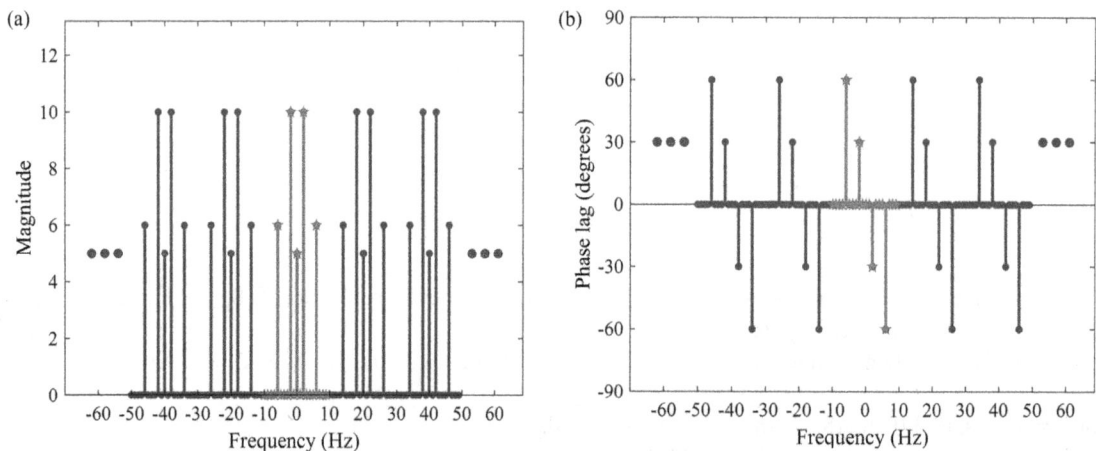

FIGURE 8.13   DFS (a) magnitude and (b) phase.

- The magnitude spectrum is symmetric in a period, while the phase spectrum is antisymmetric.
- The spectrums show the presence of complex frequency components in the signal at ±2 Hz and ±6 Hz and the dc component at 0 Hz.

8.7 FOURIER ANALYSIS OF APERIODIC DISCRETE SIGNALS (DTFT)

If a continuous signal is aperiodic, the corresponding discrete signal is also aperiodic.

Let $x_a(n)$ be an aperiodic discrete signal of finite duration, as shown in Figure 8.14a. It is a discrete version of a rectangular pulse signal. The suffix 'a' denotes the aperiodic nature of the signal. The DFS cannot be used since the signal is not periodic. But, similar to the case of the aperiodic continuous signal, $x_a(n)$ can be assumed to be a periodic signal of an infinite period. For example, Figure 8.14b shows a periodic discrete signal $x(n)$ of period T (described by N samples). With the same sampling frequency, if the period goes to infinity, then $N \to \infty$, and we can write

$$\lim_{N\to\infty} x(n) = x_a(n) \tag{8.47}$$

Therefore, the Fourier transform of $x_a(n)$ can be found as the limit of the DFS of $x(n)$ with its period approaching infinity.

The DFS of the periodic signal $x(n)$ is given by Eq. (8.34)

$$X(k) = \frac{1}{N}\sum_{n=0}^{N-1} x(n)\, e^{-ik\frac{2\pi}{N}n} \tag{8.48}$$

The quantity $\frac{2\pi}{N}$ represents the fundamental frequency of the discrete sequence with units 'radians', and let it be represented by Ω_o. (Also $\Delta t \omega_o = \Omega_o$).

When $N \to \infty$, the following changes occur:

- the frequency gap/resolution (which is equal to Ω_o) becomes infinitesimally small, and as a result, the discrete frequency $k\Omega_o$ becomes a continuous frequency variable Ω.
- The Fourier coefficient corresponding to a specific frequency becomes infinitesimally small, and hence we seek the limit of the quantity $N.X(k)$ rather than $X(k)$.
- The range of the summation over samples would be from $-\infty$ to $-\infty$.

Taking the limit of both sides of Eq. (8.48) when $N \to \infty$, and incorporating the above changes, we get

$$\lim_{N\to\infty} N.X(k) = \lim_{N\to\infty}\sum_{n=0}^{N-1} x(n)\, e^{-ik\Omega_o n} \tag{8.49}$$

In light of Eq. (8.47), we obtain

$$X_a(\Omega) = \sum_{n=-\infty}^{\infty} x_a(n)\, e^{-i\Omega n} \tag{8.50}$$

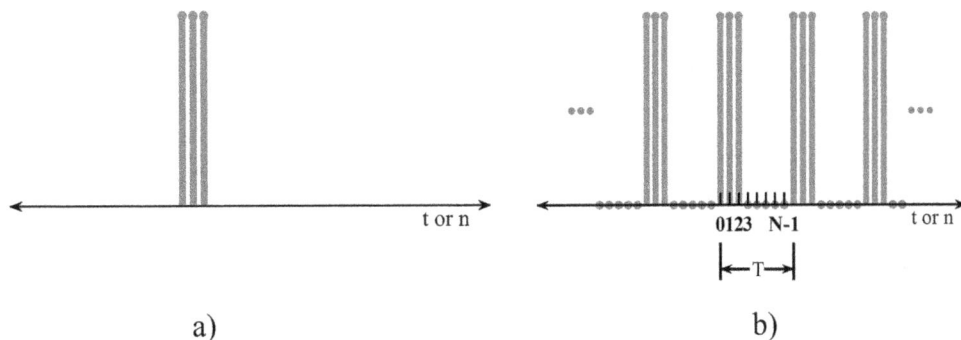

FIGURE 8.14 (a) An aperiodic discrete signal of finite duration and (b) a periodic discrete signal.

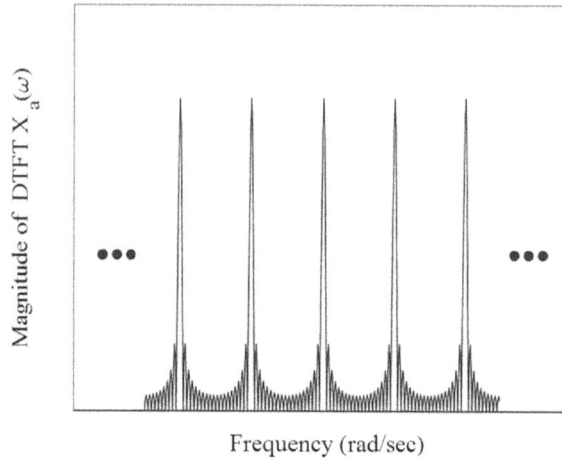

FIGURE 8.15   Plot of the magnitude of DTFT $X_a(\omega)$.

$X_a(\Omega)$ is the Fourier transform of the aperiodic discrete signal $x_a(n)$. Since it is the Fourier transform of a discrete signal, it is called the **Discrete-Time Fourier transform (DTFT)**. The following observations can be made about DTFT from Eq. (8.50).

- $X_a(\Omega)$ is continuous in frequency Ω (although the signal $x_a(n)$ is discrete).
- It is seen that $X_a(\Omega + 2\pi r) = X_a(\Omega)$ for any integer value of r, and hence the continuous spectrum is periodic in frequency Ω with the period equal to 2π.

The spectrum with frequency axis in rad/sec is given by

$$X_a(\omega) = \frac{1}{\Delta t} X_a(\Omega) \tag{8.51}$$

As an example, Figure 8.15 shows the magnitude of $X_a(\omega)$, for the signal shown in Figure 8.14a). The observations listed above are evident from these plots.

8.8   RELATIONSHIP BETWEEN DFS AND DTFT

Continuing from the previous section, we again consider $x(n)$ (periodic discrete signal) and $x_a(n)$ (aperiodic discrete signal). If the two are related by Eq. (8.47), then is there any relationship between their corresponding frequency-domain representations, i.e., $X(k)$ (which is the DFS of $x(n)$) and $X_a(\Omega)$ (which is the DTFT of $x_a(n)$)?

The signal $x_a(n)$ can be expressed as

$$x_a(n) = \begin{cases} x(n) & \text{for } 0 \leq n \leq N-1 \\ 0 & \text{otherwise} \end{cases} \tag{8.52}$$

Given the above, in the equation for DFS (Eq. 8.48), we can replace $x(n)$ with $x_a(n)$ and change the limits of the summation to $-\infty$ to ∞. In addition, by writing $\frac{2\pi}{N} = \Omega_0$, we get

$$X(k) = \frac{1}{N} \sum_{n=-\infty}^{+\infty} x_a(n)\, e^{-ik\Omega_0 n} \tag{8.53}$$

The RHS of Eq. (8.53) is nothing but the value of $\frac{1}{N} X_a(\Omega)$ at $\Omega = k\Omega_0$. Thus, Eq. (8.53) can be written as

$$X(k) = \frac{1}{N} X_a(k\Omega_0) \tag{8.54}$$

Eq. (8.54) indicates that the DFS ($X(k)$) is a sampled and scaled version of the DTFT. As the period N increases, the frequency lines in $X(k)$ become closer and in the limit when $N \to \infty$, the discrete lines in the DFS merge and form the continuous spectrum $X_a(\Omega)$.

8.9 DISCRETE FOURIER TRANSFORM (DFT)

Let us say we have an aperiodic discrete signal $x_a(n)$ and want its frequency-domain representation. To obtain this, we should ideally take its Fourier transform (using Eq. 8.50) since it is an aperiodic signal.

But instead, we can follow an alternative approach. We can construct a periodic discrete signal $x(n)$ which consists of periodically repeated copies of the signal $x_a(n)$. $x(n)$ being periodic, its frequency-domain representation ($X(k)$) can be obtained by DFS. From the result of the previous section, we know that $X(k)$ so obtained is a sampled version of the Fourier transform of $x_a(n)$ (i.e. $X_a(\Omega)$). Thus, with this approach, though, we don't get $X_a(\Omega)$, but get its sampled version.

Since $X(k)$ is periodic, only one period of it is required to represent the signal in the frequency domain, and this is called the **discrete Fourier Transform (DFT)** of the aperiodic signal $x_a(n)$. Let us represent DFT by the symbol $\bar{X}(k)$, then it is given by

$$\bar{X}(k) = \begin{cases} X(k) & \text{for} \quad 0 \le k \le N-1 \\ 0 & \text{otherwise} \end{cases} \tag{8.55}$$

Substituting, $X(k)$ from Eq. (8.34), we get

$$\bar{X}(k) = \begin{cases} \dfrac{1}{N}\displaystyle\sum_{n=0}^{N-1} x(n)\, e^{-ik\frac{2\pi}{N}n} & \text{for} \quad 0 \le k \le N-1 \\ 0 & \text{otherwise} \end{cases} \tag{8.56}$$

Since the summation in Eq. (8.56) is over 0 to N-1, $x(n)$ can be replaced by $x_a(n)$,

$$\bar{X}(k) = \begin{cases} \dfrac{1}{N}\displaystyle\sum_{n=0}^{N-1} x_a(n)\, e^{-ik\frac{2\pi}{N}n} & \text{for} \quad 0 \le k \le N-1 \\ 0 & \text{otherwise} \end{cases} \tag{8.57}$$

Eq. (8.57) is the equation for the DFT, and it is also discrete and finite, as does the time-domain signal $x_a(n)$.

If the DFT is available, then the time-domain signal can be synthesized or reconstructed using the following equation (referred to as the Inverse DFT (or IDFT)):

$$x_a(n) = \begin{cases} \displaystyle\sum_{k=0}^{N-1} \bar{X}(k) e^{ik\frac{2\pi}{N}n} & \text{for} \quad 0 \le n \le N-1 \\ 0 & \text{otherwise} \end{cases} \tag{8.58}$$

(Note that there is a factor $\dfrac{1}{N}$ in the expression of DFT (Eq. 8.57), but the expression for IDFT (Eq. 8.58) doesn't have $\dfrac{1}{N}$. In an alternative form of these expressions, the factor $\dfrac{1}{N}$ appears in the IDFT instead of the DFT.)

8.10 WHICH ONE TO USE IN PRACTICE, DFS, DTFT, OR DFT?

The signal to be measured and analyzed may be periodic or aperiodic. It may be aperiodic if it results from some transient excitation or periodic if it results from some continuous periodic excitation. Even if the actual signal is periodic, recording exactly one or an integer number of periods is difficult since the signal's frequency content may not be known. The following approach can be used for the frequency-domain representation of the measured signals.

Irrespective of the nature of the original signal, we can treat the recorded discrete signal to be a signal of finite duration and find its DFT using Eq. (8.57). The computed DFT corresponds to one of the following estimates:

- Suppose the original signal is periodic, and the recorded signal equals an integer number of signal periods. In that case, the DFT estimates the recorded signal's Fourier spectrum (i.e., DFS).
- Suppose the original signal is an aperiodic (and finite length) signal, and let the complete length of the signal is captured. In that case, the DFT gives a sampled version of the DTFT of the recorded signal.

Thus, we use the DFT for both the periodic and aperiodic signals.

8.11   RELATIONSHIPS BETWEEN THE PARAMETERS OF A DISCRETE SIGNAL AND ITS DFT

Based on the developments in the previous sections, here we review the relationships between the parameters of a discrete signal and its DFT. These relationships help choose the measurement parameters according to the requirements of the desired frequency range and resolution of the spectrum. The various parameters are

- T : length of the recorded signal in seconds
- f_s : sampling frequency in Hz; ω_s is the corresponding value in rad/sec
- N : number of discrete samples or time steps in the record T; it is also the number of frequency lines in the DFT
- Δt : sampling interval or time interval between two samples
- Δf : frequency resolution of DFT in Hz; It is the interval between any two consecutive frequency lines; $\Delta \omega$ is the corresponding value in rad/sec;
- f_{max} : the frequency of the highest frequency component in the DFT

Δt is related to f_s by

$$\Delta t = \frac{1}{f_s} \tag{8.59}$$

Δf depends on T

$$\Delta f = \frac{1}{T} \tag{8.60}$$

N is given by

$$N = \frac{T}{\Delta t} = T f_s \tag{8.61}$$

Using Eq. (8.61), Eq. (8.60) can also be written as

$$\Delta f = \frac{f_s}{N} \tag{8.62}$$

The number of frequency lines/components in the DFT is equal to N. Assuming N to be even, these components on the complex frequency axis correspond to $k = -\frac{N}{2}, \ldots -1, 0, 1, \ldots, \frac{N}{2} - 1$. The frequency of the kth component is given by

$$f_k = k \Delta f \tag{8.63}$$

Thus, the frequency lines in the DFT are at frequencies: $-\frac{N}{2} \Delta f, \ldots - \Delta f, 0, \Delta f \ldots, \left(\frac{N}{2} - 1\right) \Delta f$.

The value of f_{max} would be

$$f_{max} = \frac{N}{2} \Delta f = \frac{f_s}{2} \tag{8.64}$$

Thus, we see that the six parameters, T, f_s, N, Δt, Δf and f_{max} are related by four equations (Eqs. 8.59–8.61 and Eq. 8.64). Two of these parameters can be chosen independently in a measurement. Often, either (T, N) or (T, f_{max}) are the preferred choices.

Some valuable conclusions can be drawn from these equations.

- If the frequency resolution is to be improved, then this can be done by increasing either T or N.
- The frequency range of measurement/spectrum (i.e. f_{max}) can be increased by increasing f_s.

8.12   SINGLE-SIDED SPECTRUM

The DFT is on the complex frequency axis, which gives the spectrum in terms of positive and negative frequencies. A spectrum with positive and negative frequencies is also called a two-sided spectrum. We also note that, for a real signal, the amplitudes of a positive and the corresponding negative frequency component are related by a complex conjugate relationship. Given this, it is possible to present the spectrum only with positive frequencies, and such a spectrum is called the single-sided or one-sided spectrum. Commercial analyzers generally present the results of the DFT as a single-sided spectrum.

Assuming N to be even, let us denote the amplitudes of the complex DFT by $\overline{X}(k)$, calculated for $k = -\dfrac{N}{2}$, ..., 0, ..., $\dfrac{N}{2} - 1$. The corresponding frequencies are $-\dfrac{N}{2}\Delta f$, ..., 0,..., $\left(\dfrac{N}{2} - 1\right)\Delta f$.

Let $\overline{Y}(r)$ represent the single-sided spectrum, where $r = 0, 1, 2..., \dfrac{N}{2}$. The corresponding frequencies are $0, \Delta f, 2\Delta f..., \dfrac{N}{2}\Delta f$.

Due to the complex conjugate nature of the positive and negative frequency components, we can write

$$\left.\begin{array}{l} \left|\overline{Y}(r)\right| = 2\left|\overline{X}(r)\right| \\[2mm] \text{phase lag of } \overline{Y}(r) \ = -\tan^{-1}\dfrac{\text{Im}\left(\overline{X}(r)\right)}{\text{Re}\left(\overline{X}(r)\right)} \end{array}\right\} \quad \text{for} \ \ r = 1, 2, ..., \dfrac{N}{2} - 1 \qquad (8.65)$$

The dc component $\overline{Y}(0)$ and $\overline{Y}\left(\dfrac{N}{2}\right)$ are given by,

$$\overline{Y}(0) = \overline{X}(0) \qquad (8.66)$$

$$\overline{Y}\left(\dfrac{N}{2}\right) = \overline{X}\left(-\dfrac{N}{2}\right) \qquad (8.67)$$

Thus, Eqs. (8.65)–(8.67) enable to calculate the single-sided DFT from the complex DFT.

Example 8.2

In Example 8.1, we found the DFS of the following continuous signal:

$$x_c(t) = 5 + 20\cos(2\pi 2t - 30°) + 12\cos(2\pi 6t - 60°)$$

In this example, we find the complex and one-sided DFT of this signal. The same sampling frequency (i.e., 20 Hz) and record length (i.e., 1 second) are used. We compute the DFT using Eq. (8.57) over the range $-\dfrac{N}{2} \le k \le \dfrac{N}{2} - 1$ (which is equivalent to the range $0 \le k \le N - 1$). The discrete signal of length 1 second is used in place of $x_a(n)$ in this equation.

Figure 8.16 shows the magnitude and phase of the complex DFT. It is a two-sided spectrum. (It is the same as one period of the DFS shown in Figure 8.13). Figure 8.17 shows the magnitude and phase of the single-sided spectrum/DFT computed using Eqs. (8.64)–(8.66). It shows the amplitude and phase of the frequency components which make up the signal $x_c(t)$.

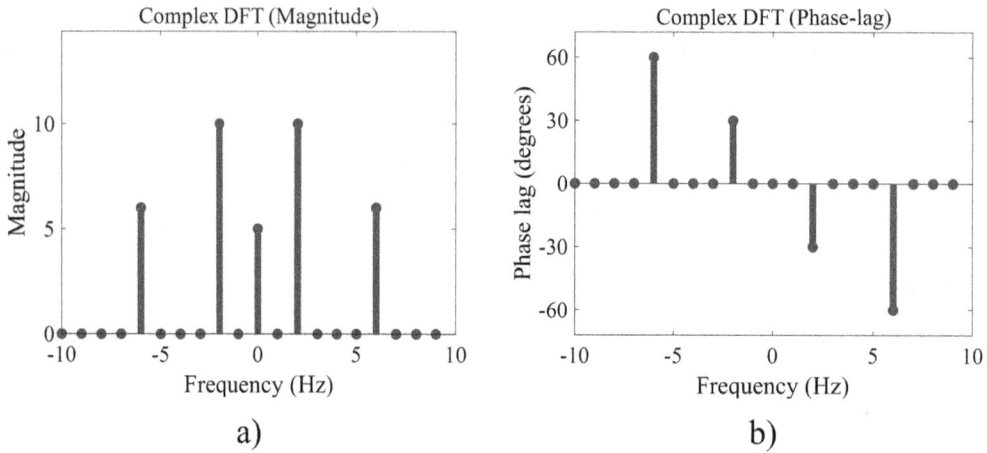

FIGURE 8.16   Complex DFT (a) magnitude and (b) phase.

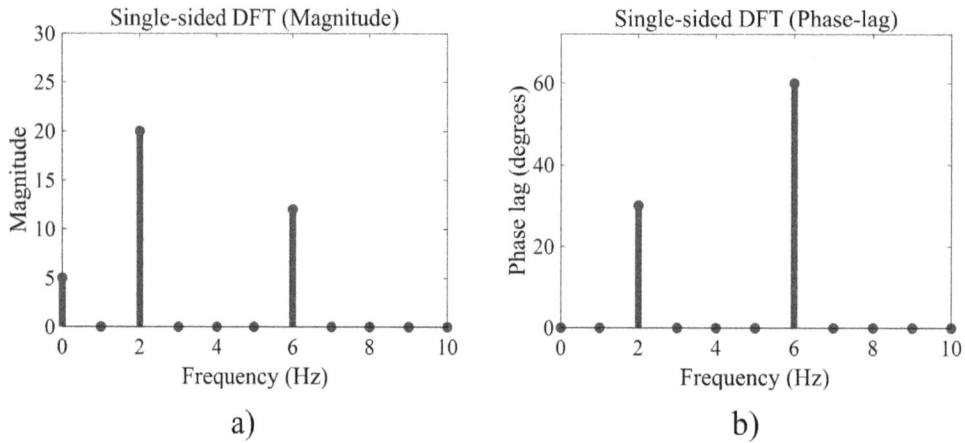

FIGURE 8.17   Single-sided DFT (a) magnitude and (b) phase.

8.13   FAST FOURIER TRANSFORM (FFT)

Determining DFT is computationally an expensive task. The **Fast Fourier Transform** (FFT) is nothing but a class of algorithms for efficient computation of the DFT of a signal. One such algorithm was first developed by Cooley and Tukey (1965). The total number of arithmetic multiplications and additions needed is a measure of the computational complexity of an algorithm. The lower this number, the lower the time needed for computation, and more is the computational efficiency of the algorithm.

It is seen that the computation time to compute the DFT of a discrete signal with N samples (which needs evaluation of RHS of Eq. 8.57) is roughly proportional to N^2. The FFT algorithms exploit the symmetry and periodicity of the DFT terms and divide DFT computation into successively smaller DFTs. The amount of computation and time FFT algorithms require to compute DFT is proportional to $N \log_2 N$.

N, the number of samples in the discrete signal, is preferably taken as a power of 2. As an example of the computational effort: if $N = 2^{12} = 4096$, then $N^2 = 2^{24} = 16777216$ and $N \log_2 N = 49152$. Thus, an FFT algorithm requires 3 orders of magnitude less computation than the direct computation of the DFT and therefore is quite efficient. All modern DSP analyzers use an FFT algorithm to compute DFT.

8.14   EFFECT OF TIME SAMPLING ON FREQUENCY-DOMAIN REPRESENTATION

We studied the Fourier analysis of discrete signals in the previous sections. We saw that the discretization or sampling of the signal leads to periodicity in the frequency domain. An important question is: Does the periodicity in the frequency domain may adversely affect the ability to determine the frequency content of the continuous signal by the Fourier analysis of the corresponding discrete signal? We will address this question in this section.

We look at the relationship between the Fourier Transforms of the continuous and the corresponding sampled signal. It gives insight into the effect of time sampling and leads to a significant result about the limit on the choice of the sampling frequency to enable a correct frequency-domain representation of a continuous signal through the Fourier transform of the corresponding discrete signal.

Let $x_c(t)$ be a continuous signal. It is periodically sampled to obtain $x_s(t)$, the sampled signal. The process of periodic sampling is mathematically represented as the product of the signal $x_c(t)$ with a signal $s(t)$ which is a train of unit impulses (the Dirac delta functions) uniformly spaced with sampling interval Δt. Thus, we have

$$x_s(t) = x_c(t).s(t) \tag{8.68}$$

Figure 8.18 shows the three signals. $x_s(t)$ is equal to $s(t)$ modulated by the signal $x_c(t)$.

Taking the Fourier transform of Eq. (8.68) and noting that the time-domain multiplication of two signals is equivalent to convolution (represented by symbol '*') of their frequency-domain counterparts, we can write

$$X_s(\omega) = \frac{1}{2\pi} X_c(\omega) * S(\omega) \tag{8.69}$$

where the capital letters represent the Fourier transforms of the corresponding time-domain signals.

Let us first find $S(\omega)$. The periodic impulse train $s(t)$, which has a period Δt, can be written as

$$s(t) = \sum_{n=-\infty}^{\infty} \delta(t - n\Delta t) \tag{8.70}$$

Since $s(t)$ is periodic, it can be represented using the Fourier series expansion (Eq. (8.19)) as

$$s(t) = \sum_{n=-\infty}^{\infty} c_n e^{in\omega_0 t} \tag{8.71}$$

where $\omega_0 = 2\pi/\Delta t$ is the fundamental frequency. It is also the rate of periodic sampling and hence is also the sampling frequency (ω_s) in rad/sec. The coefficients c_n can be obtained using Eq. (8.20) as

$$c_r = \frac{1}{\Delta t} \int_{-\Delta t/2}^{\Delta t/2} s(t)\, e^{-ir\omega_s t} dt \tag{8.72}$$

Substituting $s(t)$ from Eq. (8.70),

$$c_r = \frac{1}{\Delta t} \int_{-\Delta t/2}^{\Delta t/2} \left(\sum_{n=-\infty}^{\infty} \delta(t - n\Delta t) \right) e^{-ir\omega_s t} dt \tag{8.73}$$

Since the integration is over the duration $-\Delta t/2$ to $\Delta t/2$, we have a nonzero result of integration only for the $n = 0$ term, and therefore Eq. (8.73) reduces to

$$c_r = \frac{1}{\Delta t} \int_{-\Delta t/2}^{\Delta t/2} \delta(t)\, e^{-ir\omega_s t} dt \tag{8.74}$$

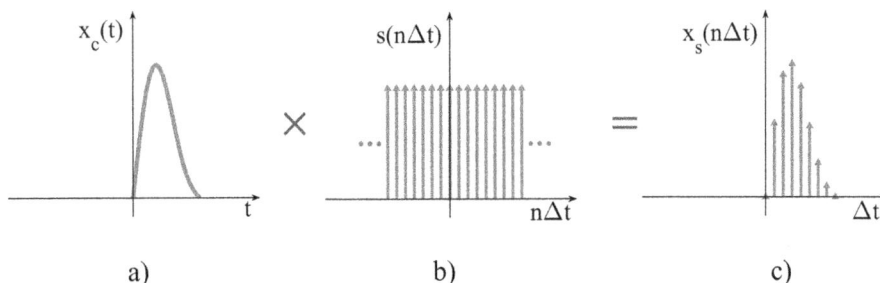

FIGURE 8.18   (a) $x_c(t)$, (b) $s(n\Delta t)$, and (c) $x_s(n\Delta t)$.

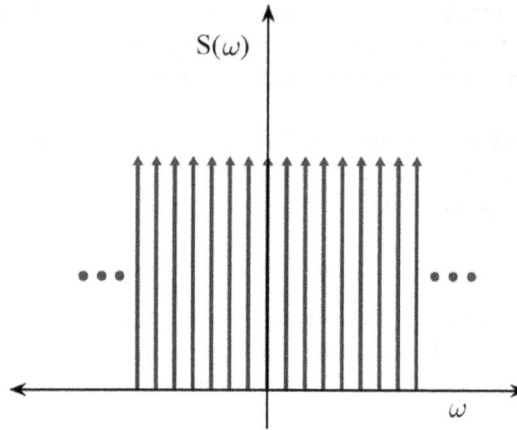

FIGURE 8.19   Fourier transform of $s(n\Delta t)$.

Using the sifting property of the delta function, which says that $\int\limits_{-\infty}^{\infty} f(t)\, \delta(t-T)dt = f(T)$, we obtain

$$c_r = \frac{1}{\Delta t} \quad \text{for} \quad -\infty \leq r \leq \infty \tag{8.75}$$

Therefore, from Eq. (8.71), the Fourier series expansion of $s(t)$ is

$$s(t) = \frac{1}{\Delta t} \sum_{n=-\infty}^{\infty} e^{in\omega_s t} \tag{8.76}$$

We know that the Fourier transform of $e^{in\omega_s t}$ is $2\pi\delta(\omega - n\omega_s)$. Using this result, the Fourier transform of Eq. (8.76) is obtained as

$$S(\omega) = \frac{2\pi}{\Delta t} \sum_{n=-\infty}^{\infty} \delta(\omega - n\omega_s) \tag{8.77}$$

Figure 8.19 shows the plot of $S(\omega)$ and we note from the plot that the Fourier transform of a periodic train of impulses is a periodic train of impulses in the frequency domain.

Now we come back to the convolution in Eq. (8.69), which can be written as

$$X_s(\omega) = \frac{1}{2\pi} \int\limits_{-\infty}^{\infty} S(g)\, X_c(\omega - g).dg \tag{8.78}$$

where g is a dummy frequency variable and has the units of ω. Substituting Eq. (8.77) in Eq. (8.78) and changing the order of integration and summation operations, we get

$$X_s(\omega) = \frac{1}{\Delta t} \sum_{n=-\infty}^{\infty} \int\limits_{-\infty}^{\infty} \delta(g - n\omega_s)\, X_c(\omega - g).dg \tag{8.79}$$

Again, making use of the sifting property of the delta function, we obtain

$$X_s(\omega) = \frac{1}{\Delta t} \sum_{n=-\infty}^{\infty} X_c(\omega - n\omega_s) \tag{8.80}$$

For clarity, we can expand the sum and write

$$X_s(\omega) = \frac{1}{\Delta t}\left(\cdots\cdots + X_c(\omega + 2\omega_s) + X_c(\omega + \omega_s) + X_c(\omega) + X_c(\omega - \omega_s) + X_c(\omega - 2\omega_s) + \cdots\cdots\right) \tag{8.81}$$

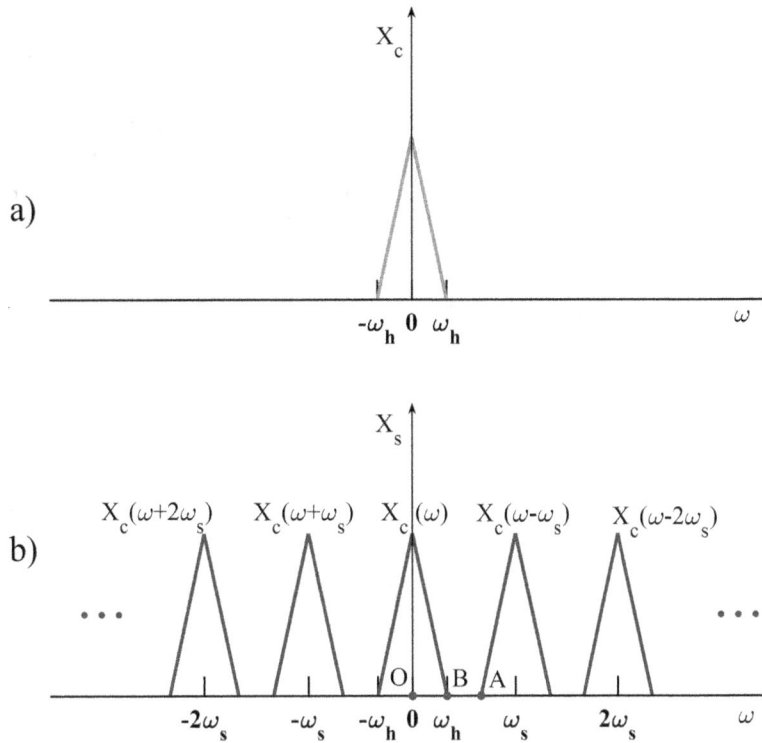

FIGURE 8.20   (a) $X_c(\omega)$, (b) $X_s(\omega)$

Eqs. (8.80) and (8.81) give the relationship between the Fourier transforms of the continuous signal and the corresponding sampled signal. Let us look at what the terms on the RHS of Eq. (8.81) represent.

- $X_c(\omega)$ is the Fourier transform of the continuous signal.
- $X_c(\omega - \omega_s)$ is $X_c(\omega)$ shifted by ω_s on the frequency axis, while $X_c(\omega - 2\omega_s)$ is $X_c(\omega)$ shifted by $2\omega_s$ on the frequency axis.
- Similarly, $X_c(\omega + \omega_s)$ and $X_c(\omega + 2\omega_s)$ are $X_c(\omega)$ shifted by $-\omega_s$ and $-2\omega_s$, respectively, on the frequency axis.

Thus, we see that $X_s(\omega)$ is a sum of $X_c(\omega)$ and its shifted versions, scaled by $\dfrac{1}{\Delta t}$. It means that $X_s(\omega)$ consists of periodically repeated copies of scaled $X_c(\omega)$ with a period equal to the sampling frequency ω_s. It is also the conclusion stated while discussing the periodicity of DFS in Section 8.6.2.

As an example, if we assume $X_c(\omega)$ to be like shown in Figure 8.20a, then $X_s(\omega)$ would be like shown in Figure 8.20b.

In this manner, we have analyzed the effect of sampling a continuous signal on the frequency-domain representation of the sampled signal.

8.15   ALIASING AND SAMPLING THEOREM

In light of the outcome in Section 8.14, a question that arises is how the highest frequency component present in $X_c(\omega)$ and the chosen sampling frequency affect $X_s(\omega)$? Let the highest frequency component in $X_c(\omega)$ has a frequency ω_h, as shown in Figure 8.20a.

It is seen from Figure 8.20b that since $\omega_s - \omega_h$ (i.e., distance OA) is more than ω_h (i.e., distance OB), we have

$$\omega_s - \omega_h > \omega_h \tag{8.82}$$

$$\omega_s > 2\omega_h \tag{8.83}$$

and the copies of $X_c(\omega)$ do not overlap, and they remain separated in $X_s(\omega)$. In this case, it is possible in principle to retrieve $x_c(t)$ by filtering $x_s(t)$ using a low-pass filter with a cut-off frequency ω_c such that $\omega_h < \omega_c < \omega_s - \omega_h$. Figure 8.21 shows the frequency response magnitude of such an ideal low-pass filter. Figure 8.22 shows the spectrum $X_{s,f}(\omega)$ obtained after low-pass filtering of $X_s(\omega)$ and applying the appropriate scaling factor. It is seen that $X_{s,f}(\omega)$ matches exactly with the spectrum of the original continuous signal $X_c(\omega)$ (shown in Figure 8.20a).

If Eq. (8.83) is not satisfied, then the copies of $X_c(\omega)$ overlap as shown in Figure 8.23a. The copies of $X_c(\omega)$ are no longer separated in $X_s(\omega)$ and therefore $X_c(\omega)$ can't be recovered by low-pass filtering. The resultant spectrum, shown in Figure 8.23b, is obtained by combining the overlapping copies of $X_c(\omega)$. Figure 8.23c shows the spectrum $X_{s,f}(\omega)$ obtained after low-pass filtering of $X_s(\omega)$ (in Figure 8.23b). $X_{s,f}(\omega)$ is not the same as the spectrum of the original continuous signal $X_c(\omega)$ (shown in Figure 8.20a). The figure shows that the high-frequency components are distorted due to the overlapping of the adjacent copies of $X_c(\omega)$ and therefore, the signal recovered by low-pass filtering is distorted. The distortion of the signal in this manner is referred to as the **aliasing** of the signal.

Nyquist sampling theorem lays down the condition on the minimum sampling frequency so that the aliasing doesn't occur. As per this theorem, in a band-limited signal $x_c(t)$, if the highest frequency component is ω_h, then $x_c(t)$ is correctly determined from its samples if

$$\omega_s > 2\omega_h \tag{8.84}$$

The frequency ω_h is called the Nyquist frequency, while $2\omega_h$ is called the Nyquist rate, which is the minimum sampling frequency to avoid aliasing. Thus, aliasing is a result of a low sampling rate, due to which the high-frequency signals appear as signals with lower frequencies.

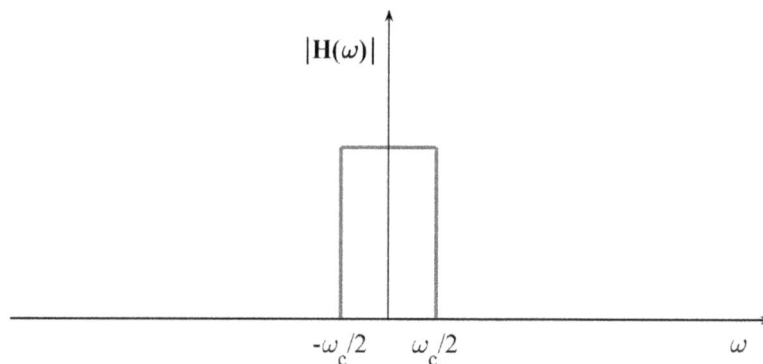

FIGURE 8.21   Frequency response magnitude of an ideal low-pass filter.

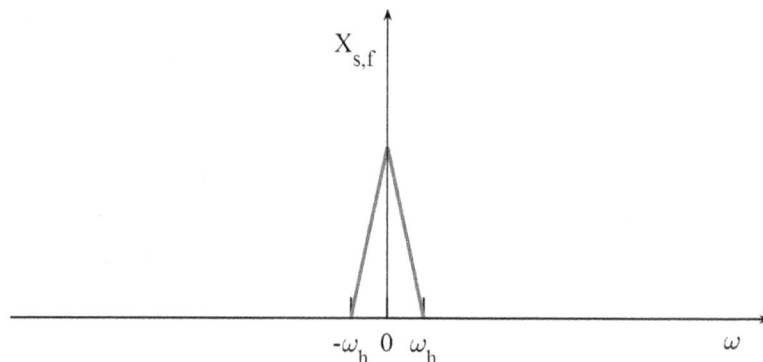

FIGURE 8.22   Frequency spectrum after low-pass filtering.

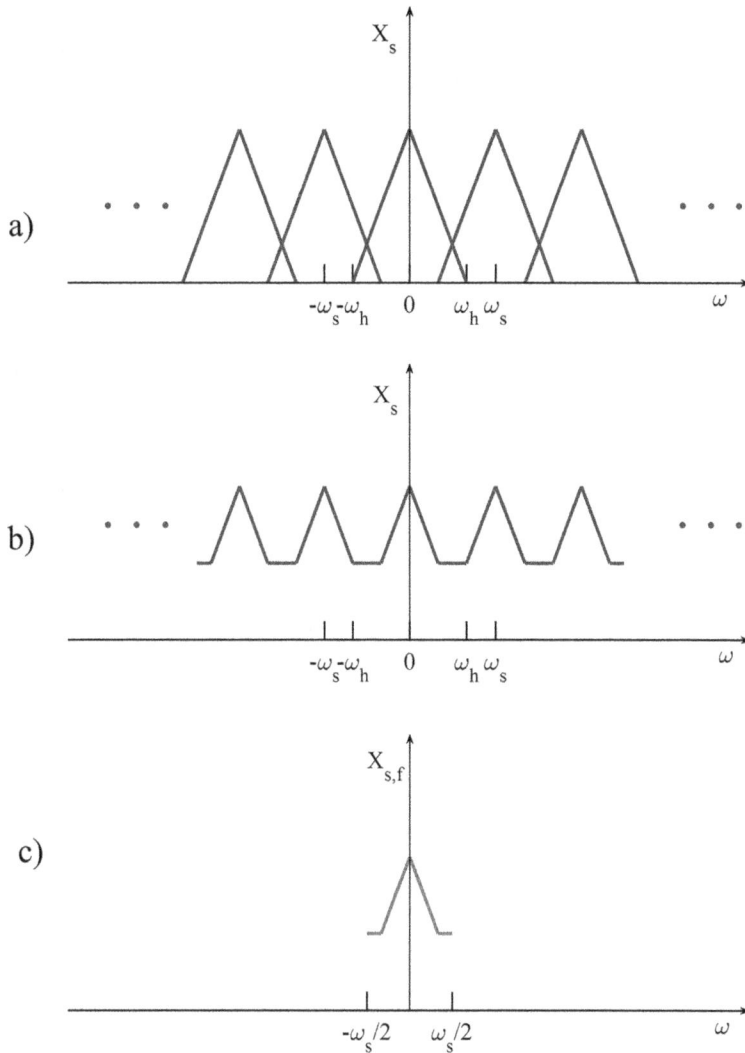

FIGURE 8.23 (a) Copies of $X_c(\omega)$, (b) $X_s(\omega)$, (c) $X_{s,f}(\omega)$.

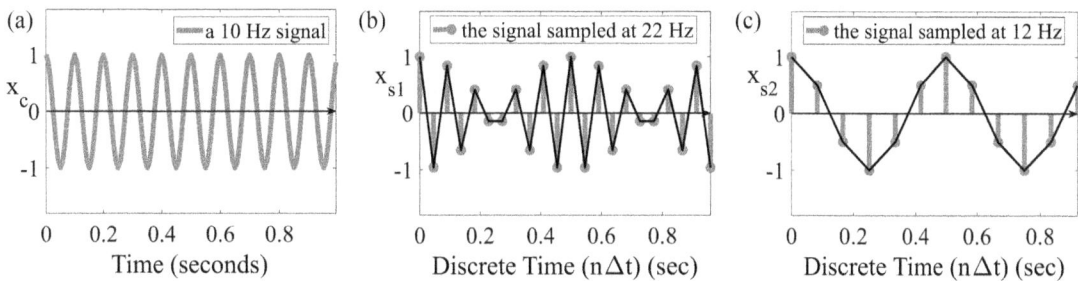

FIGURE 8.24 (a) A 10 Hz continuous sine signal, (b) the signal sampled at 22 Hz, and (c) the signal sampled at 12 Hz.

Example 8.3

Let us take an example of a harmonic signal given by $x_c(t) = 1\cos 2\pi f_o t$, with frequency $f_o = 10$ Hz, as shown in Figure 8.24a and study the effect of the sampling frequency on the resulting discrete signal.

Figure 8.24b and c show the discrete signals x_{s1} and x_{s2} obtained by sampling at 22 Hz and 12 Hz, respectively. From the time histories, it is observed that, x_{s1} represents the variations in x_c correctly, while x_{s2} varies slower than x_c. A DFT of these discrete signals reveals their frequency components at 10 Hz and 2 Hz, respectively. Thus, the frequency of x_{s1} matches with that of x_c, and thus, there is no aliasing. However, the frequency of x_{s2} does not match that of x_c. It is because the discrete signal x_{s2} is aliased since it is sampled at a frequency lower than 20 Hz, the minimum sampling frequency needed to avoid aliasing as per the Nyquist sampling theorem. It can be shown that the frequency of an aliased harmonic signal is $(f_s - f_o)$, which for x_{s2} is 2 Hz, as also correctly revealed by the DFT. Thus, the 10 Hz signal appears as a low-frequency signal of 2 Hz due to aliasing.

8.16 ANTI-ALIASING FILTER

The aliasing should be avoided to obtain a correct Fourier spectrum via DFT. One straightforward way to avoid aliasing is by satisfying the sampling theorem. The theorem can be satisfied if the frequency of the highest frequency component present in the signal is known. But in most cases, the highest frequency component present may not be known, so the sampling frequency according to the sampling theorem can't be chosen. It should be noted that the DFT can't be used to determine the highest frequency as it requires a discrete signal, which depends on the sampling frequency choice.

Another way to avoid aliasing can be by removing via low-pass filtering the frequency components beyond the frequency range of interest. However, low-pass filtering needs to be done on the analog signal before it is discretized. Such a filter is called an **anti-aliasing filter**. It is an analog filter, and the analog signal to be processed via DSP is passed through it. The filtered analog signal is then fed to the A/D converter, as shown in Figure 8.25. Figure 8.21 (given earlier) shows the frequency response of an ideal low-pass filter. It is called an ideal filter because it has a sharp cut-off frequency, i.e., for $\omega > \omega_h$ the frequency response is zero, while for $\omega \leq \omega_h$ the frequency response is constant. But in practice, the filter has a roll-off, which means the frequency response not abruptly but gradually falls to zero, as shown in Figure 8.26. The cut-off frequency is the frequency where the magnitude of the frequency response has been reduced by 3 dB. The Butterworth low-pass filter is one of the commonly used filters. The gain of the nth-order Butterworth filter is given by

$$|H(\omega)| = \frac{1}{\sqrt{1 + \left(\dfrac{\omega}{\omega_c}\right)^{2n}}} \tag{8.85}$$

The roll-off characteristics depend on the order of the filter. The higher the order, the sharper the roll-off. Commercially available spectrum or FFT analyzers have built-in anti-aliasing filters. The filter should be active during signal acquisition. If there is no built-in filter, a suitable low-pass filter should be used to prevent aliasing.

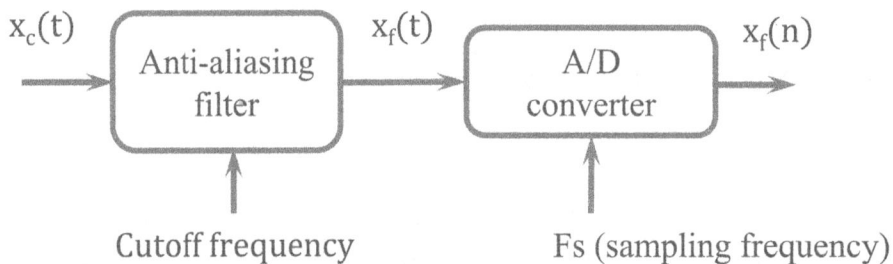

FIGURE 8.25 Anti-aliasing filter before the A/D converter.

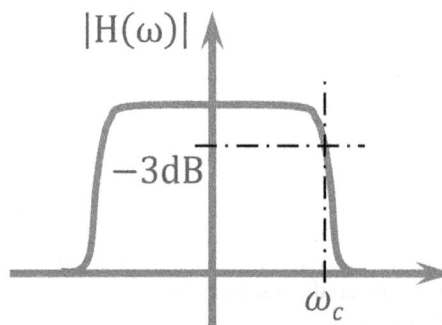

FIGURE 8.26 Low-pass filter frequency response with roll-off.

8.17   QUANTIZATION

Quantization is mapping continuous signal values into discrete values for representing them through a finite number of bits. Thus the representation of a discrete signal by a digital signal may introduce some distortion called quantization error.

The amplitudes of the samples of a discrete signal are real numbers. When these real numbers are quantized, the signal samples are quantized. The quantized samples are then represented by binary numbers to obtain the digital signal.

The sampling/discretization, quantization, and binary conversion are done by an A/D converter, as depicted in Figure 8.27. The signal is sampled by a sample-and-hold circuit that maintains its output constant for sampling time Δt. During this time, the quantizer quantizes the sample. Thus, every sample $x(n)$ is transformed or mapped to one of the predefined quantized levels $x_q(n)$ depending upon the amplitude range it belongs to. The quantized value is then converted to a binary word $x_b(n)$.

The number of quantization levels is 2^n, where n is the number of bits used for binary representation. Therefore, in a 3-bit ADC, there are eight quantization levels. Uniform quantization is the most basic quantization method in which the signal range to be quantized is divided into equal intervals.

If the input voltage range selected is $\pm V_{max}$, then the total range of the signal to be quantized is $2V_{max}$. If there are 2^n equal intervals, then the quantization interval or resolution of the ADC, or the width of each interval, is $\Delta V = 2V_{max}/2^n$. Figure 8.28 shows the correspondence between the signal ranges on the horizontal axis and the corresponding quantization levels on the vertical axis. The quantizer shown in this figure is a 3-bit uniform midrise-type quantizer. Here, the origin lies in the middle of the rise portion, and there are an even number of quantization levels. The quantization level for each range is the mid-point value of the range. The sample values are rounded off to the nearest quantization level during the quantization. Based on the location of a sample on the horizontal axis, its quantization level can be obtained from the vertical axis. The quantized samples are then converted into binary codes, as shown in Table 8.1 for a 3-bit ADC.

The signal quantization distorts the signal to some extent. The difference between the quantized and discrete signals is the quantization error or noise,

$$e(n\Delta t) = x_q(n\Delta t) - x(n\Delta t) \tag{8.86}$$

For the uniform quantizer, the quantization error or noise is in the range $-\Delta V/2 \leq e(n\Delta t) \leq \Delta V/2$. The quantization error depends on the number of quantization levels (which is related to the number of bits in the ADC). The smaller the number

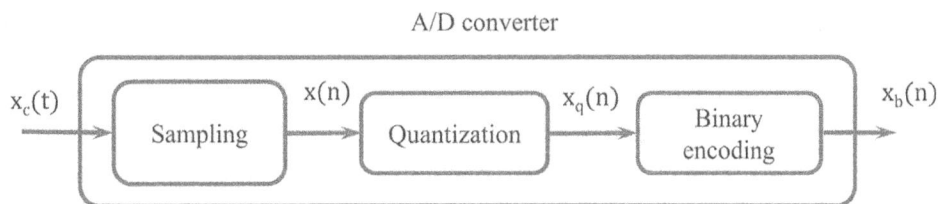

FIGURE 8.27   A/D converter.

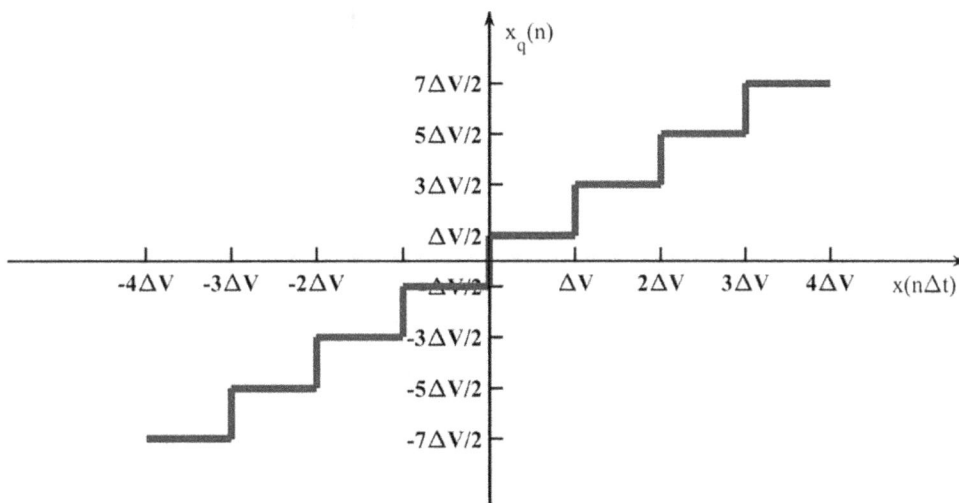

FIGURE 8.28   Signal ranges and the quantization levels.

TABLE 8.1

Conversion of Quantization Levels into 3-Bit Binary Codes

Quantization levels	$\dfrac{-7\Delta V}{2}$	$\dfrac{-5\Delta V}{2}$	$\dfrac{-3\Delta V}{2}$	$\dfrac{-\Delta V}{2}$	$\dfrac{\Delta V}{2}$	$\dfrac{3\Delta V}{2}$	$\dfrac{5\Delta V}{2}$	$\dfrac{7\Delta V}{2}$
Binary equivalent	1 0 0	1 0 1	1 1 0	1 1 1	0 0 0	0 0 1	0 1 0	0 1 1

of levels, the larger the error. A typical analyzer may have a 24-bit ADC. The quantization error also depends on the dynamic range of the measurement, as we see in Example 8.5.

Example 8.4

In this example, we study the quantization of a sinusoidal signal and the resulting quantization error.

Figure 8.29a shows the discrete sinusoidal signal. Figure 8.29b shows the quantized signal obtained using the quantizer described in Section 8.17. It is seen that the quantized signal is not the same as the discrete signal. The quantization error (found using Eq. 8.86) is shown in Figure 8.29c.

Example 8.5

In this example, we study the effect of the input voltage range of measurement on the quantization of the signal.

While taking measurements with an FFT analyzer, the input voltage range needs to be selected. One crucial consideration to keep the quantization error small is to ensure that the full selected input voltage range is utilized to measure a given signal. The input voltage range is optimally utilized if it matches the input signal level.

Let the input voltage range be set at $\pm 10\,V$; for illustration, we assume an ADC with 3 bits. Let us consider two signals $x_1(t) = 9.0\sin 2\pi 1t$ V, and $x_2(t) = 3.0\sin 2\pi 1t$ V.

Both signals have the same frequency, but the maximum value of $x_1(t)$ (9.0 V) is close to the maximum input voltage (as seen in Figure 8.30a) while the maximum value of $x_2(t)$ (3.0 V) is much less (as seen in Figure 8.31a). Due to this, the quantized signal $x_{1q}(n\Delta t)$ (Figure 8.30b) closely follows the shape of the continuous signal $x_1(t)$, while the same thing cannot be said about $x_{2q}(n\Delta t)$ (Figure 8.31b). Thus, $x_2(t)$ has poorer quantization and is said to be an 'under' signal.

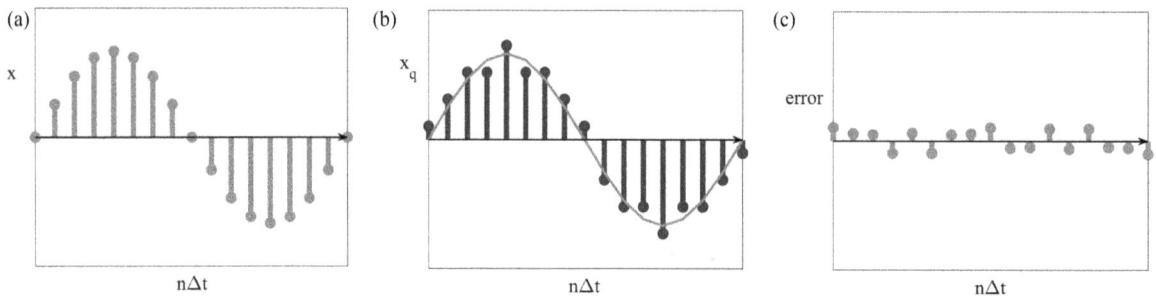

FIGURE 8.29   (a) $x(n\Delta t)$, (b) $x_q(n\Delta t)$, and (c) $e(n\Delta t)$.

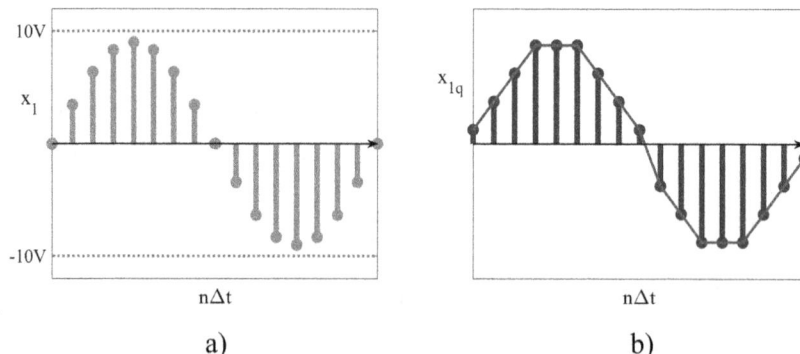

FIGURE 8.30   Optimum match between the input range and signal level (a) $x_1(n\Delta t)$ and (b) $x_{1q}(n\Delta t)$.

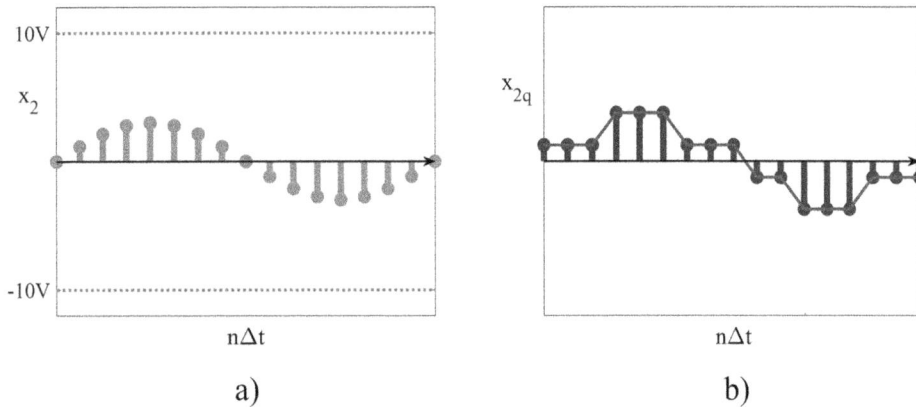

FIGURE 8.31 'Under' signal level (a) $x_2(n\Delta t)$ and (b) $x_{2q}(n\Delta t)$

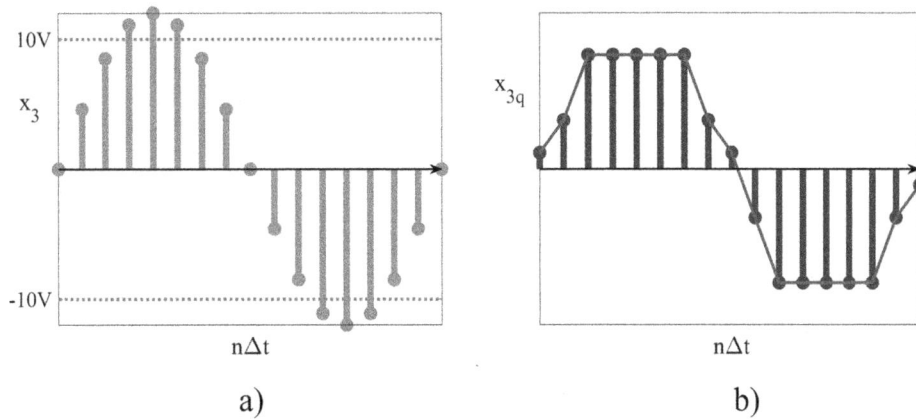

FIGURE 8.32 'Over' signal' level (a) $x_3(n\Delta t)$ and (b) $x_{3q}(n\Delta t)$.

Consider another signal $x_3(t) = 12\sin 2\pi 1t$ V. Here, the input signal level is more than the selected input voltage range (Figure 8.32a). The ADC is overloaded, the signal is clipped (Figure 8.32b), and the quantized signal is severely distorted. The analyzer generally indicates the overloading of the ADC by an 'Overload' indicator. The measurements with 'under' and 'over' signals should be rejected.

If the maximum input signal level fluctuates, choosing the input voltage range is not easy. The 'auto-range' option often available in the analyzers can be used in such cases. When this option is 'on', the analyzer automatically adjusts the input range according to the maximum input signal level.

8.18 WINDOWING

If the signal to be analyzed is finite or transient, then the complete signal can be recorded and analyzed. But if the entire length of a transient signal is not recorded, then we have only a truncated portion of the signal. Also, when the signal is not of finite length but continues indefinitely, only a finite signal record can be captured. What is the implication of analyzing a section of a signal on the frequency domain representation of the signal? We address this question in this section. The process of measuring a finite record $(x_w(t))$ from a signal continuing indefinitely $(x_c(t))$ can be viewed as the multiplication of the signal by a window function $(w(t))$, which has unit magnitude over the record length (T), while zero outside, as shown in Figure 8.33. This process of multiplication of the signal by $w(t)$ is called **windowing**. The window function in Figure 8.33 is a **rectangular window function**.

Therefore,

$$x_w(t) = x_c(t).w(t) \tag{8.87}$$

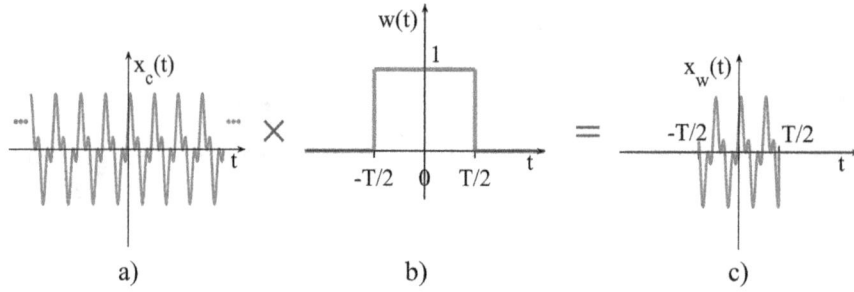

FIGURE 8.33 (a) Signal to be measured ($x_c(t)$), (b) window function ($w(t)$), and (c) recorded signal ($x_w(t)$).

Eq. (8.87) is a mathematical description of recording a finite-length signal. We would like to see how the frequency-domain representation of $x_w(t)$ compares with that of $x_c(t)$. Taking the Fourier transform of Eq. (8.87), we get

$$X_w(\omega) = \frac{1}{2\pi} X_c(\omega) * W(\omega) \tag{8.88}$$

where the capital letters represent the Fourier transforms of the corresponding time-domain signals, and the symbol '*' represents convolution operation. Eq. (8.88) can be further written as

$$X_w(\omega) = \frac{1}{2\pi} \int_{-\infty}^{\infty} W(g) \, X_c(\omega - g).dg \tag{8.89}$$

8.18.1 THE EFFECT OF WINDOWING ON A CONTINUOUS SIGNAL

Let us take an example of a continuous cosine signal to understand the effect of windowing on the frequency spectrum. Let the frequency of the signal be ω_1 rad/sec, and the signal is measured over some arbitrary length of time T by applying the window function $w(t)$. So,

$$x_c(t) = A\cos\omega_1 t \tag{8.90}$$

$$x_c(t) = \frac{A}{2}\left(e^{i\omega_1 t} + e^{-i\omega_1 t}\right) \tag{8.91}$$

Noting that the Fourier transform of $e^{\pm i\omega_1 t}$ is $2\pi\delta(\omega \mp \omega_1)$, the Fourier transform of Eq. (8.91) gives

$$X_c(\omega) = A\pi\delta(\omega - \omega_1) + A\pi\delta(\omega + \omega_1) \tag{8.92}$$

Substituting Eq. (8.92) into Eq. (8.89), we get

$$X_w(\omega) = \frac{1}{2\pi} \int_{-\infty}^{\infty} W(g) \times \left(A\pi\delta(\omega - g - \omega_1) + A\pi\delta(\omega - g + \omega_1)\right).dg \tag{8.93}$$

Using the sifting property of the delta function (mentioned before Eq. 8.75), we obtain

$$X_w(\omega) = \frac{1}{2} AW(\omega - \omega_1) + \frac{1}{2} AW(\omega + \omega_1) \tag{8.94}$$

We now work out $W(\omega)$. By definition of the Fourier transform

$$W(\omega) = \int_{-\infty}^{\infty} w(t) \, e^{-i\omega t}.dt \tag{8.95}$$

$$W(\omega) = \int_{-T/2}^{T/2} e^{-i\omega t} . dt \qquad (8.96)$$

$$W(\omega) = \left(\frac{e^{-i\omega t}}{-i\omega} \right)_{-T/2}^{T/2} \qquad (8.97)$$

which simplifies to

$$W(\omega) = T \frac{\sin \frac{\omega T}{2}}{\frac{\omega T}{2}} \qquad (8.98)$$

The RHS of Eq. (8.98) is the Fourier transform of a rectangular pulse function and $\frac{\sin \frac{\omega T}{2}}{\frac{\omega T}{2}}$ is nothing but the 'sinc' function. Figure 8.34 shows a plot of the $W(\omega)$. The plot is characterized by a main lob with side lobes on either side. The magnitude at $\omega = 0$ is T, while at $\omega = \pm 1 \frac{2\pi}{T}, \ \pm 2 \frac{2\pi}{T}, \ \pm 3 \frac{2\pi}{T}, \cdots$ it is zero. If we define $\omega_o = \frac{2\pi}{T}$ as the fundamental frequency, these latter frequencies can be represented as $\omega = \pm 1\omega_o, \pm 2\omega_o, \ \pm 3\omega_o, \cdots$.

Substituting Eq. (8.98) into Eq. (8.94), we get

$$X_w(\omega) = \frac{1}{2} AT \frac{\sin \frac{(\omega - \omega_1)T}{2}}{\frac{(\omega - \omega_1)T}{2}} + \frac{1}{2} AT \frac{\sin \frac{(\omega + \omega_1)T}{2}}{\frac{(\omega + \omega_1)T}{2}} \qquad (8.99)$$

Thus, Eq. (8.99) is the Fourier transform of the windowed signal $x_w(t)$. Figure 8.35a and b show the magnitudes of the Fourier Transforms $X_c(\omega)$ and $X_w(\omega)$, respectively. It is seen that the true signal $x_c(t)$ (which is the cosine signal) is represented in the frequency domain by impulses at $\pm\omega_1$ frequencies. On the other hand, the spectrum of the windowed signal has peak amplitudes at frequencies $\pm\omega_1$ but also has many frequency components in the neighborhood with smaller magnitudes. Thus, some signal energy can be said to have 'leaked' to the neighboring frequency components in the windowed signal. These additional frequency components in the spectrum are spurious, as they don't exist in the true spectrum and are called **leakage errors**. The leakage also causes the amplitude of the genuine frequency components to be less than their actual values.

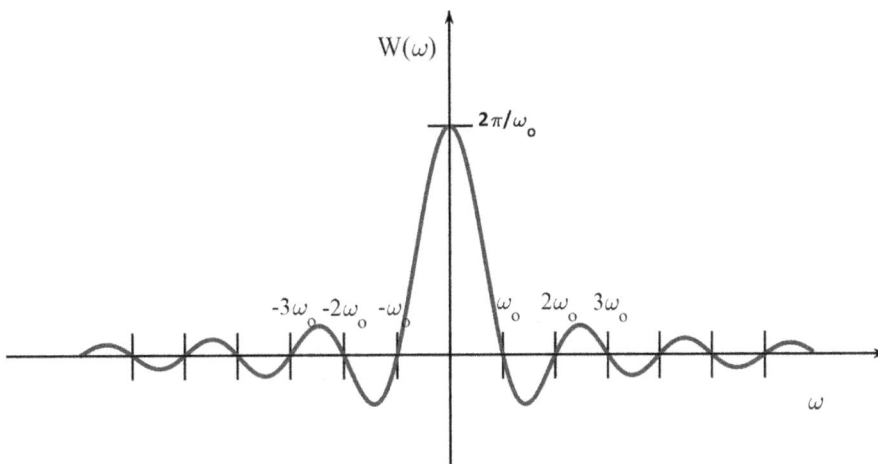

FIGURE 8.34 Fourier transform of the rectangular window function.

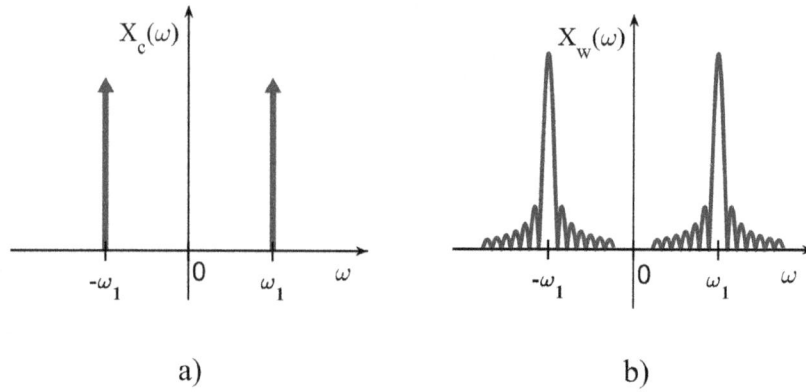

FIGURE 8.35 Fourier Transforms (a) $X_c(\omega)$ and (b) $X_w(\omega)$.

The effect of convolution of $W(\omega)$ with $X_c(\omega)$ is that the spectrum $W(\omega)$ is replicated at each frequency component of $X_c(\omega)$. The main lobe of $W(\omega)$ tends to broaden the peaks in $X_c(\omega)$ due to which the two closely spaced frequency components cannot be easily differentiated. The side lobes of $W(\omega)$ are responsible for the appearance of spurious frequency components in the neighborhood of the peaks, referred to above as the leakage error.

8.18.2 The Effect of Windowing on a Discrete Signal

The continuous signal is recorded as a discrete signal for further processing. Therefore, we now look at the effect of windowing on a discrete signal. We take the example of the cosine signal from the previous section but in a discrete form.

Let the measured signal $(x_w(n))$ is a discrete signal of length T and N samples. Thus, $x_w(n)$ can be viewed as a product of the discrete signal $x(n)$ and a discrete window function $w(n)$.

Thus,

$$x_w(n) = x(n).w(n) \tag{8.100}$$

Taking the Fourier transform of both sides, we get

$$X_w(\Omega) = X(\Omega)*W(\Omega) \tag{8.101}$$

where $X_w(\Omega)$, $X(\Omega)$ and $W(\Omega)$ are DTFTs of $x_w(n), x(n)$ and $w(n)$ respectively. We assume that there are no aliasing errors.

- $X(\Omega)$ is a continuous function and is periodic due to the discrete nature of the signal. For the discrete cosine signal, $X(\Omega)$ is a periodic train of impulses, and its one period resembles the impulses in the spectrum of the continuous cosine signal (as seen in Figure 8.35a).
- $W(\Omega)$ is also a continuous periodic function, and its one period resembles the spectrum $W(\omega)$ of the continuous window function $w(t)$ (as seen in Figure 8.34).
- After the convolution between $X(\Omega)$ and $W(\Omega)$, we get $X_w(\Omega)$, and it is also a continuous periodic function. One period of this would resemble the spectrum $X_w(\omega)$ of the continuous windowed signal $x_w(t)$ (as seen in Figure 8.35b). It would have the structure of the main and side lobes. The main lobe would be centered at the frequency of the cosine signal and would be affected by smearing, while the side lobes would lead to leakage errors.

One difficulty is that $X_w(\Omega)$ is a continuous function and cannot be determined computationally from the signal $x_w(n)$. Therefore, in practice, we find the DFT of $x_w(n)$. As explained in the section on DFT, the DFT of a discrete signal is nothing but a sampled version of the DTFT over one period. Thus, finding DFT amounts to choosing equispaced samples from the DTFT. If the frequency lines in the DFT do not coincide with the exact location of the true signal frequencies in the DTFT (like the peak frequencies of the main lobes), then the location of the peaks in the DFT would not reveal the true frequencies of the peaks in the DTFT. We take two specific examples to understand this further.

Example 8.6

In this example, we study how the record length T affects the frequency lines in the DFT and what is an ideal choice of the record length to determine the correct frequency spectrum of a periodic signal via DFT.

We take a cosine signal of frequency ω_1 rad/sec. The signal is measured over time T seconds with T equal to one period of the signal. The sampling frequency is chosen according to the sampling theorem to avoid aliasing. Let the number of samples in the record be N.

Figure 8.36 shows the DFT of the discrete windowed cosine signal $(x_w(n))$. The DFT correctly predicts the frequency content of the signal as a harmonic signal of the frequency ω_1. The reason for this is as follows.

We note that the gap between the consecutive discrete lines in the DFT is $\omega_o = \dfrac{2\pi}{T}$ and the frequencies corresponding to the various discrete lines are $-\dfrac{N}{2}\omega_o,\ \ldots, -\omega_o,\ 0,\ \omega_o, \ldots,\ \left(\dfrac{N}{2}-1\right)\omega_o$. Since T in the present example is equal to one period of the cosine signal, the discrete line in the DFT at ω_o coincides with the true frequency ω_1 of the signal. Moreover, the magnitude of the other frequency components in the DFT is zero. Because of this, the DFT correctly predicts the frequency of the signal.

The question arises as to why no windowing effect is seen in the DFT. Or, why is there no smearing and leakage? To understand this, we plot the DTFT of $x_w(n)$, after scaling the frequency axis to rad/sec, as shown in Figure 8.37. The main lobe is centered at the frequency ω_1. The discrete frequencies at which the DFT samples the DTFT are also overlaid. It is seen that the DFT draws the samples of the DTFT i) at the frequency corresponding to the peak amplitude of the main lobe and ii) at those frequencies at which the DTFT magnitude is zero. This explains why the magnitude of all the other frequency components in the DFT is zero except the one coinciding with the peaks of the main lobes in the positive and negative frequency axes. This also explains why the DFT captures the true frequency of the signal and is free of any windowing errors.

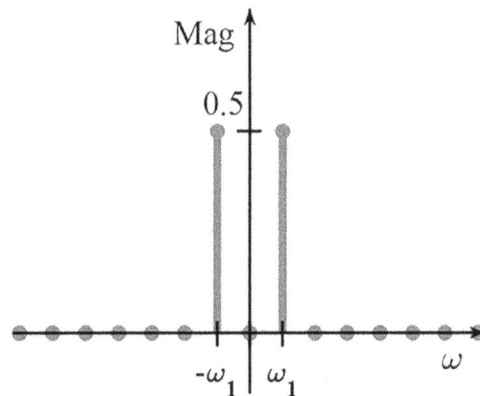

FIGURE 8.36 DFT of the discrete windowed cosine signal.

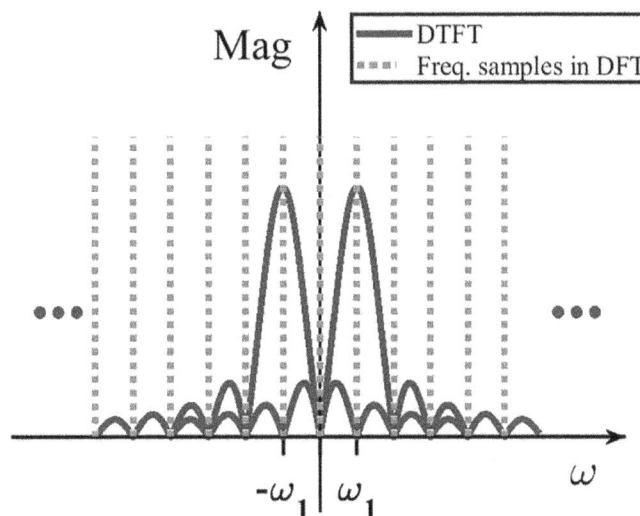

FIGURE 8.37 DTFT of $x_w(n)$ and the frequency samples in the DFT.

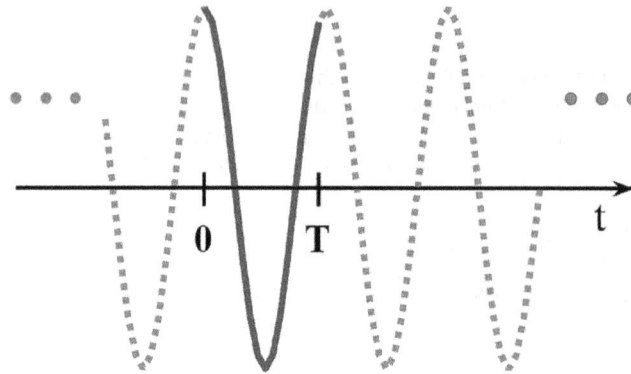

FIGURE 8.38   The signal recorded and the signal analyzed by DFT.

The above result can also be described as follows. We note that the computation of DFT is essentially computing the DFS of a hypothetical periodic signal obtained by repeating the copies of the recorded signal. The continuous line in Figure 8.38 shows the recorded length of the signal, which in the present case is one period T of the signal. The dotted line shows the hypothetical periodic signal obtained from the recorded signal. It is seen that the hypothetical periodic signal matches with the cosine signal to be analyzed, which is why we get the exact result in the DFT.

A similar result is obtained, i.e., the spectrum would be leakage-free if the record length is an integer multiple of the period of the signal. This example thus also demonstrates an ideal choice of the record length to predict the frequency spectrum of a periodic signal.

Example 8.7

In this example, we further study the effect of record length on the frequency lines in the DFT and what happens when there is a deviation from the ideal choice of the record length.

Continuing from example 8.6, let us change the record length to 1.5T. Figure 8.39 shows the DFT of the discrete windowed cosine signal $(x_w(n))$. The DFT no longer has a component at the true frequency (ω_1) of the signal, though it shows two dominating components at the adjacent neighboring frequencies. It is a result of the smearing due to the main lobe. In addition, the DFT also shows the presence of many other frequency components in the neighborhood, which are not present in the true signal. It is a result of the leakage due to the side lobes.

Why does the DFT of the recorded signal in this example suffer from these errors? Let us look at the DTFT of $x_w(n)$, as shown in Figure 8.40. The main lobe is still centered at the true signal frequency ω_1. But we see from the frequency lines that the DFT no longer draws a sample of the DTFT at the peak frequency of the main lobe and therefore misses the true frequency of the signal. The other frequencies at which it samples the DTFT are those where the DTFT has nonzero amplitudes, which appear in the DFT as spurious frequency components, representing leakage.

Like in the previous example, the hypothetical periodic signal obtained by repeating the copies of the record length 1.5T is shown in Figure 8.41 by a dotted line. It is seen that the hypothetical periodic signal, which we effectively analyze through DFT, has discontinuities after every period and is not the same as the cosine signal to be analyzed. Thus, the DFT we get is the DFT of a distorted version of the actual signal and hence does not correctly predict the frequency content of the true signal.

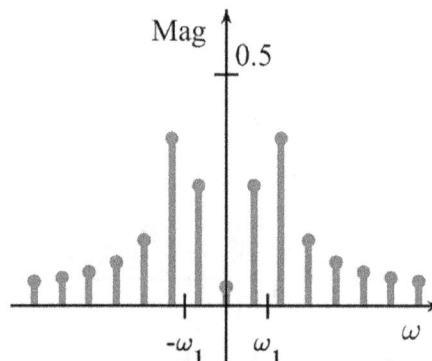

FIGURE 8.39   DFT of the discrete windowed cosine signal.

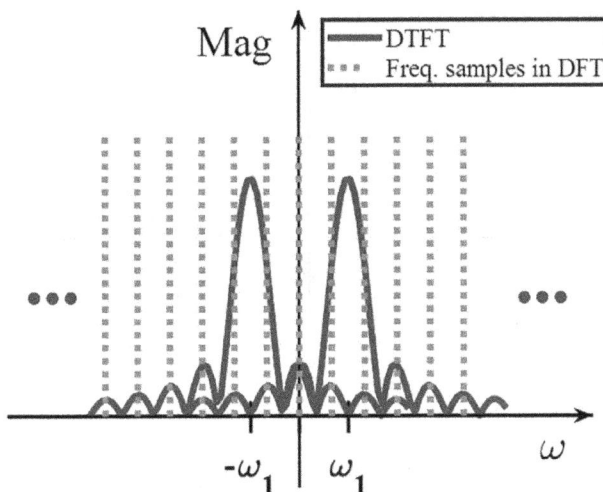

FIGURE 8.40 DTFT of $x_w(n)$ and the frequency samples in the DFT.

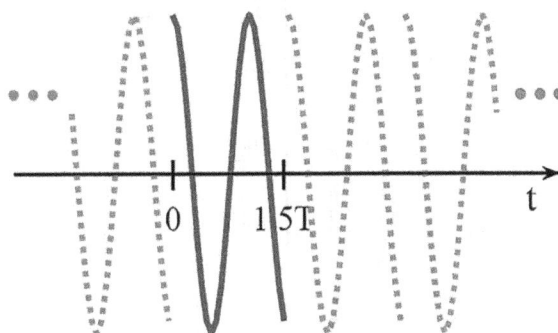

FIGURE 8.41 The signal recorded and the signal analyzed by DFT.

8.18.3 WINDOW FUNCTIONS

We saw in the previous section that the process of recording a finite-length signal from a signal continuing indefinitely could be viewed as the multiplication of the signal by a rectangular window function. We also saw that this might cause leakage errors, causing the computed spectrum to deviate from the true spectrum. The need to reduce leakage errors led to the development of window functions (referred to simply as windows). The windows are time-domain functions. The record length is multiplied by the chosen window before the record is processed to find DFT. One main characteristic of the windows is that they force the record to decay at the ends to eliminate discontinuities in the signal due to truncation.

Some of the commonly used windows are Hanning, Hamming, and Bartlett. Figure 8.42 gives details of these window functions and their plots. Note that the equations for the window functions correspond to the time limits $-\frac{T}{2} \leq t \leq \frac{T}{2}$. If the time limits are $0 \leq \tau \leq T$, then the corresponding window functions ($w(\tau)$) are obtained by substituting $t = \tau - T/2$. In Chapter 9, we study another window, called the exponential window, designed for analyzing the damped vibration response.

To get insight into the effect of window functions, let us compare the frequency spectrum of a window function, say the Hanning window, with that of the rectangular window. Figure 8.43 shows an overlay of the two spectrums.

The spectrum magnitudes are in dB. We observe the following:

- The amplitudes of the side lobes in the Hanning window are smaller than in the rectangular window. The rate of decrease of the side lobes amplitudes is also higher in the Hanning window. Due to these two characteristics, the leakage error with the Hanning window is smaller than with the rectangular window.
- The width of the main lobe of the Hanning window is larger than that of the rectangular window. Therefore, the Hanning window causes a higher smearing effect than the rectangular window and has a lesser ability to differentiate closely spaced frequency components.

(a)

$$w(t) = \begin{cases} 1 - \dfrac{|t|}{T/2} & \text{for } |t| \le \dfrac{T}{2} \\ 0 & \text{otherwise} \end{cases}$$

(b)

$$w(t) = \begin{cases} 0.5 + 0.5\cos\dfrac{2\pi t}{T} & \text{for } |t| \le \dfrac{T}{2} \\ 0 & \text{otherwise} \end{cases}$$

(c)

$$w(t) = \begin{cases} 0.54 + 0.46\cos\dfrac{2\pi t}{T} & \text{for } |t| \le \dfrac{T}{2} \\ 0 & \text{otherwise} \end{cases}$$

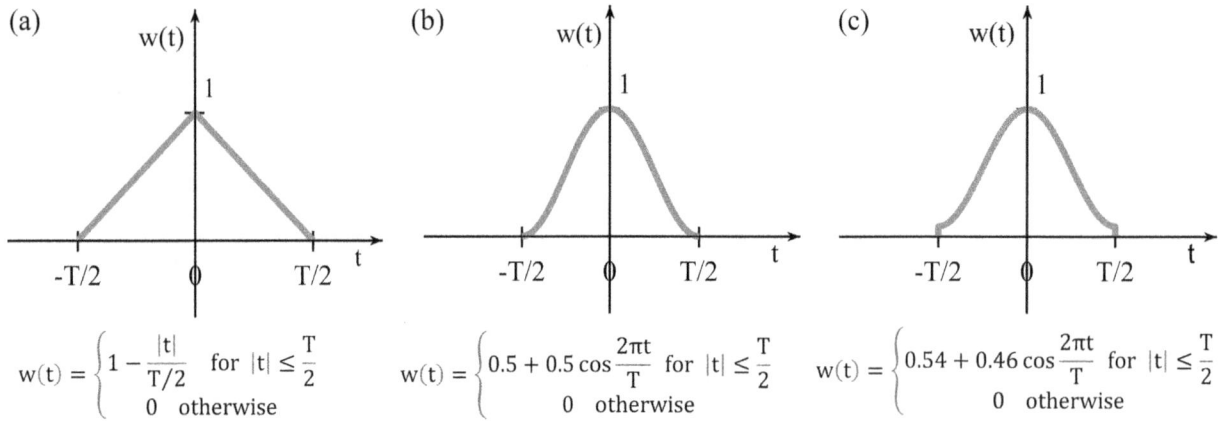

FIGURE 8.42   Window functions (a) Bartlett, (b) Hanning, and (c) Hamming.

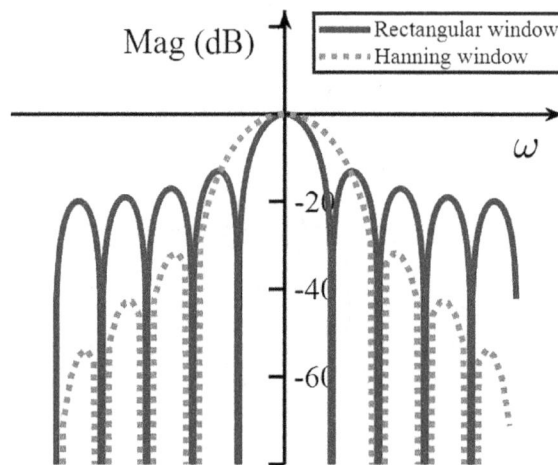

FIGURE 8.43   Fourier transform magnitudes of rectangular and Hanning windows

Therefore, in a nutshell, the use of a window function helps reduce leakage error in the following cases:

- For periodic signals, if it is not sure that the measured record length represents an integer number of the periods of the signal.
- For transient/aperiodic signals of finite length where only a partial length of the signal is recorded.
- For aperiodic signals continuing indefinitely (like random signals).

If no window function is used, it amounts to using a rectangular window.

Example 8.8

In this example, we study the effect of the Hanning window on DFT.

We calculate the DFT of the cosine signal in Example 8.7 using the Hanning window, and the same is shown in Figure 8.44. Compared with Figure 8.39 (showing the DFT with the rectangular window), it is seen that the leakage error has significantly reduced with the Hanning window.

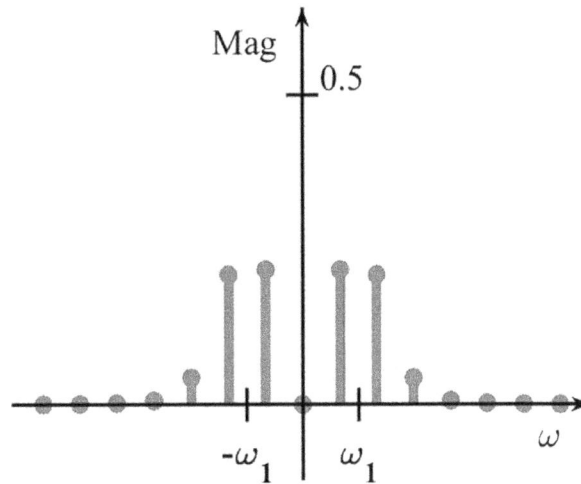

FIGURE 8.44 DFT of the cosine signal with the Hanning window.

8.19 FOURIER ANALYSIS OF RANDOM SIGNALS

Random excitation signals are used in experimental modal analysis to identify modal characteristics. A random force applied to a system causes a random vibratory response. Hence the description of the relationship between the random input, output, and the system in the frequency domain requires frequency domain representation of the random signals, and is covered in this section.

8.19.1 RANDOM SIGNALS

We studied the Fourier analysis of periodic and aperiodic/transient signals in the previous sections. Those signals are deterministic as their values can be determined or predicted precisely at any future time. However, there are signals in practice whose future values cannot be predicted precisely, and such signals are called random signals. For example, the vibration of a bicycle moving over a rough road is random since the road irregularities causing the vibration do not follow any predictable variation. The random signal is quantified using average values or other statistical measures, and the values of the signal at future instants can only be predicted in probabilistic terms.

A random signal is an aperiodic signal; hence, the Fourier series cannot be used to represent it in the frequency domain. The signal is also not finite, as it can theoretically continue forever. The infinitely long random signal also doesn't satisfy the condition below, which is necessary to be satisfied for the signal to be Fourier transformed,

$$\int_{-\infty}^{\infty} |x(t)| . dt < \infty \tag{8.102}$$

Therefore, a random signal cannot be Fourier transformed as well. However, it is seen that the autocorrelation function of a random signal satisfies the above condition, and hence its Fourier transform exists. It, therefore, allows defining the frequency-domain representation of a random signal.

We first present a classification of random processes and define the autocorrelation function in the following sections. We then present the motivation for using the autocorrelation function for the frequency-domain representation of a random signal. Then, we define the frequency-domain representation of the random signal, referred to as power spectral density, and see how it is estimated.

8.19.2 AUTOCORRELATION FUNCTION

Let us first see how any random process or phenomenon is described. A random process is described by an ensemble or a collection of sample functions of the process. A sample function is a time history of the random process. Since the process is random, just one sample function cannot characterize the whole random process. For example, in the bicycle example, one run would result in one time history or sample function $x_1(t)$. But, another run with the same bicycle on the same road would result in a time history $x_2(t)$, that would not be the same as $x_1(t)$ in a deterministic sense. Further runs lead to more such time histories. Because every time a measurement is taken, a new time history results, an infinite number of time histories are required to characterize the whole process or capture all the random variations. In practice, many time histories of sufficiently long length need to be measured to characterize the process.

Figure 8.45 shows an ensemble of sample functions $x_1(t)$, $x_2(t)$, ... describing a random process. Each sample function forms a realization of the random process.

Mean value (μ_x) is a fundamental measure of a random process. It is calculated as the mean of the instantaneous values of all the sample functions. If N is the number of sample functions, then at time $t = t_1$

$$\mu_x = \lim_{N \to \infty} \frac{1}{N} \sum_{j=1}^{N} x_j(t_1) \tag{8.103}$$

Another measure is the **autocorrelation function** $R_{xx}(\tau)$, which quantifies the average level of correlation or similarity between the values of the random signal at different instants of time. The symbol R_{xx} with x appearing twice in the suffix is used to indicate the correlation of the signal x with a displaced version of the same signal. Let $x_j(t_1)$ and $x_j(t_1 + \tau)$ are values of the jth sample function at time instants t_1 and $t_1 + \tau$ respectively, then $R_{xx}(\tau)$ is calculated as

$$R_{xx}(\tau) = \lim_{N \to \infty} \frac{1}{N} \sum_{j=1}^{N} x_j(t_1) x_j(t_1 + \tau) \tag{8.104}$$

Thus, the autocorrelation function depends on the time gap τ between the time instants.

If the mean and autocorrelation function values vary with the choice of t_1, then such a random process is said to be **non-stationary**. When these quantities don't vary with the choice of t_1 the random process is said to be **weakly stationary or stationary in a wide sense**. The mean value and autocorrelation function are the first and joint moments of the random process. Higher-order and joint moments can also be defined. If the higher-order and joint moments are also time-invariant, then the random process is said to be **strongly stationary or stationary in a strict sense.**

Note that the mean value and autocorrelation are calculated by averaging across the sample functions of the ensemble. A stationary random process in which the averages calculated for any one sample function over time are the same as the ensemble averages is said to be **ergodic**. Thus, for a **stationary ergodic process**, the mean value and autocorrelation function can also be calculated using a sample function as

$$\mu_x = \lim_{T \to \infty} \frac{1}{T} \int_0^T x_j(t).dt \tag{8.105}$$

$$R_{xx}(\tau) = \lim_{T \to \infty} \frac{1}{T} \int_0^T x_j(t).x_j(t + \tau)dt \tag{8.106}$$

Many random processes in practice behave like stationary ergodic processes. All such processes offer a significant advantage in that they can be studied by measuring only one sample function. Therefore, for the bicycle example, it would be enough to measure a time history with a single run to study its random vibrations. We assume a stationary ergodic process for further discussions related to random processes.

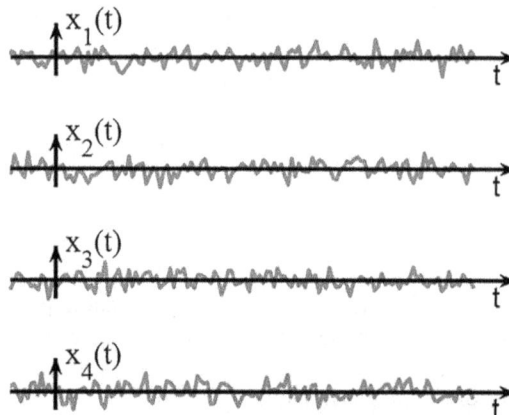

FIGURE 8.45 An ensemble of sample functions of a random process.

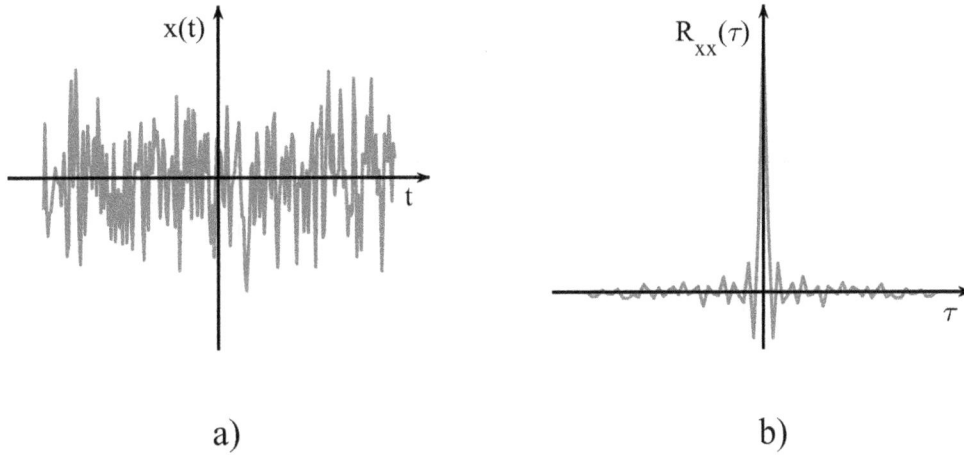

FIGURE 8.46 (a) Random signal and its (b) autocorrelation function.

Figure 8.46 shows a random signal and its autocorrelation function. The mean value of the random signal is zero, and it has no periodic components. We observe that the amplitude of the autocorrelation function reduces with the increasing values of the time gap τ. Therefore, the autocorrelation function of a random signal has a converging nature and satisfies the condition (Eq. 8.102) for the function to be absolutely integrable for the existence of the Fourier transform. Thus, while the random signal cannot be Fourier transformed, its autocorrelation function can be Fourier transformed. Note that the autocorrelation function is an even function that is $R_{xx}(-\tau) = R_{xx}(\tau)$.

8.19.3 Autocorrelation Function of Periodic Signals

We saw that the autocorrelation function could be Fourier transformed, but what about its frequency composition? Does it correlate with the frequency composition of the random signal? We determine the autocorrelation function of periodic signals to get an idea of this. The autocorrelation function is defined in the context of a random signal, but it can be calculated for deterministic signals. Let us first find the autocorrelation function of a harmonic signal. Let the signal is given by

$$x(t) = A\sin(\omega_o t + \theta) \tag{8.107}$$

By the definition of the autocorrelation function (Eq. 8.106)

$$R_{xx}(\tau) = \lim_{T \to \infty} \frac{1}{T} \int_0^T x(t).x(t+\tau)dt \tag{8.108}$$

Since the signal is harmonic, the computation of the limit in Eq. (8.108) is equivalent to simple integration over one period $(T_o = 2\pi/\omega_o)$ of the signal. Hence, we can write

$$R_{xx}(\tau) = \frac{1}{T_o} \int_0^{T_o} A\sin(\omega_o t + \theta).A\sin(\omega_o t + \theta + \omega_o \tau)dt \tag{8.109}$$

$$R_{xx}(\tau) = \frac{A^2}{T_o} \int_0^{T_o} \sin(\omega_o t + \theta) \times \left(\sin(\omega_o t + \theta)\cos\omega_o \tau + \cos(\omega_o t + \theta)\sin\omega_o \tau\right)dt \tag{8.110}$$

The integral of the second term above is zero due to the orthogonality of sin and cos functions, and hence we get

$$R_{xx}(\tau) = \frac{A^2}{T_o} \int_0^{T_o} \sin^2(\omega_o t + \theta)\cos\omega_o \tau dt \tag{8.111}$$

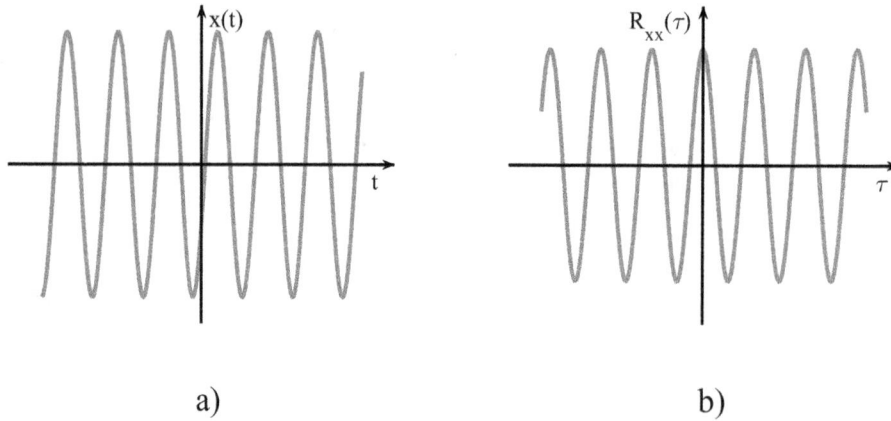

FIGURE 8.47 (a) Harmonic signal and its (b) autocorrelation function.

which finally gives

$$R_{xx}(\tau) = \frac{A^2}{2}\cos\omega_0\tau \tag{8.112}$$

Figure 8.47 show a harmonic signal and its autocorrelation function, respectively.
  We draw the following conclusions from the above result:

- The autocorrelation function of a harmonic signal is a cosine signal of the same frequency.
- The amplitude $\left(\dfrac{A^2}{2}\right)$ of the autocorrelation of a harmonic signal is the mean square value (or square of the RMS value) of the harmonic signal.
- The phase of the harmonic signal doesn't affect the autocorrelation function.

Let us now take a periodic signal, a sum of two harmonic signals of different frequencies and phases given by

$$y(t) = A\sin(\omega_1 t + \theta) + B\sin(\omega_2 t + \phi) \tag{8.113}$$

We find the autocorrelation function by following the steps above. Note that the integration should be performed over one period of signal $y(t)$, which can be worked out from the knowledge of the frequencies ω_1 and ω_2. We get the following result:

$$R_{yy}(\tau) = \frac{A^2}{2}\cos\omega_1\tau + \frac{B^2}{2}\cos\omega_2\tau \tag{8.114}$$

Let us finally take the case of a general periodic signal $x(t)$ of period T_0. Assume that its average value is zero. For a periodic signal, the autocorrelation function is given by,

$$R_{xx}(\tau) = \frac{1}{T_0}\int_0^{T_0} x(t).x(t+\tau)dt \tag{8.115}$$

The periodic signal can be represented by a sum of harmonic components using the Fourier series expansion. Let us use the trigonometric form of the Fourier series (Eq. 8.11) and substitute it in the place of $x(t)$ in Eq. (8.115),

$$R_{xx}(\tau) = \frac{1}{T_0}\int_0^{T_0}\left(\sum_{n=1}^{\infty} d_p\cos(p\omega_0 t - \theta_p)\right)\times\left(\sum_{q=1}^{\infty} d_q\cos(q\omega_0(t+\tau) - \theta_q)\right)dt \tag{8.116}$$

$$R_{xx}(\tau) = \frac{1}{T_o} \int_0^{T_o} \left(\sum_{p=1}^{\infty} d_p \cos(p\omega_o t - \theta_p) \right) \times \left(\sum_{q=1}^{\infty} d_q \left(\cos(q\omega_o t - \theta_q) \cos q\omega_o \tau + \sin(q\omega_o t - \theta_q) \sin q\omega_o \tau \right) \right) dt \quad (8.117)$$

Due to the orthogonality properties of the sine and cosine functions (Eqs. 8.3 and 8.5), the integration of various terms resulting from the product of the two series sums is zero except the integration of the product of $d_p \cos(p\omega_o t - \theta_p)$ and $d_q \cos(q\omega_o t - \theta_q) \cos q\omega_o \tau$ when p and q are equal. Therefore, Eq. (8.117) reduces to

$$R_{xx}(\tau) = \frac{1}{T_o} \int_0^{T_o} \left(\sum_{n=1}^{\infty} d_n^2 \cos^2(n\omega_o t - \theta_n) \cos n\omega_o \tau \right) dt \quad (8.118)$$

which finally leads to

$$R_{xx}(\tau) = \sum_{n=1}^{\infty} \frac{d_n^2}{2} \cos n\omega_o \tau \quad (8.119)$$

We draw the following conclusions from the result in Eq. (8.119):

- The harmonics in the periodic signal show up as cosine signals in the autocorrelation function. The cosine signal is of the same frequency as the harmonic frequency in the periodic signal.
- The amplitude of each cosine signal in the autocorrelation function is equal to the mean square value $\left(\frac{d_n^2}{2} \right)$ of the corresponding harmonic in the periodic signal.
- The phase (θ_n) of a harmonic component doesn't affect the corresponding component in the autocorrelation function.
- There is no correlation among the components with distinct frequencies.

We can derive an alternate expression for the autocorrelation function of periodic signals by using the complex form of the Fourier series (Eq. (8.19)). The final expression is given by,

$$R_{xx}(\tau) = \sum_{k=-\infty}^{\infty} c_k^* c_k e^{ik\omega_o \tau} = \sum_{k=-\infty}^{\infty} |c_k|^2 e^{ik\omega_o \tau} \quad (8.120)$$

Since, $|c_{\pm k}| = \frac{1}{2} d_n|_{n=|k|}$, the expressions in Eqs. (8.120) and (8.119) are identical.

It is seen from the above examples that the frequency components in the autocorrelation function are the same as in the original signal. Therefore, the autocorrelation function preserves the frequency composition of the signal, and hence it can be used to study its frequency composition.

8.19.4 AUTO SPECTRAL DENSITY

Given the conclusions of the previous section, the Fourier transform of the autocorrelation function is used to describe a random signal in the frequency domain and is called **Auto spectral density (ASD)**. It is also referred to as **Power spectral density (PSD)** and is commonly denoted by $S_{xx}(\omega)$. Therefore,

$$S_{xx}(\omega) = \int_{-\infty}^{\infty} R_{xx}(\tau) \, e^{-i\omega\tau} . d\tau \quad (8.121)$$

The autocorrelation function can be obtained from the PSD by the inverse Fourier transform as

$$R_{xx}(\tau) = \frac{1}{2\pi} \int_{-\infty}^{\infty} S_{xx}(\omega) \, e^{i\omega\tau} d\omega \quad (8.122)$$

Eqs. (8.121) and (8.122) are referred to as Wiener-Khinchine relations.

What does $S_{xx}(\omega)$ represent, and why is it also referred to as PSD?

We note from Eq. (8.122) that for $\tau = 0$,

$$R_{xx}(0) = \frac{1}{2\pi} \int_{-\infty}^{\infty} S_{xx}(\omega) \, d\omega \qquad (8.123)$$

The RHS of Eq. (8.123) represents the area under the PSD curve and is equal to $R_{xx}(0)$. Also, from Eq. (8.106), $R_{xx}(0)$ is obtained as

$$R_{xx}(0) = \lim_{T \to \infty} \frac{1}{T} \int_{0}^{T} x^2(t) dt = \overline{x}^2 \qquad (8.124)$$

Thus, $R_{xx}(0)$ is the mean square value $(\overline{x}^2)$ of the signal. Therefore, because of Eq. (8.123), the area under the PSD curve is nothing but the mean square value $(\overline{x}^2)$ of the signal.

The mean square value (MSV) of a signal is interpreted as a quantity proportional to the power of the signal. For example, for a voltage signal with amplitude V, the MSV is $V^2/2$. If the voltage acts across a resistance R, then the power dissipated is $V^2/2R$. Therefore, the power dissipated is proportional to the MSV of the signal. If we assume $R = 1$, then the MSV is also numerically equal to the power dissipated. Therefore, the MSV of a voltage signal is considered to represent the power of the signal. This interpretation is extended to other signals like displacement, velocity, acceleration, and force.

The autocorrelation function of a periodic signal has discrete frequency components, as seen from Eq. (8.119). This equation also shows that for $\tau = 0$, $R_{xx}(0)$ is equal to the sum of the MSV of the harmonic components in the signal. Since the MSV also represents the power of a frequency component, the $R_{xx}(0)$ represents the total power of all the frequency components.

The power is not concentrated in discrete frequencies in a random signal but is continuously distributed over a frequency band. For a random signal also, the quantity $R_{xx}(0)$ gives the MSV and hence the total power of the signal. Because of this, the RHS of Eq. (8.123), which is the area under the function $S_{xx}(\omega)$, is the total power of the signal. Since $S_{xx}(\omega)$ is a continuous function, it specifies at a frequency ω the density of the power rather than the absolute value of the power and hence is referred to as PSD. If x(t) is displacement, then the unit of $S_{xx}(\omega)$ is m^2/Hz.

Note that $S_{xx}(\omega)$ has only magnitude and no phase. It is because of the autocorrelation function, to which $S_{xx}(\omega)$ is related, which doesn't contain the phase information of various frequency components of the original signal.

8.19.5 One-Sided PSD

$S_{xx}(\omega)$ is defined over the two-sided frequency axis $(-\infty < \omega < \infty)$ as it offers mathematical convenience. From a practical viewpoint, it is more convenient to work with PSD $(G_{xx}(\omega))$ defined over the one-sided frequency axis $(0 \leq \omega < \infty)$. $G_{xx}(\omega)$ can be obtained as follows:

$$\begin{aligned} G_{xx}(\omega) &= 2S_{xx}(\omega) \quad &\text{for } 0 < \omega < \infty \text{ except } \omega \neq 0 \\ G_{xx}(\omega) &= S_{xx}(\omega) \quad &\text{for } \omega = 0 \end{aligned} \qquad (8.125)$$

Often the relationship between the one-sided and two-sided PSDs is expressed by only the first part $(G_{xx}(\omega) = 2S_{xx}(\omega))$ of the equation.

8.19.6 White Noise and Band-Limited Random Signals

We now consider autocorrelation and PSD of some important models of random signals like white noise and band-limited random signals.

a. White noise

White noise is an idealized random signal in which the power is distributed uniformly over the frequency range $-\infty < \omega < \infty$. The prefix 'white' refers to an equal contribution from the frequencies ranging from $-\infty$ to ∞. It is similar to white light, which has all the frequencies or colors in the visible spectrum.

Figure 8.48 shows two-sided PSD, a sample time history, the one-sided PSD, and the autocorrelation function of the white noise. The area under the PSD curve is infinity indicating an infinite MSV of the signal. Thus, $R_{xx}(0)$

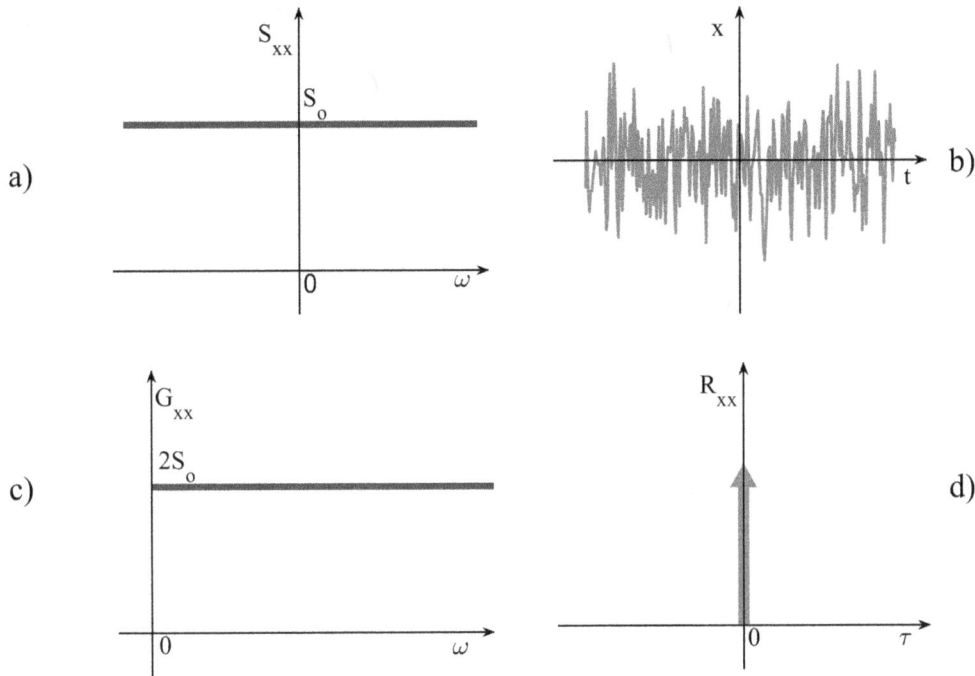

FIGURE 8.48   White noise (a) two-sided PSD, (b) a sample time history, (c) one-sided PSD, and (d) autocorrelation function.

is also infinity, and an impulse at $\tau = 0$ in the autocorrelation reflects this. At $\tau \neq 0$, $R_{xx}(\tau)$ is zero, indicating that the signal is so random that the signal values at two closely spaced instants also don't have any correlation. In practice, a signal with power distributed over an infinite frequency range can't be generated; therefore, the white noise signal is only a theoretical concept but is valuable in the study of random processes.

b. Band-limited random signal

Band-limited random signals have spectral densities limited to certain bands of frequency. The random signals in practice also have a frequency content limited mainly to specific frequency bands. Hence band-limited random signals provide idealized models to approximate random signals in practice.

A **narrowband random signal** has spectral density restricted to a narrow band, while a **wideband random signal** has spectral density distributed over a wide frequency band. The terms 'narrowband' and 'wideband' are used in a relative sense. Figure 8.49 shows PSD, a sample time history, the one-sided PSD, and the autocorrelation function for a typical narrowband random signal. Figure 8.50 shows these quantities for a typical wideband random signal.

8.19.7   ESTIMATION OF PSD OF RANDOM SIGNALS

We defined PSD as the Fourier transform of the autocorrelation function. How is it estimated from the given signal?

Before efficient DSP algorithms were developed, the estimation was done by spectral filtering of the signal. In this method, the signal is passed through a narrow band-pass filter centered at the desired frequency, and the MSV of the filtered signal is estimated, which divided by the filter's bandwidth yields an estimate of the PSD corresponding to the center frequency of the filter. The filter's center frequency is varied to find the PSD over the desired frequency range.

Another approach is to compute the autocorrelation function and find its Fourier transform via digital techniques. However, with fast and efficient digital signal processors and algorithms available, the PSD is generally estimated directly using the DFT of the measured signal, without explicitly computing the autocorrelation function. We look at the basic theory behind this approach. We first estimate the power spectrum and then see how PSD is obtained.

a. Power spectrum

Let $x(t)$ is a sample function of a random signal, and let $x(n)$ be the corresponding measured discrete signal of length T and N samples. Let $X(k)$ be the DFT of $x(n)$. (Note that taking DFT treats the signal as a periodic signal with $x(n)$ representing its one period.)

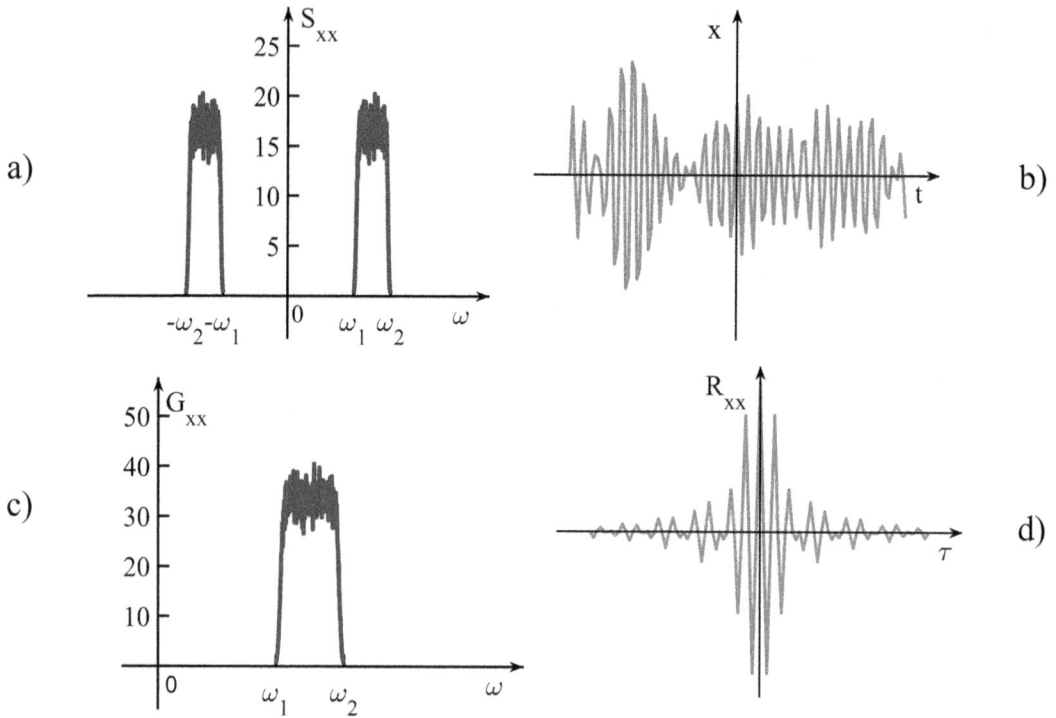

FIGURE 8.49 Narrowband random signal (a) two-sided PSD, (b) a sample time history, (c) one-sided PSD, and (d) autocorrelation function.

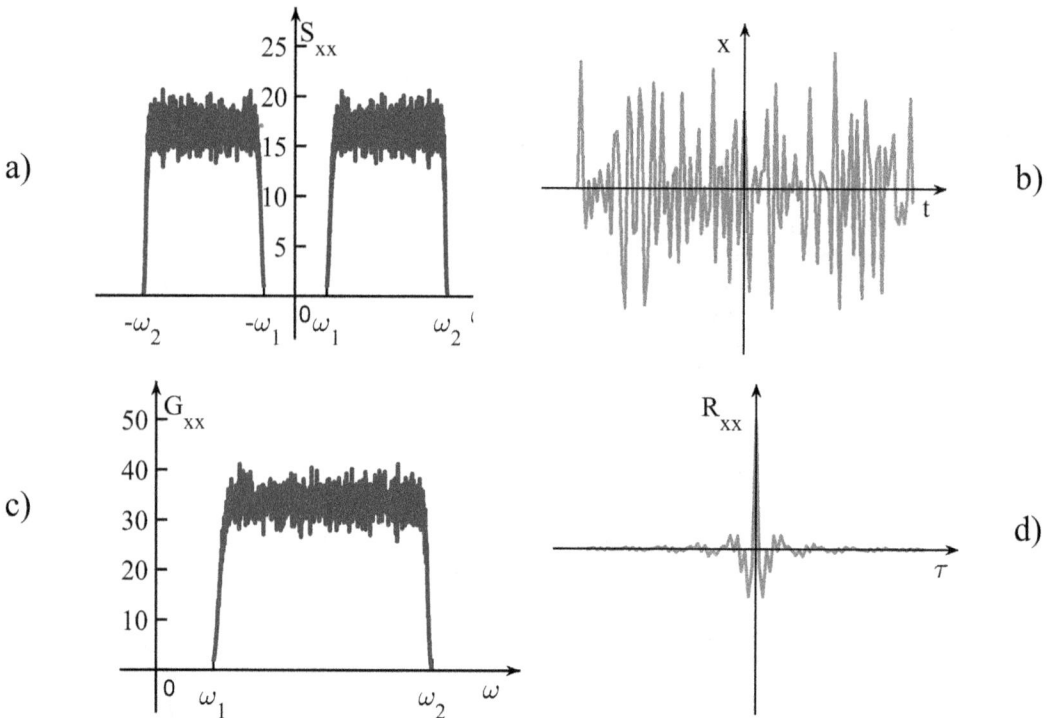

FIGURE 8.50 Wideband random signal (a) two-sided PSD, (b) a sample time history, (c) one-sided PSD, and (d) autocorrelation function.

The autocorrelation function of $x(n)$ is obtained as

$$R_{xx}(r) = \frac{1}{N} \sum_{n=0}^{N-1} x(n)x(n+r) \qquad (8.126)$$

Taking DFT of $R_{xx}(r)$

$$S_{xx}(k) = \frac{1}{N} \sum_{r=0}^{N-1} R_{xx}(r) e^{-ik\frac{2\pi}{N}r} \qquad (8.127)$$

Substituting Eq. (8.126) into (8.127), we get

$$S_{xx}(k) = \frac{1}{N} \sum_{r=0}^{N-1} \left(\frac{1}{N} \sum_{n=0}^{N-1} x(n)x(n+r) \right) e^{-ik\frac{2\pi}{N}r} \qquad (8.128)$$

$$S_{xx}(k) = \frac{1}{N} \times \sum_{r=0}^{N-1} \left(\frac{1}{N} \sum_{n=0}^{N-1} x(n) e^{ik\frac{2\pi}{N}n} x(n+r) e^{-ik\frac{2\pi}{N}n} \right) e^{-ik\frac{2\pi}{N}r} \qquad (8.129)$$

Eq. (8.129) can be written as

$$S_{xx}(k) = \left(\frac{1}{N} \sum_{n=0}^{N-1} x(n) e^{ik\frac{2\pi}{N}n} \right) \cdot \left(\frac{1}{N} \sum_{r=0}^{N-1} x(n+r) e^{-ik\frac{2\pi}{N}(n+r)} \right) \qquad (8.130)$$

Changing the variable of summation in the second bracket by letting $n+r = q$, we obtain

$$S_{xx}(k) = \left(\frac{1}{N} \sum_{n=0}^{N-1} x(n) e^{ik\frac{2\pi}{N}n} \right) \cdot \left(\frac{1}{N} \sum_{q=n}^{n+N-1} x(q) e^{-ik\frac{2\pi}{N}(q)} \right) \qquad (8.131)$$

We note that the expression in the first bracket is the complex conjugate of DFT of $x(n)$, i.e. $X^*(k)$. Since the record $x(n)$ is assumed to be a period (of N samples) of a periodic signal, the summation limits in the second bracket can be replaced with q varying from 0 to N-1. With this change, the term in the second bracket represents the DFT of $x(n)$, i.e., $X(k)$. In light of these observations, Eq. (8.131) reduces to

$$S_{xx}(k) = X^*(k)X(k) = |X(k)|^2 \qquad \text{with } k = 0, 1, 2, ..., N-1 \qquad (8.132)$$

$S_{xx}(k)$ is the DFT of the autocorrelation of the discrete signal $x(n)$. It is called **power spectrum** (or **auto spectrum**) and is an estimate of the PSD in the form of a discrete spectrum. It is also called the **periodogram** of the signal. If $x(t)$ is displacement, then the unit of the power spectrum is m^2. Therefore, the power spectrum is essentially a plot of the MSV (or power) of discrete frequency components.

[Note: In some literature, the following equation for the power spectrum is given:

$$S_{xx}(k) = \frac{1}{N} |X(k)|^2 \qquad (8.133)$$

The factor 1/N in Eq. (8.133) appears because it is based on the DFT expression, which doesn't have the factor 1/N. However, the DFT expression we have used to derive Eq. (8.132) has a factor of 1/N. Thus, either of these equations can be used to compute the power spectrum, ensuring that the corresponding expression for the DFT is used to compute $X(k)$.]

$S_{xx}(k)$ is a two-sided power spectrum, from which the one-sided power spectrum $G_{xx}(k)$ can be determined (using Eq. 8.125).

In the **Bartlett** method to compute the power spectrum, the signal is recorded over a long length and is divided into M segments, and the periodograms corresponding to individual segments are computed and averaged. It reduces the variance and improves the estimate of the power spectrum. The power spectrum estimate is

$$S_{xx}{}^{B}(k) = \frac{1}{M} \sum_{j=1}^{M} (S_{xx}(k))_j = \frac{1}{M} \sum_{j=1}^{M} X_j^*(k) X_j(k) \qquad (8.134)$$

In **Welch's estimate** of the power spectrum, M segments are constructed from the recorded signal such that the segments overlap. Each segment is then applied a window function (w(n)). Periodograms corresponding to windowed segments are then computed and averaged,

$$S_{xx}^{W}(k) = \frac{1}{M}\sum_{j=1}^{M}\frac{1}{U}\times\left|\frac{1}{N_j}\sum_{n=0}^{N_j-1}w(n)x_j(n)e^{-ik\frac{2\pi}{N}n}\right|^2 \qquad (8.135)$$

where $U = \dfrac{1}{N_j}\sum_{n=0}^{N_j-1}w^2(n)$.

b. PSD estimate

The power spectrum is discrete and gives power concentrated in discrete frequency components. It can be scaled to estimate PSD.

If ω_k is the frequency corresponding to the kth frequency component in $S_{xx}(k)$, then the PSD at this frequency can be approximated as

$$S_{xx}(\omega_k) \approx \frac{S_{xx}(k)}{\Delta f} \qquad (8.136)$$

where Δf is the frequency resolution in Hertz of the power spectrum. If the frequency resolution is improved, the frequency lines become closer. In the limit, the continuous PSD can be obtained as

$$S_{xx}(\omega) = \lim_{\Delta f \to 0} S_{xx}(\omega_k) \qquad (8.137)$$

8.19.8 CROSS-CORRELATION FUNCTION

The concept of correlation between the samples of a signal separated by a time lag can be extended to the correlation of two signals. Let $x(t)$ and $y(t)$ be two stationary ergodic random signals and let $x(t)$ and $y(t+\tau)$ be their values at time instants t and $t+\tau$ respectively. Then the cross-correlation (denoted by $R_{xy}(\tau)$) between the two signals is defined as

$$R_{xy}(\tau) = \lim_{T\to\infty}\frac{1}{T}\int_0^T x(t).y(t+\tau)dt \qquad (8.138)$$

The cross-correlation quantifies the average level of similarity between the samples of two signals separated by a time lag.

8.19.9 CROSS-CORRELATION OF PERIODIC SIGNALS

We take the example of cross-correlation between two periodic signals in this section. Let the two signals are

$$x(t) = x_1(t) + x_2(t) + x_3(t) = A_1\sin(\omega_1 t + \theta_1) + A_2\sin(\omega_2 t + \theta_2) + A_3\sin(\omega_3 t + \theta_3) \qquad (8.139)$$

$$y(t) = y_1(t) + y_2(t) = B_1\sin(\omega_1 t + \phi_1) + B_2\sin(\omega_2 t + \phi_2) \qquad (8.140)$$

Let T be the time length after which both signals repeat. Thus T can be expressed as an integer multiple of the individual periods of the two signals. Substituting $x(t)$ and $y(t)$ into Eq. (8.138), we get

$$R_{xy}(\tau) = \lim_{T\to\infty}\frac{1}{T}\int_0^T (x_1(t) + x_2(t) + x_3(t)) \times (y_1(t+\tau) + y_2(t+\tau))dt \qquad (8.141)$$

$$R_{xy}(\tau) = R_{x_1y_1}(\tau) + R_{x_2y_1}(\tau) + R_{x_3y_1}(\tau) + R_{x_1y_2}(\tau) + R_{x_2y_2}(\tau) + R_{x_3y_2}(\tau) \qquad (8.142)$$

Integration over 0 to T shows that $R_{x_2y_1}(\tau)$, $R_{x_3y_1}(\tau)$, $R_{x_1y_2}(\tau)$ and $R_{x_3y_2}(\tau)$ are zero. Evaluation of the remaining integrals gives the following result:

$$R_{xy}(\tau) = \frac{A_1 B_1}{2}\cos(\omega_1\tau + \theta_1 - \phi_1) + \frac{A_2 B_2}{2}\cos(\omega_2\tau + \theta_2 - \phi_2) \tag{8.143}$$

From the result in Eq. (8.143), we draw the following conclusions:

- The frequency components common to the two signals show up in the cross-correlation function.
- The amplitude of a frequency component in the cross-correlation function is the product of the RMS values of the component in the two signals.
- The phase of a frequency component in the cross-correlation function is the difference in the phase angles of the component in the two signals.
- The cross-correlation function preserves the relative phase between the common frequency components in the two signals.
- The cross-correlation between the frequency components of the two signals that have different frequencies is zero. In other words, any frequency component present only in one of the two signals doesn't show up in the cross-correlation function.

The above characteristics and those of the autocorrelation function are crucial for estimating the frequency response function.

8.19.10 CROSS-SPECTRAL DENSITY

Extending the definition of PSD, the **cross-spectral density (CSD)** is the Fourier transform of the cross-correlation function. It is also referred to as the **cross-power spectral density (CPSD)**.

$$S_{xy}(\omega) = \int_{-\infty}^{\infty} R_{xy}(\tau)\, e^{-i\omega\tau}.d\tau \tag{8.144}$$

The cross-correlation function is obtained from the CSD by the inverse Fourier transform

$$R_{xy}(\tau) = \frac{1}{2\pi}\int_{-\infty}^{\infty} S_{xy}(\omega)\, e^{i\omega\tau}d\omega \tag{8.145}$$

8.19.11 ESTIMATION OF THE CSD

The CSD is estimated in the same way as PSD, with the only difference that now we have two different signals. It is also generally estimated directly from the DFTs of the measured signals. We first estimate the cross-spectrum from which the CSD is estimated.

a. Cross-spectrum

Let $x(t)$ and $y(t)$ are sample functions of two stationary ergodic random processes. Let $x(n)$ and $y(n)$ be the corresponding measured discrete signals, each of length T and N samples. Let $X(k)$ and $Y(k)$ be the corresponding DFTs.

The cross-correlation between $x(n)$ and $y(n)$ is

$$R_{xy}(r) = \frac{1}{N}\sum_{n=0}^{N-1} x(n)y(n+r) \tag{8.146}$$

Cross-spectrum $(S_{xy}(k))$ is given by the DFT of $R_{xy}(r)$

$$S_{xy}(k) = \frac{1}{N}\sum_{r=0}^{N-1} R_{xy}(r)e^{-ik\frac{2\pi}{N}r} \tag{8.147}$$

By following the steps similar to those followed for the auto-spectrum, we obtain the following result:

$$S_{xy}(k) = X^*(k)Y(k) \quad \text{with } k = 0, 1, 2, \dots, N-1 \tag{8.148}$$

$S_{xy}(k)$ is a two-sided cross-spectrum, from which the one-sided cross-spectrum $G_{xy}(k)$ can be determined (using Eq. 8.125).

The Bartlett estimate of the cross-spectrum is

$$S_{xy}{}^B(k) = \frac{1}{M} \sum_{j=1}^{M} X_j^*(k) Y_j(k) \tag{8.149}$$

Welch's estimate is given by

$$S_{xy}{}^W(k) = \frac{1}{M} \sum_{j=1}^{M} \frac{1}{U} \left(\frac{1}{N_j} \sum_{n=0}^{N_j-1} w(n) x_j(n) e^{ik\frac{2\pi}{N}n} \right) \times \left(\frac{1}{N_j} \sum_{r=0}^{N_j-1} w(r) y_j(r) e^{-ik\frac{2\pi}{N}r} \right) \tag{8.150}$$

b. Estimate of CSD

We extend the process followed for defining the PSD. If ω_k is the frequency corresponding to the kth frequency component, then the CSD at this frequency can be estimated as

$$S_{xy}(\omega_k) \approx \frac{S_{xy}(k)}{\Delta f} \tag{8.151}$$

where Δf is the frequency resolution of the cross-spectrum in Hz. The continuous CSD can be found in the limit as

$$S_{xy}(\omega) = \lim_{\Delta f \to 0} S_{xy}(\omega_k) \tag{8.152}$$

Example 8.9

A signal with a frequency range of 0–1,000 Hz is to be measured. A frequency resolution of 0.5 Hz is required in the DFT. Find the time and frequency-domain parameters of the measurement.

Solution

For having resolution $\Delta f = 0.5$ Hz in the DFT: $T = \frac{1}{\Delta f} = \frac{1}{0.5} = 2$ seconds

Sampling frequency: $f_s = 2 \times 1,000 = 2,000$ Hz

Number of samples: $N = T f_s = 2 \times 2,000 = 4,000$

Time interval between two samples: $\Delta t = \frac{1}{f_s} = 0.0005$ seconds

Example 8.10

Using MATLAB® code, find the DFT of the signal given below using a one-second long record of the signal. Use a sampling frequency of 30 Hz. Also, find the single-sided DFT and plot the two DFTs.

$$x_c(t) = 4 \sin 2\pi 5t + 12 \cos 2\pi 10t$$

Solution

MATLAB program 8.1 is used to solve the problem. First, we find the complex DFT over the negative and positive frequency axes. It is converted to the single-sided DFT, as discussed in theory. The complex and single-sided DFT magnitudes are shown in Figures 8.51 and 8.52, respectively. We see that the frequencies and magnitudes in the single-sided DFT match with the given signal.

```
%##############################
%   MATLAB PROGRAM 8.1
%##############################
clear all;

%DEFINE RECORD LENGTH
T=1;

%CHOOSE SAMPLING FREQUENCY
fs=30;

%SET TIME VECTOR
Delta_t=1/fs;
t_vec=0:Delta_t:T-Delta_t;
```

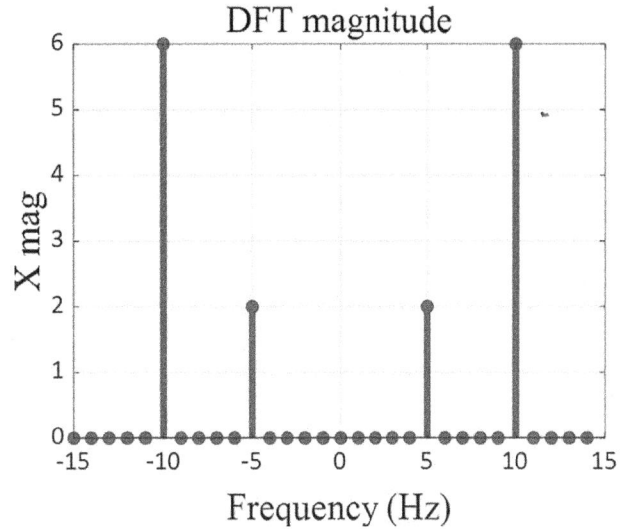

FIGURE 8.51 Complex DFT.

```
%FIND NO. OF SAMPLES
N=length(t_vec);

%SET FREQUENCY VECTORS
Delta_f=1/T;
fmax=fs/2;
f_vec_cmplx=-fmax:Delta_f:fmax-Delta_f;
f_vec_real=0:Delta_f:fmax;

%FIND DISCRETE SIGNAL
x=4*sin(2*pi*5*t_vec)+ 12*cos(2*pi*10*t_vec);

%FIND DFT (FROM -N/2 TO N/2-1)
for k=1:N
  sum=0;
  kk=k-N/2;
  for n=1:N
    sum=sum+x(n)*exp(-i*(kk-1)*2*pi*(n-1)/N);
  end
  Xk(k)=sum/N;
end

%CONVERT DFT TO SINGLE-SIDED DFT
Yk(1)=Xk(N/2+1);
Yk(2:N/2)=2*Xk(N/2+2:N);
Yk(N/2+1)=Xk(1);
%MAKE STEM PLOT OF DFT
figure;
hstem=stem(f_vec_cmplx, abs(Xk));
hstem.Color= 'b';
hstem.Marker= 'o';
hstem.MarkerSize=4;
hstem.LineWidth=4;
set(gca,'FontSize',16);
set(gca,'FontWeight','bold');
```

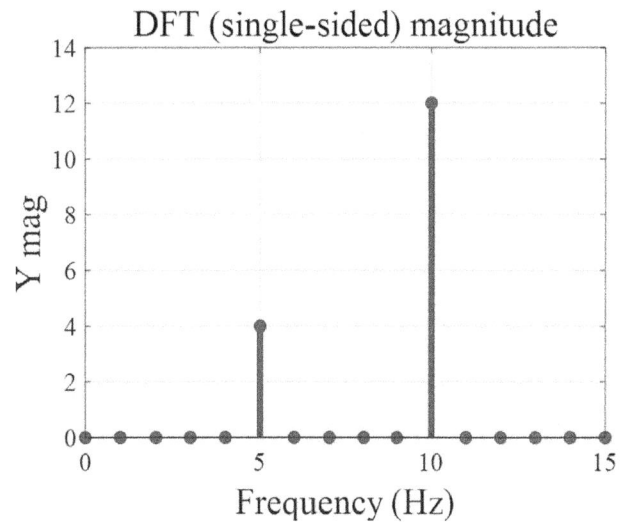

FIGURE 8.52 Single-sided DFT.

(*Continued*)

```
set(gca,'FontName','calibri');
grid on;
xlabel('Frequency (Hz)');
ylabel('X_mag');
title('DFT magnitude');
%MAKE STEM PLOT OF SINGLE-SIDED DFT
figure;
hstem=stem(f_vec_real, abs(Yk));
hstem.Color='b';
hstem.Marker='o';
hstem.MarkerSize=4;
hstem.LineWidth=4;
set(gca,'FontSize',16);
set(gca,'FontWeight','bold');
set(gca,'FontName','calibri');
grid on;
xlabel('Frequency (Hz)');
ylabel('Y mag');
title('DFT (single-sided) magnitude');
%#############################
```

Example 8.11

Find the single-sided DFT of the signal in Example 8.10 by sampling the signal at 16 Hz. Plot the DFT and comment on the result.

Solution

The code in Example 8.10 is run with a sampling frequency of 16 Hz.

Results:

Figure 8.53 shows the single-sided DFT. The frequency component at 5 Hz is correctly predicted, but the 6 Hz component is incorrect as it is nonexistent in the given signal.

Since the sampling frequency doesn't satisfy the Nyquist theorem, the 10 Hz component gets aliased and appears at a lower frequency (at 6 Hz).

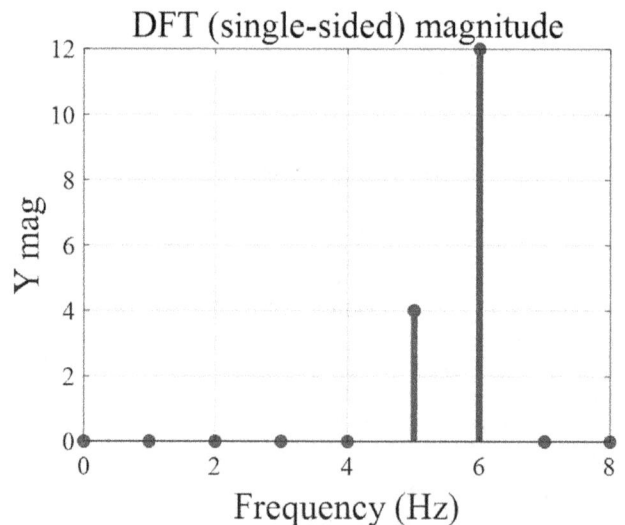

FIGURE 8.53   Single-sided DFT.

Example 8.12

Using MATLAB command 'fft()', find the DFT of the signal $x_c(t) = 10\sin 2\pi 7t + 6\cos 2\pi 13t$ using a 1.5-second length of the signal and a sampling frequency of 40 Hz. Also, find the DFT with the Hanning window and compare the DFT magnitude plots.

Solution

MATLAB program 8.2 is used to solve the problem. The program is given to find the DFT of the windowed signal (xw). The DFT of the signal without the window (i.e., signal x) is found similarly, though not shown in the code. The DFT plots (shown in Figures 8.54 and 8.55) indicate significant leakage without a window (in other words, with the rectangular window), which reduces with the Hanning window.

```
%##############################
%   MATLAB PROGRAM 8.2
%##############################
clear all;

%DEFINE RECORD LENGTH
T=1.5;

%CHOOSE SAMPLING FREQUENCY
fs=40;

%SET TIME VECTOR
Delta_t=1/fs;
t_vec=0:Delta_t:T-Delta_t;

%FIND NO. OF SAMPLES
N=length(t_vec);
```

Results:

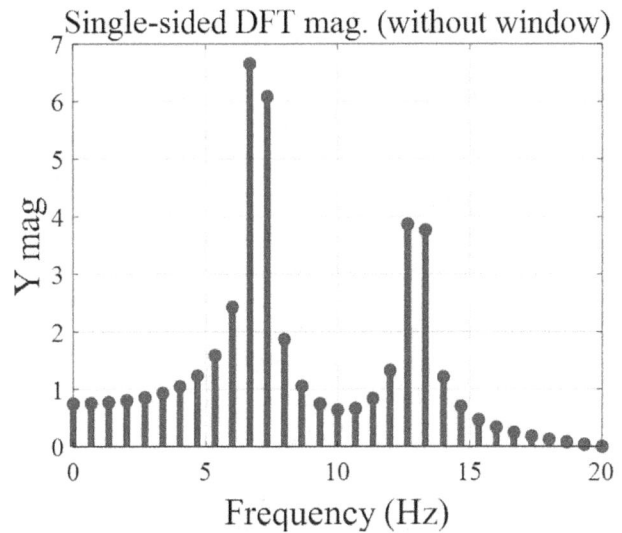

FIGURE 8.54 DFT with no window.

```
%SET FREQUENCY VECTORS
Delta_f=1/T;
fmax=fs/2;
f_vec_real=0:Delta_f:fmax;
%FIND DISCRETE SIGNAL
x=10*sin(2*pi*7*t_vec)+6*cos(2*pi*13*t_vec);

%FIND HANNING WINDOW FUNCTION
w=0.5+0.5*cos(2*pi*(t_vec-T/2)/T);

%APPLY WINDOW TO THE DISCRETE SIGNAL
xw=x.*w;

%FIND DFT WITH MATLAB COMMAND
[Xk]=fft(xw,N);
%SCALE THE DFT
Xk=Xk/N;

%FIND SINGLE-SIDED DFT
Yk(1:N/2)=2*abs(Xk(1:N/2));
Yk(N/2+1)=Xk(N/2+1);
```

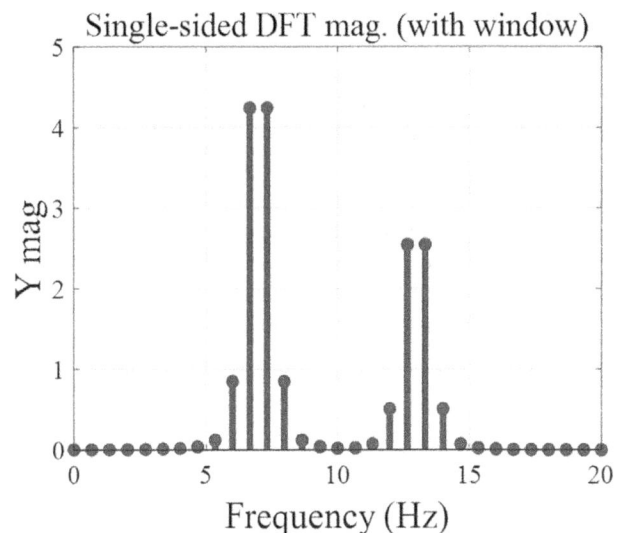

FIGURE 8.55 DFT with Hanning window.

(Continued)

```
%MAKE STEM PLOT OF SINGLE-SIDED DFT
figure;
hstem=stem(f_vec_real, abs(Yk));
hstem.Color='b';
hstem.Marker='o';
hstem.MarkerSize=4;
hstem.LineWidth=4;
set(gca,'FontSize',16);
set(gca,'FontWeight','bold');
set(gca,'FontName','calibri');
grid on;
xlabel('Frequency (Hz)');
ylabel('Y mag');
title('Single-sided DFT mag. (with window)');
%#############################
```

REVIEW QUESTIONS

1. What is the difference between harmonic and periodic signals?
2. What is the advantage of expanding a periodic function in terms of orthogonal functions, as done in the Fourier series expansion?
3. What is the Gibbs phenomenon in the Fourier series expansion of a signal with discontinuities?
4. Why is the Discrete Fourier series periodic in the frequency domain?
5. Differentiate between Discrete Fourier series, Discrete-Time Fourier transform, and Discrete- Fourier transform.
6. What is aliasing?
7. What is the sampling theorem?
8. What is an anti-aliasing filter?
9. Should an anti-aliasing filter be used before or after the discretization of the signal?
10. What is leakage? Is it possible to have a Fourier spectrum with no leakage?
11. What is windowing? Why is a window used?
12. What is quantization noise? Why is it related to the number of bits of the A/D converter?
13. How is quantization noise affected by an analyzer's maximum input voltage range?

PROBLEMS

Problem 8.1: For the signal given below, sketch the magnitude and phase of the frequency-domain representation of the signal, indicating the frequencies and the corresponding amplitudes and phases.

$$x_c(t) = 3 + 8\cos 2\pi 4t + 6\sin 2\pi 7t - 14\cos(2\pi 12t - 60^0)$$

Problem 8.2: Represent the signal in problem 8.1 by complex Fourier series. Sketch the magnitude spectrum of the complex series, indicating the frequencies and the corresponding amplitudes.

Problem 8.3: Find the time period and fundamental frequency of the following signals.

$$x_c(t) = -2 + 4\cos 5t + 6\sin 10t + 10\sin 15t$$

$$x_c(t) = 4\cos 6t + 7\sin 8t$$

Problem 8.4: A signal with a 0–500 Hz frequency range is to be measured over 4 seconds. Choose the sampling frequency and find the number of samples, time resolution of the discrete signal, and frequency resolution of DFT.

Problem 8.5: For the rectangular pulse signal shown in Figure 8.56, find the DFT of the signal using MATLAB code. Plot the single-sided DFT magnitude and phase.

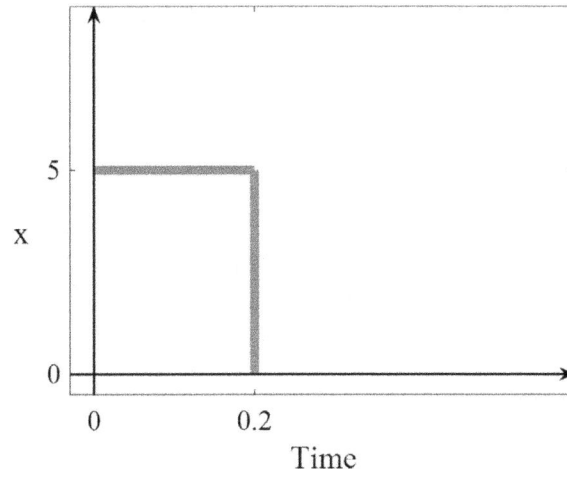

FIGURE 8.56 Rectangular pulse signal.

Problem 8.6: Using MATLAB code, find and plot the power spectrum of the following signal using its discrete version. Use a 1-second long signal and a sampling frequency of 20 Hz.

$$x_c(t) = 4\cos 6t + 7\sin 8t$$

9 FRF Measurement Using an Impact Hammer

9.1 INTRODUCTION

In Chapters 2–6, we covered the analytical approach to modal analysis, which also forms the theoretical basis for the experimental modal analysis (EMA) covered in Chapters 7–12. We have examined the characteristics of the frequency response functions (FRFs) and covered the signal processing for EMA in Chapters 7 and 8, respectively. The experimental approach to modal analysis, or modal testing, involves two major steps. The first step is the measurement of FRFs, while the second step is analyzing these FRFs to find the modal parameters. In this and the next chapter, we focus on the first step, while the second step is addressed in Chapter 11. The measurement of FRFs requires an excitation force applied to the system. Impact hammers and vibration exciters are two devices commonly used for this purpose. This chapter addresses FRF measurement using an impact hammer, while the FRF measurement using an exciter is addressed in Chapter 10.

This chapter starts by examining the basic idea behind EMA. The basic construction of an impact hammer and the role of the hammer tip are then described. Accelerometers are widely used for vibration measurement and are dealt with in detail starting from the basic principle of measuring acceleration using a seismic pickup, followed by the theory of piezoelectric accelerometers. We then analyze the role of a charge amplifier needed for charge-type accelerometers and discuss the IEPE type of accelerometers. It is followed by a discussion of the parameters influencing the selection of an accelerometer for modal analysis and the methods for mounting accelerometers. Then, the FRF estimates H1 and H2 are introduced. We look at the process of FRF measurement using an impact hammer by performing a simulation of the measurement on a cantilever structure. Test planning related to the choices regarding test points, accelerometer location, FFT/measurement parameters, and windows is discussed. In the end, we look at the calibration and boundary conditions for modal testing.

9.2 BASIC PRINCIPLE OF EMA

What is the basic principle of experimentally identifying the natural frequencies, mode shapes, and damping factors (or the eigenvalues and eigenvectors)? Can they be directly measured? We address these questions in this section.

The natural frequencies, mode shapes, and damping factors of a system/structure are also called its modal parameters or dynamic characteristics. The modal parameters depend on more fundamental properties of the structure, such as geometry/dimensions, material properties, and boundary conditions. The modal parameters can't be measured since they are abstractly related to the fundamental properties through an eigenvalue problem. However, they reveal themselves in the structure's response to dynamic forces. Thus, a knowledge of the dynamic forces and the corresponding response should, in principle, contain information about the modal parameters. Since we can apply a force to the structure and measure it, as well as we can measure the resulting response, it should be possible to estimate the modal parameters of the structure from these measurements.

We discussed in the previous chapters that the input force and the output response are related by the response model of the structure, which in the frequency domain is the FRF matrix. The element $\alpha_{jk}(\omega)$ of the FRF matrix is the displacement response at the jth DOF due to a unit amplitude harmonic force of frequency ω applied at the kth DOF. For an MDOF system with structural damping, the FRF $\alpha_{jk}(\omega)$ is related to the modal parameters by

$$\alpha_{jk}(\omega) = \sum_{r=1}^{N} \frac{\phi_{jr}\,\phi_{kr}}{\omega_r{}^2 - \omega^2 + i\eta_r\omega_r{}^2} \tag{9.1}$$

The FRF in the LHS of Eq. (9.1) can be obtained by measurement, and hence it is possible to estimate the modal parameters ω_r, η_r, ϕ_{jr} and ϕ_{kr} by fitting the expression on the RHS of the equation to the measured FRF. Curve fitting of the other FRFs can also be carried out to obtain the other modal parameters. This approach forms the basic principle of EMA for modal parameter estimation in the frequency domain.

We also studied in Chapter 6 that the impulse response function (IRF) matrix is the response model in the time domain. We discussed that, for a viscously damped system, the IRF $h_{jk}(t)$, which is the time-domain response at the jth DOF due to a unit impulse applied at the kth DOF, is given by

DOI: 10.1201/9780429454783-9

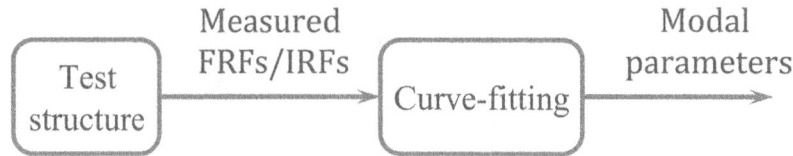

FIGURE 9.1 The basic principle of EMA.

$$h_{jk}(t) = \sum_{r=1}^{2N} \phi_{jr}\phi_{kr}e^{\bar{s}_r t} \tag{9.2}$$

Therefore, if the IRF $h_{jk}(t)$ can be obtained experimentally, then it should be possible to estimate the unknown modal parameters ϕ_{jr}, ϕ_{kr} and $\bar{s}_r$ appearing on the RHS of the equation. This approach forms the basic principle of EMA for modal parameter estimation in the time domain.

The analytical modal analysis estimates the modal parameters from a spatial model of the structure. In comparison, the experimental modal analysis (EMA) takes an inverse route and starts from the knowledge of the response model. Figure 9.1 depicts the basic idea of EMA.

9.3 SETUP FOR EMA USING IMPACT TESTING

Figure 9.2 shows a typical setup for EMA using impact testing. An impact hammer is used to apply an impact on the test structure and measure the applied force. The force applied by the hammer is measured by the force transducer attached at the front end of the hammer. The impact applied is nothing but a force pulse of short duration, referred to as an impulse. Due to the impulse excitation, the structure goes through a transient response, generally measured using an accelerometer. If the force transducer and accelerometer are of charge type, their outputs are charge signals. The charge signals are fed to charge amplifiers (shown as signal conditioners in the figure) to convert them to proportionate voltage signals. The voltage signals are measured using an FFT analyzer (or some data acquisition system). The measured voltage signals are converted into the force and acceleration signals using the force transducer and accelerometer sensitivities, respectively. These sensitivities are available from the calibration charts provided by the manufacturer. The FRF can be estimated from the measured signals using built-in functions for FRF computation generally available in the FFT analyzers. If a data acquisition system is used, the FRF must be obtained by processing the measured force and response signals.

9.4 FRF ESTIMATION USING TRANSIENT INPUT AND OUTPUT

Let the impact is applied at the kth DOF (or k^{th} test point) of the test structure and let the corresponding measured voltage signal is $V_k(t)$ volts. If S_f volt/newton is the sensitivity of the force transducer of the hammer, then the force signal is

$$F_k(t) = \frac{V_k(t)}{S_f} N \tag{9.3}$$

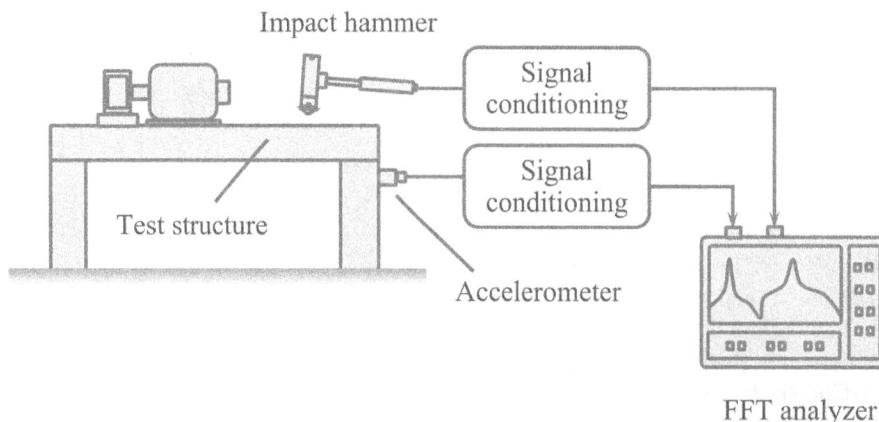

FIGURE 9.2 Setup for EMA using impact testing.

A force signal due to impact is shown in Figure 9.3a, indicating that the impact lasts only for a short time.

Similarly, let the resulting acceleration at the jth degree of freedom (or the jth test point) of the test structure is measured using an accelerometer and let the corresponding measured voltage signal is $V_j(t)$ volt. If S_a volt/m/sec^2 is the sensitivity of the accelerometer, then the acceleration signal is

$$a_j(t) = \frac{V_j(t)}{S_a} \text{ m/sec}^2 \tag{9.4}$$

An acceleration signal due to an impact is shown in Figure 9.3b, indicating the decay of vibrations.

In the previous chapters, the FRF was defined. If $F_k(\omega)$ is the amplitude of the harmonic force of frequency ω applied at the kth DOF, with no force applied at any other DOFs, and $a_j(\omega)$ is the resulting steady-state harmonic acceleration at the same frequency at the jth DOF, then the FRF (inertance $A_{jk}(\omega)$) is given by

$$A_{jk}(\omega) = \frac{a_j(\omega)}{F_k(\omega)} \tag{9.5}$$

We note that the measured force and acceleration signals (Eqs. 9.3 and 9.4) are in the time domain, while the FRF given by Eq. (9.5) requires the force and acceleration in the frequency domain. The Fourier transforms of the time domain signals give their frequency-domain representations

$$a_j(\omega) = \int_{-\infty}^{\infty} a_j(t) \, e^{-i\omega t} .dt \tag{9.6}$$

$$F_k(\omega) = \int_{-\infty}^{\infty} F_k(t) \, e^{-i\omega t} .dt \tag{9.7}$$

In practice, the frequency domain representations are computed via DFT as the measured signals are discrete. The receptance or mobility, if required, can be obtained from the inertance since they are related.

A question can be raised here: How can the FRF defined using the response to a harmonic force be obtained by applying an impulsive, non-harmonic force? We saw in Chapter 8 that a non-harmonic signal could be described by the Fourier transform as an integral sum of the harmonic signals. Hence the measurement of the FRF is possible by applying a non-harmonic force like an impulse.

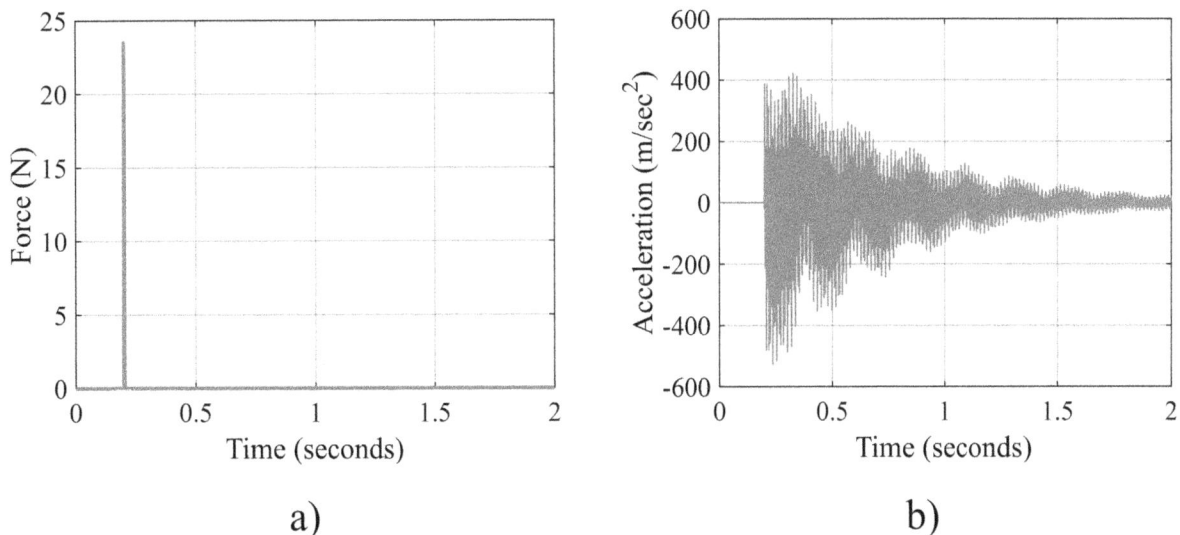

FIGURE 9.3 (a) A force signal due to impact and (b) the resulting acceleration signal.

9.5   IMPACT HAMMER

An impact hammer is an instrumented hammer that can be used to apply an impact and measure the force due to impact. Figure 9.4 shows a schematic of an impact hammer. It consists of a force transducer mounted at the front end of the hammer, with a provision to attach a tip for impact application. The material and dimensions of the tip determine its stiffness and influence the profile of the force applied. The structure's stiffness at the point of application of the impact also affects the force profile. Commercially available impact hammers come with tips of different materials (like rubber, plastic, and metal tips).

The advantage of an impact hammer is that it is easy to use and does not cause any mass loading of the structure. It is less complex than a shaker, discussed in the next chapter, and is economical. However, the amount of force that can be applied is limited, and the excitation frequency range cannot be precisely controlled.

An impact hammer uses a piezoelectric force transducer, which can be either charge or IEPE/ICP type. The charge type force transducer generates a charge proportional to the force. It requires an external charge amplifier since the charge signal is a high-impedance signal (discussed in Sections 9.7.2 and 9.8). The charge amplifier converts the signal to a proportional low-impedance voltage signal. The IEPE/ICP force transducer has a built-in charge amplifier, eliminating the need for an external charge amplifier. The sensitivities of charge type and IEPE type impact hammers are expressed in pC/N and mV/N, respectively.

Impact hammers of different capacities are commercially available to excite structures of various sizes. A miniature hammer may have a mass of 5 grams and a length of 10 cm and can generate a peak force of 200 N. The larger ones meant to excite massive structures like buildings and bridges may have a mass of 5 kg and a length of 80 cm and can generate a peak force of 20 kN.

9.6   ROLE OF THE HAMMER TIP

The material of the tip affects the tip's stiffness, affecting the duration and the amplitude of the force generated when the impact is made. Figure 9.5a shows a force profile with a rubber tip. The force profile also depends on the mass and velocity of the hammer and the mass and stiffness of the structure. Figure 9.5b shows the magnitude of the Fourier transform

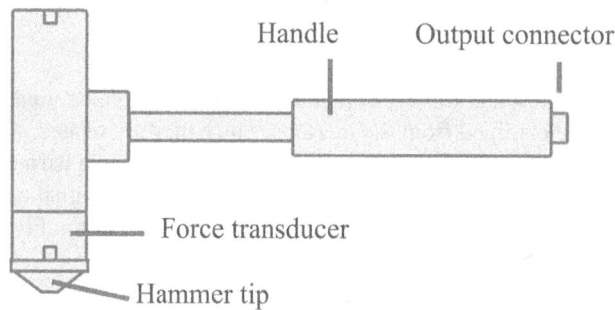

FIGURE 9.4   An impact hammer.

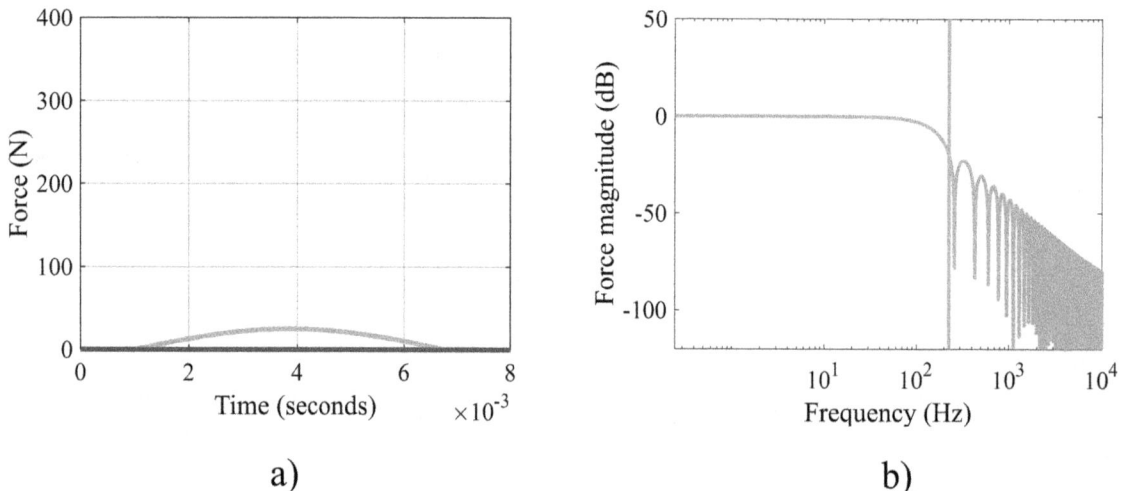

a)   b)

FIGURE 9.5   (a) Force-time history of an impact with a rubber tip and (b) the corresponding Fourier transform magnitude.

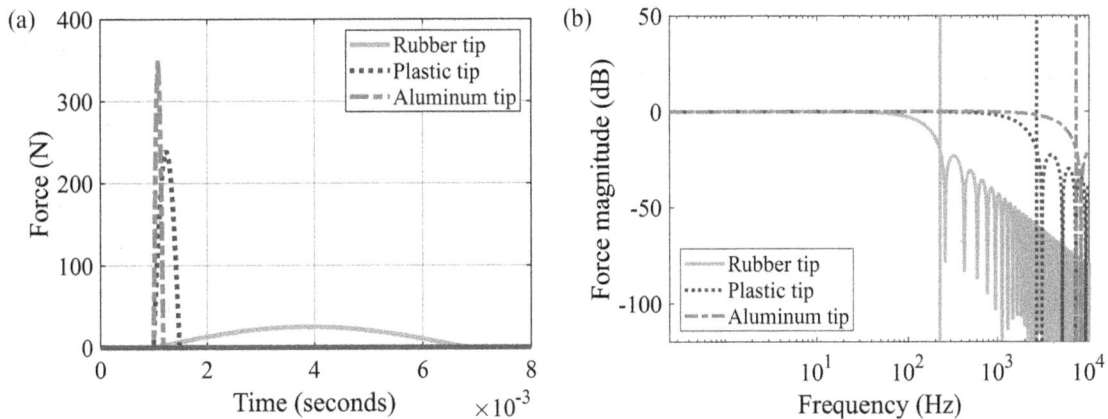

FIGURE 9.6 (a) Force-time histories of the impacts made by rubber, plastic, and aluminum tips and (b) the corresponding Fourier spectrums.

of the force on the log-frequency axis. The magnitude at the zero frequency is used as the reference value to calculate the magnitude in dB.

We observe that the side lobes of decreasing amplitudes follow the main lobe, and between every two successive lobes, the magnitude of the force is zero at some frequency. Thus, the force magnitude drops substantially beyond a certain frequency, leading to a poor signal-to-noise ratio, as the signal often also have some measurement noise.

As a thumb rule, the excitation frequency range can be taken from 0 Hz to the frequency where the force magnitude drops by 20 dB. It represents the range in which a significant portion of the input energy can be assumed to be concentrated. The upper limit of this range, shown in the figure by a vertical line, is at around 225 Hz for the impact shown.

To understand the influence of the tip on the impact, we compare the force-time histories of the impacts made with three different tips, rubber, plastic, and aluminum tips, as shown in Figure 9.6a. The magnitude of the Fourier transforms of the time histories are shown in Figure 9.6b. These graphs only indicate the impacts made with tips with a medium size hammer. The characteristics depend upon the tip material properties, dimensions, and hammer. Key observations from the two plots are:

- The impact duration is longest with the rubber tip and narrowest with the aluminum tip.
- The peak force applied by the aluminum tip is the highest (for the same hammer velocity).
- The upper limit of the excitation frequency range (shown by the vertical line at the frequency with a 20 dB drop in magnitude) is highest for the aluminum tip and lowest for the rubber tip. For the impacts shown, these frequencies for the rubber, plastic, and aluminum tips are around 225, 2,600, and 7,200 Hz, respectively.
- The narrower the impulse due to impact, the longer the excitation frequency range.

Generally, the hammer tip, which is just adequate to excite the desired frequency range, should be used. For example, using a metal tip where the desired frequency range can be excited using a rubber tip is not optimal, as a part of the excitation energy is put into a segment of the frequency range that is not under measurement. It lowers the signal level in the analysis frequency range, adversely affecting the signal-to-noise ratio.

9.7 RESPONSE MEASUREMENT

The dynamic response of a system can be described in terms of displacement, velocity, or acceleration. The question is which parameter should be measured and analyzed. The parameter with the flattest frequency spectrum is an ideal choice, as it can better describe the vibration level of a system over a wide frequency range and requires the smallest dynamic range of the measurement.

It is seen that the displacement of a system is generally significant at lower frequencies and reduces as the frequency increases. Thus, the choice of displacement as a measurement parameter requires a wider dynamic range of measurement and risks having a poor signal-to-noise ratio at higher frequencies. However, if only lower frequencies are involved, the displacement can be a preferred measurement parameter. Similarly, the acceleration of a system is generally significant at higher frequencies but smaller at lower frequencies. Thus, the choice of acceleration as a measurement parameter also requires a wider dynamic range of measurement and risks having a poorer signal-to-noise ratio at lower frequencies. But if only higher frequencies are involved, acceleration can be a preferred measurement parameter. The velocity is equally sensitive to both the frequency and displacement of the motion. In general, the range of variation of velocity over the whole frequency range is the least, requiring the smallest dynamic range of measurement and providing a better signal-to-noise ratio over a wide frequency range. Therefore, from this perspective, velocity is generally a preferred measurement parameter over a wide frequency range.

However, another consideration in the measurement parameter selection is the characteristics of the available transducer. This consideration often overrides other considerations. As a result, the acceleration is usually measured since a transducer with favorable characteristics, like a piezoelectric accelerometer, can be used for acceleration measurement. A piezoelectric accelerometer offers many advantages over displacement or velocity pickups, as listed below.

- Its weight is small. This is an essential requirement, more so in modal testing, as it ensures that the mass loading of the structure is small.
- It has a wide frequency range. We see in the next section that the higher an accelerometer's natural frequency, the higher its useful frequency range. The piezoelectric crystal enables accelerometer designs with high natural frequency and hence provides a wide frequency range of measurement.
- It has a linear response over a wide frequency range.
- It has a wide dynamic range. This makes it suitable for measuring both small and large levels of acceleration.
- The integration of the accelerometer's output is required to find the velocity and displacement. In contrast, differentiation is required to find the acceleration if velocity or displacement is measured. Integration is preferable since the differentiation is more sensitive to noise in the measured signal, adversely affecting the estimates obtained after differentiation.

Because of the above favorable characteristics, piezoelectric accelerometers are widely used for vibration measurement and modal testing.

9.7.1 SEISMIC PICKUP

The principle of acceleration measurement using a piezoelectric accelerometer is based on a seismic pickup. The seismic pickup can be designed according to the parameter to be measured, i.e., acceleration, displacement, or velocity, and the measurement frequency range.

Figure 9.7a shows the schematic of a seismic pickup. It consists of an SDOF spring-mass-dashpot system fixed to the base of the enclosed casing. The mass m is referred to as seismic mass. The seismic pickup is mounted rigidly on the test structure whose vibration is to be measured, as shown in Figure 9.7b. Figure 9.7c shows a dynamic model of the seismic pickup with the test structure. The test structure is represented as an SDOF system, with m_s, k_s, and c_s as its mass, stiffness, and damping properties. A harmonic force acts on the structure, causing vibrations y(t) and x(t) of the structure and seismic mass, respectively. Strictly speaking, the seismic pickup and test structure are coupled, requiring the solution of

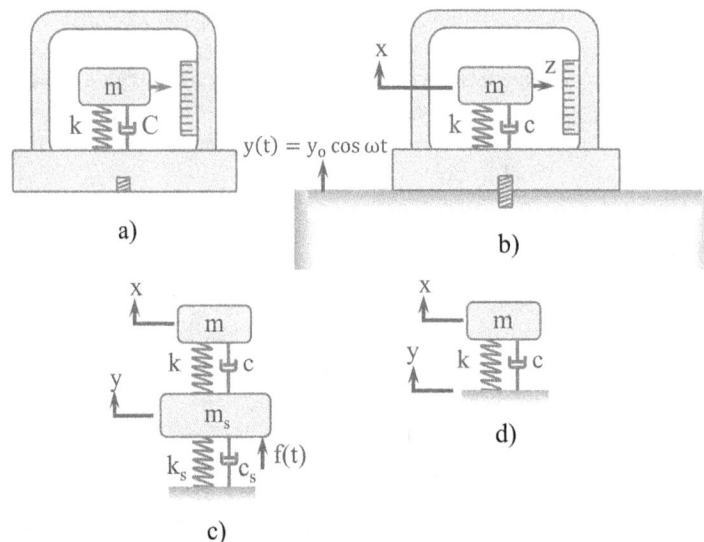

FIGURE 9.7 (a) Seismic pickup, (b) the pickup mounted on the test structure, (c) dynamic model of the pickup with the test structure, (d) simplified dynamic model.

the coupled model shown in Figure 9.7c. However, if the test structure is rigid and massive relative to the seismic pickup, the vibration y(t) of the test structure is not much affected when the seismic pickup is mounted onto it. Therefore, the seismic pickup is effectively subjected to a base excitation with a constant harmonic displacement amplitude, $y(t) = y_0 \cos \omega t$, as shown in Figure 9.7d.

The equation of motion can be written as

$$m\ddot{x} + c(\dot{x} - \dot{y}) + k(x - y) = 0 \tag{9.8}$$

If we define displacement of the seismic mass relative to its casing as z, then

$$z = x - y \tag{9.9}$$

Substituting Eq. (9.9) into Eq. (9.8), along with $y(t)$, we get

$$m\ddot{z} + c\dot{z} + kz = m\omega^2 y_0 \sin \omega t \tag{9.10}$$

The steady-state solution for z is given by $z(t) = z_0 \sin(\omega t - \phi)$, which when substituted into Eq. (9.10), gives

$$z_0 = \frac{\left(\dfrac{\omega}{\omega_n}\right)^2 y_0}{\sqrt{\left(1 - \left(\dfrac{\omega}{\omega_n}\right)^2\right)^2 + \left(2\xi\dfrac{\omega}{\omega_n}\right)^2}} \tag{9.11}$$

and

$$\phi = \tan^{-1} \frac{2\xi\dfrac{\omega}{\omega_n}}{1 - \left(\dfrac{\omega}{\omega_n}\right)^2} \tag{9.12}$$

Note that ω_n and ξ in Eqs. (9.11) and (9.12) are the undamped natural frequency and viscous damping factor of the seismic pickup and $\omega_n = \sqrt{k/m}$. The natural frequency of the pickup, assuming the test structure rigid, is called the 'mounted resonance frequency'.

Case a)
If the frequency of vibration of the structure is much larger than the natural frequency of the pickup (i.e. $\omega \gg \omega_n$), then Eq. (9.11) can be approximated as

$$z_0 \approx y_0 \tag{9.13}$$

Eq. (9.13) shows that the relative displacement amplitude of the seismic mass is approximately equal to the displacement amplitude of the structure. This approximation is valid for any value of damping in the pickup. Since the measurement of relative displacement z gives an estimate of the displacement y of the structure, the pickup in the range of vibration frequencies $\omega \gg \omega_n$ works as a displacement pickup. To satisfy the condition $\omega \gg \omega_n$, the seismic mass should be heavy, and/or the stiffness should be low. These requirements make a displacement pickup have a higher mass and size. Typical values of natural frequencies of a displacement pickup are in the range 1–6 Hz. This enables measuring displacement at low frequencies.

If the relative displacement is sensed such that the output is proportional to the rate of change of the relative displacement z, then in the range of frequencies $\omega \gg \omega_n$, the output of the pickup is proportional to the velocity of the structure. In this case, the pickup works as a velocity pickup.

Case b)
If the natural frequency of the pickup is much higher than the frequency of vibration of the structure (i.e., $\omega_n \gg \omega$), then the denominator of Eq. (9.11) approaches unity, and that equation can be approximated as

$$z_0 \approx \left(\frac{\omega}{\omega_n}\right)^2 y_0 \qquad (9.14)$$

$$z(t) = \frac{1}{\omega_n^2}\omega^2 y_0 \sin(\omega t - \phi) \qquad (9.15)$$

The quantity $\omega^2 y_0$ is the amplitude of the acceleration of the structure. Therefore, the seismic mass relative displacement is approximately equal to the acceleration of the structure scaled by the constant $1/\omega_n^2$. $z(t)$ also has a phase lag of ϕ radians, which depends on the damping factor ξ of the pickup. The seismic pickup can be used to measure acceleration, i.e., as an accelerometer, over a frequency range with an upper limit of roughly $0.2\,\omega_n$. Thus, for a wide enough frequency range for the measurement of acceleration, the natural frequency of the seismic pickup must be high. It is achieved by having a lower seismic mass (and higher stiffness), making a seismic accelerometer a lighter transducer than the displacement and velocity pickups.

9.7.2 Piezoelectric Accelerometer

We saw in the previous section that the displacement of the seismic mass relative to the casing or the base needs to be sensed to measure the acceleration. A piezoelectric accelerometer uses a piezoelectric crystal to generate an electrical signal proportional to the relative displacement of the seismic mass. Figure 9.8 shows the schematic of a piezoelectric accelerometer.

The accelerometer is shown mounted on the structure's surface, whose acceleration is to be measured. In place of the spring and dashpot in a seismic pickup, there is a disc made up of piezoelectric material attached between the seismic mass and the casing. A preloading spring is used to create an initial compression in the piezo disc so that the disc has compressive deformations for both the positive and negative values of the relative displacement z.

The property of the piezoelectric material that, when deformed, produces an electric charge is utilized in a piezoelectric accelerometer. The piezoelectric material also has a reverse effect that, when subjected to an electric field, it produces material deformation. Quartz, barium titanate, and lead zirconate titanate are some examples of piezoelectric materials.

Let us find out the output of the piezoelectric accelerometer due to the vibration y(t) of the structure on which it is mounted (Figure 9.8). The piezo disc undergoes deformation due to the relative displacement z. If the force causing deformation is F, then the charge Q produced by the piezo disc is given by

$$Q = d_{33}F \qquad (9.16)$$

d_{33} is the piezoelectric constant, which defines the charge produced on the two faces perpendicular to the orthogonal direction 3 of the crystal when a unit force is applied along the orthogonal direction 3. Its unit is picocoulomb/newton (pC/N). Direction 3 of the piezo disc is aligned in the present case with the direction of the coordinate y. If L and E are the thickness and modulus of elasticity, and A is the cross-sectional area of the piezo disc normal to direction 3, then the force F on the piezo is

$$F = z.\frac{AE}{L} \qquad (9.17)$$

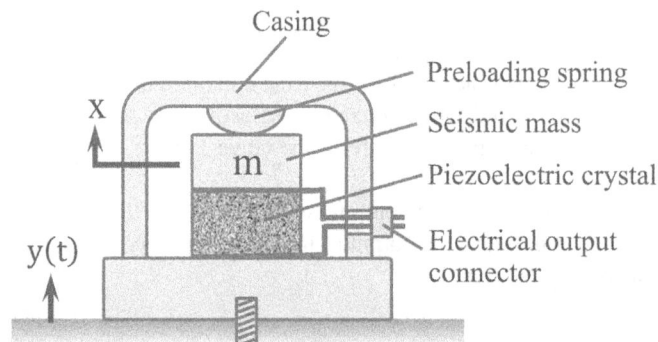

FIGURE 9.8 Piezoelectric accelerometer.

Substituting Eq. (9.17) into Eq. (9.16), we get

$$Q = d_{33} \frac{AE}{L} z \qquad (9.18)$$

From the previous section, Eq. (9.15) gives the relative displacement z. Since the damping factor of a piezoelectric accelerometer is small, the phase lag ϕ is negligible. Substituting z from Eq. (9.15) into Eq. (9.18) with zero phase lag, we get the following equation:

$$Q(t) = d_{33} \frac{AE}{L} \frac{1}{\omega_n^2} \omega^2 y_o \sin \omega t \qquad (9.19)$$

The natural frequency of the piezo-seismic mass system is approximated as

$$\omega_n = \sqrt{\frac{k_{piezo}}{m}} = \sqrt{\frac{AE/L}{m}} \qquad (9.20)$$

Substituting ω_n from Eq. (9.20) into Eq. (9.19) and noting that the acceleration of the vibrating structure is $\ddot{y}(t) = -\omega^2 y_o \sin \omega t$, we get the charge generated by the piezoelectric accelerometer

$$Q(t) = -d_{33} m \ddot{y}(t) \qquad (9.21)$$

The result in Eq. (9.21) is derived by assuming the vibration y(t) to be harmonic. If the vibration is periodic or transient, then from the Fourier series, we see that it is equivalent to a sum of harmonic components. Since the phase lag of charge generated due to each harmonic component is negligible, the total charge due to all the harmonic components remains proportional to the instantaneous acceleration of the structure. Hence, Eq. (9.21) is also valid for non-harmonic y(t).

Thus, we see that the charge signal from the piezoelectric accelerometer is proportional to the acceleration to be measured. The charge sensitivity of the accelerometer is pC/m/sec² and is proportional to the constant d_{33} of the piezoelectric material and seismic mass m. The higher the seismic mass (m) higher is the sensitivity. But the higher the mass m, the lower the natural frequency ω_n. It results in a lower frequency range since the upper limit of the range is around $0.2\omega_n$. Thus, there is a trade-off between the sensitivity and frequency range of a piezoelectric accelerometer. What is the lower limit of the frequency range? We look at it in the next section.

How is the electrical output of the piezoelectric accelerometer measured? Figure 9.9 shows an electrical model of a piezoelectric accelerometer.

It is modeled as a source of charge $Q(t)$ with the capacitance and resistance of the piezo disc, represented by C_p and R_p, respectively, placed in parallel. The generated charge creates a voltage across the opposite faces of the disc. The generated voltage is

$$V(t) = \frac{Q(t)}{C_p} \qquad (9.22)$$

If ε is the dielectric constant of the piezo material, then the capacitance of the piezo disc is given by

$$C_p = \frac{\varepsilon A}{L} \qquad (9.23)$$

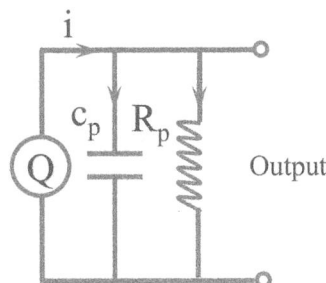

FIGURE 9.9   An electrical model of a piezoelectric accelero-meter.

Substituting Eqs. (9.23) and (9.21) into Eq. (9.22), we get

$$V(t) = -\frac{d_{33}mL}{\epsilon A}\ddot{y}(t) \qquad (9.24)$$

Thus, the voltage corresponding to the generated charge is also proportional to the acceleration to be measured. The coefficient of $\ddot{y}(t)$ represents the voltage sensitivity of the accelerometer. Can we measure this voltage by connecting the accelerometer to a voltage-measuring instrument? The output of a piezoelectric accelerometer is a high-impedance signal since the piezo disc, which is the source of the signal, has a high impedance. Relatively, a voltage-measuring instrument typically has a lower input impedance. Therefore, if the piezoelectric accelerometer output is fed to a voltage-measuring instrument, the instrument would significantly load the signal from the piezoelectric accelerometer and give an inaccurate measurement. Hence we cannot correctly measure the charge accelerometer output by directly connecting it to a voltage-measuring instrument. For correct measurement, the output of the accelerometer is first converted using a charge amplifier to a low-impedance signal before it is measured, as discussed in the next section.

9.8   CHARGE AMPLIFIER

A charge amplifier is a signal conditioning amplifier for a charge signal. It produces a low-impedance output voltage proportional to the charge. Figure 9.10 shows an electrical model of a piezoelectric accelerometer connected to a charge amplifier via a connecting cable.

The electric model of the piezoelectric accelerometer is explained in Section 9.7.2. C_b and R_b represent the capacitance and resistance of the cable, respectively. The main element in a charge amplifier is the operational amplifier (op-amp) with negative feedback with capacitance C_f and resistance R_f in the feedback path. We now determine the output voltage e_o. By Kirchhoff's current law

$$i = i_{Cp} + i_{Rp} + i_{Cb} + i_{Rb} + i_f + i_- \qquad (9.25)$$

Note that the non-inverting input (+) of the op-amp is grounded. Since the open-loop gain of the op-amp is very high, the negative feedback tends to force the potential at the inverting input (−) to follow the potential at the non-inverting input. It effectively maintains the potential at point A, almost equal to that at B. As a result, the currents i_{Cp}, i_{Rp}, i_{Cb} and i_{Rb} are negligibly small. Also, the op-amp has a very high input impedance; as a result, the current drawn by it is very small. Thus, i_- is also negligible. Therefore, Eq. (9.25) can be approximated as

$$i = i_f \qquad (9.26)$$

$$i = i_{Cf} + i_{Rf} \qquad (9.27)$$

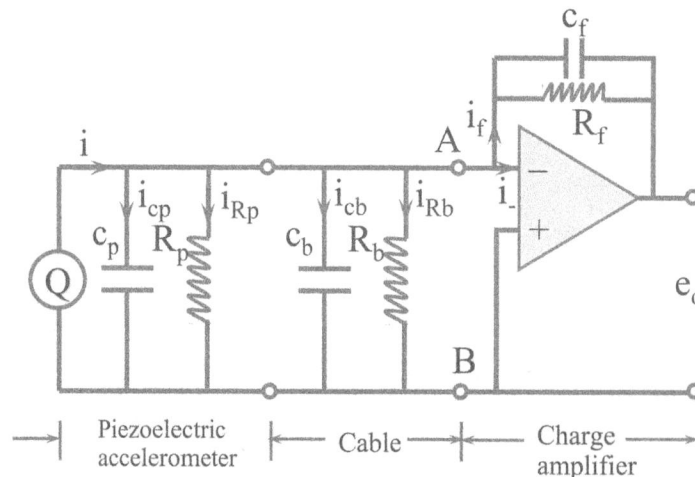

FIGURE 9.10   An electrical model of a piezoelectric accelerometer connected to a charge amplifier.

$$\frac{dQ}{dt} = -\frac{e_o}{R_f} - C_f \frac{de_o}{dt} \tag{9.28}$$

Taking the Laplace transform of Eq. (9.28) and assuming zero initial conditions on the charge and output voltage, we get the following equation after simplification:

$$\frac{e_o(s)}{Q(s)} = -\frac{s\,R_f}{1 + R_f C_f s} \tag{9.29}$$

The resistance R_f is chosen very high to make the time constant $R_f C_f$ large, and hence Eq. (9.29) can be approximated as

$$\frac{e_o(s)}{Q(s)} = -\frac{1}{C_f} \tag{9.30}$$

The Laplace inverse gives

$$e_o(t) = -\frac{Q(t)}{C_f} \tag{9.31}$$

Thus, the amplifier's output voltage is proportional to the charge generated. The output impedance of an op-amp is very small, and hence the voltage signal is a low-impedance signal and can be measured using a voltage-measuring instrument like an oscilloscope or FFT, which have a much higher input impedance.

It is noted from Eq. (9.31) that the output voltage is dependent only on the capacitance in the feedback path. That is because the high input impedance of the op-amp causes nearly the whole of the charge to accumulate on the capacitor C_f. It offers the advantage that the capacitance of the cable does not affect the output. Therefore, longer cables can be used without affecting the voltage sensitivity of the accelerometer-amplifier system.

We now substitute Eq. (9.21) into Eq. (9.31), which gives

$$e_o(t) = \frac{d_{33}m}{C_f}\ddot{y}(t) \tag{9.32}$$

Thus, the charge amplifier voltage output is proportional to the acceleration of the structure on which the accelerometer is mounted. The voltage sensitivity of the accelerometer-amplifier system has units V/m/sec² and is dependent on the piezo-electric constant, the seismic mass, and the feedback capacitance of the charge amplifier.

9.8.1 ACCELEROMETER RESPONSE TO STATIC INPUTS

Can a piezoelectric accelerometer be used to measure static acceleration or zero-frequency acceleration? To check this, let a unit step acceleration input is applied at t=0, as shown in Figure 9.11a, to a piezoelectric accelerometer connected to a charge amplifier. In the Laplace domain, it can be expressed as

$$\ddot{y}(s) = 1/s \tag{9.33}$$

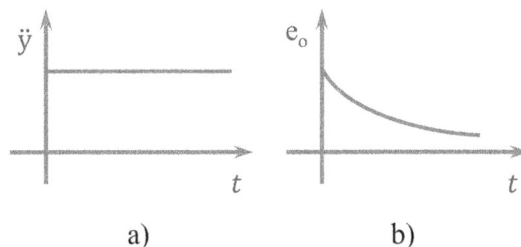

a)   b)

FIGURE 9.11   (a) A unit step acceleration input, (b) the resulting voltage output of the charge amplifier.

Taking the Laplace transform of Eq. (9.21), substituting Eq. (9.33) into it, and then substituting $Q(s)$ so obtained into Eq. (9.29), we get

$$e_o(s) = -\frac{s\,R_f}{1+R_f C_f s} \times -\frac{d_{33}m}{s} \tag{9.34}$$

$$e_o(s) = \frac{d_{33}m}{C_f}\left(\frac{1}{s+\dfrac{1}{R_f C_f}}\right) \tag{9.35}$$

Taking the Laplace inverse leads to

$$e_o(t) = \frac{d_{33}m}{C_f}e^{-t/R_f C_f} \tag{9.36}$$

Eq. (9.36) is the voltage output of the charge amplifier to a unit static acceleration input, which is plotted in Figure 9.11b. The voltage response decays with time, eventually becoming zero. That happens because the static charge accumulated on the feedback capacitor C_f gradually leaks through the feedback resistance R_f. The time constant of this process is equal to $R_f C_f$. Theoretically, if R_f is infinity, the output voltage does not decay. In practice, R_f is finite, causing decay of voltage. Even for low frequencies, the decay rate is fast enough to prevent accurate measurement. Hence, static or quasi-static accelerations cannot be measured using piezoelectric accelerometers. Note that the time constant, which affects the lower frequency limit, can be controlled by changing R_f or C_f.

We can also verify the static response by finding the frequency response of the accelerometer-amplifier system. Take the Laplace transform of Eq. (9.21), and substitute it into Eq. (9.29) to obtain the transfer function between the voltage output and the acceleration input. Substituting $s = i\omega$ into the transfer function, we get the frequency response as

$$\frac{e_o(i\omega)}{\ddot{y}(i\omega)} = \frac{i\omega\,R_f\,d_{33}m}{1+i\omega R_f C_f} \tag{9.37}$$

The magnitude of the frequency response is

$$\left|\frac{e_o(i\omega)}{\ddot{y}(i\omega)}\right| = \frac{\omega\,d_{33}m}{\sqrt{\omega^2 C_f^2 + \dfrac{1}{R_f^2}}} \tag{9.38}$$

It is seen that for static acceleration input, or $\omega = 0$, the frequency response magnitude is zero. The result matches the steady-state value of $e_o(t)$ from Eq. (9.36) (obtained when $t \to \infty$). For a dynamic input, i.e., input with a nonzero ω, and with a high R_f, the magnitude of the frequency response can be approximated as $\dfrac{d_{33}m}{C_f}$. (This result matches the voltage amplitude from Eq. (9.32) for a harmonic acceleration input of unit amplitude.)

9.9   IEPE PIEZOELECTRIC ACCELEROMETERS

The output of a charge-type piezoelectric accelerometer is a high-impedance signal and needs to be conditioned using a charge amplifier before it can be measured, as discussed in the previous sections. The piezoelectric accelerometers based on IEPE (integrated electronics piezoelectric) standard have a built-in electronics/charge amplifier in the form of a small integrated circuit chip. The output from such an accelerometer, called IEPE accelerometer, is in the form of a low-impedance voltage signal and can be measured using a voltage-measuring device. These types of accelerometers are available under different trademarks like ICP, DeltaTron, Piezotron, and ISOTRON.

An IEPE accelerometer does not require an external charge amplifier for signal conditioning, but the built-in amplifier requires a power supply. Most FFT analyzers have constant current drive inputs, meaning they can supply the necessary power (2 to 4 mA current) to the transducers with built-in preamplifiers when connected to them. If the FFT analyzer doesn't have constant current drive inputs, then an external signal conditioner capable of delivering the necessary power must be used between the accelerometer and the analyzer. Though the accelerometers with built-in amplifiers are more convenient to use, the gain of amplification cannot be changed, which is possible with an external charge amplifier. The built-in amplifier also reduces the permissible maximum temperature up to which the accelerometer can be used without causing damage to the built-in preamplifier. So measurement under elevated temperatures may require a charge-type accelerometer.

9.10 MOUNTING OF ACCELEROMETERS

An accelerometer can be mounted on a structure in various ways, like stud mounting, magnetic mounting, adhesive mounting and cementing stud mounting or the acceleration may be sensed using a probe. The mounting type determines the stiffness of the connection between the accelerometer and the test structure. It affects the resonance frequency of the accelerometer with the mounting and the usable frequency range. The mounting with the highest contact stiffness provides the highest effective frequency range. Figure 9.12 shows five methods of mounting an accelerometer.

a. Stud mounting

It requires a drilled and tapped hole on the surface of the test structure, and the accelerometer is attached through a stud. The stud mounting gives the highest usable frequency range compared to other mounting methods and allows for measurement at the exact desired location on the test structure. As it needs a tapped hole on the surface, this method cannot be used where any damage to the surface is unacceptable.

b. Magnetic mounting

In this method, the accelerometer is secured to the surface via a magnetic disc. The magnet mounting is fast and doesn't damage the surface, but the surface should be ferromagnetic. The magnet mass adds to the mass loading of the structure. The frequency range of the accelerometer with this mounting is lesser than that with stud mounting.

c. Adhesive mounting

In this method, the accelerometer is secured to the surface using a thin layer of adhesive. The surface must be smooth and clean before the adhesive is applied to ensure a strong bond. This method can be used for permanent mounting of the accelerometer if the stud mounting is not possible. The frequency range of the accelerometer with this mounting is lesser than that with the magnetic mount. Another option is to glue a cementing stud to the surface with an adhesive and then secure the accelerometer to the stud. This mounting allows easy removal of the accelerometer as compared to adhesive mounting.

d. Using a hand probe

A hand probe can be used as a quick way of initially assessing the vibration. The probe is a long thin rod with a provision to attach the accelerometer at its one end. The probe is hand-held perpendicular to the surface, and its tip is positioned at the desired location on the surface. This method allows fast measurement and is especially

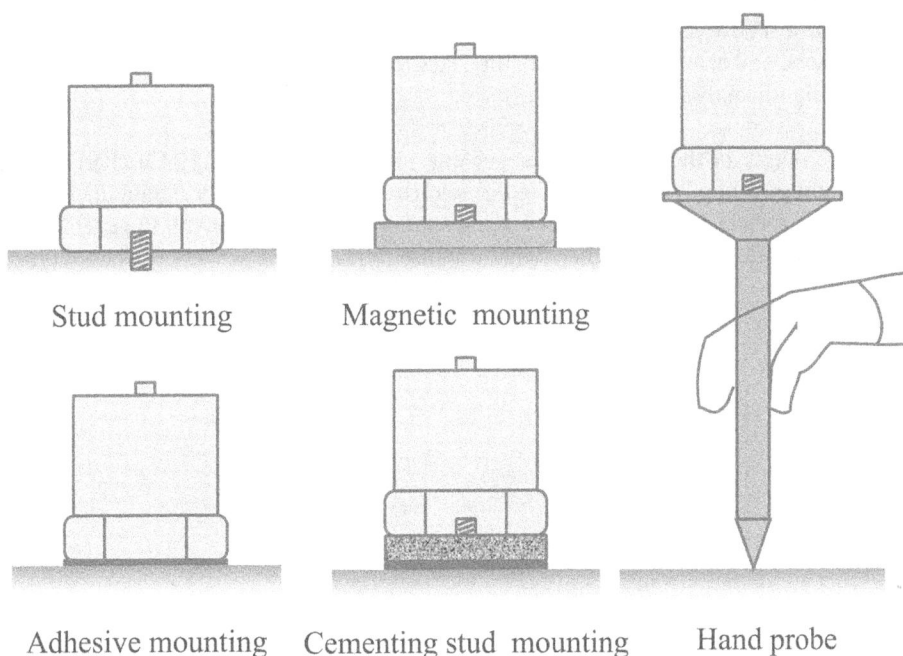

FIGURE 9.12 Methods of mounting an accelerometer.

useful when the measurement location does not permit or has access to using other mounting methods. The useful frequency range is the smallest compared to other mounting methods.

9.11 ACCELEROMETER SELECTION

The following characteristics need to be considered in selecting a piezoelectric accelerometer for modal testing:

a. Uniaxial or triaxial

A uniaxial accelerometer allows acceleration measurement in one direction. However, a triaxial accelerometer allows simultaneous acceleration measurement in three perpendicular directions. A triaxial measurement may be of interest in cases where the vibration is predominant in more than one direction. The triaxial accelerometer is a better option than three uniaxial accelerometers, as it would generally cause less mass loading and be more convenient for measurement.

b. IEPE or charge type

An IEPE accelerometer is convenient and economical as it doesn't need an external charge amplifier. However, a charge-type accelerometer may be necessary for measurement under elevated temperatures.

c. Design

The design refers to how the piezoelectric crystal is arranged with the seismic mass. The type of arrangement determines the loading on the crystal when subjected to acceleration. Figure 9.13 shows two designs. In the compression design, the crystal is subjected to a compressive load, while in the shear design, it is subjected to a shear load. 'm' represents the seismic mass, and the arrows in the figure represent the direction of polarization of the piezoelectric crystals. The piezoelectric materials also generate charge due to temperature changes; a phenomenon called the pyroelectric effect. The faces of the crystal on which this charge appears are perpendicular to the polarization direction. In the compression design, these are also the faces on which the charge due to crystal deformation appears. Therefore, the charge signal is corrupted if there are temperature changes. In the shear design, the acceleration input generates a shear load on the crystal, and the charge develops on the faces on which the shear load acts. These faces differ from the ones where the charge develops due to temperature changes. Thus, the accelerometers based on shear designs have the advantage that they are free from measurement errors due to temperature changes.

d. Weight

The weight of the accelerometer is a critical consideration. An accelerometer needs to be mounted on the structure where acceleration is to be measured. Since the structure, while vibrating, must accelerate the accelerometer mass as well, the accelerometer causes massloading of the structure. Therefore, the acceleration of the structure-accelerometer system may be (slightly) different from the acceleration of the structure alone. The smaller the accelerometer mass, the better it is. But how much a large mass is tolerable depends on the mass of the structure under test. The accelerometers are available in many sizes weighing, typically 60 grams (meant for machines, compressors, and heavier machines) to tiny ones (called miniature accelerometers) weighing as low as 2 grams (meant for significantly lighter structures).

e. Sensitivity

Sensitivity is the output of the accelerometer per unit input. The sensitivity of a charge-type piezoelectric accelerometer is expressed in pC/m/sec^2 or pC/g ('g' referring to 9.8 m/sec^2 acceleration). The sensitivity of an IEPE (or ICP type) piezoelectric accelerometer is expressed in mV/m/sec^2 or mV/g. In general, an accelerometer with higher sensitivity is desired for a higher signal-to-noise ratio. But the sensitivity chosen must also consider

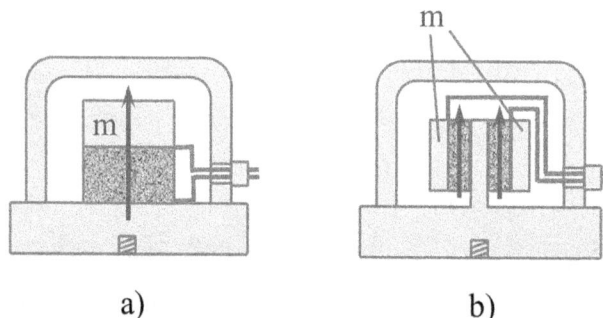

a)   b)

FIGURE 9.13   (a) Compression design. (b) Shear design.

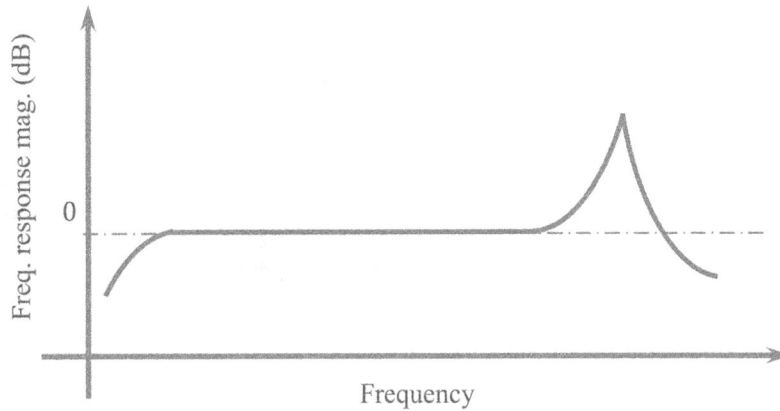

FIGURE 9.14 Frequency response of a piezoelectric accelero-meter.

the dynamic range of acceleration to be measured. For example, if the measurement involves very low and very high levels of acceleration, then choosing an accelerometer with very high sensitivity may be fine for measuring very low levels of acceleration, but measuring the very high levels of acceleration may cause overloading of the input end of the analyzer. Therefore, an accelerometer sensitive enough but suitable for measuring the dynamic range of acceleration should be chosen. A lower sensitivity allows a higher dynamic range of measurement. It should be noted that, as studied in Section 9.7.2, an accelerometer with a higher sensitivity has a higher mass, causing higher mass loading. It also has a smaller frequency range.

f. Frequency range

The frequency range of an accelerometer is the range in which the frequency response of the accelerometer is linear. Figure 9.14 shows the frequency response of a piezoelectric accelerometer. The magnitude of the frequency response (given by the ratio of voltage output and acceleration input) in dB is plotted against frequency on a log-axis. The resonance peak limits the upper cut-off frequency of the accelerometer since the frequency response is nonlinear around the resonance and beyond it. We also studied in Section 9.8.1 that the piezoelectric accelerometer has a low-frequency limit determined by the time constant of the amplifier circuit. The useful frequency range of the accelerometer is the linear region between the low and high-frequency bounds. The frequency response looks almost constant in this range but deviates somewhat. The permitted deviation (5% or 10%) determines the frequency range. For a typical accelerometer, the frequency range could be from 1 to 6,000 Hz, with a 10% deviation from the reference sensitivity.

g. Dynamic range

The frequency range relates to the signal frequency, while the dynamic range relates to the signal amplitude. The dynamic range of an accelerometer refers to the range of acceleration over which its output is proportional to the acceleration input. The lower limit of the dynamic range is limited by the inherent noise in the measuring system, while the upper limit is governed by the accelerometer design. Beyond the upper limit, the output becomes nonlinear. The lower and upper absolute limits of the dynamic range of a typical charge-type accelerometer are $2 \times 10^{-3} \, \text{m/sec}^2$ and $2 \times 10^{4} \, \text{m/sec}^2$ respectively, giving a dynamic range of 140 dB.

h. Transverse sensitivity

A piezoelectric accelerometer subjected to an acceleration normal to its measurement axis also generates some charge. This charge per unit of acceleration in the direction normal to the measurement axis is called transverse sensitivity. It is generally specified as % of the sensitivity along the measurement axis. Preferably it should not be more than 5%.

9.12 LASER DOPPLER VIBROMETER (LDV)

LDV is a laser-based transducer for the measurement of vibratory response. It is based on the principle that when the light reflects from a moving surface, then the frequency of the light changes, called Doppler shift, and since the amount of shift is proportional to the velocity of the reflecting surface, the velocity of the surface can be estimated from the measurement of Doppler shift.

The velocity of vibrating surfaces in practice is small; hence, the Doppler shift as a fraction of the frequency of the light is small and limits the possibility of direct estimation of the Doppler shift from spectral analysis of the reflected light. To deal with this, the LDV, which has a laser emitting a beam of coherent light, splits the emitted beam into two beams, referred to as the reference and measuring beams. The measuring beam is incident at the measurement point on the vibrating surface and suffers reflection. The reflected beam frequency changes due to the Doppler shift, while the reference beam frequency is unchanged. These two beams are made to interfere, and the Doppler shift is determined from the result of interference. The velocity of the surface is then estimated from the Doppler shift, representing the component of the velocity along the incident beam.

Scanning LDV, or SLDV, automates the vibration measurement over discrete test points on a planer surface. It is an extension of the single-point LDV technique in which the measuring beam is moved automatically from one test point to the next. A 3D-SLDV measures vibration at a point on a surface in three directions and allows the construction of the complete velocity vector at that point. It uses three SLDVs oriented differently in space and directs their measuring beams at the test point.

LDV/SLDV has the following advantages:

- It is a non-contact method of vibration measurement and doesn't cause mass loading of the structure. The measurements, therefore, are free from any bias error due to the mass loading.
- Measuring from a considerable distance (typically up to 100m) is possible. It is because the laser beam, being coherent, can travel over a considerable distance.
- The size of the laser spot is narrow (typically 12 μm with LDV located 20cm from the surface) and enables measurement precisely at the desired point. It also helps to obtain a good spatial resolution.
- Measurement on surfaces with high temp or in a harsh environment is possible.

Limitations of the LDV are:

- The surface should be reflective and reasonably smooth or polished. A reflective tape may need to be pasted at the point of measurement.
- There must be a line of sight. The laser should have a clear path to reach the measurement point on the test structure.
- There can be a problem of speckle noise due to the diffusion of the laser beam from the rough surface, causing a drop in the output signal.

9.13 ESTIMATION OF FRF FROM AUTO AND CROSS SPECTRUMS

We saw in Section 9.4 that the FRF can be estimated as the ratio of the Fourier transforms of the force and response time histories. Representing the FRF by H(ω), we can write

$$H(\omega) = \frac{X(\omega)}{F(\omega)} \tag{9.39}$$

But often, there is some random noise in the measured force and response signals. Averaging can be used to reduce random noise. In this section, we look at two FRF estimates based on auto and cross-spectrums that can reduce the effect of measurement noise using averaging.

9.13.1 ESTIMATE H₁

Note that F(ω) and X(ω) are, in general, complex. Let us multiply and divide the RHS of Eq. (9.39) by F*(ω), which is the complex conjugate of F(ω). It gives

$$H(\omega) = \frac{F^*(\omega)\,X(\omega)}{F^*(\omega)\,F(\omega)} \tag{9.40}$$

We recall from Chapter 8 that the numerator is the cross-spectrum between the force and response, while the denominator is the auto-spectrum of the force. If the quantities F(ω) and X(ω) are one-sided spectrums, the numerator and the denominator are the one-sided cross- and auto-spectrums, respectively. We get

$$H(\omega) = \frac{G_{fx}(\omega)}{G_{ff}(\omega)} \tag{9.41}$$

Thus, the FRF can also be computed from the auto and cross spectrums. The FRF estimate in Eq. (7.41) is referred to as estimate H_1. Hence,

$$H_1(\omega) = \frac{G_{fx}(\omega)}{G_{ff}(\omega)} \tag{9.42}$$

9.13.2 ESTIMATE H_2

Alternatively, if we multiply and divide the RHS of Eq. (9.39) by $X^*(\omega)$, which is the complex conjugate of $X(\omega)$, we get

$$H(\omega) = \frac{X^*(\omega)\, X(\omega)}{X^*(\omega)\, F(\omega)} \tag{9.43}$$

The numerator and denominator are auto and cross-spectrums, respectively, and hence we can write

$$H(\omega) = \frac{G_{xx}(\omega)}{G_{xf}(\omega)} \tag{9.44}$$

The FRF estimate in Eq. (9.44) is referred to as estimate H_2, Therefore

$$H_2(\omega) = \frac{G_{xx}(\omega)}{G_{xf}(\omega)} \tag{9.45}$$

9.13.3 SPECTRUM AVERAGING

In the ideal case, when there is no noise in the signals $x(t)$ and $f(t)$, the FRF obtained from Eq. (9.39), and the estimates H_1 and H_2 are all equal to the true FRF (H_{true}) (which is unknown). But the estimates would deviate from the true FRF if there is any noise.

The effect of uncorrelated random noise on the FRF estimates can be reduced by spectrum averaging. For averaging, we repeat the measurement at a point many times, each time measuring the force and response and estimating auto and cross-spectrums. If there are M measurements, we find the averaged auto and cross-spectrums estimates as

$$G_{xx}(\omega) = \frac{1}{M} \sum_{j=1}^{M} X_j^*(\omega) X_j(\omega) \tag{9.46}$$

$$G_{ff}(\omega) = \frac{1}{M} \sum_{j=1}^{M} F_j^*(\omega) F_j(\omega) \tag{9.47}$$

$$G_{fx}(\omega) = \frac{1}{M} \sum_{j=1}^{M} F_j^*(\omega) X_j(\omega) \tag{9.48}$$

It can be shown that

$$G_{xf}(\omega) = G_{fx}^*(\omega) \tag{9.49}$$

Once the average auto and cross-spectrums are obtained from Eqs. (9.46) to (9.49), the estimates H_1 and H_2 are obtained by Eqs. (9.42) and (9.45).

Note that the spectrum averaging approach cannot be used to reduce random noise in $X(\omega)$ and $F(\omega)$ since different measurements correspond to different phase angles, and their averaging tends to zero for a large number of averages. On the other hand, the averaging is meaningful for the auto-spectrum since it has only magnitude and no phase, and also for the cross-spectrum, since it contains the relative phase between the response and the force, which is invariant over the averages.

9.13.4 EFFECT OF NOISE ON H_1 AND H_2

We now study how the noise in the input and output signals affects the estimates H_1 and H_2.

a. Noise only in the output

Let there is random noise in the output but no noise in the input. The noise in the output could result from the response generated due to extraneous excitation sources, like ambient sources.

The extraneous excitation is uncorrelated with the applied force, and therefore the corresponding responses are also uncorrelated. Hence, the response to extraneous excitation appears like uncorrelated random noise in the output. The output noise could also be due to random noise in the measurement system. If $\bar{x}(t)$ is the measured output with noise, $n(t)$ is the uncorrelated random noise in the output and $f(t)$ is the measured input, then we can write

$$\bar{x}(t) = x(t) + n(t) \tag{9.50}$$

$$\bar{X}(\omega) = X(\omega) + N(\omega) \tag{9.51}$$

The cross-spectrum is

$$G_{f\bar{x}}(\omega) = \frac{1}{M} \sum_{j=1}^{M} F_j^*(\omega) \bar{X}_j(\omega) \tag{9.52}$$

Substituting $\bar{X}_j(\omega)$ from Eq. (9.51), we get

$$G_{f\bar{x}}(\omega) = \frac{1}{M} \sum_{j=1}^{M} \left(F_j^*(\omega) X_j(\omega) + F_j^*(\omega) N_j(\omega) \right) \tag{9.53}$$

$$G_{f\bar{x}}(\omega) = G_{fx}(\omega) + G_{fn}(\omega) \approx G_{fx}(\omega) \tag{9.54}$$

In Eq. (9.54), $G_{fn}(\omega)$ tends to zero for large values of M since the noise is random and uncorrelated with force. The auto-spectrum $G_{ff}(\omega)$ is noise-free since f(t) is noise-free.

Therefore, the estimate H_1 gives

$$H_1(\omega) = \frac{G_{f\bar{x}}(\omega)}{G_{ff}(\omega)} = \frac{G_{fx}(\omega)}{G_{ff}(\omega)} = H_{true} \tag{9.55}$$

Thus, the estimate H_1 is unaffected by the random uncorrelated noise in the output and gives the true FRF. What about H_2?

The auto-spectrum $G_{\bar{x}\bar{x}}(\omega)$ is

$$G_{\bar{x}\bar{x}}(\omega) = \frac{1}{M} \sum_{j=1}^{M} \bar{X}_j^*(\omega) \bar{X}_j(\omega) \tag{9.56}$$

Substituting $\bar{X}_j(\omega)$ from Eq. (9.51), we get

$$G_{\bar{x}\bar{x}}(\omega) = \frac{1}{M} \sum_{j=1}^{M} \left(X_j^*(\omega) X_j(\omega) + X_j^*(\omega) N_j(\omega) + N_j^*(\omega) X_j(\omega) + N_j^*(\omega) N_j(\omega) \right) \tag{9.57}$$

Therefore,

$$G_{\bar{x}\bar{x}}(\omega) = G_{xx}(\omega) + G_{xn}(\omega) + G_{nx}(\omega) + G_{nn}(\omega) \approx G_{xx}(\omega) + G_{nn}(\omega) \tag{9.58}$$

In Eq. (9.58), the terms $G_{xn}(\omega)$ and $G_{nx}(\omega)$ tend to zero for large values of M, since the noise is random and uncorrelated with the response. However, there is a component $G_{nn}(\omega)$ due to the noise which contaminates the auto-spectrum of the response. From Eq. (9.49), we can write

$$G_{\bar{x}f}(\omega) = G_{f\bar{x}}^*(\omega) \tag{9.59}$$

Using the results in Eqs. (9.54) and (9.49), Eq. (9.59) becomes

$$G_{\bar{x}f}(\omega) = G_{fx}^*(\omega) = G_{xf}(\omega) \tag{9.60}$$

Therefore, H_2 is obtained as

$$H_2(\omega) = \frac{G_{\overline{xx}}(\omega)}{G_{\overline{x}f}(\omega)} = \frac{G_{xx}(\omega) + G_{nn}(\omega)}{G_{xf}(\omega)} \tag{9.61}$$

$$H_2(\omega) = \frac{G_{xx}(\omega)}{G_{xf}(\omega)} + \frac{G_{nn}(\omega)}{G_{xf}(\omega)} = H_{true} + \frac{G_{nn}(\omega)}{G_{xf}(\omega)} \tag{9.62}$$

Thus, the estimate H_2 is affected by the random uncorrelated noise in the output. $G_{nn}(\omega)$ being the auto-spectrum is always positive, and hence $H_2 \geq H_{true}$. From the analysis, it is clear that H_1 is the preferred estimate of the FRF when there is uncorrelated random noise only in the output.

b. Noise only in the input

Let there is random noise in the input but no noise in the output. The noise in the input could result from the measurement noise. If $x(t)$ is the measured output, $\overline{f}(t)$ is the measured input, and $q(t)$ is the uncorrelated noise in the input, then we can write

$$\overline{f}(t) = f(t) + q(t) \tag{9.63}$$

$$\overline{F}(\omega) = F(\omega) + Q(\omega) \tag{9.64}$$

By proceeding like in the previous case, we obtain H_1 as

$$H_1(\omega) = \frac{G_{\overline{f}x}(\omega)}{G_{\overline{ff}}(\omega)} = \frac{G_{fx}(\omega)}{G_{ff}(\omega) + G_{qq}(\omega)} \tag{9.65}$$

$$H_1(\omega) = \frac{G_{fx}(\omega)}{G_{ff}(\omega)}\left(\frac{G_{ff}(\omega)}{G_{ff}(\omega) + G_{qq}(\omega)}\right) = H_{true}\left(1 - \frac{G_{qq}(\omega)}{G_{ff}(\omega) + G_{qq}(\omega)}\right) \tag{9.66}$$

Thus, the estimate H_1 is affected by the random uncorrelated noise in the input. Both $G_{qq}(\omega)$ and $G_{ff}(\omega)$ being auto-spectrums are always positive, and hence $H_1 \leq H_{true}$.

The estimate H_2 is obtained as

$$H_2(\omega) = \frac{G_{xx}(\omega)}{G_{x\overline{f}}(\omega)} = \frac{G_{xx}(\omega)}{G_{xf}(\omega)} = H_{true} \tag{9.67}$$

Thus, the estimate H_2 gives the true FRF and is the preferred estimate of the FRF when there is uncorrelated random noise only in the input.

c. Noise in both the input and output

In this case, both the measured input and output are contaminated by random noise. We obtain H_1 as

$$H_1(\omega) = \frac{G_{\overline{f}x}(\omega)}{G_{\overline{ff}}(\omega)} = \frac{G_{fx}(\omega)}{G_{ff}(\omega) + G_{qq}(\omega)} = H_{true}\left(1 - \frac{G_{qq}(\omega)}{G_{ff}(\omega) + G_{qq}(\omega)}\right) \tag{9.68}$$

While H_2 is obtained as

$$H_2(\omega) = \frac{G_{\overline{xx}}(\omega)}{G_{\overline{x}\overline{f}}(\omega)} = \frac{G_{xx}(\omega) + G_{nn}(\omega)}{G_{xf}(\omega)} = H_{true} + \frac{G_{nn}(\omega)}{G_{xf}(\omega)} \tag{9.69}$$

Thus, both H_1 and H_2 are affected by the uncorrelated random noise, and deviate from the true FRF. For this general case also, we note that

$$H_1 \leq H_{true} \leq H_2 \tag{9.70}$$

Thus, the estimates H_1 and H_2 form the lower and upper bounds of the true FRF.

9.14   COHERENCE FUNCTION

Is it possible to assess the quality of the measured FRF? From the previous section, we see that the estimates H_1 and H_2 form the lower and upper bounds of the true FRF, and as a result, the ratio H_1/H_2 is always less than or equal to one. This ratio, therefore, can serve as an indicator of noise in the FRF and is called the coherence function. The coherence function is given by

$$\gamma^2(\omega) = \frac{H_1(\omega)}{H_2(\omega)} \tag{9.71}$$

$$\gamma^2(\omega) = \frac{G_{\bar{f}\bar{x}}(\omega)G_{\bar{x}\bar{f}}(\omega)}{G_{\bar{x}\bar{x}}(\omega)G_{\bar{f}\bar{f}}(\omega)} = \frac{\left|G_{\bar{x}\bar{f}}(\omega)\right|^2}{G_{\bar{x}\bar{x}}(\omega)G_{\bar{f}\bar{f}}(\omega)} \tag{9.72}$$

Note that the coherence at any frequency is bounded by $0 \le \gamma^2(\omega) \le 1.0$.

For the case of random uncorrelated noise in the input and output, using the results in the previous section, we can write

$$\gamma^2(\omega) = \frac{\left|G_{xf}(\omega)\right|^2}{\left(G_{xx}(\omega) + G_{nn}(\omega)\right)\left(G_{ff}(\omega) + G_{qq}(\omega)\right)} \tag{9.73}$$

Eq. (9.73) indicates that the coherence at a frequency is less than one if there is noise at that frequency. The noise in the input, output, or both essentially breaks the linear relationship between the measured input and output. Because of this, the two estimates of the FRF tend to differ. The coherence function quantifies the degree of linear relationship between the output and the input. Deviation from a linear relationship may occur due to noise or nonlinearity in the structure, making coherence less than one.

9.15   FRF MEASUREMENT USING IMPACT TESTING

In this section, we go through the steps involved in measuring FRFs using impact testing. We present these steps by simulating the test on a cantilever structure. The objective is to see how the force and response signals are measured and processed and the role and effect of various measurement parameters on the estimated FRF. This understanding is necessary to make an appropriate choice of parameters for measurement. This phase, where various decisions related to the test are made, is also referred to as test planning.

Figure 9.15a shows a cantilever beam under test. The beam is made of steel and has a rectangular cross-section. The beam's length, width, and thickness are 0.5, 0.05, and 0.003 m, respectively. This beam is used further in all the remaining chapters for simulating shaker testing, modal parameter estimation, phase resonance testing, OMA, and modal testing applications.

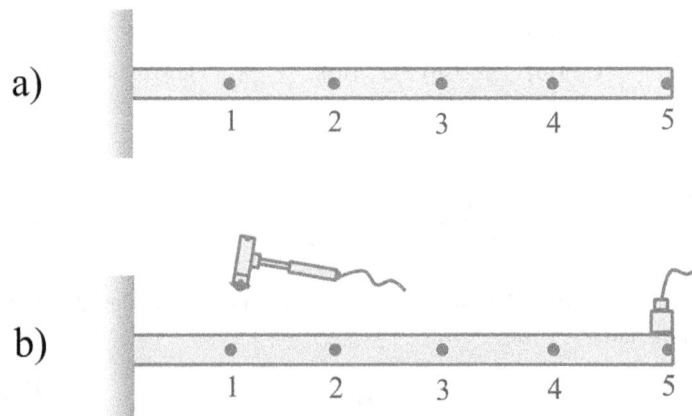

FIGURE 9.15   (a) Cantilever structure used for simulation. (b) Impact test.

9.15.1 Choice of the Frequency Range of Measurement

The frequency range of measurement should be the range in which the dynamic response of the structure is to be studied, which, in turn, depends on the frequencies of the dynamic forces acting on the structure. If the objective is to identify the modal model of the structure, the measurement frequency range should correspond to the frequency range for which the model needs to be used. We assume that a modal model of the structure is required for the 0–400 Hz frequency range.

9.15.2 Choice of Measurement Points

The number of test points for FRF measurement depends on the test objective. If the objective is to get an idea of the structure's resonant frequencies and modal damping factors, then the knowledge of the FRFs at a few points can provide that information. But if the objective is to identify the modal model of the structure, then it also requires identifying the mode shapes. In that case, the number of test points should be enough to describe mode shapes to avoid spatial aliasing. Spatial aliasing refers to an insufficient number of points or data to represent a curve or surface in a geometric space. Due to spatial aliasing, a mode with a complicated shape is represented incorrectly by a simpler shape.

High-frequency modes have more complicated mode shapes than low-frequency modes, and hence the high-frequency mode shapes govern the choice of the number of test points. But since the experimental mode shapes become known only after the modal testing, one way to decide the number of test points for modal testing could be based on the estimates of the mode shapes from an analytical or numerical model of the structure. The chosen points should not be concentrated in a local region but distributed such that the distribution captures the mode shapes over the whole structure.

A prior analytical model reveals four modes in the 0–400 Hz range for the cantilever structure. The cantilever's fundamental mode has no node, but the number of nodes increases by one per mode. Therefore, we need at least five test points to describe the first four modes without causing spatial aliasing. Of course, taking a higher number of test points would better describe the mode shapes, but we take five test points for simulating the FRF measurement.

9.15.3 FRFs to be Measured at the Test Points

Once the number of test points is decided, the relevant question is the FRFs to be measured at these test points. There are two issues: the DOFs at each test point for which the FRFs must be measured and whether the full FRF matrix must be measured.

In structures made up of beams, plates, shells, or a combination of these, the translational and rotational DOFs are the relevant DOFs at a point. However, only translational DOFs are generally measured due to the lack of suitable and reliable transducers for measuring rotational DOFs. For the cantilever structure, let us assume that we are interested in in-plane modes of vibration. Hence, the FRFs corresponding to the translational DOFs in the vertical direction at the five test points are measured.

Do we need to measure the complete FRF matrix corresponding to translational DOFs at these five test points? To answer this, let us look at the relationship between an FRF and the modal parameters, which for an MDOF system with structural damping is

$$\alpha_{jk}(\omega) = \sum_{r=1}^{N} \frac{\phi_{jr}\,\phi_{kr}}{\omega_r^2 - \omega^2 + i\eta_r\omega_r^2} \tag{9.74}$$

We note from the RHS of Eq. (9.74) that α_{jk} contains information about the

- natural frequencies of all the modes,
- damping factors of all the modes, and
- the product of the jth and k^{th} elements of all the mode shapes.

Thus, an FRF has information about the natural frequencies and damping factors of all the modes, but a particular FRF, like α_{jk}, has information about only the product of the jth and k^{th} elements of a mode shape for all the modes.

If we fix the jth DOF as the response coordinate and measure the FRFs with excitation at all the DOFs, then the measured FRFs contain information about all the elements of the mode shapes. These FRFs comprise the jth row of the FRF matrix, as shown in Figure 9.16a. Hence we conclude that the measurement of any one row of the FRF matrix is sufficient to estimate the modal model of the system corresponding to the chosen test DOFs.

a)
$$[\alpha]_{p\times p} = \begin{bmatrix} \alpha_{11} & \alpha_{12} & \cdots & \alpha_{1p} \\ \alpha_{21} & \alpha_{22} & \cdots & \alpha_{2p} \\ \cdots & \cdots & \cdots & \cdots \\ \boxed{\alpha_{j1}} & \boxed{\alpha_{j2}} & \boxed{\cdots} & \boxed{\alpha_{jp}} \\ \cdots & \cdots & \cdots & \cdots \\ \alpha_{p1} & \alpha_{p2} & \cdots & \alpha_{pp} \end{bmatrix}$$

b)
$$[\alpha]_{p\times p} = \begin{bmatrix} \alpha_{11} & \alpha_{12} & \cdots & \alpha_{1k} & \cdots & \alpha_{1p} \\ \alpha_{21} & \alpha_{22} & \cdots & \alpha_{2k} & \cdots & \alpha_{2p} \\ \cdots & \cdots & \cdots & \cdots & \cdots & \cdots \\ \alpha_{p1} & \alpha_{p2} & \cdots & \alpha_{pk} & \cdots & \alpha_{pp} \end{bmatrix}$$

FIGURE 9.16   (a) j^{th} row of the FRF matrix, (b) k^{th} column of the FRF matrix.

To measure any one row of the FRF matrix, say the jth row, the response transducer is fixed at the jth test DOF. The FRF α_{jk} is measured by applying force at the kth test DOF and the process is repeated for all the values of k. This approach is convenient to follow with an impact hammer as the location of the force application can be easily shifted to the desired test DOF. It is called the **Roving hammer test** as the hammer is moved from one test point to the other with the location of the response transducer fixed during the test.

Alternatively, if we fix the excitation coordinate at the kth DOF and measure the FRFs with the response at all the test DOFs, then these FRFs also contain information about all the elements of the mode shapes. The measured FRFs, comprise the kth column of the FRF matrix, as shown in Figure 9.16b. Hence we conclude that the measurement of any one column of the FRF matrix is also sufficient to estimate the modal model of the system corresponding to the chosen test DOFs.

To measure any one column of the FRF matrix, say the kth column, the force is applied to the kth test DOF. The FRF α_{jk} is measured by measuring the response at the jth test DOF and the process is repeated for all the values of j. This approach is preferred when using a vibration exciter (to be discussed in the next chapter), as relocating the exciter to a new location is inconvenient. The easier option is to shift the location of the response transducer from one test DOF to another. This approach is called the **Roving accelerometer test**.

In this section, we perform a roving hammer test for the cantilever structure. Since there are five test points, we must measure five FRFs corresponding to any row of the FRF matrix.

9.15.4   CHOICE OF RESPONSE TRANSDUCER LOCATION

In a roving hammer test, the location of the response transducer/accelerometer is fixed. The location of the accelerometer should be such that all the modes in the analysis frequency range are observable. For this to happen, the accelerometer location should not coincide with any of the nodes of the modes. But this requires a knowledge of the mode shapes yet to be identified from the test. If available, an analytical/numerical model of the structure can be used in this regard. Otherwise, an initial measurement of FRFs with different accelerometer locations can be carried out to choose a suitable location. The test points where the response is likely to be small, like those close to fixed boundaries, should be avoided. For the cantilever structure, the accelerometer is fixed at test point 5, as shown in Figure 9.15b.

9.15.5   CHOICE OF TIME AND FREQUENCY DOMAINS PARAMETERS

For the FRF measurement, we need to decide the parameters of the time and frequency domains. These parameters are: the frequency range of the FRF, defined by the upper limit f_{max}, sampling frequency f_s, record length T, time resolution Δt, number of samples in the time domain N_t, and frequency resolution Δf. From the relationships between these parameters presented in Chapter 8, we know that we can choose any two parameters while the remaining get defined based on these choices.

For FRF measurement on the cantilever structure, the analysis frequency range f_{max} is chosen (equal to 400 Hz). The other parameter often chosen is the number of samples in the time domain (N_t). In the FFT analyzers, the choice of N_t needs to be made from a set of values expressible as 2^n, since it makes the FFT computation efficient.

The other parameters can be evaluated from the values of f_{max} and N_t as follows.

- As per the Nyquist theorem, the sampling frequency must be at least twice of f_{max}. As discussed in Chapter 8, the signal is first passed through an anti-aliasing filter (a low-pass filter) to avoid aliasing. The frequency response of an anti-aliasing filter doesn't have a sharp cut-off, but has a roll-off, which is compensated by taking the sampling frequency $2.56\ f_{max}$.

- The cut-off frequency of the anti-aliasing filter is equal to $\dfrac{f_s}{2}$.

- The sampling time is $\Delta t = 1/f_s$.
- The record length T depends on the choice of N_t which is influenced by the following considerations:
  - If the response signal decays completely in time T, then the windowing of the response signal can be avoided, which means that no window is required. T can be increased by choosing a higher N_t.
  - T longer than necessary may make noise more dominant in the latter part of the signal, adversely affecting the FRF quality.
  - Higher is N_t, higher is T, and hence smaller is Δf (or better is the frequency resolution).
  - But the higher the N_t, the higher the processing time and the time required for the test.

Table 9.1 summarizes the relationships between the time and frequency domain parameters. Table 9.2 shows the choices for the number of samples in the time domain typically available in FFT analyzers. The corresponding number of frequency lines in the frequency domain is also shown.

We make an initial choice of 2,048 time samples for the modal testing of the cantilever structure, giving a record length of 2 seconds and a frequency resolution of 0.5 Hz. We can revise the choice, if necessary, once we know the decay time of the transient response. Table 9.3 gives the values of all the time and frequency domain parameters based on the choices of $f_{max} = 400$ Hz and $N_t = 2,048$.

TABLE 9.1

Relationship between the Time and Frequency Domain Parameters

Frequency range (Hz)	f_{max}
Number of time samples	N_t
Sampling frequency (Hz)	$f_s = 2.56\ f_{max}$
Cut-off frequency of the anti-aliasing filter (Hz)	$f_c = \dfrac{f_s}{2}$
Time resolution or sample time (seconds)	$\Delta t = \dfrac{1}{f_s}$
Record length (seconds)	$T = N_t \Delta t$
Frequency resolution (Hz)	$\Delta f = \dfrac{1}{T}$
Number of frequency lines in the spectrum	$N_f = \dfrac{f_{max}}{\Delta f}$

TABLE 9.2

Choices for the number of samples (N_t) in the time domain typically available in FFT analyzers and the Corresponding Number of Frequency Lines (N_f)

N_t	256	512	1,024	2,048	4,096	8,192	16,384
N_f	100	200	400	800	1,600	3,200	6,400

TABLE 9.3

Measurement Parameters for the Modal Testing of the Cantilever Structure

f_{max}	N_t	f_s	f_c	T	Δt	Δf	N_f
400 Hz	2048	1,024 Hz	512 Hz	2 seconds	0.0009765 seconds	0.5 Hz	800

9.15.6 OTHER MEASUREMENT SETTINGS IN THE FFT ANALYZER

The analyzer must have at least two channels for modal testing since the force and response must be simultaneously measured to estimate the FRF. There are some other measurement settings necessary in an FFT analyzer. The relevant settings are briefly discussed below.

- Input voltage: It is the maximum voltage at the input. It should ideally be just a little more than the maximum signal voltage. It ensures that the dynamic range of the A/D converter is utilized fully, keeping the quantization noise to a minimum (as discussed in Chapter 8). The 'Autoranging' option in the analyzer can be used initially to detect the maximum signal voltage and set the input voltage accordingly.
- AC/DC coupling: Coupling should be set to AC when the DC component in the input is not to be measured.
- CCLD input: Should be set on when using an IEPE/ICP type of transducer to enable the necessary current to be delivered to the built-in preamplifier in the transducer.
- Trigger: Trigger is the condition when satisfied starts the measurement. The signal level and slope are parameters of the trigger condition. The channel with the force signal is generally chosen as the source of the trigger signal.
- Pre-trigger delay: It is the time difference between the fulfillment of the trigger condition and the start of the measurement. A negative delay is often used to start the measurement a few samples before the trigger condition is satisfied.
- Averages: Represents the desired number of spectrum averages.
- Sensitivity and units: The sensitivity of the transducers and the engineering units for displaying the signals (like m, m/sec, m/sec^2, or N) should be fed into the analyzer.
- Window: The type of window for the input signals should be selected. The window functions generally available are: Rectangular, Exponential, Force, Hanning, Hamming, Flat-top, or user-defined.
- Output or data type: Represents the quantity computed from the measured signals. Some choices are Fourier spectrum (mag and phase), Auto-spectrum, Cross-spectrum, Frequency response (FRF) (mag and phase), Autocorrelation function, Cross-correlation function, and Coherence function.
- Mathematical Processing: Whether the input signal or spectrum is to be integrated or differentiated. If the inertance is measured, this option can be used to obtain the receptance or mobility by choosing the integration option and specifying the number of integrations.

9.15.7 FRF MEASUREMENT

In this section, we study the process of measuring the force and response signals and estimating the FRFs by impact testing. We do this by simulating this process on our cantilever structure. As mentioned earlier, the accelerometer is fixed at test point 5 to conduct a roving hammer test.

The manufacturer's data sheet for an impact hammer may give information on the frequency range that can be excited using a particular hammer tip. Alternatively, whether a particular tip is suitable for exciting the given structure in the desired frequency range can be verified. Let us use a rubber tip and verify its suitability. The measurement parameters used are given in Table 9.3. We use a small pre-trigger delay. In this section, the FRF is computed as a ratio of the DFTs of the acceleration and force signals, while the FRF computation using spectrum averaging is considered in the next section.

We first measure the point FRF, which means the excitation is applied at the location of the accelerometer. Let us first look at the role of the anti-aliasing filter, as it is the first element through which the signals pass before processing. Figure 9.17a shows the magnitude of the point inertance at test point 5 (i.e., FRF A$_{55}$) measured without an anti-aliasing filter. For comparison, the true/correct estimate of the FRF is overlaid in part (b) of the figure. We see spurious peaks in the measured FRF and distortion at other frequencies. It is due to the aliasing of the time signals since the sampling frequency of 1,024 Hz is good enough only up to a frequency range of 512 Hz. The frequencies in the time signal beyond this range get aliased, causing the measured FRF to deviate from the true FRF.

Therefore, to avoid aliasing, we use an anti-aliasing filter (a Butterworth analog filter of order 4) with a cut-off frequency of 512 Hz. Thus, the accelerometer and hammer outputs are passed through the anti-aliasing filter before discretization and measurement. Figure 9.18 shows the time history of the force applied at test point 5. Ideally, we expect no signal outside the duration of the force pulse. However, the force signal may contain some random measurement noise, as seen in the figure in the zoomed view of a portion of the time history beyond the impact force pulse. Since the noise is not a part of the force applied, it can be removed by applying a rectangular window around the impact force pulse. Figure 9.19 shows a narrow **rectangular force window**. The window's location should coincide with the force pulse and width enough to cover the impact force duration without truncating the physical force signal. Figure 9.20 shows the force signal after applying the force window, indicating that the noise is eliminated from a significant portion of the signal, though any noise inside the window is still there.

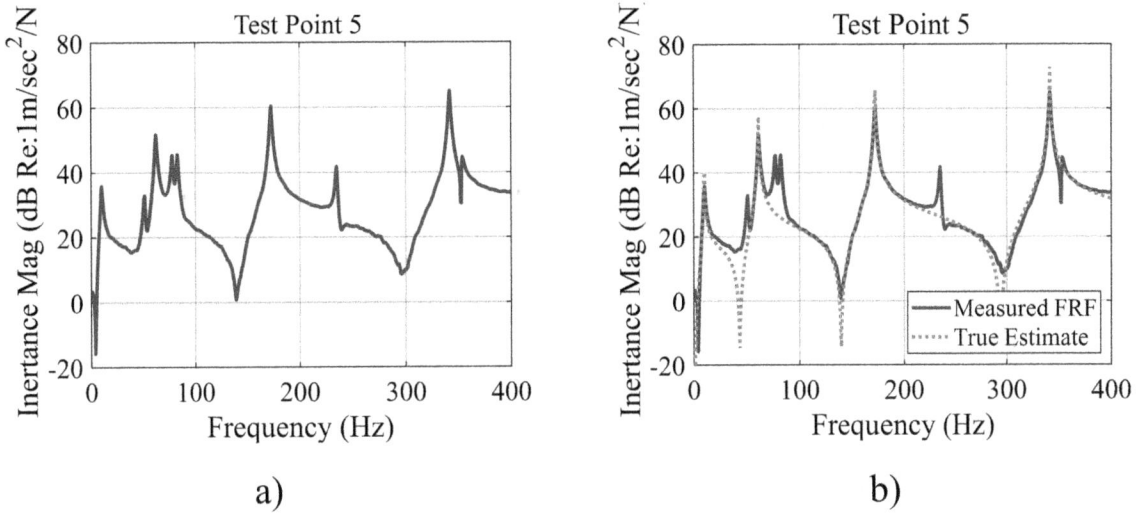

FIGURE 9.17   (a) FRF measurement without an anti-aliasing filter, (b) overlay with the true estimate.

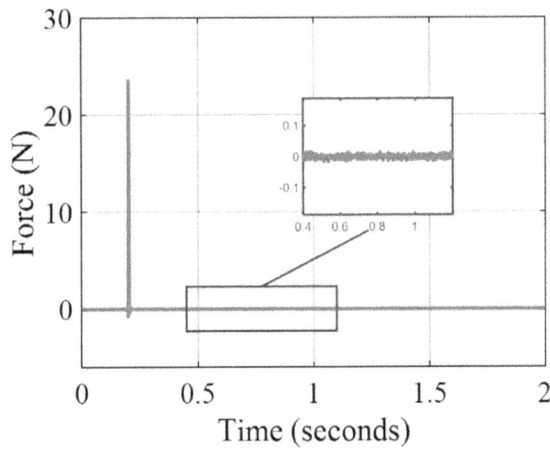

FIGURE 9.18   Time history of the impact force.

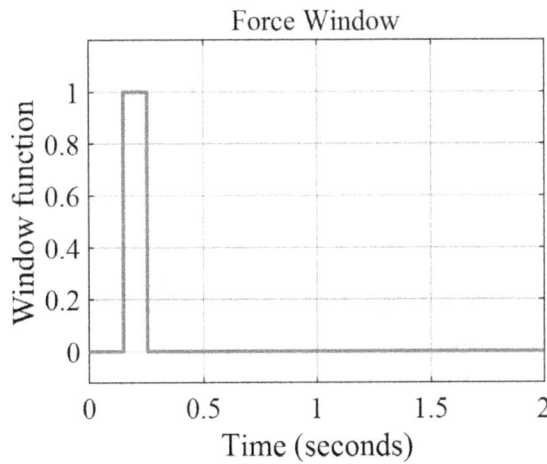

FIGURE 9.19   Rectangular force window.

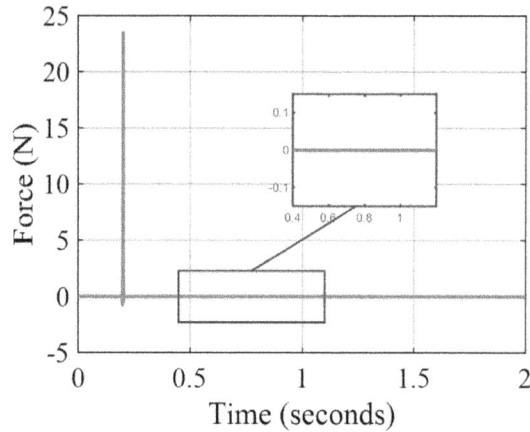

FIGURE 9.20   Windowed force signal.

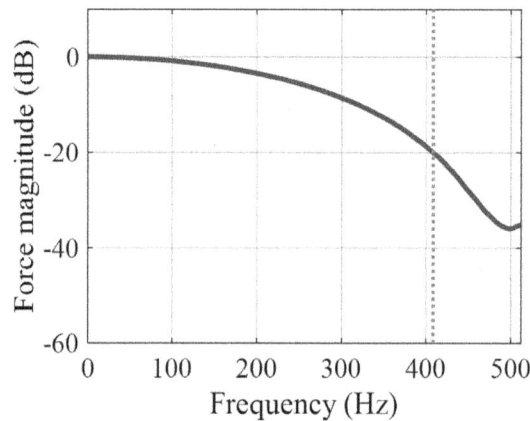

FIGURE 9.21   Magnitude of the DFT of the windowed force signal.

Figure 9.21 shows the magnitude of the DFT of the windowed force signal. The frequency where the amplitude drops by 20 dB is 409 Hz, shown by a vertical line. Thus, the impact produced with the rubber tip can excite frequencies up to 400 Hz (the chosen analysis frequency range); hence, we use this tip for FRF measurement.

Figure 9.22 shows the time history of the resulting acceleration at test point 5, while Figure 9.23 shows the plot of the magnitude of its DFT. We observe that the acceleration signal has not decayed completely within the chosen record length of 2 seconds. It results in signal truncation, causing leakage in the frequency domain, as seen in the DFT plot and the zoomed view. The leakage also affects the inertance estimate computed from this signal, as seen in Figure 9.24.

There are two ways to deal with signal truncation and the resulting leakage. We either use a window function to force the signal to decay completely in the chosen record length or increase the record length. Let us first explore using a window function, an exponential window function, as shown in Figure 9.25. The **exponential window function** is described by

$$w(t) = w_0 e^{-\beta t} \qquad (9.75)$$

where the variables w_0 and β depend upon the parameters b and t_0 of the window function. The acceleration signal is multiplied by the window function $w(t)$. Figure 9.26 shows the time history of the windowed acceleration signal at test point 5. Due to the window function, the signal is forced to decay completely in the record length. It significantly reduces the leakage, as seen from the FRF plot shown in Figure 9.27, though there is some noise due to the noise in the signals, as seen in the zoomed view. We should note that the window function adds some artificial damping to the response; hence, the damping identified from the measured FRFs may be overestimated. A correction can be made to the damping estimates based on the time constant of the exponential window.

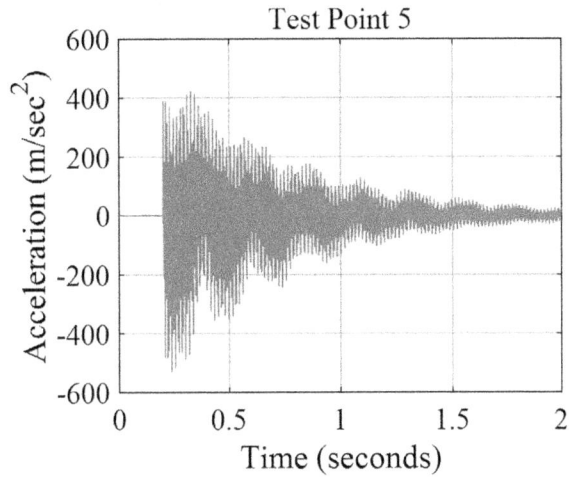

FIGURE 9.22   Time history of the acceleration at test point 5.

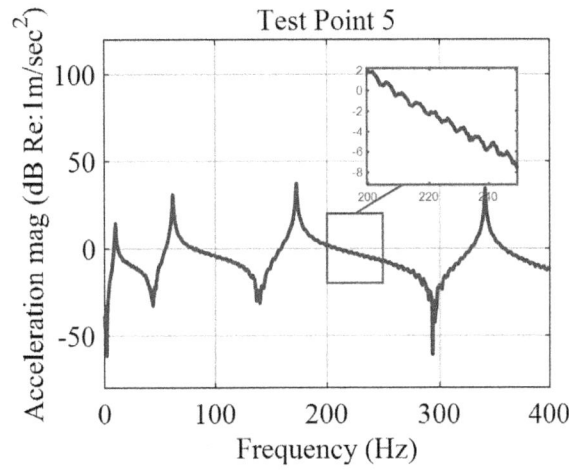

FIGURE 9.23   Magnitude of the DFT of the acceleration at test point 5.

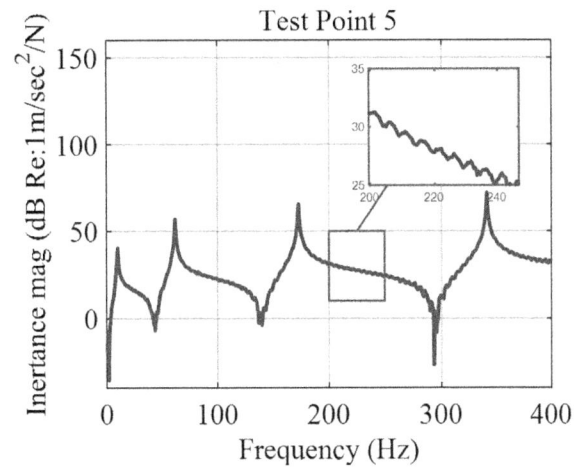

FIGURE 9.24   Magnitude of inertance A_{55}.

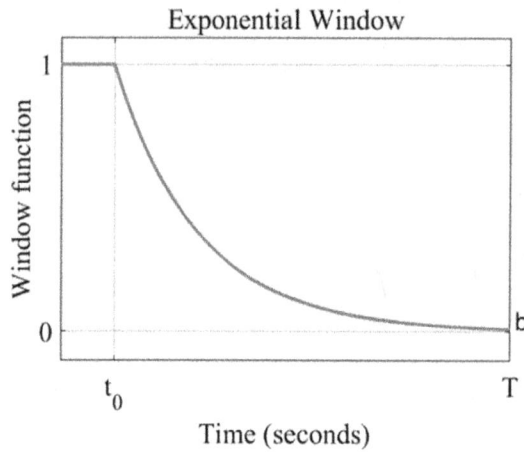

FIGURE 9.25   Exponential window function.

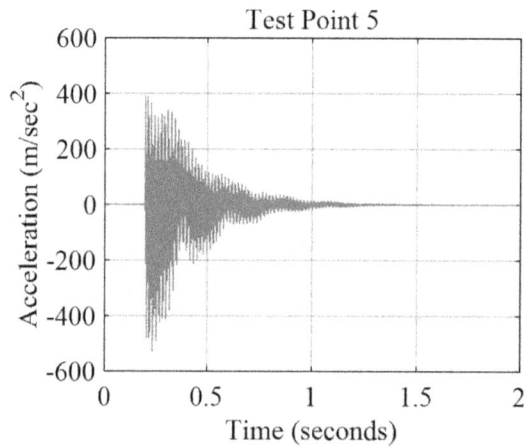

FIGURE 9.26   Time history of the acceleration at test point 5 after windowing.

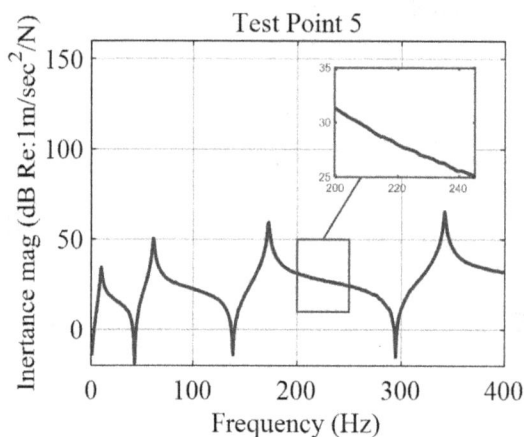

FIGURE 9.27   Magnitude of inertance A_{55}.

Next, we explore the second option to deal with the truncation of the signal and increase the number of samples to 4,096. It changes the record length to 4 seconds while the frequency resolution and frequency lines change to 0.25 Hz and 1,600, respectively. The other parameters don't change. Figure 9.28 shows the acceleration signal at test point 5, and we observe that the signal has decayed entirely in 4 seconds. Hence, no window function is used. Figure 9.29 shows the magnitude and phase of inertance A_{55}. The estimated FRF has no leakage, though some noise is present due to the noise in the

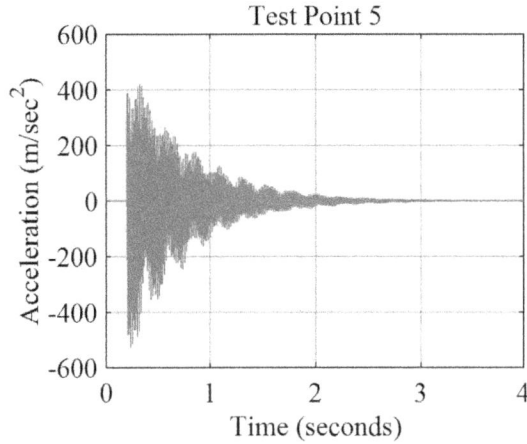

FIGURE 9.28   Acceleration Time history at test point 5.

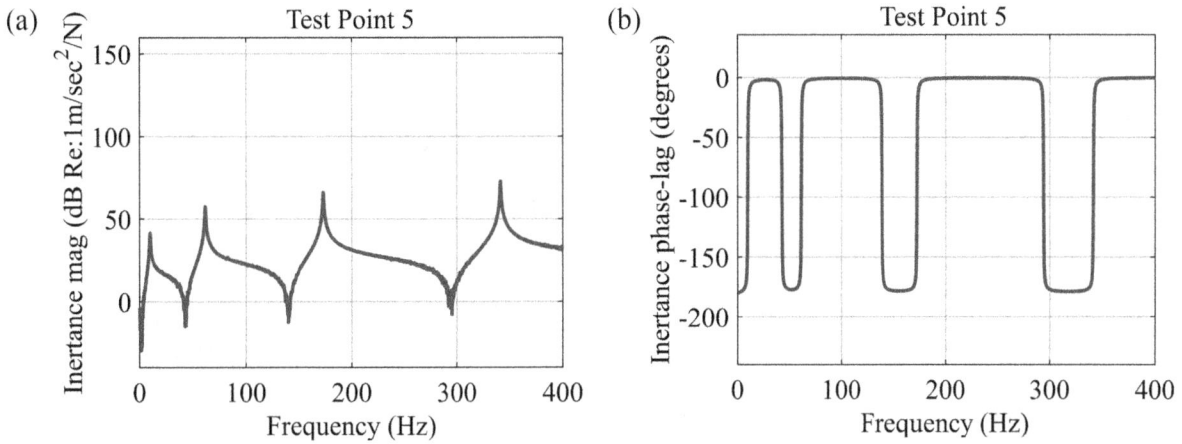

FIGURE 9.29   (a) Magnitude and (b) phase of inertance A_{55}.

measured signals. Also, the FRF peaks are sharper than those in Figure 9.27, thus indicating a lesser level of damping in the structure. It is a correct reflection of the damping since the FRF is free from the artificial damping introduced using an exponential window, as was the case with the FRF in Figure 9.27.

We now measure the FRFs at other test points (1–4) by applying impacts at those points and using a record length of 4 seconds. Figure 9.30 shows the magnitude of the inertances so obtained (A_{51}, A_{52}, A_{53} and A_{54}). It completes the measurement of one row of the FRF matrix corresponding to the translational DOFs at the five test points. Curve fitting of the measured FRFs to estimate modal parameters is dealt with in Chapter 11.

Let us finally check what happens if we use a rubber tip to measure the FRF over a range beyond 400 Hz, say up to 800 Hz. We use the same record length (4 seconds), so windowing is not required. However, the time samples need to be increased to 8,192. For these choices, the values of the anti-aliasing filter cut-off frequency and sampling frequency become 1,024 Hz and 2,048 Hz, respectively. Figure 9.31 shows the magnitudes of the DFT of the force and the measured FRF A_{55}. We observe that beyond 400 Hz, the force is very small or even zero at some frequencies. In this range, the signal-to-noise ratio of the force signal is poor, which makes the signal-to-noise ratio of the response signal also poor since the response is the result of the force. It makes the FRF estimate also inaccurate in this range. These results justify the thumb rule that the effective frequency range of excitation by the hammer is the frequency where the force has dropped substantially (say by 20 dB).

9.15.8   FRF Measurement with Spectrum Averaging

In this section, we look at the process of FRF measurement with spectrum averaging. To measure A_{55}, we apply an impact at test point 5, measure force and the acceleration signals, and compute the auto and cross-spectrums. This process is repeated M times, and the averaged estimates of the auto and cross-spectrums are obtained (using Eqs. 9.46–9.49), which are then used to find the FRF.

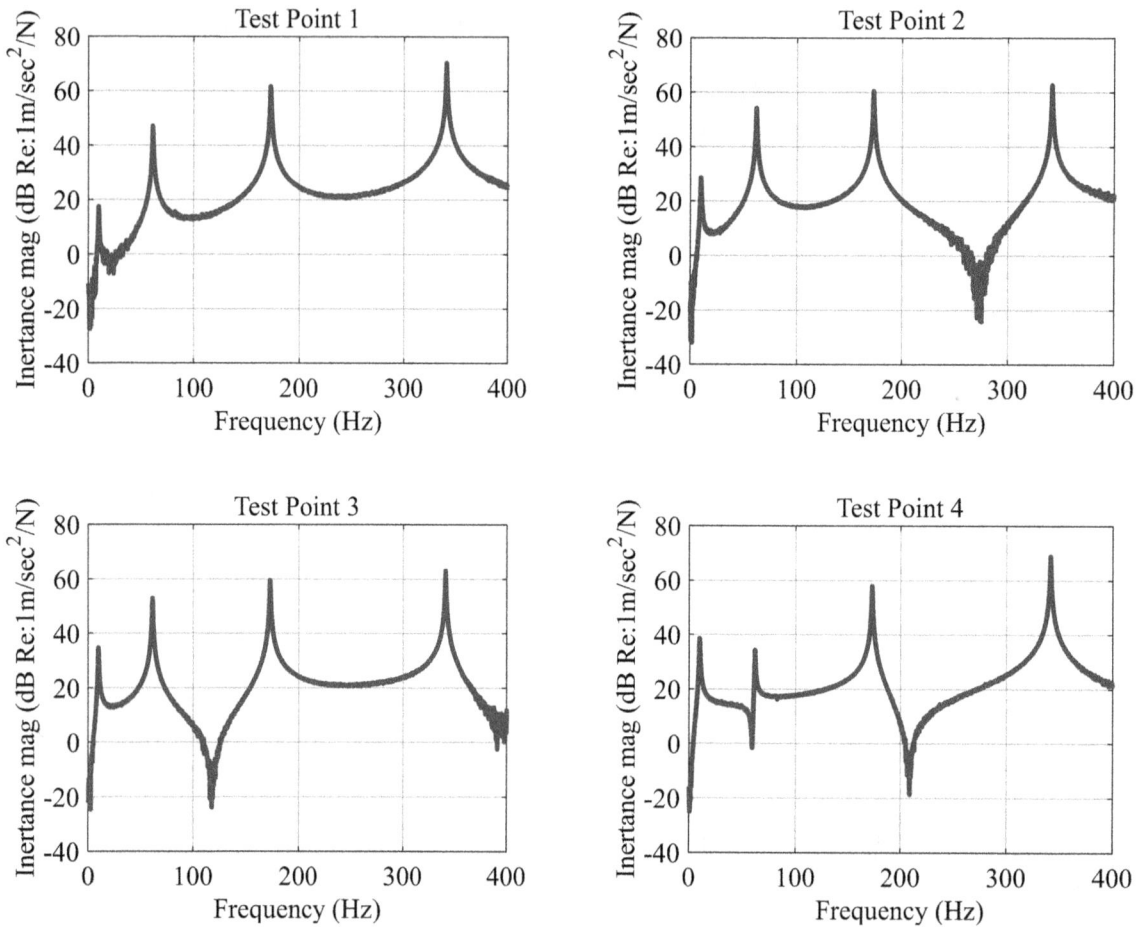

FIGURE 9.30 Magnitude of inertances A_{51}, A_{52}, A_{53} and A_{54}.

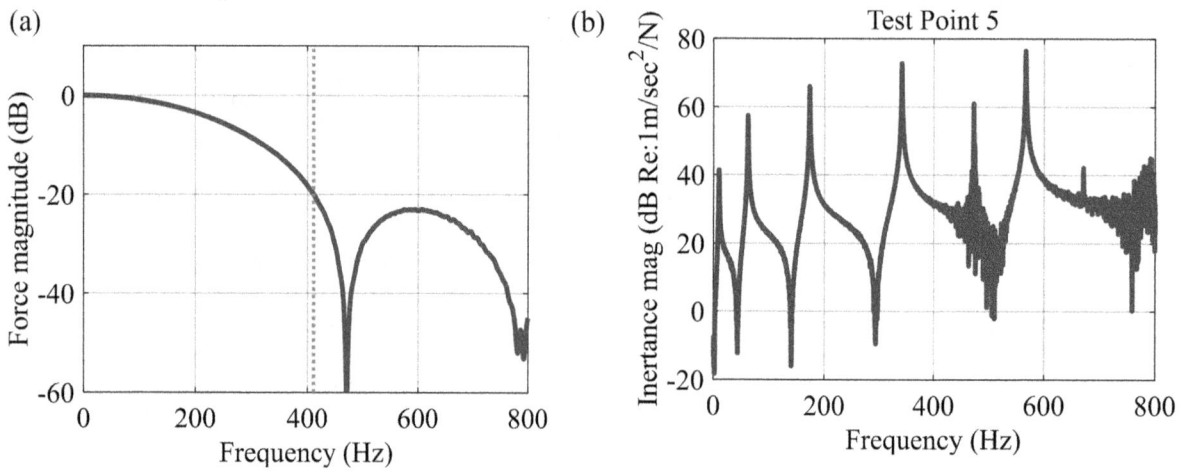

FIGURE 9.31 Magnitudes of (a) DFT of the impact force and (b) FRF A_{55}.

Figure 9.32a and b show the auto-spectrum and the cross-spectrum magnitudes at test point 5 obtained using four averages (M=4). Figure 9.33a and b show the estimate H_1 of the FRF A_{55} and the coherence function, respectively.

The FRF has lesser noise than the direct estimate shown in Figure 9.29a. The estimate H_1 is based on spectrum averaging and can reduce the random noise in the FRF estimate. Part (b) of the figure shows that the coherence is poor at the antiresonance frequencies due to the poor signal-to-noise ratio at these frequencies. Otherwise, the coherence is excellent and indicates that the FRF estimate is accurate. Figure 9.34 shows the H_1 estimate obtained with 2, 4, 8, and 16 averages. With an increase in the averages, the random noise in the FRF reduces.

In practice, spectrum averaging should be used carefully, ensuring no variations in the location and direction of the impacts over the averages. Minor variations in the strength of the impact should not be an issue if the structure is excited within the range of its linear behavior. If there is not much noise, then the FRF can be measured without the spectrum averaging.

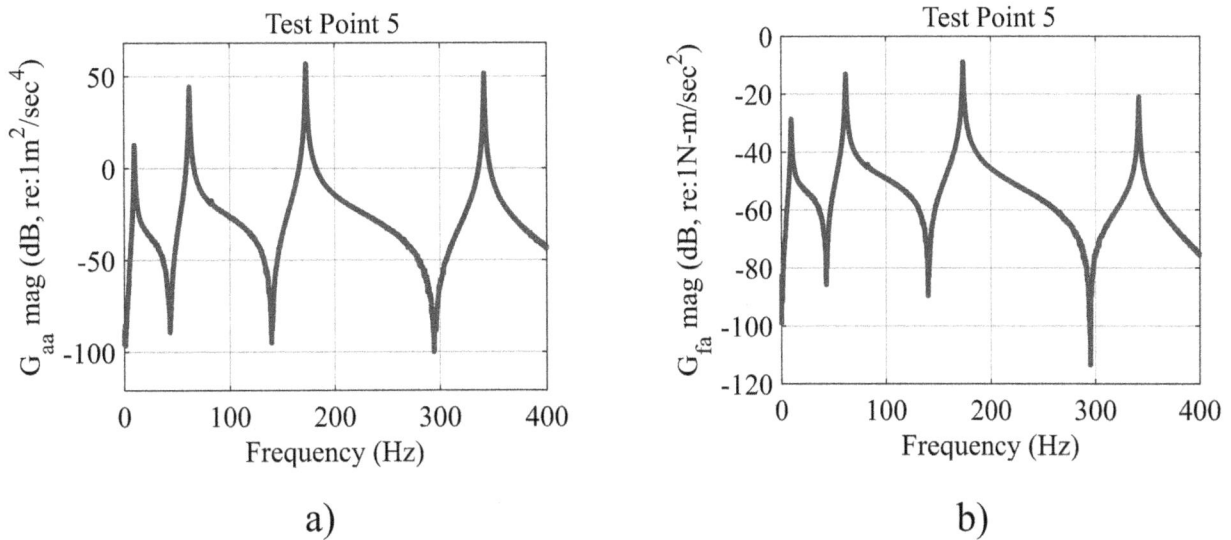

a)

b)

FIGURE 9.32 (a) Auto-spectrum of the acceleration and (b) the magnitude of the cross-spectrum between the force and acceleration signals at test point 5.

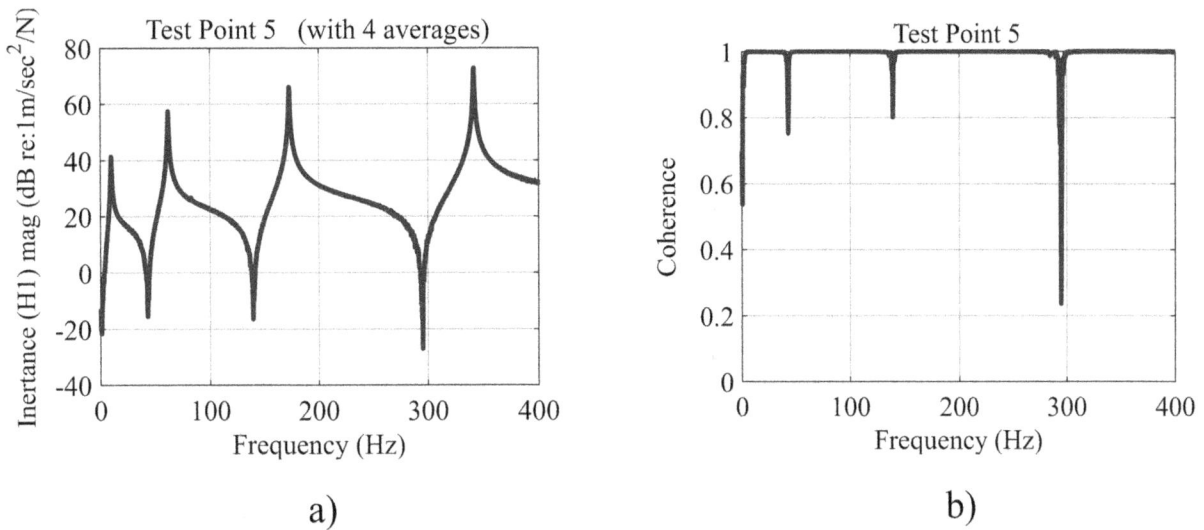

a)

b)

FIGURE 9.33 (a) Magnitude of the H1 estimate of FRF A_{55} and (b) coherence function.

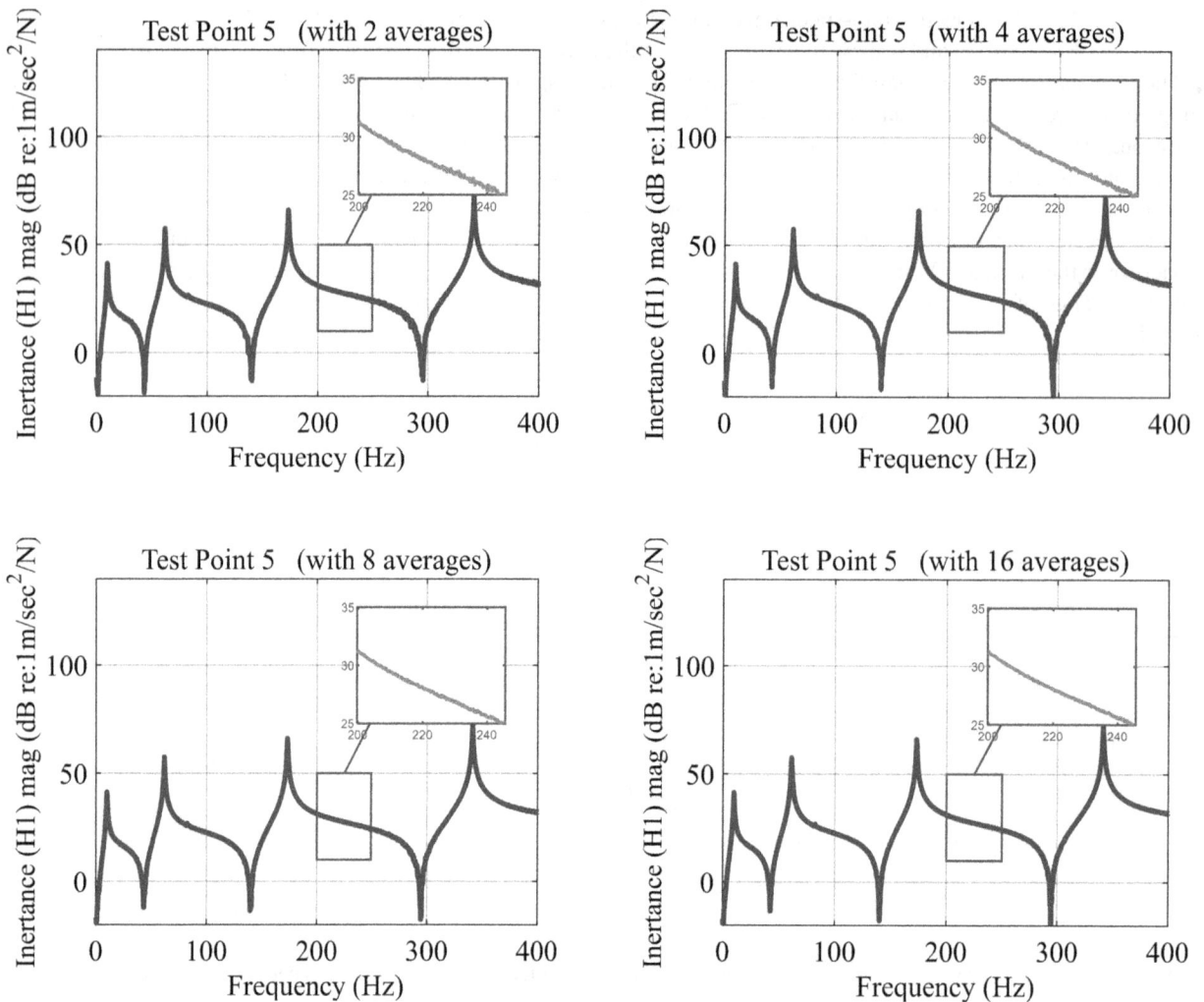

FIGURE 9.34 FRF estimate H1 obtained with 2, 4, 8, and 16 averages.

9.16   FRF VALIDITY CHECKS

Some checks can be performed to ascertain whether the measured FRF is a valid estimate or not.

a. Reciprocity check

We studied in Chapter 2 that a linear system follows reciprocity, which makes

$$A_{jk} = A_{kj} \tag{9.76}$$

that is, if the response and excitation points are interchanged, the FRF remains unaltered. Thus, for a linear structure, Eq. (9.76) is a valuable equation to check the validity of the measured FRFs. The equation can also be used if we seek a linear approximation of a nonlinear structure. There must be no change in the test setup when the measurements with the interchanged position of the force and response points are taken.

b. Linearity check

If the structure is linear, the FRF doesn't change with the amplitude of the force. Therefore, comparing the FRFs measured with different magnitudes of the excitation force can indicate whether the structure is linear.

c. Repeatability check

A valid measurement must be repeatable. An FRF measured repeatedly must give identical results if the test conditions are unchanged. Noise in the measured signals or nonlinearities in the system may make the measurements nonrepeatable. Excitations originating from other sources and influencing the test may also make measurements nonrepeatable.

9.17 BOUNDARY CONDITIONS FOR MODAL TESTING

The boundary conditions under which the given structure must be tested is another decision that needs to be made as part of the test planning. The boundary conditions for modal testing of a component or an assembly depend on the purpose for which the modal test results are to be used.

9.17.1 TESTING WITH THE ACTUAL BOUNDARY CONDITIONS

In many cases, modal testing is carried out to know the system's natural frequencies to avoid the resonant operation of the system. It requires accurate knowledge of the system's natural frequencies. Also, if the objective of the modal test is to identify a modal model for response simulation to dynamic forces, then for this application also, the modal model must correspond to the system with the actual boundary conditions for the predictions to be accurate.

In such cases, the modal testing of the system needs to be carried out with the same boundary conditions as those existing during the system's operation. It is because the boundary conditions influence the dynamic characteristics, and deviations from the actual boundary conditions may lead to inaccurate estimates of natural frequencies and other modal parameters.

One approach to ensure actual boundary conditions is to perform testing in situ. For many structures like those encountered in civil engineering, like bridges and buildings, there is no option other than to perform testing in situ. In many cases, in situ testing may not be feasible due to constraints of size, physical access, resources required, or environmental conditions.

The second approach is to simulate the actual boundary conditions of the structure in a laboratory setup. However, recreating identical boundary conditions is challenging.

9.17.2 FREE-FREE AND FIXED BOUNDARY CONDITIONS

Another vital application of modal testing is validating and updating numerical models. If the model of the system corresponding to its operational state is to be validated, then at some point, the modal test for the system with the actual boundary conditions would have to be carried out. But in many cases, the system model validation is preceded by the validation of its components and subassemblies. In these cases, the subassembly or the component may be removed from the main assembly for validation. The idea is first to validate the models of the components in isolated conditions, and after validation, the component models are assembled for the next or system level validation.

Two boundary conditions or supports frequently used for component and subassembly testing are: free-free support and fixed or grounded support. These boundary conditions can be easily incorporated into a numerical or FE model and are widely used.

A further application of testing with free-free or fixed supports is in the coupled structural analysis, which involves combining models of the subassemblies and components of an assembled structure to build the model of the complete structure. In the hybrid approach, the individual component/subassembly models may come from an experimental route like modal testing or an analytical route like an FE model. The experimental model derived from modal testing may be obtained by testing under free-free or fixed boundary conditions and can be incorporated into the system model. Another instance where the free-free boundary condition is useful is in the experimental identification of the inertia properties. The inertia properties are associated with the rigid body modes of a body. By modal testing of the body under the free-free boundary conditions, the inertia properties of the body can be identified. This approach is particularly useful for bodies with complex and irregular shapes.

How the free-free and fixed boundary conditions are realized? An ideal free-free boundary condition refers to no constraint to the motion of the body. A body in space can have independent motions in six different ways, three translations and three rotations. These are referred to as rigid body modes of the body. The body is not subjected to any deformation or strain in these modes. Since no elastic restoring forces act on the body while having motion in rigid body modes, the natural frequencies of rigid body modes are equal to zero. In practice, a body cannot be held in space without support; hence, an ideal free-free boundary condition cannot be realized. To approximate free-free boundary conditions, the body is supported on soft or highly compliant supports, like foam, or is suspended from a rigid ceiling using soft rubber or elastic bands. The soft supports apply small elastic restoring forces during the motion, making the natural frequencies of the rigid body modes nonzero. The supports should be soft enough to keep the rigid modes' natural frequencies as small as possible so that they have a negligible effect on the flexural or strain modes of the body. The farther these two sets of natural frequencies are, the less the interference and the modal parameters of the flexible modes obtained by testing with soft supports would be accurate estimates of the modal parameters under the true/ideal free-free boundary conditions. The soft supports also dissipate some energy when the test structure vibrates during testing. Therefore, the damping factors identified from a free-free test may be overestimated.

Fixed boundary conditions constrain all the degrees of freedom at some points on the test structure. In practice, this boundary condition is not easy to realize since the supporting structure used for fixing or clamping would have a finite stiffness and inertia, causing deviation from a truly fixed boundary condition.

9.18 CALIBRATION

The sensitivities of the force and response transducers must be known for converting their electrical output to the relevant engineering units. The manufacturer generally provides this data in a calibration chart with the transducer. However, with time, due to handling and environmental conditions, the sensitivity may change, and the measurement would be in error if the correct value of the sensitivity is not known. The use of erroneous sensitivity would make FRF magnitude inaccurate. Inaccurate FRF magnitude may lead to scaling errors in the modes shapes identified from them. Using inaccurate FRFs for response simulation would also lead to erroneous response predictions.

It calls for the periodic calibration of a transducer to determine the sensitivity by measuring the transducer's output to a known input. Individual calibration of the accelerometer and the force transducer requires transducer-specific calibration systems. However, a simpler method can be followed for FRF measurement in which the whole system is calibrated rather than the individual transducers. It is often referred to as ratio calibration. In this method, the FRF is measured on a system with the known FRF. The comparison of the measured FRF with the known FRF is used to calculate a 'calibration factor' to be applied to the measured FRF to obtain a 'corrected' or 'calibrated' estimate.

A typical setup for calibrating the system using the above method is shown in Figure 9.35. A heavy lumped mass m is suspended such that it can freely oscillate in the vertical plane. The system is like a pendulum and behaves like an SDOF system. If L is the length of the strings and assuming an equivalent viscous damping c for the energy dissipation in the system, the equation of motion for the small amplitude of motion can be written as

$$m_e \ddot{x} + c\dot{x} + m_e \frac{g}{L} x = F(t) \tag{9.77}$$

where m_e is the equivalent mass of the pendulum, which includes the lumped mass and the accelerometer mass. The equivalent string mass can be neglected as it is much smaller than the lumped mass. The inertance is given by

$$|A_t(\omega)| = \frac{\frac{1}{m_e}\frac{\omega^2}{\omega_n^2}}{\sqrt{\left(1-\frac{\omega^2}{\omega_n^2}\right)^2 + \left(2\xi\frac{\omega}{\omega_n}\right)^2}} \tag{9.78}$$

Figure 9.36 shows the plot of $|A_t(\omega)|$. It is observed that for $\omega \gg \omega_n$ the inertance is nearly constant (called the mass line studied in Chapter 7), and the ordinate of the line is

$$|A_t(\omega)| \approx \frac{1}{m_e} \tag{9.79}$$

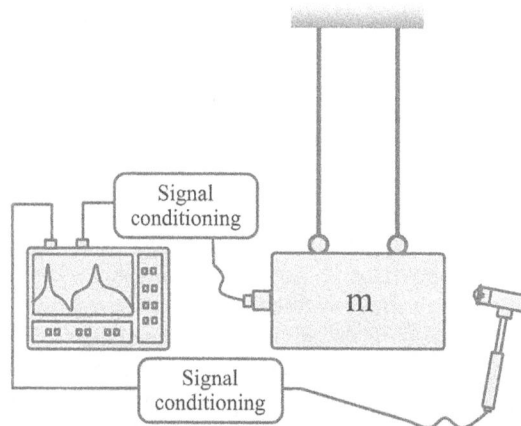

FIGURE 9.35   Setup for the calibration of the measurement system.

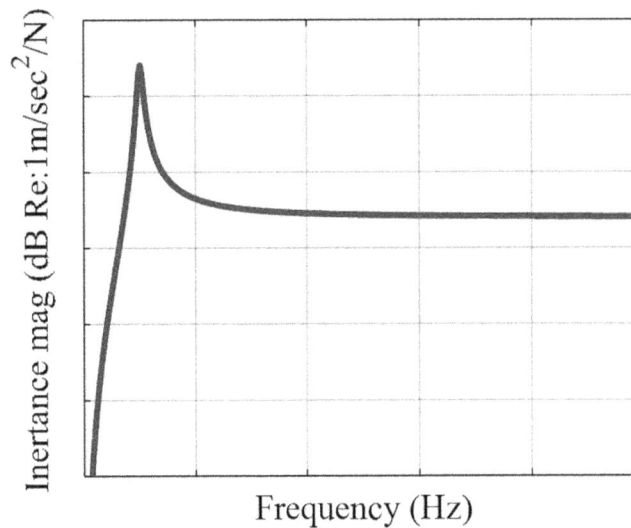

FIGURE 9.36 Plot of the magnitude of the inertance A_t.

Since the equivalent mass m_e is known accurately, the inertance A_t in the range $\omega \gg \omega_n$ is accurately known from Eq. (9.78). The inertance is measured using the setup shown in Figure 9.35. Let A_m be the absolute magnitude of the measured FRF in the range $\omega \gg \omega_n$ where the inertance becomes nearly constant. The calibration or correction factor is calculated as

$$C_f = \frac{A_t}{A_m} \tag{9.80}$$

The FRF measured on the given test structure is multiplied by the factor C_f to obtain a calibrated estimate of the measured FRF. Note that the above method takes care of changes in the sensitivities of the force and response transducers and any scaling errors in the measurement.

REVIEW QUESTIONS

1. What is an IEPE/ICP accelerometer?
2. Why is a charge amplifier needed for signal conditioning with a charge-type accelerometer?
3. Can a piezoelectric accelerometer be used to measure zero or low-frequency acceleration signals?
4. What determines the low and high-frequency limits of a piezoelectric accelerometer?
5. Why should a piezoelectric accelerometer have a high natural frequency?
6. What is the difference between compression-type and shear-type piezoelectric accelerometers?
7. How the nature of the hammer tip influences the frequency range of excitation using an impact hammer?
8. Define the estimates H_1 and H_2 to compute an FRF.
9. What is the coherence function? What could be the reasons for coherence to be less than one?
10. Why is the coherence function likely to be poor at antiresonance frequencies?
11. How can the system for FRF measurement be calibrated?
12. For FRF measurement, how the measurement parameters should be changed to increase the frequency resolution by a factor of 2?
13. How can free-free boundary conditions for a structure be simulated for modal testing?
14. What is the motivation for testing under free-free boundary conditions?
15. What do you mean by reciprocity, linearity, and repeatability checks that are used to ascertain the validity of the measured FRFs?

10 FRF Measurement Using Shaker Excitation

10.1 INTRODUCTION

In Chapter 9, we studied FRF measurement using impact excitation. If the structure is large and heavy, the impact hammer may not be suitable to impart sufficient energy for vibration. A stronger impact may excite nonlinearities and cause local damage to the structure. Excitation using a vibration shaker may be suitable in this situation. Another advantage of shaker testing is that the excitation frequency range can be precisely controlled and tailored as per the requirement, which is not possible with impact excitation. In this chapter, we study FRF measurement using shaker excitation.

We first look at the basic measurement system using a shaker (also called an exciter), followed by the construction and principle of operation of an electromagnetic shaker. The shaker needs to be attached to the structure to be excited. We study through a lumped parameter model how the shaker interacts with the dynamics of the structure resulting in force dropout at some frequencies, and how the interaction affects the measured FRF in voltage and current modes of operation. We then briefly study the sensors like force transducer and impedance head used in shaker testing. Like the impact testing in Chapter 9, the FRF measurement with a shaker is simulated on the cantilever structure. The FRF measurement is presented using various excitation signals such as random, periodic random, burst random, sine sweep, and sine chirp.

10.2 EXPERIMENTAL SETUP FOR FRF MEASUREMENT USING A SHAKER

Figure 10.1 shows a typical setup for FRF measurement on a structure using a shaker. In this example, the structure is suspended using soft elastic bands to simulate the free-free boundary conditions. A force transducer is mounted on the structure at the excitation point to measure the excitation force. The shaker is attached to the force transducer via a thin rod called a stinger. The stinger is attached to the shaker table at one end and to the force transducer at the other. The shaker is driven by a signal generator and a power amplifier. The purpose of the signal generator is to generate a voltage signal with the desired frequency range and characteristics. Since the signal generator output is a low-power signal, a power amplifier is used to amplify and supply the necessary power to the shaker. The force generated by the shaker is transmitted to the structure through the stinger and force transducer, making the structure vibrate. The structure's response is measured using an accelerometer. The output of the transducers may require signal conditioning, depending upon the type of the transducers, i.e., the charge or IEPE type (as discussed in Chapter 9). The FRF is then estimated from the measured force and response signals.

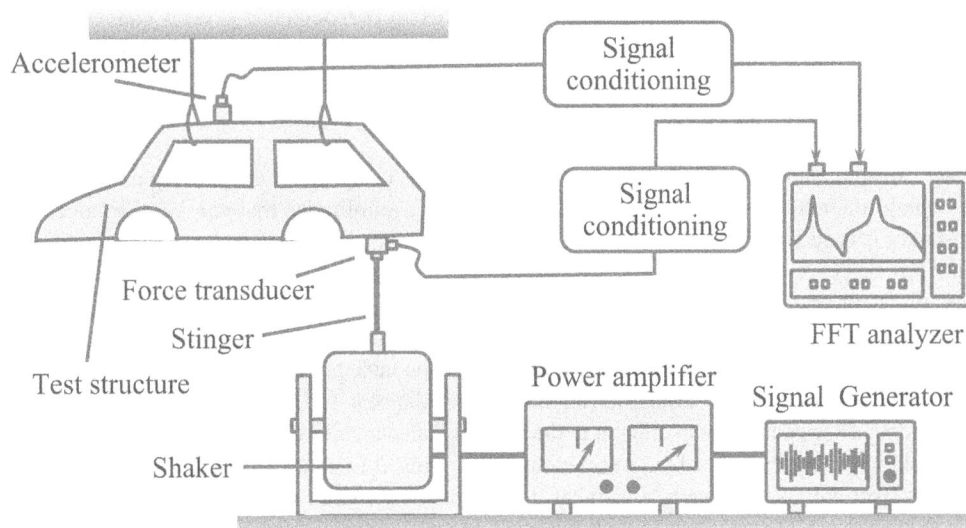

FIGURE 10.1 Experimental setup for FRF measurement using a shaker.

DOI: 10.1201/9780429454783-10

10.3 VIBRATION SHAKER/EXCITER

A vibration shaker or exciter allows the generation of an oscillating or dynamic force/displacement and is used in many applications such as modal testing, vibration acceptance testing, material properties identification, and fatigue life assessment.

Some structures in their service life have to operate in environments with vibration and shocks. For ex., a satellite structure should sustain the dynamic loads encountered during launch. Similarly, electrical/electronic instruments mounted on-board ground and aerial vehicles must function without failure in the vibratory environment they have to operate. A shaker is used to simulate the vibratory environment experienced by a system or product. It allows laboratory testing of the product to verify its ability to sustain the vibratory environment it has to face in practice without failure. Acceptance testing is often done for specific vibration levels, so a vibration controller, based on feedback control, is used with the shaker. It compares the actual and desired vibration levels and regulates the input to the shaker to ensure that the actual level follows the desired vibration level.

In modal testing, the objective is to identify the system from the knowledge of the applied input and the corresponding output. The type and magnitude of the input used are guided mainly by modal testing considerations and generally have nothing to do with the forces that the structure under test is subjected to in practice. For the same reason, it is also not necessary to maintain any specific levels of vibration as is the case for acceptance testing, and hence the vibration controller is not required with the shaker for modal testing.

There are several types of vibration shakers, such as mechanical, hydraulic, electromagnetic, and piezoelectric.

- Mechanical vibration shakers are either direct-drive or reaction types. The direct-drive design of the shaker utilizes the slider-crank mechanism, in which the slider makes up the vibration table, and the crank is driven by a motor. This shaker generates a constant amplitude displacement for the permissible loads on the table and frequencies of operation. The reaction-type design has two eccentric masses rotating in opposite directions, and the excitation force arises from the centrifugal forces on the masses. In this design, controlling the excitation force at any frequency is difficult as it requires a change in the unbalance. The mechanical shakers are limited to low-frequency excitations.
- Hydraulic shakers generate excitation force through high-pressure oil delivered by a pump. The shaker uses a pilot or servo valve to control the direction of the flow of the high-pressure oil to a power cylinder. The idea is to alternately switch the high-pressure oil flow to the two sides of the power cylinder piston. This action generates an alternating force on the piston connected to the mounting table of the shaker. The pilot valve is electrically actuated and needs only a little power for its operation. The frequency of the input electrical signal actuating the valve controls the frequency of the force generated on the table. Since the valve needs to be operated at the excitation frequency, the valve dynamics limit the higher excitation frequencies. Hence, hydraulic shakers are used at relatively lower frequencies than electrodynamic shakers, though larger excitation forces can be generated at low frequencies. Since the excitation force emanates from the oil pressure, hydraulic shakers can also generate static forces, which may be a requirement in many cases.
- The electromagnetic shaker is described in the next section.
- A piezoelectric shaker consists of a stack of piezoelectric actuators. It is based on the inverse piezoelectric effect by which the application of an electric voltage across a piezoelectric crystal causes it to extend/contract. They can generate small forces over a wide frequency range.

10.4 ELECTROMAGNETIC SHAKER

An electromagnetic shaker, also known as an electrodynamic shaker, is a popular and widely used shaker for modal and vibration testing due to the extensive range of frequencies and force amplitudes that can be generated. The principle of working of this shaker is based on the fact that a current-carrying conductor/coil placed in a magnetic field experiences a force (the Lorentz force). The force varies at the same frequency as the current in the coil. So, an alternating force can be generated by passing an alternating current through the coil.

We now look at the basic construction of an electromagnetic shaker. Figure 10.2 shows a schematic arrangement of this type of shaker. The armature coil is wound on a spider which, along with the table at the top, makes the armature. The armature is supported on the shaker body by a flexible suspension so that the impedance to the motion of the armature is small. The shaker body can be turned about a horizontal axis to orient it in the vertical plane to apply the force at an angle.

The magnetic field is generated either by a permanent magnet (used in small shakers) or by electromagnets (used in large shakers). The shaker with electromagnets also has a field coil. The shaker shown in the figure has a permanent magnet (N-S). Two pole pieces of soft iron are used to direct and concentrate the magnetic field into the small gap between the two pole pieces. The pole pieces have a much higher permeability than air; thus, they help to increase the flux density and reduce leakage. As the figure shows, the armature coil is positioned in this small gap cutting the magnetic lines of forces.

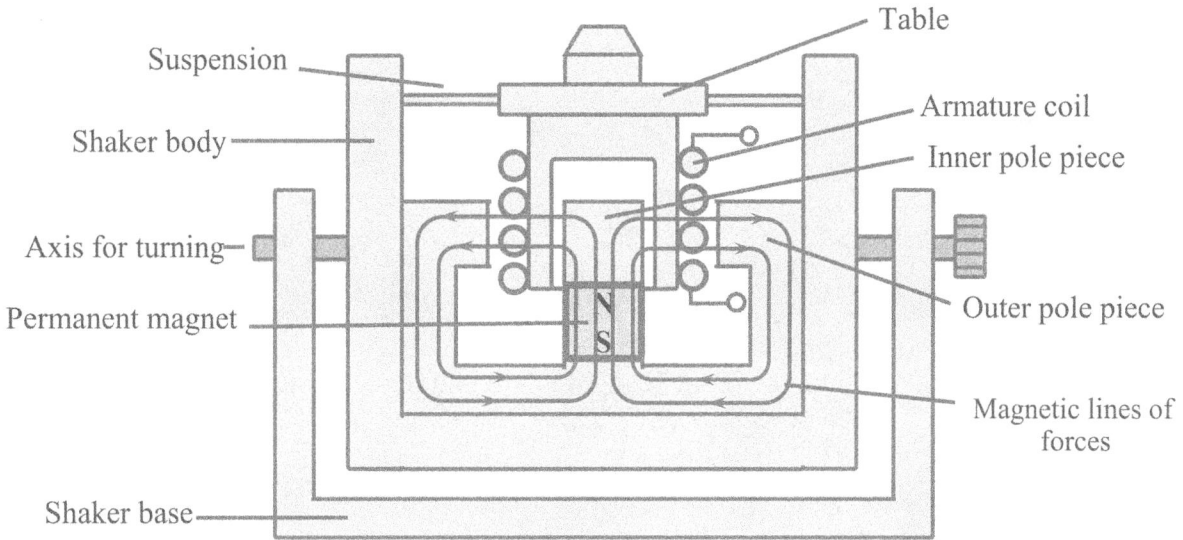

FIGURE 10.2 Schematic arrangement of an electromagnetic shaker.

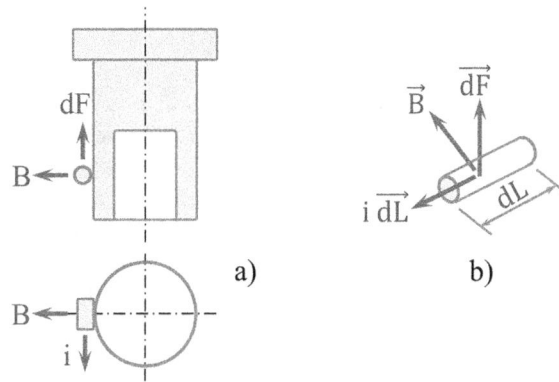

FIGURE 10.3 (a) The front and top views of a small length dL of the coil. (b) The directions of the magnetic field and current through length Dl.

A force acts on the coil when the current passes through it. Consider a small length dL of the coil, as shown in Figure 10.3. Also shown are the directions of the magnetic field (B) and the current (i) through dL. The magnetic field direction is radially outwards and perpendicular to the element dL having current i. The cross-product of the current-element and magnetic field vectors is the force generated and is given by

$$\overrightarrow{dF} = i\ \overrightarrow{dL} \times \vec{B} \qquad (10.1)$$

Thus, the direction of the force is perpendicular to the plane formed by the vectors $\vec{B}$ and $\overrightarrow{dL}$ as shown in the figure, and causes the armature to move axially. Integrating the equation over the length of the armature coil, we get the total force on the coil, whose magnitude is

$$F = B\ i\ L \qquad (10.2)$$

10.5 ATTACHMENT AND SUPPORT OF SHAKER

The shaker needs to be attached to the structure for excitation. The requirement is that the force is applied only along the degree of freedom we intend to excite. A widely used method is to connect the two through a so-called stinger. The stinger is a thin rod that is axially stiff but flexible in the transverse direction. This ensures that while the force along the stinger length is transmitted faithfully, the resistance to the motion of the structure at the point of attachment is least along the other DOFs. Typically, there is a collet chuck at the shaker end to grip the stinger, while at the structure end, the stinger is connected to the force transducer mounted on the structure through a threaded connection.

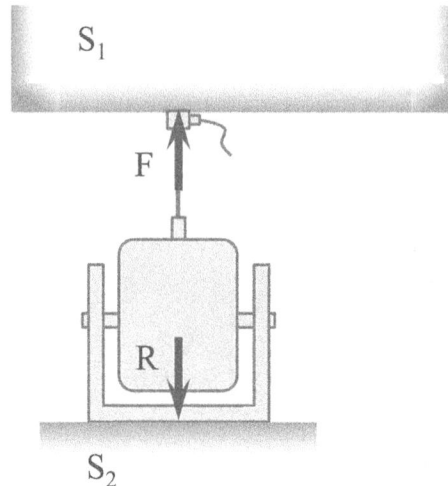

FIGURE 10.4   Forces generated by the shaker.

The alignment of the stinger and shaker with the attachment point on the structure is crucial. This alignment essentially means that with the shaker placed in the desired position and orientation, the shaker's axis of vibration, hypothetically extended in space, passes through the point of attachment. There should not be any misalignment; otherwise, the stinger exerts forces and moments along other DOFs, distorting the FRFs.

Careful consideration should also be given to support the shaker. It can be placed directly on a rigid floor, supported on a rigid platform, or suspended from a rigid ceiling. This decision is influenced by the point of excitation on the structure. Note that if the support is not rigid enough, its displacement influences the displacement and force transmitted to the structure. The reaction forces from the shaker should also not disturb the structure under test. Figure 10.4 shows a shaker supported on the surface S_2 and exciting the structure S_1. The shaker applies the force F on S_1 generating a reaction force R on S_2. For the correct measurement of FRF, the reaction force R on S_2 should not be transmitted to S_1 in any way. There should not be any compliant path between S_1 and S_2. Otherwise, the measured FRF will be invalid as there would be additional response due to the transmitted force, and the structure would also be subjected to forces at more than one DOFs, violating the definition of the FRF.

10.6   SHAKER-STRUCTURE INTERACTION MODEL

As the shaker is attached to the structure, we effectively have a coupled system consisting of the structure and the exciter. Thus, the force generated by the exciter excites the coupled system and not the structure alone. Since we want to identify the dynamic characteristics of the structure and not of the coupled system, we must study the influence of the shaker properties on the dynamics of the coupled system. It would help us figure out the measurements needed to identify the dynamic characteristic of the structure and the effects of the shaker properties on those measurements. In this section, we present a lumped parameter electromechanical model of the coupled system, as shown in Figure 10.5, and determine the FRFs of the system. The structure and the shaker are represented by SDOF models. The power amplifier can be operated either in the voltage mode or in the current mode, and let G_v and G_i be the gains of the power amplifier in these two modes, respectively.

We use the following symbols:

- m_s, k_s and c_s. represent the mass, stiffness, and viscous damping coefficient of the structure, respectively.
- m_e, k_e and c_e. represent the mass, stiffness, and viscous damping coefficient of the exciter, respectively. (m_e is the mass of the armature, as that is the moving element, and k_e is the stiffness of the suspension).
- R and L represent the resistance and inductance of the armature coil.
- F represents the force generated by the coil.
- V is the voltage output of the signal generator.
- E and i represent the voltage and current outputs of the power amplifier, respectively
- e_b represents the back EMF generated by the coil

10.6.1   Voltage Mode Operation

In the voltage mode of operation, the amplifier output voltage E is held constant, while the output current may vary with the excitation frequency. The output voltage is proportional to the input voltage V through gain G_v. Hence,

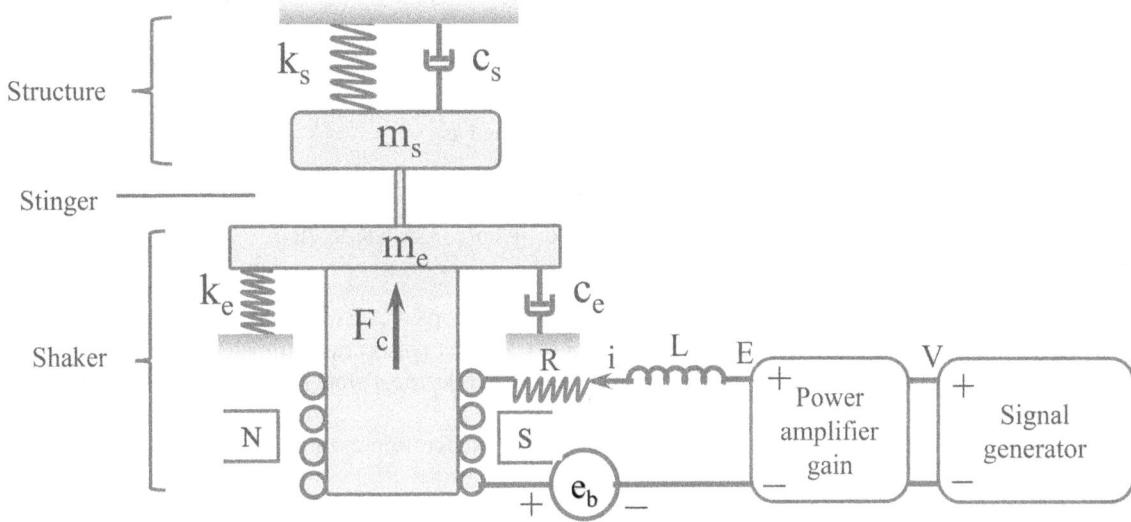

FIGURE 10.5 A lumped parameter electro-mechanical model of the coupled shaker-structure system.

$$E(t) = G_v \, V(t) \tag{10.3}$$

According to Faraday's law, whenever the magnetic flux linked to a coil changes with time, a voltage called the back EMF (e_b) is induced in the coil. As per Lenz's law, the back EMF acts opposite to the voltage that induces the current in the coil. In the present case, we have a coil moving in the magnetic field, due to which the flux linked to the coil changes with time. As a result, a back EMF is induced in the armature circuit, as shown in the figure. The back EMF is proportional to the velocity of the coil and is given by

$$e_b(t) = B \, L \, \dot{x} = K_b \, \dot{x} \tag{10.4}$$

where K_b is the back EMF constant of the shaker. Applying Kirchhoff's voltage law to the armature circuit, we get

$$E(t) = Ri + L\frac{di}{dt} + e_b \tag{10.5}$$

The force on the coil is given by Eq. (10.2) and can be written as

$$F(t) = K_f \, i \tag{10.6}$$

where K_f is the force constant of the shaker. Since the stinger is assumed rigid, the masses m_e and m_s vibrate as a single unit, and the equation of motion can be written as

$$(m_s + m_e)\ddot{x} + (c_s + c_e)\dot{x} + (k_s + k_e)x = F(t) \tag{10.7}$$

The time constant of the armature circuit is generally much smaller than that of the mechanical system, and hence the term $L\frac{di}{dt}$ in Eq. (10.5) can be dropped for a simplified model,

$$E(t) \cong Ri + e_b \tag{10.8}$$

Substituting i from Eq. (10.8) into Eq. (10.6) and also making use of Eq. (10.3), we get

$$F(t) = \frac{K_f}{R} \left(G_v \, V(t) - K_b \, \dot{x} \right) \tag{10.9}$$

Substituting $F(t)$ from Eq. (10.9) into Eq. (10.7) gives after simplification

$$(m_s + m_e)\ddot{x} + \left(c_s + c_e + \frac{K_f K_b}{R} \right)\dot{x} + (k_s + k_e)x = \frac{K_f}{R} G_v \, V(t) \tag{10.10}$$

Thus, Eq. (10.10) is the governing equation between the input voltage and the output response of the structure under the voltage mode operation of the power amplifier. We note that the term $\dfrac{K_f K_b}{R}$, called electromagnetic damping, due to the back EMF, also contributes to the damping of the coupled system.

For a harmonic input voltage, the FRF between V and x is obtained as

$$\frac{x(\omega)}{V(\omega)} = \frac{K_f G_v/R}{(k_s+k_e)-(m_s+m_e)\omega^2 + j\omega(c_s+c_e+K_f K_b/R)} \tag{10.11}$$

The peaks in $x(\omega)/V(\omega)$ correspond to the natural frequencies of the coupled electro-mechanical shaker-structure system. Therefore, using the FRFs based on the measured voltage input would lead to inaccurate estimates of the structure's dynamic characteristics, as these FRFs are affected by the shaker's mechanical and electrical properties. The same is true with the FRFs based on the measured current.

Let us also determine the forces shared by the structure and the exciter. Representing these by F_s and F_e respectively, we can write

$$F = F_s + F_e \tag{10.12}$$

Further, the motion of the structure under the action of F_s is governed by

$$m_s\ddot{x} + c_s\dot{x} + k_s x = F_s(t) \tag{10.13}$$

The corresponding FRF is

$$\frac{x(\omega)}{F_s(\omega)} = \frac{1}{k_s - m_s\omega^2 + j\omega c_s} \tag{10.14}$$

The peaks in $x(\omega)/F_s(\omega)$ correspond to the natural frequencies of the structure alone. Therefore, the FRFs based on the force acting on the structure should be used to identify the structure's dynamic characteristics. These FRFs can be found by measuring the force acting on the structure.

The FRF between V and F_s is obtained by dividing Eq. (10.11) by Eq. (10.14)

$$\frac{F_s(\omega)}{V(\omega)} = \frac{(K_f G_v/R)(k_s - m_s\omega^2 + j\omega c_s)}{(k_s+k_e)-(m_s+m_e)\omega^2 + j\omega(c_s+c_e+K_f K_b/R)} \tag{10.15}$$

The peaks in $F_s(\omega)/V(\omega)$ correspond to the natural frequencies of the coupled electro-mechanical shaker-structure system. But, the valleys in this FRF correspond to the structure's natural frequencies, which means that the force acting on the structure becomes very small at the structure's natural frequencies, a phenomenon referred to as **'force dropout'**. It happens because, at its natural frequencies, the structure can vibrate naturally and hence offers a very little impedance, due to which the force it shares drops. The force dropout leads to a poor signal-to-noise ratio at the structure's natural frequencies and may adversely affect the FRF estimate at these frequencies.

Proceeding similarly, the FRF between V and F_e is

$$\frac{F_e(\omega)}{V(\omega)} = \frac{(K_f G_v/R)(k_e - m_e\omega^2 + j\omega c_e)}{(k_s+k_e)-(m_s+m_e)\omega^2 + j\omega(c_s+c_e+K_f K_b/R)} \tag{10.16}$$

10.6.2 Current Mode Operation

In the current mode of operation, the amplifier output current i is held constant, while the output voltage may vary with the excitation frequency. The output current is proportional to the input voltage V through gain G_i. Hence,

$$i(t) = G_i V(t) \tag{10.17}$$

The equations corresponding to the armature circuit (Eq. 10.8), back EMF (Eq. 10.4), and force (Eq. 10.6) remain the same. Substituting i from Eq. (10.17) into Eq. (10.6), we get

$$F(t) = K_f G_i \ V(t) \tag{10.18}$$

Substituting $F(t)$ from Eq. (10.18) into Eq. (10.7) gives

$$(m_s + m_e)\ddot{x} + (c_s + c_e)\dot{x} + (k_s + k_e)x = K_f G_i \ V(t) \tag{10.19}$$

Thus, Eq. (10.19) is the governing equation between the input voltage and the output response of the structure under the current mode of operation of the power amplifier. We note that the back EMF doesn't contribute to the damping since, in this mode, the amplifier compensates the back EMF by adjusting the output voltage to maintain the constant current.

For a harmonic input voltage, the FRF between V and x is obtained as

$$\frac{x(\omega)}{V(\omega)} = \frac{K_f G_i}{(k_s + k_e) - (m_s + m_e)\omega^2 + j\omega(c_s + c_e)} \tag{10.20}$$

The other FRFs are

$$\frac{x(\omega)}{F_s(\omega)} = \frac{1}{k_s - m_s\omega^2 + j\omega c_s} \tag{10.21}$$

$$\frac{F_s(\omega)}{V(\omega)} = \frac{K_f G_i (k_s - m_s\omega^2 + j\omega c_s)}{(k_s + k_e) - (m_s + m_e)\omega^2 + j\omega(c_s + c_e)} \tag{10.22}$$

and,

$$\frac{F_e(\omega)}{V(\omega)} = \frac{K_f G_i (k_e - m_e\omega^2 + j\omega c_e)}{(k_s + k_e) - (m_s + m_e)\omega^2 + j\omega(c_s + c_e)} \tag{10.23}$$

Compared to the voltage mode, the current mode doesn't add electromagnetic damping; consequently, the force dropout is slightly less.

10.7   DYNAMIC EFFECTS OF SHAKER-STRUCTURE INTERACTION

In the previous section, we looked at a lumped parameter model of the shaker-structure system. Using this model, we now study the dynamic effects of the shaker-structure interaction in this section. Table 10.1 gives the properties of the structure and shaker used in the simulation.

The undamped natural frequencies of the structure and the combined structure-exciter system are 128.9 and 123 Hz, respectively.

We first use the voltage mode of operation. Let the signal generator voltage is 5 V and $G_v = 1.0$ V/V. Figure 10.6 shows the magnitudes of the FRFs $x(\omega)/V(\omega)$ and $x(\omega)/F_s(\omega)$. Note that the frequency of the peak in $x(\omega)/F_s(\omega)$ correctly represents the natural frequency of the structure (128.9 Hz), but the peak in the FRF $x(\omega)/V(\omega)$ corresponds to the natural frequency of the combined structure-exciter system (123 Hz).

TABLE 10.1

Properties of the Structure and Shaker Used in the Simulation

$m_s = 2.5$ kg	$m_e = 0.25$ kg
$k_s = 1{,}640{,}000$ N/m	$k_e = 4{,}100$ N/m
$c_s = 50$ N−sec/m	$c_e = 14$ N−sec/m
$K_b = 6.5$ V−sec/rad	$R = 4\ \Omega$
$K_f = 6.5$ N/a	

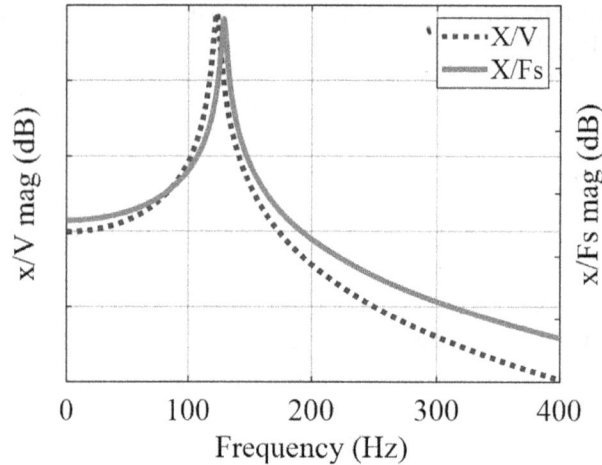

FIGURE 10.6   Overlay of FRFs $x(\omega)/V(\omega)$ and $x(\omega)/F_s(\omega)$.

Figure 10.7 shows the magnitudes of $F_s(\omega)/V(\omega)$ and $x(\omega)/F_s(\omega)$. As discussed in the previous section, we observe from the plot of $F_s(\omega)/V(\omega)$ that the force on the structure undergoes 'force dropout', i.e., drops to a low value at the structure's natural frequency, as discussed in the previous section.

Let us compare $x(\omega)/V(\omega)$ in voltage and current modes. Since these FRFs are associated with different gains, we compare the following scaled FRFs. For the voltage mode, we write the following from Eq. (10.11):

$$\frac{x(\omega)}{K_f G_v V(\omega)/R} = \frac{1}{(k_s + k_e) - (m_s + m_e)\omega^2 + j\omega(c_s + c_e + K_f K_b/R)} \tag{10.24}$$

Similarly, for the current mode of operation, we write the following from Eq. (10.20):

$$\frac{x(\omega)}{K_f G_i V(\omega)} = \frac{1}{(k_s + k_e) - (m_s + m_e)\omega^2 + j\omega(c_s + c_e)} \tag{10.25}$$

Figure 10.8 shows an overlay of the FRFs in Eqs. (10.24) and (10.25). In the voltage mode, the peak is less sharp due to the contribution of electromagnetic damping.

Similarly, we compare the following scaled versions of the forces acting on the structure for the voltage and current modes obtained from Eqs. (10.15) and (10.22), respectively:

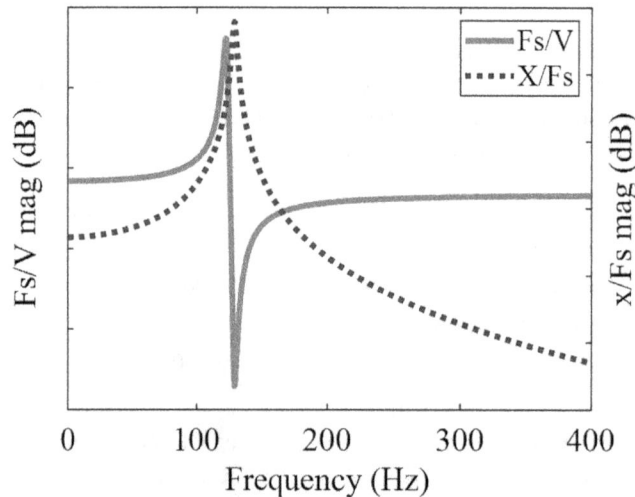

FIGURE 10.7   FRFs $F_s(\omega)/V(\alpha$ and $x(\omega)/F_s(\omega)$.

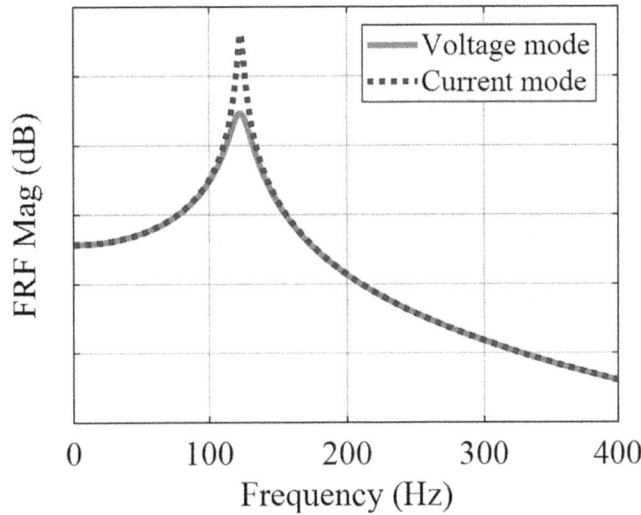

FIGURE 10.8 Overlay of the FRFs for voltage and current modes.

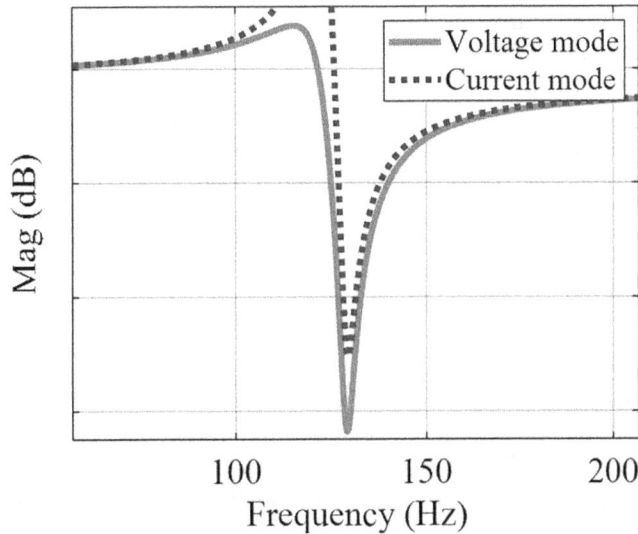

FIGURE 10.9 Force dropout in voltage and current modes.

$$\frac{F_s(\omega)}{K_f G_v V(\omega)/R} = \frac{\left(k_s - m_s\omega^2 + j\omega c_s\right)}{\left(k_s + k_e\right) - \left(m_s + m_e\right)\omega^2 + j\omega\left(c_s + c_e + K_f K_b/R\right)} \quad (10.26)$$

$$\frac{F_s(\omega)}{K_f G_i V(\omega)} = \frac{\left(k_s - m_s\omega^2 + j\omega c_s\right)}{\left(k_s + k_e\right) - \left(m_s + m_e\right)\omega^2 + j\omega\left(c_s + c_e\right)} \quad (10.27)$$

Figure 10.9 shows the FRFs in Eqs. (10.26) and (10.27), indicating that in the current mode of operation, the force dropout is slightly less than that in the voltage mode.

10.8 SHAKER SPECIFICATIONS

The specifications relevant to selecting a shaker are:

- Force rating: It is the peak excitation force the exciter can apply. The force required depends on the mass of the structure to be excited. Though a large force of excitation is not required for modal testing, it should be large enough to provide a good signal-to-noise ratio for both the force and response signals.

- Stroke length: It is the peak-to-peak displacement of the armature of the exciter. A higher value is necessary for excitation in the low-frequency range.
- Frequency range: It is the frequency range in which the shaker's frequency response is linear. The suspension mode of the shaker table determines the lower limit, while the coil resonance mode (in which the coil moves out of phase with the table) determines the upper limit of the frequency range. The shaker's frequency range should cover the frequency range of FRF measurement.
- Moving mass: It is the mass of the armature/moving parts of the exciter. It should be small compared to the structure's mass for smaller force dropouts at the structure's resonances.
- Cooling: Forced cooling may be necessary if the exciter is to be used over extended periods.
- Stinger attachment: A suitable arrangement for the attachment of a stinger must be there.
- Power amplifier: A matching power amplifier is required to drive the exciter.

10.9   FORCE TRANSDUCER

A piezoelectric force transducer is often used to measure the force acting on the structure under test. Figure 10.10 shows a schematic arrangement of a piezoelectric force transducer. It consists of a pair of piezoelectric discs sandwiched between two metallic pieces. Tapped holes are provided in the top part and base for attaching the transducer to the structure and shaker. This arrangement ensures that the force transmitted by the shaker to the structure passes through the piezoelectric discs. As a result, the deformation of the discs and generated charge are proportional to the force transmitted. The measurement of the charge or the corresponding voltage after signal conditioning provides an estimate of the force transmitted. The arrangement of a piezoelectric force transducer is different from that of a piezoelectric accelerometer in which there is a seismic mass, and the deformation of the piezoelectric crystal is proportional to the relative displacement between the seismic mass and the base. The piezoelectric force transducers, like accelerometers, are available as charge or IEPE types.

It is noted that the measured force consists of the force transmitted to the structure and the force required to accelerate the mass of the transducer part between the structure and the piezoelectric discs. Keeping the mass of this part small is an essential consideration in the design of the piezoelectric force transducers so that the measured force is closer to the force that excites the structure.

10.10   MASS CANCELLATION

It is observed that some part of the force transmitted to the structure is used to accelerate the accelerometer mass, which should be accounted for to obtain a correct estimate of FRF. Therefore, the mass loading by the accelerometer affects not only the measured response but also the measured force.

The FRF estimate can be corrected by mass cancellation, in which the effect of the accelerometer mass and mass of the transducer part between the structure and the piezoelectric discs on the measured FRF is nullified. Let the force transducer and accelerometer be mounted at locations j and k, respectively. Let f_j^m and f_j^s represent the measured force and the force exciting the structure at location j, respectively, and let $\ddot{x}_k$ represent the acceleration at location k. Let m_c represent the correction mass for this pair of locations. Therefore, we can write

$$f_j^m(t) = f_j^s(t) + m_c\,\ddot{x}_k(t) \qquad (10.28)$$

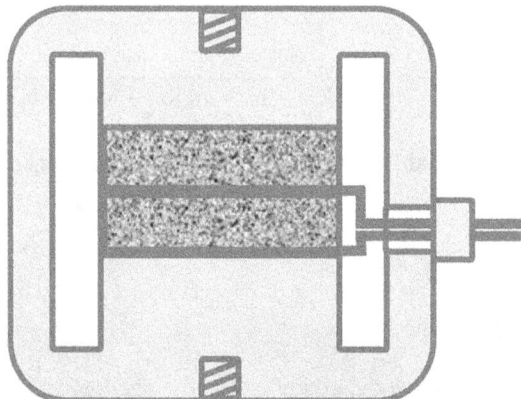

FIGURE 10.10   Piezoelectric force transducer.

Taking the Fourier transform, we get

$$F_j^m(\omega) = F_j^s(\omega) + m_c \, \ddot{X}_k(\omega) \tag{10.29}$$

Dividing by $F_j^s(\omega)$ and noting that the desired estimate of the inertance based on the force exciting the structure is $A_{kj}^s(\omega) = \ddot{X}_k(\omega)/F_j^s(\omega)$, we get

$$F_j^m(\omega)/F_j^s(\omega) = 1 + m_c \, A_{kj}^s(\omega) \tag{10.30}$$

The FRF estimate based on the measured force can be written as, $A_{kj}^m(\omega) = \ddot{X}_k(\omega)/F_j^m(\omega)$. Substituting for $F_j^m(\omega)$ from Eq. (10.30) and simplifying, we get

$$A_{kj}^s(\omega) = \frac{A_{kj}^m(\omega)}{1 - m_c \, A_{kj}^m(\omega)} \tag{10.31}$$

Eq. (10.31) gives the corrected estimate of the inertance based on the force exciting the structure. 'mc' is taken as the sum of the mass of the transducer part between the structure and the piezoelectric discs and a fraction of the accelerometer mass. For point FRFs, the complete mass of the accelerometer is effective, while for the transfer FRFs, only a portion of the accelerometer mass is effective, and its estimate depends on the relative locations of the accelerometer and force transducer. Thus, improving the FRF estimate through mass cancellation is more complex for the transfer FRFs.

10.11 IMPEDANCE HEAD

Measurement of the point FRF requires the response measured precisely at the same point where the force is applied. For beam and plate-like structures, it may sometimes be possible to accomplish this by mounting the response and force transducers on the opposite faces. But when this is not possible, an impedance head can be used. As shown in Figure 10.11, an impedance head combines a force transducer and an accelerometer in a single housing. It allows simultaneous measurement of force and acceleration at the same point. The impedance head is attached to the structure at the desired location through the driving surface. This puts the force transducer side of the impedance head closer to the structure, which ensures that the measured force closely represents the force exciting the structure. The impedance head is attached to the shaker at the other end through a stinger. The impedance head's design ensures that the stiffness of the connection between the accelerometer and the force transducer is extremely high so that the measured acceleration accurately represents the acceleration at the driving point and the phase difference between the acceleration and force signals is negligible.

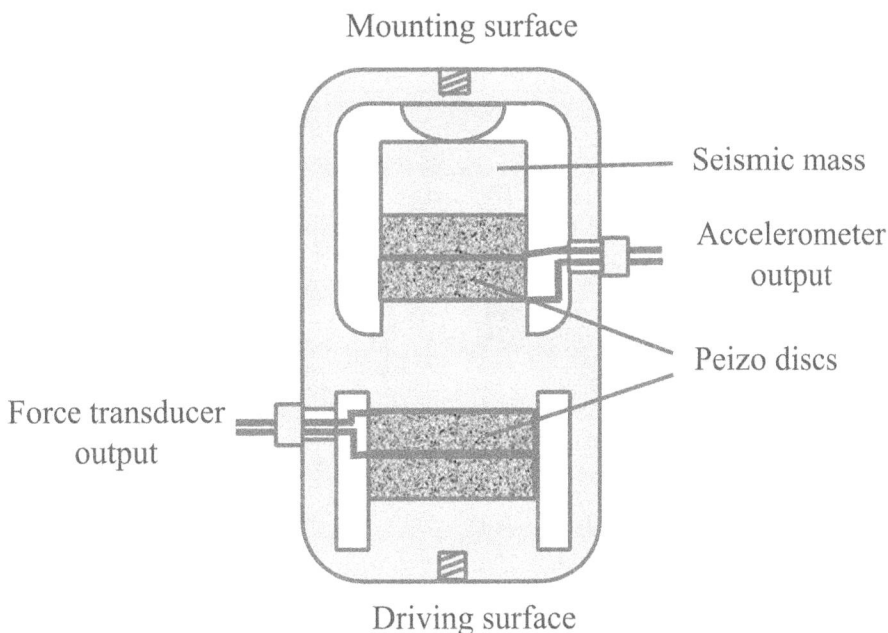

FIGURE 10.11 Impedance head.

10.12 FRF MEASUREMENT USING SHAKER TESTING WITH RANDOM EXCITATION

In this section, we go through the steps involved in shaker testing for measuring FRFs. We present these steps by simulating the test on the cantilever structure used in Chapter 9 for impact testing. A random excitation force is used to measure the FRFs in this section, and other excitation signals are discussed in the next sections.

In Chapter 9, we have discussed the choice of the frequency range of measurement, measurement points, and time/frequency domain parameters. We use the same measurement parameters as in hammer testing. To recall, we used $f_{max} = 400$ Hz and $N_t = 4,096$. Table 10.2 gives details of the time and frequency domain parameters based on these choices.

With shaker excitation, we perform a roving accelerometer test as discussed in Section 9.14 and measure a column of the FRF matrix. We also discussed the choice of the response transducer location in a roving hammer test. Similar considerations apply to choosing the location for the excitation force in a roving accelerometer test. We assume the exciter to be attached at test point 5.

The exciter used for shaker testing has the properties given in Table 10.1, and the power amplifier is assumed to be in voltage mode with a gain of $G_v = 1.0$ V/V. The signal generator is used to generate a random voltage signal of 5 Vrms amplitude. A small uncorrelated random noise is simulated in the input and output.

The random signals are described in the frequency domain by auto and cross PSDs, and the auto and cross-spectrums are their discrete counterparts, as discussed in Chapter 8. Based on the auto and cross-spectrums of the random input and output signals, the FRF can be obtained using the H_1 or H_2 estimate.

Figure 10.12 shows the time histories of the excitation force on the structure and the resulting acceleration at test point 5. From these signals, the auto spectrum of the force ($G_{ff}(\omega)$) and cross-spectrum ($G_{fa}(\omega)$) between the force and acceleration are computed, as shown in Figure 10.13. Figure 10.14 shows the inertance (H_1) and the coherence function. Since the random signal doesn't satisfy the periodicity assumption, leakage is seen in the auto and cross-spectrums and the FRF. It is indicated by poor coherence, especially at and near the resonance frequencies.

To reduce the leakage, we use a window function, a Hanning window, as shown in Figure 10.15. The windowed force and the acceleration signal at test point 5 are shown in Figure 10.16, which are then used to compute the force auto spectrum and the cross-spectrum, as shown in Figure 10.17. The inertance (H_1) and coherence function obtained are shown in Figure 10.18. We see that with the window function, there is a significant improvement in the FRF and the coherence function.

Figure 10.19a show the auto-spectrum of the voltage input to the exciter; it is uniform over the measurement frequency range (0–400 Hz). However, the auto spectrum of the excitation force, as shown in part (b) of the figure with the inertance

TABLE 10.2

Measurement Parameters for Shaker Testing of the Cantilever Structure

f_{max}	N	f_s	f_c	T	Δt	Δf	N_f
400 Hz	4,096	1,024 Hz	512 Hz	4 seconds	0.0009765 seconds	0.25 Hz	1,600

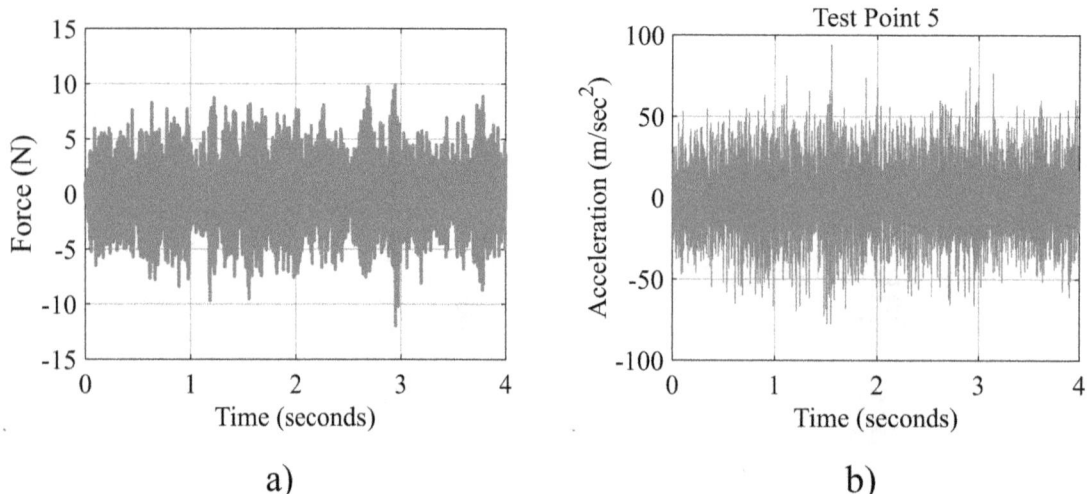

FIGURE 10.12 Time histories of (a) force on the structure and (b) acceleration at test point 5.

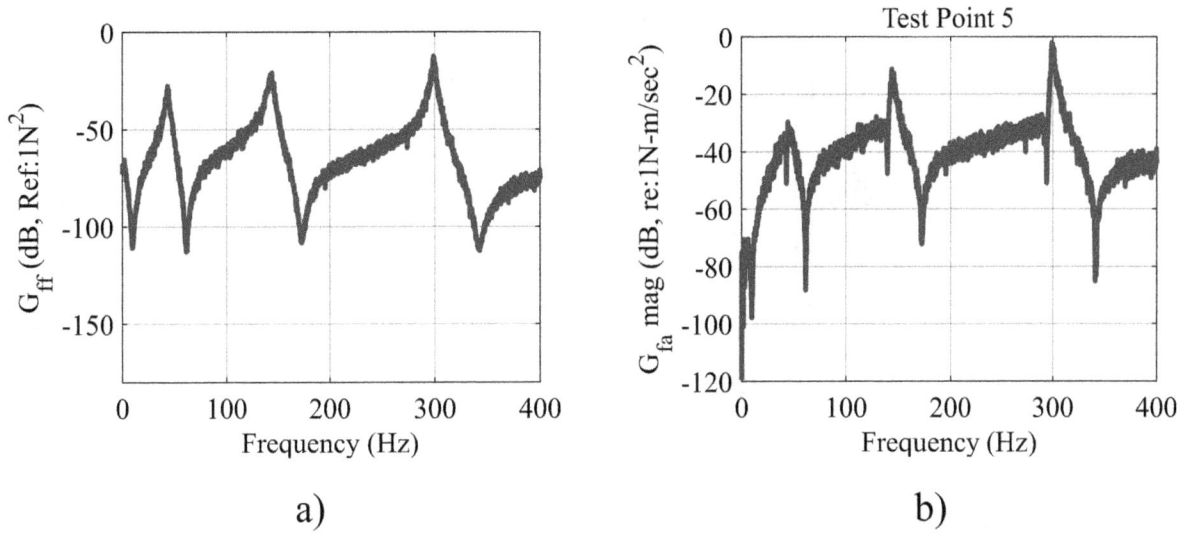

FIGURE 10.13 (a) Auto spectrum of the force ($G_{ff}(\alpha)$ and (b) cross-spectrum ($G_{fa}(\alpha)$ between the force and acceleration.

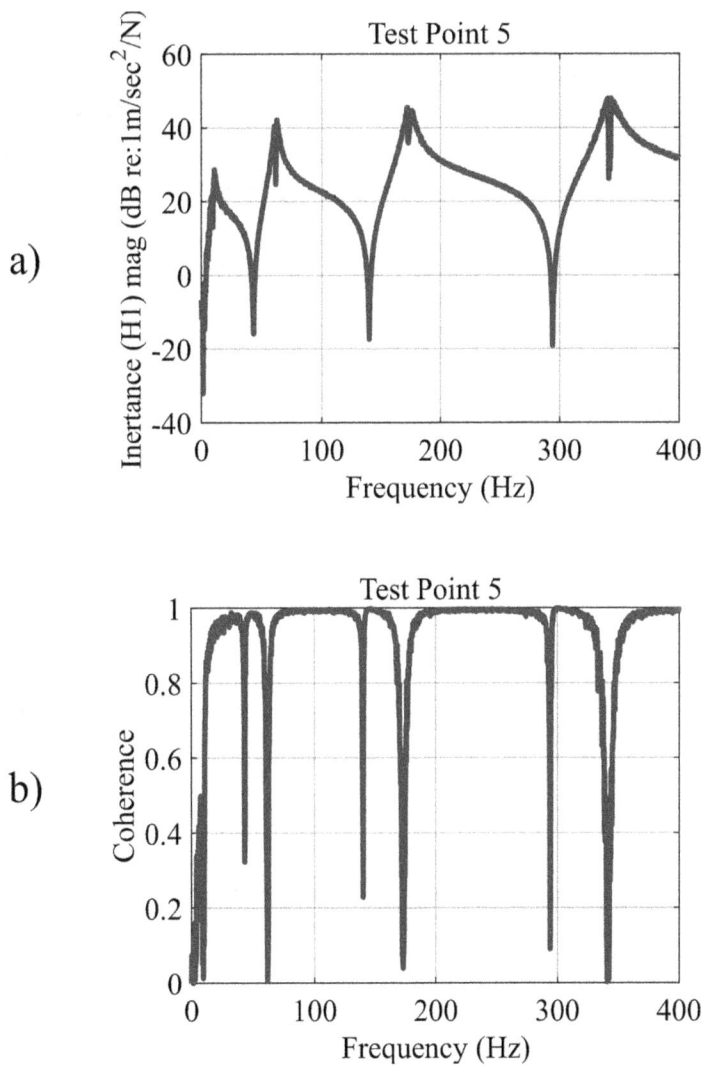

FIGURE 10.14 Random excitation- (a) inertance (H1) and (b) coherence at test point 5.

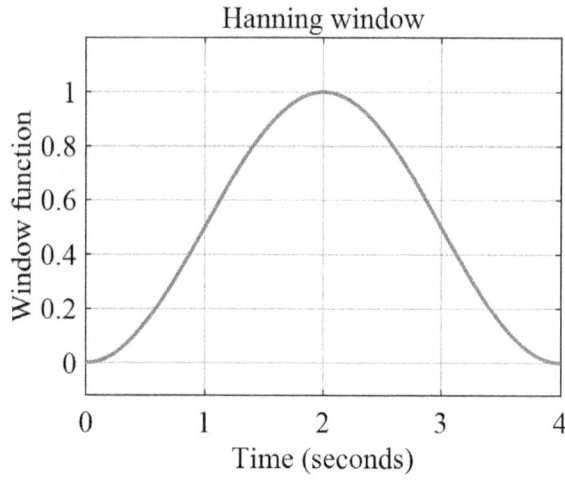

FIGURE 10.15   Hanning window.

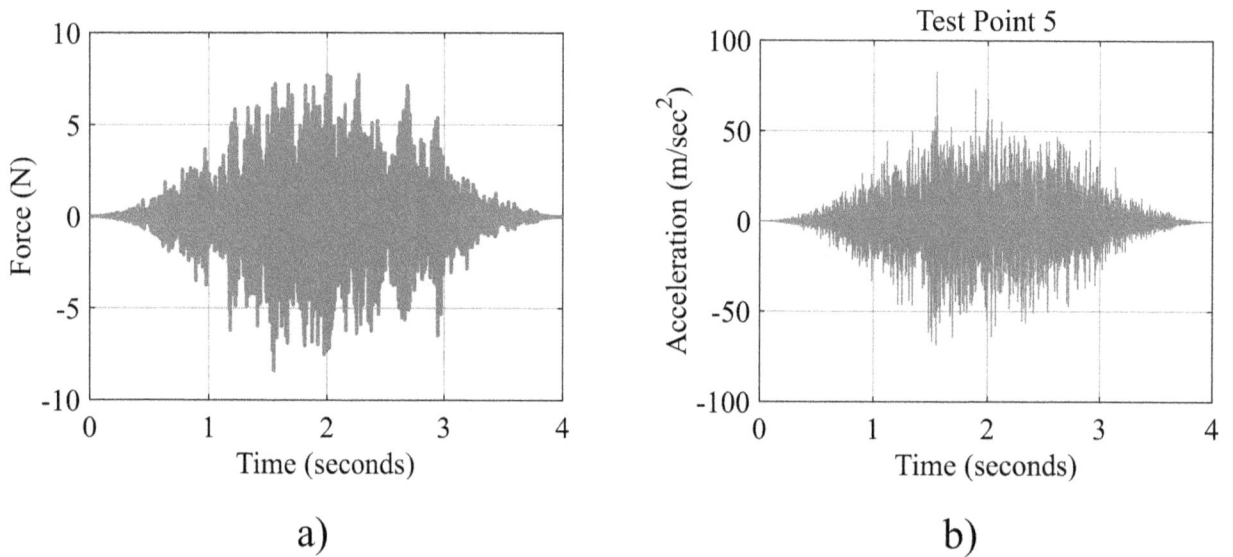

a)

b)

FIGURE 10.16   After windowing: (a) force and (b) acceleration signals at test point 5.

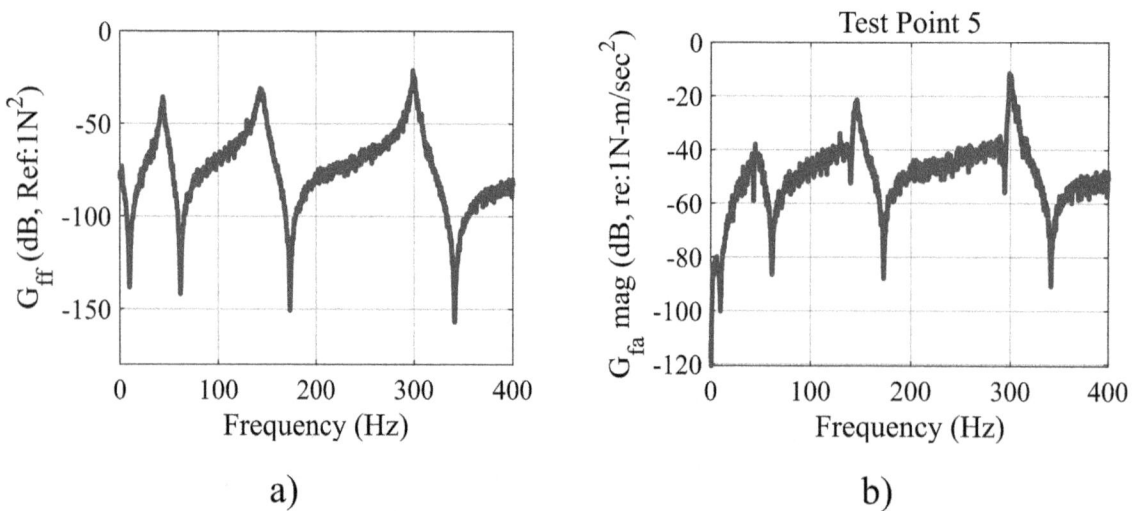

a)

b)

FIGURE 10.17   After windowing- (a) force auto spectrum $(G_{ff}(\omega))$ and (b) cross-spectrum $(G_{fa}(\omega))$ between the force and acceleration.

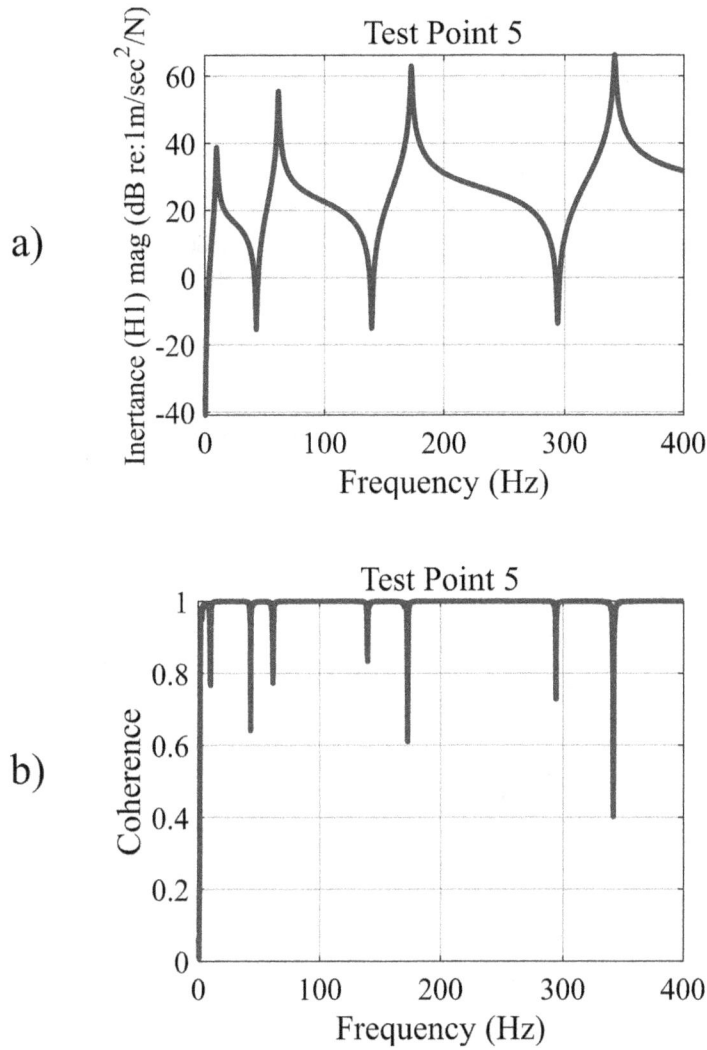

FIGURE 10.18 After windowing- (a) inertance (H1) and (b) coherence at test point 5.

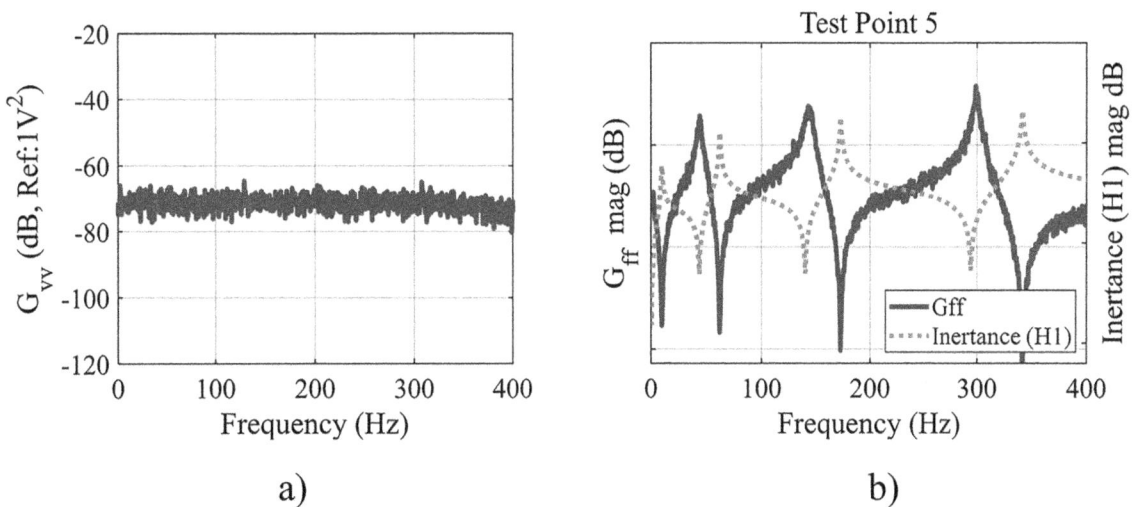

FIGURE 10.19 (a) Auto spectrum of the voltage input to the exciter and (b) auto spectrum of the force transmitted to the structure overlaid on the inertance.

overlaid onto it, shows force dropouts at the resonance frequencies of the structure. The force dropouts lead to a poor signal-to-noise ratio at these frequencies and are reflected in lower coherence values. The signal-to-noise ratio is poor at antiresonance frequencies also due to the smaller response and is responsible for smaller coherence.

A random signal is an aperiodic signal continuing indefinitely; hence, it has the disadvantage that a finite signal length suffers from leakage in the frequency domain and requires a window function to reduce the leakage. Moreover, several averages are often required to get an acceptable estimate of the FRF. Thus, the overall time of shaker testing with random excitation can be high. However, the FRF measurement with random excitation offers one advantage. Since the excitation force is random, the structure is excited over a broad range of force amplitudes, and any nonlinearities in the structure are averaged out. As a result, the estimated FRF corresponds to a linear approximation of the structure.

10.13 FRF MEASUREMENT USING OTHER RANDOM EXCITATION SIGNALS

In the previous section, we looked at the FRF measurement using a random excitation signal. It was noted that a suitable window function is necessary with this excitation signal. While the windows help reduce the leakage, they also distort the force and response time histories and affect the FRF estimates. In this section, we look at some variants of the random excitation signals developed to address the drawbacks of pure random excitation.

10.13.1 Pseudo-Random Excitation

Let a signal is generated by periodically repeating a block of length T of a random signal. The generated signal has the property that a record of length T or an integer multiple of T drawn from the signal is free of leakage and hence does not need any window function. Such a signal is called a pseudo-random signal. It is so named as it is not truly a random signal. The pseudo-random signal is periodic and hence has a discrete spectrum.

For shaker testing, the pseudo-random signal is synthesized by prescribing the frequency domain parameters of the signal. The main steps for synthesizing the signal are as follows.

a. Let T be the record length of measurement.
b. Let f_{max} be the frequency range of measurement.
c. Calculate the frequency resolution: $\Delta f = 1/T$.
d. Calculate the number of discrete frequency components: $N_f = f_{max}/\Delta f$.
e. Calculate the frequencies of the discrete frequency components: $f_k = k\Delta f$, $k = 0, 1, 2,...,N_f - 1$.
f. Let the amplitudes (A_k) of all the frequency components be equal. Let $A_k = V_0$, $k = 1, 2,...,N_f - 1$. Let the dc component amplitude is zero (i.e., $A_0 = 0$).
g. Generate random phase angles (θ_k) of the frequency components such that $-\pi \leq \theta_k \leq \pi$, $k = 0, 1, 2,...,N_f - 1$.
h. Synthesize the pseudo-random voltage signal by the inverse FFT:

$$V(t) = \sum_{k=0}^{N_f-1} A_k \cos(2\pi f_k t - \theta_k) \qquad (10.32)$$

The synthesized pseudo-random voltage signal is fed to the power amplifier to drive the exciter. For a linear system, the period and frequency content of the steady-state system response is the same as that of the pseudo-random excitation signal. Since the steady-state response needs to be measured for FRF estimation, the force and response should be measured after the transient response has died out. The same synthesized signal is used for excitation over all the averages.

We now simulate FRF measurement on the cantilever structure using pseudo-random excitation. Figure 10.20 shows the synthesized pseudo-random voltage signal. Figure 10.21 shows the time histories of the force on the structure and the resulting acceleration at test point 5. Since the excitation is periodic, the steady-state response after the decay of the transient part will also be periodic. In practice, we can simply wait some time to let the transient decay and then measure the steady-state response. However, to show the transient response's existence and detect the onset of the steady-state in the simulations presented in this and the next few sections, the last block of data is subtracted from all the blocks. The result of subtraction is shown in Figure 10.22. We observe no transient part beyond the first block of data. Thus, the steady-state can be said to have reached after the first block, and the steady-state measurements can be taken.

Since no leakage is expected, no window is used. A small number of averages are used to reduce the effect of random measurement noise on the FRFs. Figure 10.23 shows the estimated inertance and coherence function, which are found to be good.

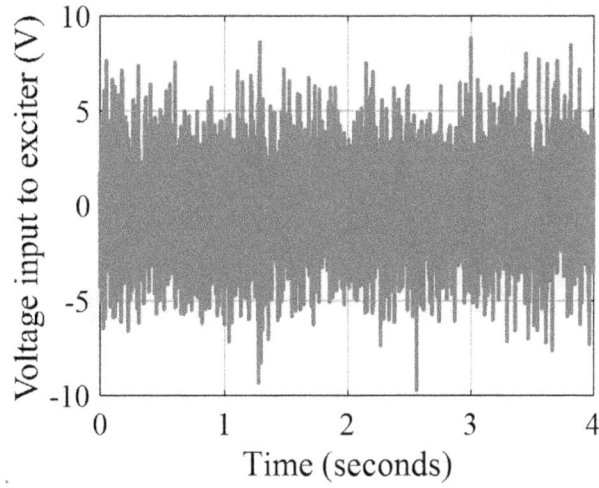

FIGURE 10.20   Synthesized pseudo-random voltage signal.

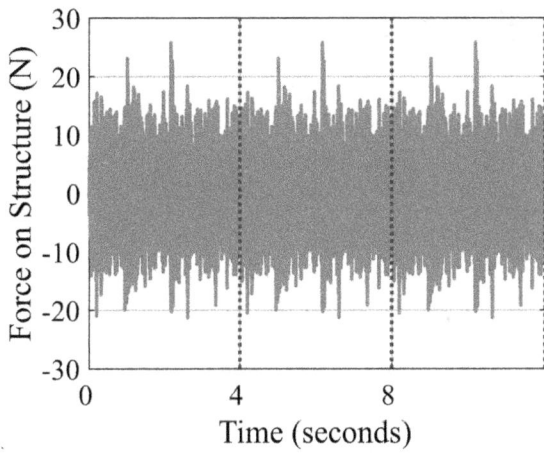

a)

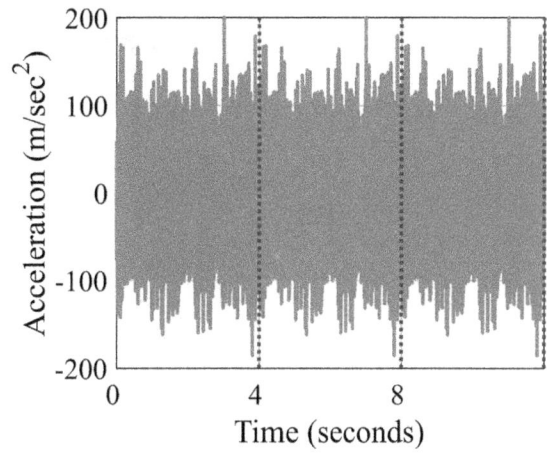

b)

FIGURE 10.21   With pseudo-random excitation, time histories of (a) force on the structure and (b) acceleration at test point 5.

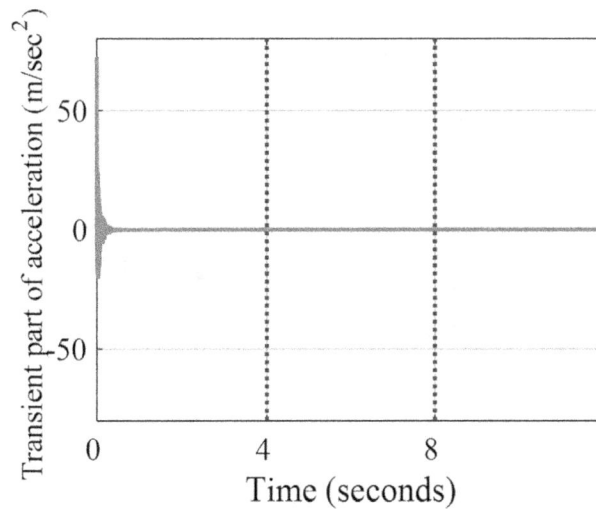

FIGURE 10.22   With pseudo-random excitation, the time history after the last block of data is subtracted from all the blocks.

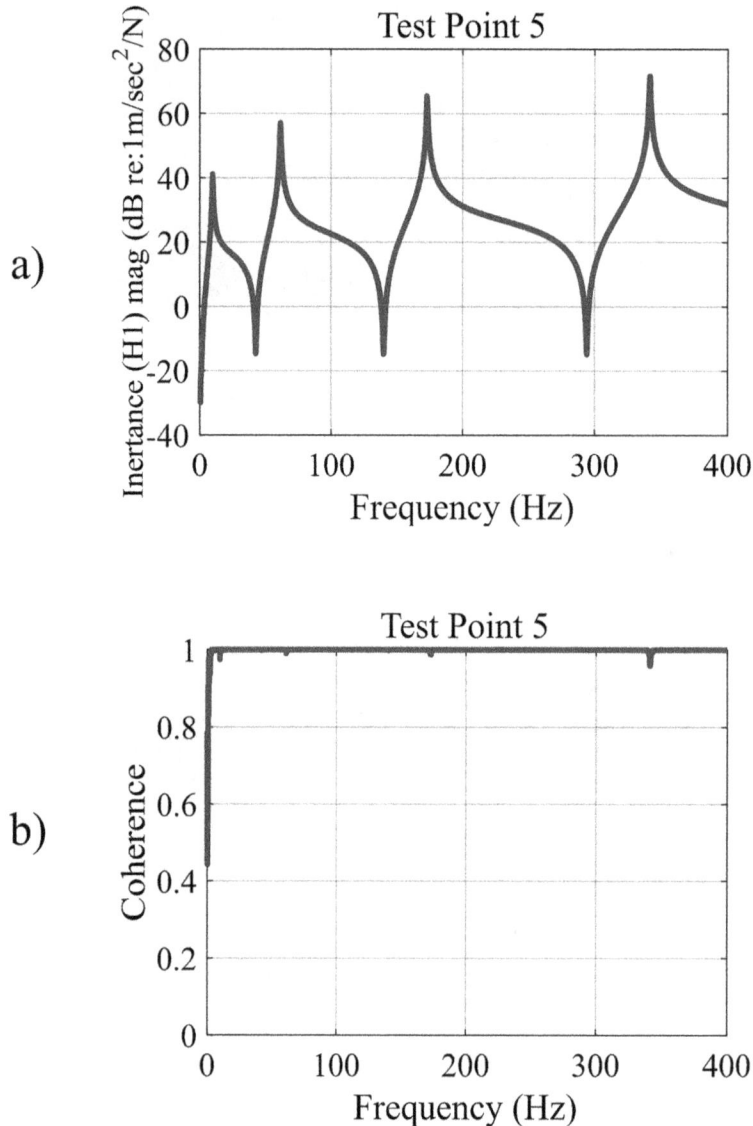

FIGURE 10.23 With pseudo-random excitation- (a) inertance (H1) and (b) coherence at test point 5.

One disadvantage of the pseudo-random signal is that if there are any nonlinearities, it doesn't yield an FRF corresponding to a linear approximation of the structure. It is because the same synthesized signal is used over various averages, exciting the structure with the same force amplitude and preventing the possible averaging of the nonlinear response in the absence of a broader range of excitation.

10.13.2 PERIODIC RANDOM EXCITATION

This excitation is obtained by making a slight modification to the pseudo-random excitation. Instead of using the same synthesized signal for each average, the periodic random excitation uses the synthesized signals generated using a random amplitude over the averages. The periodic random signal, therefore, excites the structure with a broader range of force amplitudes. As a result, any nonlinearities are averaged out, yielding an FRF estimate corresponding to a linear approximation of the structure. Moreover, since the signal is periodic for every average, the leakage is avoided, and no window is needed. The signal is synthesized as follows.

- We follow steps (a) to (h) in Section 10.13.1 to synthesize the periodic random signal with one change that for each average, the amplitude V_0 in step (f) is chosen randomly. For each average, we wait for the steady state to reach and measure the force and response for length T and estimate the FRF.

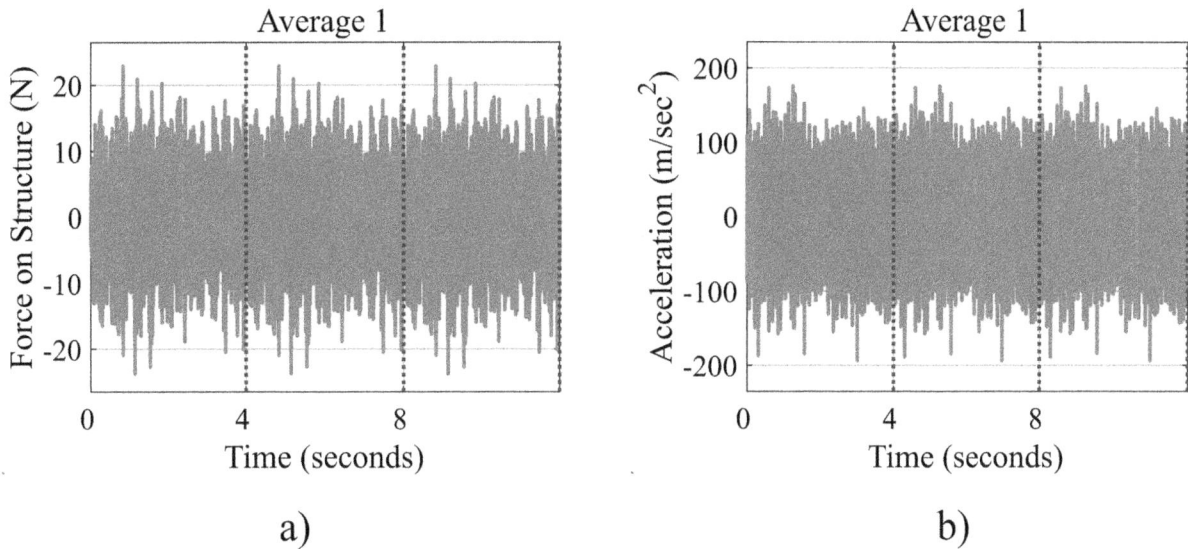

FIGURE 10.24 With periodic random excitation (for the first average), time histories of (a) force on the structure and (b) acceleration at test point 5.

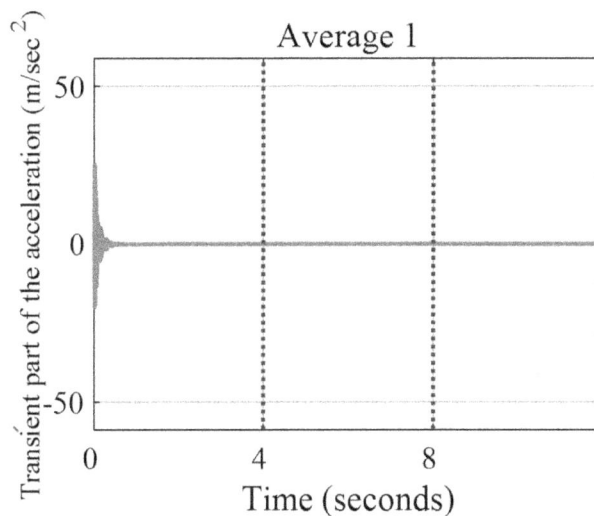

FIGURE 10.25 With periodic random excitation (for the first average), acceleration time history after the last block of data is subtracted from all the blocks.

For the first average, Figure 10.24 shows the time histories of the force on the structure and the resulting acceleration at test point 5. Three blocks of data, each 4 seconds long, are shown. Similar to the previous section, for detecting the onset of the steady state, the last block is subtracted from all the blocks. The signal after subtraction is shown in Figure 10.25. We observe no transient part beyond the first block of data, so the steady-state measurements are taken after that.

For the second average, Figure 10.26 shows the time histories of the force on the structure and resulting acceleration at test point 5. Note that the force amplitude is different than in the first average. Figure 10.27 shows the transient part of the response, and we measure the force and response after the transient has decayed. This way, the measurements for all the averages are taken, and the FRF is estimated.

10.13.3 Burst Random Excitation

The pseudo and periodic random excitations are periodic; hence, the energy is distributed over discrete frequency components and is a disadvantage compared to pure random excitation. Burst random excitation consists of a random signal making up only a portion of the record length T, with no input in the remaining part of the record length.

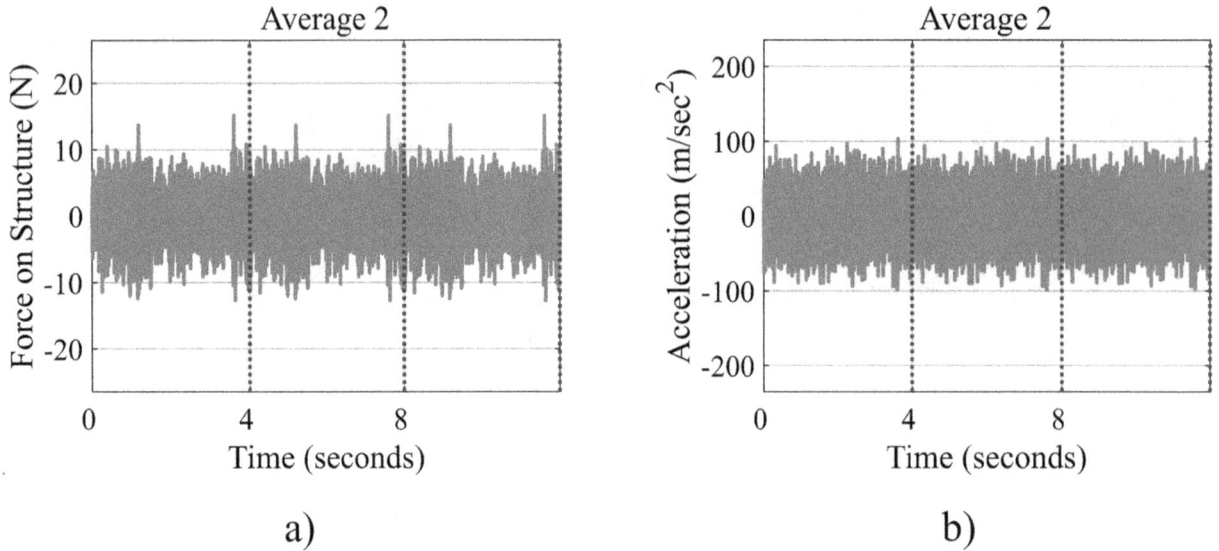

FIGURE 10.26 With periodic random excitation (for the second average), time histories of (a) force on the structure and (b) acceleration at test point 5.

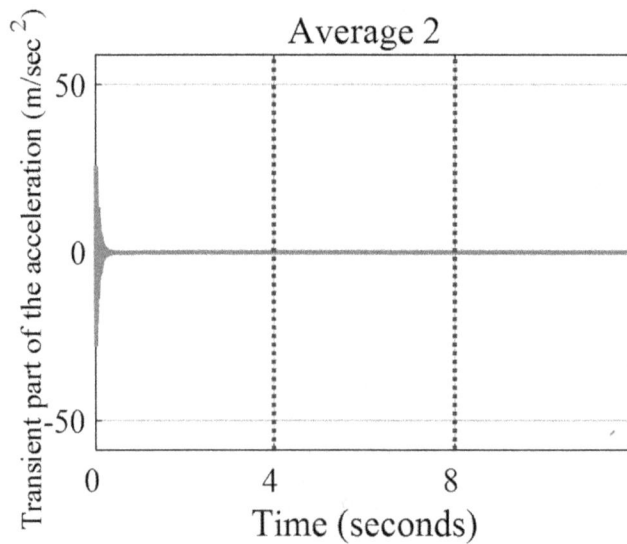

FIGURE 10.27 With periodic random excitation (for the second average), acceleration time history after the last block of data is subtracted from all the blocks.

When this excitation is applied, the response starts decaying once the input becomes zero. If the response decays completely within the record length T, then there is no leakage, and no window is required. This observation forms the basic idea of the burst random excitation. Since the non-zero part of the excitation signal is random, the energy is distributed continuously over the measurement frequency range. For each average, the burst part of the excitation signal is different, and hence this excitation also allows obtaining a linearized estimate of the FRF in case the structure has nonlinearities. Thus, the burst random excitation combines the advantages of random, pseudo-random, and periodic random excitations. But the burst random excitation, like the random excitation, requires a more significant number of averages to get a good quality FRF.

We now simulate the FRF measurement on the cantilever structure using the burst random excitation. Figure 10.28 shows the time histories of the force on the structure and the resulting acceleration at test point 5. We see that once the force becomes zero, the response decays completely within the record length of the measurement. Figure 10.29 shows the inertance and coherence function obtained with just 2 averages, while Figure 10.30 shows the results with 20 averages. With an increase in the number of averages, there is a significant improvement in the coherence and FRF.

a)   b)

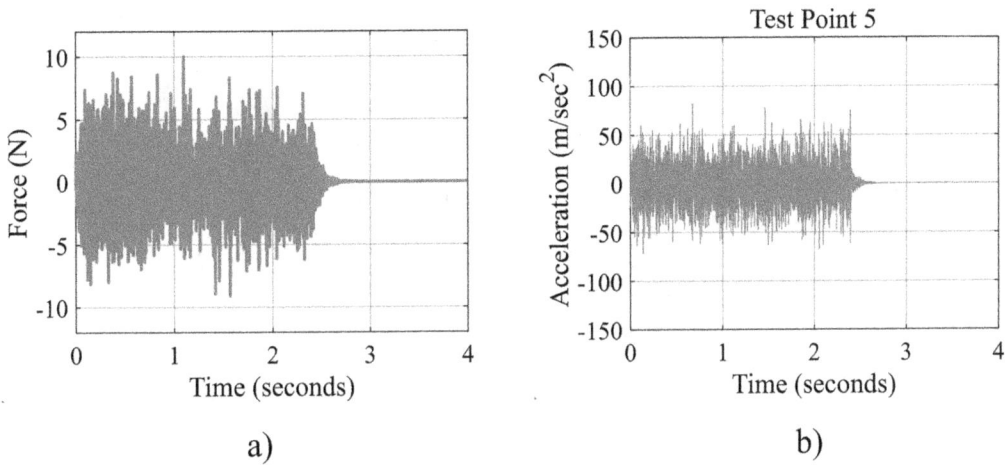

FIGURE 10.28   With burst random excitation, time histories of (a) force on the structure and (b) acceleration at test point 5.

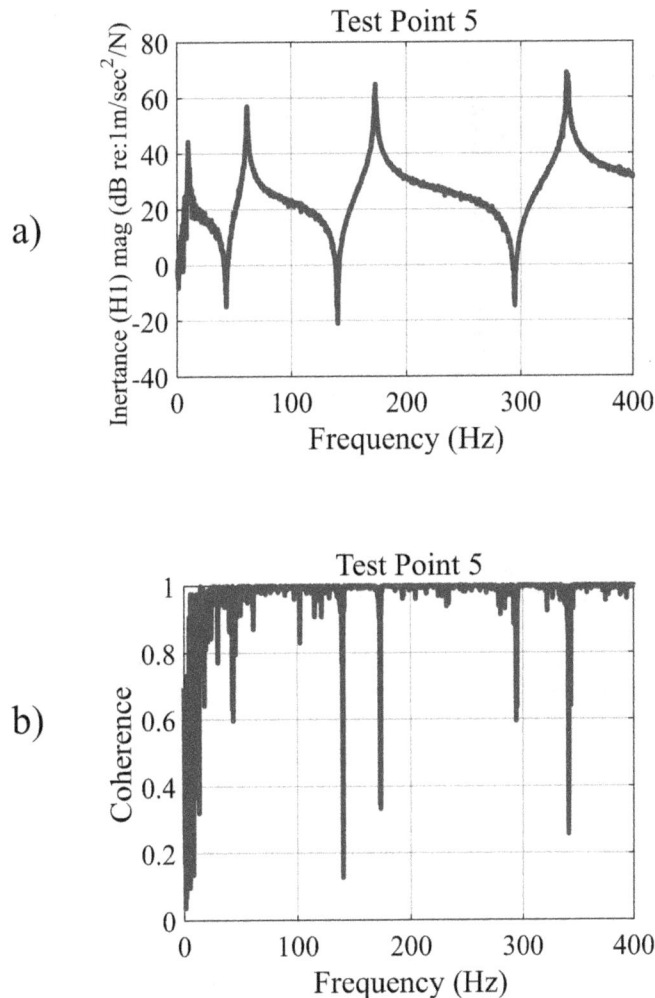

a)

b)

FIGURE 10.29   Burst random excitation (with 2 averages): (a) inertance (H1) and (b) coherence at test point 5.

10.14   FRF MEASUREMENT USING SINE EXCITATION

Unlike random excitation signals, a sine signal is a deterministic excitation signal. The sine sweep signal is a sinusoidal signal whose instantaneous frequency is swept continuously from one end to the other end of the analysis frequency range. We first briefly discuss the stepped sine and the traditional slow sweep technique and then look at the more commonly used sweep technique, the sine chirp technique.

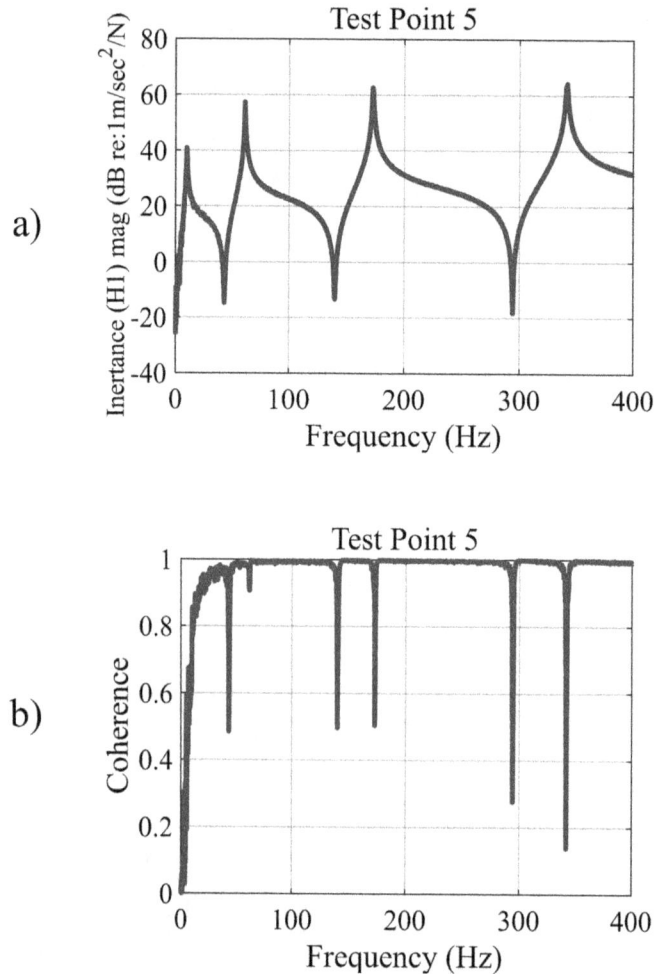

FIGURE 10.30   Burst random excitation (with 20 averages): (a) inertance (H1) and (b) coherence at test point 5.

10.14.1   STEPPED SINE

In stepped sine excitation, a sinusoidal force of the desired frequency is applied. Once the transient response has decayed, the excitation force and the steady-state response are measured to find the magnitude and phase of the FRF at the current frequency. The frequency is stepped to the next frequency, and the measurement and FRF estimation is carried out. This process is repeated to cover the whole analysis frequency range. The total time required for a stepped sine test is generally quite large. But the advantage is that the record length can be chosen equal to an integer number of signal periods since the signal frequency is precisely known. As a result, the leakage can be eliminated, and the windowing can be avoided.

The sine excitation signals are advantageous if nonlinearities are to be detected or quantified since, with this excitation, it is possible to control precisely the amplitude and frequency of the input to the system. But this excitation signal is not helpful if a linearized model of the system in the presence of any nonlinearities is required.

10.14.2   SLOW SINE SWEEP

In this technique, the frequency of the sine excitation signal is swept continuously but slowly through the frequency range. The sweep rate is kept slow enough to ensure that the steady-state condition can be approximately reached, and the measurements can be taken to obtain the frequency response at the current frequency. The overall time for the test is generally very large, and the accuracy of the measured FRF depends on the sweep rate.

10.14.3   SINE CHIRP

The sine chirp technique is a rapid sine sweep technique. Linear and exponential sweeps are two commonly used methods of sweeping the frequency range. In the linear sweep, the instantaneous frequency of the signal is changed linearly with time; therefore, the sweep rate, which is the time rate of change of frequency, is constant. For the exponential sweep, the instantaneous frequency of the signal is varied exponentially with time.

We now derive the equation for the sine signal with the linear sweep. Consider a sinusoidal signal,

$$V(t) = V_0 \sin \theta(t) \tag{10.33}$$

where $\theta(t)$ is the instantaneous phase of the sine signal and V_0 is the signal amplitude. The instantaneous frequency of the signal is given by

$$\omega(t) = \frac{d\theta(t)}{dt} \tag{10.34}$$

Let the instantaneous frequency is swept from ω_1 to ω_2 over the time 0 to T. For the linear sweep, the instantaneous frequency $\omega(t)$ can be written as

$$\omega(t) = a\,t + b \tag{10.35}$$

where a and b are constants and are obtained as

$$a = \frac{\omega_2 - \omega_1}{T} \tag{10.36}$$

$$b = \omega_1 \tag{10.37}$$

Substituting Eq. (10.35) into (10.34) and integrating with time gives,

$$\int (a\,t + b)\,dt = \int d\theta(t) \tag{10.38}$$

$$\theta(t) = \frac{1}{2} a\,t^2 + b\,t + c \tag{10.39}$$

where c is a constant of integration. If at $t = 0$ the phase of the sine signal is β, then from the above equation, we get $c = \beta$. Substituting Eq. (10.39) into (10.33), we get the sine sweep signal

$$V(t) = V_0 \sin\left(\frac{1}{2} a\,t^2 + b\,t + \beta \right) \tag{10.40}$$

No specific frequencies can be associated with the whole length of the sine sweep signal. It is interesting to look at the frequency domain composition of the sweep signal. Using Eq. (10.40), we generate a sine sweep signal of amplitude 2 V with the frequency swept from 100 to 300 Hz in 1 second (and with $\beta = 0$). Figure 10.31 shows the time history of the signal (on a log time axis) and the magnitude of its DFT. The DFT magnitude is predominant over the swept frequency band, and therefore the sweep signal can be used to excite the structure in this frequency band. The sine chirp signal can be used for FRF measurement either as a periodic or as a transient excitation signal.

a. Sine chirp as a periodic excitation signal

We generate a periodic signal with the sine chirp signal of length T as its period. The generated periodic signal is used to excite the structure under test. We should measure the response and excitation force under the steady-state of the vibration when the transient response has decayed completely. One advantage of this signal is that under the steady state, the force and response are periodic, satisfying the periodicity requirement of the Fourier series. Hence, the FRF is free of leakage, and no window is required. Averaging can be used to reduce the effect of any random measurement noise on the FRF.

We now perform the FRF measurement on our cantilever structure using the periodic sine chirp signal. We use the same measurement parameters as in Table 10.2. Since the measurement frequency range is 0–400 Hz and $T = 4$ seconds, we generate a sine sweep signal with its frequency swept from 0 to 400 Hz in 4 seconds (and with $\beta = 0$). The amplitude of the signal is 5 V. The signal is shown in Figure 10.32, on a linear time axis.

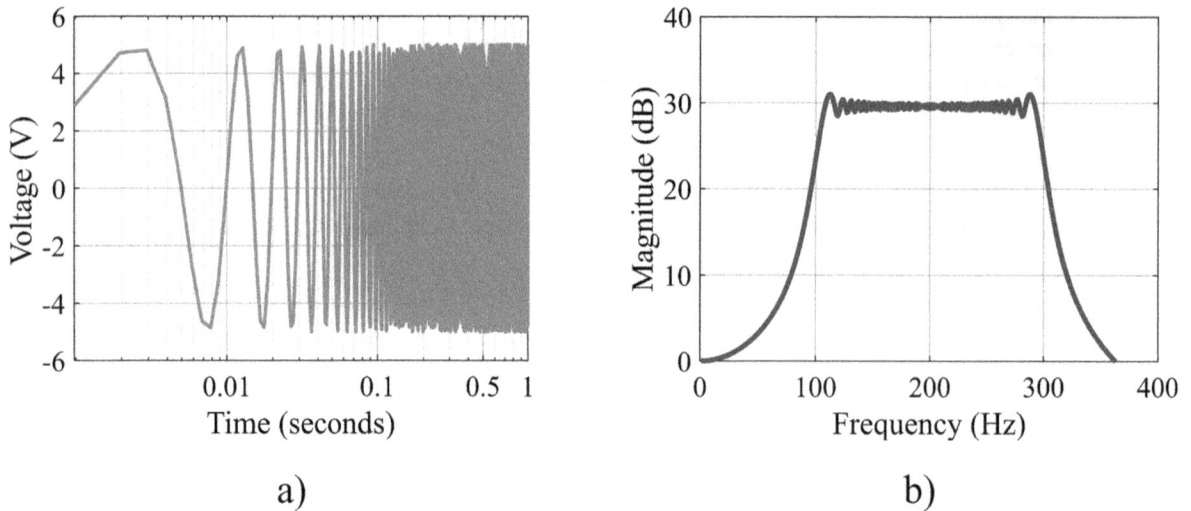

a)   b)

FIGURE 10.31   Sweep signal (a) time history and (b) DFT magnitude.

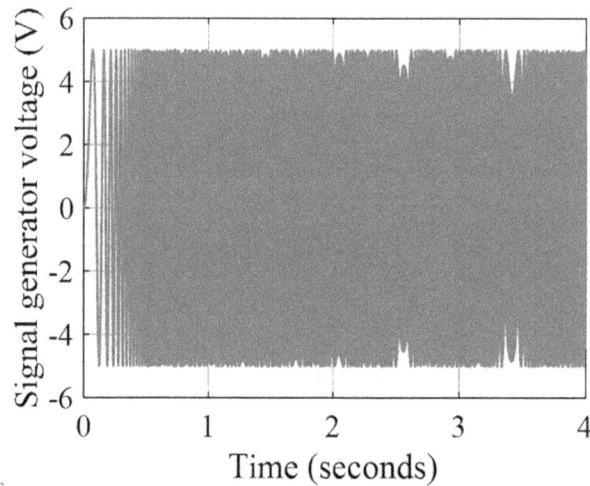

FIGURE 10.32   Sine chirp signal.

Figure 10.33 shows the time histories of the force on the structure and the resulting acceleration at test point 5. It shows three blocks of data, 4 seconds each. We observe that though the voltage amplitude is constant, the amplitude of the force applied fluctuates a lot in a block. It is a result of the interaction between the structure and the shaker; as the excitation frequency sweeps through the structure's resonances, the force transmitted to the structure drops (due to the phenomenon of force dropout we studied earlier in the chapter). The acceleration also shows many fluctuations resulting from the frequency sweeping through resonances and antiresonances of the structure.

Similar to simulations in the previous sections, to assess the transient part of the response, the last block of data in Figure 10.33b is subtracted from all the blocks in this figure, and the resulting time history is shown in Figure 10.34. It is seen that there is no transient part beyond the first block of data, and hence the steady-state measurements are taken after this block.

Figure 10.35 shows the inertance and coherence function estimated without using any window. A small number of averages were used to reduce the effect of random noise on the FRF. The quality of the FRF is good, as indicated by good coherence. The drop in coherence at resonances is due to force dropout at these frequencies.

b. Sine chirp as a transient excitation signal

The sine chirp signal is used as a transient excitation signal similar to the burst random excitation. The sine chirp excitation signal is generated over the initial part τ of the record length T, and the signal is zero over the length $T - \tau$. When such excitation is applied, the response decays once the input becomes zero at $t = \tau$. The time T is selected such that the response decays completely within this time. This ensures that there is no leakage and

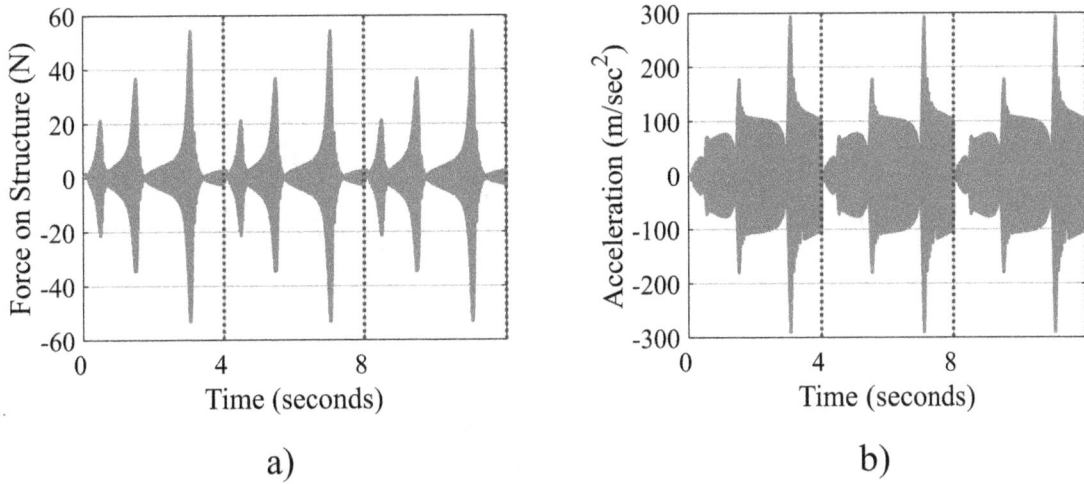

FIGURE 10.33   With periodic sine chirp excitation, time histories of (a) force on the structure and (b) acceleration at test point 5.

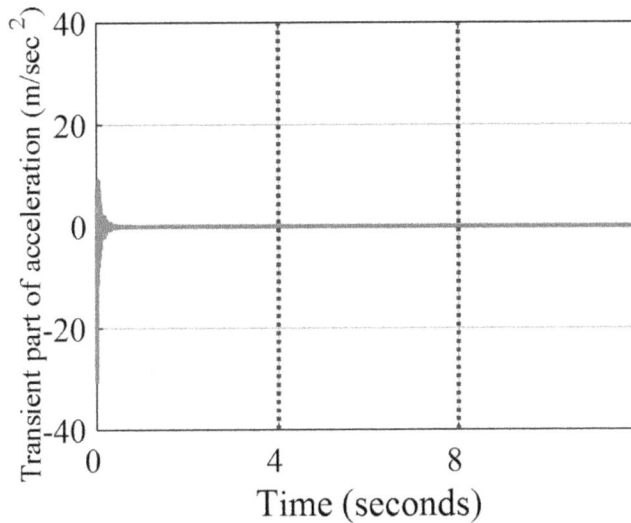

FIGURE 10.34   With periodic sine chirp excitation, acceleration time history after the last block of data is subtracted from all the blocks.

no window is required. The force and responses over length T are measured, and the FRF is estimated. Averaging is done to reduce the effect of any random measurement noise on the FRF.

For the FRF measurement on our cantilever structure, we generate a transient sine sweep signal with its frequency swept from 0 to 400 Hz in 2.4 seconds (and with $\beta = 0$), while the signal is zero over the remaining 1.6 seconds of the record length. The signal amplitude is 5 V. The generated signal is shown in Figure 10.36.

Figure 10.37 shows the time histories of the force on the structure and the resulting acceleration at test point 5. Like in the case of periodic sine chirp excitation, the amplitude of the force signal fluctuates a lot. The acceleration decays completely within the record length, as desired. Figure 10.38 shows the inertance and coherence function estimated without using any window and are found to be good.

10.15   FRF MEASUREMENT USING MULTI-REFERENCE TESTING

In the previous sections and Chapter 9, we looked at FRF measurement that can be classified as either Single-input-Single-output (SISO), or Single-input-Multi-output (SIMO) measurement. Applying force using a hammer or a shaker at one DOF and measuring the resulting response at a DOF forms a SISO measurement. If the resulting response is measured simultaneously at multiple DOFs, it forms a SIMO measurement.

But in either of these cases, we get a row or a column of the FRF matrix at the end of the test. These measurements are also referred to as single reference testing.

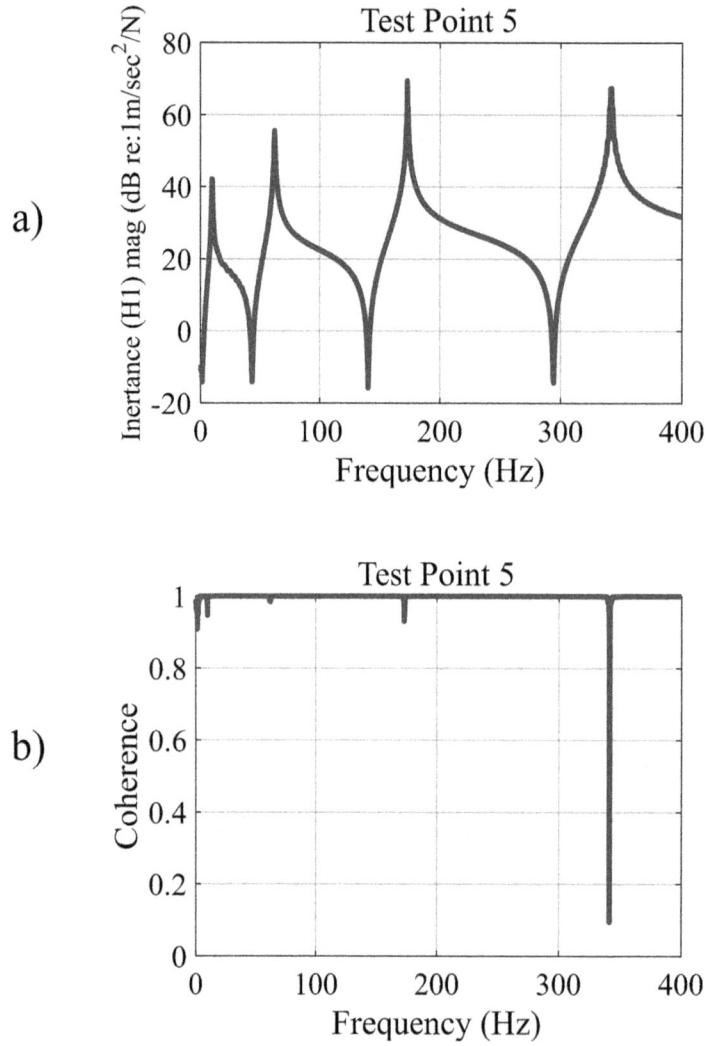

FIGURE 10.35 Periodic sine chirp excitation: (a) inertance (H1) and (b) coherence at test point 5.

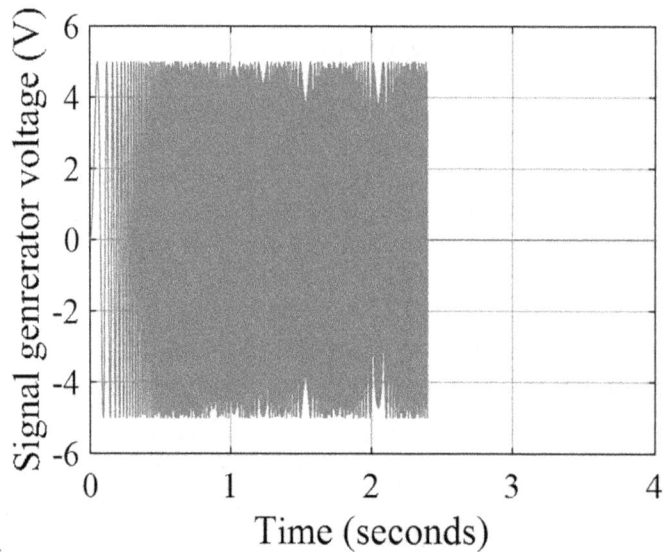

FIGURE 10.36 Transient sine chirp signal.

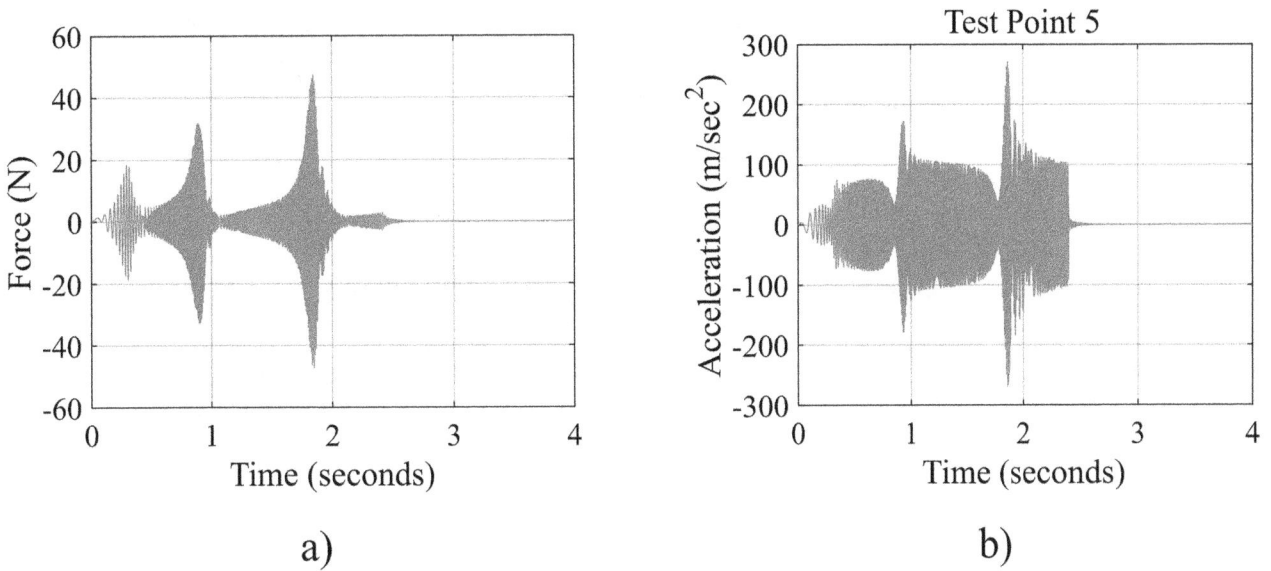

FIGURE 10.37 With transient sine chirp excitation, time histories of (a) force on the structure and (b) acceleration at test point 5.

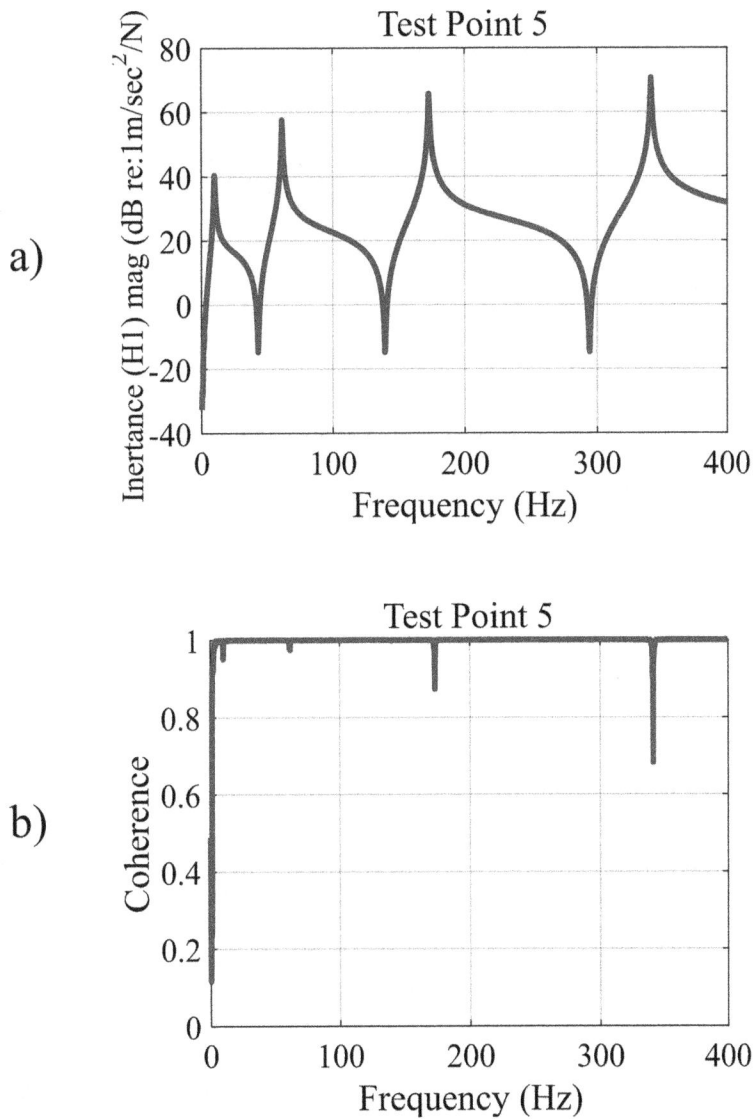

FIGURE 10.38 Transient sine chirp excitation: (a) inertance (H1) and (b) coherence at test point 5.

As discussed in Sections 9.15.3, 9.15.4, and 10.12, if the reference DOF could be chosen appropriately, then one row or one column of the FRF matrix is enough to find the modal parameters of the modes of interest. However, in some situations, one row or column may not be enough to find accurately the modal parameters of all the modes of interest, as discussed later in the section. Multi-reference testing was developed to deal with these situations.

a. Multi-reference testing

Multi-reference testing involves measuring FRFs with more than one reference DOFs. In a Multi-reference hammer test, more than one DOFs are fixed as response coordinates (which act as reference DOFs), and the FRFs are measured with excitation applied by the hammer at all the DOFs. Such a test gives as many rows of the FRF matrix as the number of response coordinates.

In a multi-reference shaker test, shaker excitations are applied at more than one DOFs (which act as reference DOFs), and the FRFs are measured with response measurements at all the DOFs. Such a test gives as many columns of the FRF matrix as the number of excitation coordinates. Multi-reference shaker test is also referred to as multi-input modal test. If the response is measured at one of the DOFs at a time, then it forms a multi-input-single-output (MISO) measurement, and if the responses at multiple DOFs are measured simultaneously, then it forms a multi-input-multi-output (MIMO) measurement and is referred to as MIMO testing.

b. Need for Multi-reference testing

The choice of reference DOF is important in a SISO or SIMO test. Reference DOF is the DOF corresponding to the fixed response transducer location in a roving hammer test and to the fixed shaker location in a roving accelerometer test.

The reference DOF should not be coincident with any of the nodes of the modes of interest. In addition, the modes of interest must be sufficiently visible from the reference location, as otherwise, these may be masked in the FRFs by any noise or other dominant modes. In cases where this cannot be ensured, single reference testing will not be able to identify such modes accurately.

Some structures may have modes in orthogonal directions. Excitation in one of the orthogonal directions would excite the modes having a predominant displacement in that direction, but other modes may not be excited well, and such modes can't be accurately identified. For example, the modes of a simple beam with displacement predominant in the horizontal plane can't be accurately identified by excitation in the vertical plane due to poor coupling between the two directions.

Symmetrical structures may have repeated roots, that is, the existence of more than one mode with the same natural frequency. Multi-reference tests can be used to identify modes having the same natural frequencies.

Another problem with single reference testing could be in the testing of large structures (such as an airplane, naval structures, and machine tools). Excitation at one point may not provide sufficient energy for the vibration of different parts of the structure. As a result, the response at DOFs far away from the excitation DOF may not be enough leading to a poor signal-to-noise ratio and hence inaccurate FRF estimates. To address this, if a higher excitation level is used, it may excite the nonlinearities, adversely affecting the FRF estimates. Here, a multi-reference shaker test with more than one shaker exciting the structure simultaneously can help realize a sufficient vibrational energy over the whole structure and give a more accurate FRF estimate.

c. FRF estimation in multi-reference testing

In multi-reference hammer testing, the response at the fixed response DOFs can be simultaneously measured, giving consistent FRFs corresponding to both the reference DOFs. The FRFs can be estimated in the usual manner as done in a single reference hammer test.

In multi-reference shaker testing, there can be two approaches. In the first approach, the shaker excitation is applied at one of the reference DOFs and the responses at all the DOFs are measured. The FRFs are estimated in the usual manner as done in a single reference shaker test, yielding the column of the FRF matrix corresponding to the reference DOF. This process is repeated by shifting the shaker to other reference DOFs, yielding corresponding columns of the FRF matrix. Such an approach may have to be followed if only one shaker is available. But it may not yield a consistent set of FRFs as the combined shaker-structure system is not identical when the FRF measurements for two different shaker locations are taken.

The second multi-reference shaker testing approach is to simultaneously apply (uncorrelated) excitation at all the reference DOFs. The responses are measured at all the DOFs either sequentially or simultaneously, depending upon the availability of response transducers. This approach gives a consistent set of FRFs and is expected to yield a more accurate estimate of the modal parameters than the first approach. The FRFs with simultaneous multiple excitations can be estimated as follows.

Let there be 'm' reference DOFs where the shaker excitations are applied, and let there be 'n' response DOFs. The force and response measurements might contain some noise. Depending on the presence of noise, either in input or in output or in both, an optimal estimate of the FRF to minimize the effect of noise can be found. Here, we look at an estimate based on the assumption that noise is in the output but not in the input. Representing the noise in the response at the jth DOF by $N_j(\omega)$, and the FRF by $H_{jk}(\omega)$, we can write

$$x_1(\omega) = H_{11}(\omega)\, f_1(\omega) + H_{12}(\omega)\, f_2(\omega) + \ldots + H_{1m}(\omega)\, f_m(\omega) + N_1(\omega)$$

$$x_2(\omega) = H_{21}(\omega)\, f_1(\omega) + H_{22}(\omega)\, f_2(\omega) + \ldots + H_{2m}(\omega)\, f_m(\omega) + N_2(\omega)$$

$$\ldots\ldots\ldots\ldots\ldots\ldots\ldots\ldots\ldots\ldots\ldots\ldots\ldots\ldots\ldots\ldots\ldots\ldots\ldots\ldots\ldots$$

$$x_n(\omega) = H_{n1}(\omega)\, f_1(\omega) + H_{n2}(\omega)\, f_2(\omega) + \ldots + H_{nm}(\omega)\, f_m(\omega) + N_n(\omega) \tag{10.41}$$

Eq. (10.41) can be written in the matrix form as

$$\{x(\omega)\}_{n\times1} = [H(\omega)]_{n\times m}\ \{f(\omega)\}_{m\times1} + \{N(\omega)\}_{n\times1} \tag{10.42}$$

The estimate of the FRFs based on the least square error is

$$\left[\hat{H}(\omega)\right]_{n\times m} = \ [G_{xf}(\omega)]_{n\times m}\ [G_{ff}(\omega)]_{m\times m}^{-1} \tag{10.43}$$

where $[G_{xf}(\omega)]$ is the matrix of cross-spectrums between output responses and force inputs, while $[G_{ff}(\omega)]$ is the matrix of cross-spectrums between force inputs.

It is noted that FRF estimation using Eq. (10.43) needs computation of the inverse of the input cross-spectrum matrix. Due to this, the input signals used to drive the m shakers must be uncorrelated so that the cross-spectrum matrix is not rank-deficient and its inverse can be computed. At frequencies coinciding with the structure's natural frequencies, there is force drop out, and the noise dominates the input, making the FRF estimates inaccurate.

The estimate $\left[\hat{H}(\omega)\right]$ contains m columns of the FRF matrix at n response DOFs. These FRFs can be analyzed using a poly-reference curve-fitting algorithm to find the modal parameters.

REVIEW QUESTIONS

1. What is the working principle of the operation of an electromagnetic exciter?
2. What determines the usable frequency range of operation of an electromagnetic exciter?
3. What is the role of a signal generator and power amplifier in a shaker system?
4. What is a stinger? Can we use a thick rod as a stinger?
5. What is force dropout, and what are its consequences on FRF measurement?
6. For modal testing of a machine tool, can we place the shaker on the machine's table to measure its FRFs?
7. Explain how the accelerometer mass interferes with the correct measurement of the force and response.
8. What is a burst excitation signal? What are its advantages?
9. Why doesn't a pseudo-random signal require windows for FRF measurement?
10. Why is a periodic-random signal better suited to obtaining a linearized approximation of FRFs than a pseudo-random signal?
11. Why is a stepped sine signal preferred for identifying and quantifying nonlinearities in a system?
12. Why FRF measurement with a random excitation signal requires windows?
13. What is Multi-reference testing?
14. What are the situations where Multi-reference testing may be needed?
15. Why the input signals must be uncorrelated in Multi-reference shaker testing?

11 Modal Parameter Estimation Methods

11.1 INTRODUCTION

We studied the measurement of FRFs on a structure through impact and shaker testing in Chapters 9 and 10. The next step is to estimate the modal parameters of the structure from these measured FRFs. Estimating modal parameters from the FRFs is also referred to as **curve-fitting** since the basic philosophy of modal parameter estimation is to fit a known theoretical relationship between the FRF and the modal parameters to the measured data.

The curve-fitting methods are well-advanced and established, and many commercial software packages are available that allow curve-fitting through many alternative approaches. Most methods operate in a black-box mode, which means that once the measured data is given as the input, they yield the modal parameters without requiring much user intervention. However, a basic understanding of the modal parameter estimation from the measured data and the various issues involved is necessary to enable a correct and optimum usage of the available tools and make decisions about the acceptability and validity of the results. This chapter aims to give details of some of the curve-fitting approaches to provide insight into the parameter estimation process and to bring about the steps involved in these methods. The emphasis is on building a fundamental understating of these methods.

This chapter starts by discussing the peak-picking method, the most straightforward approach, to estimate the modal parameters from FRFs. SDOF frequency domain approaches like circle-fit and line-fit are considered next. MDOF curve-fitting with the rational fraction polynomial method is then covered, followed by its extension to global curve-fitting. The concept of residuals is discussed. Curve-fitting in the time domain using the complex exponential method, least-squares complex exponential method, the Ibrahim time-domain (ITD) method, and the Eigensystem realization algorithm (ERA) is presented. Curve-fitting on a cantilever structure using simulated FRFs is carried out to provide insight into the working of various methods.

11.2 CLASSIFICATION OF THE CURVE-FITTING METHODS

The curve-fitting methods can be classified as follows:

- SDOF and MDOF methods: Whether one mode or more than one mode is curve-fitted at a time.
- Frequency domain and time-domain methods: Whether the curve-fitting is done in the frequency domain or in the time domain.
- Local and global methods: Whether the curve-fitting of one FRF (referred to as single input single output method (SISO)) or all the FRFs (referred to as single input multiple output method (SIMO)) is carried out.
- Single reference and multi/poly-reference methods: Whether curve-fitting is done on the FRFs obtained by excitation at one point or multiple points (referred to as multiple input multiple output method (MIMO)).

11.3 ANALYTICAL FORMS OF FRF FOR CURVE-FITTING

Let us briefly review the FRF relationships that can be fit to the measured FRF. The choice of damping model for curve-fitting is necessary, and the structural/hysteretic or viscous damping models can be chosen. Often, the damping in practical systems occurs due to multiple sources. For example, in structural assemblies, while the individual parts may predominantly have structural damping, the joints, and interfaces may have coulomb damping due to friction. The presence of lubrication, hydrodynamic bearings, and drag due to the surrounding air may be more appropriately represented by a viscous damping model. If viscous effects are negligible, the structural damping model may be preferred. But one disadvantage of this model is that the transient time-domain response analysis to arbitrary forces is problematic.

If the structural damping model is used, then the receptance relationship used for curve-fitting is

$$\alpha_{jk}(\omega) = \sum_{r=1}^{N} \frac{\phi_{jr}\,\phi_{kr}}{\omega_r^2 - \omega^2 + i\eta_r\omega_r^2} = \sum_{r=1}^{N} \frac{{}_rA_{jk}}{\omega_r^2 - \omega^2 + i\eta_r\omega_r^2} \tag{11.1}$$

where ${}_rA_{jk} = \phi_{jr}\,\phi_{kr}$ is referred to as the modal constant for the rth mode of FRF $\alpha_{jk}(\omega)$. If the inertance is measured then it can be converted to receptance or mobility as needed.

DOI: 10.1201/9780429454783-11

In Chapter 4, the transfer function of an SDOF system with viscous damping was shown to be

$$H(s) = \frac{x(s)}{F(s)} = \frac{1}{\left(ms^2 + cs + k\right)} \tag{11.2}$$

which can also be written as

$$H(s) = \frac{1/m}{\left(s^2 + 2\xi\omega_n s + \omega_n^2\right)} \tag{11.3}$$

The denominator in Eq. (11.3) is the characteristic polynomial, and its roots are called the **poles** of the system. Since the denominator is of order 2, there are two roots, and the system has two poles. Assuming the system to be underdamped, poles are complex conjugate and are given by

$$s_1, \ s_1^* = -\xi\omega_n \pm i\omega_n\sqrt{1-\xi^2} = -\xi\omega_n \pm i\omega_d \tag{11.4}$$

Representing the RHS of Eq. (11.3) by the sum of partial fractions, we get

$$H(s) = \frac{R}{s - s_1} + \frac{R^*}{s - s_1^*} \tag{11.5}$$

where R and R^* are complex conjugate residues. They are obtained from Eqs. (11.3) and (11.5) as

$$R = \frac{1}{2im\omega_d} \quad \text{and} \quad R^* = -\frac{1}{2im\omega_d} \tag{11.6}$$

Substituting $s = i\omega$ into Eq. (11.5) leads to the **pole-residue** form of the FRF

$$H(i\omega) = \frac{R}{i\omega - s_1} + \frac{R^*}{i\omega - s_1^*} \tag{11.7}$$

We can also substitute $s = i\omega$ into Eq. (11.2), leading to an alternative expression

$$H(i\omega) = \frac{1}{k - m\omega^2 + i\omega c} \tag{11.8}$$

For an MDOF system with viscous damping, the receptance was derived in Chapter 6. Using the notation $H_{jk}(i\omega)$ for the FRF, we can write that equation as

$$H_{jk}(i\omega) = \sum_{r=1}^{N}\left(\frac{_rR_{jk}}{i\omega - s_r} + \frac{_rR_{jk}^*}{i\omega - s_r^*}\right) \tag{11.9}$$

where

$$_rR_{jk} = \phi_{jr}\ \phi_{kr} \quad \text{and} \quad _rR_{jk}^* = \phi_{jr}^*\ \phi_{kr}^* \tag{11.10}$$

Eq. (11.9) is the pole-residue form of the FRF for an MDOF system with viscous damping. s_r and s_r^* are complex conjugate poles, while $_rR_{jk}$ and $_rR_{jk}^*$ are the corresponding residues of the rth mode in H_{jk}. Thus, each modal contribution in an FRF is associated with a pair of complex conjugate poles and residues. If this data for all the modes in all the FRFs is estimated, then the natural frequencies, modal viscous damping factors, and modes shapes can be obtained using Eqs. (11.4) and (11.10).

11.4   PEAK-PICKING METHOD

The peak-picking method is the most straightforward approach to modal parameter estimation. It is a frequency domain method based on the SDOF assumption. Let us understand the basic idea and procedure of this method. The system is assumed to have structural damping.

a. This method assumes that the FRF in a narrow frequency band around a resonance peak is contributed by that mode alone. Thus, the contributions of all modes other than the mode being analyzed are neglected. To analyze the rth mode, we can write Eq. (11.1) as

$$\alpha_{jk}(\omega) = \frac{{}_rA_{jk}}{\omega_r^2 - \omega^2 + i\eta_r\omega_r^2} + \sum_{q=1,\,\neq r}^{N} \frac{{}_qA_{jk}}{\omega_q^2 - \omega^2 + i\eta_q\omega_q^2} \qquad (11.11)$$

The peak-picking method assumes the second term on the RHS to be negligible in the band around the rth mode, and hence

$$\alpha_{jk}(\omega) \approx \frac{{}_rA_{jk}}{\omega_r^2 - \omega^2 + i\eta_r\omega_r^2} \qquad (11.12)$$

At $\omega = \omega_r$, the denominator magnitude is minimum, and hence the FRF magnitude should be maximum. Therefore, in this method, the estimate of natural frequency ω_r is taken as the frequency where the FRF magnitude is maximum.

We should note that the peak frequency and the natural frequency coincide for an SDOF system with structural damping but not for a system with viscous damping, though the difference is negligible for lower levels of damping. For MDOF systems, irrespective of the nature of damping, the peak and natural frequencies are not coincident due to the contributions of multiple modes. The difference between them depends on the damping level and the modal coupling.

b. The modal loss factor for the mode is estimated using the **half-power bandwidth (HPB)**. HPB is the width of the frequency band where the FRF magnitude drops by a factor of $1/\sqrt{2}$ (or by 3dB). (The factor arises from the fact that a change in voltage by a factor of $1/\sqrt{2}$ changes the electrical power by a factor of one-half.) HPB can be obtained by solving the following equation for ω:

$$\left|\alpha_{jk}(\omega)\right| = \frac{1}{\sqrt{2}}\left|\alpha_{jk}(\omega_r)\right| \qquad (11.13)$$

Substituting Eq. (11.12) into Eq. (11.13) and simplifying, we get

$$\eta_r\omega_r^2 = \pm\left(\omega_r^2 - \omega^2\right) \qquad (11.14)$$

which leads to

$$\eta_r\omega_r^2 = \omega_r^2 - \omega_a^2 \quad \text{and} \quad \eta_r\omega_r^2 = \omega_b^2 - \omega_r^2 \qquad (11.15)$$

Adding the two equations in (11.15) and assuming that $\omega_r \cong (\omega_b + \omega_a)/2$, we get the following expression:

$$\eta_r = \frac{\omega_b - \omega_a}{\omega_r} \qquad (11.16)$$

where the numerator on the RHS is nothing but the HPB. The frequencies ω_a and ω_b can be obtained from the measured FRF corresponding to half-power points, and the loss factor for the rth mode can be determined from Eq. (11.16).

c. Eq. (11.12) written at $\omega = \omega_r$ yields the following expression for the modal constant

$$_rA_{jk} = i\eta_r\omega_r^2\alpha_{jk}(\omega_r) \qquad (11.17)$$

Since ω_r and η_r are obtained in the previous steps, the modal constant for the rth mode can be obtained from Eq. (11.17).

d. Using steps (a)–(c), the natural frequency, loss factor, and modal constant are obtained for all the peaks ($r = 1, 2, \ldots, m$) in the FRF $\alpha_{jk}(\omega)$. This completes modal parameter estimation for one FRF.

e. If one row of the FRF matrix is recorded, then we apply steps (a)–(d) to all the FRFs $\alpha_{jk}(\omega)$ with k = 1, 2, …, n.

f. In general, the estimates of ω_r and η_r obtained from various FRFs would have some variance. But since the natural frequency and damping factor are global characteristics of the structure, as they don't depend on the excitation and response locations, we obtain a global estimate by averaging the estimates obtained from various FRFs. The modal constants estimates can also be revised in step (c) using these global estimates.

g. Once we have modal constants for all the FRFs, the mass-normalized mode shapes can be obtained. We demonstrate this process to find the rth mode shape. The modal constants for the driving point FRF are critical to find the mode shapes. When one row of the FRF matrix is measured, the driving point FRF is α_{jj} and the modal constant for the rth mode is

$$_r A_{jj} = \phi_{jr}\ \phi_{jr} \tag{11.18}$$

which yields

$$\phi_{jr} = \sqrt{_r A_{jj}} \tag{11.19}$$

For α_{jk} the modal constant for the rth mode is

$$_r A_{jk} = \phi_{jr}\ \phi_{kr} \tag{11.20}$$

Since ϕ_{jr} is known from Eq. (11.19), we get

$$\phi_{kr} = \frac{_r A_{jk}}{\phi_{jr}} \tag{11.21}$$

Thus, Eq. (11.21) can be used to find all the elements of the rth mode shape, thus yielding the complete mode shape

$$\{\phi\}_r = \left\{\phi_{1r}\quad \phi_{2r}\quad \cdots\quad \phi_{nr}\right\}^T \tag{11.22}$$

Example 11.1: Modal Parameter Estimation Using the Peak-Picking Method

We apply the peak-picking method to analyze the receptances of the cantilever structure, considered in Chapters 9 and 10 for impact and shaker testings. Figure 9.15a in Chapter 9 shows five test points on the cantilever structure. We assume that the fifth row of the FRF matrix corresponding to translational DOFs at these test points is available.

Let us analyze one of the peaks, say, the third peak, in the FRFs.

Step 1: Start with the point FRF α_{55}. Locate the frequency at which the FRF magnitude is maximum. From Figure 11.1, this frequency is 173.0 Hz, and hence, $\omega_3 = 2\pi \times 173$ rad/sec.

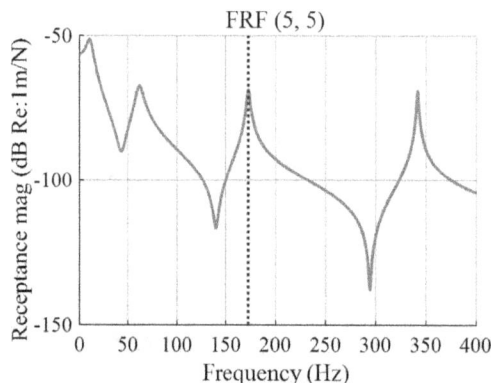

FIGURE 11.1   Frequency corresponding to the maximum FRF magnitude.

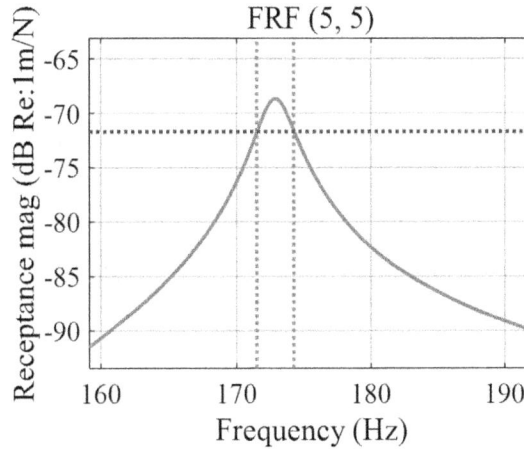

FIGURE 11.2 Frequencies corresponding to the half-power points.

Step 2: Locate the half-power points as shown in the zoomed view in Figure 11.2. The horizontal line corresponds to a 3 dB drop in the FRF magnitude from the maximum value. The frequencies corresponding to the two vertical lines drawn at these two points are the half-power frequencies. The modal loss factor calculated using Eq. (11.16) is

$$\eta_3 = \frac{\omega_b - \omega_a}{\omega_3} = \frac{174.25 - 171.5}{173} = 0.015896$$

Step 3: Steps 1 and 2 are repeated for the third peak in the other four FRFs.

Step 4: The natural frequencies and modal loss factors from the five FRFs are averaged. These average estimates are

$$\bar{\omega}_3 = 2\pi\, 173 \text{ rad/sec} \quad \text{and} \quad \bar{\eta}_3 = 0.015914$$

Step 5: Modal constant for the third peak is now calculated for the point FRF α_{55}. Using Eq. (11.17), and noting that $\alpha_{55}(\bar{\omega}_3) = -5.56 \times 10^{-5} - 0.00036\,i$, we get

$$_3A_{55} = i\bar{\eta}_3\bar{\omega}_3{}^2\alpha_{55}(\bar{\omega}_3) = 6.812 - 1.043\,i$$

Similarly, the modal constants for the third peak in the other four FRFs are calculated.

Step 6: We now compute mode shapes as explained in point (g) of Section 11.4. The mass-normalized mode shape for mode 3 is obtained as

$$\{\phi\}_3 = \left\{ \begin{array}{ccccc} 1.57 + 0.22i & 1.36 + 0.24i & -1.23 - 0.12i & -1.02 - 0.18i & 2.61 - 0.19i \end{array} \right\}^\mathrm{T}$$

For comparison, the correct modal parameters for mode 3 are as follows. The correct natural frequency and modal loss factor are $2\pi \times 172.84$ rad/sec, 0.015963, respectively, while the correct mode shape is

$$\{\phi\}_3 = \left\{ \begin{array}{ccccc} 1.59 + 0.006i & 1.38 + 0.03i & -1.24 - 0.029i & -1.04 - 0.0003i & 2.63 - 0.02i \end{array} \right\}^\mathrm{T}$$

11.5 CIRCLE-FIT METHOD

The circle-fit method, also based on the SDOF assumption, utilizes the property that the Nyquist plot of the receptance of an SDOF system with structural damping is a circular arc. For fitting a viscous damping model, it utilizes the property that the mobility of the system traces a circular arc in the Nyquist plane. These properties were derived in Sections 7.6.2 and 7.6.3 of Chapter 7.

a. Basic idea of the method

Consider an MDOF system with structural damping. To analyze the rth mode in the receptance, we write Eq. (11.1) as

$$\alpha_{jk}(\omega) = \frac{_rA_{jk}}{\omega_r{}^2 - \omega^2 + i\eta_r\omega_r{}^2} + \sum_{q=1,\,\neq r}^{N} \frac{_qA_{jk}}{\omega_q{}^2 - \omega^2 + i\eta_q\omega_q{}^2} \tag{11.23}$$

It is represented as

$$\alpha_{jk}(\omega) = {}_r C_{jk}(\omega) + {}_r B_{jk}(\omega) \qquad (11.24)$$

The term ${}_r B_{jk}(\omega)$ is called the residual and represents the contribution of all the modes other than mode r to the FRF. The term ${}_r C_{jk}(\omega)$ represents the contribution of mode r. Since ${}_r C_{jk}(\omega)$ represents the contribution of a single mode, it is like the FRF of an SDOF system; hence, its plot on the Nyquist plane traces a circular arc. However, the Nyquist plot of $\alpha_{jk}(\omega)$ would not trace a circular arc since it is contributed by more than one mode.

If we consider a frequency band around the rth peak, then in this band ${}_r C_{jk}(\omega)$ makes a dominant contribution to the FRF as compared to the contribution ${}_r B_{jk}(\omega)$. If ${}_r B_{jk}(\omega)$ is constant in the band, i.e., does not vary with ω, then the Nyquist plot of $\alpha_{jk}(\omega)$ over the frequency band would trace a circular arc in the Nyquist plane since the term ${}_r B_{jk}(\omega)$ would only translate the plot of ${}_r C_{jk}(\omega)$ in the Nyquist plane. But, in reality, ${}_r B_{jk}(\omega)$ is not constant but varies with ω to some extent; therefore, the Nyquist plot of $\alpha_{jk}(\omega)$ in the frequency band would not be a perfectly circular arc, though it can be so approximated if the modes are well separated, and the frequency band is narrow enough. This observation is also clear from Figure 11.3, which shows for an MDOF system, the Nyquist plots of ${}_r C_{jk}(\omega)$, ${}_r B_{jk}(\omega)$ and $\alpha_{jk}(\omega)$ over a narrow frequency band around the rth mode. It is seen that the circular nature of ${}_r C_{jk}(\omega)$ plot over the narrow band is not much distorted by ${}_r B_{jk}(\omega)$ as it is much smaller than ${}_r C_{jk}(\omega)$. Given this observation, the modal parameters of the rth mode can be obtained by fitting a circular arc to the Nyquist plot of $\alpha_{jk}(\omega)$ over a narrow frequency band around the rth peak. This conclusion forms the basic principle of modal parameter estimation using the circle-fit method.

b. Circle-fitting

The first step in this method is to choose a narrow frequency band around the peak to be analyzed in the given FRF α_{jk}. Let x and y denote the real$(\alpha_{jk}(\omega))$ and imag$(\alpha_{jk}(\omega))$ parts, respectively. The equation of the circle with center (x_c, y_c) and radius R is given by

$$(x - x_c)^2 + (y - y_c)^2 = R^2 \qquad (11.25)$$

It can be simplified as

$$x^2 + y^2 + ax + by + c = 0 \qquad (11.26)$$

where a, b, and c are related to (x_c, y_c) and R by the following relationships:

$$x_c = -\frac{a}{2}; \quad y_c = -\frac{b}{2} \quad \text{and} \quad R = \sqrt{\left(\frac{a}{2}\right)^2 + \left(\frac{b}{2}\right)^2 - c} \qquad (11.27)$$

For fitting the circle, let m number of frequency points are chosen in the frequency band. Let the value of (x, y) at these frequency points be (x_i, y_i), with $i = 1, 2, ..., m$.

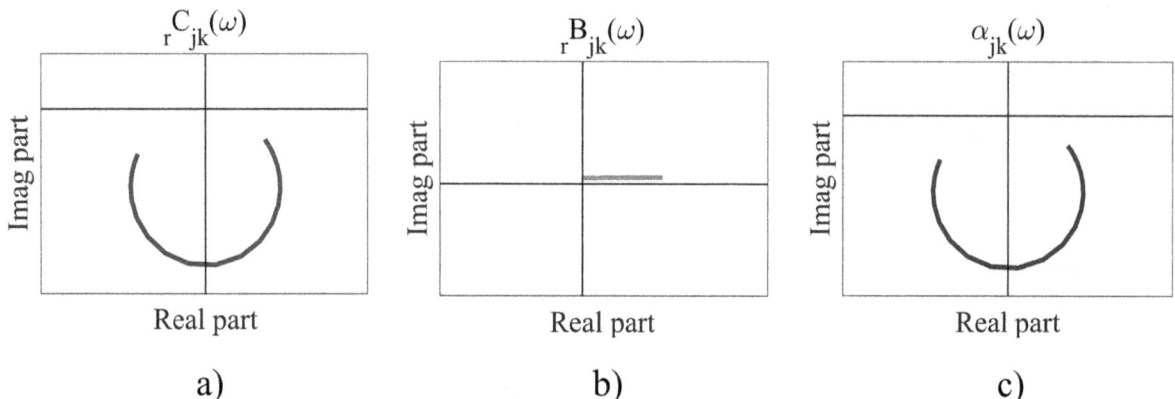

FIGURE 11.3 Nyquist plots of ${}_r C_{jk}(\omega)$, ${}_r B_{jk}(\omega)$ and $\alpha_{jk}(\omega)$ over a narrow frequency band around the rth mode.

If these points lie on a circle, they must satisfy Eq. (11.26). Therefore, if the point (x_i, y_i) is substituted in this equation, then the LHS magnitude represents the error in the circle-fit. Hence, the total error at all the frequency points is defined as

$$E(a,b,c) = \sum_{i=1}^{m} \left(x_i^2 + y_i^2 + ax_i + by_i + c \right)^2 \tag{11.28}$$

For the minimum value of the error function, we should have

$$\frac{\partial E}{\partial a} = 0; \quad \frac{\partial E}{\partial b} = 0 \quad \text{and} \quad \frac{\partial E}{\partial c} = 0; \tag{11.29}$$

Applying conditions in Eq. (11.29) to the error function and simplifying, we get the following matrix equation:

$$\begin{bmatrix} \sum x_i^2 & \sum x_i y_i & \sum x_i \\ \sum x_i y_i & \sum y_i^2 & \sum y_i \\ \sum x_i & \sum y_i & m \end{bmatrix} \begin{Bmatrix} a \\ b \\ c \end{Bmatrix} = \begin{Bmatrix} -\sum x_i z_i^2 \\ -\sum y_i z_i^2 \\ -\sum z_i^2 \end{Bmatrix} \tag{11.30}$$

where the summation is over m frequencies, and

$$z_i^2 = x_i^2 + y_i^2 \tag{11.31}$$

Matrix equation (11.30) represents three simultaneous linear equations in three unknowns, and its solution gives the coefficients $\{a, \ b, \ c\}^T$, which, when substituted in Eq. (11.27), gives the circle's center and radius. It provides a least-squares error fit to the frequency points in the band.

c. Estimation of natural frequency

Once a circle is fitted to the Nyquist plot of the FRF over a narrow frequency band around the rth peak, the center (x_c, y_c) and radius (R) of the circle are known. But it is not straightforward to determine the modal parameters directly from this information as the two are related by a nonlinear relationship. The estimate of the natural frequency from the circle-fit is based on the concept of the maximum sweep rate proposed by Kennedy and Pancu [51]. The sweep rate is the rate of change of angle subtended at the center of the circle with frequency. For obtaining the sweep rate, it is enough to analyze only the term $_r C_{jk} (\omega)$ since it is the term that represents the circular arc. We know that

$$_r C_{jk} (\omega) = \frac{_r A_{jk}}{\omega_r^2 - \omega^2 + i \eta_r \omega_r^2} \tag{11.32}$$

Expressing $_r C_{jk} (\omega)$ as

$$_r C_{jk} (\omega) = _r A_{jk} \cdot _r C_{jk}^u (\omega) \tag{11.33}$$

where

$$_r C_{jk}^u (\omega) = \frac{1}{\omega_r^2 - \omega^2 + i \eta_r \omega_r^2} \tag{11.34}$$

The trace generated by $_r C_{jk}^u (\omega)$ is also a circular arc and is referred to as a unit modal circle since it is based on a unit modal constant. The only difference between the circular arcs corresponding to $_r C_{jk} (\omega)$ and $_r C_{jk}^u (\omega)$ is the role played by the modal constant $_r A_{jk}$. Since $_r A_{jk}$ is generally complex, it contracts or expands the size of the circle traced by $_r C_{jk}^u (\omega)$ and rotates it in the Nyquist plane to produce the circle due to $_r C_{jk} (\omega)$. Therefore, the angle subtended at the center of the circular arc can also be obtained by analyzing the plot of $_r C_{jk}^u (\omega)$.

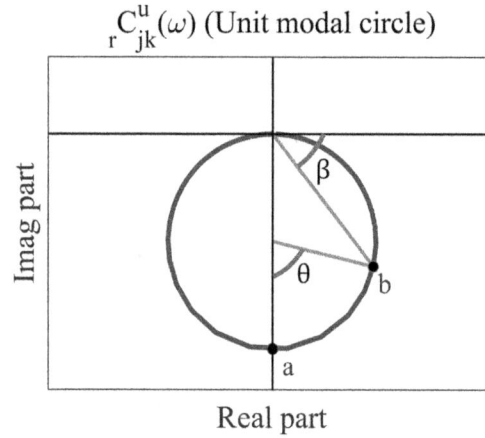

FIGURE 11.4   Plot of $_rC_{jk}^u(\omega)$ in the Nyquist plane.

Figure 11.4 shows the plot of $_rC_{jk}^u(\omega)$ (the unit modal circle) in the Nyquist plane. θ is the angle subtended at the center of the circular arc by the arc ab. Point b corresponds to frequency ω, while point 'a' is the point of intersection of the circle with the imaginary axis. It is noted that point 'a' corresponds to the frequency $\omega = \omega_r$ since at this frequency, the real part of $_rC_{jk}^u(\omega)$ is zero. From the figure, we can write

$$\tan\beta = \frac{\left|\text{Im}\left(_rC_{jk}^u(\omega)\right)\right|}{\text{Re}\left(_rC_{jk}^u(\omega)\right)} = \frac{\eta_r\omega_r^2}{\omega_r^2 - \omega^2} \tag{11.35}$$

Since

$$\theta = 2\left(\frac{\pi}{2} - \beta\right) \tag{11.36}$$

we get

$$\tan\frac{\theta}{2} = \frac{\omega_r^2 - \omega^2}{\eta_r\omega_r^2} \tag{11.37}$$

$$\omega^2 = \omega_r^2 - \eta_r\omega_r^2\tan\frac{\theta}{2} \tag{11.38}$$

Differentiating ω^2 with θ, and taking its reciprocal, we get

$$\frac{d\theta}{d\omega^2} = \frac{-2\eta_r\omega_r^2}{\left(\omega_r^2 - \omega^2\right)^2 + \left(\eta_r\omega_r^2\right)^2} \tag{11.39}$$

The quantity $\dfrac{d\theta}{d\omega^2}$ is referred to as sweep rate, and it is observed from Eq. (11.39) that it is maximum when $\omega = \omega_r$. This observation provides the basis for locating the natural frequency on the circle. Since the circle has been fitted, it is easy to calculate the sweep rate at different frequencies, and the natural frequency can be determined based on the maximum sweep rate criterion.

d. Estimation of modal loss factor

Eq. (11.37) provides the relation between the loss factor η_r and the angle θ subtended by frequency points ω and ω_r on the circle. If we choose two frequencies, ω_a and ω_b, in the band, one each on either side of ω_r, then we can write

$$\tan\frac{\theta_a}{2} = \frac{\omega_a^2 - \omega_r^2}{\eta_r\omega_r^2} \quad \left(\text{for } \omega_a > \omega_r\right) \tag{11.40}$$

$$\tan\frac{\theta_b}{2} = \frac{\omega_r^2 - \omega_b^2}{\eta_r\omega_r^2} \quad (\text{for } \omega_b < \omega_r) \tag{11.41}$$

Subtracting the two equations gives the following equation from which the loss factor can be computed:

$$\eta_r = \frac{\omega_a^2 - \omega_b^2}{\omega_r^2\left(\tan\dfrac{\theta_a}{2} + \tan\dfrac{\theta_b}{2}\right)} \tag{11.42}$$

e. Estimation of modal constant

The modal constant can be estimated by making use of the relationship between $_rC_{jk}^u(\omega)$ and $_rC_{jk}(\omega)$ given by Eq. (11.33). At $\omega = \omega_r$, we can write

$$_rC_{jk}(\omega_r) = {}_rA_{jk}\cdot{}_rC_{jk}^u(\omega_r) \tag{11.43}$$

Each quantity in (11.43) is a complex quantity. Writing these quantities in the polar form leads to

$$\left|{}_rC_{jk}(\omega_r)\right| = \left|{}_rA_{jk}\right|\cdot\left|{}_rC_{jk}^u(\omega_r)\right| \tag{11.44}$$

$\left|{}_rC_{jk}^u(\omega_r)\right|$ is nothing but the diameter of the unit modal circle, obtained from Eq. (11.34) as

$$\left|{}_rC_{jk}^u(\omega_r)\right| = \frac{1}{\eta_r\omega_r^2} \tag{11.45}$$

Eq. (11.44) states that the factor $\left|{}_rA_{jk}\right|$ amplifies the diameter of the unit modal circle to produce $\left|{}_rC_{jk}(\omega_r)\right|$, which is nothing but the diameter (2R) of the circle fitted in step (a). Therefore, the magnitude of the modal constant is obtained as

$$\left|{}_rA_{jk}\right| = 2R\eta_r\omega_r^2 \tag{11.46}$$

From Eq. (11.43), we also have

$$\angle{}_rA_{jk} = \angle{}_rC_{jk}(\omega_r) - \angle{}_rC_{jk}^u(\omega_r) \tag{11.47}$$

Thus, the phase angle of $_rA_{jk}$ can be obtained from the phase angles of $_rC_{jk}(\omega_r)$ and $_rC_{jk}^u(\omega_r)$. $_rC_{jk}^u(\omega_r)$ is aligned with the negative imaginary axis, and hence the phase of the modal constant is equal to the phase angle of $_rC_{jk}(\omega_r)$ measured from the negative imaginary axis.

Example 11.2: Modal Parameter Estimation Using the Circle-Fit Method

We apply the circle-fit procedure to a transfer receptance $\alpha_{15}(\omega)$ of the cantilever structure.

Figure 11.5 shows the Nyquist plot of this FRF over the frequency range 0–400 Hz covering four modes. Each loop in the Nyquist plot is dominated by a particular mode. Let us apply the circle-fit procedure to the third peak in the FRF.

Step 1: We select a frequency band around the third peak, as shown in Figure 11.6. There are 21 frequency points in the band. The band should be wide enough to have enough points for circle-fitting. But too wide a band should not be used as the assumption of constancy of contribution of other modes may not be valid.

Figure 11.7 shows the Nyquist plot of $\alpha_{15}(\omega)$ in the frequency band. The circle is fitted to the frequency points using Eq. (11.30). It requires setting up the LHS coefficient matrix and the RHS vector of this equation using the real and imaginary parts of the FRF at the frequency points. Figure 11.8 shows the circle fitted to the frequency points. The center and the radius of the fitted circle are: (xc, yc) = (-4.962×10^{-8}, -1.112×10^{-4}) and R = 1.113×10^{-4}.

Step 2:

We calculate from the fitted circle the angle swept by a pair of consecutive frequency points at the circle's center. The coordinates of the frequency points and the circle's center are used to find the angle subtended by each pair of points. The natural frequency is estimated as the average of the pair's frequency points with the maximum swept angle. Based on this principle, the natural frequency of mode 3 is obtained as $\omega_3 = 2\pi \times 172.87$ rad/sec.

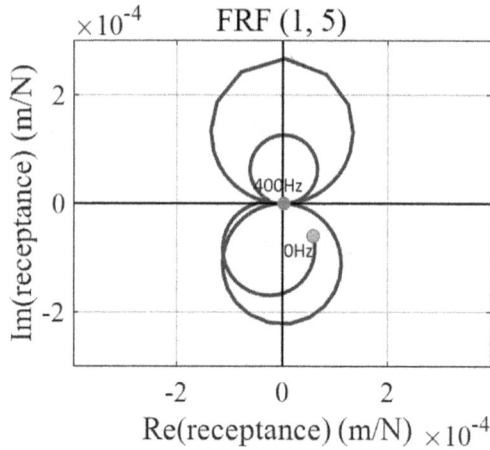

FIGURE 11.5 Nyquist plot of FRF $\alpha_{15}(\omega)$.

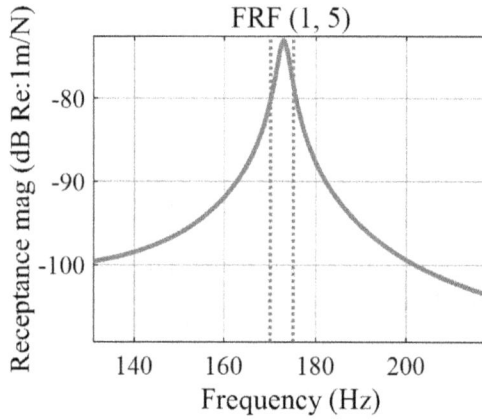

FIGURE 11.6 Frequency band around the third peak.

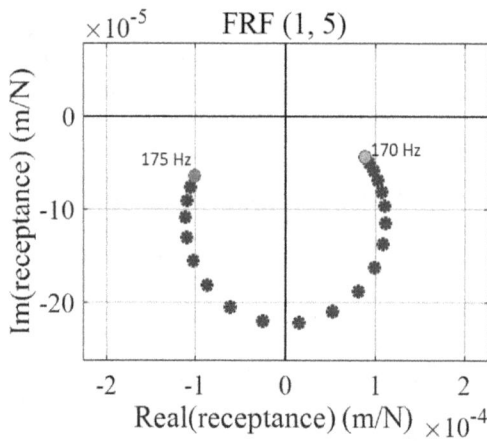

FIGURE 11.7 Nyquist plot of $\alpha_{15}(\omega)$ in the selected band.

Step 3:

Figure 11.9 shows the fitted circle with the estimated natural frequency. To estimate the modal loss factor, we choose one frequency point on either side of the natural frequency, as shown in the figure. The frequency points and the computed angles are

$$(\omega_a, \ \omega_b) = (2\pi \times 171.5, \quad 2\pi \times 174) \ \text{rad/sec}$$

$$(\theta_a, \ \theta_b) = (80.3°, \quad 87.9°)$$

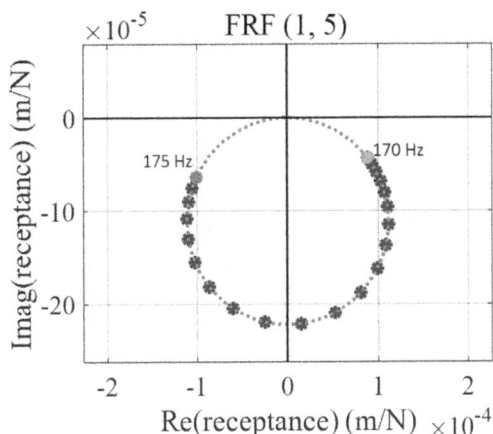

FIGURE 11.8 The circle fitted to the frequency points.

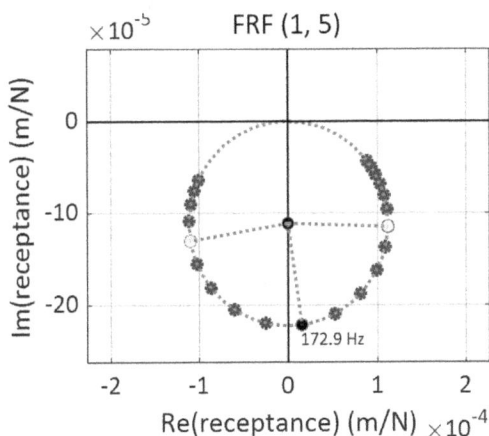

FIGURE 11.9 The frequency points chosen for estimating the modal loss factor.

Using Eq. (11.42), we get

$$\eta_3 = \frac{\omega_a{}^2 - \omega_b{}^2}{\omega_3{}^2\left(\tan\dfrac{\theta_a}{2} + \tan\dfrac{\theta_b}{2}\right)} = 0.0159$$

Step 4:
The magnitude of the modal constant is obtained using Eq. (11.46)

$$\left|{}_3A_{15}\right| = 2R\eta_3\omega_3{}^2 = 2 \times 1.113 \times 10^{-4} \times 0.0159 \times \left(2\pi \times 172.87\right)^2 = 4.175$$

From Figure 11.9, the angle of the line joining the center with the point ω_3 (the natural frequency) with the negative imaginary axis, which is the phase of the modal constant, is found as $\angle_r A_{15} = -7.82°$.
Step 5:
Steps 1–4 are repeated for the other three peaks in the FRF.
Step 6: All the other FRFs are analyzed. The mode shapes can be obtained from the modal constants by the procedure followed in Example 11.1.

11.6 LINE-FIT METHOD

The line-fit method, developed by Dobson [33], is based on fitting a straight line to the inverse of an FRF. Hence, it is also called the inverse FRF method. Since the FRF-inverse has units of dynamic stiffness, the method is also referred to as the dynamic stiffness method. The line-fit method is a frequency domain method based on the SDOF assumption and was presented assuming a structural damping model.

a. Basic idea of the method

In light of Eqs. (11.23) and (11.24), we can write

$$\alpha_{jk}(\omega) = \frac{{}_rA_{jk}}{\omega_r^2 - \omega^2 + i\eta_r\omega_r^2} + {}_rB_{jk}(\omega) \tag{11.48}$$

In this method, the residual term ${}_rB_{jk}(\omega)$, representing the contribution of all modes other than mode r to the FRF, is assumed to be negligible. Hence, Eq. (11.48) becomes

$$\alpha_{jk}(\omega) \approx \frac{{}_rA_{jk}}{\omega_r^2 - \omega^2 + i\eta_r\omega_r^2} \tag{11.49}$$

Representing the inverse of $\alpha_{jk}(\omega)$ by $z_{jk}(\omega)$, we get

$$z_{jk}(\omega) = \frac{1}{\alpha_{jk}(\omega)} \approx \frac{\omega_r^2 - \omega^2 + i\eta_r\omega_r^2}{{}_rA_{jk}} \tag{11.50}$$

Since the modal constant ${}_rA_{jk}$ is, in general, complex, let ${}_rA_{jk} = c + id$. Express Eq. (11.50) in terms of the real and imaginary parts

$$z_{jk}(\omega) = z_R + iz_I \tag{11.51}$$

where z_R and z_I are

$$z_R = \mathrm{Real}\left(z_{jk}(\omega)\right) = \frac{\left(\omega_r^2 - \omega^2\right)c + \eta_r\omega_r^2 d}{c^2 + d^2} \tag{11.52}$$

$$z_I = \mathrm{Imag}\left(z_{jk}(\omega)\right) \frac{-\left(\omega_r^2 - \omega^2\right)d + \eta_r\omega_r^2 c}{c^2 + d^2} \tag{11.53}$$

It is seen that z_R and z_I are linear functions of ω^2 and can be expressed as

$$z_R = m_R\omega^2 + q_R \tag{11.54}$$

$$z_I = m_I\omega^2 + q_I \tag{11.55}$$

Comparing Eqs. (11.54)–(11.55) with Eqs. (11.52)–(11.53), we get

$$m_R = \frac{-c}{c^2 + d^2} \tag{11.56}$$

$$q_R = \frac{c\omega_r^2 + \eta_r\omega_r^2 d}{c^2 + d^2} \tag{11.57}$$

$$m_I = \frac{d}{c^2 + d^2} \tag{11.58}$$

$$q_I = \frac{-d\omega_r^2 + \eta_r\omega_r^2 c}{c^2 + d^2} \tag{11.59}$$

Solving Eqs. (11.56)–(11.59), ω_r, η_r, c and d are obtained as

$$\eta_r = \frac{m_R\, q_I - m_I\, q_R}{m_R\, q_R + m_I\, q_I} \tag{11.60}$$

$$\omega_r = \sqrt{\frac{q_R}{\eta_r m_I - q_R}} \qquad (11.61)$$

$$c = \frac{-m_R}{m_R^{\,2} + m_I^{\,2}} \qquad (11.62)$$

$$d = \frac{m_I}{m_R^{\,2} + m_I^{\,2}} \qquad (11.63)$$

From the above equations, the following observations are made:

- Eq. (11.54) represents a straight line between z_R and ω^2 with m_R and q_R being the slope and intercept of the line. Therefore, a straight line can be fitted to the values of z_R and ω^2. For the line-fit, the frequency points should be chosen from a narrow frequency band around the rth peak so that the approximation of Eq. (11.48) by (11.49) is reasonably satisfied.
- Similarly, Eq. (11.55) represents a straight line between z_I and ω^2 with m_I and q_I being the slope and intercept of the line. A straight line can be fitted to the values of z_R and ω^2.
- Once the slopes and intercepts of the two lines are found from line-fits, the modal parameters ω_r, η_r, c and d for the rth mode can be obtained from Eqs. (11.60)–(11.63).

To get an idea of what the inverse FRF looks like, we plot the FRF magnitude and the real and imaginary parts of the inverse of the FRF (i.e., the quantities z_R and z_I) as a function of ω^2 as shown in Figure 11.10. It is seen that around the peaks in the FRF magnitude plot, the real and imaginary parts of the inverse FRF have straight-line-like variations. However, away from the resonance regions, the variation is not linear (as the contributions of more than one mode become comparable).

b. Fitting of the straight lines

The first step in this method is to choose a narrow frequency band around the peak to be analyzed in the given FRF. Let us consider in the band the line-fit between z_R and ω^2. The equation to be fitted (Eq. 11.54) is rewritten below.

$$m_R \omega^2 + q_R - z_R = 0 \qquad (11.64)$$

We choose m number of frequencies (ω_i, with $i = 1, 2, \ldots, m$) in the band around the peak to be analyzed. Let $\left(\omega_i^{\,2}, z_{R,i}\right)$ represent the values of $\left(\omega^2, z_R\right)$ at frequency ω_i. If $\left(\omega_i^{\,2}, z_{R,i}\right)$ is substituted into the LHS of Eq. (11.64), then the LHS magnitude represents the error in the line-fit. The total error at all the frequency points is defined as

$$E\left(m_R, q_R\right) = \sum_{i=1}^{m} \left(m_R \omega_i^{\,2} + q_R - z_{R,i}\right)^2 \qquad (11.65)$$

Inverse FRF and FRF

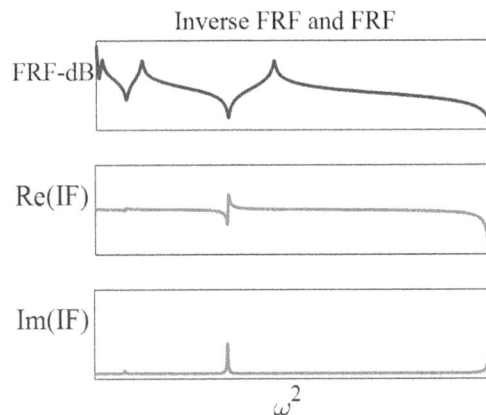

FIGURE 11.10   FRF magnitude and the real and imaginary parts of the inverse of the FRF as a function of ω^2.

For the minimum value of the error function, we should have

$$\frac{\partial E}{\partial m_R} = 0 \quad \text{and} \quad \frac{\partial E}{\partial q_R} = 0 \tag{11.66}$$

Applying the conditions in Eq. (11.66) to the error function and simplifying, we get the following matrix equation:

$$\begin{bmatrix} \sum \omega_i^4 & \sum \omega_i^2 \\ \sum \omega_i^2 & m \end{bmatrix} \begin{Bmatrix} m_R \\ q_R \end{Bmatrix} = \begin{Bmatrix} \sum \omega_i^2 z_{R,i} \\ \sum z_{R,i} \end{Bmatrix} \tag{11.67}$$

where the summation is over m frequencies.

Matrix equation (11.67) represents two simultaneous linear equations in two unknowns, and its solution gives the slope and intercept (m_R, q_R) of the line.

By proceeding in the same way as mentioned above, the following matrix equation is obtained to fit a straight line between z_I and ω^2, whose solution gives the slope and the intercept (m_I, q_I) of the line.

$$\begin{bmatrix} \sum \omega_i^4 & \sum \omega_i^2 \\ \sum \omega_i^2 & m \end{bmatrix} \begin{Bmatrix} m_I \\ q_I \end{Bmatrix} = \begin{Bmatrix} \sum \omega_i^2 z_{I,i} \\ \sum z_{I,i} \end{Bmatrix} \tag{11.68}$$

As mentioned before, once the slopes and intercepts of the two lines are known, the modal parameters ω_r, η_r, c and d for the rth mode can be obtained from Eqs. (11.60) to (11.63). The modal constant is obtained as $_r A_{jk} = c + id$.

We note that though the method is simple to apply, the residuals are neglected. Neglecting residuals may be acceptable for well-separated modes, but the results may not be accurate for the systems with close modes or modes with higher modal coupling.

Example 11.3: Modal Parameter Estimation Using the Line-Fit Method

In this example, we apply the line-fit procedure to our cantilever structure to analyze its transfer receptance α_{15}. We analyze the third peak in the FRF.

Step 1: We first select a frequency band around the third peak. There are 11 frequency points in the selected band. The comments made in the circle-fit procedure to choose the band also apply in this case. Figure 11.11 shows the plot of Z_R in the selected band.

A straight line is fitted between Z_R and ω^2 to the frequency points using Eq. (11.67). It requires setting up the LHS coefficient matrix and RHS vector of the equation using the real part of the inverse FRF at the frequency points. The slope and intercept of the line are found to be $(m_R, q_R) = (-2.383 \times 10^{-1}, 2.811 \times 10^5)$.

Figure 11.12 shows an overlay of Z_R and the line-fit.

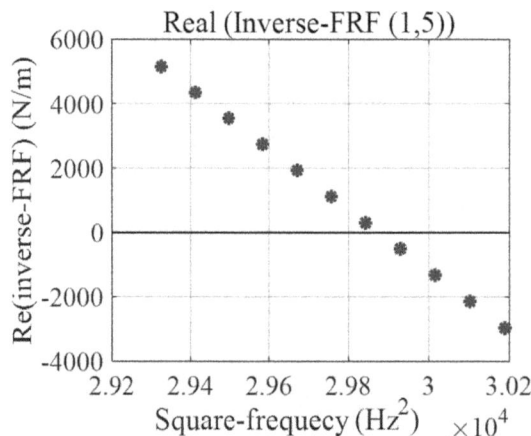

FIGURE 11.11 Plot of Z_R in the selected band.

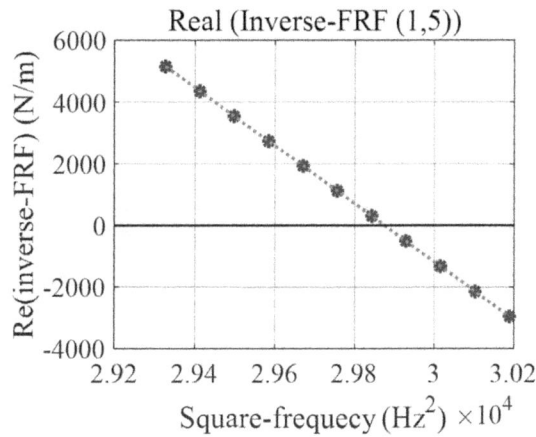

FIGURE 11.12 Line-fit of Z_R.

Similarly, a straight line is fitted to the imaginary part of the inverse FRF using Eq. (11.68). The slope and intercept of
the line are found to be $(m_I, q_I) = (-5.655 \times 10^{-5},\ 4.557 \times 10^3)$.

Figures 11.13 and 11.14 show the plot of Z_I in the selected band and its overlay with the fitted line, respectively.

Step 2:

The modal parameters are obtained from Eqs. (11.60) to (11.63) as given below using the slopes and intercepts of
the two lines.

$$\omega_3 = 2\pi \times 172.84 \text{ rad/sec}; \qquad \eta_3 = 0.0159; \qquad {}_3A_{15} = c + id = 4.194 - i\,0.0009$$

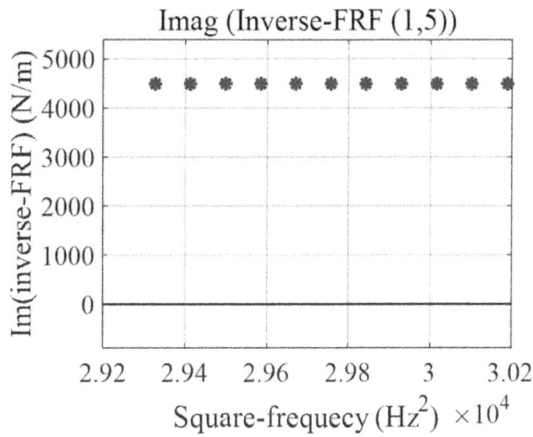

FIGURE 11.13 Plot of Z_I in the selected band.

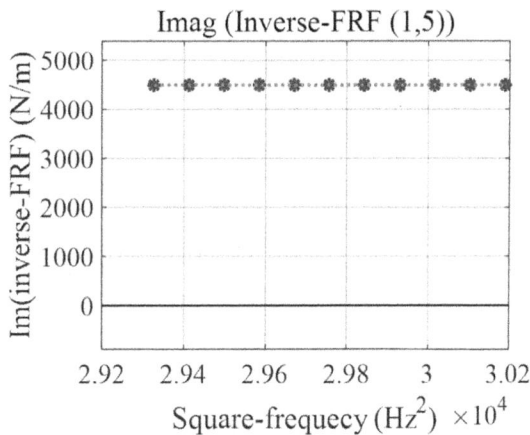

FIGURE 11.14 Line-fit of Z_I.

Step 3:
Steps 1–2 are repeated for the other three peaks in the FRF.
Step 4: Other FRFs are analyzed as mentioned above. The mode shapes are obtained as explained in Example 11.1.

11.7   MODIFIED LINE-FIT METHOD

The line-fit method, described in Section 11.6, neglects the residual term affecting the accuracy of the estimated modal parameters. The modified line-fit method, developed by Dobson [32], is formulated by considering the residual term.

a. Basic idea of the method

In this method, a fixing frequency Ω is chosen in the frequency band to approximately account for the residual term. From Eq. (11.48), the FRF at the fixing frequency can be written as

$$\alpha_{jk}(\Omega) = \frac{{}_r A_{jk}}{\omega_r^2 - \Omega^2 + i\eta_r\omega_r^2} + {}_r B_{jk}(\Omega) \tag{11.69}$$

If the residual term is assumed to be constant, then the subtraction of Eq. (11.69) from (11.48) cancels out the residual term leading to

$$\alpha_{jk}(\omega) - \alpha_{jk}(\Omega) = \frac{{}_r A_{jk}}{\omega_r^2 - \omega^2 + i\eta_r\omega_r^2} - \frac{{}_r A_{jk}}{\omega_r^2 - \Omega^2 + i\eta_r\omega_r^2} \tag{11.70}$$

Writing ${}_r A_{jk} = c + id$, and taking the inverse of Eq. (11.70) and simplifying, we get

$$\frac{\omega^2 - \Omega^2}{\alpha_{jk}(\omega) - \alpha_{jk}(\Omega)} = \frac{c - id}{c^2 + d^2}\left(\left(\omega_r^2 - \omega^2\right)\left(\omega_r^2 - \Omega^2\right) - \eta_r^2\omega_r^4 + i\eta_r\omega_r^2\left(2\omega_r^2 - \omega^2 - \Omega^2\right)\right) \tag{11.71}$$

Represent the LHS by Δ, that is

$$\Delta = \frac{\omega^2 - \Omega^2}{\alpha_{jk}(\omega) - \alpha_{jk}(\Omega)} \tag{11.72}$$

It is seen that the real and imaginary parts of the RHS of Eq. (11.71) are linear functions of ω^2. Hence, we can write

$$\mathrm{Re}(\Delta) = g_R\omega^2 + h_R \tag{11.73}$$

$$\mathrm{Im}(\Delta) = g_I\omega^2 + h_I \tag{11.74}$$

It is also clear from Eq. (11.71) that the slopes g_R and g_I and intercepts h_R and h_I depend on the choice of the fixing frequency and are a linear function of Ω^2. Hence, we can express them as

$$g_R = n_R\Omega^2 + u_R \tag{11.75}$$

$$g_I = n_I\Omega^2 + u_I \tag{11.76}$$

From Eq. (11.71), we obtain the following relationships between the slopes and intercepts in Eqs. (11.75) and ((11.76) and the modal parameters:

$$P = \frac{n_I}{n_R} \text{ and } Q = \frac{u_I}{u_R} \tag{11.77}$$

$$\eta_r = \frac{Q - P}{PQ + 1} \tag{11.78}$$

$$\omega_r^2 = -\frac{u_R}{n_R}\left(\frac{1}{1 - P\eta_r}\right) \tag{11.79}$$

$$c = \frac{n_R}{n_R{}^2 + n_I{}^2} \tag{11.80}$$

$$d = -cP \tag{11.81}$$

The main steps in the modified line-fit method are as follows:
- We perform line-fits in Eqs. (11.73) and (11.74) for several choices of fixing frequency $\Omega = \Omega_s$, with $s = 1, 2, \ldots, n$ (such that $\Omega_s \neq \omega_i$, to avoid singularity in the denominator of Eq. 11.71). It gives a set of values of g_R and g_I and the corresponding intercepts.
- With these values, we perform line-fits corresponding to Eqs. (11.75) and (11.76) to obtain n_R and u_R and n_I and u_I.
- Eqs. (11.77)–(11.81) are then solved for the modal parameters.

b. Line-fit between $\text{Re}(\Delta)$ and ω^2 and $\text{Im}(\Delta)$ and ω^2

We choose a narrow frequency band around the peak to be analyzed in the given FRF and m number of frequencies ω_i, with $i = 1, 2, \ldots, m$ in this band. We also choose a set of fixing frequencies, Ω_s, with $s = 1, 2, \ldots, n$, such that, $\Omega_s \neq \omega_i$.

Rewriting Eq. (11.60) after taking $\Omega = \Omega_s$

$$g_R \omega^2 + h_R - \text{Re}(\Delta)_s = 0 \tag{11.82}$$

Let $\left(\omega_i{}^2, \text{Re}(\Delta)_{s,i}\right)$ represent the values of $\left(\omega^2, \text{Re}(\Delta)_s\right)$ at the frequency ω_i. If $\left(\omega_i{}^2, \text{Re}(\Delta)_{s,i}\right)$ is substituted into the LHS of Eq. (11.69), then the LHS magnitude represents the error in the line-fit. Hence, the total error at all the frequency points is defined as

$$E(g_R, h_R) = \sum_{i=1}^{m} \left(g_R \omega_i{}^2 + h_R - \text{Re}(\Delta)_{s,i}\right)^2 \tag{11.83}$$

For the minimum value of the error function, we should have

$$\frac{\partial E}{\partial g_R} = 0 \quad \text{and} \quad \frac{\partial E}{\partial h_R} = 0 \tag{11.84}$$

Applying the conditions in Eq. (11.84) to the error function and simplifying, we get the following matrix equation:

$$\begin{bmatrix} \sum \omega_i^4 & \sum \omega_i^2 \\ \sum \omega_i^2 & m \end{bmatrix} \left\{ \begin{array}{c} g_R \\ h_R \end{array} \right\} = \left\{ \begin{array}{c} \sum \left(\omega_i^2 \text{Re}(\Delta)_{s,i}\right) \\ \sum \text{Re}(\Delta)_{s,i} \end{array} \right\} \tag{11.85}$$

where the summation is over m frequencies.

Matrix equation (11.85) represents two simultaneous linear equations in two unknowns, and its solution gives the slope and intercept (g_R, h_R) of the line for one choice of the fixing frequency. The line-fit is repeated for all selected values of Ω_s. This yields n values of (g_R, h_R).

By proceeding similarly, we make a line-fit between $\text{Im}(\Delta)$ and ω^2 using the following matrix equation. We get n values of (g_I, h_I) for the line-fits corresponding to n values of Ω_s.

$$\begin{bmatrix} \sum \omega_i^4 & \sum \omega_i^2 \\ \sum \omega_i^2 & m \end{bmatrix} \left\{ \begin{array}{c} g_I \\ h_I \end{array} \right\} = \left\{ \begin{array}{c} \sum \left(\omega_i^2 \text{Im}(\Delta)_{s,i}\right) \\ \sum \text{Im}(\Delta)_{s,i} \end{array} \right\} \tag{11.86}$$

c. Line-fit between g_R and Ω^2 and g_I and Ω^2

We now fit a straight line to the set of values of g_R and Ω^2. Let us rewrite Eq. (11.75) as

$$n_R\Omega^2 + u_R - g_R = 0 \qquad (11.87)$$

Let $\left(\Omega_i^2, g_{R,i}\right)$ represent the values of $\left(\Omega^2,\ g_R\right)$ at fixing frequency Ω_i. The total error in the line-fit is defined

$$E(n_R, u_R) = \sum_{i=1}^{n}\left(n_R\Omega_i^2 + u_R - g_{R,i}\right)^2 \qquad (11.88)$$

For the minimum value of the error function, we should have

$$\frac{\partial E}{\partial n_R} = 0 \quad \text{and} \quad \frac{\partial E}{\partial u_R} = 0 \qquad (11.89)$$

Applying the conditions in Eq. (11.89) to the error function and simplifying, we get the following matrix equation:

$$\begin{bmatrix} \sum \Omega_i^4 & \sum \Omega_i^2 \\ \sum \Omega_i^2 & n \end{bmatrix}\begin{Bmatrix} n_R \\ u_R \end{Bmatrix} = \begin{Bmatrix} \sum\left(\omega_i^2 g_{R,i}\right) \\ \sum g_{R,i} \end{Bmatrix} \qquad (11.90)$$

where the summation is over n.

Matrix equation (11.90) represents two simultaneous linear equations in two unknowns, and its solution gives the slope and intercept (n_R, u_R) of the line.

By proceeding similarly, we fit a straight line between g_I and Ω^2. The matrix equation that needs to be solved to obtain (n_I, u_I) is,

$$\begin{bmatrix} \sum \Omega_i^4 & \sum \Omega_i^2 \\ \sum \Omega_i^2 & n \end{bmatrix}\begin{Bmatrix} n_I \\ u_I \end{Bmatrix} = \begin{Bmatrix} \sum\left(\omega_i^2 g_{I,i}\right) \\ \sum g_{I,i} \end{Bmatrix} \qquad (11.91)$$

Once (n_R, u_R) and (n_I, u_I) are available, solutions of Eqs. (11.77)–(11.81) gives the model parameters for the rth mode.

Example 11.4: Modal Parameter Estimation Using the Modified Line-Fit Method

In this example, we apply the modified line-fit procedure to our cantilever structure to analyze its transfer receptance α_{15}. We analyze the third peak in the FRF.

Step 1: Select a frequency band around the third peak. Also, choose fixing frequencies Ω_s, with s = 1, 2, ..., n, such that $\Omega_s \neq \omega_i$. For each Ω_s, we perform a line-fit between $\text{Re}(\Delta)$ and ω^2 using Eq. (11.85).
Figure 11.15 shows the plots of $\text{Re}(\Delta)$ versus ω^2 for various fixing frequencies Ω_s, with the corresponding line-fits. This yields 'n' values of (g_R, h_R).
Similarly, for each Ω_s, we fit a straight line between $\text{Im}(\Delta)$ and ω^2 using Eq. (11.86). Figure 11.16 shows the plots of $\text{Re}(\Delta)$ versus ω^2 for various fixing frequencies Ω_s, with the corresponding line-fits. This yields 'n' values of (g_I, h_I).
Step 2: We now fit a straight line to the set of values of g_R and Ω^2 using Eq. (11.90). Figure 11.17 shows the plot of g_R versus Ω^2 with the corresponding line-fit. The slope and intercept of the line are obtained as $(n_R, u_R) = (0.2382,\ -281037)$
Similarly, a straight line is fitted to the set of values of g_I and Ω^2 using Eq. (11.91). Figure 11.18 shows the plot of g_I versus Ω^2 with the corresponding lie fit. The slope and intercept of the line are obtained as $(n_I, u_I) = (0.0005,\ -5171.02)$
Step 3: Using the values of (n_R, u_R) and (n_I, u_I), the modal parameters are obtained from Eqs. (11.77-11.81) as

$$\omega_3 = 2\pi \times 172.84 \text{ rad/sec}; \quad \eta_3 = 0.0159; \quad _3A_{15} = c + id = 4.196 - i\,0.01$$

Step 4:
Steps 1–4 are repeated for the other three peaks in the FRF.

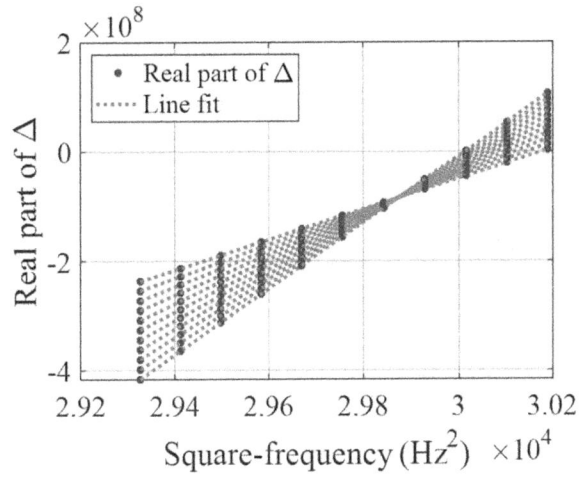

FIGURE 11.15 $\mathrm{Re}(\Delta)$ versus ω^2 for various fixing frequencies Ω_s and the corresponding line-fits.

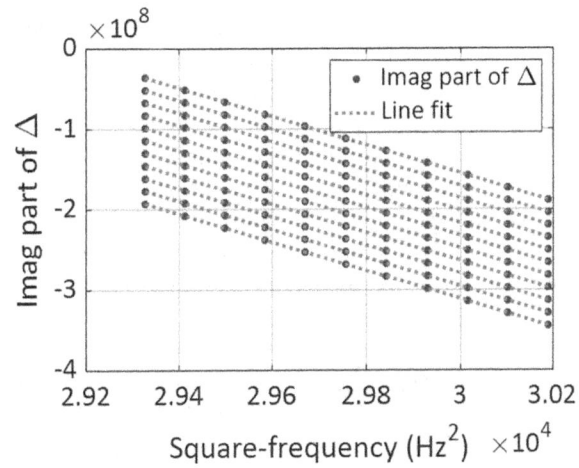

FIGURE 11.16 $\mathrm{Im}(\Delta)$ versus ω^2 for various fixing frequencies Ω_s and the corresponding line-fits.

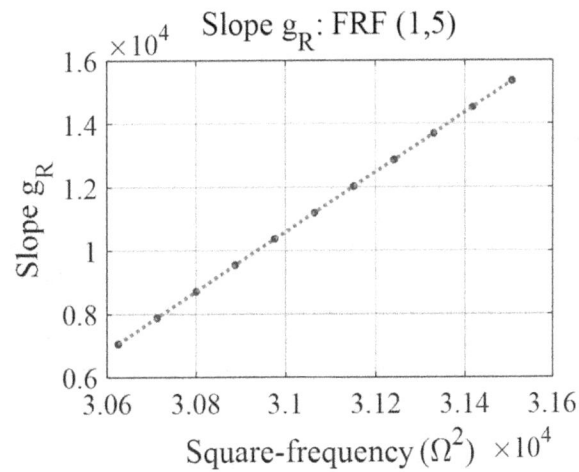

FIGURE 11.17 g_R versus Ω^2 with the corresponding line-fit.

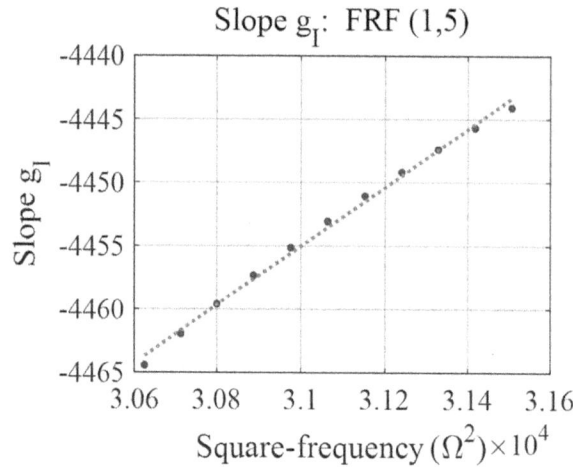

FIGURE 11.18 g_I versus Ω^2 with the corresponding line-fit.

11.8 RESIDUALS

Modal parameter estimation in the frequency domain is performed in a selected frequency band to identify the modes in the frequency band. However, it should be noted that the FRF is contributed by the modes inside and outside the selected band. The contribution of the modes outside the band is called **residual**. If the residual is not accounted for in the modal parameter estimation, then the residual contribution is attributed to the modes in the analysis frequency band, which adversely affects the accuracy of the estimated modal parameters.

In the previous sections, we discussed that the SDOF methods of modal parameter estimation select a frequency band covering only one mode at a time. The residual contribution is either neglected, like in the peak-picking and line-fit methods or is approximated as a constant, like in the circle-fit and modified line-fit methods.

Let us look further at the nature of the residual contribution and how it can be approximated for modal parameter extraction. The receptance of a system with structural damping is given by

$$\alpha_{jk}(\omega) = \sum_{r=1}^{N} \frac{{}_rA_{jk}}{\omega_r^2 - \omega^2 + i\eta_r\omega_r^2} \tag{11.92}$$

Let us say the frequency band of analysis contains the modes from m_1 to m_2. Eq. (11.92) can then be expressed as

$$\alpha_{jk}(\omega) = \sum_{r=1}^{m_1-1} \frac{{}_rA_{jk}}{\omega_r^2 - \omega^2 + i\eta_r\omega_r^2} + \sum_{r=m_1}^{m_2} \frac{{}_rA_{jk}}{\omega_r^2 - \omega^2 + i\eta_r\omega_r^2} + \sum_{r=m_2+1}^{N} \frac{{}_rA_{jk}}{\omega_r^2 - \omega^2 + i\eta_r\omega_r^2} \tag{11.93}$$

$$= \alpha_{jk}^{L}(\omega) + \alpha_{jk}^{Band}(\omega) + \alpha_{jk}^{H}(\omega)$$

where $\alpha_{jk}^{Band}(\omega)$ is the contribution of the modes inside the band. $\alpha_{jk}^{L}(\omega)$ and $\alpha_{jk}^{H}(\omega)$ are the contributions of the outside modes from the low and high-frequency sides of the band and are referred to as low and high-frequency residuals, respectively. Figure 11.19a shows the magnitudes of $\alpha_{jk}(\omega)$ and $\alpha_{jk}^{Band}(\omega)$ for a structure. The lower and upper limits of the frequency band covering modes 2 and 3 are shown by two vertical lines. The difference between the two curves inside the band is nothing but the total residual contribution. The plot shows that ignoring the residual contribution during curve-fitting in the analysis band amounts to forcing the fit of $\alpha_{jk}^{Band}(\omega)$ onto $\alpha_{jk}(\omega)$, which distorts the modal parameter estimates.

Figure 11.19b shows the magnitude of $\alpha_{jk}^{L}(\omega)$ on a logarithmic frequency axis. It is seen that the low-frequency residual in the frequency band can be approximated as the mass line of the receptance of an SDOF system. The receptance based on the mass line of an SDOF system can be written (from Chapter 7) as

$$\alpha_{jk}^{L}(\omega) \approx -\frac{1}{M_{jk}\omega^2} \tag{11.94}$$

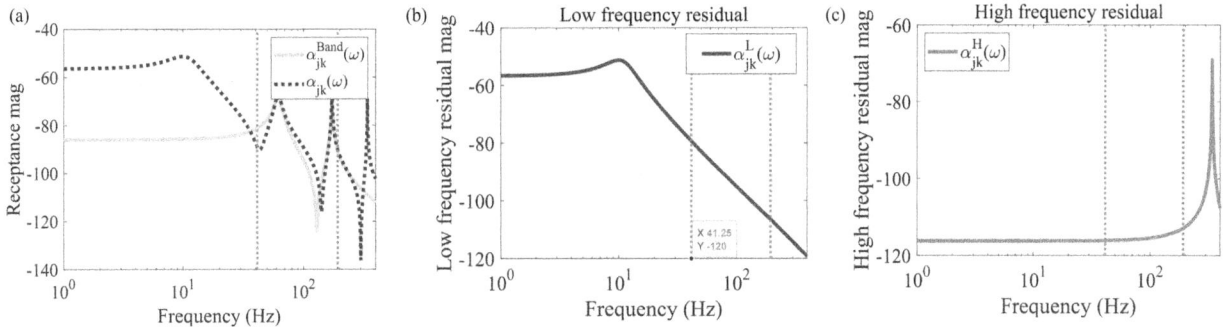

FIGURE 11.19 Magnitudes of a) $\alpha_{jk}(\omega)$ and $\alpha_{jk}^{Band}(\omega)$ b) $\alpha_{jk}^{L}(\omega)$ and c) $\alpha_{jk}^{H}(\omega)$.

Figure 11.19c shows the magnitude of $\alpha_{jk}^{H}(\omega)$ on a logarithmic frequency axis. It is seen that the high-frequency residual in the frequency band can be approximated as the stiffness line of the receptance of an SDOF system. The receptance based on the stiffness line of an SDOF system can be written (from Chapter 7) as

$$\alpha_{jk}^{H}(\omega) \approx \frac{1}{K_{jk}} \tag{11.95}$$

Thus, Eqs. (11.94) and (11.95) provide approximations to the unknown low and high-frequency residuals. In general, more than one out-of-band mode makes up the residual terms, but in these equations, the residuals have been approximated as SDOF contributions. With these approximations, Eq. (11.93) can be written as

$$\alpha_{jk}(\omega) = -\frac{1}{M_{jk}\omega^2} + \sum_{r=m_1}^{m_2} \frac{{}_r A_{jk}}{\omega_r^2 - \omega^2 + i\omega_r\omega_r^2} + \frac{1}{K_{jk}} \tag{11.96}$$

Therefore, in addition to the unknowns related to the modes m_1 to m_2, there are two more unknowns, M_{jk} and K_{jk}, the low and high-frequency residual coefficients, respectively, to be determined in the curve-fitting.

Another approach is to model the residuals as pseudo-modes, where each residual is modeled as a pseudo-mode contribution described by $1/\left(K_{jk} - M_{jk}\omega^2\right)$. In that case, there are four more unknowns.

Apart from the fact that the modeling of residuals is necessary for accurate modal parameter estimation, the residuals are also valuable to accurately regenerate or reconstruct the corresponding FRF using the identified modal parameters. It allows a more accurate dynamic response prediction to given forces. The low and high-frequency residuals for an inertance FRF can be similarly worked out using the stiffness and mass line characteristics of the inertance studied in Chapter 7.

11.9 RATIONAL FRACTION POLYNOMIAL METHOD

The methods considered in the previous sections were SDOF methods, which performed curve-fitting of an FRF mode-by-mode. This approach has the shortcoming that the contribution of out-of-band modes is not correctly accounted for, affecting the accuracy of the estimated modal parameters. The rational fraction polynomial (RFP) method (developed by Richardson and Formenti [92]) is an MDOF curve-fitting method and can simultaneously fit multiple peaks in the FRF. The MDOF curve-fitting ensures that the modal contributions of all the considered modes are accounted for. RFP is a method in the frequency domain.

The curve-fitting of FRFs one at a time is referred to as the 'local' approach. In the 'global' approach, all the FRFs are processed simultaneously to obtain a consistent estimate of the modal parameters. This section considers the local RFP, while the global RFP is considered after that.

RFP is based on expressing an FRF of the system in the form of a rational fraction, a ratio of two polynomials. The formulation of the method is based on the viscous damping model. From Section 11.3, we know that the FRF of an MDOF system with viscous damping can be written as

$$H_{jk}(\omega) = \sum_{r=1}^{N} \left(\frac{{}_r R_{jk}}{i\omega - s_r} + \frac{{}_r R_{jk}^*}{i\omega - s_r^*} \right) \tag{11.97}$$

As mentioned before, Eq. (11.97) is the pole-residue form with s_r and s_r^* being the complex conjugate poles of the rth mode, and $_rR_{jk}$ and $_rR_{jk}^*$ being the corresponding residues for the FRF H_{jk}. Since the natural frequencies, modal viscous damping factors, and mode shapes are related to the poles and residues, RFP aims to estimate the poles and residues from the measured FRFs.

Note that the FRF in Eq. (11.97) is in the partial fraction form. However, RFP writes this equation in the rational fraction form, which (after dropping the subscript 'jk' for brevity) is given by

$$H(\omega) = \frac{\sum_{k=0}^{m} a_k (i\omega)^k}{\sum_{k=0}^{n} b_k (i\omega)^k} \tag{11.98}$$

m and n in Eq. (11.98) are the upper limits of the summations, which for the FRF in Eq. (11.97) are m = 2N − 1 and n = 2N. With the FRF in the rational fraction form, the coefficients a_k and b_k are the primary unknowns and not the poles and residues. However, the poles and residues are related to these coefficients and can be found once the coefficients are determined.

11.9.1 RFP FORMULATION USING ORDINARY POLYNOMIALS

The numerator and denominator polynomials in Eq. (11.98) are in the form of ordinary polynomials (the polynomials involving the increasing power of the independent variable ($i\omega$ in the present case) as the basis functions). We first look at the RFP formulation based on Eq. (11.98), and an alternative formulation is considered in the next section.

In Eq. (11.98), we can choose one of the coefficients arbitrarily as unity, which is the same result obtained by dividing the numerator and denominator by that coefficient. Given this, we rewrite Eq. (11.98), with the coefficient of the highest degree term in the denominator chosen as unity,

$$H(\omega) = \frac{\sum_{k=0}^{m} a_k (i\omega)^k}{\sum_{k=0}^{n-1} b_k (i\omega)^k + (i\omega)^n} \tag{11.99}$$

Let $H_m(\omega)$ represent the corresponding measured FRF. The difference between the curve-fit and the measured FRF is

$$\Delta H(\omega) = H(\omega) - H_m(\omega) \tag{11.100}$$

Substituting Eq. (11.99) into (11.100), we get

$$\Delta H(\omega) = \frac{\sum_{k=0}^{m} a_k (i\omega)^k - H_m(i\omega)\left(\sum_{k=0}^{n-1} b_k (i\omega)^k + (i\omega)^n\right)}{\sum_{k=0}^{n} b_k (i\omega)^k} \tag{11.101}$$

The numerator of Eq. (11.101) ideally should be zero, as this makes $\Delta H(\omega)$ zero, and, therefore, can be taken as representing the error in the curve-fit at frequency ω. Hence,

$$e(\omega) = \sum_{k=0}^{m} a_k (i\omega)^k - H_m(\omega)\left(\sum_{k=0}^{n-1} b_k (i\omega)^k + (i\omega)^n\right) \tag{11.102}$$

Using Eq. (11.102), the error $e(\omega)$ at discrete frequencies $\omega = \omega_j$, with j = 1, 2, …, L can be written down, and the L equations so obtained can be combined and represented by the following matrix equation:

$$\{E\} = [P]\{a\} - [T]\{b\} - \{w\} \tag{11.103}$$

The vectors and matrices in Eq. (11.103) are

$$\{E\} = \begin{Bmatrix} e(\omega_1) \\ e(\omega_2) \\ \vdots \\ e(\omega_L) \end{Bmatrix}, \quad \{w\} = \begin{Bmatrix} H_m(\omega_1)(i\omega_1)^n \\ H_m(\omega_2)(i\omega_2)^n \\ \vdots \\ H_m(\omega_L)(i\omega_L)^n \end{Bmatrix} \tag{11.104}$$

$$[P] = \begin{bmatrix} (i\omega_1)^0 & (i\omega_1)^1 & \cdots & (i\omega_1)^m \\ (i\omega_2)^0 & (i\omega_2)^1 & \cdots & (i\omega_2)^m \\ \vdots & \vdots & \cdots & \vdots \\ (i\omega_L)^0 & (i\omega_L)^1 & \cdots & (i\omega_L)^m \end{bmatrix}, \quad \{a\} = \begin{Bmatrix} a_0 \\ a_1 \\ \vdots \\ a_m \end{Bmatrix} \tag{11.105}$$

$$[T] = \begin{bmatrix} H_m(\omega_1)(i\omega_1)^0 & H_m(\omega_1)(i\omega_1)^1 & \cdots & H_m(\omega_1)(i\omega_1)^{n-1} \\ H_m(\omega_2)(i\omega_2)^0 & H_m(\omega_2)(i\omega_2)^1 & \cdots & H_m(\omega_2)(i\omega_2)^{n-1} \\ \vdots & \vdots & \cdots & \vdots \\ H_m(\omega_L)(i\omega_L)^0 & H_m(\omega_L)(i\omega_L)^1 & \cdots & H_m(\omega_L)(i\omega_L)^{n-1} \end{bmatrix}, \quad \{b\} = \begin{Bmatrix} b_0 \\ b_1 \\ \vdots \\ b_{n-1} \end{Bmatrix} \tag{11.106}$$

The objective function, defined as the sum of squares of the errors, is

$$J = \{E\}^T \{E\} \tag{11.107}$$

The objective function J is a quadratic function of the unknown polynomial coefficients. The partial derivative of J with each unknown coefficient must be zero at its minimum value. Hence

$$\frac{\partial J}{\partial \{a\}} = 2[P]^H[P]\{a\} - 2\mathrm{Re}\left([P]^H[T]\{b\}\right) - 2\mathrm{Re}\left([P]^H\{w\}\right) = 0 \tag{11.108}$$

$$\frac{\partial J}{\partial \{b\}} = 2[T]^H[T]\{b\} - 2\mathrm{Re}\left([T]^H[P]\{a\}\right) - 2\mathrm{Re}\left([T]^H\{w\}\right) = 0 \tag{11.109}$$

where superscript 'H' represents Hermitian transpose.

Equations (11.108) and (11.109) can be written in the matrix form as

$$\begin{bmatrix} [P]^H[P] & -\mathrm{Re}\left([P]^H[T]\right) \\ -\mathrm{Re}\left([T]^H[P]\right) & [T]^H[T] \end{bmatrix} \times \begin{Bmatrix} \{a\} \\ \{b\} \end{Bmatrix} = \begin{Bmatrix} \mathrm{Re}\left([P]^H\{w\}\right) \\ \mathrm{Re}\left([T]^H\{w\}\right) \end{Bmatrix} \tag{11.110}$$

Eq. (11.110) is a set of simultaneous linear equations whose solution yields $\{a\}$ and $\{b\}$. Once the coefficient vector $\{b\}$ is available, the roots of the denominator polynomial of Eq. (11.99) are found using any root-finding procedure. These are nothing but the poles of the system. From the knowledge of the poles and the coefficient vector $\{a\}$, the residues corresponding to the poles are obtained by applying the pole-residue theorem. From the knowledge of the poles, it is straightforward to obtain the natural frequencies and modal viscous damping factors, while the elements of the mode shapes are obtained from the residues. All these steps are considered in more detail in the next section.

One advantage of the RFP formulation is that it requires the solution of linear equations to find the primary unknowns compared to the solution of a nonlinear problem for the MDOF curve-fit based on the partial fraction form. Finding the roots of the denominator polynomial to obtain the poles does require solving a nonlinear equation, but it is much less demanding, as well-established root-finding methods are available.

It is experienced that the solution of matrix equation (11.110) is difficult as the coefficient matrix in this equation is found to be ill-conditioned. Due to ill-conditioning, a small error in the equations, like a truncation error, results in a significant error in the solution. To overcome this difficulty, the RFP was reformulated using orthogonal polynomials presented in the following section.

11.9.2   RFP Formulation Using Orthogonal Polynomials

a. Basic idea of the method

The RFP was formulated using orthogonal polynomials to address the numerical difficulties in solving the resulting equations obtained using ordinary polynomials. Two polynomials are orthogonal if the integration of their product over the domain is zero. It is found that the condition of the final equations is significantly improved when the orthogonal polynomials are used, which makes the equations amenable to an accurate solution.

Let $\phi_k(i\omega)$, with $k = 1, 2, ..., m$ be a set of orthogonal polynomials. Note that in the rational fraction form of the FRF based on the ordinary polynomials, the basis functions forming the polynomials were $(i\omega)^k$, with $k = 1, 2, ..., m$ and were not orthogonal. These functions cause ill-conditioning due to the increasing powers of ω with k. Instead, $\phi_k(i\omega)$ are used as the basis functions for the numerator polynomial. Similarly, let $\theta_k(i\omega)$, with $k = 1, 2, ..., n$ be another set of orthogonal polynomials used as the basis functions for the denominator polynomial.

With this, Eq. (11.99), in the form of orthogonal polynomials, is written as

$$H(\omega) = \frac{\sum_{k=0}^{m} c_k \phi_k(i\omega)}{\sum_{k=0}^{n-1} d_k \theta_k(i\omega) + \theta_n(i\omega)} \tag{11.111}$$

As in the previous section, the coefficient of the highest degree polynomial in the denominator is chosen as unity. Once the orthogonal polynomials are generated, as presented below, the coefficients c_k and d_k are the unknowns of the curve-fitting problem. These coefficients then need to be transformed into ordinary polynomial coefficients, from which the poles and residues are obtained, leading to the modal parameters.

b. Generation of orthogonal polynomials

The generation of orthogonal polynomials is done using the Forsythe method. Richardson and Formenti [92] developed a modified approach utilizing the Hermitian symmetry of the FRFs and the properties of the orthogonal polynomials.

Let us, as before, choose some frequency points $\omega = \omega_j$, with $j = 1, 2, ..., L$ in the frequency band selected for curve-fitting. Let us represent the value of the kth orthogonal polynomial at a frequency ω_j by symbol $P_{j,k}$. $P_{j,k}$ therefore, can represent either $\phi_k(\omega_j)$ or $\theta_k(\omega_j)$.

$P_{j,k}$ is computed using the following recursive relations:

$$P_{j,k} = (i)^k R_{j,k} \tag{11.112}$$

$$R_{j,-1} = 0 \tag{11.113}$$

$$R_{j,0} = \frac{1}{\sqrt{2\sum_{j=1}^{L} q_j}} \tag{11.114}$$

$$S_{j,k} = \omega_j R_{j,k-1} - V_{k-1} R_{j,k-2} \quad \text{for } k = 1,2,... \tag{11.115}$$

where

$$V_{k-1} = 2\sum_{j=1}^{L}\left(\omega_j R_{j,k-1} R_{j,k-2} q_j\right) \tag{11.116}$$

$$R_{j,k} = \frac{S_{j,k}}{D_k} \tag{11.117}$$

$$D_k = \sqrt{2 \sum_{j=1}^{L} \left(S_{j,k}\right)^2 q_j} \tag{11.118}$$

- Equations (11.112)–(11.118) are used to generate numerator orthogonal polynomials $\phi_k(\omega)$, for $k = 1, 2, \ldots, m$ at frequency points $\omega = \omega_j$, with $j = 1, 2, \ldots, L$. The weighting function q_j is taken as

$$q_j = 1 \tag{11.119}$$

- Equations (11.112)–(11.118) are also used to generate denominator orthogonal polynomials $\theta_k(\omega)$, for $k = 1, 2, \ldots, n$ at frequency points $\omega = \omega_j$, with $j = 1, 2, \ldots, L$. The weighting function q_j is taken as

$$q_j = \left| H_m\left(\omega_j\right) \right|^2 \tag{11.120}$$

It is seen that the orthogonal polynomials so generated satisfy the following orthogonality relationships:

$$\sum_{j=1}^{L} \left(\phi_k\left(\omega_j\right)\right)^H \phi_s\left(\omega_j\right) = \begin{cases} 0 & \text{for } k \neq s \\ 0.5 & \text{for } k = s \end{cases} \tag{11.121}$$

$$\sum_{j=1}^{L} \left(\theta_k\left(\omega_j\right)\right)^H \left| H_m\left(\omega_j\right) \right|^2 \theta_s\left(\omega_j\right) = \begin{cases} 0 & \text{for } k \neq s \\ 0.5 & \text{for } k = s \end{cases} \tag{11.122}$$

c. Formulation of the equations

Starting from the rational fraction form in Eq. (11.111) and following the steps similar to those in Eqs. (11.100)–(11.103), we get

$$\{E_o\} = [P_o]\{c\} - [T_o]\{d\} - \{w_o\} \tag{11.123}$$

where subscript 'o' indicates that the corresponding quantities are based on orthogonal polynomials. The various vectors and matrices in Eq. (11.123) are given by

$$\{E_o\} = \begin{Bmatrix} e_o(\omega_1) \\ e_o(\omega_2) \\ \vdots \\ e_o(\omega_L) \end{Bmatrix}, \quad \{w_o\} = \begin{Bmatrix} H_m(\omega_1)\theta_n(\omega_1) \\ H_m(\omega_2)\theta_n(\omega_2) \\ \vdots \\ H_m(\omega_L)\theta_n(\omega_L) \end{Bmatrix} \tag{11.124}$$

$$[P_o] = \begin{bmatrix} \phi_0(\omega_1) & \phi_1(\omega_1) & \cdots & \phi_m(\omega_1) \\ \phi_0(\omega_2) & \phi_1(\omega_2) & \cdots & \phi_m(\omega_2) \\ \vdots & \vdots & \cdots & \vdots \\ \phi_0(\omega_L) & \phi_1(\omega_L) & \cdots & \phi_m(\omega_L) \end{bmatrix}, \quad \{c\} = \begin{Bmatrix} c_0 \\ c_1 \\ \vdots \\ c_m \end{Bmatrix} \tag{11.125}$$

$$[T_o] = \begin{bmatrix} H_m(\omega_1)\theta_0(\omega_1) & H_m(\omega_1)\theta_1(\omega_1) & \cdots & H_m(\omega_1)\theta_{n-1}(\omega_1) \\ H_m(\omega_2)\theta_0(\omega_2) & H_m(\omega_2)\theta_1(\omega_2) & \cdots & H_m(\omega_2)\theta_{n-1}(\omega_2) \\ \vdots & \vdots & \cdots & \vdots \\ H_m(\omega_L)\theta_0(\omega_L) & H_m(\omega_L)\theta_1(\omega_L) & \cdots & H_m(\omega_L)\theta_{n-1}(\omega_L) \end{bmatrix}, \quad \{d\} = \begin{Bmatrix} d_0 \\ d_1 \\ \vdots \\ d_{n-1} \end{Bmatrix} \tag{11.126}$$

Following the steps similar to those in Eqs. (11.107)–(11.110), we get the following set of linear equations:

$$
\begin{bmatrix}
[P_o]^H[P_o] & -\mathrm{Re}\left([P_o]^H[T_o]\right) \\
-\mathrm{Re}\left([T_o]^H[P_o]\right) & [T_o]^H[T_o]
\end{bmatrix}
\begin{Bmatrix}
\{c\} \\
\{d\}
\end{Bmatrix}
=
\begin{Bmatrix}
\mathrm{Re}\left([P_o]^H\{w_o\}\right) \\
\mathrm{Re}\left([T_o]^H\{w_o\}\right)
\end{Bmatrix}
\tag{11.127}
$$

It is seen that due to the orthogonality properties of the orthogonal polynomials

$$
[P_o]^H[P_o] = \frac{1}{2}[I]
\tag{11.128}
$$

$$
[T_o]^H[T_o] = \frac{1}{2}[I]
\tag{11.129}
$$

and

$$
\mathrm{Re}\left([T_o]^H\{w_o\}\right) = \{0\}
\tag{11.130}
$$

Let

$$
[X] = -\mathrm{Re}\left([P_o]^H[T_o]\right) \text{ and } \{G\} = \mathrm{Re}\left([P_o]^H\{w_o\}\right)
\tag{11.131}
$$

After making substitutions from Eqs. (11.128) to (11.131), Eq. (11.127) takes the following form:

$$
\begin{bmatrix}
0.5[I] & [X] \\
[X]^T & 0.5[I]
\end{bmatrix}
\begin{Bmatrix}
\{c\} \\
\{d\}
\end{Bmatrix}
=
\begin{Bmatrix}
\{G\} \\
\{0\}
\end{Bmatrix}
\tag{11.132}
$$

Using the second part of Eq. (11.132), $\{c\}$ is eliminated from the first part, leading to

$$
\left([I] - 4[X]^T[X]\right)\{d\} = -4[X]^T\{G\}
\tag{11.133}
$$

The first part of Eq.(11.132) is

$$
\{c\} = \frac{1}{2}\left(\{G\} - [X]\{d\}\right)
\tag{11.134}
$$

Eq. (11.133) is first solved for the denominator coefficients, after which Eq. (11.120) yields the numerator coefficients.

d. Determination of ordinary polynomial coefficients

The next step is to convert the orthogonal polynomial coefficients into ordinary polynomial coefficients. To perform this, we develop a transformation matrix as follows.

To obtain ordinary polynomial coefficients $\{a\}$ from $\{c\}$, we write the numerator polynomial at m+1 number of frequency points using the orthogonal and ordinary numerator polynomial coefficients and equate them as follows:

$$
\begin{Bmatrix}
(i\omega_1)^0 & (i\omega_1)^1 & \cdots & (i\omega_1)^m \\
(i\omega_2)^0 & (i\omega_2)^1 & \cdots & (i\omega_2)^m \\
\vdots & \vdots & \cdots & \vdots \\
(i\omega_{m+1})^0 & (i\omega_{m+1})^1 & \cdots & (i\omega_{m+1})^m
\end{Bmatrix}
\begin{Bmatrix}
a_0 \\
a_1 \\
\vdots \\
a_m
\end{Bmatrix}
=
\begin{bmatrix}
\phi_0(\omega_1) & \phi_1(\omega_1) & \cdots & \phi_m(\omega_1) \\
\phi_0(\omega_2) & \phi_1(\omega_2) & \cdots & \phi_m(\omega_2) \\
\vdots & \vdots & \cdots & \vdots \\
\phi_0(\omega_{m+1}) & \phi_1(\omega_{m+1}) & \cdots & \phi_m(\omega_{m+1})
\end{bmatrix}
\begin{Bmatrix}
c_0 \\
c_1 \\
\vdots \\
c_m
\end{Bmatrix}
\tag{11.135}
$$

Representing the coefficient matrices as $[A_1]$ and $[A_2]$, we get

$$[A_1]\{a\} = [A_2]\{c\} \tag{11.136}$$

The numerator ordinary polynomial coefficients vector is obtained as

$$\{a\} = [A_1]^{-1}[A_2]\{c\} \tag{11.137}$$

To obtain $\{b\}$ from $\{d\}$, we write the denominator polynomial at n+1 number of frequency points using the orthogonal and ordinary denominator polynomial coefficients and equate them as follows:

$$
\begin{bmatrix}
(i\omega_1)^0 & (i\omega_1)^1 & \dots & (i\omega_1)^m \\
(i\omega_2)^0 & (i\omega_2)^1 & \dots & (i\omega_2)^m \\
\vdots & \vdots & \dots & \vdots \\
(i\omega_{n+1})^0 & (i\omega_{n+1})^1 & \dots & (i\omega_{n+1})^m
\end{bmatrix}
\begin{Bmatrix}
b_0 \\ b_1 \\ \vdots \\ b_n
\end{Bmatrix}
=
\begin{bmatrix}
\theta_0(\omega_1) & \theta_1(\omega_1) & \dots & \theta_{n+1}(\omega_1) \\
\theta_0(\omega_2) & \theta_1(\omega_2) & \dots & \theta_{n+1}(\omega_2) \\
\vdots & \vdots & \dots & \vdots \\
\theta_0(\omega_{n+1}) & \theta_1(\omega_{n+1}) & \dots & \theta_{n+1}(\omega_{n+1})
\end{bmatrix}
\begin{Bmatrix}
d_0 \\ d_1 \\ \vdots \\ 1
\end{Bmatrix} \tag{11.138}
$$

Representing the coefficient matrices as $[B_1]$ and $[B_2]$, we get

$$[B_1]\{b\} = [B_2]\{d\} \tag{11.139}$$

The denominator ordinary polynomial coefficients vector is obtained as

$$\{b\} = [B_1]^{-1}[B_2]\{d\} \tag{11.140}$$

e. Computation of poles

Once the denominator ordinary polynomial coefficients ($\{b\}$) are known, the poles are obtained by solving the characteristics equation given by

$$\sum_{k=0}^{n} b_k(i\omega)^k = b_0(i\omega)^0 + b_1(i\omega)^2 + \dots + b_n(i\omega)^n = 0 \tag{11.141}$$

The roots of Eq. (11.141) are the poles of the system and can be obtained using any standard root solver. The poles are represented by s_r, with r = 1, 2, ...n and are in complex conjugate pairs.

f. Computation of residues

Computation of residue corresponding to each pole can be done using the pole-residue theorem. Since the poles are known, the FRF (now written using the full notation H_{jk}) can be expressed in the partial fraction form. From Eq. (11.97), we can write,

$$H_{jk}(i\omega) = \sum_{r=1}^{n} \frac{{}_rR_{jk}}{i\omega - s_r} \tag{11.142}$$

Note that on the RHS in Eq. (11.142), we have the sum over n poles rather than over n/2 complex conjugate pairs (as in Eq. 11.97). To find residue ${}_qR_{jk}$ of the qth pole, s_q, we multiply both sides by $(i\omega - s_q)$ and take the limit $i\omega$ approaching s_q. It isolates the term corresponding to the qth pole on the RHS, and we get

$${}_qR_{jk} = \lim_{i\omega \to s_r} H_{jk}(i\omega)(i\omega - s_q) \tag{11.143}$$

which can be written as

$${}_qR_{jk} = \frac{\sum_{k=0}^{m} {}_qa_k(s_q)^k}{\prod_{u=1,\ \neq q}^{n}(s_q - s_u)} \tag{11.144}$$

g. Computation of modal parameters

Natural frequencies and damping factors are obtained from the poles by the following relationships:

$$\omega_r = \sqrt{\left(\text{Re}(s_r)\right)^2 + \left(\text{Imag}(s_r)\right)^2} \tag{11.145}$$

$$\xi_r = \frac{-\text{Re}(s_r)}{\omega_r} \tag{11.146}$$

We note from Eq. (11.10) that

$$_r R_{jk} = \phi_{jr}\ \phi_{kr} \tag{11.147}$$

Once we have residues for all the FRFs, the mass-normalized mode shapes can be obtained using Eq. (11.147) and the procedure described in Section 11.4.

h. Accounting residuals in RFP

We saw in Section 11.8 that the residuals could be approximated by low and high-frequency residuals or through pseudo-modes. The discussion in that section was about the systems with structural damping, but the concepts presented also apply to the systems with viscous damping.

Residuals can be easily accounted for in curve-fitting using the RFP method. To account for the residuals in RFP, we should look at how the residual approximation terms affect the FRF in the rational fraction form. Adding a residual term essentially alters the degree of the numerator and/or denominator polynomial.

For example, let us include the low-frequency residual term in the FRF expression in the rational fraction form. It gives

$$H(\omega) = \frac{1}{M_{jk}(i\omega)^2} + \frac{\sum_{k=0}^{m} a_k (i\omega)^k}{\sum_{k=0}^{n} b_k (i\omega)^k} \tag{11.148}$$

If we combine the two terms on the RHS, the degree of the numerator and denominator polynomials go up by 2. Therefore, the low-frequency residual can be accounted for during the curve-fitting with RFP by increasing the variables m and n each by 2. Thus, by appropriately increasing the variables m and n, the residuals can be accounted for in curve-fitting by the RFP method.

i. Curve-fitting of inertance using RFP

The degree of the numerator of the inertance is two more than that of the receptance. Hence increasing the number of coefficients (m) in the numerator by two, the curve-fitting of the inertance using RFP can be carried out.

Example 11.5: Modal Parameter Estimation Using RFP Method

In this example, we apply the RFP method to our cantilever structure to analyze its driving point receptance H_{55}. We perform an MDOF curve-fit covering the first three peaks/modes in the FRF.

Step 1: We start by selecting the frequency band covering the first three peaks and choosing several frequency points in this band, as shown in Figure 11.20. The frequency points should be more than the number of unknown coefficients and be higher to force a least-squares fit.
Step 2: We now choose the degree of the numerator and denominator polynomials. They depend on the number of modes we expect in the frequency band. In many cases, counting the number of peaks in the band may be good enough to guess the number of modes, but it may not be straightforward in cases involving close modes, higher damping levels, and random noise. More sophisticated methods like mode indicator functions (discussed later in this chapter and Chapter 12) can be used in such cases. The stabilization diagram (to be discussed in later sections) is another alternative.
In the present case, from the FRF, it is clear that there are three modes. In addition, let us also include the high-frequency residual to approximate the contribution of modes beyond the upper limit of the frequency band in the form of a pseudo-mode and hence increase the degree of both the polynomials by 2. Therefore, the degree of the denominator polynomial is $n = 2\times3+2 = 8$, and the degree of the numerator polynomial is $m = 2\times3-1+2 = 7$.

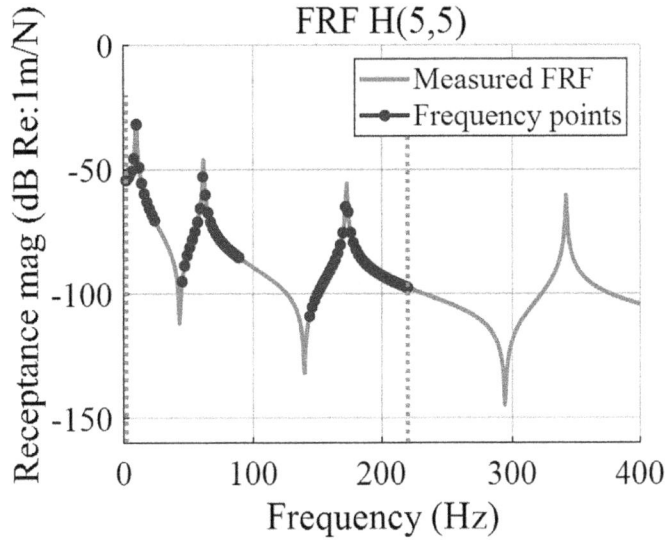

FIGURE 11.20 Selected frequency band and frequency points covering the first three peaks.

Step 3: We generate the orthogonal polynomials at the chosen frequency points using the recursive relations in Section 11.9.2(b). The numbers of the denominator and numerator orthogonal polynomials are $n+1$ and $m+1$, i.e., 9 and 8, respectively.

Step 4: The matrices $[X]$ and $\{G\}$ are set up using the orthogonal polynomials using the equations given in Section 11.9.2(c). Then Eqs. (11.133) and (11.134) are solved to find the denominator coefficients ($\{d\}$) and the numerator coefficients ($\{c\}$).

Step 5: The ordinary polynomial coefficients, $\{b\}$ and $\{a\}$, are computed using Eqs. (11.137) and (11.140).

Step 6: Poles and residues are now obtained. In this example, the function 'residue' in MATLAB® is used. The function requires the numerator and denominator polynomial coefficients as input. It gives the poles, the corresponding residues, and the direct term, if any, as the outputs. The direct term is zero when the degree of the numerator polynomial is less than the degree of the denominator polynomial, as in the present case.

The poles and residues obtained are shown in Table 11.1. Note that there are eight poles since the degree of the denominator polynomial is 8. The first three pairs of complex conjugate poles correspond to the three modes in the frequency band. The last pole pair is the high-frequency residual.

Step 7: From the poles with positive imaginary parts, the corresponding natural frequencies, and modal viscous damping factors are computed using Eqs. (11.145) and (11.146). These are:

$$\omega_1 = 2\pi\, 9.81 \text{ rad/sec} \quad \text{and} \quad \xi_1 = 0.0283$$

$$\omega_2 = 2\pi\, 61.54 \text{ rad/sec} \quad \text{and} \quad \xi_2 = 0.00452$$

$$\omega_3 = 2\pi\, 172.86 \text{ rad/sec} \quad \text{and} \quad \xi_3 = 0.00161$$

Step 8: We regenerate the FRF using the identified poles and residues. Figure 11.21 shows an overlay of the regenerated and the 'measured' point FRF, which shows an excellent fit in the frequency band in which curve-fitting was done, fitting three modes simultaneously.

TABLE 11.1
Poles and Residues

S. No.	Pole	Residue
1, 2	$-1.7489 \pm 61.65i$	$-1.52e\text{-}06 \pm 0.0554i$
3, 4	$-1.7486 \pm 386.69i$	$4.33e\text{-}07 \pm 0.0088i$
5, 6	$-1.7502 \pm 1086.12i$	$1.42e\text{-}07 \pm 0.0031i$
7, 8	$-4.7988 \pm 2371.46i$	$0.0001 \pm 0.0029i$

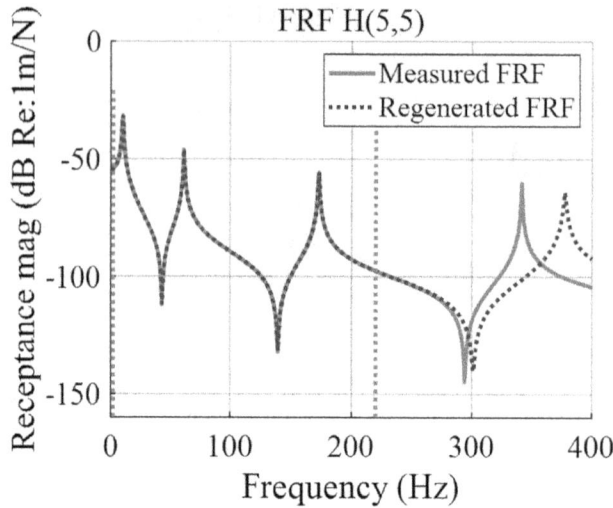

FIGURE 11.21   Overlay of the regenerated and the 'measured' FRF.

11.10   GLOBAL RFP METHOD

a. Basic idea of the method

The RFP presented in Section 11.9 is a local method as it is used to analyze one FRF at a time. When each measured FRF is analyzed separately, the estimates of the natural frequencies and damping factors obtained from different FRFs may have some variance. Local methods or single FRF methods rely on an averaging process to get a single estimate. However, such an estimate doesn't ensure an optimal fit of all the FRFs.

Fundamentally, the natural frequencies and damping factors don't change with the change in the location of the excitation force or response and hence are the global properties of the system. On the other hand, the residues or modal constants depend on the force and response locations and hence are the local properties. Therefore, to obtain consistent estimates, the estimation of the natural frequencies and damping factors should utilize all the FRFs in the estimation process. The basic approach through which the global formulation of RFP achieves this is a two-step process ([93]). It is observed from Eqs. (11.133) and (11.134) that the estimation of the denominator polynomial coefficients is decoupled from that of the numerator polynomial coefficients. Since the natural frequencies and damping factors depend only on the denominator polynomial, the curve-fitting problem can be formulated using all the FRFs first to estimate only the denominator coefficients. It forms the first step of the global RFP. Once this is done, each FRF is analyzed individually to estimate the corresponding residues in the second step.

b. Formulation

Let us say we have measured the p number of FRFs belonging to a column of the FRF matrix.

We write Eq. (11.133) for the jth measured FRF, after adding a subscript 'j', as

$$\left([I] - 4[X_j]^T[X_j]\right)\{d\} = -4[X_j]^T\{G_j\} \tag{11.149}$$

Eq. (11.140) is used to transform the denominator orthogonal polynomial coefficients into ordinary polynomial coefficients. This equation for the jth measured FRF is

$$[B_{1,j}]\{b\} = [B_{2,j}]\{d\} \tag{11.150}$$

Substituting $\{d\}$ from Eq. (11.149) into Eq. (11.150) leads to,

$$[S_j]\{b\} = \{u_j\} \tag{11.151}$$

where

$$[S_j] = \left([I] - 4[X_j]^T[X_j]\right)[B_{2,j}]^{-1}[B_{1,j}] \quad \text{and} \quad \{u_j\} = -4[X_j]^T\{G_j\} \tag{11.152}$$

Eq. (11.151) is set up for all the p FRFs. All the resulting equations have a common unknown vector and are combined as

$$
\begin{bmatrix} [S_1] \\ [S_2] \\ \vdots \\ [S_p] \end{bmatrix} \{b\} = \begin{Bmatrix} \{u_1\} \\ \{u_2\} \\ \vdots \\ \{u_p\} \end{Bmatrix}
$$

(11.153)

Eq. (11.153) is solved using the least-squares method. The resulting estimate of $\{b\}$ is a single consistent estimate of the denominator ordinary polynomial coefficient vector for all the p FRFs.

To estimate the numerator polynomial coefficient vector for an FRF, the use of Eq. (11.134) needs the denominator orthogonal polynomial coefficient vector ($\{d\}$), which is no longer estimated. Instead, the equations are formulated directly from the rational fraction form of the FRF (Eq. (11.111)) written in the following form:

$$
H(\omega) = \sum_{k=0}^{m} g(i\omega)\, c_k \phi_k (i\omega)
$$

(11.154)

where $g(i\omega)$ is the inverse of the denominator and can be calculated since the denominator ordinary polynomial coefficient vector ($\{b\}$) is known. $g(i\omega)$ is given by

$$
g(i\omega) = \frac{1}{\sum_{p=0}^{n-1} d_p \theta_p (i\omega) + \theta_n (i\omega)} = \frac{1}{\sum_{p=0}^{n} b_p (i\omega)^p}
$$

(11.155)

Further, let us define

$$
z_k (i\omega) = g(i\omega)\phi_k (i\omega)
$$

(11.156)

The error in the FRF fit is

$$
e(\omega) = \Delta H(\omega) = H(\omega) - H_m (\omega)
$$

(11.157)

Substituting Eqs. (11.154) and (11.156) into Eq. (11.157), writing the resulting equation at frequencies $\omega = \omega_j$, with $j = 1, 2, \ldots, L$, and combining all these equations gives the following equation in matrix form:

$$
\{E\} = [z]\{c\} - \{v\}
$$

(11.158)

The various vectors and matrices in Eq. (11.158) are

$$
\{E\} = \begin{Bmatrix} e(\omega_1) \\ e(\omega_2) \\ \vdots \\ e(\omega_L) \end{Bmatrix}, \quad \{v\} = \begin{Bmatrix} H_m(\omega_1) \\ H_m(\omega_2) \\ \vdots \\ H_m(\omega_L) \end{Bmatrix}
$$

(11.159)

$$
[z] = \begin{bmatrix} z_0(\omega_1) & z_1(\omega_1) & \cdots & z_m(\omega_1) \\ z_0(\omega_2) & z_1(\omega_2) & \cdots & z_m(\omega_2) \\ \vdots & \vdots & \cdots & \vdots \\ z_0(\omega_L) & z_1(\omega_L) & \cdots & z_m(\omega_L) \end{bmatrix}, \quad \{c\} = \begin{Bmatrix} c_0 \\ c_1 \\ \vdots \\ c_m \end{Bmatrix}
$$

(11.160)

Minimizing the objective function $J = \{E\}^T \{E\}$ gives the following least-squares solution to the numerator orthogonal polynomial coefficients of the FRF:

$$\{c\} = \left([z]^H [z]\right)^{-1} [z]^H \{v\} \tag{11.161}$$

We convert the vector $\{c\}$ to the ordinary polynomial coefficient vector ($\{a\}$) using Eq. (11.137).

The above procedure to find the numerator ordinary polynomial coefficients is applied to all the FRFs. Roots of the denominator ordinary polynomial give a consistent set of poles applicable to all the FRFs, while the residues for each FRF are obtained from the corresponding numerator ordinary polynomial and the common denominator ordinary polynomial.

Example 11.6: Modal Parameter Estimation Using Global RFP

We apply the global RFP method to the cantilever structure to analyze the receptances at five test points. We perform an MDOF curve-fit covering the first three peaks/modes in the FRFs. Some random measurement noise is also simulated in the 'measured' FRFs.

Step 1: The choices regarding the frequency band, number of frequency points in the band, the degrees of the numerator and denominator polynomials to be fitted, and the residual are kept the same as in Example 11.5.

Step 2: We setup the coefficient matrix and the RHS vector of Eq. (11.153) and solve for the denominator ordinary polynomial coefficients vector ($\{b\}$).

Step 3: For each FRF, the numerator ordinary polynomial coefficients vector ($\{a\}$) is determined using Eqs. (11.161) and (11.137).

Step 4: Using the MATLAB function 'residue', the poles and residues are obtained for each numerator ordinary polynomial coefficients vector paired with the common denominator ordinary polynomial coefficients vector. The poles are identical for each FRF. The poles and the residues for the five FRFs (corresponding to the poles with a positive imaginary part) are given in Table 11.2.

Step 5: The natural frequencies and damping factors are computed from the poles, and the normalized mode shapes are computed from the residues. These, along with the correct values, are given in Table 11.3 and are found to be in good agreement.

Step 6: The FRFs are regenerated using the estimated poles and residues. Overlays of the point FRF (H_{55}) and a transfer FRF (H_{15}) with the corresponding 'measured' FRFs are shown in Figures 11.22 and 11.23, respectively. It is observed that the global RFP method has yielded a good least-squares fit of the FRFs in the frequency band, even in the presence of noise.

11.11 COMPLEX EXPONENTIAL METHOD

a. Basic idea of the method

The complex exponential method (CEM) is a modal parameter estimation method in the time domain. It is one of the earliest modal analysis methods and was developed by Prony. The time-domain estimation relies on the fact

TABLE 11.2

Poles (with Positive Real Part) and the Corresponding Residues Identified Using Global RFP

	Poles		
	Mode 1	Mode 2	Mode 3
	−2.341+61.4j	−1.507+387.04j	−1.159+1086.48j
	Residues		
FRF	**Mode 1**	**Mode 2**	**Mode 3**
H_{15}	−0.00017−0.0043 i	0.00016+0.0026 i	−6.5e-5−0.0018 i
H_{25}	−0.00057−0.0152 i	0.00043+0.0066 i	−9.7e-5−0.0018 i
H_{35}	−0.00134−0.0336 i	0.00028+0.00489 i	3.5e-5+0.0015 i
H_{45}	−0.0018−0.0521 i	−8.54e-5−0.00066 i	1.69e-5+0.001 i
H_{55}	−0.0027−0.0684 i	−0.0007−0.0099 i	−0.00011−0.0034 i

TABLE 11.3

Modal Parameters Identified Using Global RFP and Its Comparison with Exact Values

	Mode 1		Mode 2		Mode 3	
	Global RFP	Exact	Global RFP	Exact	Global RFP	Exact
Natural Freq. (Hz)	9.78	9.81	61.6	61.54	172.91	172.86
Damping factor	0.0381	0.0283	0.0038	0.0045	0.00106	0.0016
Mode shape $\{\phi\}$	$0.011-0.011\,i$	$0.01-0.01\,i$	$-0.017+0.018\,i$	$-0.02+0.02\,i$	$0.021-0.022i$	$0.024-0.024\,i$
	$0.04-0.04\,i$	$0.038-0.038\,i$	$-0.045+0.048\,i$	$-0.045+0.045\,i$	$0.021-0.022\,i$	$0.021-0.021i$
	$0.089-0.092\,i$	$0.076-0.076\,i$	$-0.033+0.035\,i$	$-0.039+0.039\,i$	$-0.019+0.019\,i$	$-0.018+0.018\,i$
	$0.138-0.143\,i$	$0.12-0.12\,i$	$0.004-0.0051\,i$	$0.004-0.004\,i$	$-0.012+0.012\,i$	$-0.015+0.015\,i$
	$0.181-0.188\,i$	$0.166-0.166\,i$	$0.068-0.073\,i$	$0.066-0.066\,i$	$0.04-0.04\,i$	$0.039-0.039\,i$

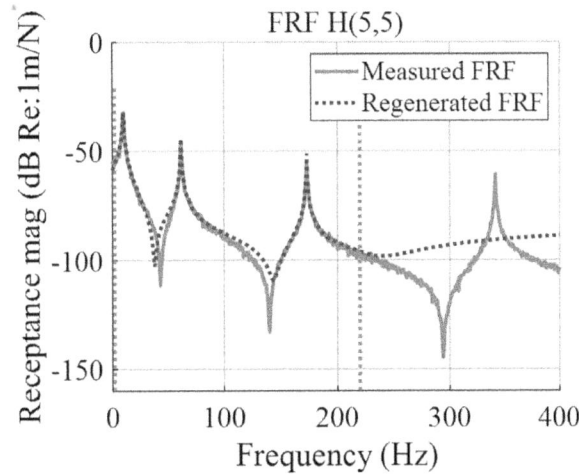

FIGURE 11.22 Overlay of the point FRF (H_{55}) with the corresponding 'measured' FRF.

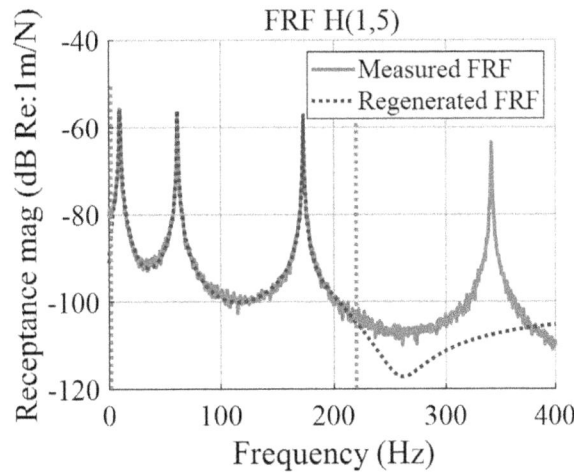

FIGURE 11.23 Overlay of a transfer FRF (H_{15}) with the corresponding 'measured' FRF.

that the system's impulse response function (IRF) consists of contributions from all the system modes, and hence an analysis of the IRFs of a system can yield information on its modal parameters. In Chapter 6, the IRF $h_{jk}(t)$ of a viscously damped system was derived from the fundamental principles. $h_{jk}(t)$ is the time-domain response at the jth DOF due to a unit impulse applied at the kth DOF and is given by

$$h_{jk}(t) = \sum_{r=1}^{N} \left(\psi_{jr}\psi_{kr}e^{s_r t} + \psi_{jr}^{*}\psi_{kr}^{*}e^{s_r^* t} \right) \qquad (11.162)$$

In light of Eq. (11.10), Eq. (11.162) can be written in terms of residues as

$$h_{jk}(t) = \sum_{r=1}^{N} {}_rR_{jk}e^{s_r t} + {}_rR_{jk}^{*}e^{s_r^{*}t} \tag{11.163}$$

For a compact representation, Eq. (11.163) can be written as follows (remembering that the 2N terms consist of N complex conjugate pole-residue pairs):

$$h_{jk}(t) = \sum_{r=1}^{2N} {}_rR_{jk}e^{s_r t} \tag{11.164}$$

In practice, an IRF is obtained by inverse DFT of the corresponding measured FRF since direct measurement of the IRF is faced with the difficulty of applying an ideal impulsive force (in the form of the Dirac-delta function). It is noted that the IRF and the unknown parameters, i.e., the poles and residues, are related through a nonlinear relationship (Eq. 11.164). The basic idea of CEM is resolving this nonlinearity and converting the nonlinear curve-fitting problem in the time domain into simpler problems that are less difficult to solve.

It should also be noted that there is nothing like an SDOF modal parameter estimation method in the time domain as we have in the frequency domain. It is because all the modes contribute to the IRF at all times. Therefore, all the system modes present in the impulse response must be identified, always necessitating an MDOF curve fit. It is unlike the frequency domain, where the modal contribution of a mode can be considered dominant in a small frequency band around the corresponding peak in the FRF. Hence the mode can be approximately identified by an SDOF fit.

b. Formulation of the method

The IRF from the measurement is available in the form of its values at discrete times. If Δt is the sampling time, then at $t = 0\Delta t, 1\Delta t, 2\Delta t, \ldots, L\Delta t$, we have the IRF values $h_{jk}(0\Delta t)$, $h_{jk}(1\Delta t)$, $h_{jk}(2\Delta t)$, $\ldots h_{jk}(L\Delta t)$.

At each instant, Eq. (11.164) must hold true. Hence, we can write that equation at various instants as

$$h_{jk}(0\Delta t) = {}_1R_{jk}e^{s_1 0\Delta t} + {}_2R_{jk}e^{s_2 0\Delta t} + \ldots + {}_{2N}R_{jk}e^{s_{2N} 0\Delta t}$$

$$h_{jk}(1\Delta t) = {}_1R_{jk}e^{s_1 1\Delta t} + {}_2R_{jk}e^{s_2 1\Delta t} + \ldots + {}_{2N}R_{jk}e^{s_{2N} 1\Delta t}$$

$$\cdots \quad \cdots \quad \cdots \quad \cdots \quad \cdots \quad \cdots \quad \cdots \quad \cdots \quad \cdots \quad \cdots \tag{11.165}$$

$$\cdots \quad \cdots \quad \cdots \quad \cdots \quad \cdots \quad \cdots \quad \cdots \quad \cdots \quad \cdots \quad \cdots$$

$$h_{jk}(L\Delta t) = {}_1R_{jk}e^{s_1 L\Delta t} + {}_2R_{jk}e^{s_2 L\Delta t} + \ldots + {}_{2N}R_{jk}e^{s_{2N} L\Delta t}$$

Exponential terms like $e^{s_r(k\Delta t)}$ on the RHS are a nonlinear function of the poles. The first simplification is achieved by defining new variables as follows.

$$v_1 = e^{s_1\Delta t}, \quad \therefore e^{s_1 0\Delta t} = (v_1)^0, \quad e^{s_1 1\Delta t} = (v_1)^1, \quad \ldots, \quad e^{s_1 L\Delta t} = (v_1)^L$$

$$v_2 = e^{s_2\Delta t}, \quad \therefore e^{s_2 0\Delta t} = (v_2)^0, \quad e^{s_2 1\Delta t} = (v_2)^1, \quad \ldots, \quad e^{s_2 L\Delta t} = (v_2)^L$$

$$\cdots \quad \cdots \quad \cdots \quad \cdots \quad \cdots \quad \cdots \quad \cdots \quad \cdots \quad \cdots \quad \cdots \tag{11.166}$$

$$v_r = e^{s_r\Delta}, \quad \therefore e^{s_r 0\Delta t} = (v_r)^0, \quad e^{s_r 1\Delta t} = (v_r)^1, \quad \ldots, \quad e^{s_r L\Delta t} = (v_r)^L$$

$$\cdots \quad \cdots \quad \cdots \quad \cdots \quad \cdots \quad \cdots \quad \cdots \quad \cdots \quad \cdots \quad \cdots$$

Substituting Eqs. (11.166) into Eqs. (11.165), we get

$$h_{jk}\left(0\Delta t\right) = {}_1R_{jk}\left(v_1\right)^0 + {}_2R_{jk}\left(v_2\right)^0 + \ldots + {}_{2N}R_{jk}\left(v_{2N}\right)^0$$

$$h_{jk}\left(1\Delta t\right) = {}_1R_{jk}\left(v_1\right)^1 + {}_2R_{jk}\left(v_2\right)^1 + \ldots + {}_{2N}R_{jk}\left(v_{2N}\right)^1$$

$$\ldots \quad \ldots \quad \ldots \quad \ldots \quad \ldots \quad \ldots \quad \ldots \quad \ldots \quad \ldots \quad \ldots \quad \ldots \quad (11.167)$$

$$\ldots \quad \ldots \quad \ldots \quad \ldots \quad \ldots \quad \ldots \quad \ldots \quad \ldots \quad \ldots \quad \ldots$$

$$h_{jk}\left(L\Delta t\right) = {}_1R_{jk}\left(v_1\right)^L + {}_2R_{jk}\left(v_2\right)^L + \ldots + {}_{2N}R_{jk}\left(v_{2N}\right)^L$$

In Eq. (11.167), there are nonlinear terms due to the powers of the unknown quantities v_r and their product with unknown residues. The next simplification carried out was to multiply each equation in (11.167) by a scalar β_p, with 'p' being the time index of the equation, and then to sum the resulting equations. The multiplication by β_p gives

$$\beta_0 h_{jk}\left(0\Delta t\right) = \beta_0 \left({}_1R_{jk}\left(v_1\right)^0 + {}_2R_{jk}\left(v_2\right)^0 + \ldots + {}_{2N}R_{jk}\left(v_{2N}\right)^0\right)$$

$$\beta_1 h_{jk}\left(1\Delta t\right) = \beta_1 \left({}_1R_{jk}\left(v_1\right)^1 + {}_2R_{jk}\left(v_2\right)^1 + \ldots + {}_{2N}R_{jk}\left(v_{2N}\right)^1\right)$$

$$\ldots \quad \ldots \quad \ldots \quad \ldots \quad \ldots \quad \ldots \quad \ldots \quad \ldots \quad \ldots \quad \ldots \quad \ldots \quad (11.168)$$

$$\ldots \quad \ldots \quad \ldots \quad \ldots \quad \ldots \quad \ldots \quad \ldots \quad \ldots \quad \ldots \quad \ldots$$

$$\beta_L h_{jk}\left(L\Delta t\right) = \beta_L \left({}_1R_{jk}\left(v_1\right)^L + {}_2R_{jk}\left(v_2\right)^L + \ldots + {}_{2N}R_{jk}\left(v_{2N}\right)^L\right)$$

Summing all the equations in (11.168) and collecting together all the terms with a common residue gives

$$\beta_0 h_{jk}\left(0\Delta t\right) + \beta_1 h_{jk}\left(1\Delta t\right) + \ldots + \beta_L h_{jk}\left(L\Delta t\right)$$

$$= {}_1R_{jk}\left(\beta_0\left(v_1\right)^0 + \beta_1\left(v_1\right)^1 + \ldots + \beta_L\left(v_1\right)^L\right)$$

$$+ {}_2R_{jk}\left(\beta_0\left(v_2\right)^0 + \beta_1\left(v_2\right)^1 + \ldots + \beta_L\left(v_2\right)^L\right) \qquad (11.169)$$

$$\ldots \quad \ldots \quad \ldots \quad \ldots \quad \ldots \quad \ldots \quad \ldots \quad \ldots$$

$$+ {}_{2N}R_{jk}\left(\beta_0\left(v_{2N}\right)^0 + \beta_1\left(v_{2N}\right)^1 + \ldots + \beta_L\left(v_{2N}\right)^L\right)$$

If all β_p could be selected such that the expression in the bracket associated with each residue is made zero, then we have

$$\beta_0\left(v_1\right)^0 + \beta_1\left(v_1\right)^1 + \ldots + \beta_L\left(v_1\right)^L = 0$$

$$\beta_0\left(v_2\right)^0 + \beta_1\left(v_2\right)^1 + \ldots + \beta_L\left(v_2\right)^L = 0$$

$$\ldots \quad \ldots \quad \ldots \quad \ldots \quad \ldots \quad \ldots \quad \ldots \quad \ldots \quad \ldots \quad (11.170)$$

$$\ldots \quad \ldots \quad \ldots \quad \ldots \quad \ldots \quad \ldots \quad \ldots \quad \ldots$$

$$\beta_0\left(v_{2N}\right)^0 + \beta_1\left(v_{2N}\right)^1 + \ldots + \beta_L\left(v_{2N}\right)^L = 0$$

Due to Eqs. (11.170), the LHS of Eq. (11.169) should be equal to zero, and hence, we get

$$\beta_0 h_{jk}(0\Delta t) + \beta_1 h_{jk}(1\Delta t) + \ldots + \beta_L h_{jk}(L\Delta t) = 0 \qquad (11.171)$$

So, after getting Eqs. (11.170) and (11.171), what is the gain? We see that $\beta_p{}^s$ are additional unknowns, but they are related by a linear equation (Eq. 11.171), which can be solved easily. Every equation in Eqs. (11.170) is an independent equation and represents a polynomial in v_r of degree L. The solution of a polynomial is essentially a root-finding procedure and is well-established. Therefore, the roots of any of the Eqs. (11.170) gives information about the poles of the system since v_r is related to pole s_r (Eq. 11.152). Once all v_r are known, Eq. (11.167) gives a set of simultaneous linear equations in unknown residues that can be solved easily.

Now we look at how $\beta_p{}^s$ are obtained from Eq. (11.171). Since the equation is homogenous, we arbitrarily assume $\beta_L = 1$, (which is equivalent to dividing the whole equation by β_L), giving us

$$\beta_0 h_{jk}(0\Delta t) + \beta_1 h_{jk}(1\Delta t) + \ldots + \beta_{L-1} h_{jk}((L-1)\Delta t) = h_{jk}(L\Delta t) \qquad (11.172)$$

Many more equations like Eq. (11.172) can be written by displacing the samples of the IRF by discrete time steps. Since there are L unknowns (β_0 to β_{L-1}), we need at least L equations. However, we use more equations than L to enable a least-squares error solution to reduce the effect of any measurement noise. Therefore, we get

$$\beta_0 h_{jk}(0\Delta t) + \beta_1 h_{jk}(1\Delta t) + \ldots + \beta_{L-1} h_{jk}((L-1)\Delta t) = h_{jk}(L\Delta t)$$

$$\beta_0 h_{jk}(1\Delta t) + \beta_1 h_{jk}(2\Delta t) + \ldots + \beta_{L-1} h_{jk}(L\Delta t) = h_{jk}((L+1)\Delta t)$$

$$\cdots \quad \cdots \quad \cdots \quad \cdots \quad \cdots \quad \cdots \quad \cdots \quad \cdots \quad \cdots \quad \cdots \quad \cdots \quad \cdots \qquad (11.173)$$

$$\cdots \quad \cdots \quad \cdots \quad \cdots \quad \cdots \quad \cdots \quad \cdots \quad \cdots \quad \cdots \quad \cdots \quad \cdots \quad \cdots$$

$$\beta_0 h_{jk}(m\Delta t) + \beta_1 h_{jk}((m+1)\Delta t) + \ldots + \beta_{L-1} h_{jk}((m+L-1)\Delta t) = h_{jk}((L+m)\Delta t)$$

Eq. (11.173) is written in compact form as

$$[G]\{\beta\} = \{q\} \qquad (11.174)$$

where

$$[G] = \begin{bmatrix} h_{jk}(0\Delta t) & h_{jk}(1\Delta t) & \cdots & h_{jk}((L-1)\Delta t) \\ h_{jk}(1\Delta t) & h_{jk}(2\Delta t) & \cdots & h_{jk}(L\Delta t) \\ \vdots & \vdots & \cdots & \vdots \\ h_{jk}(m\Delta t) & h_{jk}((m+1)\Delta t) & \cdots & h_{jk}((m+L-1)\Delta t) \end{bmatrix}$$

$$\qquad (11.175)$$

$$\{\beta\} = \begin{Bmatrix} \beta_0 \\ \beta_1 \\ \vdots \\ \beta_{L-1} \end{Bmatrix} \text{ and } \{q\} = \begin{Bmatrix} h_{jk}(L\Delta t) \\ h_{jk}((L+1)\Delta t) \\ \vdots \\ h_{jk}((L+m)\Delta t) \end{Bmatrix}$$

The least-squares estimate of $\{\beta\}$ is given by

$$\{\beta\} = \left([G]^T [G]\right)^{-1} [G]\{u\} \qquad (11.176)$$

We now find the roots of one of the polynomials in Eq. (11.170), which is rewritten below.

$$\beta_0 (v_1)^0 + \beta_1 (v_1)^1 + \ldots + \beta_L (v_1)^L = 0 \tag{11.177}$$

It gives L roots of v_1. For the jth root (denoted by $v_{1,j}$), the jth pole (in the light of Eq. 11.166) is obtained as

$$s_j = \frac{1}{\Delta t} \log_e v_{1,j} \quad \text{with} \quad j = 1, 2, \ldots, L \tag{11.178}$$

The natural frequency and viscous damping factor corresponding to the pole are obtained using Eqs. (11.145) and (11.146) given earlier. We note that to extract 2N complex modes, L should be equal to 2N so that Eq. (11.177) is of degree 2N, and its solution yields 2N roots. But, in practice, the number of modes that make up the measurement is generally not known, and therefore a mode order solution strategy based on the stabilization diagram (to be discussed later) can be used.

Once the poles are known, the unknown residues are obtained from Eq. (11.167), a set of simultaneous linear equations. Similar equations written at a few more time instants can be added to pose the problem as a least-squares problem. The resulting equations can be written in a compact matrix form, as given below.

$$[W]\{R\} = \{u\} \tag{11.179}$$

where

$$[W] = \begin{bmatrix} e^{s_1 0 \Delta t} & e^{s_2 0 \Delta t} & \ldots & e^{s_{2N} 0 \Delta t} \\ e^{s_1 1 \Delta t} & e^{s_2 1 \Delta t} & \ldots & e^{s_{2N} 0 \Delta t} \\ \vdots & \vdots & \ldots & \vdots \\ e^{s_1 (n-1) \Delta t} & e^{s_2 (n-1) \Delta t} & \ldots & e^{s_{2N} (n-1) \Delta t} \end{bmatrix},$$

$$\{R\} = \begin{Bmatrix} {}_1 R_{jk} \\ {}_2 R_{jk} \\ \vdots \\ {}_{2N} R_{jk} \end{Bmatrix} \quad \text{and} \{u\} = \begin{Bmatrix} h_{jk}(0\Delta t) \\ h_{jk}(1\Delta t) \\ \vdots \\ h_{jk}((n-1)\Delta t) \end{Bmatrix} \tag{11.180}$$

The least-squares estimate of $\{R\}$ is given by

$$\{R\} = \left([W]^T [W]\right)^{-1} [W]\{u\} \tag{11.181}$$

In this way, the poles and residues corresponding to IRF h_{jk} are obtained in the time domain by the CEM. The process is repeated for the other IRFs. The natural frequencies and viscous damping factors for various modes corresponding to various IRFs can be averaged to obtain a single global estimate, and the residues can be processed to obtain the mode shapes, as discussed earlier in this chapter.

11.12   STABILIZATION DIAGRAM

One of the variables to be decided in an MDOF curve-fit is the number of poles that should be identified. The number of poles is also referred to as the model order, as it equals the degree of freedom in state space. Several approaches have been suggested to determine the model order. In the frequency domain, counting the number of peaks in the FRF can serve the purpose but may lead to an incorrect assessment in cases where close and/or repeated modes exist, damping is higher, or when the measured data is noisy, resulting in false peaks. Moreover, some peaks might be absent in an FRF if the response/force location coincides with a node of the modes. Another approach is to overlay all the measured FRFs. It helps detect the peaks that consistently appear in all the FRFs and prevent any peak from being missed.

Another method to determine the model order is based on what is called the stabilization diagram. Suppose the true model order is 2N. In the absence of knowledge of the true model order, the idea is to perform curve-fitting with a high model order (say $2M_1$) and find the poles and residues. Assuming $2M_1 > 2N$, the curve-fitting identifies some extra poles. These extra poles are computational poles, which have no relation to the system's dynamics but provide more 'computational space' to the method to obtain a closer data fit. The curve-fitting is repeated with a decreasing model order. It may be done up to a lower value of model order (say $2M_2$, assuming $2M_2 < 2N$). In the end, we get poles and residue corresponding to various model orders from $2M_1$ to $2M_2$. The plot of the natural frequencies on the x-axis with the model orders on the y-axis gives the stabilization diagram for natural frequencies.

To determine the model order from the stabilization diagram, we must watch the natural frequencies that are stabilizing when the model order is reduced from a higher value to a lower value. The natural frequencies corresponding to computational poles fluctuate with the model order, don't stabilize and may disappear from subsequent runs. On the contrary, the natural frequencies corresponding to the poles of the system fluctuate less and stabilize along vertical lines at the true values of the natural frequencies as the model order is reduced. The true model order is that for which the true natural frequencies have stabilized and there are no computational poles.

Further validation of the model order can be done from stabilization diagrams for the damping factors and residues. In the example given next, we use the stabilization diagram to identify the true model order.

Example 11.7: Modal Parameter Estimation Using the CEM

In this example, we apply the CEM to our cantilever structure to analyze its driving point IRF h_{55} (t). We use the stabilization diagram to identify the true model order.

Step 1: The first step is to obtain the IRF. It is obtained as the inverse DFT of the corresponding 'measured' FRF $H_{55}(\omega)$. $H_{55}(\omega)$ is the line spectrum measured over the 0–400 Hz frequency range. It is defined over the positive frequency axis. To obtain the inverse DFT, presented in Chapter 8, we first find the corresponding two-sided FRF (denoted by $\bar{H}_{55}(\omega)$). $\bar{H}_{55}(\omega)$ can be obtained using the following equations:

$$\bar{H}_{55}(\omega) = H_{55}(\omega) \quad \omega \geq 0$$

$$\bar{H}_{55}(\omega) = \left(H_{55}(|\omega|)\right)^* \quad \omega < 0$$

Figure 11.24 shows the magnitude of two-sided FRF $\bar{H}_{55}(\omega)$, which is a line spectrum. Figure 11.25 shows the inverse DFT of $\bar{H}_{55}(\omega)$, i.e. $h_{55}(t)$. The IRF is 4 seconds long and has 4,096 samples.

Step 2: We choose arbitrarily a high model order, say, $2M_1 = 20$ for curve-fitting. It means that the algorithm would try to fit $2M_1$ complex conjugate poles and residue (or M_1 physical modes) to the IRF.

Step 3: The next step is to determine the vector $\{\beta\}$ using Eq. (11.174). To set up this equation, we need to choose the number of samples in one record and the number of such records displaced by discrete time steps. The number of samples in one record should be $2M_1 + 1$. The number of displaced records used is 790. With this, the least-squares solution of $\{\beta\}$ is obtained from Eq. (11.176).

Step 4: Eq. (11.177) is solved to find roots using the command 'roots()' in MATLAB. The input to this command is the coefficient vector in the polynomial, which is nothing but the vector $\{\beta\}$. The output is a vector containing $2M_1$ roots. Eq. (11.178) is then solved to find the poles sj, with j = 1, 2, ..., $2M_1$.

Step 5: The next step is to determine the residue vector $\{R\}$ using Eq. (11.179). We need to decide the number of IRF samples to be used. The minimum number of samples required is $2M_1$, since there are an equal number of residues to be determined. But we take many more (equal to 409 in this example) to impose a least-squares error fit. The least-squares estimate of $\{R\}$ is obtained from Eq. (11.181).

Step 6: This completes the curve-fitting process for the chosen model order.

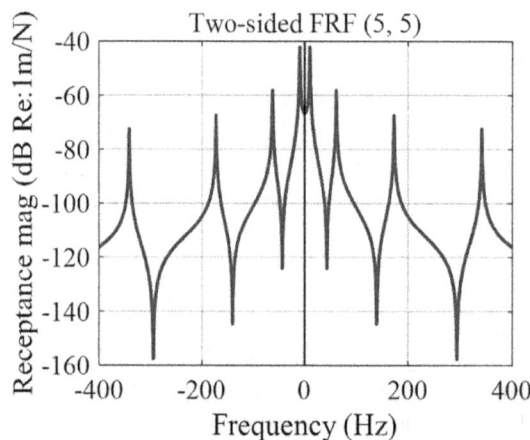

FIGURE 11.24   The magnitude of two-sided FRF $\bar{H}_{55}(\omega)$.

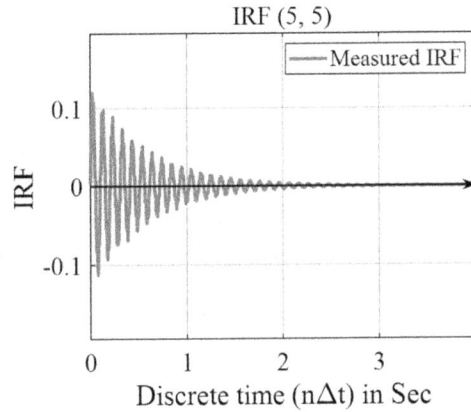

FIGURE 11.25 Inverse DFT of $\bar{H}_{55}(\omega)$.

Step 7: However, to assess the true model order, we make the stabilization diagram. To plot the stabilization diagram, we repeat steps 2–5 for $2M_1$ = 18, 16, 14, 12, 10, 8, 6, 4 and 2

Step 8: Figure 11.26 shows the stabilization diagram for the natural frequencies. We observe that as the model order reduces, true modes stabilize, while the false or computational modes fluctuate and gradually disappear. For model order 10, one identified natural frequency is close to zero 0 Hz. It is unlikely to be a true natural frequency as the structure under test has no rigid body modes. On reducing the model order from 10 to 8, the four dots remain stable, indicating that they should be the true natural frequencies. A further reduction in the model order to six knocks off one true mode. Overlays of the regenerated and 'measured' IRFs for model orders 10, 8, and 6 are shown in Figure 11.27. The lack of fit in the overlay for model order 6 is clear evidence that a true mode is being removed away by this choice of model order. For model order 8, the regenerated IRF matches perfectly with the 'measured' IRF. This observation, along with the observations from the stabilization diagram, indicates that the true model order of the 'measured' IRF is 8. It corresponds to four physical modes.

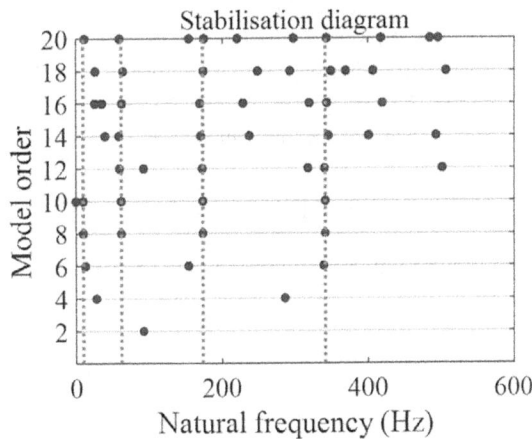

FIGURE 11.26 Stabilization diagram for the natural frequencies.

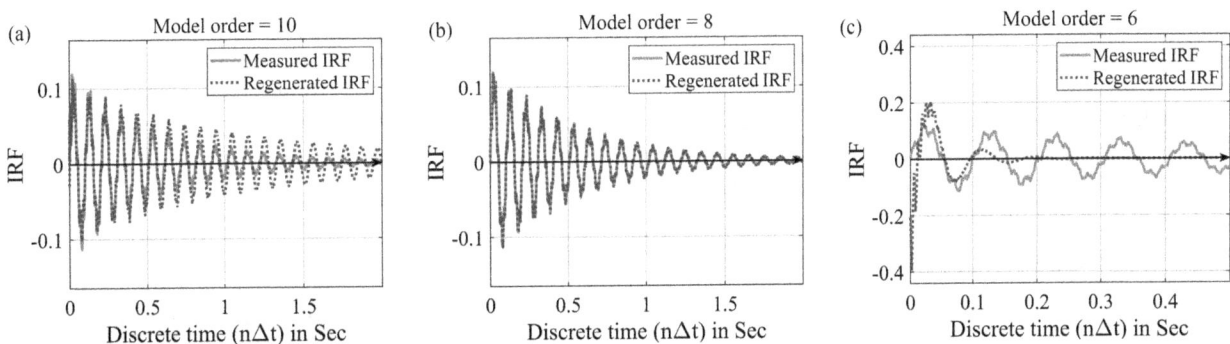

FIGURE 11.27 Overlays of the regenerated and 'measured' IRF for model orders 10, 8, and 6.

TABLE 11.4
Poles and Residues

S. No.	Pole	Residue
1	$-1.723+61.67\,i$	$-0.0002-0.0551\,i$
2	$-1.722-61.67\,i$	$-0.0002+0.0551\,i$
3	$-1.816+386.76\,i$	$-0.0001-0.0089\,i$
4	$-1.815-386.76\,i$	$-0.0001+0.0089\,i$
5	$-1.734-1086.12i$	$7.0e-07+0.0031\,i$
6	$-1.734+1086.12\,i$	$6.3e-07-0.0031\,i$
7	$-1.746-2145.63\,i$	$-1.9e-06+0.0016\,i$
8	$-1.746+2145.63\,i$	$-1.9e-06-0.0016\,i$

Step 9: The poles and residues obtained based on model order 8 are given in Table 11.4.

Step 10: From the poles with positive imaginary parts, the corresponding natural frequencies and modal viscous damping factors are computed using Eqs. (11.145) and (11.146), and these are,

$$\omega_1 = 2\pi\,9.81\ \text{rad/sec} \quad \text{and} \quad \xi_1 = 0.0279$$

$$\omega_2 = 2\pi\,61.55\ \text{rad/sec} \quad \text{and} \quad \xi_2 = 0.00469$$

$$\omega_3 = 2\pi\,172.86\ \text{rad/sec} \quad \text{and} \quad \xi_3 = 0.00159$$

$$\omega_4 = 2\pi\,341.48\ \text{rad/sec} \quad \text{and} \quad \xi_3 = 0.000814$$

11.13   LEAST-SQUARES COMPLEX EXPONENTIAL METHOD

This method is an extension of the CEM to analyze all the measured IRFs simultaneously. It is a global time-domain method and yields a consistent global estimate of the poles of the system.

Let us say we have p number of measured FRFs belonging to a column of the FRF matrix. We find the corresponding IRFs by the procedure explained in Example 11.7.

For the jth measured IRF, we write Eq. (11.174), after adding subscript 'j', as

$$\left[G_j\right]\{\beta\} = \left\{q_j\right\} \tag{11.182}$$

Eq. (11.182) is written for each of the p IRFs, and all such equations have a common unknown vector $\{\beta\}$. These equations are combined as

$$\begin{bmatrix} [G_1] \\ [G_2] \\ \vdots \\ [G_p] \end{bmatrix} \{\beta\} = \begin{Bmatrix} \{q_1\} \\ \{q_2\} \\ \vdots \\ \{q_p\} \end{Bmatrix} \tag{11.183}$$

A least-squares estimate of $\{\beta\}$ is obtained from Eq. (11.183). Next, we find the roots of the polynomial in Eq. (11.177), which are nothing but the values of v_r. The poles are obtained from Eq. (11.178), giving a single consistent global estimate of the poles corresponding to all the IRFs. Now, for each IRF, Eq. (11.179) is set up, and the corresponding residues are obtained from Eq. (11.181). Once the poles and residues are known, the natural frequencies, viscous damping factors, and mode shapes can be obtained, as discussed in earlier sections.

11.14   SINGULAR VALUE DECOMPOSITION

Singular value decomposition (SVD) is a powerful tool for numerical analysis. Like eigenvalue decomposition (EVD), SVD decomposes a matrix into a product of three matrices with special properties. SVD applies even to a rectangular matrix, while EVD applies only to a square matrix.

SVD of matrix $[A]$ of order $n \times m$ is given by

$$[A] = [U] \ [S] \ [V]^T \atop {n \times m} \quad {n \times n} \quad {n \times m} \quad {m \times m}$$ (11.184)

- $[U]$ is a matrix of left singular vectors of $[A]$. Columns of $[U]$ are eigenvectors of $[A][A]^T$ and form an orthonormal basis of the row space of $[A]$. $[U]$ is orthonormal and hence satisfies

$$[U][U]^T = [I] \quad \text{or} \quad [U]^{-1} = [U]^T$$ (11.185)

- Similarly, $[V]$ is a matrix of right singular vectors of $[A]$. Columns of $[V]$ are eigenvectors of $[A]^T[A]$ and form an orthonormal basis of the column space of $[A]$. $[V]$ is also orthonormal and hence satisfies

$$[V][V]^T = [I] \quad \text{or} \quad [V]^{-1} = [V]^T$$ (11.186)

- $[S]$ is a matrix of singular values of $[A]$ and is of the same order as that of $[A]$. $[S]$ is a diagonal matrix with singular values on the main diagonal. Singular values are also the eigenvalues of $[A][A]^T$ and $[A]^T[A]$.

Some important applications of SVD are summarized below.

- The rank of a matrix can be determined from $[S]$. The rank is equal to the number of nonzero singular values.
- The condition of a matrix can be assessed from $[S]$. The condition number is calculated as the ratio of the largest and the smallest singular values. A condition number equal to unity represents the best condition of the matrix, while a high value of the condition number represents an ill-conditioned matrix. An ill-conditioned matrix arising in an inverse problem amplifies the truncation and round-off errors leading to significant errors in the estimation.
- SVD is also used to solve least-squares problems. All inverse problems, including the modal parameter estimation being studied in the present chapter, involve a least-squares problem at some stage, and SVD can be used to solve this problem robustly. The relative magnitudes of the singular values provide a valuable means to determine the system order. The existence of very small singular values in the presence of very large singular values may be indicative of some noise in the data. Truncating such relatively small singular values may help reduce the effect of the noise in the estimation process.

11.15 IBRAHIM TIME-DOMAIN (ITD) METHOD

The ITD method (Ibrahim and Mikulcik [48]) is one of the first few methods introduced for modal parameter identification. It is a time-domain identification method that simultaneously processes the responses at all the measured DOFs and hence is also a global method. ITD is based on free-response decays and therefore doesn't require the knowledge of the inputs to the system. We first present the formulation of the method with complete measured data and then consider the form of the final equations for the practical case of incomplete measured data.

a. Formulation with complete data

The equation of motion for free vibration of an MDOF system with viscous damping is given by

$$[M]\{\ddot{x}\} + [C]\{\dot{x}\} + [K]\{x\} = \{0\}$$ (11.187)

The free vibration response of the system can be written as

$$\{x\} = \sum_{r=1}^{N} A_r \{\phi\}_r e^{s_r t} + A_r^* \{\phi\}_r^* e^{s_r^* t}$$ (11.188)

where A_r and A_r^* are complex conjugate arbitrary constants, s_r and s_r^* are complex conjugate eigenvalues and $\{\phi\}$ and $\{\phi\}_r^*$ are the corresponding normalized complex conjugate eigenvectors for the rth mode. Representing the sum in Eq. (11.188) through a single term, we can write

$$\{x\} = \sum_{r=1}^{2N} A_r \{\phi\}_r e^{s_r t} = \sum_{r=1}^{2N} \{\psi\}_r e^{s_r t}$$ (11.189)

where, $\{\psi\}_r$ is the rth unscaled eigenvector.

We assume that the measured data is complete, that is, the responses are available at all the N DOFs. It is noted that since the eigenvalues and eigenvectors are unknowns, the RHS of Eq. (11.189) is nonlinear, and hence the following procedure is used to simplify the solution.

At time instant $t = t_1$, Eq. (11.189) can be written as

$$\begin{Bmatrix} x_1(t_1) \\ x_2(t_1) \\ \cdots \\ x_N(t_1) \end{Bmatrix} = \begin{Bmatrix} \psi_{11} & \psi_{12} & \cdots & \psi_{12N} \\ \psi_{21} & \psi_{22} & \cdots & \psi_{22N} \\ \cdots & \cdots & \cdots & \cdots \\ \psi_{N1} & \psi_{N2} & \cdots & \psi_{N2N} \end{Bmatrix} \begin{Bmatrix} e^{s_1 t_1} \\ e^{s_2 t_1} \\ \cdots \\ e^{s_{2N} t_1} \end{Bmatrix} \tag{11.190}$$

We write a similar equation at time instants $t = t_2, t_3, \ldots, t_L$, and we combine all these equations in a single matrix equation leading to

$$\begin{bmatrix} x_1(t_1) & x_1(t_2) & \cdots & x_1(t_L) \\ x_2(t_1) & x_2(t_2) & \cdots & x_2(t_L) \\ \cdots & \cdots & \cdots & \cdots \\ x_N(t_1) & x_N(t_2) & \cdots & x_N(t_L) \end{bmatrix}$$

$$= \begin{bmatrix} \psi_{11} & \psi_{12} & \cdots & \psi_{12N} \\ \psi_{21} & \psi_{22} & \cdots & \psi_{22N} \\ \cdots & \cdots & \cdots & \cdots \\ \psi_{N1} & \psi_{N2} & \cdots & \psi_{N2N} \end{bmatrix} \begin{bmatrix} e^{s_1 t_1} & e^{s_1 t_2} & \cdots & e^{s_1 t_L} \\ e^{s_2 t_1} & e^{s_2 t_2} & \cdots & e^{s_2 t_L} \\ \cdots & \cdots & \cdots & \cdots \\ e^{s_{2N} t_1} & e^{s_{2N} t_2} & \cdots & e^{s_{2N} t_L} \end{bmatrix} \tag{11.191}$$

Eq. (11.191) is represented as

$$\underset{N \times L}{[X_0]} = \underset{N \times 2N}{[\psi]} \ \underset{2N \times L}{[\Lambda]} \tag{11.192}$$

We shift forward all the time instants by $p\Delta t$, where p is an integer and Δt is the time resolution of the samples in the free decays. Therefore, at time instant $t = t_1 + p\Delta t$, Eq. (11.189) can be written as

$$\left\{ x(t_1 + p\Delta t) \right\} = \sum_{r=1}^{2N} \{\psi\}_r \, e^{s_r(t_1 + p\Delta t)} = \sum_{r=1}^{2N} \{\psi\}_r \, e^{s_r p\Delta t} e^{s_r t_1} = \sum_{r=1}^{2N} \{\bar{\psi}\}_r \, e^{s_r t_1} \tag{11.193}$$

where

$$\{\bar{\psi}\}_r = \{\psi\}_r \, e^{s_r p\Delta t} \tag{11.194}$$

Writing equations like (11.193) at time instants $t = t_2 + p\Delta t, t_3 + p\Delta t, \ldots, t_L + p\Delta t$, and then combining all these equations in a single matrix equation, we get the following equation similar to Eq. (11.192):

$$\underset{N \times L}{[X_p]} = \underset{N \times 2N}{[\bar{\psi}]} \ \underset{2N \times L}{[\Lambda]} \tag{11.195}$$

Combining Eqs. (11.192) and (11.195), we get

$$\underset{2N \times L}{\begin{bmatrix} [X_0] \\ [X_p] \end{bmatrix}} = \underset{2N \times 2N}{\begin{bmatrix} [\psi] \\ [\bar{\psi}] \end{bmatrix}} \underset{2N \times L}{[\Lambda]} \tag{11.196}$$

which is represented as

$$\underset{2N\times L}{[X_1]} = \underset{2N\times 2N}{[\psi_1]} \underset{2N\times L}{[\Lambda]} \tag{11.197}$$

We now generate another set of equations similar to Eq. (11.197), by shifting all the time instants used in writing Eq. (11.197) by time step qΔt, where q is an integer. We can directly write down the resulting equation by noting the result in Eq. (11.194), as per which the time shifting leads to the multiplication of the columns of $[\psi]$ and $[\overline{\psi}]$ by factor $e^{s_r q\Delta t}$. It gives rise to a diagonal matrix $[\alpha]$ as shown in the equation below,

$$\underset{2N\times L}{[X_2]} = \underset{2N\times 2N}{[\psi_2]} \underset{2N\times L}{[\Lambda]} = \underset{2N\times 2N}{[\psi_1]} \underset{2N\times 2N}{[`\alpha`]} \underset{2N\times L}{[\Lambda]} \tag{11.198}$$

where

$$[`\alpha`] = \left[`e^{s_r q\Delta t}` \right] \tag{11.199}$$

From Eq. (11.197),

$$\underset{2N\times L}{[\Lambda]} = \underset{2N\times 2N}{[\psi_1]^{-1}} \underset{2N\times L}{[X]_1} \tag{11.200}$$

Substituting Eq. (11.200) into Eq. (11.198), we get

$$\underset{2N\times L}{[X_2]} = \underset{2N\times 2N}{[\psi_1]} \underset{2N\times 2N}{[`\alpha`]} \underset{2N\times 2N}{[\psi_1]^{-1}} \underset{2N\times L}{[X_1]} \tag{11.201}$$

which can be represented as

$$\underset{2N\times 2N}{[B]} \underset{2N\times L}{[X_1]} = \underset{2N\times L}{[X_2]} \tag{11.202}$$

where

$$\underset{2N\times 2N}{[B]} = \underset{2N\times 2N}{[\psi_1]} \underset{2N\times 2N}{[`\alpha`]} \underset{2N\times 2N}{[\psi_1]^{-1}} \tag{11.203}$$

In Eq. (11.202), $[X_1]$ and $[X_2]$ are known as they depend on the measured free decay responses, while the matrix $[B]$ is unknown. Post-multiplying by $[X_1]^T$ or $[X_2]^T$ leads to two different estimates, and hence an average of the two estimates is taken, giving the so-called double least-squares estimate as

$$[B] = \frac{1}{2}\left([X_2][X_1]^T \left([X_1][X_1]^T \right)^{-1} + [X_2][X_2]^T \left([X_1][X_2]^T \right)^{-1} \right) \tag{11.204}$$

Once $[B]$ is known, from Eq. (11.203), we can write

$$\underset{2N\times 2N}{[B]} \underset{2N\times 2N}{[\psi_1]} = \underset{2N\times 2N}{[\psi_1]} \underset{2N\times 2N}{[`\alpha`]} \tag{11.205}$$

or

$$\underset{2N\times 2N}{[B]} \underset{2N\times 1}{\{\psi_1\}_r} = \alpha_r \underset{2N\times 1}{\{\psi_1\}_r} \tag{11.206}$$

Eq. (11.206) is an eigenvalue problem whose solution gives 2N eigenvalues (i.e., values of α_r, for r = 1, 2, .., 2N) and the corresponding eigenvectors (i.e. values of $\{\psi_1\}_r$).

The pole s_r is obtained from the eigenvalue α_r (which, in general, is complex) as follows. Let

$$\alpha_r = \beta_r + i\gamma_r \tag{11.207}$$

and

$$s_r = a_r + ib_r \tag{11.208}$$

From the relationship in Eq. (11.199), and making use of Eq. (11.204), we get

$$\alpha_r = e^{s_r q \Delta t} = e^{a_r q \Delta t} . e^{i b_r q \Delta t} \tag{11.209}$$

The complex conjugate of Eq. (11.209) gives

$$\alpha_r^* = e^{a_r q \Delta t} . e^{-i b_r q \Delta t} \tag{11.210}$$

Multiplying Eqs. (11.209) and (11.210) and simplifying, we get

$$|\alpha_r|^2 = e^{2 a_r q \Delta t} \tag{11.211}$$

$$a_r = \frac{1}{2 q \Delta t} \log_e \left(\beta_r^2 + \gamma_r^2 \right) \tag{11.212}$$

From Eqs. (11.209) and (11.211), we obtain

$$b_r = \frac{1}{q \Delta t} \tan^{-1} \frac{\gamma_r}{\beta_r} \tag{11.213}$$

Thus, Eqs. (11.212) and (11.213) give the real and imaginary parts of the poles. As explained in earlier sections, the natural frequencies and modal viscous damping factors can be obtained from the poles. The first half of $\{\psi_1\}_r$ gives $\{\psi\}_r$ (r^{th} unscaled mode shape of the system).

b. Formulation with incomplete data

In practice, the measured data is incomplete, that is, the number of measured DOFs (denoted by n) is smaller than the DOF (N) of the system. Let us see how the equations obtained in the previous section, to be set up and solved, change because of this.

The order of the matrices in Eq. (11.202) changes leading to

$$\underset{2n \times 2n}{[B]} \; \underset{2n \times L}{[X_1]} = \underset{2n \times L}{[X_2]} \tag{11.214}$$

With changes in matrix orders, matrix $[B]$ is obtained using Eq. (11.204). EVP in Eq. (11.206) takes the following form:

$$\underset{2n \times 2n}{[B]} \; \underset{2n \times 1}{\{\psi_1\}_r} = \alpha_r \underset{2n \times 1}{\{\psi_1\}_r} \tag{11.215}$$

and its solution gives 2n eigenvalues and eigenvectors. The poles are obtained from Eqs. (11.212) and (11.213).

It is suggested that the size of the matrix $[B]$ can be increased to $2N \times 2N$ by having more 'pseudo' measurements, i.e., more equations like (11.195), obtained by considering the time-shifted records. It enables estimating all the modes present in the data.

Since the number of modes present in the data is not known in practice, the number of measurements 'n' can sometimes be more than the number of modes. In that case, the estimate of n modes obtained using the eigenvalue problem (Eq. (11.215)) yields some computational modes along with the physical modes. The noise modes can be distinguished from the physical modes by repeating the identification procedure but with all the records displaced by a common time step. Physical modes are not expected to change, while the noise modes would change significantly. A modal confidence factor is proposed to differentiate the physical and noise modes.

Example 11.8: Modal Parameter Estimation Using the ITD Method

We apply the ITD method to our cantilever structure using simulated measured data in this example.

Step 1: The free decay responses can be obtained in several ways. In the present case, they are found by applying a light impact at the cantilever tip. Let the objective is to identify the first four modes. Therefore, we need four or more response points to ensure that the order of the matrix [B] is 8 or higher, enabling the identification of 8 or more complex conjugate modes (or four or more physical modes). Because of this, let us measure the responses at five test points using the sensors mounted at those points. Thus, n = 5. The input force is not measured. The simulated measured data at a sampling frequency of 2,048 Hz is used. Figure 11.28 shows the 'measured' free decays at two test points.

Step 2: We first estimate the matrix [B] using Eq. (11.204). To do this, we need to construct the matrices $[X_1]$ and $[X_2]$. The matrix $[X_1]$ is given by

$$\underset{2n \times L}{[X_1]} = \left[\begin{array}{c} [X_0] \\ [X_p] \end{array} \right]_{2N \times L}$$

The no. of instants (L) chosen should be more than 2n. The matrix $[X_0]$ (which is the LHS of Eq. (11.191)) is first constructed. The matrix $[X_p]$ is then set up using the responses obtained by shifting forward by $p\Delta t$ all the instants used to set up $[X_0]$. We take p = 1.

The matrix $[X_2]$ is then set up using the responses obtained by shifting forward by $q\Delta t$ all the instants used in setting up $[X_1]$. We take q=2.

[B] is now computed.

Step 3: The next step is to solve the eigenvalue problem in [B] defined by Eq. (11.215). It gives 2n eigenvalues and eigenvectors.

Step 4: The poles are obtained from Eqs. (11.212) and (11.213). The natural frequencies and modal viscous damping factors are obtained from the poles and are given in Table 11.5. The first half of the eigenvector $\{\psi_1\}_r$ gives $\{\psi\}_r$ (the rth unscaled mode shape of the system). The estimated five unscaled mode shapes are listed in Table 11.6. The modal properties of the first four identified modes are found to be in reasonable agreement with the corresponding exact values. The fifth identified mode is closer to mode 5 though not accurately identified.

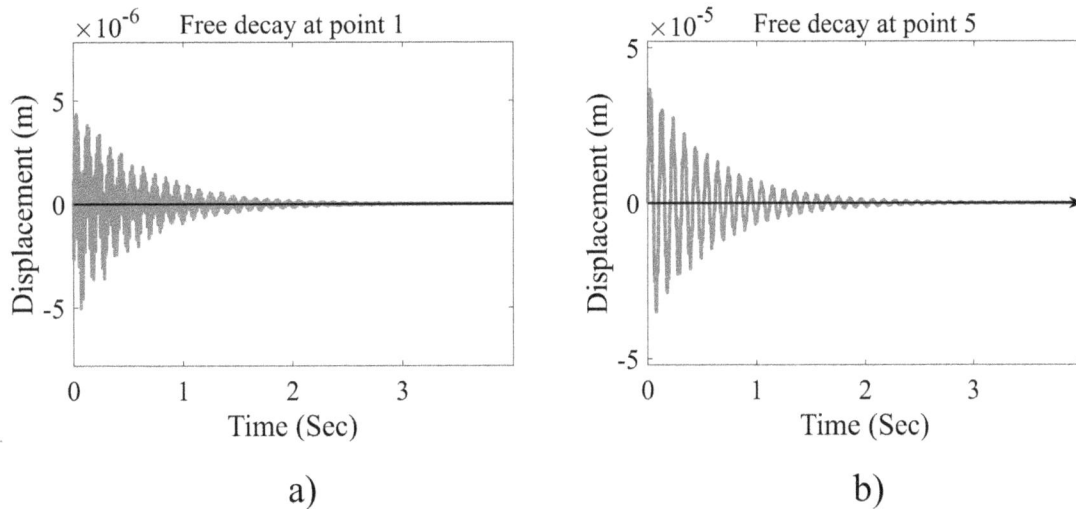

a)   b)

FIGURE 11.28   Free decays at two test points.

TABLE 11.5

Estimated Natural Frequencies and Modal Viscous Damping Factors Using ITD

Mode No.	1	2	3	4	5
Natural frequency (Hz)	9.79	61.32	172.61	337.58	457.19
Damping factor	0.026956	0.004649	0.001569	0.000278	0.000635

TABLE 11.6
Estimated Unscaled Mode Shapes

Mode Shape 1	Mode Shape 2	Mode Shape 3	Mode Shape 4	Mode Shape 5
$-0.033 - 0.0001\,i$	$-0.154 + 0.0001\,i$	$0.3 + 0.0034\,i$	$-0.366 - 0.0078\,i$	$0.286 - 0.0001\,i$
$-0.121 + 2.08e\text{-}5i$	$-0.350 - 0.0016\,i$	$0.264 + 0.001\,i$	$0.151 + 0.0031\,i$	$-0.305 + 0.0002\,i$
$-0.243 - 5.79e\text{-}5\,i$	$-0.303 \text{-} 0.001\,i$	$-0.234 - 0.001\,i$	$0.163 + 0.0053\,i$	$0.304 - 0.0003\,i$
$-0.382 - 0.0002\,i$	$0.033 + 0.0007\,i$	$-0.199 - 0.0005\,i$	$-0.313 - 0.0075\,i$	$-0.254 + 0.0001\,i$
-0.527	0.51	0.495	0.469	0.409

11.16 EIGENSYSTEM REALIZATION ALGORITHM (ERA)

The ERA developed by Juang and Pappa [50] is a time-domain multiple-input-multiple-output (MIMO) algorithm for modal parameter identification. Multi-input excitation has the advantage that repeated eigenvalues, and close modes can be identified. For example, symmetrical structures may have repeated or close eigenvalues, in which case, the use of single excitation along one orthogonal direction may not be able to excite the modes along other orthogonal directions. It can be resolved by applying additional excitations in other suitable directions.

The ERA is based on the system realization approach used in control theory, in which a state-space model of the system is identified. The modal parameters are then computed from the state-space model.

a. Continuous-time state-space model

Since the algorithm is based on a discrete state-space model of the system, let us first develop a state-space model in continuous time and then discretize it. The equation of motion for the forced vibration of an MDOF system with viscous damping is given by

$$[M]\{\ddot{x}\} + [C_v]\{\dot{x}\} + [K]\{x\} = \{f\} \tag{11.216}$$

Note that in this section, we use the symbol $[C_v]$ to represent the viscous damping matrix since the symbol $[C]$ is used below to represent the output matrix.

Let us choose the displacements and velocities at all the DOFs as state variables and denote them by the vectors $\{y_1\}$ and $\{y_2\}$, respectively. Thus,

$$\{y_1\} = \{x\} \tag{11.217}$$

$$\{y_2\} = \{\dot{x}\} \tag{11.218}$$

Differentiating Eqs. (11.217–11.218) once with time and using Eqs. (11.216)–(11.218), we get

$$\{\dot{y}_1\} = \{\dot{x}\} = \{y_2\} \tag{11.219}$$

$$\{\dot{y}_2\} = \{\ddot{x}\} = -[M]^{-1}[K]\{y_1\} - [M]^{-1}[C_v]\{y_2\} + [M]^{-1}\{f\} \tag{11.220}$$

Eqs. (11.219) and (11.220) can be combined and represented as

$$\underset{2N\times1}{\{\dot{y}\}} = \underset{2N\times2N}{[A_c]} \underset{2N\times1}{\{y\}} + \underset{2N\times N}{[B_{cf}]} \underset{N\times1}{\{f\}} \tag{11.221}$$

where

$$\{y\} = \left\{ \begin{array}{c} \{y_1\} \\ \{y_2\} \end{array} \right\}, \; [A_c] = \left[\begin{array}{cc} [0] & [I] \\ -[M]^{-1}[K] & -[M]^{-1}[C_v] \end{array} \right], \; \text{and } [B_{cf}] = \left[\begin{array}{c} [0] \\ [M]^{-1} \end{array} \right] \tag{11.222}$$

Assuming nonzero input forces at q DOFs, the force vector $\{f\}_{N\times 1}$ after dropping the null elements can be modified and represented by $\{u\}_{q\times 1}$. Similarly, matrix $[B_{cf}]_{2N\times N}$ after dropping the columns corresponding to DOFs with zero forces, is modified to $[B_c]_{2N\times q}$. With these changes, Eq. (11.221) is modified to

$$\underset{2N\times 1}{\{\dot{y}\}} = \underset{2N\times 2N}{[A_c]}\ \underset{2N\times 1}{\{y\}} + \underset{2N\times q}{[B_c]}\ \underset{q\times 1}{\{u\}} \tag{11.223}$$

In Eq. (11.223), $\{y\}$ is the state vector and $[A_c]$ and $[B_c]$ are the state and input matrices, respectively. Subscript 'c' indicates association with continuous time. If $\{z\}$ is a vector of the p outputs, then it can be related to the state vector by

$$\underset{p\times 1}{\{z\}} = \underset{p\times 2N}{[C]}\ \underset{2N\times 1}{\{y\}} \tag{11.224}$$

where [C] is the output matrix. Eqs. (11.223) and (11.224) are referred to as state and output equations, respectively, and together they form the state-space model of the system.

If the initial value of the state vector at $t = t_0$ is $\{y(t_0)\}$, then the solution of the state equation (Eq. 11.223) is given by

$$\{y(t)\} = e^{[A_c](t-t_0)}\{y(t_0)\} + \int_{t_0}^{t} e^{[A_c](t-\tau)}[B_c]\{u(\tau)\}\ d\tau \tag{11.225}$$

where τ is a dummy time variable.

b. Discrete-time state-space model

Let us obtain the state-space model with time as a discrete variable. If Δt is the sampling time, then the discrete time-instants can be represented by $t = 0\Delta t,\ 1\Delta t,\ 2\Delta t,\ ..., k\Delta t,\ (k+1)\Delta t,....$ Let us denote two consecutive discrete time-instants as

$$\begin{aligned} t(k) &= k\Delta t \\ t(k+1) &= (k+1)\Delta t \end{aligned} \tag{11.226}$$

If we treat $t(k)$ as the initial time t_0 and $t(k+1)$ as the future time t, and assume that $\{u(\tau)\}$ is constant over the time step, then making these substitutions in Eq. (11.225), we get

$$\{y((k+1)\Delta t)\} = e^{[A_c]\Delta t}\{y(k\Delta t)\} + \left(\int_{k\Delta t}^{(k+1)\Delta t} e^{[A_c]((k+1)\Delta t-\tau)}[B_c]d\tau\right)\{u(k\Delta t)\} \tag{11.227}$$

To simplify, we transform the variable of integration by assuming $\gamma = \tau - k\Delta t$, where γ is another dummy time variable. With this, Eq. (11.227) is transformed to

$$\{y((k+1)\Delta t)\} = e^{[A_c]\Delta t}\{y(k\Delta t)\} + \left(\int_{0}^{\Delta t} e^{-[A_c](\Delta t-\gamma)}[B_c]d\gamma\right)\{u(k\Delta t)\} \tag{11.228}$$

Let

$$[A] = e^{[A_c]\Delta t} \tag{11.229}$$

and

$$[B] = \int_{0}^{\Delta t} e^{-[A_c](1-\gamma)}[B_c]d\gamma = e^{-[A_c]\Delta t}[A_c]^{-1}\left(e^{[A_c]\Delta t} - [I]\right)[B_c] \tag{11.230}$$

Here $[A]$ and $[B]$ are state and input matrices in discrete-time space. Hence Eq. (11.228) and the output equation (Eq. 11.224) together form the discrete state-space model and can be written (after dropping Δt for brevity) as

$$\underset{2N \times 1}{\{y(k+1)\}} = \underset{2N \times 2N}{[A]} \ \underset{2N \times 1}{\{y(k)\}} + \underset{2N \times q}{[B]} \ \underset{q \times 1}{\{u(k)\}} \tag{11.231}$$

$$\underset{p \times 1}{\{z(k)\}} = \underset{p \times 2N}{[C]} \ \underset{2N \times 1}{\{y(k)\}} \tag{11.232}$$

c. Free pulse response

The ERA is based on free decay responses. The formulation is developed using free pulse responses as the outputs. To find the free pulse response for the first input location, we consider a unit impulse applied at the first input location with no force at the other input locations. Thus, we define the input vector as

$$\underset{q \times 1}{\{u(k)\}} = \underset{q \times 1}{\{\delta_1(k)\}} \tag{11.233}$$

such that,

$$\left. \begin{array}{l} \{\delta_1(0)\} = \left\{ \begin{array}{cccccc} 1 & 0 & 0 & \cdots & 0 \end{array} \right\}^T_{q \times 1} \\[2em] \{\delta_1(k)\} = \left\{ \begin{array}{cccccc} 0 & 0 & 0 & \cdots & 0 \end{array} \right\}^T_{q \times 1} \text{ for } k > 0 \end{array} \right\} \tag{11.234}$$

Substituting Eqs. (11.233) and (11.234) into Eqs. (11.231) and (11.232), and assuming that the system is at rest at $t = 0$, i.e., $y\{0\} = \{0\}$, we can find the state and the output vectors for various discrete times, as shown in Table 11.7.

By induction, we can write the state vector and free pulse response due to unit impulse at the first input location as

$$\begin{aligned} \{y(k)\} &= [A]^{k-1}[B]\{\delta_1(0)\} \\ \{z(k)\} &= [C][A]^{k-1}[B]\{\delta_1(0)\} \end{aligned} \tag{11.235}$$

TABLE 11.7

Computation of State and the Output Vectors at Discrete Times

k=0	$\{y(1)\} = [B]\{\delta_1(0)\}$ $\{z(k)\} = \{0\}$
k=1	$\{y(2)\} = [A]\{y(1)\} + [B]\{\delta_1(1)\} = [A][B]\{\delta_1(0)\}$ $\{z(1)\} = [C]\{y(1)\} = [C][B]\{\delta_1(0)\}$
k=2	$\{y(3)\} = [A]\{y(2)\} + [B]\{\delta_1(2)\} = [A]^2[B]\{\delta_1(0)\}$ $\{z(2)\} = [C]\{y(2)\} = [C][A][B]\{\delta_1(0)\}$
k=3	$\{y(4)\} = [A]\{y(3)\} + [B]\{\delta_1(3)\} = [A]^3[B]\{\delta_1(0)\}$ $\{z(3)\} = [C]\{y(3)\} = [C][A]^2[B]\{\delta_1(0)\}$

To find the free pulse response for the second input location, we consider a unit impulse applied at the second input location with no force at the other input locations. Thus, we have

$$\{u(k)\}_{q\times 1} = \{\delta_2(k)\}_{q\times 1} \tag{11.236}$$

such that,

$$\left.\begin{aligned}
\{\delta_2(0)\} &= \left\{\begin{array}{ccccc} 0 & 1 & 0 & \cdots & 0 \end{array}\right\}_{q\times 1}^{T} \\
\{\delta_2(k)\} &= \left\{\begin{array}{ccccc} 0 & 0 & 0 & \cdots & 0 \end{array}\right\}_{q\times 1}^{T} \quad \text{for } k > 0
\end{aligned}\right\} \tag{11.237}$$

Repeating the procedure as given in Table 11.6, we get the free pulse response due to unit impulse at the second input location as

$$\begin{aligned}
\{y(k)\} &= [A]^{k-1}[B]\{\delta_2(0)\} \\
\{z(k)\} &= [C][A]^{k-1}[B]\{\delta_2(0)\}
\end{aligned} \tag{11.238}$$

In this way, the pulse response for the remaining input locations can be obtained. Combining pulse responses corresponding to all q input locations stacked as columns, we get the pulse response matrix $([Z])$ as

$$[Z(k)] = \left[\begin{array}{cccc} [C][A]^{k-1}[B]\{\delta_1(0)\} & [C][A]^{k-1}[B]\{\delta_2(0)\} & \cdots & [C][A]^{k-1}[B]\{\delta_q(0)\} \end{array}\right] \tag{11.239}$$

$$[Z(k)] = [C][A]^{k-1}[B]\left[\begin{array}{cccc} \{\delta_1(0)\} & \{\delta_2(0)\} & \cdots & \{\delta_q(0)\} \end{array}\right] = [C][A]^{k-1}[B][I] \tag{11.240}$$

$$\underset{p\times q}{[Z(k)]} = \underset{p\times 2N}{[C]} \underset{2N\times 2N}{[A]^{k-1}} \underset{2N\times q}{[B]} \tag{11.241}$$

d. System realization

From Eq. (11.241) we see that the matrix of free pulse responses $([Z(k)])$ at any instant is related to the matrices $[A]$, $[B]$ and $[C]$ that form the state-space model. The process of identifying these matrices from $[Z(k)]$ is called system realization. Value of $[Z(k)]$ at just one instant would not be enough to identify the three matrices; hence, we construct the block Hankel matrix using the free pulse response at different instants, as explained below.

We write down the equation for $[Z(k)]$ at 's' discrete times $k\Delta t$, $(k+1)\Delta t,\ldots, (k+s-1)\Delta t$. These pulse response matrices are arranged column-wise to make the first row of the Hankel matrix. Next, the second row of the Hankel matrix is obtained by using the pulse response matrices at discrete times obtained by displacing by a unit time step the discrete times used for the first row. The second row, therefore, corresponds to the 's' discrete times $(k+1)\Delta t$, $(k+2)\Delta t,\ldots, (k+s)\Delta t$. In this manner, 'r' rows are built with each row obtained by displacing the immediate previous row by one time-step. Thus, the Hankel matrix is given by

$$\underset{pr\times qs}{[H(k-1)]} = \left[\begin{array}{cccc}
[Z(k)] & [Z(k+1)] & \cdots & [Z(k+s-1)] \\
[Z(k+1)] & [Z(k+2)] & \cdots & [Z(k+s)] \\
\cdots & \cdots & \cdots & \cdots \\
[Z(k+r-1)] & [Z(k+r)] & \cdots & [Z(k+r+s-2)]
\end{array}\right]_{pr\times qs} \tag{11.242}$$

Substituting Eq. (11.241) into Eq. (11.242), we get

$$
[H(k-1)] =
\begin{bmatrix}
[C][A]^{k-1}[B] & [C][A]^{k}[B] & \cdots & [C][A]^{k+s-2}[B] \\
[C][A]^{k}[B] & [C][A]^{k+1}[B] & \cdots & [C][A]^{k+s-1}[B] \\
\cdots & & \cdots & \cdots \\
[C][A]^{k+r-2}[B] & [C][A]^{k+r-1}[B] & \cdots & [C][A]^{k+r+s-3}[B]
\end{bmatrix}
\tag{11.243}
$$

$$
[H(k-1)] =
\begin{bmatrix}
[C] \\
[C][A] \\
\vdots \\
[C][A]^{r-1}
\end{bmatrix}
[A]^{k-1}
\begin{bmatrix}
[B] & [A][B] & \cdots & [A]^{s-1}[B]
\end{bmatrix}
\tag{11.244}
$$

$$
\underset{pr\times qs}{[H(k-1)]} = \underset{pr\times 2N}{[Q]} \; \underset{2N\times 2N}{[A]^{k-1}} \; \underset{2N\times qs}{[W]}
\tag{11.245}
$$

where $[Q]$ and $[W]$ are the observability and controllability matrices and are given by

$$
[Q] =
\begin{bmatrix}
[C] \\
[C][A] \\
\vdots \\
[C][A]^{r-1}
\end{bmatrix}
\text{ and } [W] =
\begin{bmatrix}
[B] & [A][B] & \cdots & [A]^{s-1}[B]
\end{bmatrix}
\tag{11.246}
$$

The objective of ERA is to seek a minimum realization of the state-space matrices. There are an infinite number of realizations of these matrices that can reproduce the measured responses, but the one having a minimum order of the matrices is the true representation of the system. ERA uses SVD to obtain a minimum order of the state space matrices.

From Eq. (11.245),

$$
\text{at } k=1 : \underset{pr\times qs}{[H(0)]} = \underset{pr\times 2N}{[Q]} \; \underset{2N\times 2N}{[A]^{0}} \; \underset{2N\times qs}{[W]} = \underset{pr\times 2N}{[Q]} \; \underset{2N\times qs}{[W]}
\tag{11.247}
$$

Taking SVD of $[H(0)]$,

$$
\underset{pr\times qs}{[H(0)]} = \underset{pr\times pr}{[U]} \; \underset{pr\times qs}{[S]} \; \underset{qs\times qs}{[V]^{T}}
\tag{11.248}
$$

The zero or near-zero singular values can be dropped from the singular values matrix since they don't represent the physical modes (as only 2N singular values are expected to be nonzero). The left and right singular vectors corresponding to the zero/near-zero singular values are also dropped. After these changes, Eq. (11.248) becomes

$$
\underset{pr\times qs}{[H(0)]} = \underset{pr\times 2N}{[U]} \; \underset{2N\times 2N}{[S]} \; \underset{2N\times qs}{[V]^{T}}
\tag{11.249}
$$

which can further be written as

$$
\underset{pr\times qs}{[H(0)]} = \underset{pr\times 2N}{[U]} \; \underset{2N\times 2N}{[S]^{1/2}} \; \underset{2N\times 2N}{[S]^{1/2}} \; \underset{2N\times qs}{[V]^{T}}
\tag{11.250}
$$

Comparing Eqs. (11.247) and (11.250), we can write

$$\underset{pr \times 2N}{[Q]} = \underset{pr \times 2N}{[U]} \; \underset{2N \times 2N}{[S]^{1/2}} \tag{11.251}$$

$$\underset{2N \times qs}{[W]} = \underset{2N \times 2N}{[S]^{1/2}} \; \underset{2N \times qs}{[V]^{T}} \tag{11.252}$$

From Eq. (11.245), we can write

$$\text{at } k = 2: \quad [H(1)] = [Q][A]^{1}[W] \tag{11.253}$$

Substituting $[Q]$ and $[W]$ from Eqs. (11.251) and (11.252) into Eq. (11.253), we get

$$[H(1)] = [U] \; [S]^{1/2} \; [A]^{1} \; [S]^{1/2} \; [V]^{T} \tag{11.254}$$

which leads to

$$[A] = \left([U] \; [S]^{1/2}\right)^{+} \; [H(1)] \; \left([S]^{1/2} \; [V]^{T}\right)^{+} \tag{11.255}$$

The superscript '+' represents the pseudo-inverse. On simplification and noting the properties of SVD that $[U]^{+} = [U]^{T}$ and $[V]^{T+} = [V]$, Eq. (11.255) becomes

$$\underset{2N \times 2N}{[A]} = \underset{2N \times 2N}{[S]^{-1/2}} \; \underset{2N \times pr}{[U]^{T}} \; \underset{pr \times qs}{[H(1)]} \; \underset{qs \times 2N}{[V]} \; \underset{2N \times 2N}{[S]^{-1/2}} \tag{11.256}$$

Eq. (11.256) thus gives the minimum realization of the state matrix $[A]$. To find the matrices $[B]$ and $[C]$, we proceed as follows. The matrix of free pulse responses given by Eq. (11.241) can be related to the Hankel matrix $[H(k-1)]$ (Eq. 11.243) as follows:

$$\underset{p \times q}{[Z(k)]} = \underset{p \times pr}{[E_{p}]^{T}} \; \underset{pr \times qs}{[H(k-1)]} \; \underset{qs \times q}{[E_{q}]} \tag{11.257}$$

where the matrices $[E_{p}]$ and $[E_{q}]$ are given by

$$\underset{p \times pr}{[E_{p}]^{T}} = \left[\underset{p \times p}{[I]} \; \underset{p \times p}{[0]} \; \cdots \; \underset{p \times p}{[0]} \right] \tag{11.258}$$

$$\underset{qs \times q}{[E_{q}]} = \begin{bmatrix} \underset{q \times q}{[I]} \\ \underset{q \times q}{[0]} \\ \vdots \\ \underset{q \times q}{[0]} \end{bmatrix} \tag{11.259}$$

On substituting $[H(k-1)]$ from Eq. (11.245) into Eq. (11.257), we get

$$[Z(k)] = [E_{p}]^{T} \; [Q][A]^{k-1}[W] \; [E_{q}] \tag{11.260}$$

On substituting $[Q]$ and $[W]$ from Eqs. (11.251) and (11.252) into Eq. (11.260), we obtain

$$[Z(k)] = [E_{p}]^{T} \; [U] \; [S]^{1/2} \; [A]^{k-1} \; [S]^{1/2} \; [V]^{T} \; [E_{q}] \tag{11.261}$$

Comparing Eq. (11.261) with Eq. (11.241), we can write

$$\underset{p\times 2N}{[C]} = \underset{p\times pr}{\left[E_p\right]^T} \; \underset{pr\times 2N}{[U]} \; \underset{2N\times 2N}{[S]^{1/2}} \tag{11.262}$$

$$\underset{2N\times q}{[B]} = \underset{2N\times 2N}{[S]^{1/2}} \; \underset{2N\times qs}{[V]^T} \; \underset{qs\times q}{\left[E_q\right]} \tag{11.263}$$

Thus, Eqs. (11.262) and (11.263), respectively, give the minimum realization of the output matrix $[C]$ and the input matrix $[B]$.

e. Modal parameter computation

The eigenvalues and eigenvectors of the realized state matrix are determined to compute the modal parameters. Let these be represented by s_r^d and $\{\psi\}_r^d$, respectively, with $r = 1, 2, .., 2N$ The superscript 'd' indicates association with the discrete-time state matrix. The following is the relationship between the frequencies in discrete and continuous time spaces:

$$s_r^d = e^{s_r \Delta t} \tag{11.264}$$

Therefore, the eigenvalue (or pole) of the system in continuous time is given by

$$s_r = \frac{1}{\Delta t} \log_e s_r^d \tag{11.265}$$

The natural frequencies and modal viscous damping factors are obtained from the poles. The continuous-time eigenvector is obtained from the discrete-time eigenvector using the following relationship:

$$\{\psi\}_r = [C]\{\psi\}_r^d \tag{11.266}$$

Example 11.9: Modal Parameter Estimation Using the ERA

We apply the ERA to our cantilever structure using simulated measured data in this example. In Section 11.16, the ERA formulation was presented using free pulse responses, which are nothing but impulse responses. These can be obtained by measuring FRFs and then taking their inverse Fourier transforms. However, in this example, we use free decay responses instead of impulse responses (though it would yield only the unscaled mode shapes).

Step 1: In this example, though there are no repeated or close modes, to demonstrate the multi-input nature of the ERA, we use a multi-input excitation to measure the free decays and perform modal parameter estimation.

There are five test points on the cantilever structure. We apply band-limited white noise excitation (over 0–400 Hz) at test points 2 and 5. The excitation is applied simultaneously at these two points for 4 seconds, after which the excitation is stopped. Figure 11.29 shows the excitation forces applied at points 2 and 5, and Figure 11.30 shows the resulting responses at test points 1 and 5. It is seen that from 0 to 4 seconds, there is a random forced response, while after 4 seconds, there is free decay response. A dashed vertical line demarcates these two regions in the figure. The free decay responses at all five test points are measured simultaneously using the sensors mounted at these points. Figure 11.31 shows the measured free decay responses at test points 1 and 5.

Step 2: The next step is to obtain the Hankel matrix [H(0)] using Eq. (11.242). It requires [Z(k)] and choice of variables 'r' and 's'. In the present case, the inputs are acting simultaneously, and the resulting responses are measured at five points. Therefore q=1 and p=5. Thus, [Z(k)] is of order 5×1. We take r=164 and s=130, leading to [H(0)] of order 650×154.

Step 3: [H(1)] is then built using the responses obtained by shifting all the time instants used to set [H(0)] by 1 Δt.

Step 4: Find the SVD of [H(0)] using the MATLAB command 'svd()'. It gives the matrices [U], [S] and [V]. To determine the minimum order of the system, we look at the singular values. Figure 11.32 shows a plot of the top 50 singular values. It is seen that the first eight singular values are significant, and the remaining values are wholly or nearly zero. Thus, the system order is 2N=8. The matrices of the singular values and the singular vectors are truncated to retain only the first eight singular values and the corresponding singular vectors.

Step 5: Using Eq. (11.256), we compute the realized state matrix [A].

Step 6: We solve the eigenvalue problem in [A]. It gives s_r^d and $\{\psi\}_r^d$, with $r = 1, 2, .., 8$.

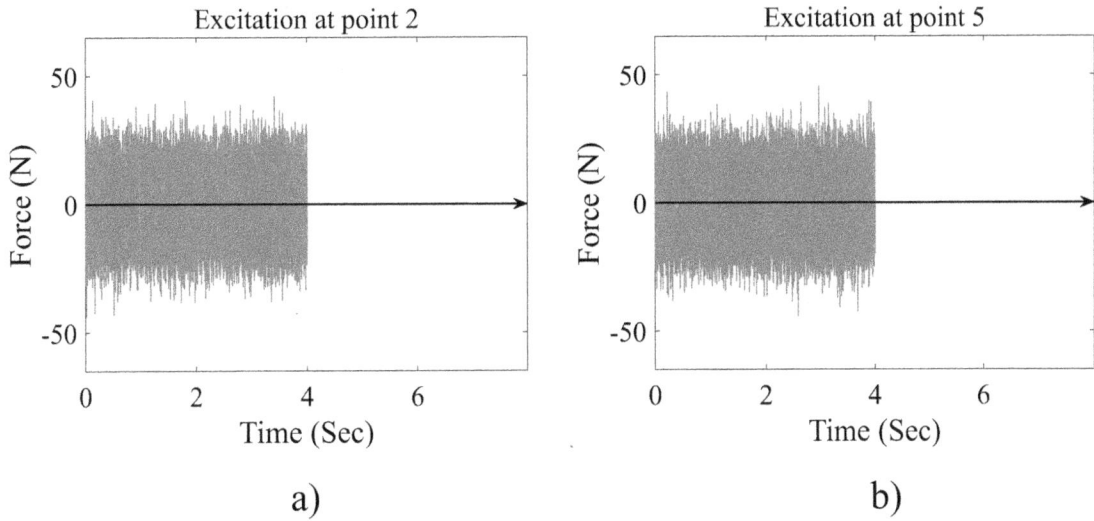

FIGURE 11.29   Excitation forces applied at points 2 and 5.

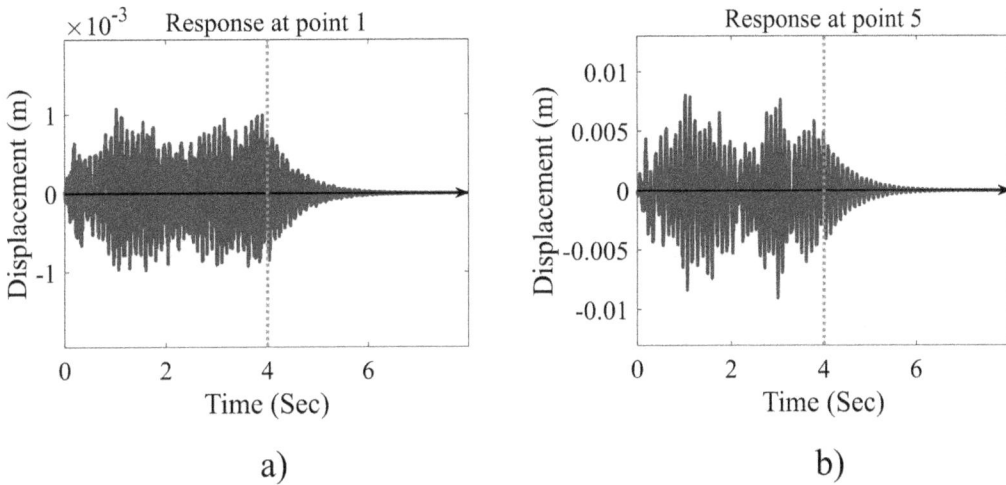

FIGURE 11.30   Responses at test points 1 and 5.

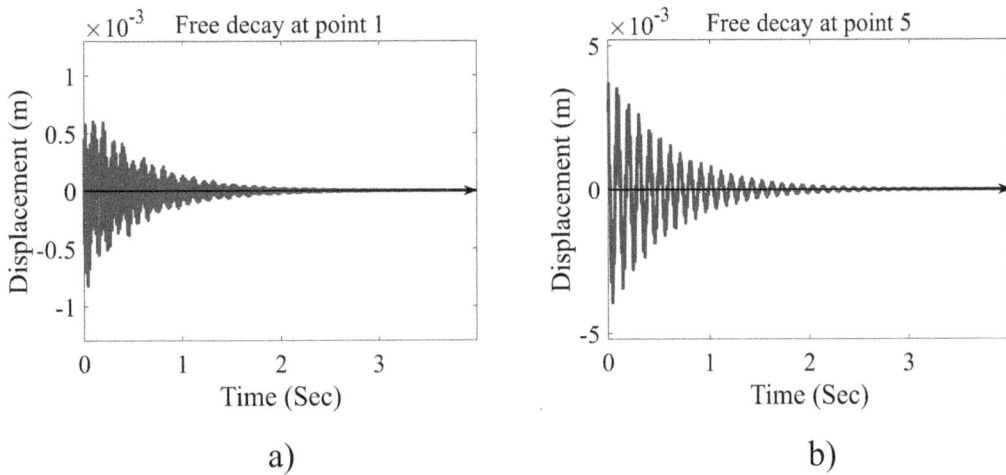

FIGURE 11.31   Measured free decay responses at test points 1 and 5.

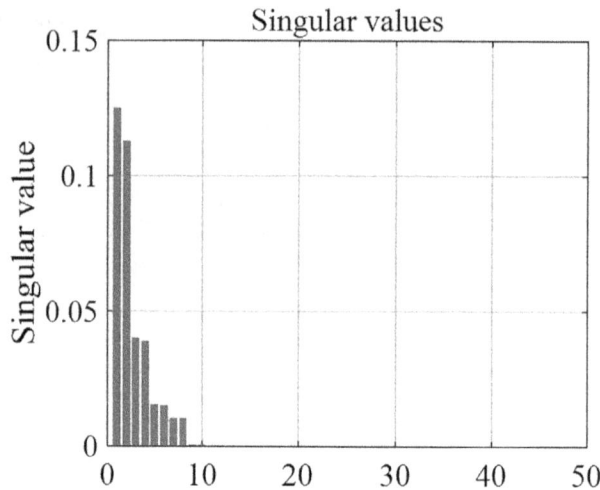

FIGURE 11.32 Plot of the singular values.

TABLE 11.8
Complex Conjugate Poles and the Corresponding Natural Frequencies and Modal Viscous Damping Factors Identified Using ERA

Mode No.	Complex Conjugate Poles	Natural Frequency (Hz)	Damping Factor
1	$-1.74993 \pm 61.651j$	9.81	0.028373
2	$-1.74998 \pm 386.7j$	61.54	0.004525
3	$-1.75068 \pm 1,086.13j$	172.86	0.001612
4	$-1.75043 \pm 2,145.63j$	341.48	0.000816

TABLE 11.9
Unscaled Mode Shapes Identified Using ERA

Mode Shape 1	Mode Shape 2	Mode Shape 3	Mode Shape 4
$-0.0013 + 0.0008i$	$0.0035 - 0.0022i$	$0.0051 - 0.0001i$	$0.0042 + 0.0027i$
$-0.0048 + 0.0031i$	$0.0080 - 0.0052i$	$0.0044 - 0.0001i$	$-0.0017 - 0.0011i$
$-0.0097 + 0.0062i$	$0.0069 - 0.0045i$	$-0.004 + 0.0001i$	$-0.0018 - 0.0012i$
$-0.0153 + 0.0097i$	$-0.0008 + 0.0005i$	$-0.0033 + 9.3e\text{-}5i$	$0.0036 + 0.0023i$
$-0.0212 + 0.0134i$	$-0.0117 + 0.0076i$	$0.0084 - 0.0002i$	$-0.0055 - 0.0036i$

Step 7: We use Eqs. (11.265) and (11.266) to find poles S_r and the corresponding eigenvectors $\{\psi\}_r$ in physical space. The complex conjugate poles and the corresponding natural frequencies and modal viscous damping factors for the four identified physical modes are given in Table 11.8, while the corresponding eigenvectors/mode shapes are given in Table 11.9. All the identified modal parameters match closely with their correct values.

11.17 COMPLEX MODE INDICATOR FUNCTION (CMIF)

Mode indicators functions (MIFs) are used to detect the presence of modes. They help to determine the model order of the system as that information is needed for curve-fitting, as seen in the previous sections. Several MIFs have been introduced to detect the normal modes, and these are covered in Chapter 12, which presents normal mode testing (or phase resonance testing). The complex mode indicator function (CMIF) (Shih and Hammond [98]), however, detects complex modes and gives an estimate of the damped natural frequencies. CMIF is a plot of the eigenvalues of the 'normal' FRF matrix (denoted by $\left[H(i\omega)\right]^{N}$), computed from the measured FRF matrix $\left(\left[H(i\omega)\right]\right)$. The 'normal' FRF matrix is defined as

$$\left[H(i\omega)\right]^{N} = \left[H(i\omega)\right]^{H}\left[H(i\omega)\right] \qquad (11.267)$$

Taking SVD of $\left[H(i\omega)\right]$

$$\left[H(i\omega)\right] = \left[U(i\omega)\right]\left[S(i\omega)\right]\left[V(i\omega)\right]^{H} \qquad (11.268)$$

and substituting Eq. (11.268) into Eq. (11.267), we get

$$\left[H(i\omega)\right]^{N} = \left[V(i\omega)\right]\left[S^{2}(i\omega)\right]\left[V(i\omega)\right]^{H} \qquad (11.269)$$

From Eq. (11.269), we conclude that the eigenvalues of the 'normal' FRF matrix are nothing but the square of the singular values of the FRF matrix $\left[H(i\omega)\right]$.

To plot CMIF, the eigenvalues of $\left[H(i\omega)\right]^{N}$ are computed at each frequency, and the eigenvalues are plotted as a function of the frequency. The peaks in the CMIF plot show the modes, and the corresponding frequencies provide estimates of the damped natural frequencies. Structures with symmetries often have repeated or closely located poles, and multi-reference testing is needed to detect and identify such poles. CMIF can detect repeated poles from the FRF matrix from a multi-reference test.

REVIEW QUESTIONS

1. What is meant by curve-fitting?
2. What is the basic philosophy of an SDOF curve-fitting method?
3. Under what situations an SDOF curve-fit may lead to significant errors in the estimated modal parameters?
4. What is the principle of the circle-fit method?
5. What is the principle of the line-fit method?
6. What are residuals?
7. What are low and high-frequency residuals?
8. Does there exist an SDOF curve-fitting method in the time domain?
9. What is the RFP method? How are residuals accounted for using this method?
10. What is a stabilization diagram, and how is it used to determine the model order?
11. How is a system with minimum order realized in Eigensystem Realization Algorithm?
12. Under what conditions single-input modal testing may be inadequate, needing multi-input testing?

12 Phase Resonance Testing

12.1 INTRODUCTION

In Chapters 9–11, we studied the measurement of FRFs and the estimation of the modal parameters from them using curve-fitting methods. In this chapter, we study another approach to modal parameter identification called phase resonance testing (also referred to as normal mode testing, or appropriated force testing). We first define phase resonance testing and derive the condition for phase resonance and excitation of normal modes. This chapter then introduces the methods to determine appropriation force vectors and presents mode indicator functions (MIFs). It is followed by a simulation of phase resonance testing of a 3-DOF system and the cantilever structure considered in the previous chapters. In the end, we compare the phase resonance and curve-fitting approaches.

12.2 OBJECTIVE OF PHASE RESONANCE TESTING

Phase resonance testing is the excitation and measurement of the normal modes of a structure. The undamped modes of vibration of a structure are called the normal modes. Every structure in practice has damping, evident from the fact that the free vibration due to an initial disturbance decays with time. The undamped modes represent the vibration modes of the structure if the damping is hypothetically removed from it. The motivation to experimentally identify the normal modes originates from the need to validate and, if necessary, update the analytical/numerical models.

A numerical model yields undamped modes of the structure, as damping is either not modeled or modeled as proportional damping due to the lack of an accurate model. Even if the model is damped, finding the undamped modes from a numerical model is straightforward, as it can be done by ignoring the damping matrix and solving the eigenvalue problem based on only the mass and stiffness matrices. Hence, the experimentally identified normal modes are needed to enable a direct and consistent comparison with the numerical model modes. There has been considerable interest in phase resonance testing in the aircraft industry due to the reliable model validation necessary for accurate aeroelastic and stability limit speed predictions.

In phase resonance testing, the idea is to apply the harmonic forces of the same frequency at multiple locations using vibration shakers such that the structure vibrates in a normal mode. The question addressed is what the frequency, magnitude, and phase of the excitation forces should be so that a normal mode is excited. To realize a normal mode, the frequency of the excitation forces on the structure needs to be equal to the undamped natural frequency of that mode, and the forces need to be tuned such that the resulting response is entirely governed by the corresponding normal mode shape. We see in the next section that this is achieved if the condition for phase resonance is satisfied. Thus, **phase resonance testing** is about making a structure vibrate in a normal mode and then determining the modal parameters of that mode by direct measurement. However, the curve-fitting approach discussed in Chapter 11 is about computationally estimating the modal parameters from the measured FRFs. The FRFs contain the contribution of all the structure modes, and hence the estimation process mathematically separates the contribution of various modes to determine the modal parameters of individual modes. Hence, modal testing with curve fitting is also called **phase separation testing**. Phase separation testing generally yields complex modes.

12.3 PHASE RESONANCE CONDITION

In this section, we determine the condition for the vibration of a system in a normal mode. Let us consider the vibration of an MDOF system with viscous damping subjected to harmonic forces. The governing equation of motion of the system in matrix form with a harmonic excitation force vector can be written as

$$[M]\{\ddot{x}\} + [C]\{\dot{x}\} + [K]\{x\} = \{F_0\}e^{i\omega t} \tag{12.1}$$

The steady-state response is given by

$$\{x(t)\} = \{x_0\}e^{i\omega t} \tag{12.2}$$

Let us assume that the vector of force amplitudes, $\{F_0\}$, is a real vector and study the implications of this on the response. Thus, the forces at various DOFs are in or out of phase. It makes the force vector a monophase vector. However, the vector of resulting displacement amplitudes, $\{x_0\}$, is, in general, a complex vector and can be broken into its real and imaginary parts as

$$\{x_0\} = \{x_{0R}\} + i\{x_{0I}\} \tag{12.3}$$

Substituting, Eqs. (12.2)–(12.3) into Eq. (12.1), and separating and equating the real and imaginary parts on the two sides, we get the following two equations:

$$\left([K] - \omega^2[M]\right)\{x_{0R}\} - \omega[C]\{x_{0I}\} = \{F_0\} \tag{12.4}$$

$$\left([K] - \omega^2[M]\right)\{x_{0I}\} + \omega[C]\{x_{0R}\} = \{0\} \tag{12.5}$$

Let us investigate the state of vibration of the system when the real part of its response at all the points is zero, i.e.,

$$\{x_{0R}\} = \{0\} \tag{12.6}$$

Substituting Eq. (12.6) into Eq. (12.5) gives

$$\left([K] - \omega^2[M]\right)\{x_{0I}\} = \{0\} \tag{12.7}$$

Eq. (12.7) is nothing but the eigenvalue problem of the corresponding undamped system.

Thus, if the force vector is mono-phased and its frequency is equal to an undamped natural frequency of the system, then the real part of the response will be zero, and its imaginary part will match the corresponding undamped mode shape. This effectively excites an undamped mode. Under these conditions, the excitation forces are entirely balanced by the damping forces of the system, as revealed by the following equation obtained by substituting Eq. (12.6) into Eq. (12.4).

$$-\omega[C]\{x_{0I}\} = \{F_0\} \tag{12.8}$$

Let us also determine the response of the system. Substituting Eqs. (12.3) and (12.6) into Eq. (12.2) gives

$$\{x(t)\} = i\{x_{0I}\}e^{i\omega t} \tag{12.9}$$

Comparing Eq. (12.9) with the excitation force vector $\{F_0\}e^{i\omega t}$ indicates that the phase difference between the responses and excitation forces is 90°. This state of vibration is also called a quadrature response or response being in the quadrature with force.

Thus, Eq. (12.9) must hold if Eq. (12.6) is satisfied and vice versa. That is, if the real part of the response vector is zero, then the total response vector is in quadrature with the force vector. This is the condition for excitation of a normal mode of the system and is referred to as the **phase resonance condition**.

Note that no assumption regarding the type of damping distribution, i.e., proportional or nonproportional, was made in arriving at the above conclusions. Therefore, for both these cases, a normal mode can be excited by choosing a monophase force vector, and the phase resonance condition would be satisfied by the structure when vibrating in a normal mode. Further, by proceeding as above, a similar analysis for a system with structural damping can be made, and it can be shown that the conclusions drawn above are also valid for such a system.

In the following sections, we look at determining the force vector for the excitation of normal modes.

12.4   EXCITATION OF NORMAL MODES

Having looked at the conditions that signify forced vibration in a normal mode, the question arises as to how the excitation of a normal mode can be realized. It can be realized by applying a specific pattern of the excitation forces (also called an **appropriated force vector**) at a frequency equal to the undamped natural frequency of that mode. Theoretically, the undamped natural frequencies can be obtained by solving the eigenvalue problem in Eq. (12.7), and the corresponding necessary force patterns (or the force vectors) can also be obtained from Eq. (12.8). However, these equations are based on a mathematical model of the structure and are not useful in exciting a normal mode experimentally.

In this section, we look at the basic principle of force vector estimation, and in the later sections, we look at some methods based on that. The force vector estimation methods can be classified as direct and iterative. We study some direct methods and briefly discuss the principle of the iterative method.

12.4.1 Basic Principle of Force Vector Estimation

The force vector for the excitation of a normal mode can be related to the FRF matrix. Since the FRFs can be measured on the structure under test, this relationship provides a fundamental basis for experimentally estimating the required force vector. The steady-state displacement response is given by

$$\{x(\omega)\} = [\alpha]\{F(\omega)\} \tag{12.10}$$

where $[\alpha]$ is the receptance matrix of the system. Eq. (12.10) in terms of the steady-state amplitudes is

$$\{x_0\} = [\alpha]\{F_0\} \tag{12.11}$$

The FRF matrix is complex and can be written in terms of its real and imaginary parts as

$$[\alpha] = [\alpha_R] + i[\alpha_I] \tag{12.12}$$

Substituting Eqs. (12.3) and (12.12) into Eq. (12.11), and separating and equating the real and imaginary parts on the two sides, and noting that the force amplitude vector is real, we get the following two equations:

$$\{x_{0R}\} = [\alpha_R]\{F_0\} \tag{12.13}$$

$$\{x_{0I}\} = [\alpha_I]\{F_0\} \tag{12.14}$$

We studied in the last section that when a normal mode is excited, the real part of the response is zero (Eq. 12.6), and hence from Eq. (12.13), we get

$$[\alpha_R]\{F_0\} = \{0\} \tag{12.15}$$

Eq. (12.15) provides a fundamental relation for estimating the force vector required for the excitation of normal modes. This equation is homogenous, and for a solution to exist, the determinant of the coefficient matrix must be zero,

$$\det([\alpha_R]) = 0 \tag{12.16}$$

Thus, all the frequencies at which the determinant of the real part of the FRF matrix is zero are the undamped natural frequencies of the system (Asher [4]). Let r^{th} undamped natural frequency is denoted by $\omega_{u,r}$. The subscript 'u' denotes association with undamped mode. $\omega_{u,r}$ is substituted back into Eq. (12.15) to find the force vector and let it be $\{F_0\}_{u,r}$. $\{F_0\}_{u,r}$ is essentially the force vector that would excite the rth undamped/normal mode.

The structure under test can be excited with the force vector $\{F_0\}_{u,r}$, and the resulting responses under the steady state can be measured. Further tuning of the force vector may be necessary if the phase resonance condition is not satisfied due to inaccuracies in the force vector estimate. After the phase resonance condition is satisfied, the response vector can be measured, and it represents the rth normal mode shape. Alternatively, $\{F_0\}_{u,r}$ can be substituted in Eq. (12.14), with the resulting response vector giving an estimate of the rth normal mode shape. Thus,

$$\{\psi\}_{u,r} = [\alpha_I]\{F_0\}_{u,r} \tag{12.17}$$

12.5 EXTENDED ASHER'S METHOD OF FORCE VECTOR ESTIMATION

In practice, due to limitations on the number of exciters that can be used, the number of excitation points (denoted by e) is far less than the number of points at which the responses are measured (denoted by m). In such a situation, the measured FRF matrix is rectangular (of order $m \times e$), and hence the condition in Eq. (12.16) cannot be applied as that requires a square FRF matrix. Even for a square FRF matrix, Eq. (12.16) may not be satisfied exactly at any frequency due to measurement noise. In these cases, the condition of phase resonance or modal tuning cannot be achieved perfectly. Instead, tuning in a least-square sense is sought by solving the following minimization problem (Ibanez [46]):

$$\min \quad \lambda = \frac{\left\|\{x_{0R}\}\right\|^2}{\left\|\{F_0\}\right\|^2} \tag{12.18}$$

where $\|\cdot\|^2$ denotes the square of the norm of a vector. Since at phase resonance, the real part of the response must ideally be zero, the minimization problem in Eq. (12.18) seeks to achieve this by minimizing the square of the norm of the real part of the response weighted by the square of the norm of the force vector. λ can be written as

$$\lambda = \frac{\{x_{0R}\}^T \{x_{0R}\}}{\{F_0\}^T \{F_0\}} \tag{12.19}$$

Substituting Eq. (12.13) into (Eq. (12.19), we obtain

$$\lambda = \frac{\{F_0\}^T [A]\{F_0\}}{\{F_0\}^T \{F_0\}} \tag{12.20}$$

where

$$[A] = [\alpha_R]^T [\alpha_R] \tag{12.21}$$

From Eq. (12.20),

$$\{F_0\}^T [A]\{F_0\} = \lambda \{F_0\}^T \{F_0\} \tag{12.22}$$

or,

$$[A]\{F_0\} = \lambda \{F_0\} \tag{12.23}$$

Eq. (12.23) is an eigenvalue problem and is solved at each frequency (ω). Since our interest lies in minimizing λ, we look for minimum values of the lowest eigenvalue over the frequency range. The excitation frequencies corresponding to the minimum eigenvalues are the undamped natural frequencies. The eigenvectors corresponding to these eigenvalues are the force vectors that excite the normal modes. The normal mode shapes are then obtained, as explained in Section 12.4.1.

12.6 REAL MODE INDICATOR FUNCTION (RMIF)

In this method, the problem of determining normal modes is formulated as follows. The response is a monophase response if the real part of the response is proportional to its imaginary part. That is,

$$\{x_{0R}\} = \lambda \{x_{0I}\} \tag{12.24}$$

The following minimization problem is solved to find λ for obtaining a monophase response.

$$\min \quad J = \left\|\{x_{0R}\} - \lambda \{x_{0I}\}\right\|^2 \tag{12.25}$$

which can be written as

$$\min \quad J = \left(\{x_{0R}\} - \lambda \{x_{0I}\}\right)^T \left(\{x_{0R}\} - \lambda \{x_{0I}\}\right) \tag{12.26}$$

Applying the necessary condition for minimum, $dJ/d\lambda = 0$, we get

$$\{x_{0I}\}^T \{x_{0R}\} = \lambda \{x_{0I}\}^T \{x_{0I}\} \tag{12.27}$$

Substituting Eqs. (12.13) and (12.14) into Eq. (12.27), and simplifying, we obtain

$$[\alpha_I]^T [\alpha_R]\{F_0\} = \lambda [\alpha_I]^T [\alpha_I]\{F_0\} \tag{12.28}$$

Eq. (12.28) is a generalized eigenvalue problem and is solved at each frequency ω. The monophase response corresponding to $\lambda = 0$ satisfies the phase resonance condition and hence represents a normal mode. Therefore, the eigenvalues obtained through the solution of (12.28) can be plotted against frequency ω, and the zero crossings of the plots represent the undamped natural frequencies. Since the plot of the eigenvalue versus frequency allows identifying a real or normal mode, it is called a real mode indicator function (RMIF). The eigenvectors corresponding to zero eigenvalues are the force vectors that excite the normal modes. The normal mode shapes can then be obtained, as explained in Section 12.4.1. Note that RMIF can also be applied to rectangular FRF matrices.

12.7   MULTIVARIATE MODE INDICATOR FUNCTION (MMIF)

In this method (Williams et al. [106]), the problem of determining normal modes is formulated as the following minimization problem:

$$\min \quad \lambda = \frac{\left\|\{x_{0R}\}\right\|^2}{\left\|\{x_{0R}\} + i\{x_{0I}\}\right\|^2} \tag{12.29}$$

λ can be written as

$$\lambda = \frac{\{x_{0R}\}^T \{x_{0R}\}}{\{x_{0R}\}^T \{x_{0R}\} + \{x_{0I}\}^T \{x_{0I}\}} \tag{12.30}$$

Substituting Eqs. (12.13) and (12.14) into (Eq. (12.30), we obtain

$$\lambda = \frac{\{F_0\}^T [A]\{F_0\}}{\{F_0\}^T ([A] + [B])\{F_0\}} \tag{12.31}$$

where

$$[A] = [\alpha_R]^T [\alpha_R] \tag{12.32}$$

$$[B] = [\alpha_I]^T [\alpha_I] \tag{12.33}$$

Simplification of Eq. (12.31) leads to

$$[A]\{F_0\} = \lambda ([A] + [B])\{F_0\} \tag{12.34}$$

Eq. (12.34) is an eigenvalue problem and is solved at each frequency ω, with λ and $\{F_0\}$ being the eigenvalue and eigenvector, respectively. Since our interest is to minimize λ, we look for the lowest eigenvalue at each frequency. The plot of the lowest eigenvalue against ω is referred to as the multivariate mode indicator function (MMIF). The excitation frequencies corresponding to the local minimum values in this plot are the undamped natural frequencies. The eigenvectors corresponding to the eigenvalues at these local minimum points are the force vectors that excite the corresponding normal modes. The normal mode shapes can then be obtained, as explained in Section 12.4.1. MMIF can also be used for rectangular FRF matrices. It is called a multivariate function, as the plot of the other eigenvalues allows the detection of repeated normal modes, if any.

12.8   ESTIMATION OF OTHER MODAL PROPERTIES

In the previous sections, we studied the methods for estimating force vectors (or appropriated force vectors) to excite normal modes. We also studied how the undamped natural frequencies and the corresponding (unscaled) normal mode shapes can be obtained. In this section, we look at how the modal damping factor and the modal mass of the modes can be estimated to get a complete description of the modal properties

One approach to finding these properties is removing the appropriated force excitation vector applied to the structure once the normal mode shape has been measured. The resulting free decay response of the structure is used to estimate the modal damping factor and modal mass. If proportional damping is there, then the structure's response would continue to be purely in the normal mode that was excited. Hence, the modal damping factor for that mode can be found by logarithmic decrement from the free decay assuming viscous damping. The modal damping coefficient, modal mass, and modal stiffness for the mode can also be estimated from the free decay response. The mass-normalization of the normal modes can be done using the estimated modal masses.

If the damping is nonproportional, then the structure's response after removing the force vector would not be purely in the normal mode that was excited. This is because the cross-modal damping forces due to nonproportional damping would also excite other modes. Therefore, using logarithmic decrement would be invalid but, if used, would not give an accurate estimate. An MDOF curve fit in the time domain can be carried out to find the modal damping factors and modal masses (Naylor et al. [75]).

The other approach that has been used to find the damping factor is to perform a sine sweep measurement over a narrow frequency range around the identified natural frequency.

12.9   BASIC STEPS IN PHASE RESONANCE TESTING

The steps followed in phase resonance testing are listed below.

a. Decide the number of response transducers and exciters to be used. Due to cost and other constraints, often the number of exciters is not many. However, it should be enough to excite the normal modes of interest. At least more than one exciter is needed. The choice of location of the exciters is also crucial as it determines the ability to excite the desired modes.
b. Measure all the FRFs between the response and excitation DOFs.
c. Determine an estimate of the undamped natural frequencies and the corresponding force appropriation vectors in the measurement frequency range (using one of the methods presented in the previous sections).
d. In normal mode testing, one mode at a time is excited and measured. Let us say the rth normal mode, identified in step (c), is to be excited. The corresponding estimated force appropriation vector is applied to the structure using the exciters. The frequency of excitation is the estimated undamped natural frequency for the mode. However, it may not necessarily tune the rth normal mode exactly. It may be because the estimates obtained in (c) are inaccurate due to measurement noise, inaccuracies in the measured FRFs, inadequate excitation points, or the finite frequency resolution of the FRFs. Therefore, in practice, a tuning procedure is used to fine-tune the frequency and forces to satisfy the phase resonance condition. For tuning, the outputs of the response and force transducers are used to determine the relative phase between the responses and the forces to check the phase resonance condition. If the condition is not satisfied, then the inputs to the exciters are varied iteratively through a closed loop control to minimize some index that quantifies deviation from the phase resonance condition. We should note that the tuning imposes the phase resonance condition for only those DOFs where the responses are measured. It doesn't guarantee the satisfaction of the resonance condition over the whole structure.
e. Once the rth normal mode is excited, the measured responses give the normal mode shape. The other modal parameters, like modal damping factor and modal mass, are estimated as discussed in Section 12.8.
f. Steps (d) and (e) are repeated for other undamped natural frequencies identified in step (c).

12.10   PHASE RESONANCE TESTING SIMULATION

In this section, we perform numerical simulations of the phase resonance testing on some simple systems. Three studies are performed, two on a 3-DOF system and one on our cantilever structure. The objective of these simulations is to use the principles and methods studied in the previous sections to get further insight into the excitation of normal modes and see how it allows a direct measurement of the modal parameters of normal modes.

12.10.1   3-DOF SYSTEM WITH PROPORTIONAL VISCOUS DAMPING

A 3-DOF system with proportional viscous damping is considered. The mass, stiffness, and damping matrices of the system are

$$[M] = \begin{bmatrix} 5 & 0 & 0 \\ 0 & 7.5 & 0 \\ 0 & 0 & 3.75 \end{bmatrix},$$

$$[K] = \begin{bmatrix} 8,000 & -4,000 & 0 \\ -4,000 & 8,000 & -4,000 \\ 0 & -4,000 & 8,000 \end{bmatrix}$$

$$[C] = 0.002[K] + 1.2[M] = \begin{bmatrix} 22 & -8 & 0 \\ -8 & 25 & -8 \\ 0 & -8 & 20.5 \end{bmatrix}$$

a. We assume that the excitation is applied and the responses are measured at all the DOFs. Thus, the full FRF matrix of the system (of order 3×3) is measured.

b. We use RMIF and MMIF methods to determine the undamped natural frequencies and the corresponding force appropriation vectors. Since the full FRF matrix is known, we also use the determinant method.

c. Figure 12.1 shows the plot of the determinant of the FRF matrix versus frequency. The frequencies corresponding to the zero crossings of the curve are the undamped natural frequencies. The plot has three zero crossings, marked by dots. The determinant method cannot be used in practice as the full FRF matrix is generally unavailable.

To obtain the RMIF, we solve the eigenvalue problem given by Eq. (12.28) at each frequency. Figure 12.2 shows the plot of the three eigenvalues versus frequency. The frequencies corresponding to the zero crossings of the eigenvalue curves, marked by dots in the figure, are the undamped natural frequencies. The eigenvectors corresponding to the zero eigenvalues give the force appropriation vectors.

We calculate MMIF by solving the eigenvalue problem given by Eq. (12.34) at each frequency. Figure 12.3 shows the MMIF (i.e., the plot of the lowest eigenvalue versus frequency). The frequencies corresponding to the local minimums of the MMIF (marked by dots in the figure) are the undamped natural frequencies. The eigenvectors corresponding to the eigenvalues at these local minimum points give the force appropriation vectors.

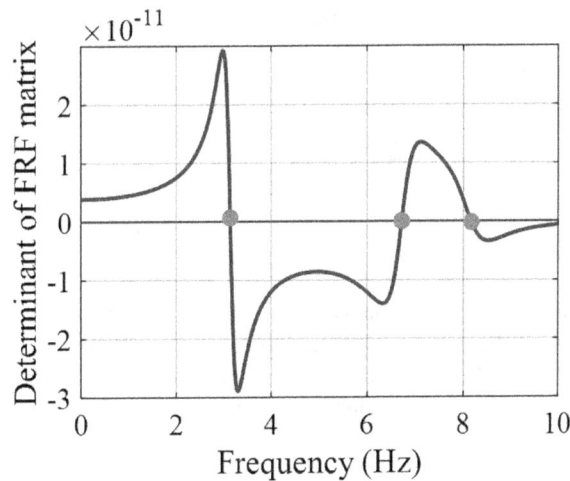

FIGURE 12.1   Plot of the determinant of the FRF matrix versus frequency.

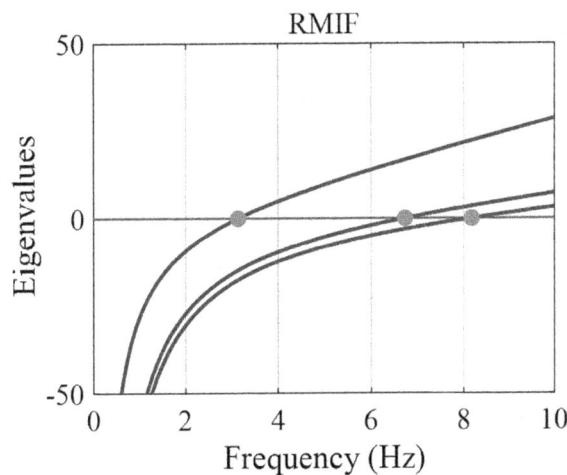

FIGURE 12.2   Real mode indicator function.

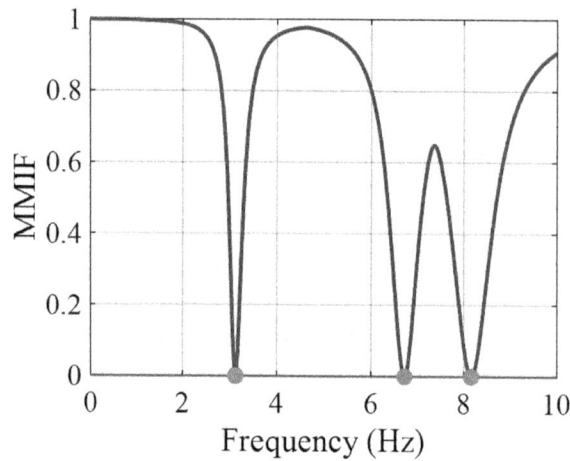

FIGURE 12.3   Multivariate mode indicator function.

TABLE 12.1
Undamped Natural Frequencies and the Corresponding Force Appropriation Vectors

	Identified by RMIF			Identified by MMIF		
Undamped natural frequency (Hz)	3.14	6.72	8.17	3.14	6.72	8.16
Force appropriation vector (N)	$\begin{Bmatrix} 6.532 \\ 10 \\ 5.547 \end{Bmatrix}$	$\begin{Bmatrix} -10 \\ 3.422 \\ 5.203 \end{Bmatrix}$	$\begin{Bmatrix} -4.817 \\ 9.254 \\ -10 \end{Bmatrix}$	$\begin{Bmatrix} 6.532 \\ 10 \\ 5.547 \end{Bmatrix}$	$\begin{Bmatrix} -10 \\ 3.422 \\ 5.216 \end{Bmatrix}$	$\begin{Bmatrix} 4.818 \\ -9.254 \\ 10 \end{Bmatrix}$

Table 12.1 gives the undamped natural frequencies and the corresponding force appropriation vectors identified by RMIF and MMIF. The correct values of the undamped natural frequencies in Hz calculated from the system matrices are 3.136, 6.719, and 8.16. The undamped natural frequencies identified by both methods are nearly equal to the corresponding correct values. The force appropriation vectors given in the table are scaled with the maximum force value equal to 10.

d. As an example, we excite one of the normal modes, mode 2, identified using MMIF. The corresponding force appropriation vector is used to excite the system at a frequency equal to the undamped natural frequency of mode 2, and the resulting responses are measured. From Table 12.1, the force appropriation vector for mode 2 is $\{F\} = \{-10 \quad 3.422 \quad 5.216\}^T$, and the undamped natural frequency is 6.72 Hz. Figure 12.4 shows the time histories of the excitation forces and the responses at the three DOFs. The responses initially show a transient phase, where the response builds up, after which the steady state is reached.

Now the question is, does the state of vibration of the system under the steady state satisfy the phase resonance condition as that is the evidence of the normal mode vibration? To verify this, we look at the relative phase of the force and response signals. Figure 12.5 shows the time histories of the excitation forces and the responses over a short duration. The forces and responses at a common instant are marked by stems in the figure. It is seen that the displacements at the three DOFs make a monophase vector since all the displacements reach their mean or zero value simultaneously. The figure also confirms the monophase nature of the force vector, as all three excitation forces reach their absolute maximum values simultaneously. We also observe that when the forces are at their absolute maximum values, the displacements are at the mean values. Thus, the two have a 90° phase, and the phase resonance condition is satisfied. These observations show that the force appropriation vector estimated from the MMIF excites the normal mode.

Let us compute the modal forces and modal responses to get further insight into why a pure normal mode is excited. Figure 12.6 shows the time histories of the modal forces and modal responses for the three modes. The modal forces for modes 1 and 3 are zero; hence, the corresponding modal responses are also zero. The modal force for only mode 2 is nonzero, causing a response in mode 2 alone.

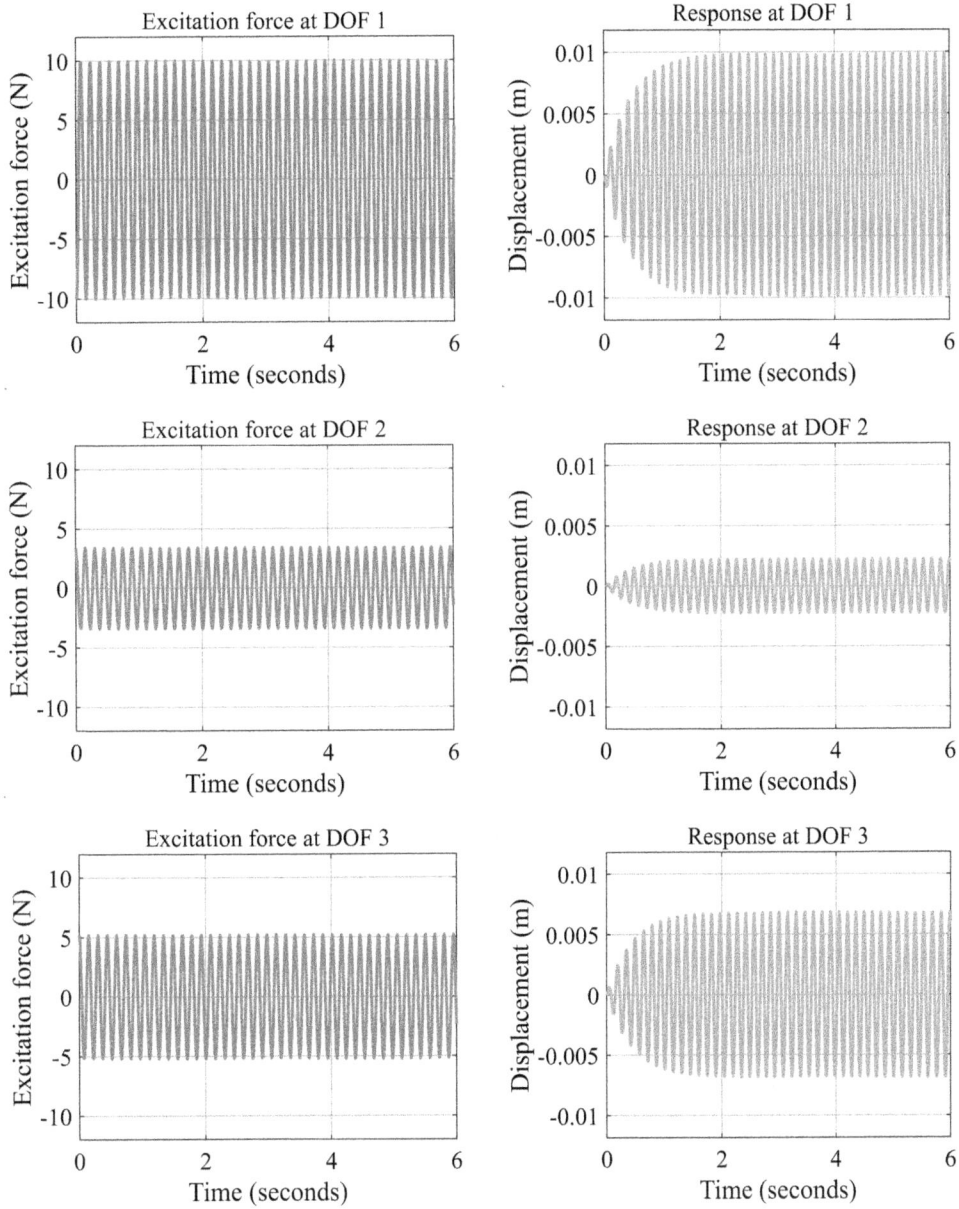

FIGURE 12.4 Time histories of the excitation forces and the responses at the three DOFs.

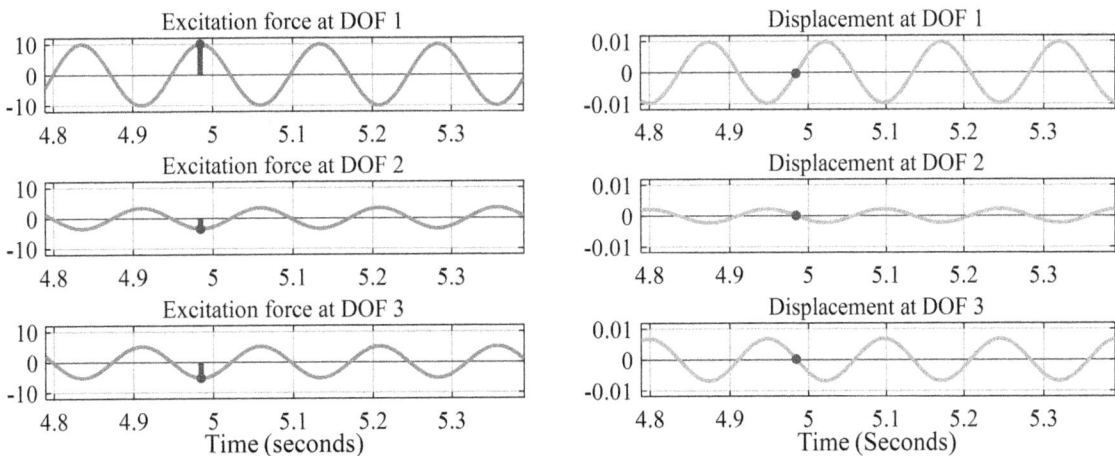

FIGURE 12.5 Time histories of the excitation forces and the responses over a short duration.

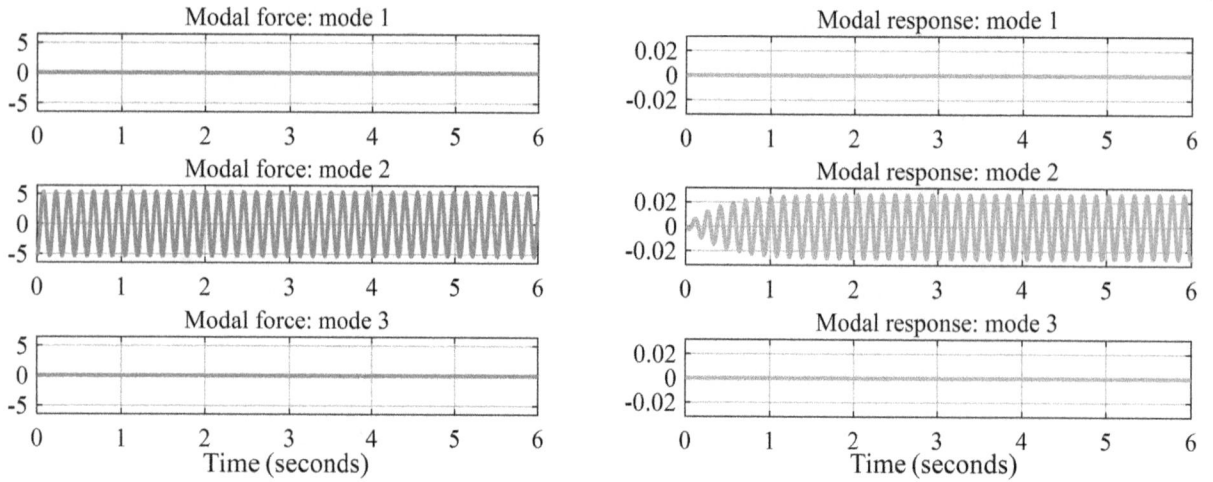

FIGURE 12.6   Time histories of the modal forces and modal responses.

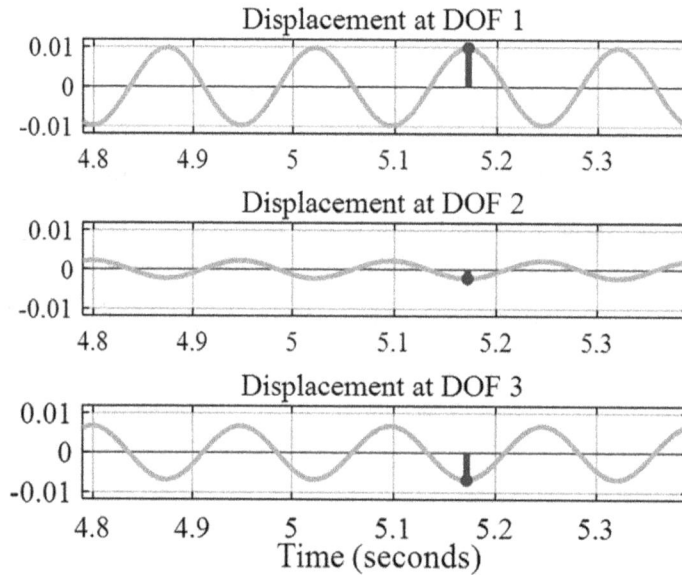

FIGURE 12.7   Measurement of response at all DOFs.

e. Once the desired normal mode is excited, the responses can be measured to get the mode shape. Figure 12.7 shows the measurements taken at all DOFs. All measurements must correspond to the same phase of the vibration. Because of this, the responses at a common instant are measured (marked by stems). It gives the mode shape

for mode 2 as $\{\psi\}_X = \{0.00994 \quad -0.00226 \quad -0.0069\}^T$.

   The experimental mode shape is now compared with the analytical estimate of the corresponding undamped mode shape, $(\{\psi\}_A)$, obtained from the system matrices. Both the eigenvectors are normalized such that the maximum element is unity. These two mode shapes are: $\{\psi\}_X = \{1 \quad -0.2281 \quad -0.6938\}^T$ and $\{\psi\}_A = \{1 \quad -0.22816 \quad -0.69375\}^T$. Thus, the mode shape obtained from the phase resonance test is found to be an accurate estimate of the undamped mode shape of the system.

f. To measure the modal damping factor of the mode, we remove the excitation forces from all the DOFs to get free decay response. Figure 12.8 shows the force and response time histories at DOF 1 when the excitation forces are cut off. Since the system has proportional damping, the free decay response is purely in mode 2. Using the logarithmic decrement, the modal damping factor is found to be $\xi_X = 0.05641$. It is in good agreement with the corresponding estimate obtained from the system matrices ($\xi_A = 0.056431$).

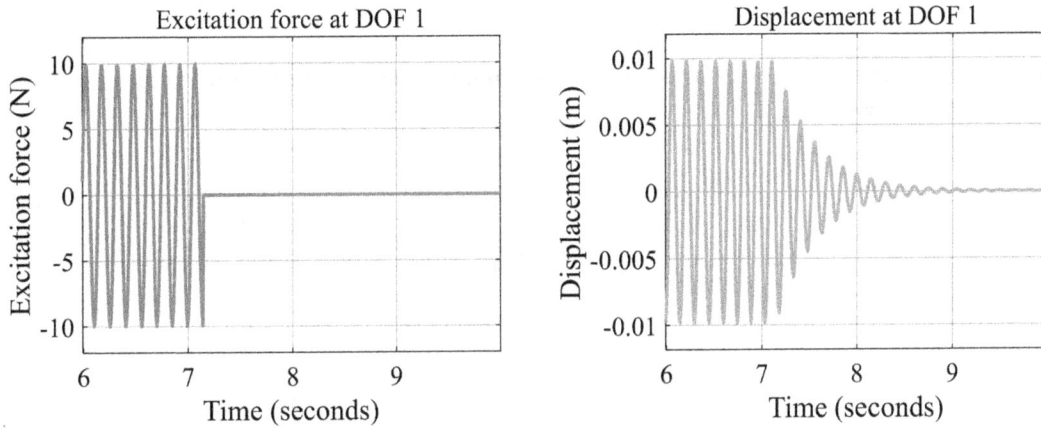

FIGURE 12.8   Force and response time histories at DOF 1 when the excitation forces are cut off.

12.10.2   3-DOF SYSTEM WITH NONPROPORTIONAL VISCOUS DAMPING

The damping matrix of the 3-DOF system in the last section is replaced by the nonproportional viscous damping matrix given below.

$$[C] = \begin{bmatrix} 12 & -12 & 0 \\ -12 & 24 & 0 \\ 0 & 0 & 12 \end{bmatrix}$$

a. For this case as well, the full FRF matrix of the system is assumed to be measured. We use only the MMIF method to detect normal mode frequencies.

We calculate the MMIF, as in the previous section, by solving the eigenvalue problem given by Eq. (12.34). Figure 12.9 shows the plot of the MMIF. The frequencies corresponding to the local minimum values of MMIF (marked by dots in the figure) are the undamped natural frequencies. The eigenvectors corresponding to the eigenvalues at these local minimum points give the force appropriation vectors.

Table 12.2 gives the details of the undamped natural frequencies identified using MMIF. These agree well with the corresponding correct values of the undamped natural frequencies in the previous section. The force appropriation vectors shown in the table are scaled with a maximum value of 10. The force vectors are different from those in the previous section (the proportional damping case). It is due to changes in the magnitude and distribution of the damping forces.

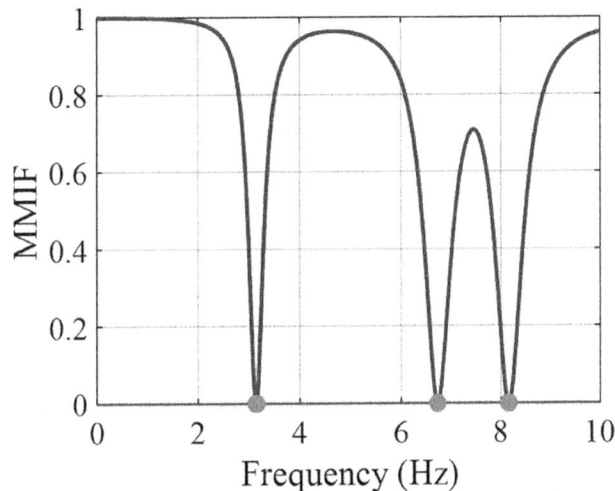

FIGURE 12.9   Multivariate mode indicator function.

TABLE 12.2

Undamped Natural Frequencies and the Corresponding Force Appropriation Vectors Identified by the MMIF Method

Undamped natural frequency (Hz)	3.14	6.72	8.16
Force appropriation vector (N)	$\begin{Bmatrix} -1.703 \\ -10 \\ -7.384 \end{Bmatrix}$	$\begin{Bmatrix} -8.321 \\ 10 \\ 4.791 \end{Bmatrix}$	$\begin{Bmatrix} 6.468 \\ -10 \\ 7.860 \end{Bmatrix}$

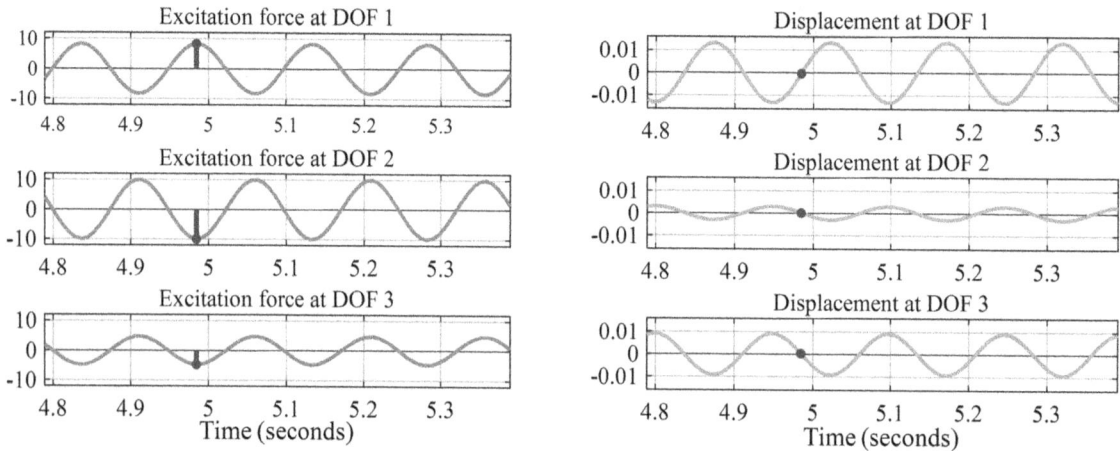

FIGURE 12.10 Time histories of the excitation forces and the responses over a short duration.

b. For example, let us consider the excitation of normal mode 2. The system is excited with the force appropriation vector $\{F\} = \{-8.321 \quad 10 \quad 4.791\}^T$ at the undamped natural frequency of mode 2 (6.72 Hz). The resulting responses are measured.

   Let us first check whether the phase resonance condition is satisfied or not. Figure 12.10 shows the time histories of the excitation forces and the responses over a short duration in the steady state. The forces and responses in the figure at an instant are marked by stems. For this case also, the observations in the proportional case in the previous section are applicable. The force and displacement vectors are monophase, and when the forces are at their maximum values, the displacements are at their mean values. Thus, the two have a phase difference of 90° and hence the phase resonance condition is satisfied. It shows that the phase resonance condition can be achieved by force appropriation even for a system with nonproportional damping.

   Let us compute the modal forces and responses to gain insight into normal mode excitation. Figure 12.11 shows the time histories of the modal excitation forces and modal responses. Unlike the proportional case, the modal forces for modes 1 and 3 are no longer zero, even though it is only mode 2 that is excited. It is because due to nonproportional damping, there is modal coupling between various normal modes, which gives rise to modal damping forces in modes 1 and 3. The nonzero modal forces for modes 1 and 3 are produced to counter the modal damping forces in these modes. We also note from the figure that modal responses for modes 1 and 3 are zero under the steady state but are nonzero in the transient phase. Therefore in the transient phase, the response is multi-modal, but as the steady state is achieved, the response is purely in the second normal mode.

c. Once the desired normal mode is excited, we measure the responses as shown in Figure 12.12 (marked by stems). It gives the second mode shape: $\{\psi\}_X = \{0.01346 \quad -0.00307 \quad -0.00934\}^T$. The mode shape after normalization (with the maximum element unity) is $\{\psi\}_X = \{1 \quad -0.2281 \quad -0.6940\}^T$. It matches well with the analytical estimate of the corresponding undamped mode shape given in the previous section.

d. If we remove the excitation forces from all the DOFs, we get free decay responses, as shown in Figure 12.13. The free decay response is not purely in mode 2 since the modal responses in modes 2 and 3 are nonzero in the transient phase (as discussed in step (c)). Since they are small, we use logarithmic decrement to find an approximate value of the damping factor for mode 2, yielding $\xi_X = 0.04053$. This is in good agreement with the correct value obtained from the system matrices, $\xi_A = 0.04035$.

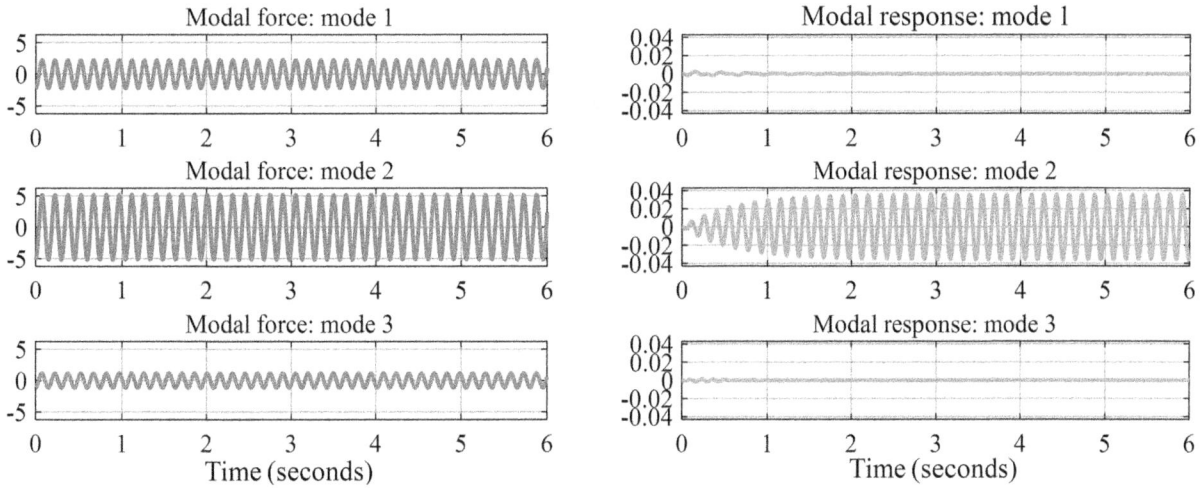

FIGURE 12.11 Time histories of the excitation forces and the responses over a short duration.

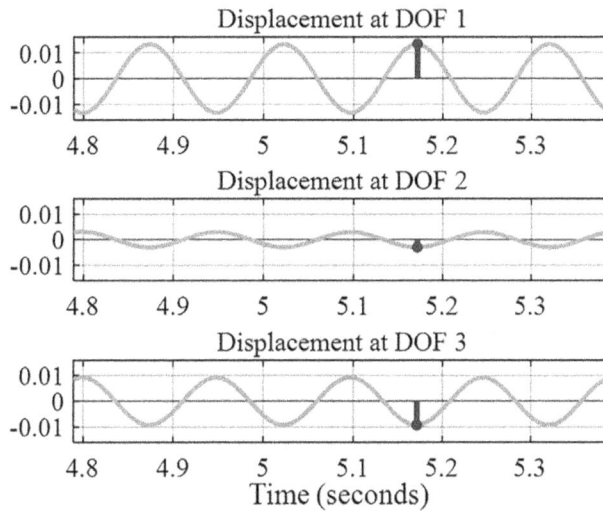

FIGURE 12.12 Measurement of response at all DOFs.

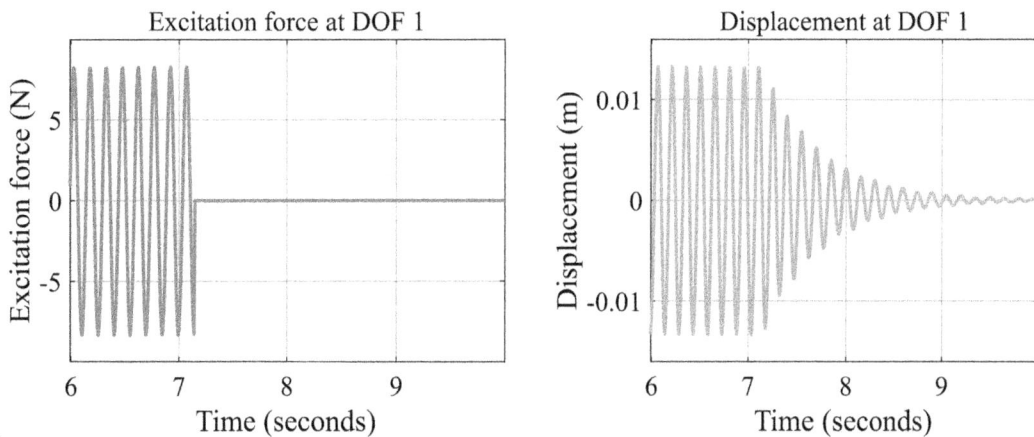

FIGURE 12.13 Force and response time histories at DOF 1 when the excitation forces are cut off.

12.10.3 CANTILEVER STRUCTURE

In this section, we simulate phase resonance testing of the cantilever structure (shown in Figure 9.15a) we have used in the previous chapters. Proportional viscous damping is simulated.

    a. In the previous two sections, we assumed that the complete FRF matrix was measured. But in practice, the response and excitation points are limited. For the cantilever, we assume that the responses in transverse directions are measured at five test points, and the excitation is applied in the transverse direction at only two test points (1 and 5).

    b. The measured FRF matrix is a rectangular matrix of order 5×2. The frequency range is 0–400 Hz.

    c. We use the MMIF method for identifying undamped natural frequencies and the corresponding force appropriation vectors. Figure 12.14 shows the MMIF.

        Table 12.3 gives the undamped natural frequencies identified using MMIF. These agree well with the corresponding exact values of the undamped natural frequencies of the cantilever, which (in Hz) are 9.816, 61.545, 172.863, and 341.488. The table also gives the force appropriation vectors corresponding to each undamped natural frequency. It should be noted that since there are only two excitation forces, the identified force appropriation vector as given in the table may not, in general, be able to enforce the phase resonance condition fully at all the response points.

    d. As an example, consider the excitation of the first normal mode of the cantilever. The system is excited with the force appropriation vector $\{F\} = \{10 \quad -7.027\}^T$ with its frequency equal to the undamped natural frequency of the first mode (9.82 Hz). The resulting responses are measured. Figure 12.15 shows the time histories of the excitation forces and the responses over a short duration in the steady state. The forces and responses at a time instant are marked by stems in the figure. We see that the responses and forces have a phase difference of 90°, and hence the phase resonance condition is satisfied.

        Figure 12.16 shows the time histories of the modal excitation forces and modal responses for the first four modes. We see that the modal forces in modes other than the first mode are also nonzero, a result of only two excitation forces being used. However, the modal response is dominant in mode 1, indicating that nearly a pure normal mode is excited.

    e. The mode shape is found by measuring the responses, as shown in Figure 12.17 (marked by stems). It gives

$$\{\psi\}_X = \left\{ 0.0639 \quad 0.2299 \quad 0.4612 \quad 0.7255 \quad 1.0 \right\}^T.$$ It matches well with the corresponding analytical estimate of

the undamped mode shape given by $\{\psi\}_A = \left\{ 0.06387 \quad 0.22988 \quad 0.46113 \quad 0.72547 \quad 1.0 \right\}^T$.

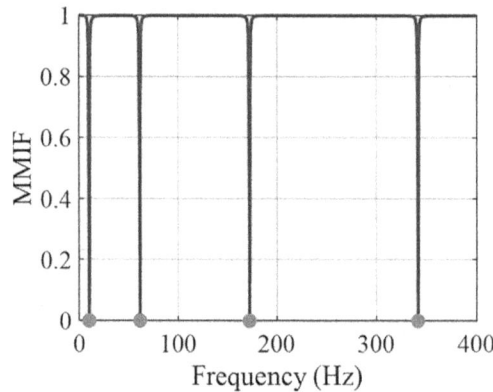

FIGURE 12.14　Multivariate mode indicator function.

TABLE 12.3

Undamped Natural Frequencies and the Corresponding Force Appropriation Vectors Identified by the MMIF Method

Undamped natural frequency (Hz)	9.82	61.55	172.86	341.49
Force appropriation vector (N)	$\left\{ \begin{array}{c} 10 \\ -7.027 \end{array} \right\}$	$\left\{ \begin{array}{c} 10 \\ -0.524 \end{array} \right\}$	$\left\{ \begin{array}{c} -2.037 \\ -10 \end{array} \right\}$	$\left\{ \begin{array}{c} -10 \\ -9.16 \end{array} \right\}$

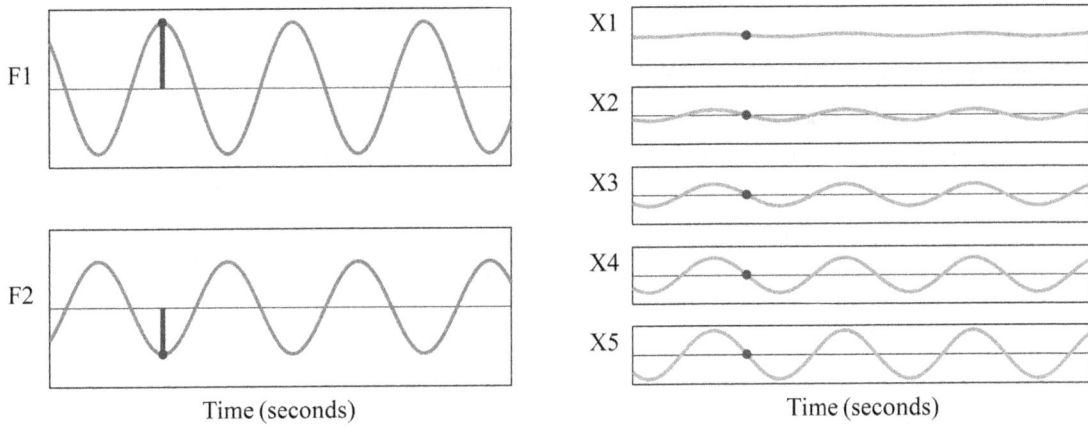

FIGURE 12.15 Time histories of the excitation forces and the responses over a short duration.

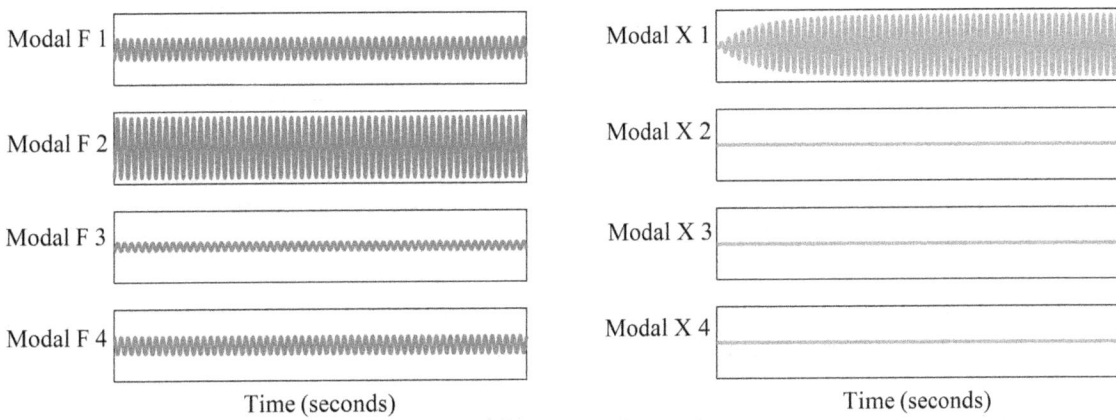

FIGURE 12.16 Time histories of the excitation forces and the responses over a short duration.

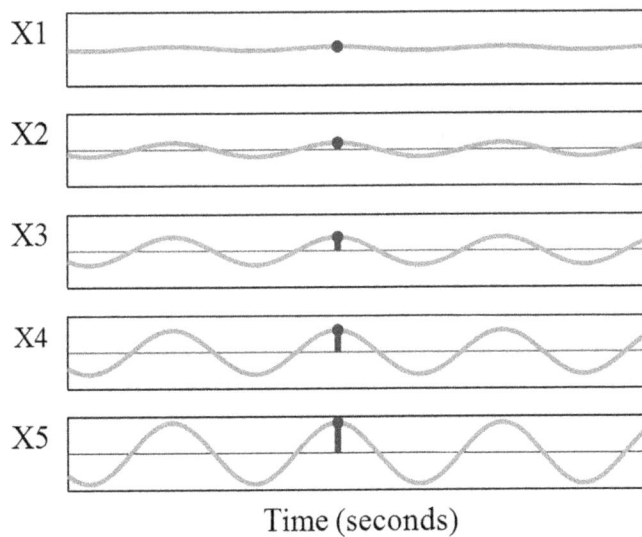

FIGURE 12.17 Measurement of response at all DOFs.

TABLE 12.4

Comparison of Phase Resonance and Phase Separation Techniques

	Phase Resonance Testing	Phase Separation Testing
1	Involves excitation of structure at one frequency for identifying a normal mode	Most methods, like impulse and random excitations, involve excitation over a broad frequency range covering all the modes in the range
2	Requires excitation at multiple points to isolate a mode	In many cases, single-point excitation may be enough; a multi-input test may be necessary in some cases
3	Modal parameters of individual normal modes are directly measured	Modal parameters of modes are estimated via curve fitting by separating the individual modal contributions mathematically
4	Yields normal modes directly; these can be directly used for comparison with FE model modes	Generally yields complex modes; modes need to be realized for comparison with FE model modes;
5	Provides highly accurate modal parameters; also good at identifying close modes since it relies on the physical excitation of a mode	Accuracy of modal parameters depends on the accuracy of curve fitting; good at identifying well-separated modes; the accuracy may be an issue if close modes are there;
6	Requires relatively more time	Is relatively faster
7	The cost of instrumentation for the test is higher; it requires a greater number of transducers and exciters; also requires a system for mode tuning; it is preferred where accuracy is of more concern;	The cost of instrumentation for the test is relatively lower; with the time for the test also being lower, it is preferred in most situations

f. By removing the excitation forces, we get free decay response, which corresponds to mode 1 as the contribution of other modes is negligible. From the free decay, the viscous damping factor obtained using the logarithmic decrement is $\xi_X = 0.02839$. The corresponding estimate from the analytical model of the cantilever is $\xi_A = 0.028374$.

Thus, even with a rectangular FRF matrix, the normal mode can be excited reasonably well with phase resonance testing, and the properties of a normal or undamped mode can be directly measured.

12.11　COMPARISON OF PHASE RESONANCE AND PHASE SEPARATION TECHNIQUES

Table 12.4 compares the main features of these two modal testing methods.

REVIEW QUESTIONS

1. What is a normal mode?
2. What is the phase resonance condition?
3. What is the objective of phase resonance testing?
4. How is phase resonance testing different from phase separation testing?
5. What is the force appropriation vector?
6. Show that a monophase vector with a frequency equal to the undamped natural frequency can excite the corresponding normal mode if the real part of the response is zero.

13 Operational Modal Analysis

13.1 INTRODUCTION

Operational modal analysis (OMA) identifies the modal parameters of a vibrating system from its response to ambient or operational excitation. It is different from experimental modal analysis (EMA), in which an artificial excitation is applied, often in a laboratory setup, and the input to the system is also measured along with the output to estimate the modal parameters. OMA is a relatively recent development compared to EMA.

The objective of this chapter is to present the basic idea of OMA, how the OMA deals with the lack of knowledge about the input to the system, and modal parameter identification by OMA in the time and frequency domains. Estimation of free decays using correlation functions and random decrement is also presented.

13.2 BASIC IDEA OF OMA

13.2.1 MOTIVATION AND PRINCIPLE

Most structures/systems in mechanical engineering such as automotive vehicles, aircraft, industrial equipment, and machine tools, to name a few, are routinely tested by EMA in laboratory conditions, which involves the application of an artificial excitation force. However, applying an artificial excitation may be difficult for many structures in civil engineering, like bridges and buildings. In EMA, the structure should be excited to have a sufficient response at all points to ensure a good signal-to-noise ratio. It is also necessary that the forces applied are not too large to achieve this as that may excite nonlinearities and damage the structure locally. Because of these constraints and economic considerations, the artificial excitation of large and massive structures, as mentioned above, is not practicable. Large and massive structures in other engineering areas may also pose similar difficulties.

It is noted that a structure like a building is exposed to ambient excitation due to wind. The excitation, in general, is random, exciting the modes within the bandwidth of the excitation. Therefore, the resulting response contains information about the excited modes. The ambient excitation is natural and difficult to quantify and hence is an unmeasured quantity, though the resulting responses can be measured easily. The objective of OMA is to measure and analyze these ambient responses and estimate the modal parameters of the modes excited and within the frequency range of measurement. It is seen that when the ambient excitation is white, the correlation functions of the responses consist of a sum of decaying sinusoids at the damped natural frequencies of the system. Thus, the correlation functions are similar to the impulse response functions or free decays and can be used to identify the modal parameters. Since the OMA is based on the measurement of only system outputs, it is also called **output-only modal analysis**.

13.2.2 MODELING OF AMBIENT EXCITATION

OMA assumes that the ambient excitation is white. The white noise excitation has uniform power spectral density throughout the frequency range and excites all the modes in the bandwidth of the excitation. Therefore, if the signal-to-noise ratio is good, then a meaningful measurement can be made, and its analysis can yield modal parameters.

In practice, the ambient excitation may not be purely white but may consist of a combination of random and deterministic components. For example, a bridge, along with the wind loading, which may be approximated as white, may also be subjected to excitations from the vehicles moving over the bridge, which may be deterministic. Therefore, the ambient response in the frequency domain would have peaks corresponding to not only the bridge natural frequencies excited by the random wind loading but also at frequencies of the excitations from the vehicles.

The situation is depicted in Figure 13.1. Part (a) of the figure shows that the actual ambient excitation, which may not necessarily be a pure white noise excitation, acts on the system producing an ambient response. Part (b) of the figure shows the implications of the OMA assumption that the excitation is white. Analyzing the ambient response by treating it as being generated by white noise excitation effectively adds a hypothetical 'mathematical filter' before the system. The filter modifies the hypothetical white noise input into the actual ambient excitation, which acts on the system. Therefore, OMA effectively identifies the modal parameters of the combined system (indicated by the outer rectangle in the figure), which includes the hypothetical filter. The output of the identified system contains the poles associated with the physical modes of the system but also the non-physical modes or poles related to the 'mathematical filter'. Therefore, one of the challenges in OMA is distinguishing between the physical and non-physical modes.

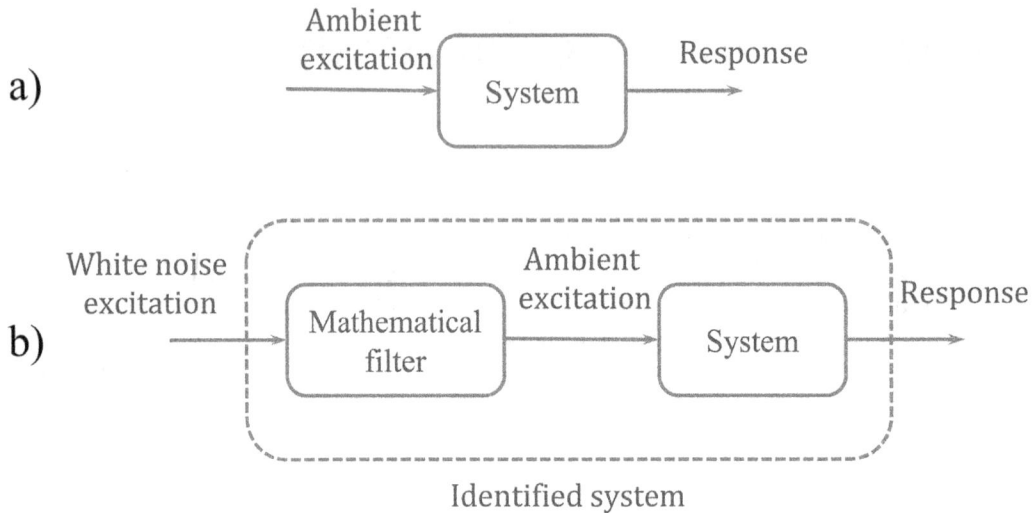

FIGURE 13.1 (a) The actual system with ambient excitation, (b) the identified system in OMA with white noise excitation.

13.2.3 REFERENCE MEASUREMENTS

Another distinct aspect of OMA is the need for reference measurements. In OMA, since the input excitation causing the response is unmeasured, the question arises as to how the ambient responses at various points on the structure measured at different times can be processed and compared to find the relative phase and magnitude. An accurate estimate of the relative phase and magnitude of the motions at various points is necessary to estimate mode shapes.

Since the input causing the responses is measured in EMA, the FRFs at various points can be computed and used for modal parameter estimation. The FRFs essentially normalize the responses to a common input, making them comparable across the test points and allowing estimation of the relative phase and magnitude necessary for mode shape identification.

In OMA, the challenge is to normalize the measured responses with the inputs being unmeasured. The response measurement in OMA can proceed in two ways: the responses at all the test points are measured simultaneously or in batches. If the responses at all the test points on the structure are measured simultaneously, then there is no problem as the responses are comparable since they correspond to the same input and time.

But when the number of response points is large, and the available sensors and instrumentation are limited, a simultaneous measurement of responses at all the points is not possible. In such a case, the only option is to perform response measurement in batches using the available hardware. The responses measured in batches correspond to motion at different times.

How the responses measured at different times can be processed? There are two issues. One is that the phase difference between the motions at the test points on the structure cannot be obtained from the responses at those points. Second, the relative magnitude of the responses at those points cannot be obtained as they might correspond to different input excitation levels. These problems are addressed in OMA by having some measurements common to all the batches, called reference measurements. There are roving and reference sensors in a batch of measurements. The responses measured in one batch make up a dataset. The sensors whose locations change from one dataset to another are called roving sensors, while the sensors whose locations are fixed over all datasets are called reference sensors.

OMA uses correlation functions to process the responses. We studied in Chapter 8 that the cross-correlation function captures the relative phase between two signals. Thus, the cross-correlation of roving and reference responses in a dataset gives the relative phase and magnitude of the responses. The reference measurements at points common to all the datasets make it possible to normalize and compare the cross-correlations across the datasets. The reference measurements, therefore, provide a basis to estimate mode shapes by normalization across the datasets.

We illustrate the idea of the reference and roving sensors with an example. Figure 13.2 shows a beam structure in which the transverse vibration modes in the vertical plane are of interest. Let there be n test points. Sensor A is the reference sensor. The location of the reference sensor should not ideally coincide with any nodes of the modes of interest so that all the excited modes can be identified. Node 1 is an acceptable choice for the reference sensor location in the present example since no nodes exist at that point. In general, however, the location of the nodes may not be known a priori, and the remedy could be to use more than one reference sensor as it may help ensure that the modes are observable from at least one of them. In the figure, Sensor B is a roving sensor used to measure the responses at test points 2 to n. Each roving sensor measurement has a corresponding reference measurement at test point 1. The total time to complete the measurements at all the test points can be reduced by increasing the number of roving sensors.

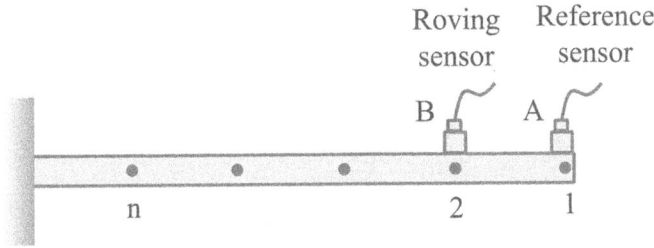

FIGURE 13.2 Reference and roving sensors on a beam structure.

13.3 CROSS-CORRELATION OF THE SYSTEM OUTPUT UNDER WHITE NOISE EXCITATION

The natural excitation technique (NExT) was developed by James et al. [49] for identifying modal parameters from in-situ measurements of responses. The technique was developed in the context of modal testing of wind turbines subjected to ambient excitations. The method involved the estimation of the correlation functions of the measured responses and their subsequent use as free decays with the eigensystem realization algorithm (ERA) to find modal parameters. They derived the auto and cross-correlation functions of the outputs for a white noise input and showed that the correlation functions contain information about the modal parameters of the system. That development is reviewed in this section.

The equation of motion of an MDOF system with viscous damping subjected to excitation forces is given by

$$[M]\{\ddot{x}\}+[C]\{\dot{x}\}+[K]\{x\}=\{F\} \tag{13.1}$$

If the modes are assumed real (or the damping is proportional), Eq. (13.1) can be decoupled using the mass-normalized modes by substituting $\{x(t)\}=[\phi]\{y(t)\}$. Pre-multiplying by $[\phi]^T$ and using the orthogonality properties, we get

$$[I]\{\ddot{y}\}+[`2\xi_r\omega_r`]\{\dot{y}\}+[`\omega_r^2`]\{y\}=[\phi]^T\{F\} \tag{13.2}$$

The equation of motion for the rth mode in modal coordinates can be written as

$$\ddot{y}_r+2\xi_r\omega_r\dot{y}_r+\omega_r^2 y_r=\{\phi\}_r^T\{F\} \tag{13.3}$$

It yields the modal response in the rth mode

$$y_r(t)=\int_{-\infty}^{t} h_r(t-\tau)\{\phi\}_r^T\{F(\tau)\}\ d\tau \tag{13.4}$$

The response in physical coordinates is

$$\{x(t)\}=\sum_{r=1}^{N}\{\phi\}_r\ y_r(t) \tag{13.5}$$

$$\{x(t)\}=\sum_{r=1}^{N}\{\phi\}_r\left(\int_{-\infty}^{t} h_r(t-\tau)\{\phi\}_r^T\{F(\tau)\}\ d\tau\right) \tag{13.6}$$

Let F_k is a random force acting at the kth DOF, with no force at the other DOFs. Substituting the force into Eq. (13.6) gives

$$\{x(t)\}=\sum_{r=1}^{N}\{\phi\}_r\ \phi_{kr}\left(\int_{-\infty}^{t} h_r(t-\tau)F_k(\tau)\ d\tau\right) \tag{13.7}$$

The output at the ith DOF is

$$x_i(t)=\sum_{r=1}^{N}\phi_{ir}\phi_{kr}\left(\int_{-\infty}^{t} h_r(t-\tau)F_k(\tau)\ d\tau\right) \tag{13.8}$$

Similarly, the output at the jth DOF (after taking the index of summation as s and the variable of integration as σ) can be written as

$$x_j(t) = \sum_{s=1}^{N} \phi_{js}\phi_{ks} \left(\int_{-\infty}^{t} h_s(t-\sigma)F_k(\sigma)\,d\sigma \right) \tag{13.9}$$

The cross-correlation between the outputs at the ith and j^{th} DOFs (with input at k^{th} DOF) is

$$R_{ijk}(T) = E\left(x_i(t+T).x_j(t) \right) \tag{13.10}$$

Substituting Eqs. (13.8) and (13.9) into Eq. (13.10), and noting that only the force is random, we get

$$R_{ijk}(T) = \sum_{r=1}^{N}\sum_{s=1}^{N} \phi_{ir}\phi_{kr}\phi_{js}\phi_{ks} \left(\int_{-\infty}^{t}\int_{-\infty}^{t+T} h_r(t+T-\tau)h_s(t-\sigma)\,E\left(F_k(\tau)F_k(\sigma) \right)\,d\sigma\,d\tau \right) \tag{13.11}$$

If F_k is a white noise excitation, then, we can write

$$E\left(F_k(\tau)F_k(\sigma) \right) = \alpha_k\delta(\tau-\sigma) \tag{13.12}$$

where α_k is a scalar. Substituting Eq. (13.12) into Eq. (13.11), we obtain

$$R_{ijk}(T) = \sum_{r=1}^{N}\sum_{s=1}^{N} \alpha_k\phi_{ir}\phi_{kr}\phi_{js}\phi_{ks} \left(\int_{-\infty}^{t}\left(\int_{-\infty}^{t+T} h_r(t+T-\tau)h_s(t-\sigma)\,\delta(\tau-\sigma)\,d\sigma \right)d\tau \right) \tag{13.13}$$

The inner integral can be evaluated using the sifting property of the delta function (studied in Chapter 8), which leads to

$$R_{ijk}(T) = \sum_{r=1}^{N}\sum_{s=1}^{N} \alpha_k\phi_{ir}\phi_{kr}\phi_{js}\phi_{ks} \left(\int_{-\infty}^{t} h_r(t+T-\tau)h_s(t-\tau)\,d\tau \right) \tag{13.14}$$

Introducing transformation $\lambda = t - \tau$, we get

$$R_{ijk}(T) = \sum_{r=1}^{N}\sum_{s=1}^{N} \alpha_k\phi_{ir}\phi_{kr}\phi_{js}\phi_{ks} \left(\int_{0}^{\infty} h_r(\lambda+T)h_s(\lambda)\,d\lambda \right) \tag{13.15}$$

We know that $h_r(t)$ is the impulse response of the system in the rth mode and is given by

$$h_r(t) = \frac{e^{-\xi_r\omega_r t}}{\omega_{dr}}\sin\omega_{dr}t \tag{13.16}$$

where ω_{dr} is the damped natural frequency of the rth mode. After substituting $h_r(t)$ into Eq. (13.15), and simplifying, the equation can be written in the following form:

$$R_{ijk}(T) = \sum_{r=1}^{N} \left(G_{ijk,r}\,e^{-\xi_r\omega_r T}\sin\omega_{dr}T + H_{ijk,r}\,e^{-\xi_r\omega_r T}\cos\omega_{dr}T \right) \tag{13.17}$$

where

$$\left\{ \begin{array}{c} G_{ijk,r} \\ H_{ijk,r} \end{array} \right\} = \sum_{s=1}^{N} \frac{\alpha_k\phi_{ir}\phi_{kr}\phi_{js}\phi_{ks}}{\omega_{dr}\omega_{ds}} \int_{0}^{\infty} e^{-(\xi_r\omega_r + \xi_s\omega_s)\lambda}\sin\omega_{ds}\lambda \left\{ \begin{array}{c} \sin\omega_{dr}\lambda \\ \cos\omega_{dr}\lambda \end{array} \right\} d\lambda \tag{13.18}$$

The following conclusions are drawn from Eqs. (13.17) and (13.18):

- The two terms in Eq. (13.17) are damped sinusoids at the damped natural frequencies, with their decay rates governed by the modal damping factors.
- The coefficients $G_{ijk,r}$ and $H_{ijk,r}$ of the sinusoids don't depend on the time variable (T) of the cross-correlation function. They are functions of the modal parameters.

Thus, it is found that the cross-correlation between the outputs of a system subjected to a white noise input is a sum of damped sinusoids. The same conclusion is drawn when the above analysis is extended to the case where the white noise input acts at multiple DOFs (Brincker and Ventura [20]). This outcome forms the basis of OMA as it indicates that if the excitation is white, then the information about the modal parameters is contained in the cross-correlation of the outputs, and hence these parameters can be estimated from the cross-correlation estimates. Since estimating the cross-correlations requires only the knowledge of the outputs, the modal analysis can be carried out without the knowledge of the excitation forces.

We also note from Eqs. (13.17) and (13.18) that there is a scaling factor α_k (associated with the magnitude of the white noise input) appearing in the cross-correlation. This means that the cross-correlation of the measured ambient responses is dependent on the strength of the input excitation. Since the input forces are unmeasured in OMA, the scaling factor in the cross-correlations is unknown. The consequence is that the mode shapes estimated in OMA are unscaled, i.e., associated with an unknown arbitrary scaling factor.

13.4 ESTIMATION OF FREE DECAYS FOR OMA

Free decay is the free vibration response of a system to initial disturbance. The free decay is governed by the system's modal parameters, apart from the initial conditions, and hence can be used for modal parameter identification. For an SDOF system subjected to stationary Gaussian white noise excitation, the autocorrelation of the output is proportional to the free decay response to initial conditions. In OMA, it is assumed that the correlation functions are proportional to the free decay responses. Therefore, the first step in OMA is to obtain free decay from the measured ambient responses. It is done by estimating either the correlation or random decrement functions.

13.4.1 CORRELATION FUNCTIONS

We have defined the auto and cross-correlation of signals in Chapter 8 and looked at its various properties. One important property is that the cross-correlation preserves the relative phase between the frequency components of the two signals.

The cross-correlation between two discrete signals, $x_i(n)$ and $x_j(n)$, each N sample long, can be computed as

$$R_{x_i x_j}(m) = \frac{1}{N} \sum_{n=1}^{N} x_i(n) x_j(n+m) \quad \text{for} \quad -(N-1) \le m \le N-1 \quad (13.19)$$

The expression in Eq. (13.19) gives a biased cross-correlation estimate due to some terms in the sum on the RHS becoming zero. The unbiased estimate is given by

$$R_{x_i x_j}(m) = \frac{1}{N-|m|} \sum_{n=1}^{N-1-|m|} x_i(n) x_j(n+m) \quad \text{for} \quad -(N-1) \le m \le N-1 \quad (13.20)$$

The unbiased cross-correlation can also be computed using the MATLAB® command 'xcorr()' by using the normalization option 'unbiased'.

If there are 'p' measured ambient responses, then a matrix of correlation functions $[R_{xx}]$ can be defined as

$$[R_{xx}] = \begin{bmatrix} R_{x_1 x_1} & R_{x_1 x_2} & \dots & R_{x_1 x_p} \\ R_{x_2 x_1} & R_{x_2 x_2} & \dots & R_{x_2 x_p} \\ \dots & \dots & \dots & \dots \\ R_{x_p x_1} & R_{x_p x_2} & \dots & R_{x_p x_p} \end{bmatrix}_{p \times p} \quad (13.21)$$

The various elements of the matrix $[R_{xx}]$ can be computed from Eq. (13.20). One or more columns of $[R_{xx}]$ or the full correlation matrix can be used for OMA, treating them as free decays.

13.4.2 RANDOM DECREMENT (RD)

a. Definition

The random decrement (RD) was introduced by Cole [25] as a method of identification of structures under ambient loadings. This method defines the RD function, which can be treated as the free-decay response. Let $x(t)$ be the ambient response of a system, and let it be stationary and ergodic. RD is an average of the time segments picked up from the signal based on some triggering condition. Random decrement is defined as

$$D_{xx}(\tau) = E\left[x(t+\tau) \mid T_{x(t)} \right] \quad (13.22)$$

From the above definition, random decrement is the expected value of the time segment $x(t+\tau)$ (with τ representing the length of the segment) picked out from the signal $x(t)$ when $x(t)$ satisfies some chosen condition (referred to as the triggering condition and denoted by $T_{x(t)}$) at time t. The symbol $D_{xx}(\tau)$ denotes RD of length τ, with the first subscript denoting the signal from which the time segments are picked (signal $x(t)$ in the present case) and the second subscript representing the signal used to evaluate the triggering condition and the corresponding time instants (again signal $x(t)$ in the present case).

The random decrement function $D_{xx}(\tau)$ is called an auto RD function. Suppose we use another signal $y(t)$ to evaluate the triggering condition and the corresponding time instants, while the time segments for averaging are selected from the signal $x(t)$. The random decrement function estimated in this manner is called a cross RD function and is denoted by $D_{xy}(\tau)$.

b. Estimation of RD function

$D_{xx}(\tau)$ can be estimated from a sampled signal as

$$D_{xx}(\tau)=\frac{1}{N}\sum_{j=1}^{N}x(t_j+\tau)\mid T_{x(t_j)} \tag{13.23}$$

where N represents the number of trigger points at which the triggering condition is satisfied. Several triggering conditions have been proposed to calculate the RD. The level-crossing triggering condition given by the following equation checks whether the signal is equal to a certain chosen level x_0.

$$T_{x(t_j)} \Rightarrow \quad x(t_j)=x_0 \tag{13.24}$$

Due to the discrete nature of the signal, the level-crossing condition is not satisfied at a sufficient number of points. It prevents an accurate estimation of the RD. A more useful triggering condition is the positive point triggering condition, where it is checked whether the signal lies within a small narrow range around the trigger level. It is given by

$$T_{x(t_j)} \Rightarrow \quad x_0-\epsilon \leq x(t_j)\leq x_0+\epsilon \tag{13.25}$$

where ϵ is a small positive number. Few other triggering conditions can be used, like based on the slope/velocity or a combination of the trigger level with slope/velocity. However, the positive point triggering condition is widely used in practice.

We now look at the steps involved in implementing Eq. (13.23) to find the RD. We use the positive point triggering condition. Figure 13.3 shows a random response signal $x(t)$. The chosen trigger level x_0 is shown by a horizontal line. The RD function is estimated as follows:
- Find the time instants at which the trigger line intersects $x(t)$. Let these be t_1, t_2, ..., t_N, as shown in the figure. (The trigger line visibly intersects the response at several points. But note that the signal is discrete, and the triggering condition is evaluated for discrete samples of the signal using a narrow range around the trigger level.)
- Choose the desired length of the RD function. Let it is τ.
- Pick out the time segments of length τ starting at the times t_1, t_2, ..., t_N, as shown in other plots in Figure 13.3.
- Take the average of the selected time segments to estimate the RD (shown in Figure 13.4).
     The cross RD function for the sampled signals is estimated as

$$D_{xy}(\tau)=\frac{1}{N}\sum_{j=1}^{N}x(t_j+\tau)\mid T_{y(t_j)} \tag{13.26}$$

c. Interpretation of RD function
Vandiver et al. [103] developed a mathematical basis for the RD and showed that for an LTI system subjected to a zero mean, stationary, Gaussian random input, the RD of the output, assuming a level triggering condition, is proportional to the autocorrelation function of the output. Thus,

$$D_{xx}(\tau)=\left(\frac{x_0}{R_{xx}(0)}\right)R_{xx}(\tau) \tag{13.27}$$

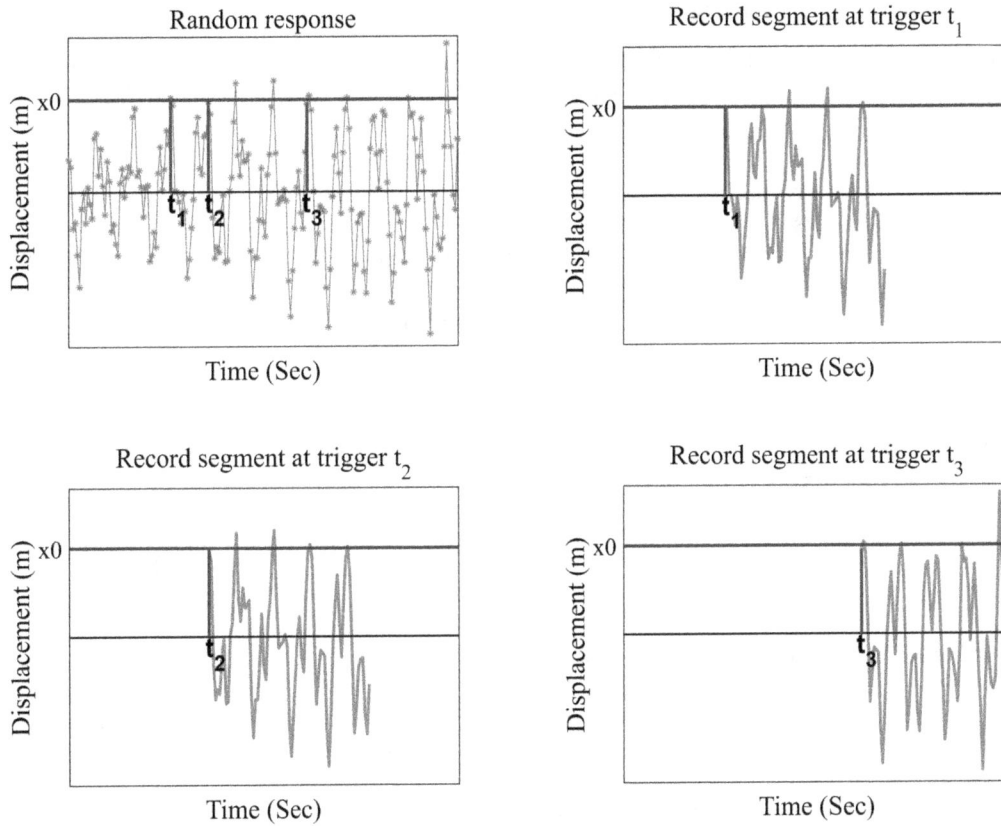

FIGURE 13.3 Computation of random decrement.

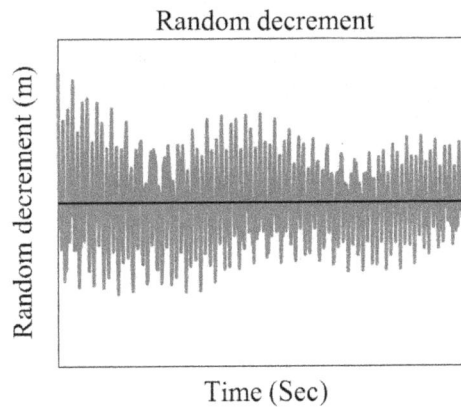

FIGURE 13.4 Random decrement.

Since autocorrelation of the output of such a process represents a free decay, the RD also represents the free decay response of the system. The free decay interpretation of the RD can be understood intuitively for an SDOF system subjected to zero mean, stationary Gaussian random force. The response of the system is a linear combination of the response due to initial displacement, initial velocity, and a random component due to the random force. If time segments of the response satisfying the trigger condition $x(t_j) = x_0$ are picked up and averaged, then the response component due to initial velocity (i.e., slope) would average out to zero because the velocity associated with each time segment would be random. Similarly, the component due to the random force would also average out to zero, being random. It leaves only the response due to initial displacement x_0, as that is identical for all the segments. Thus, the RD with level-crossing condition would be equal to the free decay due to initial displacement x_0.

13.5 OMA IN THE TIME DOMAIN

The time-domain methods for EMA, like the Ibrahim time-domain method (ITD) and the eigensystem realization method (ERA), described in Chapter 11, were formulated with free decay responses. OMA can be carried out with these methods as described below.

- Choose the reference and roving sensors locations and measure the ambient responses, as explained in Section 13.2.3 for different datasets.
- For each dataset, compute free decays corresponding to the measured responses by estimating either the correlation or the RD functions as explained in Section 13.4.
- Using the free decays of each dataset, identify the corresponding modal parameters by the ITD or ERA, as explained in Chapter 11.
- Find estimates of the natural frequencies and damping factors, taking an average of these values identified in each dataset.
- Merge the mode shapes by normalizing part mode shapes of each dataset to obtain the mode shape of the whole system (as explained in Example 13.1 below).

ARMA (autoregressive moving average) and SSI (stochastic subspace identification) are other time-domain methods for OMA.

Example 13.1: OMA in the Time Domain by ERA

In this example, we perform OMA on our cantilever structure by ERA using the correlation functions. We assume that the structure is subjected to ambient excitation in the form of stationary white noise. Simulated measured data is used.

Step 1: We consider 5 test points on the cantilever structure and designate point 5 (which is at the tip) as the location of the reference sensor for all the datasets/measurements. We use one roving sensor, moved sequentially from test point 1 to 5. Thus, we have two signals in each dataset, one each from the reference and roving sensors. The signals are measured with a sampling frequency of 1,024 Hz with a record length of 24 seconds. Table 13.1 summarizes the measured ambient responses in various datasets. The subscript corresponds to the test point, while the superscript (in the reference signals) corresponds to the dataset. Figure 13.5 shows the first 8 seconds of the reference and roving sensor signals in datasets 1 and 5.

Step 2: For dataset 1, we compute the correlation functions between the roving and reference sensor signals (i.e., $y_1(t)$ and $x_5^1(t)$, respectively) using 'xcorr' command (with the 'unbiased' option) in MATLAB. We obtain the following correlation function matrix.

$$\left[R_{y_1x_5^1} \right] = \begin{bmatrix} R_{y_1y_1}(t) & R_{y_1x_5^1}(t) \\ R_{x_5^1y_1}(t) & R_{x_5^1x_5^1}(t) \end{bmatrix}$$

Figure 13.6 shows the estimated correlation functions $R_{y_1x_5^1}(t)$ and $R_{x_5^1x_5^1}(t)$ over a length of 1 second.

Step 3: The correlation function matrix of dataset 1 is now used as a matrix of free decays from which the Hankel matrices in ERA are constructed. Further steps to be followed for applying ERA are the same as those described in Example 11.9 in Chapter 11. Four modes are identified, and we get the corresponding natural frequencies and modal viscous damping factors. We also get part mode shapes corresponding to test points 1 and 5.

Step 4: We repeat steps 2 and 3 for the other datasets (2 to 5).

Step 5: Table 13.2 shows the estimates of natural frequencies found in each dataset and the average values over the datasets. Table 13.3 give these results for modal damping factors. The estimates are close to their correct values.

TABLE 13.1

Summary of the Measured Ambient Responses in Various Datasets

Dataset		1	2	3	4	5
Reference sensor	Location	5	5	5	5	5
	Signal measured	$x_5^1(t)$	$x_5^2(t)$	$x_5^3(t)$	$x_5^4(t)$	$x_5^5(t)$
Roving sensor	Location	1	2	3	4	5
	Signal measured	$y_1(t)$	$y_2(t)$	$y_3(t)$	$y_4(t)$	$y_5(t)$

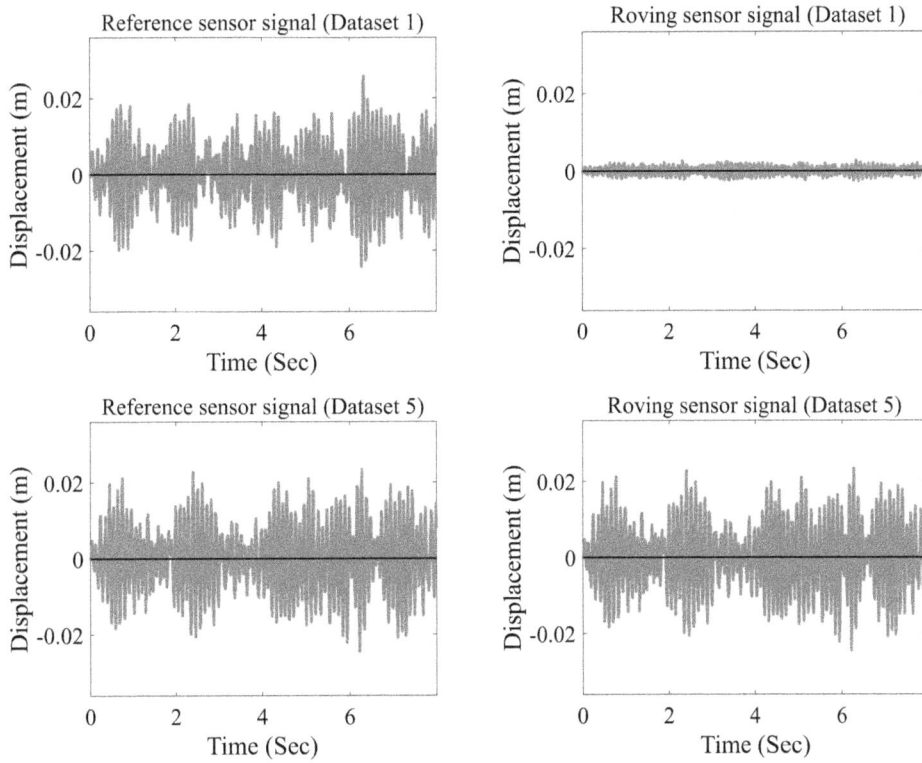

FIGURE 13.5 The reference and roving sensor signals corresponding to datasets 1 and 5.

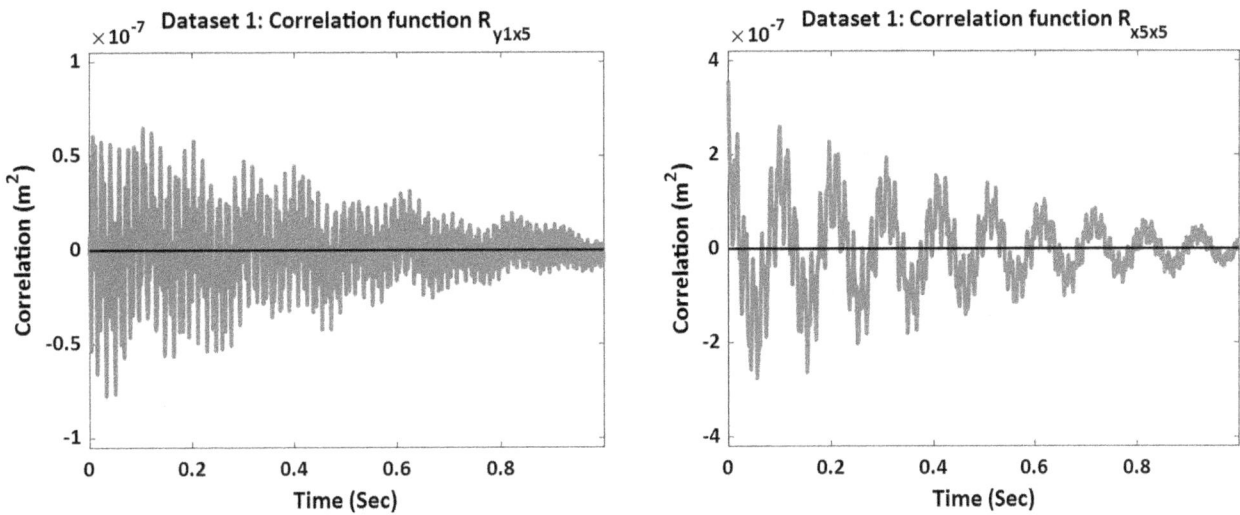

FIGURE 13.6 The estimated correlation functions $R_{y_1 x_5^1}(\tau)$ and $R_{x_5^1 x_5^1}(\tau)$.

TABLE 13.2
Natural Frequencies Found from Each Dataset and the Average Values (Hz)

Dataset	Mode 1	Mode 2	Mode 3	Mode 4
1	9.65	61.62	172.91	341.41
2	9.74	61.58	172.83	341.39
3	9.82	61.54	172.84	341.60
4	9.86	61.56	172.88	341.56
5	9.82	61.54	172.82	341.45
Average value	9.78	61.57	172.86	341.48

TABLE 13.3

Modal Viscous Damping Factors Found from Each Dataset and the Average Values

Dataset	Mode 1	Mode 2	Mode 3	Mode 4
1	0.03464	0.00440	0.00124	0.00038
2	0.03962	0.00610	0.00121	0.00069
3	0.03065	0.00359	0.00118	0.00072
4	0.02074	0.00509	0.00138	0.00098
5	0.03484	0.00417	0.00122	0.00094
Average value	0.03210	0.00467	0.00124	0.00074

TABLE 13.4

Identified Unscaled Mode Shapes

Mode Shape 1	Mode Shape 2	Mode Shape 3	Mode Shape 4
$0.06381 - 8.91e\text{-}05i$	$-0.30157 + 0.00089i$	$0.6043 - 4.35e\text{-}05i$	$-0.75887 - 0.0008i$
$0.22967 - 0.00055i$	$-0.68044 + 0.0088i$	$0.52667 + 0.00235i$	$0.31354 + 0.00313i$
$0.46158 + 0.00019i$	$-0.58655 + 0.0044i$	$-0.47735 + 0.00296i$	$0.33273 - 0.00233i$
$0.72576 - 0.00064i$	$0.07232 - 0.00011i$	$-0.39321 + 0.00478i$	$-0.64875 - 0.00809i$
1.0	1.0	1.0	1.0

Step 6: The part mode shapes obtained for datasets are normalized and merged/assembled. Let us see how the part mode shapes corresponding to the first natural frequency are normalized and assembled. The part mode shape of the first mode in dataset 1 is normalized with the element corresponding to the reference location. In this example, we normalize by dividing the part mode shape by the element corresponding to the reference location. It is done for all the datasets, and then the part mode shapes of all the datasets are assembled to obtain the complete mode shape of the first mode. The procedure is repeated for other modes in the datasets. Table 13.4 shows the mode shapes obtained in this manner.

Example 13.2: Estimation of RDs

In this example, we compute the RDs for the ambient response data of the cantilever structure in Example 13.1. The computed RDs for different datasets can be used as free decays for OMA using ERA by following the steps in Example 13.1.

Step 1: We use the positive point triggering condition and implement it with a trigger level of $x_0 = 3.972 \times 10^{-4}$ and $\epsilon = 5\%$ of x_0.

Step 2: The length of the RD also needs to be fixed, and we take it as 512 samples. It corresponds to a length of 0.5 second.

Step 3: We compute the random decrements of each dataset. For dataset 1, we have random decrements $D_{y_1 y_1}(\tau), D_{y_1 x_5^1}(\tau), D_{x_5^1 y_1}(\tau)$ and $D_{x_5^1 x_5^1}(\tau)$. Figure 13.7 shows the estimated random decrements.

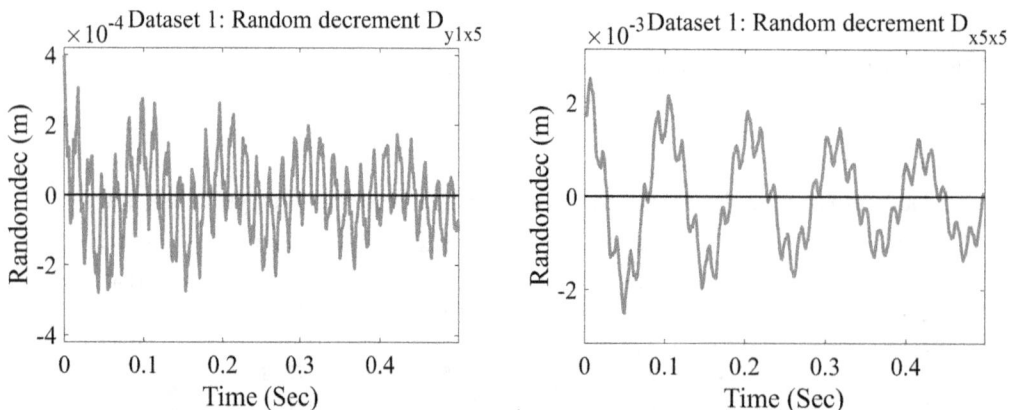

FIGURE 13.7   The estimated random decrements $D_{y_1 x_5^1}(\tau)$ and $D_{x_5^1 x_5^1}(\tau)$.

13.6   OMA IN THE FREQUENCY DOMAIN

OMA can be carried out in the frequency domain by analyzing the power spectral densities (PSDs) obtained by the Fourier transform of the correlation functions. The most straightforward approach to analyzing PSDs is the classical frequency-domain approach ([12]). It is like the peak picking approach in EMA. We briefly review this approach in this section.

Let $\{x(t)\}$ be the vector of ambient responses measured in a dataset. This approach assumes that at a resonance peak in the PSD, the response is dominated by the contribution of only the corresponding mode, and the contributions of all other modes are negligible. Therefore, the frequency corresponding to a resonance peak is taken as the estimate of the natural frequency. The mode shape is obtained as follows. The correlation matrix of the measured responses is

$$[R_{xx}(\tau)] = E\left(\{x(t)\}\{x(t+\tau)\}^T\right) \tag{13.28}$$

At the rth peak in the PSD, the response is dominated by mode r. Thus, the response can be approximated as

$$\{x(t)\} = \{\phi\}_r \, y_r(t) \tag{13.29}$$

where $y_r(t)$ is the modal response in the rth mode. Substituting Eq. (13.29) into (13.28), we get

$$[R_{xx}(\tau)] = E\left(\{\phi\}_r \, y_r(t) y_r(t+\tau)\{\phi\}_r^T\right) \tag{13.30}$$

$$[R_{xx}(\tau)] = \{\phi\}_r \, E(y_r(t) y_r(t+\tau))\{\phi\}_r^T \tag{13.31}$$

$$[R_{xx}(\tau)] = R_{y_r y_r}(\tau)\{\phi\}_r \{\phi\}_r^T \tag{13.32}$$

In Eq. (13.32) $R_{y_r y_r}(\tau) = E(y_r(t) y_r(t+\tau))$ is the autocorrelation of the rth mode modal response. Taking the Fourier transform of Eq. (13.32) leads to

$$[G_{xx}(\omega)] = G_{y_r y_r}(\omega)\{\phi\}_r \{\phi\}_r^T \tag{13.33}$$

where $[G_{xx}(\omega)]$ is the PSD matrix. If there are n measurements in the dataset, then the PSD matrix can be written as

$$[G_{xx}(\omega)]_{n \times n} = \left[G_{y_r y_r}(\omega)\phi_{1r}\begin{Bmatrix}\phi_{1r}\\\phi_{2r}\\\vdots\\\phi_{nr}\end{Bmatrix} G_{y_r y_r}(\omega)\phi_{2r}\begin{Bmatrix}\phi_{1r}\\\phi_{2r}\\\vdots\\\phi_{nr}\end{Bmatrix} \cdots G_{y_r y_r}(\omega)\phi_{nr}\begin{Bmatrix}\phi_{1r}\\\phi_{2r}\\\vdots\\\phi_{nr}\end{Bmatrix} \right]_{n \times n} \tag{13.34}$$

We see that each column of the PSD matrix is a scaled version of the rth mode shape. The same also holds for any row of the matrix. Thus, from the measured ambient responses, the correlation matrix can be estimated, whose Fourier transform gives the PSD matrix. Any column or row of the PSD matrix at the rth natural frequency is an estimate of the rth mode shape, though the scaling factor is unknown. The modal damping of the mode can be obtained by half-power bandwidth.

The frequency domain approach discussed above can be used when the modes are well separated. There are more refined approaches, like the frequency domain decomposition (FDD) method (Brincker et al. [21]) and enhanced FDD method (Brincker et al. [19]), based on the SVD of the PSD matrix, that can deal more effectively with close modes.

Example 13.3: OMA in the Frequency Domain

In this example, we perform OMA on the cantilever structure using the classical frequency domain approach described in Section 13.6. We use the ambient response data of Example 13.1.

Step 1: The ambient responses measured in five datasets are available as detailed in Table 13.1.

Step 2: For dataset 1, we compute the correlation functions matrix $\left(\left[R_{y_1 x_5^1}(\tau)\right]\right)$ between the roving and reference sensor signals (i.e. $y_1(t)$ and $x_5^1(\tau)$ respectively) using 'xcorr' command (with the 'unbiased' option) in MATLAB. Figure 13.8 shows the estimated correlation functions $R_{y_1 x_5^1}(\tau)$ and $R_{x_5^1 x_5^1}(\tau)$.

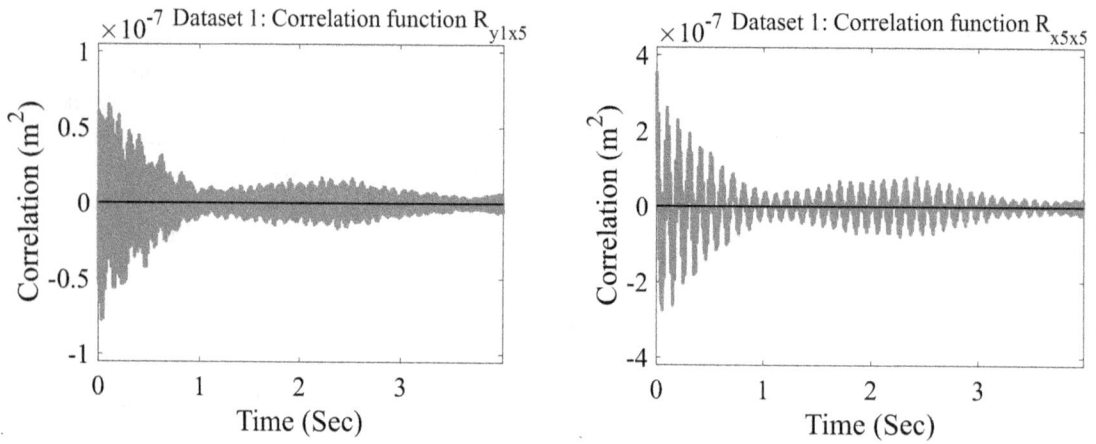

FIGURE 13.8 The estimated correlation functions $R_{y_1x_5^1}(\tau)$ and $R_{x_5^1x_5^1}(\tau)$.

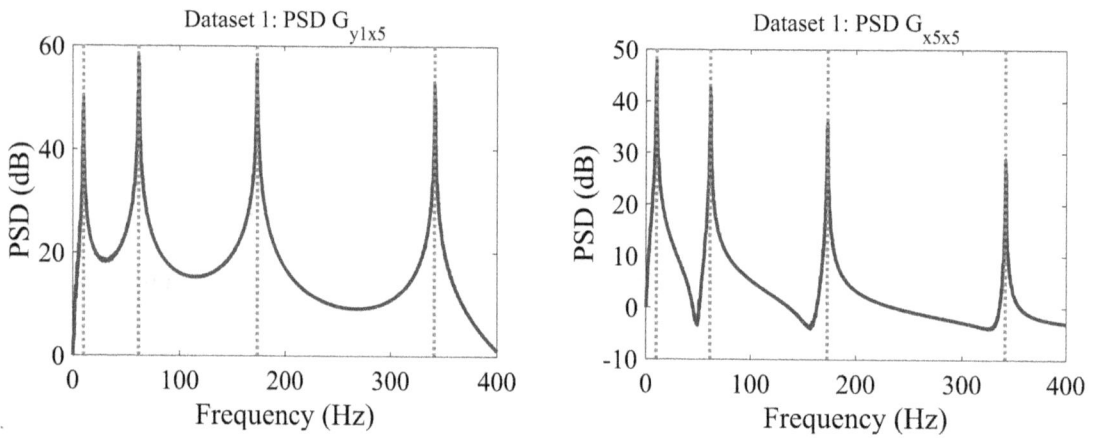

FIGURE 13.9 PSDs $G_{y_1x_5^1}(f)$ and $G_{x_5^1x_5^1}(f)$.

Step 3: For OMA in the frequency domain, we need to find the PSD matrix by the Fourier transform of the correlation functions. Therefore, we choose the length of the estimated correlation function such that the entire length of the correlation function is nearly covered without signal truncation. In this example, a length of 4 seconds is found to be adequate. Further, an exponential window is applied to the last small portion of the estimated correlation function to reduce leakage errors.

We compute the DFT of each element of $\left[R_{y_1x_5^1}(\tau)\right]$. It gives two-sided discrete PSDs, which are converted to one-sided discrete PSDs. Figure 13.9 shows the discrete PSDs $G_{y_1x_5^1}(f)$ and $G_{x_5^1x_5^1}(f)$. The frequency resolution of the PSD is 0.25 Hz.

Step 4: We apply the classical frequency domain method to each peak, obtaining an estimate of the natural frequencies. Any column or row of the PSD matrix at the peak frequency gives the corresponding part mode shape for the dataset.

Step 5: We repeat steps 2 and 4 for the other datasets (2 to 5).

Step 6: The average estimates of the natural frequencies for the four identified modes from the five datasets in Hz are 10.0, 61.5, 173.0, and 341.25.

Step 8: The part mode shapes obtained for each dataset are merged, as explained in Example 13.1. Table 13.5 shows the mode shapes after merging the part mode shapes of the individual datasets. They are found to be close to the correct mode shapes.

TABLE 13.5
Identified Unscaled Mode Shapes

Mode Shape 1	Mode Shape 2	Mode Shape 3	Mode Shape 4
$0.0653+0.0065i$	$-0.3032-0.0006i$	$0.6073-7.17e-5i$	$-0.7584+0.0053i$
$0.2303-0.0017i$	$-0.6835+0.0028i$	$0.5309-0.0142i$	$0.3139-0.1107i$
$0.4609+0.0025i$	$-0.5921+0.0139i$	$-0.4861+0.0062i$	$0.3582-0.0418i$
$0.7257-0.0009i$	$0.0431-0.1444i$	$-0.4631-0.004i$	$-0.6735+0.0448i$
1.0	1.0	1.0	1.0

13.7   HARMONICS DETECTION IN OMA

OMA assumes white noise ambient excitation. But in practice, the ambient excitation may also contain deterministic excitation forces. Mechanical systems often have harmonic excitation due to rotating parts like shafting, gears, pulleys, and prime movers like engines and motors. In such cases, the output response is not only due to the stochastic ambient excitation but also to the harmonic excitation forces acting on the structure.

Analyzing these responses using OMA leads to modes containing the structural or physical modes and harmonic modes. The physical modes are the true vibration modes, while the harmonic modes only describe the deflection shapes of the structure corresponding to the frequencies of the harmonic forces (called **operational deflection shapes (ODS)**) and are not the modes of vibration of the structure.

The presence of harmonics or harmonic modes presents two difficulties. Firstly, these may be wrongly construed as structural modes and thus may lead to an incorrect picture of the modal parameters of the structure. Secondly, its presence may also cause interference in modal parameter estimation. Hence, it is necessary to distinguish between the harmonics and true structural modes and remove them from the responses, if possible.

Some approaches suggested to distinguish harmonics from physical modes are as follows.

- An identified mode is regarded as a harmonic mode if the estimated value of the damping factor for the mode is zero or nearly zero.
- Brincker et al. [18] proposed an approach based on the statistical properties of the output signal. It is based on the idea that the probability density function (PDF) of the response of a structural mode to a random input is a Gaussian distribution, while the PDF of a harmonic signal has a two-peak bath-tub-like characteristic.
- Another approach proposed utilizes random decrement (Modak [70]). It is seen that the amplitude of the structural mode components in the random decrement decay with time, while those of harmonic components stay constant. The random decrement of the ambient response can be divided into two (or more) blocks of data, and the FFTs of the blocks are computed. The amplitudes of the peaks in the FFTs of the successive blocks are compared. The peaks with a constant magnitude in the FFTs correspond to harmonic frequencies, while those with decreasing amplitudes correspond to the natural frequencies of the structural modes.

13.8   ADVANTAGES AND LIMITATIONS OF OMA

- OMA is performed on the system under in-situ conditions, and hence the modal parameters of the structure corresponding to the actual boundary conditions can be obtained. The prevailing environmental conditions are also taken care of. However, simulating actual boundary and environmental conditions is challenging in EMA.
- In OMA, the estimated modal parameters correspond to the actual magnitude of the operational forces acting on the system. In EMA, the forces applied are small and may not correspond to the forces the system is subjected to during operation. The identified modal parameters in EMA may not accurately depict the system behavior at the operating loads if any nonlinearities are present.
- OMA can be performed without disrupting the operation of the system.
- OMA is especially useful for large structures, like many in civil engineering, where the application of artificial forces, as needed for EMA, can be difficult and uneconomical.
- In OMA, the estimated mode shapes are unscaled, while EMA yields the scaled mode shapes. A modal model with unscaled mode shapes doesn't make a complete dynamic model of the system. For some applications, like response simulation to external forces, the knowledge of the scaled mode shapes is necessary. Some methods are proposed to obtain scaling factors in OMA but require additional tests.
- OMA requires the presence of some white noise excitation. Many systems, especially in mechanical engineering, have strong harmonic excitation forces and may have only a little random excitation. OMA is not helpful in such cases.

REVIEW QUESTIONS

1. What is operational modal analysis? Under what conditions is the OMA suitable for modal analysis?
2. Why are the reference measurements required in OMA?
3. What is random decrement? What does it represent?
4. How are the part mode shapes of different datasets assembled in OMA?
5. Why does OMA not yield mass-normalized mode shapes?
6. What are the advantages of OMA over EMA?
7. What are harmonic modes?
8. Discuss the approaches for detecting harmonics in OMA briefly.

14 Applications of Experimental Modal Analysis

14.1 INTRODUCTION

In many situations, the primary objective of the experimental modal analysis (EMA) is to know the natural frequencies of the system to avoid its resonant operation. Knowledge of a system's natural frequencies and mode shapes is also helpful in troubleshooting any associated vibration and noise problems. Since EMA also yields a dynamic model of the system in the form of the modal model, there are other applications for which EMA can be used, as we see in this chapter. First, we look at why the modal model obtained from EMA is incomplete and then discuss the possibility of identifying the spatial model from the modal model. We then consider EMA applications such as response simulation, structural modification, coupling of structures, and force identification. Another application, the finite element model validation and updating is covered separately in Chapter 15.

14.2 INCOMPLETE MODEL

Compared to a numerical/FE model, the modal model obtained from EMA is not complete and is an 'incomplete model' of the system for the following reasons.

    a. We know that the number of rows in the eigenvector matrix obtained from EMA is equal to the number of DOFs at which the FRFs are measured. But this number is generally less than the DOF of the numerical/FE model. While the FRFs at the DOFs on a structure's surface can be measured, those inside the body or situated at inaccessible locations cannot be measured. Also, for 1D and 2D structures (made up of beams, plates, and shells), the rotational DOFs generally are not measured due to a lack of simple and robust measurement transducers. Thus, the number of measured FRFs (say, n) is less than the DOF of the spatial model (say, N), and hence the eigenvectors obtained from EMA are incomplete. This is often referred to as the **coordinate incompleteness** of the modal data.

    b. The number of eigenvalues or natural frequencies (which is also the number of columns in the eigenvector matrix) obtained from EMA is equal to the number of modes identified. The modes identified in EMA depend on the measurement frequency range and the modes in that range. The measurement frequency range is limited and is dictated by the range in which the dynamic behavior of a given structure is to be studied. If it is desired to extract all the modes (i.e., N), then that requires exciting a vast frequency range as many of these modes would have a very high frequency. However, measuring all the modes is difficult due to limits on the capacities of the excitation system, response transducers, and measurement systems. Therefore, the number of identified modes from EMA (say, m) is less than the DOF N of the spatial model. This is often referred to as the **modal incompleteness** of the modal data.

Thus, $\left[\begin{smallmatrix} \cdot \\ \lambda_r \end{smallmatrix}\right]_{m \times m}$ and $\left[\phi\right]_{n \times m}$ are the eigenvalue and eigenvector matrices obtained from EMA and form an incomplete set of modal data, as compared to the numerical/FE model.

14.3 CAN WE IDENTIFY THE SPATIAL MODEL FROM THE EXPERIMENTAL MODAL MODEL?

We studied in earlier chapters that a spatial model is a description of the structure in the physical domain, and it consists of mass, stiffness, and damping matrices. The spatial model description is helpful as it allows incorporating any design changes or modifications directly in the physical domain to study their effect on the system response. Can the spatial model be determined from the modal model identified via EMA? Theoretically, it should be possible to do so using the orthogonality properties that provide the relationship between these two models. These relationships, in the case of an N DOF undamped system are

DOI: 10.1201/9780429454783-14

$$[\phi]^T_{N \times N} \ [M]_{N \times N} \ [\phi]_{N \times N} = [I]_{N \times N} \tag{14.1}$$

$$[\phi]^T_{N \times N} \ [K]_{N \times N} \ [\phi]_{N \times N} = [\diagdown \lambda r \diagdown]_{N \times N} \tag{14.2}$$

If we know the complete eigenvalue and mass-normalized eigenvector matrices, i.e. $[\diagdown \lambda r \diagdown]$ and $[\phi]$, then the mass and stiffness matrices can be obtained from Eqs. (14.1) and (14.2) as

$$[M]_{N \times N} = [\phi]^{-T}_{N \times N} [\phi]^{-1}_{N \times N} \tag{14.3}$$

$$[K]_{N \times N} = [\phi]^{-T}_{N \times N} [\diagdown \lambda r \diagdown]_{N \times N} [\phi]^{-1}_{N \times N} \tag{14.4}$$

However, we saw in Section 14.2 that, in practice, the modal model obtained from EMA is incomplete. Therefore, the eigenvector matrix obtained from EMA is rectangular and cannot be inverted as required to estimate the spatial matrices using Eqs. (14.3) and (14.4).

One may enforce the condition that m = n by changing either the frequency range of measurement or the number of measured DOFs, allowing the inverse to be computed and the spatial matrices to be estimated. But these matrices would not generally be an accurate representation of the system due to the coordinate and modal incompleteness since m and n, even if equal, would be much smaller than N. Thus, estimating a spatial model of a practical system from the modal model identified from a modal test is not feasible.

14.4   RESPONSE SIMULATION

In the design cycle of a system or a product, one of the considerations is to reduce the extent of physical tests that might be required. One such requirement is to know whether the system based on the proposed design can sustain the dynamic loads expected to act on the system during its operation and service. The modal model identified from EMA on a prototype of the system can be used to answer these questions, as the model can be used to determine the system's response to the dynamic forces. It eliminates the physical tests required to validate the designs.

Let us say we have identified the modal model of the system by EMA assuming viscous damping. It consists of 'm' pairs of complex conjugate eigenvalues and eigenvectors. Each eigenvector is known at the 'n' DOFs of the system, the DOFs where the FRFs were measured for EMA. The response can be predicted if the dynamic forces act at one or more of these DOFs. The first step is to determine the receptance matrix for the n DOFs. It can be obtained using the following relationship:

$$[\alpha(\omega)]_{n \times n} = [\phi]_{n \times m} \left[\diagdown \frac{1}{i\omega - s_r} \diagdown \right]_{m \times m} [\phi]^T_{m \times n} + [\phi]^*_{n \times m} \left[\diagdown \frac{1}{i\omega - s_r^*} \diagdown \right]_{m \times m} [\phi]^{*T}_{m \times n} \tag{14.5}$$

The force vector $\{f(\omega)\}$ is constructed for the given dynamic forces. At frequency ω, the steady-state response at n DOFs is given by

$$\{x(\omega)\}_{n \times 1} = [\alpha(\omega)]_{n \times n} \{f(\omega)\}_{n \times 1} \tag{14.6}$$

If the forces are periodic or transients, then they can be first transformed to the frequency domain by DFT, and then Eq. (14.6) can be used to find the steady-state response at each frequency ω of the DFT. The responses at all the frequencies are added algebraically to find the response to the periodic or transient forces. If the forces are random, then a similar process can be followed by using PSDs.

14.5   DYNAMIC DESIGN

Dynamic design refers to the design of a product or a system to ensure that it has acceptable dynamic characteristics and/ or acceptable dynamic response to the dynamic forces that are expected to act during its operation and service.

The acceptable dynamic characteristics may relate to the specification of acceptable natural frequencies, mode shapes, FRFs, and damping factors. It could be the absence of natural frequencies in the specified band/s of frequencies, favorable mode shapes, the requirement of nodes of mode shapes at specific locations, or the specification of resonances, and

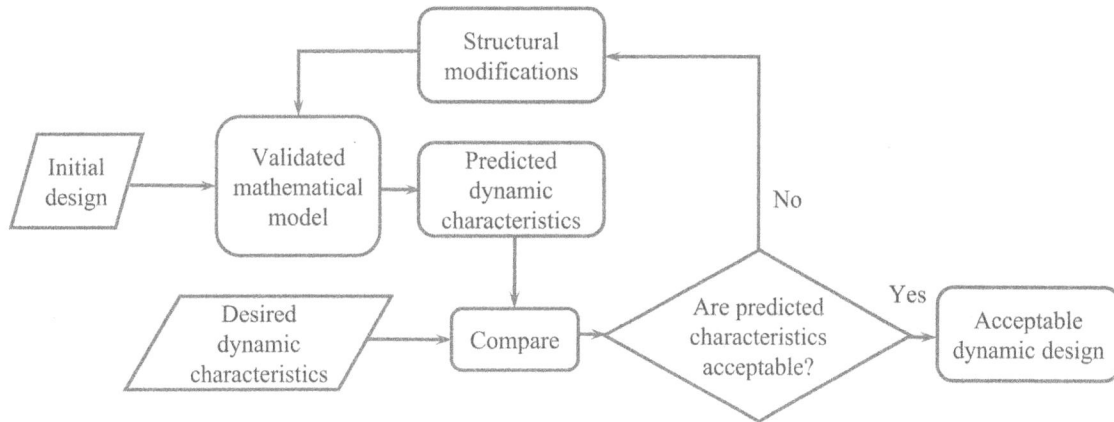

FIGURE 14.1 Dynamic design.

antiresonances of FRFs. The acceptable dynamic response could be in terms of permissible displacement, velocity, or acceleration levels.

Generally, the dynamic design starts after the basic mechanical design of the product has been accomplished in terms of its configuration, dimensions and geometry, materials, and choice of manufacturing processes. The basic mechanical design is based on failure criteria such as static strength, fatigue strength, rigidity, fracture, wear, creep, or other relevant criteria. It is difficult to impose the dynamic design requirements at the initial/basic design stage since the dynamic characteristics are undefined until the dimensions, material properties, boundary conditions and manufacturing process are fixed.

Once the basic design is complete, the dynamic design can be carried out. The dynamic design aims to achieve the desired specifications by making minor structural modifications such that the basic design is not compromised. It ensures that the failure criteria used in the basic design are not violated. Typical modifications include adding lumped masses, springs, stiffeners, dashpots, or damping layers.

How can the necessary design modifications be determined? Since the initial design is available, it is possible to construct a dynamic mathematical model to evaluate the dynamic performance and introduce modifications, if necessary. A model-based approach is preferable since a test-based approach to test the effectiveness of any proposed modification would be uneconomical and time-consuming and would not be amenable to achieving an optimum design. The problem of dynamic design using a mathematical model can be approached as a direct problem or an inverse problem.

Figure 14.1 shows a flow chart depicting the generic steps followed for the dynamic design using a direct approach. In this approach, the dynamic characteristics predicted by the mathematical model of the product are compared with the user-specified desired dynamic characteristics. If the comparison indicates that the predicted dynamic characteristics are not acceptable, a structural modification is selected and simulated in the mathematical model. The process is iterated till an acceptable dynamic design is achieved.

In the inverse approach, the problem is formulated with the structural modifications as the unknowns to minimize an objective criterion quantifying the difference between the predicted and the desired dynamic characteristics. The inverse problem is more challenging to formulate and solve.

The mathematical model used for dynamic design can be a lumped parameter or an FE model of the system. These models are generally validated using modal test data before they are used. The mathematical model can also be a modal model identified via EMA, as it can be used to predict the modified dynamic characteristics as presented in the next section.

14.6 LOCAL STRUCTURAL MODIFICATION

Local structural modification is a method to predict the natural frequencies and mode shapes of the modified structure due to a local/small structural modification using the modal model of the unmodified or base structure (Luk and Mitchell [60]). This method is also called modal synthesis or dual-modal space method.

14.6.1 FORMULATION OF THE METHOD

The equation of motion for the free vibration of an undamped system (referred as unmodified structure) is

$$[M]\{\ddot{x}\} + [K]\{x\} = \{0\} \tag{14.7}$$

Assuming the free vibration response as

$$\{x\} = \{\phi\} \cos \lambda t \tag{14.8}$$

we obtain the following eigenvalue problem:

$$\left([K] - \lambda^2 [M]\right)\{\phi\} = \{0\} \tag{14.9}$$

Let $\left[\,`\lambda_r^2\,`\right]$ and $[\phi]$ be the corresponding eigenvalue and eigenvector matrices, respectively. If $[\Delta M]$ and $[\Delta K]$ represent the changes in the mass and stiffness matrices because of a structural modification, then the equation of motion for the modified structure can be written as

$$[M + \Delta M]\{\ddot{y}\} + [K + \Delta K]\{y\} = \{0\} \tag{14.10}$$

We transform the response of the modified structure to the modal space of the unmodified structure.

$$\{y\} = [\phi]\{q\} \tag{14.11}$$

where $\{q\}$ is the modal response of the modified structure in the modal space of the unmodified structure. Substituting Eq. (14.11) into (14.10), post-multiplying the resulting equation by $[\phi]^T$, and using the orthogonality properties of the eigenvectors of the unmodified structure, we get

$$\left([I] + [\phi]^T [\Delta M][\phi]\right)\{\ddot{q}\} + \left(\left[\,`\lambda_r^2\,`\right] + [\phi]^T [\Delta K][\phi]\right)\{q\} = \{0\} \tag{14.12}$$

Assuming the free vibration response as

$$\{q\} = \{\psi\} \cos \beta t \tag{14.13}$$

we obtain the following eigenvalue problem corresponding to Eq. (14.12):

$$\left(\left(\left[\,`\lambda_r^2\,`\right] + [\phi]^T [\Delta K][\phi]\right) - \beta^2 \left([I] + [\phi]^T [\Delta M][\phi]\right)\right)\{\psi\} = \{0\} \tag{14.14}$$

The solution to the eigenvalue problem gives eigenvalues $\left(\left[\,`\beta_r^2\,`\right]\right)$ and eigenvectors $\left([\psi]\right)$ of the modified structure. The eigenvectors $[\psi]$ are in the modal space of the unmodified structure. By substituting $\{q\} = \{\psi\}_r$ in Eq. (14.11), we get the corresponding eigenvector in the physical coordinates. Therefore, if we represent the eigenvector matrix of the modified structure in the physical space by $[z]$, then it is given by

$$[z] = [\phi][\psi] \tag{14.15}$$

Eq. (14.15) shows that the modified structure's eigenvectors are linear combinations of the unmodified structure's eigenvectors. Let's look at the implications of solving the eigenvalue problem of the modified structure (Eq. 14.14).

- The eigenvalue problem needs eigenvalues and eigenvectors of the unmodified structure, which can be obtained from the EMA of the unmodified structure. It also needs a spatial description of the mass and stiffness changes (i.e. $[\Delta M]$ and $[\Delta K]$) due to a structural modification. Notably, the mass and stiffness matrices of the unmodified structure are not required. Thus, the eigenvalue problem in (14.14) can be solved using the experimental modal model, and the modal data of the modified structure can be predicted using the unmodified structure's modal model. This is the basic principle of the local structural modification method.
- Let n and m represent the numbers of measured DOFs and identified modes, respectively. Thus, the order of the matrices $\left[\,`\lambda_r^2\,`\right]$ and $[\phi]$ obtained from EMA of the unmodified structure are m × m and n × m, respectively. Because of this, $[\Delta M]$ and $[\Delta K]$ each must be of order n × n, as evident from Eq. (14.14). Thus, any structural modification that affects the DOFs other than the n measured DOFs cannot be described accurately.
- The eigenvalue problem (Eq. (14.14)) considers only m modes of the unmodified structure, i.e., those obtained through EMA, and hence is based on a truncated modal basis. The matrices $[\phi]^T [\Delta K][\phi]$ and $[\phi]^T [\Delta M][\phi]$ are not diagonal; hence, the truncation of the unmodified structure's modes adversely affects the accuracy of the estimates of the modal data of the modified structure.

Though the method above is described for an undamped system, it can be extended to damped systems to predict the effects of damping modifications.

14.6.2 Typical Structural Modifications

a. Lumped mass modification

Let a lumped mass m_0 is added at the jth DOF, belonging to the set of n measured DOFs. The matrix $[\Delta K]$ is a matrix of all zeros since a lumped mass practically does not affect the stiffness. Matrix $[\Delta M]$ is a matrix of zeros with the jth diagonal element equal to m_0 and is given by

$$[\Delta M]_{n \times n} = \begin{bmatrix} 0 & \cdots & \cdots & 0 \\ \cdots & \cdots & \cdots & \cdots \\ \cdots & \cdots & m_0 & \cdots \\ 0 & \cdots & \cdots & 0 \end{bmatrix} \tag{14.16}$$

The lumped mass modification represents a unit rank modification, as $[\Delta M]$ has a unit rank. The rank is unity because $[\Delta M]$ can be generated with one vector, as shown in the equation below.

$$[\Delta M]_{n \times n} = m_0 \begin{Bmatrix} 0 \\ \cdots \\ 1 \\ 0 \end{Bmatrix}_{n \times 1} \begin{Bmatrix} 0 \\ \cdots \\ 1 \\ 0 \end{Bmatrix}_{1 \times n}^{T} \tag{14.17}$$

b. Lumped stiffness modification

Let a lumped spring of stiffness k_0 is added between the ith and j^{th} DOFs, both belonging to the set of n measured DOFs. In this case, the matrix $[\Delta M]$ is a matrix of all zeros if the mass of the spring is negligible compared to the structural mass. The matrix $[\Delta K]$ is

$$[\Delta K]_{n \times n} = \begin{bmatrix} 0 & \cdots & \cdots & 0 \\ \cdots & k_0 & -k_0 & \cdots \\ \cdots & -k_0 & k_0 & \cdots \\ 0 & \cdots & \cdots & 0 \end{bmatrix} \tag{14.18}$$

The lumped stiffness modification also represents a unit rank modification, as $[\Delta K]$ has a unit rank. The rank is unity because $[\Delta K]$ can be generated with one vector, as shown in the equation below.

$$[\Delta K]_{n \times n} = k_0 \begin{Bmatrix} 0 \\ 1 \\ -1 \\ 0 \end{Bmatrix}_{n \times 1} \begin{Bmatrix} 0 \\ 1 \\ -1 \\ 0 \end{Bmatrix}_{1 \times n}^{T} \tag{14.19}$$

c. Beam modification

Beam modification is a more complex modification than a spring. It could be in the form of a stiffener incorporated into the structure between two points. The addition of a beam/stiffener adds both mass and stiffness to the structure, and hence both $[\Delta M]$ and $[\Delta K]$ are non-zero. A stiffener offers both the reaction forces and moments and has mass and rotational inertia. Thus, $[\Delta M]$ and $[\Delta K]$ for a stiffener have translational and rotational DOFs and can be described using the FE matrices for beam elements. Since in EMA, the rotational DOFs are generally unmeasured, the experimental eigenvectors $([\phi])$ don't have rotational DOFs, and this makes the orders of the matrices $[\phi]$ and $[\Delta M]$ and $[\Delta K]$ inconsistent. One approach to resolving this inconsistency is expanding the experimental eigenvectors to include the rotational DOFs (Avitabile et al. [9]).

14.7 COUPLED STRUCTURAL ANALYSIS

In coupled structural analysis, also referred to as sub-structuring, the mathematical model of an assembled structure is obtained by combining the models of the individual components or substructures of the assembly. This approach can be used in the following situations.

- A reduced model of the assembled structure is required to reduce the computational effort and time. The individual sub-component models can be reduced by performing modal reduction by retaining only limited and relevant modes and dropping the modes beyond the frequency range of interest. It reduces the size of the sub-component modal models but retains the description over all the physical DOFs. The reduced sub-component modal models can be coupled to derive a reduced model of the assembled structure.
- There may be a situation where the individual sub-component models are derived from different techniques, and the model of the assembled structure is required. For example, some component models are derived from FEM and others from EMA.
- All sub-component models are derived through EMA in the form of modal models, and the modal model of the assembled structure is required.

This section looks at a class of coupling methods known as the modal coupling methods (also called component mode synthesis (CMS) methods). In these methods, the modal model of the assembled structure is synthesized from the modal models of the substructures. We look at two formulations of these methods: the fixed-interface and the free-interface methods. The fixed-interface method can be used to couple the numerical models of the substructures but is not amenable to combining the modal models obtained through EMA. The free-interface method can be used even with experimentally identified modal models.

14.7.1 MODAL COUPLING USING THE FIXED-INTERFACE METHOD

Let us consider a coupled structure with substructures A and B, as shown in Figure 14.2. A and B are coupled at connection/coupling/interface DOFs, while the remaining DOFs of the two substructures are their internal DOFs, as shown. The subscripts 'i' and 'c' are used to denote the internal and coupling DOFs, respectively. The objective is to combine the individual modal models of A and B under uncoupled states to obtain the model of the coupled structure.

The basic idea behind the fixed-interface method is to couple the modal models of substructures A and B with fixed boundary conditions at the coupling DOFs (as shown in Figure 14.3).

The equation of motion for substructure A, when not coupled to substructure B, can be written as

$$\left[M^A \right] \left\{ \ddot{x}^A \right\} + \left[K^A \right] \left\{ x^A \right\} = \left\{ f^A \right\} \tag{14.20}$$

Eq. (14.20) after partitioning into internal and coupling DOFs becomes

$$\begin{bmatrix} M_{ii}^A & M_{ic}^A \\ M_{ci}^A & M_{cc}^A \end{bmatrix} \left\{ \begin{matrix} \ddot{x}_i^A \\ \ddot{x}_c^A \end{matrix} \right\} + \begin{bmatrix} K_{ii}^A & K_{ic}^A \\ K_{ci}^A & K_{cc}^A \end{bmatrix} \left\{ \begin{matrix} x_i^A \\ x_c^A \end{matrix} \right\} = \left\{ \begin{matrix} f_i^A \\ f_c^A \end{matrix} \right\} \tag{14.21}$$

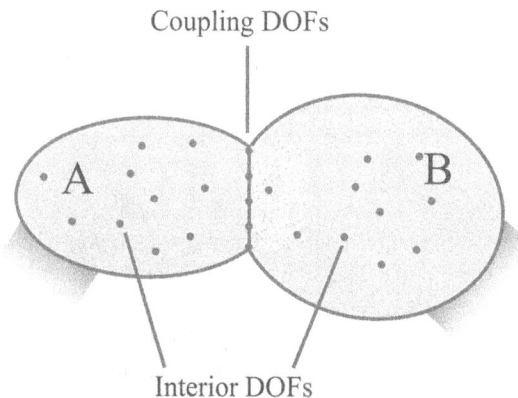

FIGURE 14.2   Coupled structure with substructures A and B.

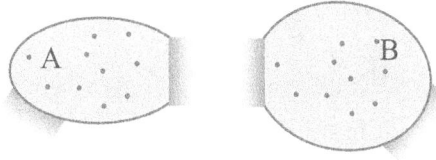

FIGURE 14.3 Substructure A and B with fixed boundary conditions at the coupling DOFs.

If the external forces on the internal DOFs are zero (i.e. $f_i^A = 0$), and the coupling DOFs are fixed (i.e. $x_c^A = 0$), as shown in Figure 14.3, then from Eq. (14.21) we obtain

$$\left[M_{ii}^A\right]\left\{\ddot{x}_i^A\right\}+\left[K_{ii}^A\right]\left\{x_i^A\right\}=\{0\} \tag{14.22}$$

Eq. (14.22) is nothing but the equation of motion for free vibration of substructure A fixed at the coupling DOFs. The solution of the corresponding eigenvalue problem gives the fixed-interface modal model of A. Let $\left[\,^{\diagdown}\lambda^A_{\diagdown}\right]$ and $\left[\phi_i^A\right]$ be the corresponding eigenvalue and eigenvector matrices, respectively. The displacement at the internal DOFs with the fixed interface can be expressed in the modal domain as

$$\left\{x_i^A\right\}=\left[\phi_i^A\right]\left\{q^A\right\} \tag{14.23}$$

Eq. (14.22) describes the motion at the internal DOFs of substructure A when substructure A is fixed at the coupling DOFs. To describe the motion at the internal DOFs, without the fixed boundary, we must add the motion at the internal DOFs, which occurs due to the motion at the coupling DOFs. An approximation is obtained by considering force equilibrium at the internal DOFs of substructure A by neglecting the inertia forces. With this simplification, we obtain the following equation from the first part of Eq. (14.21):

$$\left[K_{ii}^A\right]\left\{x_i^A\right\}+\left[K_{ic}^A\right]\left\{x_c^A\right\}\cong\{0\} \tag{14.24}$$

$$\left\{x_i^A\right\}=-\left[K_{ii}^A\right]^{-1}\left[K_{ic}^A\right]\left\{x_c^A\right\} \tag{14.25}$$

$$\left\{x_i^A\right\}=\left[D^A\right]\left\{x_c^A\right\} \tag{14.26}$$

where

$$\left[D^A\right]=-\left[K_{ii}^A\right]^{-1}\left[K_{ic}^A\right] \tag{14.27}$$

Eq. (14.26) statically relates the displacements at the internal and coupling DOFs. For this reason, the columns of $\left[D^A\right]$ are referred to as static modes, which linearly combine to produce the displacement at the internal DOFs.

Summing the displacement in Eqs. (14.23) and (14.26) gives the total displacement at the internal DOFs of substructure A

$$\left\{x_i^A\right\}=\left[\phi_i^A\right]\left\{q^A\right\}+\left[D^A\right]\left\{x_c^A\right\} \tag{14.28}$$

Therefore, the displacement vector of all the DOFs of substructure A is obtained as

$$\left\{x^A\right\}=\left\{\begin{array}{c}x_i^A\\x_c^A\end{array}\right\}=\left[\begin{array}{cc}\phi_i^A & D^A\\0 & I\end{array}\right]\left\{\begin{array}{c}q^A\\x_c^A\end{array}\right\}=\left[T^A\right]\left\{\begin{array}{c}q^A\\x_c^A\end{array}\right\} \tag{14.29}$$

Substituting Eq. (14.29) into Eq. (14.21) (with $f_i^A = 0$), and pre-multiplying by $\left[T^A\right]^T$ gives

$$\left[T^A\right]^T\left[\begin{array}{cc}M_{ii}^A & M_{ic}^A\\M_{ci}^A & M_{cc}^A\end{array}\right]\left[T^A\right]\left\{\begin{array}{c}\ddot{q}^A\\\ddot{x}_c^A\end{array}\right\}+\left[T^A\right]^T\left[\begin{array}{cc}K_{ii}^A & K_{ic}^A\\K_{ci}^A & K_{cc}^A\end{array}\right]\left[T^A\right]\left\{\begin{array}{c}q^A\\x_c^A\end{array}\right\}=\left[T^A\right]^T\left\{\begin{array}{c}0\\f_c^A\end{array}\right\} \tag{14.30}$$

Simplifying and using the orthogonality properties leads to

$$
\begin{bmatrix} I^A & \overline{M_{ic}^A} \\ \overline{M_{ci}^A} & M_{cc}^A \end{bmatrix} \begin{Bmatrix} \ddot{q}^A \\ \ddot{x}_c^A \end{Bmatrix} + \begin{bmatrix} \lambda^A & 0 \\ 0 & K_{cc}^A \end{bmatrix} \begin{Bmatrix} q^A \\ x_c^A \end{Bmatrix} = \begin{Bmatrix} 0 \\ f_c^A \end{Bmatrix}
\tag{14.31}
$$

By following the above steps for substructure B, we get a similar equation

$$
\begin{bmatrix} I^B & \overline{M_{ic}^B} \\ \overline{M_{ci}^B} & M_{cc}^B \end{bmatrix} \begin{Bmatrix} \ddot{q}^B \\ \ddot{x}_c^B \end{Bmatrix} + \begin{bmatrix} \lambda^B & 0 \\ 0 & K_{cc}^B \end{bmatrix} \begin{Bmatrix} q^B \\ x_c^B \end{Bmatrix} = \begin{Bmatrix} 0 \\ f_c^B \end{Bmatrix}
\tag{14.32}
$$

Note that the two substructures are still separate. When coupled, they must satisfy the compatibility conditions related to displacement continuity and force equilibrium at the coupling DOFs. These conditions are

$$
\{x_c^A\} = \{x_c^B\} = \{x_c\}
\tag{14.33}
$$

$$
\{f_c^A\} + \{f_c^B\} = \{0\}
\tag{14.34}
$$

We substitute the condition in Eq. (14.33) into Eqs. (14.31) and (14.32) and then impose the condition (14.34). The three equations we get can be combined into the following matrix equation

$$
\begin{bmatrix} I^A & 0 & \overline{M_{ic}^A} \\ 0 & I^B & \overline{M_{ic}^B} \\ \overline{M_{ci}^A} & \overline{M_{ci}^B} & \overline{M_{cc}^A} + \overline{M_{cc}^B} \end{bmatrix} \begin{Bmatrix} \ddot{q}^A \\ \ddot{q}^B \\ \ddot{x}_c \end{Bmatrix} + \begin{bmatrix} \lambda^A & 0 & 0 \\ 0 & \lambda^B & 0 \\ 0 & 0 & \overline{K_{cc}^A} + \overline{K_{cc}^B} \end{bmatrix} \begin{Bmatrix} q^A \\ q^B \\ x_c \end{Bmatrix} = \begin{Bmatrix} 0 \\ 0 \\ 0 \end{Bmatrix}
\tag{14.35}
$$

The solution to the eigenvalue problem corresponding to Eq. (14.35) gives the eigenvalues $(\lceil \beta \rfloor)$ and eigenvectors $([\psi])$ of the coupled structure.

However, the computed eigenvectors $[\psi]$ correspond to the coordinates in Eq. (14.35). The eigenvectors of the coupled structure in physical space are obtained as follows. Eq. (14.28) and the corresponding equation for substructure B can be combined and written as

$$
\begin{Bmatrix} x_i^A \\ x_i^B \\ x_c \end{Bmatrix} = \begin{bmatrix} \phi_i^A & 0 & D^A \\ 0 & \phi_i^B & D^B \\ 0 & 0 & I \end{bmatrix} \begin{Bmatrix} q^A \\ q^B \\ x_c \end{Bmatrix}
\tag{14.36}
$$

Therefore, the eigenvectors of the coupled structure in physical space, denoted by the matrix $[z]$, are given by

$$
[z] = \begin{bmatrix} \phi_i^A & 0 & D^A \\ 0 & \phi_i^B & D^B \\ 0 & 0 & I \end{bmatrix} [\psi]
\tag{14.37}
$$

The following observations are made from the formulation of the fixed-interface method.

- We note from Eq. (14.35) that the eigenvalue problem for the coupled structure requires the modal model of the substructures and also some other matrices that require the knowledge of the stiffness and mass matrices of these substructures. Therefore, the fixed-interface method is not amenable to combining the modal models obtained from EMA but can be used with the numerical models of the substructures.
- It is noted that the method can be implemented with a limited number of modes for each substructure. This aspect can be used to obtain a reduced coupled structure model, making the coupled structure analysis computationally efficient.

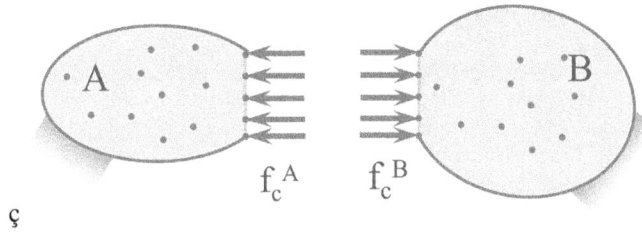

FIGURE 14.4   Substructures A and B with free boundary conditions at the coupling DOFs.

14.7.2   MODAL COUPLING USING THE FREE-INTERFACE METHOD

In this method also, the objective is to build the modal model of the coupled structure (shown in Figure 14.2) from the modal models of the substructures. The basic idea behind the free interface method is to couple the modal models of substructures A and B with free boundary conditions at the coupling DOFs (as shown in Figure 14.4). Note that the interconnection forces at the coupling DOFs act on both structures. The equation of motion for substructure A shown in the figure, assuming that the external forces on the internal DOFs are zero (i.e. $f_i^A = 0$), can be written as

$$\left[M^A\right]\left\{\ddot{x}^A\right\} + \left[K^A\right]\left\{x^A\right\} = \left\{f^A\right\} = \left\{\begin{array}{c} 0 \\ f_c^A \end{array}\right\} \tag{14.38}$$

Let $\left[\diagdown\lambda^A\diagdown\right]$ and $\left[\phi_i^A\right]$ be the matrices of 'mA' eigenvalues and eigenvectors, respectively, for substructure A with free boundary conditions at the coupling DOFs. Consider the modal transformation given by

$$\left\{x^A\right\} = \left[\phi^A\right]\left\{q^A\right\} = \left[\begin{array}{c} \phi_i^A \\ \phi_c^A \end{array}\right]\left\{q^A\right\} \tag{14.39}$$

where $\left\{q^A\right\}$ is the modal response vector. Substituting Eq. (14.39) into Eq. (14.38), pre-multiplying by $\left[\phi^A\right]^T$ and using the orthogonality properties, we get

$$\underset{mA\times mA}{\left[I^A\right]} \ \underset{mA\times 1}{\left\{\ddot{q}^A\right\}} + \underset{mA\times mA}{\left[\diagdown\lambda^A\diagdown\right]} \ \underset{mA\times 1}{\left\{q^A\right\}} = \underset{mA\times nc}{\left[\phi_c^A\right]^T} \ \underset{nc\times 1}{\left\{f_c^A\right\}} \tag{14.40}$$

where 'nc' represents the number of coupling DOFs. If 'mB' is the number of eigenvalues and eigenvectors for substructure B with free boundary conditions at the coupling DOFs, then the equation for substructure B corresponding to Eq. (14.40) can be written as

$$\underset{mB\times mB}{\left[I^B\right]} \ \underset{mB\times 1}{\left\{\ddot{q}^B\right\}} + \underset{mB\times mB}{\left[\diagdown\lambda^B\diagdown\right]} \ \underset{mB\times 1}{\left\{q^B\right\}} = \underset{mB\times nc}{\left[\phi_c^B\right]^T} \ \underset{nc\times 1}{\left\{f_c^B\right\}} \tag{14.41}$$

Eqs. (14.40) and (14.41) can be combined, leading to the following matrix equation

$$\left[\begin{array}{cc} I^A & 0 \\ 0 & I^B \end{array}\right]\left\{\begin{array}{c} \ddot{q}^A \\ \ddot{q}^B \end{array}\right\} + \left[\begin{array}{cc} \lambda^A & 0 \\ 0 & \lambda^B \end{array}\right]\left\{\begin{array}{c} q^A \\ q^B \end{array}\right\} = \left[\begin{array}{cc} \left[\phi_c^A\right]^T & 0 \\ 0 & \left[\phi_c^B\right]^T \end{array}\right]\left\{\begin{array}{c} f_c^A \\ f_c^B \end{array}\right\} \tag{14.42}$$

Eq. (14.42) does not yet represent the equation for the coupled structure since the conditions of displacement continuity and equilibrium of coupling forces are not yet imposed.

When coupled, the displacement continuity at the coupling DOFs of substructures A and B requires that,

$$\underset{nc\times 1}{\left\{x_c^A\right\}} = \underset{nc\times 1}{\left\{x_c^B\right\}} \tag{14.43}$$

Considering the modal transformations for the two substructures

$$\{x_c^A\} = [\phi_c^A]\{q^A\} \quad\quad (14.44)$$

$$\{x_c^B\} = [\phi_c^B]\{q^B\} \quad\quad (14.45)$$

Substituting Eqs. (14.44) and (14.45) into Eq. (14.43), we get

$$[\phi_c^A]\{q^A\} - [\phi_c^B]\{q^B\} = \{0\} \quad\quad (14.46)$$

$$\left[[\phi_c^A] \quad -[\phi_c^B] \right] \begin{Bmatrix} \{q^A\} \\ \{q^B\} \end{Bmatrix} = \{0\} \quad\quad (14.47)$$

Eq. (14.47) relates the (mA+mB) modal coordinates by 'nc' linear equations. Thus, the coupled structure would effectively have (mA+ mB-nc) modal coordinates or modes. Assuming nc < mA, we can partition $[\phi_c^A]$ and $\{q^A\}$ and write Eq. (14.47) as

$$\left[[\phi_c^A]_s \quad [\phi_c^A]_r \quad -[\phi_c^B] \right] \begin{Bmatrix} \{q^A\}_s \\ \{q^A\}_r \\ \{q^B\} \end{Bmatrix} = \{0\} \quad\quad (14.48)$$

The above equation can be solved for $\{q^A\}_s$ yielding

$$\{q^A\}_s = -[\phi_c^A]_s^{-1}[\phi_c^A]_r\{q^A\}_r + [\phi_c^A]_s^{-1}[\phi_c^B]\{q^B\} \quad\quad (14.49)$$

Therefore, the modal response vector can be written as

$$\begin{Bmatrix} q^A \\ q^B \end{Bmatrix} = \begin{Bmatrix} \{q^A\}_s \\ \{q^A\}_r \\ \{q^B\} \end{Bmatrix} = \begin{bmatrix} -[\phi_c^A]_s^{-1}[\phi_c^A]_r & [\phi_c^A]_s^{-1}[\phi_c^B] \\ I & 0 \\ 0 & I \end{bmatrix} \begin{Bmatrix} \{q^A\}_r \\ \{q^B\} \end{Bmatrix} = [T]\{Q\} \quad\quad (14.50)$$

where $[T]$ is the transformation matrix and $\{Q\}$ is the vector of modal coordinates of the coupled structure. These are given by

$$[T] = \begin{bmatrix} -[\phi_c^A]_s^{-1}[\phi_c^A]_r & [\phi_c^A]_s^{-1}[\phi_c^B] \\ I & 0 \\ 0 & I \end{bmatrix} \text{ and } \{Q\} = \begin{Bmatrix} \{q^A\}_r \\ \{q^B\} \end{Bmatrix} \quad\quad (14.51)$$

Thus, Eq. (14.50) is the result of imposing the displacement continuity at the coupling DOFs. Now we substitute Eq. (14.50) into Eq. (14.42) and pre-multiply by $[T]^T$ leading to

$$[T]^T \begin{bmatrix} I^A & 0 \\ 0 & I^B \end{bmatrix} [T]\{\ddot{Q}\} + [T]^T \begin{bmatrix} \lambda^A & 0 \\ 0 & \lambda^B \end{bmatrix} [T]\{Q\} = [T]^T \begin{bmatrix} \left[\phi_c^A\right]^T & 0 \\ 0 & \left[\phi_c^B\right]^T \end{bmatrix} \begin{Bmatrix} f_c^A \\ f_c^B \end{Bmatrix} \qquad (14.52)$$

We note that for the equilibrium of the coupling forces, we must have

$$\{f_c^A\} + \{f_c^B\} = \{0\} \qquad (14.53)$$

It is found that the RHS of Eq. (14.52) after simplification and given Eq. (14.53) is a null vector. Therefore, Eq. (14.52) becomes

$$[M]\{\ddot{Q}\} + [K]\{Q\} = \{0\} \qquad (14.54)$$

where, $[M]$ and $[K]$ are the mass and stiffness matrices of the coupled structure in the modal domains of substructures A and B under the free-interface boundary condition and with the coupling constraints imposed. They are given by

$$[M] = [T]^T[T] \quad \text{and} \quad [K] = [T]^T \begin{bmatrix} \lambda^A & 0 \\ 0 & \lambda^B \end{bmatrix} [T] \qquad (14.55)$$

The solution to the eigenvalue problem corresponding to Eq. (14.54) gives the eigenvalues ($\lceil \beta_\searrow \rceil$) and eigenvectors ($[\psi]$) of the coupled structure.

However, the computed eigenvectors $[\psi]$ correspond to the coordinates in Eq. (14.54). The eigenvectors of the coupled structure in physical space are obtained as follows. Eq. (14.39) and the corresponding equation for substructure B can be combined and written as

$$\begin{Bmatrix} x^A \\ x^B \end{Bmatrix} = \begin{bmatrix} \phi^A & 0 \\ 0 & \phi^B \end{bmatrix} \begin{Bmatrix} q^A \\ q^B \end{Bmatrix} \qquad (14.56)$$

Therefore, given Eq. (14.50), the eigenvectors of the coupled structure in physical space, denoted by the matrix $[z]$, are given by,

$$[z] = \begin{bmatrix} \phi^A & 0 \\ 0 & \phi^B \end{bmatrix} [T][\psi] \qquad (14.57)$$

The following observations are made from the formulation of the free-interface method.

- From Eq. (14.54), it is seen that setting up the eigenvalue problem for the coupled structure requires only the modal models of the substructures with the free interface. These can be obtained by EMA on the substructures with free boundary conditions at the coupling DOFs. Hence, the free-interface method is amenable to using experimentally identified modal models to obtain the modal model of the coupled structure.
- It is noted that the eigenvectors of the substructures must include all the DOFs at the interface. It means that when the beams or plate-like substructures are to be coupled, the rotational DOFs at the interface also need to be measured. Since the rotational DOF measurements are generally unavailable, a finite-difference approximation based on translational measurements is one possibility. The other could be the expansion of the measured mode shapes, but it requires a numerical model of the substructure.
- The method can be implemented with limited modes for each substructure. This aspect allows for obtaining a reduced model of the coupled structure, making the analysis of the coupled structure computationally efficient.

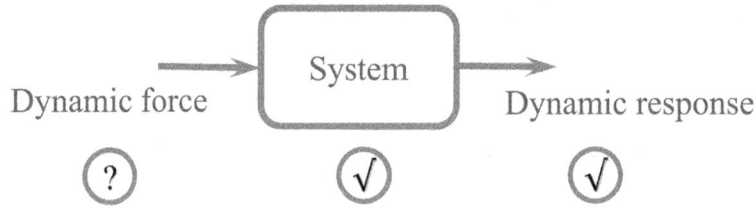

FIGURE 14.5 The problem of force identification.

14.8 FORCE IDENTIFICATION

Knowledge of the dynamic forces a system or a structure is subjected to during its operation is vital to predicting its dynamic response. It is also a necessary input for the design of the system. The direct measurement of the operational forces by force transducers, load cells, or strain gauges is difficult due to transducer mounting issues and associated modifications that may have to be made in the system. Instead, the dynamic forces can be measured indirectly by identifying them from the measurement of the system response to the dynamic forces and FRFs. Figure 14.5 depicts the problem of force identification through a block diagram. It is an inverse problem as the unknown input is to be identified from the knowledge of the system and output. In this section, we present two force identification methods.

14.8.1 FRF-BASED METHOD

This method is based on the input-output relationship in the frequency domain. In terms of the receptance matrix, this relationship is

$$\{x(\omega)\} = [\alpha(\omega)]\{F(\omega)\} \tag{14.58}$$

The objective is to estimate $\{F(\omega)\}$ from the above equation using the measured values of $\{x(\omega)\}$ and $[\alpha(\omega)]$.

Let us say the forces at 'q' DOFs are to be identified, and the responses are measured at 'n' DOFs. With these variables, Eq. (14.58) can be written as

$$\{x(\omega)\}_{n\times1} = [\alpha(\omega)]_{n\times q}\{F(\omega)\}_{q\times1} \tag{14.59}$$

Thus, we need to measure FRFs between the q force and n response locations. It yields the FRF matrix $[\alpha(\omega)]_{n\times q}$. It is desired that $n > q$ so that Eq. (14.59) is an overdetermined set of equations, and a unique solution can be obtained by a least-square error approach. If the FRF matrix is full rank, then the least-square solution of Eq. (14.59) is given by

$$\{F(\omega)\} = \left([\alpha(\omega)]^H[\alpha(\omega)]\right)^{-1}[\alpha(\omega)]^H\{x(\omega)\} \tag{14.60}$$

Note that Eq. (14.60) needs to be solved at each frequency ω. A common issue faced in the inverse problems is related to the ill-conditioning of the coefficient matrix, which is the matrix $[\alpha(\omega)]$ in the present case. If $[\alpha(\omega)]$ is ill-conditioned, the inverse $\left([\alpha(\omega)]^H[\alpha(\omega)]\right)^{-1}$ may yield incorrect results, and the forces cannot be correctly identified. $[\alpha(\omega)]$ may be ill-conditioned if the number of modes (m) contributing to the FRF is less than the number of forces (q) to be identified. At frequencies equal to or close to the natural frequencies also, $[\alpha(\omega)]$ may be ill-conditioned as only one mode dominates the FRF. This is especially the case with lightly damped structures and structures with well-separated modes. Regularization techniques can be used to solve the ill-conditioned equations (as discussed in section 15.17 of Chapter 15).

14.8.2 MODAL COORDINATE TRANSFORMATION METHOD

The modal coordinate transformation method is based on transforming the equations of motion to the modal domain. The equation of motion for a system with proportional viscous damping is given by

$$[M]\{\ddot{x}\} + [C]\{\dot{x}\} + [K]\{x\} = \{F(t)\} \tag{14.61}$$

Considering the modal transformation using the mass-normalized eigenvectors of the system, we have

$$\{x\} = [\phi]\{y\} \tag{14.62}$$

Substituting Eq. (14.62) into Eq. (14.61), pre-multiplying by $[\phi]^T$ and using orthogonality properties, we get

$$[I]\{\ddot{y}\} + \left[\,^\backprime 2\xi_r\omega_{r\backprime}\right]\{\dot{y}\} + \left[\,^\backprime \omega_r^2\,_\backprime\right]\{y\} = \{f_r(t)\} \tag{14.63}$$

where, $\{f_r\}$ is the vector of modal forces.

$$\{f_r\} = [\phi]^T\{F\} \tag{14.64}$$

Thus, Eq. (14.63) represents uncoupled modal equations relating modal responses to the corresponding modal forces. For the harmonic forces at frequency ω, Eq. (14.63) can be written as

$$\{y(\omega)\} = \left[\,^\backprime\left(\omega_r^2 - \omega^2 + 2\xi_r\omega_r\omega i\right)_\backprime\right]^{-1}\{f_r(\omega)\} \tag{14.65}$$

Substituting Eq. (14.64) into Eq. (14.65) and then substituting the resulting equation into Eq. (14.62), we get

$$\{x(\omega)\} = [\phi]\left[\,^\backprime\left(\omega_r^2 - \omega^2 + 2\xi_r\omega_r\omega i\right)_\backprime\right]^{-1}[\phi]^T\{F(\omega)\} \tag{14.66}$$

Eq. (14.66) could have also been written directly from Eq. (14.58) by substituting the modal series form of the FRF matrix in that equation. If n, q, and m represent the numbers of responses, forces to be identified, and modes, respectively, Eq. (14.66) can be written as

$$\underset{n\times 1}{\{x(\omega)\}} = \underset{n\times m}{[\phi]}\ \underset{m\times m}{\left[\,^\backprime\left(\omega_r^2 - \omega^2 + 2\xi_r\omega_r\omega i\right)_\backprime\right]^{-1}}\underset{m\times q}{[\phi]^T}\underset{q\times 1}{\{F(\omega)\}} \tag{14.67}$$

$$\underset{q\times 1}{\{F(\omega)\}} = \underset{q\times m}{[\phi]^{+T}}\ \underset{m\times m}{\left[\,^\backprime\left(\omega_r^2 - \omega^2 + 2\xi_r\omega_r\omega i\right)_\backprime\right]}\underset{m\times n}{[\phi]^+}\underset{n\times 1}{\{x(\omega)\}} \tag{14.68}$$

The superscript '+' represents the pseudo-inverse of the matrix. Eq. (14.68) can be used to find the unknown forces using the modal model identified from EMA and the measured responses.

The solution of Eq. (14.68) can be broken down into three steps.

- Estimate the modal responses from Eq. (14.62) in the frequency domain.

$$\underset{m\times 1}{\{y(\omega)\}} = \underset{m\times n}{[\phi]^+}\underset{n\times 1}{\{x(\omega)\}} \tag{14.69}$$

It is desired that n>m to ensure over-determinacy so that the estimates of the modal responses are meaningful.
- Using Eq. (14.65) compute the modal forces from the estimated modal responses.

$$\underset{m\times 1}{\{f_r(\omega)\}} = \underset{m\times m}{\left[\,^\backprime\left(\omega_r^2 - \omega^2 + 2\xi_r\omega_r\omega i\right)_\backprime\right]}\ \underset{m\times 1}{\{y(\omega)\}} \tag{14.70}$$

- Finally, using Eq. (14.64), estimate the unknown physical forces from the estimated modal forces.

$$\underset{q\times 1}{\{F(\omega)\}} = \underset{q\times m}{[\phi]^{+T}}\underset{m\times 1}{\{f_r(\omega)\}} \tag{14.71}$$

It is desired that m>q to ensure over-determinacy.

Compared to the FRF-based method, the modal coordinate transformation method has the advantage that the pseudo-inverses need to be computed only twice, and the computed inverses are independent of the frequency. However, the method requires a prior modal analysis of the FRFs to obtain the modal model.

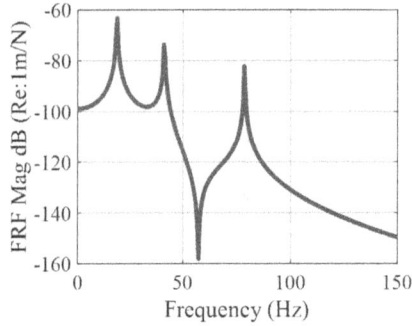

FIGURE 14.6   Measured FRF.

Example 14.1

A rotating machine is to be installed on a structure at a certain location (location 1). The machine's operating speed is 3,000 RPM, and it has a rotating unbalance of 0.002 kg-m. Predict the response at location 2 on the structure due to the unbalance force when the machine is installed. The receptance between these two locations is given in Figure 14.6.

Solution

Frequency of the force: $f = 3,000/60 = 50$ Hz

The amplitude of the force: $F_0 = me\omega^2 = 0.002 \times (2\pi 50)^2 = 197.39$ N
The FRF magnitude at 50 Hz from the plot is −116 dB.
The absolute value of FRF: $\alpha = 10^{-116/20} = 1.584 \times 10^{-6}$ m/N
The amplitude of the response: $X_0 = \alpha F_0 = 1.584 \times 10^{-6} \times 197.39 = 0.00031$ m

Example 14.2

The mass and stiffness matrices of a 2-DOF system are given below. An excitation force at a frequency of 49 Hz acts at the second DOF. Determine the stiffness to be added between the first DOF and ground to make the response zero at the second DOF.

MASS MATRIX :		STIFFNESS MATRIX :	
2	0	130,000	−80,000
0	5	−80,000	105,000

Solution

The response at the second DOF due to harmonic force at the second DOF can be made zero if the antiresonance frequency of FRF α_{22} coincides with the frequency of the harmonic force. Let stiffness k is added between the first DOF and the ground. The new stiffness matrix is [130,000 + k − 80,000; − 80,000 105,000].

Based on the discussion in Chapter 7, the antiresonance frequency of the FRF α_{22} can be obtained by solving the EVP in the matrices $[K]^{22}$ and $[M]^{22}$, which are obtained by deleting the second row and second column of $[K]$ and $[M]$. The EVP gives

$$(130,000 + k) - 2\omega^2 = 0$$

Substituting $\omega = 2\pi \times 49$, we get $k = 59,575$ N/m.

Example 14.3

The natural frequencies and mass-normalized mode shapes of a 3-DOF system are given below. A spring with a stiffness of 20,000 N/m is added between the second and third DOFs. Find the natural frequencies and mass-normalized mode shapes of the modified system by the local structural modification method. Use a MATLAB® program.

Natural frequencies in Hz	Mass-normalized mode shapes
5.94 13.02 24.86	[−0.3277 0.4272 −0.4584
	−0.5431 0.3844 0.7465
	−0.3501 −0.3490 −0.0750]

Solution

MATLAB program 14.1 is used to solve the problem.

```
%##############################          Results:
%   MATLAB PROGRAM 14.1
%##############################          NF_Hz:
clear all;
                                         33.3359   6.3546   17.4474
%DEFINE NATURAL FREQUENCIES
Nf_vec_Hz=[5.94 13.02 24.86];            Evec_Phy:

%SYSTEM DOF                              −0.2014  −0.2775   0.6184
N=length(Nf_vec_Hz);                      0.8878  −0.4520   0.0862
                                         −0.1808  −0.4005  −0.2386
%DEFINE EIGENVECTOR MATRIX
phi=[−0.3277   0.4272  −0.4584
    −0.5431   0.3844   0.7465
    −0.3501  −0.3490  −0.0750];

%DEFINE MASS MATRIX IN MODAL SPACE
m=eye(N,N);

%DEFINE STIFFNESS MODIFICATION MATRIX
k0=20000;
dk=zeros(N,N);
dk(2,2)=k0;
dk(2,3)=-k0;
dk(3,2)=-k0;
dk(3,3)=k0;

%DEFINE STIFFNESS MATRIX IN MODAL SPACE
k=lamda+phi'*dk*phi;

%SOLVE EVP
[Evec,Eval]=eig(k,m);

%FIND NATURAL FREQUENCY IN rad/sec
NF_rad_sec=sqrt(real(diag(Eval)));

%FIND NATURAL FREQUENCY IN Hz
NF_Hz=NF_rad_sec/(2*pi)

%FIND MODAL MASS MATRIX IN MODAL SPACE
mr=Evec.'*m*Evec;

%FIND NORMALIZED EIGENVECTORS
for p=1:N
  Evec_Nor(:,p)=Evec(:,p)/sqrt(mr(p,p));
end

%FIND EIGENVECTORS IN PHYSICAL SPACE
Evec_Phy=phi*Evec_Nor
%##############################
```

REVIEW QUESTIONS

1. Why is the modal model obtained by EMA said to be an incomplete model?
2. Is it possible to identify the spatial model from the modal model obtained by EMA?
3. Explain the principle of response simulation based on the experimental modal model.
4. What is the need for structural modification in the dynamic design of a system?
5. Why is adding lumped mass or stiffness said to be unit rank modifications?
6. What are the difficulties in predicting the effects of beam modification by the local modification method?
7. What is the need for force identification methods?
8. Why is force identification an inverse problem?
9. Explain the principle of force identification based on an experimental modal model of the system.
10. What is the fixed-interface method of coupled structural analysis? Can it be used to obtain the coupled models from the experimentally obtained modal models of the substructures?
11. What is the free-interface method of coupled structural analysis? Can it be used to obtain the coupled models from the experimentally obtained modal models of the substructures?

PROBLEMS

Problem 14.1: An electric motor mounted (at point 1) on a cantilever-type bracket is running at 1440 RPM and generates an excitation force due to unbalance. The RMS value of acceleration measured at some other point (point 2) on the bracket is 0.17 g. Figures 14.7 and 14.8 show the point-receptance at point 1 and cross-receptance between 1 and 2, respectively.

(a) Find the amplitude of the excitation force due to the motor. (b) Find the magnitude of the unbalance in the motor. (c) What is the RMS value of acceleration at the motor location?

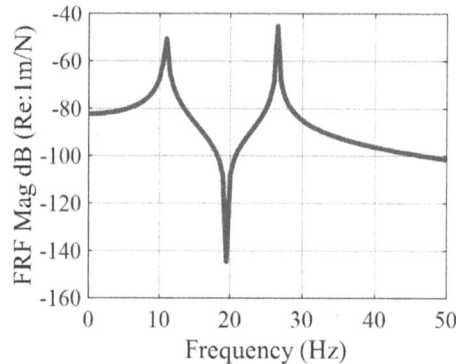

FIGURE 14.7 Point-receptance at point 1.

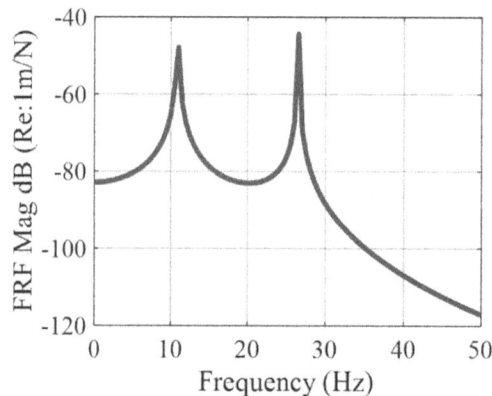

FIGURE 14.8 Cross-receptance between 1 and 2.

Problem 14.2: The natural frequencies and mass-normalized mode shapes of a 2-DOF system are given below. A mass of 1.2 kg is added at the first DOF. Find the natural frequencies and mass-normalized mode shapes of the modified system by the local structural modification method.

Natural frequencies in Hz	Mass-normalized mode shapes
10.12 24.63	
	[−0.5245 −0.4742
	−0.6706 0.7418]

Problem 14.3: The eigenvalue and mass-normalized eigenvector matrix of a 2-DOF system are given below. A spring with a stiffness of 6,400 N/m is added between the first and second DOFs. (a) Find the eigenvalue and mass-normalized eigenvector matrix of the modified system using the local structural modification method. (b) Can you justify the results?

Eigenvalue matrix:	Mass-normalized eigenvector matrix
[4,000 0	[−0.5 −0.5
0 20,000]	−0.5 0.5]

Problem 14.4: The mass and stiffness matrices of a 3-DOF system are given below. An excitation force at a frequency of 45.9 Hz acts at the first DOF. Determine the stiffness to be added between the third DOF and ground to make the response zero at the first DOF.

MASS MATRIX :			STIFFNESS MATRIX :		
2	0	0	260,000	−140,000	0
0	1	0	−140,000	220,000	−80,000
0	0	4	0	−80,000	180,000

15 Finite Element Model Validation and Updating

15.1 INTRODUCTION

The finite element method (FEM) is widely used for the modal and dynamic analysis of vibrating systems. This chapter presents another application of modal testing: finite element model validation and updating, which is undertaken to ensure that the analysis results and design decisions based on the FE model are reliable. We first discuss the need for FE model validation and look at some standard correlation tools used for this purpose. Model reduction and expansion methods are presented, which may be required to resolve the inconsistency between the FE model and modal test data. Then we look at the motivation for FE model updating and classify updating methods, which is followed by a description of direct and iterative updating methods based on modal and FRF data. Issues of parameter selection and ill-conditioning in model updating are discussed. Examples of model updating using simulated data are presented to demonstrate the steps involved in each method described in this chapter. In the end, the need for stochastic FE model updating is introduced.

15.2 NEED FOR FE MODEL VALIDATION

FEM is a well-established technique for the modeling and simulation of dynamic systems. We presented this technique in Chapter 3. An FE model is a spatial model of the system and offers the following advantages for modal analysis and dynamic design as compared to a test-based or experimental approach:

- The model can be as detailed and refined as necessary, with virtually no limit on the degree of the details incorporated and model size, both limited only by the available computational resources.
- It is easy to test the effect of any modifications in the system design or parameters, because the potential modifications can be directly incorporated into the model.
- Alternative designs can be evaluated and compared economically to achieve an optimal design.

However, before an FE model is used for modal analysis and design, it needs to be ascertained that the FE model predictions are accurate enough. Can there be any sources of modeling inaccuracies in a FE model that may make its predictions inaccurate? To answer this question, we look at various phases of the FE modeling process and see if any uncertainties might exist. We break down the whole process broadly into the following phases.

a. Solid modeling

This phase involves defining the system's geometry, and a solid model of the system is built. If it is an assembled structure, the individual parts are modeled, followed by their assembly. The geometry can generally be accurately modeled based on the available drawings, but uncertainties may arise if geometric simplifications or idealizations are made to model complex geometries.

b. Material properties and models

- The relevant material properties inputs to the model are generally based on the standard material databases. These values represent the statistical average estimates obtained through material testing, and variations are possible, leading to inaccuracies.
- Joints, such as bolted, welded, riveted, or adhesive, are approximated, in many cases, by simplified models to save modeling time and effort and reduce the solution time. Joints may affect the dynamics considerably, and simplifying their modeling may lead to inaccuracies.
- The modeling of the interfaces in the assemblies and the contact models employed may be inaccurate and could be the source of uncertainty.
- Accurate damping modeling is essential if the FE model has to predict the dynamic response. Often a proportional damping model is assumed and may not represent the damping characteristics accurately. The values of the damping proportionality constants also remain uncertain.

c. Boundary conditions

Often ideal boundary condition like fixed, free or simple support is assumed. The actual boundary condition may be different, leading to inaccuracies.

d. FE mesh

If the mesh is not sufficiently fine, it may lead to discretization errors. However, it can be avoided by ensuring that the mesh satisfies the convergence requirements and quality checks.

e. Model solution

The model solvers are well advanced, and the model can be solved accurately with the high computational power available.

To summarize, a structural dynamic FE model may have uncertainties related to modeling joints, interfaces, boundary conditions, damping, and material properties. The extent of the uncertainties, if any, varies from system to system, as different systems may have different degrees of complexity. For a given system, any given model is not unique; hence, the extent of the uncertainties may also differ in different models of the same system. Thus, the model needs to be checked to know whether it can be considered accurate enough for the task for which it is to be used. The process of confirming that the model can accurately predict the system's behavior for the intended application is called **model validation**.

In industrial practice, **verification and validation (V&V)** of a computational model are essential steps to be carried out before the model is used for its intended application. The verification of a model involves checking the correctness of the codes and algorithms used and ensuring that the discretization error is small. The verification is done to confirm that the computational model is a good numerical approximation of the continuous mathematical model of the physical system. The next step is the validation of the model. It is based on comparing the model's prediction with those obtained from the relevant tests conducted on the system.

15.3 REFERENCE DATA FOR FE MODEL VALIDATION

Reference data is required against which the model's predictions can be compared for FE model validation. EMA results can be used as reference data. It may be argued that the results of any experimental test, like EMA, generally have some measurement noise and, therefore, how can the results of an experimental test be used as reference data to check the accuracy of the predictions of an FE model?

While it is true that the results of EMA may have some measurement noise, the noise is random and can be reduced by statistical averaging. Since the test is conducted on the structure (or its prototype), it is an experimental observation of the actual dynamics and includes the effects of all the structural details, including joints, interfaces, material properties, boundary conditions, and damping. If precautions are taken to take error-free measurements, then the output of the EMA can be considered more accurate though less precise (due to any random noise) than the FE model predictions. The FE model predictions are precise but are likely to be less accurate due to the modeling inaccuracies that may occur, as discussed in Section 15.2. Given these observations, the modal test results are used as reference data for validating FE models for dynamic analysis and design.

15.4 FE MODEL CORRELATION

An objective comparison of the model and experimental results is necessary for FE model validation. FE model correlation is a comparison of the FE model predictions with the results of EMA on a sample or prototype of the structure. The comparison quantifies the degree to which the FE model predictions correlate with the corresponding data from the test. The correlation also indicates the error in the FE model predictions with the test data treated as a reference. Since the data from the EMA is available in the form of natural frequencies, mode shapes, damping factors, and FRFs, this data is computed from the FE model for correlation. In the following sections, we look at how the correlation can be quantified.

15.5 CORRELATION OF MODE SHAPES

15.5.1 PAIRING OF MODE SHAPES

An acceptable correlation between the FE model and experimental natural frequencies is critical in many situations. However, pairing the FE model and experimental mode shapes is necessary before the correlation between the natural frequencies can be quantified. It is because we need to compare the natural frequencies from the FE model and test, which correspond to the same mode of vibration of the system. Such mode pairs are called correlated mode pairs (CMPs).

Just listing the natural frequencies from the two sets in ascending orders may not guarantee the correct pairing for the following reasons:

- The FE model inaccuracies may shift various natural frequencies by different amounts, shuffling their order compared to the experiment. It is more likely to happen when close modes are there.
- Some modes in the experiment might be missing. For example, for a planar frame, the test might have been conducted to measure only the in-plane modes, while the FE model may be based on a model capturing even the out-of-plane modes. The other reason could be that one or more modes are not excited in the EMA and hence are unmeasured.

Therefore, the first step to enable a comparison of the modes is to perform the mode shape correlation between the test and FE model modes. An initial assessment of the correlation of the mode shapes can be done visually by comparing the plots or animations of the mode shapes from the two sets. But often, a quantification and an objective assessment of the correlation are desired, and this can be done using the mode shape correlation indices such as orthogonality check, pseudo-orthogonality check, or MAC, as discussed in the following sections.

15.5.2 ORTHOGONALITY CHECK

Mass or cross orthogonality check (Targoff [100]) is based on the fact that if the FE model closely represents the structure, then the measured mode shapes must be orthogonal to the analytical mass matrix. However, for applying this check, the coordinate incompleteness of the measured mode shapes needs to be resolved since the number of elements in measured mode shapes is generally less than the size of the FE model mass matrix. This requires that either the analytical mass matrix be reduced to the size of the measured mode shapes or the measured mode shapes be expanded to the size of the analytical mass matrix. (The mode shape expansion and model reduction are presented in the subsequent sections). Another issue is that the FE model is generally undamped or uses a proportional damping model, and hence the FE mode shapes are real, whereas the experimental mode shapes, in general, are complex. Hence, undamped/real mode shapes need to be estimated from the complex mode shapes to apply the orthogonality check. This process is often referred to as the realization of mode shapes.

If $[\phi]_A$ and $[\phi]_X$ represent the FE model (with the subscript A denoting the analytical/FE model) and measured mode shape matrices (with the subscript X denoting the experiment/ measured), respectively, and assuming that the measured mode shapes have been expanded and realized, then the cross-orthogonality matrix ($[COM]$) is calculated as

$$\underset{N\times m}{[COM]} = \underset{N\times N}{[\phi]_A^T} \; \underset{N\times N}{[M]} \; \underset{N\times m}{[\phi]_X} \tag{15.1}$$

If the correlation is perfect, and assuming that the two sets of the mode shapes are mass normalized, then the elements of COM would either be zero or one. A value of one indicates the corresponding analytical and experimental modes as correlated, forming a CMP. The orthogonality is generally not perfect, causing COM elements to deviate from one or zero. Therefore, the analytical-experimental mode pair corresponding to the element with the maximum magnitude in each column of the COM matrix may be taken as a CMP.

The pseudo-orthogonality check (Avitabile and O'callahan [8]) makes use of system equivalent reduction/expansion process (SEREP) to either expand the measured modes or reduce the mass matrix to implement the orthogonality check.

15.5.3 MODAL ASSURANCE CRITERION

Modal assurance criterion (MAC) (Allemang and Brown [3]) quantifies the correlation between an FE model and experimental mode shape as

$$\text{MAC}\left(\{\phi\}_A, \{\phi\}_X\right) = \frac{\left|\{\phi\}_A^H \{\phi\}_X\right|^2}{\left(\{\phi\}_A^H \{\phi\}_A\right)\left(\{\phi\}_X^H \{\phi\}_X\right)} \tag{15.2}$$

where the superscript 'H' represents Hermitian transpose. To calculate the MAC, we use only those DOFs in the FE mode shapes that correspond to the DOFs in the experimental mode shapes. The MAC value ranges from 0.0 to 1.0. A MAC value close to one indicates a good correlation, while a value close to zero indicates a poor correlation. MAC values between all pairs of FE and experimental mode shapes are computed and can be presented as a MAC matrix. The mode-shape pair with the highest MAC value in each column/row represents a CMP. A 3D plot of the MAC matrix visually shows the mode pairs that have a good correlation.

MAC is quite popular for mode shape correlation as no expansion or reduction is required, making it easier to compute and implement. Since MAC is not based on mass orthogonality, one consequence is that the MAC values for uncorrelated mode pairs would not generally be zero, which may sometimes make distinguishing between the correlated and uncorrelated mode pairs difficult.

15.5.4   Coordinate Modal Assurance Criterion

The coordinate modal assurance criterion (CoMAC) (Lieven and Ewins [58]) quantifies the correlation at a degree of freedom between all pairs of the correlated FE and experimental mode shapes. CoMAC at the jth DOF is defined as

$$\text{CoMAC}(j) = \frac{\sum_{r=1}^{m} |\phi_{A,jr} \phi_{X,jr}|^2}{\left(\sum_{r=1}^{m} (\phi_{A,jr})^2\right)\left(\sum_{r=1}^{m} (\phi_{X,jr})^2\right)} \tag{15.3}$$

where $\phi_{A,jr}$ and $\phi_{X,jr}$ represent the jth elements of the analytical and experimental mode shapes of the rth CMP. A plot of CoMAC against the degrees of freedom visually shows the DOFs having poor mode shape correlation.

15.6   CORRELATION OF THE NATURAL FREQUENCIES

Once the CMPs are identified from the mode shape correlation, the natural frequencies of the analytical and experimental modes in each mode pair can be compared, and the difference can be quantified. One method could be plotting the analytical natural frequencies against experimental ones. If all the points lie on a straight line at 45^0 from the origin, it represents a perfect correlation between the two sets of natural frequencies. The more the deviation of the points from this line, the higher the error in the analytical natural frequencies. The error in each analytical natural frequency can be quantified by finding either the absolute error or % error with the corresponding experimental natural frequency treated as the reference.

15.7   CORRELATION OF THE FRFs

One straightforward way of comparing the FRFs is to overlay the magnitude/phase of the corresponding analytical and experimental FRFs. The FRF correlation can also be quantified by calculating the error between the corresponding FRFs from the two sets or defining an assurance criterion (like MAC).

15.8   MODEL REDUCTION

The primary motivation in the development of the model reduction techniques was to reduce the computational time and cost of solving a large-size model. However, in the context of FE model correlation, model reduction may be required to achieve consistency between the order of the experimental mode shapes and the FE model matrices. Some FE model updating methods may also need the consistency of the two datasets, as we see in later sections. Model reduction achieves this by reducing the order of the analytical model to the order of the experimental mode shapes. In this section, we look at some popular model reduction methods.

15.8.1   Guyan Reduction

Guyan reduction ([41]) is a static reduction technique. It divides the total DOFs into two sets, master DOFs (represented by subscript 'm') and slave DOFs (represented by subscript 's'). The master DOFs are retained in the reduced model, while the slave DOFs are related to the master DOFs and are eliminated.

The equation of motion for the forced vibration of an undamped system is

$$[M]\{\ddot{x}\} + [K]\{x\} = \{F\} \tag{15.4}$$

Partitioning the displacement vector into the master and slave DOFs, we get

$$\{x\} = \left\{ \begin{array}{c} x_m \\ x_s \end{array} \right\} \tag{15.5}$$

Substituting Eq. (15.5) into Eq. (15.4) and partitioning the mass and stiffness matrices and force vector, we obtain

$$\begin{bmatrix} M_{mm} & M_{ms} \\ M_{sm} & M_{ss} \end{bmatrix} \begin{Bmatrix} \ddot{x}_m \\ \ddot{x}_s \end{Bmatrix} + \begin{bmatrix} K_{mm} & K_{ms} \\ K_{sm} & K_{ss} \end{bmatrix} \begin{Bmatrix} x_m \\ x_s \end{Bmatrix} = \begin{Bmatrix} F_m \\ F_s \end{Bmatrix} \qquad (15.6)$$

The equilibrium equation corresponding to the slave DOFs is

$$[M_{sm}]\{\ddot{x}_m\} + [M_{ss}]\{\ddot{x}_s\} + [K_{sm}]\{x_m\} + [K_{ss}]\{x_s\} = \{F_s\} \qquad (15.7)$$

If the slave DOFs are chosen such that the external forces at these DOFs are zero, then $\{F_s\} = \{0\}$. If it is also assumed that the inertia forces at the slave DOFs are negligible, then Eq. (15.7) reduces to

$$[K_{sm}]\{x_m\} + [K_{ss}]\{x_s\} \cong \{0\} \qquad (15.8)$$

or,

$$\{x_s\} = -[K_{ss}]^{-1}[K_{sm}]\{x_m\} \qquad (15.9)$$

Eq. (15.9) relates the displacement at the slave DOFs to those at the master DOFs. Substituting $\{x_s\}$ from Eq. (15.9) into Eq. (15.5), we get

$$\{x\} = \begin{bmatrix} [I] \\ -[K_{ss}]^{-1}[K_{sm}] \end{bmatrix} \{x_m\} = [T]\{x_m\} \qquad (15.10)$$

where $[T]$ is the transformation matrix between $\{x\}$ and $\{x_m\}$ given by

$$[T] = \begin{bmatrix} [I] \\ -[K_{ss}]^{-1}[K_{sm}] \end{bmatrix} \qquad (15.11)$$

Substituting Eq. (15.10) into (15.4) and pre-multiplying by $[T]^T$, we get

$$[M_R]\{\ddot{x}_m\} + [K_R]\{x_m\} = \{F_R\} \qquad (15.12)$$

where $[M_R]$ and $[K_R]$ are the reduced mass and stiffness matrices given by

$$[M_R] = [T]^T[M][T]$$
$$[K_R] = [T]^T[K][T] \qquad (15.13)$$

Since the external forces at the slave DOFs are zero, the force vector for the reduced model simplifies to

$$\{F_R\} = [T]^T\{F\} = \{F_m\} \qquad (15.14)$$

We see that Eq. (15.8), which leads to the transformation matrix, represents the equilibrium of the elastic forces at the slave DOFs. Since it is also the static equilibrium equation for these DOFs, the Guyan reduction is also called the static reduction/condensation technique. It should be noted that while the static condensation is an exact reduction for a statics problem, it is not exact for a dynamics problem, as the inertia forces are neglected. Therefore, the natural frequencies and mode shapes of the reduced model are not the same as that of the original model; the higher the frequency, the higher the difference.

15.8.2 Dynamic Reduction

Dynamic reduction (Paz [84]) is an improvement over the Guyan reduction technique. In this technique, the transformation matrix is derived such that it produces an exact response at the chosen frequency ω_0. If we consider Eq. (15.7), the equilibrium equation corresponding to the slave DOFs with no external forces at these DOFs, and consider the inertia forces at a frequency ω_0, then we get

$$-\omega_0^2[M_{sm}]\{X_m\} - \omega_0^2[M_{ss}]\{x_s\} + [K_{sm}]\{x_m\} + [K_{ss}]\{x_s\} = \{0\} \tag{15.15}$$

It yields

$$\{x_s\} = -\left([K_{ss}] - \omega_0^2[M_{ss}]\right)^{-1}\left([K_{sm}] - \omega_0^2[M_{sm}]\right)\{x_m\} \tag{15.16}$$

Substituting $\{x_s\}$ from Eq. (15.16) into Eq. (15.5), we obtain

$$\{x\} = \begin{bmatrix} [I] \\ -\left([K_{ss}] - \omega_0^2[M_{ss}]\right)^{-1}\left([K_{sm}] - \omega_0^2[M_{sm}]\right) \end{bmatrix}\{x_m\} = [T]\{x_m\} \tag{15.17}$$

where $[T]$ is the transformation matrix given by

$$[T] = \begin{bmatrix} [I] \\ -\left([K_{ss}] - \omega_0^2[M_{ss}]\right)^{-1}\left([K_{sm}] - \omega_0^2[M_{sm}]\right) \end{bmatrix} \tag{15.18}$$

Substituting $[T]$ from Eq. (15.18) into Eq. (15.13) gives the reduced matrices as per the dynamic reduction technique. We note that,

- if ω_0 is zero, then the transformation matrix in the dynamic reduction reduces to that in the Guyan reduction.
- the model reduction is exact at ω_0, but inexact at other frequencies as the inertia forces are not accounted for exactly.

15.8.3 System Equivalent Reduction/Expansion Process (SEREP)

SEREP (O'Callahan et al. [78]) is a reduction technique that preserves in the reduced model the natural frequencies and mode shapes of the selected modes of the original model. This characteristic of the method is an advantage over the Guyan and dynamic reduction methods.

The equation of motion for the free vibration of an MDOF undamped system is

$$[M]\{\ddot{x}\} + [K]\{x\} = \{0\} \tag{15.19}$$

Let us say 'p' number of system modes is to be preserved in the reduced model. Consider transformation to modal coordinates using these p modes

$$\{x\}_{N\times1} = [\phi]_{N\times p}\ \{y\}_{p\times1} \tag{15.20}$$

where $[\phi]$ is the matrix of the p mode-shapes and $\{y\}$ is the vector of p modal coordinates. Let 'a' and 'd' denote the numbers of master DOFs (the DOFs in the reduced model) and slave DOFs (the DOFs eliminated), respectively. Partitioning the displacement vector and mode shape matrix in Eq. (15.20) into the master and slave DOFs, we get

$$\begin{Bmatrix} x_m \\ x_s \end{Bmatrix} = \begin{bmatrix} \phi_m \\ \phi_s \end{bmatrix}\{y\} \tag{15.21}$$

The equation corresponding to the master DOFs can be written as

$$\{x_m\}_{a\times1} = [\phi_m]_{a\times p}\ \{y\}_{p\times1} \tag{15.22}$$

From Eq. (15.22), the estimate of the modal coordinates is

$$\{y\}_{p\times1} = [\phi_m]^+_{p\times a}\ \{x_m\}_{a\times1} \tag{15.23}$$

where $[\phi_m]^+$ is the pseudo-inverse of $[\phi_m]$.

Equation (15.23) forms the basis for model reduction. This equation expresses the modal response in p modes in terms of the physical responses at master DOFs. The response at slave DOFs is now obtained using the estimated modal responses. Substituting Eq. (15.23) into Eq. (15.20), we get

$$\underset{N\times 1}{\{x\}} = \underset{N\times p}{[\phi]} \underset{p\times a}{[\phi_m]^+} \underset{a\times 1}{\{x_m\}} = \underset{N\times a}{[T]} \underset{a\times 1}{\{x_m\}} \tag{15.24}$$

where $[T]$ is the transformation matrix given by

$$\underset{N\times a}{[T]} = \underset{N\times p}{[\phi]} \underset{p\times a}{[\phi_m]^+} = \underset{N\times p}{\begin{bmatrix} \phi_m \\ \phi_s \end{bmatrix}} \underset{p\times a}{[\phi_m]^+} \tag{15.25}$$

Substituting Eq. (15.24) into Eq. (15.19) and pre-multiplying by $[T]^T$, we get

$$\underset{a\times a}{[M_R]}\{\ddot{x}_m\} + \underset{a\times a}{[K_R]}\{x_m\} = \{0\} \tag{15.26}$$

where $[M_R] = [T]^T[M][T]$ and $[K_R] = [T]^T[K][T]$ are the reduced mass and stiffness matrices. Equation (15.22) should be overdetermined for a meaningful estimate of the modal responses and hence $p \le a$. It is also noted that while the static and dynamic condensation methods don't require the solution of the eigenvalue problem of the complete system model, the SEREP requires that to compute the reduced model matrices.

15.9   MODE SHAPE EXPANSION

The mode shape expansion involves expanding the size of the experimental mode shapes to the size of the analytical model matrices. The SEREP can also be used for mode shape expansion. The method requires only the analytical mode shapes for expansion. The basic idea of the method is to determine the linear combination of the analytical mode shapes that best represents the experimental mode shape and then use that combination of the analytical mode shapes to obtain the unmeasured part of the experimental mode shape.

Let us use the subscripts 'm' and 'u' to denote the parts of the mode shape corresponding to measured and unmeasured DOFs. Therefore, the experimental mode shape after expansion, including the measured and unmeasured parts, is denoted as

$$\{\phi\}_X = \begin{Bmatrix} \phi_m \\ \phi_u \end{Bmatrix}_X \tag{15.27}$$

Expression of the rth experimental mode shape at the measured DOFs by a linear combination of the 'p' analytical mode shapes can be written as

$$\{\phi_m\}_X = \sum_{j=1}^{p} c_j \{\phi_m\}_{A,j} = [\phi_m]_A \{c\} \tag{15.28}$$

The coefficient vector is estimated as

$$\{c\} = [\phi_m]_A^+ \{\phi_m\}_X \tag{15.29}$$

The unmeasured part of the experimental mode shape is obtained by linearly combining, using $\{c\}$, the 'p' analytical mode shapes corresponding to the unmeasured DOFs. Hence,

$$\{\phi_u\}_X = [\phi_u]_A \{c\} = [\phi_u]_A [\phi_m]_A^+ \{\phi_m\}_X \tag{15.30}$$

Therefore, the expanded r^{th} experimental mode shape is given by

$$\{\phi\}_X = \begin{Bmatrix} \{\phi_m\}_X \\ \{\phi_u\}_X \end{Bmatrix} = \begin{bmatrix} [I] \\ [\phi_u]_A [\phi_m]_A^+ \end{bmatrix} \{\phi_m\}_X \tag{15.31}$$

15.10 FE MODEL UPDATING

We studied in the previous sections that a structural dynamic FE model needs to be validated, which requires quantifying the correlation between the FE and modal test data. We also looked at some correlation measures and indices to quantify the correlation.

Once the correlation is quantified, the next step is to check whether the correlation meets the criterion of acceptable correlation. The application for which the FE model is to be used needs to be considered while deciding the acceptable level of correlation. For example, suppose the purpose is to predict the response to a broadband excitation. In that case, the limit on the acceptable error in the predicted natural frequencies need not be so stringent since the predicted response is expected to be less sensitive to the perturbation in the natural frequencies. But suppose the objective is to predict the possibility of the resonant operation of a system with periodic forces acting on the system. In that case, the limit on the acceptable error in the predicted natural frequencies must be stringent to enable accurate prediction of the resonant condition.

If the FE model satisfies the criterion of the acceptable correlation, then the FE model can be used with confidence for the intended application. But, if it doesn't, then what could be the remedy? If the EMA has been carried out with due precautions, avoiding any errors, then the unacceptable error in the correlation can be attributed to the inaccuracies in the FE model. We have discussed the possible sources of inaccuracies in an FE model in Section 15.2. Therefore, the question arises is it possible to improve the FE model predictions by addressing these inaccuracies in any way? This question has been addressed extensively in the last four/five decades leading to the development of several techniques to improve the correlation of FE model with modal test data. It essentially involves adjusting the chosen parameters of the model to improve the model's correlation with the test data, and this process is known as **Finite element model updating** or **model updating**. Model updating is also referred to as **Model adjustment, Model tuning, Model improvement, or Model calibration**.

Figure 15.1 shows the process of FE model updating via a flowchart. FE model updating is typically carried out as part of the design cycle of a product or a system. An FE model of the system can be constructed based on an initial design. EMA is conducted on a prototype of the system to obtain the reference data for model validation. If the correlation between the FE model predictions and test results is not acceptable, then the FE model is updated to make the correlation acceptable.

15.11 UPDATING PARAMETERS SELECTION

Once it is established that the FE model needs to be updated to improve its correlation with the test data, the next question is what parameters to update in the model. The direct matrix updating methods don't allow a choice of the parameters to be updated; hence, what to update is not an issue with those methods. In practice generally iterative updating methods are used as they permit a choice of the parameters to be updated, and with these methods, the parameters to be updated needs to be decided.

Ideally, the FE model should be updated to reduce the modeling inaccuracies. If the measured data is complete and the updating parameters represent all the modeling inaccuracies, then the updating can potentially reduce the modeling inaccuracies and correctly identify the parameters. However, many challenges are faced in parameter selection, as discussed below.

- The test data is generally incomplete and limited, which constraints the number of parameters that can be adjusted. Model updating is an inverse problem since the unknown parameters of the system are identified from the knowledge of the output data. The problem must be posed as an overdetermined problem to ensure more information about the system than the number of unknown system parameters. An overdetermined problem helps to adjust the parameters to their realistic and meaningful values. The constraint on the number of parameters prohibits parameterizing all the perceived sources of modeling inaccuracies, small or big, adversely affecting the quality of the updated model.

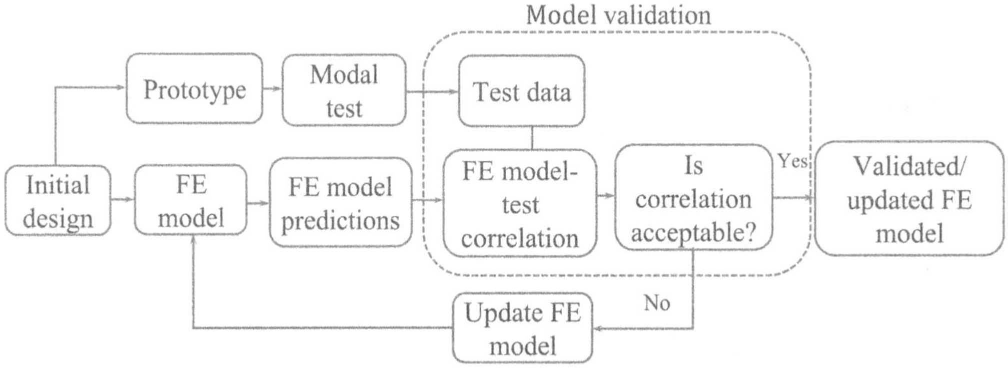

FIGURE 15.1 FE model updating.

- The modeling inaccuracies in a given model are unknown. Some error localization methods are developed, and their success is demonstrated for specific problems. But no reliable methods are available applicable to a general problem to identify or localize the inaccuracies.
- In many cases, suitable parameters associated with the perceived modeling inaccuracy may not exist. The model may have to be modified so that the missing feature can be updated.
- There is no way to verify the correctness of the selected parameters. If the selected parameters contain some parameters which don't need updating, then updating may distort the model.

Due to the above constraints, parameter sensitivities (which is the rate of change of data to be correlated with the parameters) are also used in parameter selection. Updating sensitive parameters needs smaller changes to achieve the desired level of correlation, ensuring that the model is perturbed less from the initial model. However, it has the undesirable consequence that the sensitive parameters may be adjusted more than those in error. Therefore, to ensure physically meaningful adjustment during updating, the strategy should be to choose those parameters that represent the modeling inaccuracies and are also sensitive enough to the data to be correlated. It also needs to be ensured that no two or more parameters have an identical or nearly identical effect on the data, as this makes the identification problem ill-conditioned. In such cases, eliminating or grouping such parameters can be explored. Thus, combined use of the error localization methods, correlation indices, knowledge of the parameter sensitivities to the data, and engineering judgment about the perceived modeling errors in the model of a given system must be made for parameter selection.

Physical parameters of the model are widely used as updating parameters as the changes can be physically interpreted and related to the modeling inaccuracies. Some of the commonly used parameters for updating are:

- material properties (such as modulus of elasticity, shear modulus, Poisson's ratio, and density)
- geometric properties (such as thickness, width, length, area of cross-section, second moment of area)
- flexural rigidity, torsional rigidity
- boundary conditions and joints (modeled by lumped linear and torsional springs and dashpots)
- lumped masses to account for uncertain masses
- damping matrix proportionality constants
- mass, stiffness, and damping submatrix correction factors
- joint parameters, offset parameters

In some cases, the distribution of the % error in various natural frequencies may give a clue about the parameters that might be in error in the FE model. For example, in beams and plates, the error in some global parameters like the material property (say, modulus of elasticity or density or both) may have a uniform effect on different natural frequencies. But the localized modeling inaccuracies, like those associated with joints and boundary conditions, may have a nonuniform effect.

15.12   CLASSIFICATION OF MODEL UPDATING METHODS

The model updating methods can be classified as given below.

a. Direct and iterative methods: The research in model updating was initiated in the aircraft industry during the 1970s and 80s to have accurate models to meet stringent demands on safety and performance. The majority of the development in this period was based on the formulation of the problem as a minimization problem in which the entries in the mass and stiffness matrices were the candidates that could be directly adjusted. Hence, this class of methods is known as direct matrix updating methods.

Iterative updating methods are based on an alternative formulation in which an index of correlation between the FE model and test data is minimized. Most methods linearize the index to obtain linear equations to be solved. Due to linear approximation, the parameters are iteratively updated. The parameters to be updated can be chosen in these methods, which provides better control over the changes made during updating.

b. The direct matrix updating methods are based on the modal data. But the iterative methods can also be classified as those based on modal or FRF data. The FRF-based iterative methods can be further classified as equation error or output error methods.

c. Updating methods can also be classified as those that update only the stiffness and mass matrices and those that also update the damping matrix. The first type of method ideally requires undamped modal or FRF data, whereas the second type is based on complex modal or FRF data.

d. The methods updating all three matrices, mass, stiffness, and damping, can also be classified as either single-stage or two-stage methods. The first type of method updates all three matrices in a single stage, whereas the second type updates mass and stiffness matrices in the first stage, followed by updating the damping matrix in the second stage.

e. Model updating methods can also be classified based on the approach used to solve the updating problem, i.e., using a local or global optimization approach. Methods using genetic algorithms, swarm optimization, and neural networks have also been developed.

15.13   DIRECT MATRIX UPDATING

In the direct matrix updating, the elements of the uncertain matrices are changed optimally by treating the data assumed to be correct as a reference. Several direct methods with different choices of reference data are developed. In this section, we consider one such method (Berman and Nagy [14]) in detail. The method consists of two steps, and Sections 15.13.1 and 15.13.2 present updating formulation in these two steps, respectively. The symbols used are given in Table 15.1.

15.13.1   DIRECT MATRIX UPDATING OF THE MASS MATRIX

In the first step of the method, the mass matrix is updated, treating measured eigenvectors as reference data. Due to modeling errors, the measured eigenvectors are generally not orthogonal to the analytical mass matrix. So, let

$$\underset{m \times N}{[\phi]^{T}} \ \underset{N \times N}{[M]} \ \underset{N \times m}{[\phi]} = \underset{m \times m}{\left[\bar{M}\right]} \tag{15.32}$$

But, the measured eigenvectors must be orthogonal to the updated mass matrix, and hence

$$\underset{m \times N}{[\phi]^{T}} \ \underset{N \times N}{\left[M^{U}\right]} \ \underset{N \times m}{[\phi]} = \underset{m \times m}{[I]} \tag{15.33}$$

which, taking note of the symbols used, can also be written as

$$\underset{m \times N}{[\phi]^{T}} \ \underset{N \times N}{[\Delta M]} \ \underset{N \times m}{[\phi]} = \underset{m \times m}{[I]} - \underset{m \times m}{\left[\bar{M}\right]} \tag{15.34}$$

The direct mass matrix updating problem is formulated by seeking a minimum change in the mass matrix subject to the constraint that the measured eigenvectors are orthogonal to the updated mass matrix (given by Eq. 15.34). Thus, the constrained minimization problem is

$$\text{Minmimise}: \ J = [W][\Delta M][W]$$

$$\text{subject to}: \ \ [\phi]^{T}[\Delta M] \ [\phi] - [I] + \left[\bar{M}\right] = [0] \tag{15.35}$$

where W is a weighting matrix and is taken as

$$[W] = [M]^{-1/2} \tag{15.36}$$

TABLE 15.1
Symbols Used in Direct Matrix Updating

N: DOF of the FE/Analytical Model	m: Number of Measured Modes
$\underset{N \times N}{[M]}$: Analytical mass matrix	$\underset{N \times N}{[K]}$: Analytical stiffness matrix
$\underset{N \times N}{\left[M^{U}\right]}$: Updated analytical mass matrix	$\underset{N \times N}{\left[K^{U}\right]}$: Updated analytical stiffness matrix
$[\Delta M] = \left[M^{U}\right] - [M]$ (Change in the mass matrix due to updating)	$[\Delta K] = \left[K^{U}\right] - [K]$ (Change in the stiffness matrix due to updating)
$\underset{N \times m}{[\phi]}$: Matrix of the measured eigenvectors expanded to the size of the analytical matrices	$\underset{m \times m}{[\Lambda]}$: Diagonal matrix of the measured eigenvalues
$\dfrac{\partial}{\partial[A]}$: Represents partial derivative with each element of the matrix $[A]$	

The equivalent unconstrained problem is obtained by defining the Lagrangian, which is the sum of the objective function and all the constraints multiplied by the corresponding Lagrange multipliers. Therefore, we have

$$\text{Minimize}: L = J + Q = \sum_{k=1}^{N} \sum_{j=1}^{m} \left(\left([W][\Delta M][W]\right)_{kj} \right)^2 + \sum_{k=1}^{N} \sum_{j=1}^{m} \lambda_{kj} \left([\phi]^T [\Delta M] \, [\phi] - [I] + [\bar{M}] \right)_{kj} \qquad (15.37)$$

The unknowns in the problem (15.37) are elements of $[\Delta M]$ and the Lagrange multipliers λ_{kj}. The necessary conditions for the minimum of L are that the first-order partial derivatives of L with all unknowns are zero. For ease in presenting the equations, we find the derivatives of J and Q separately. The partial derivative of J with $[\Delta M]$ is obtained as follows.

$$\frac{\partial J}{\partial \Delta M_{rs}} = \frac{\partial}{\partial \Delta M_{rs}} \left(\sum_{k=1}^{N} \sum_{j=1}^{m} \left(\left([W][\Delta M][W]\right)_{kj} \right)^2 \right) \qquad (15.38)$$

$$\frac{\partial J}{\partial \Delta M_{rs}} = \frac{\partial}{\partial \Delta M_{rs}} \left(\sum_{k=1}^{N} \sum_{j=1}^{m} \left(\sum_{i=1}^{N} W_{ki} \sum_{p=1}^{N} \Delta M_{ip} W_{pj} \right)^2 \right) \qquad (15.39)$$

$$\frac{\partial J}{\partial \Delta M_{rs}} = \sum_{k=1}^{N} \sum_{j=1}^{m} 2 W_{rk}^{T} \left([W][\Delta M][W]\right)_{kj} W_{js}^{T} \qquad (15.40)$$

Hence, we get

$$\frac{\partial J}{\partial [\Delta M]} = 2[W]^{T}[W][\Delta M][W][W]^{T} \qquad (15.41)$$

Due to Eq. (15,36), $[W]^{T}[W] = [W][W]^{T} = [M]^{-1}$, and hence

$$\frac{\partial J}{\partial [\Delta M]} = 2[M]^{-1}[\Delta M][M]^{-1} \qquad (15.42)$$

Now, we find the partial derivative of Q with $[\Delta M]$.

$$\frac{\partial Q}{\partial \Delta M_{rs}} = \frac{\partial}{\partial \Delta M_{rs}} \left(\sum_{k=1}^{N} \sum_{j=1}^{m} \lambda_{kj} \left([\phi]^T [\Delta M] \, [\phi] - [I] + [\bar{M}] \right)_{kj} \right) \qquad (15.43)$$

$$\frac{\partial Q}{\partial \Delta M_{rs}} = \frac{\partial}{\partial \Delta M_{rs}} \left(\sum_{k=1}^{N} \sum_{j=1}^{m} \lambda_{kj} \left(\left(\sum_{i=1}^{N} \phi_{ki}^{T} \sum_{p=1}^{N} \Delta M_{ip} \phi_{pj} \right) - I_{kj} + \bar{M}_{kj} \right) \right) \qquad (15.44)$$

$$\frac{\partial Q}{\partial \Delta M_{rs}} = \sum_{k=1}^{N} \sum_{j=1}^{m} \phi_{rk} \lambda_{kj} \phi_{js}^{T} \qquad (15.45)$$

$$\frac{\partial Q}{\partial [\Delta M]} = [\phi] \, [\lambda] \, [\phi]^{T} \qquad (15.46)$$

Using the results in Eqs. (15.42) and (15.46), we get the first necessary condition as

$$\frac{\partial L}{\partial [\Delta M]} = 2[M]^{-1}[\Delta M][M]^{-1} + [\phi] \, [\lambda] \, [\phi]^{T} = [0] \qquad (15.47)$$

Now, we find the partial derivatives with $[\lambda]$.

$$\frac{\partial J}{\partial [\lambda]} = [0] \tag{15.48}$$

$$\frac{\partial Q}{\partial \lambda_{rs}} = \frac{\partial}{\partial \lambda_{rs}} \left(\sum_{k=1}^{N} \sum_{j=1}^{m} \lambda_{kj} \left([\phi]^T [\Delta M] [\phi] - [I] + [\bar{M}] \right)_{kj} \right) \tag{15.49}$$

$$\frac{\partial Q}{\partial \lambda_{rs}} = \left([\phi]^T [\Delta M] [\phi] - [I] + [\bar{M}] \right)_{rs} \tag{15.50}$$

$$\frac{\partial Q}{\partial [\lambda]} = [\phi]^T [\Delta M] [\phi] - [I] + [\bar{M}] \tag{15.51}$$

Using the results in Eqs. (15.48) and (15.51), we obtain the second necessary condition

$$\frac{\partial L}{\partial [\lambda]} = [\phi]^T [\Delta M] [\phi] - [I] + [\bar{M}] = [0] \tag{15.52}$$

The matrix equations (15.47) and (15.52) can be solved to find the unknowns. From Eq. (15.47), we get

$$[\Delta M] = -\frac{1}{2} [M] [\phi] [\lambda] [\phi]^T [M] \tag{15.53}$$

Substituting Eq. (15.53) into Eq. (15.52) and using Eq. (15.32) yields

$$[\lambda] = -2 [\bar{M}]^{-1} \left([I] - [\bar{M}] \right) [\bar{M}]^{-1} \tag{15.54}$$

Substituting Eq. (15.54) into Eq. (15.53), we get the solution for $[\Delta M]$. The updated mass matrix is given by

$$\left[M^U \right] = [M] + [M] [\phi] [\bar{M}]^{-1} \left([I] - [\bar{M}] \right) [\bar{M}]^{-1} [\phi]^T [M] \tag{15.55}$$

15.13.2 Direct Matrix Updating of the Stiffness Matrix

In the second step, the stiffness matrix is updated using the knowledge of the updated mass matrix. This step of the Berman and Nagy method is the same as that followed by Baruch and Bar-Itzhack [10].

The direct stiffness matrix updating problem is formulated by seeking a minimum change in the stiffness matrix subject to the constraint that the updated stiffness and mass matrices satisfy the eigenvalue problem for all the measured eigenvalues and eigenvectors. In addition, the updated stiffness matrix must be symmetric. Thus, the constrained minimization problem is defined as

$$\text{Minimise}: \quad J = \frac{1}{2} [W][\Delta K][W]$$

$$\text{subject to}: \quad \left[K^U \right] [\phi] = \left[M^U \right] [\phi] [\Lambda] \tag{15.56}$$

$$\left[K^U \right] = \left[K^U \right]^T$$

where W is a weighting matrix and is taken as

$$[W] = \left[M^U \right]^{-1/2} \tag{15.57}$$

Since the FE model stiffness matrix is symmetric, the symmetricity constraint of the updated stiffness matrix is equivalent to

$$[\Delta K] = [\Delta K]^T \tag{15.58}$$

We construct the equivalent unconstrained problem by defining the Lagrangian as

$$\text{Minimize}: L = J + Q + G = \sum_{k=1}^{N}\sum_{j=1}^{m}\frac{1}{2}\left(([W][\Delta K][W])_{kj}\right)^2 + 2\sum_{k=1}^{N}\sum_{j=1}^{m}\lambda_{kj}\begin{pmatrix}[\Delta K][\phi]+[K]\;[\phi]\\-[M^U][\phi][\Lambda]\end{pmatrix}_{kj} + \sum_{k=1}^{N}\sum_{j=1}^{m}\eta_{kj}\left([\Delta K]-[\Delta K]^T\right)_{kj}$$

(15.59)

The unknowns in the problem (15.59) are elements of $[\Delta K]$ and the Lagrange multipliers λ_{kj} and λ_{kj}. The necessary conditions for the minimum of L are that the first-order partial derivatives of L with all unknowns are zero. For ease of presenting the equations, we first find the derivatives of J, Q, and G separately. The partial derivative of J with $[\Delta K]$ is obtained as follows.

$$\frac{\partial J}{\partial \Delta K_{rs}} = \frac{\partial}{\partial \Delta K_{rs}}\left(\sum_{k=1}^{N}\sum_{j=1}^{m}\frac{1}{2}\left(([W][\Delta K][W])_{kj}\right)^2\right)$$

(15.60)

$$\frac{\partial J}{\partial \Delta K_{rs}} = \frac{\partial}{\partial \Delta K_{rs}}\left(\sum_{k=1}^{N}\sum_{j=1}^{m}\frac{1}{2}\left(\sum_{i=1}^{N}W_{ki}\sum_{p=1}^{N}\Delta K_{ip}W_{pj}\right)^2\right)$$

(15.61)

$$\frac{\partial J}{\partial \Delta K_{rs}} = \sum_{k=1}^{N}\sum_{j=1}^{m}W_{rk}^T\left([W][\Delta K][W]\right)_{kj}W_{js}^T$$

(15.62)

Hence, we get

$$\frac{\partial J}{\partial[\Delta K]} = [W]^T[W][\Delta K][W][W]^T$$

(15.63)

Due to Eq. (15,57), $[W]^T[W]=[W][W]^T=[M^U]^{-1}$, and hence

$$\frac{\partial J}{\partial[\Delta K]} = [M^U]^{-1}[\Delta K][M^U]^{-1}$$

(15.64)

Now, we find the partial derivative of Q with $[\Delta K]$.

$$\frac{\partial Q}{\partial \Delta K_{rs}} = \frac{\partial}{\partial \Delta K_{rs}}\left(\sum_{k=1}^{N}\sum_{j=1}^{m}2\lambda_{kj}\begin{pmatrix}[\Delta K][\phi]+[K]\;[\phi]\\-[M^U][\phi][\Lambda]\end{pmatrix}_{kj}\right)$$

(15.65)

$$\frac{\partial Q}{\partial \Delta K_{rs}} = \sum_{j=1}^{m}2\lambda_{rj}\phi_{js}^T$$

(15.66)

$$\frac{\partial Q}{\partial[\Delta K]} = 2\;[\lambda][\phi]^T$$

(15.67)

Now, we find the partial derivative of G with $[\Delta K]$.

$$\frac{\partial G}{\partial \Delta K_{rs}} = \frac{\partial}{\partial \Delta K_{rs}}\left(\sum_{k=1}^{N}\sum_{j=1}^{m}\Delta_{kj}\left([\Delta K]-[\Delta K]^T\right)_{kj}\right)$$

(15.68)

$$\frac{\partial G}{\partial \Delta K_{rs}} = \frac{\partial}{\partial \Delta K_{rs}}\left(\sum_{k=1}^{N}\sum_{j=1}^{m}\eta_{kj}\left(\Delta K_{kj}-\Delta K_{kj}^T\right)\right)$$

(15.69)

$$\frac{\partial G}{\partial \Delta K_{rs}} = \eta_{rs}-\eta_{rs}^T$$

(15.70)

$$\frac{\partial G}{\partial[\Delta K]} = [\eta] - [\eta]^{T} \tag{15.71}$$

Using the results in Eqs. (15.64), (15.67) and (15.71), we get the first necessary condition as

$$\frac{\partial L}{\partial[\Delta K]} = [M^{U}]^{-1}[\Delta K][M^{U}]^{-1} + 2[\lambda][\phi]^{T} + [\eta] - [\eta]^{T} = [0] \tag{15.72}$$

Regarding the partial derivatives with $[\lambda]$, we note that the derivative of only Q is non-zero, which is obtained as

$$\frac{\partial Q}{\partial \lambda_{rs}} = \frac{\partial}{\partial \lambda_{rs}} \left(\sum_{k=1}^{N} \sum_{j=1}^{m} \lambda_{kj} \left(\left[K^{U} \right][\phi] - \left[M^{U} \right][\phi][\Lambda] \right)_{kj} \right) \tag{15.73}$$

$$\frac{\partial Q}{\partial \lambda_{rs}} = \left(\left[K^{U} \right][\phi] - \left[M^{U} \right][\phi][\Lambda] \right)_{rs} \tag{15.74}$$

$$\frac{\partial Q}{\partial[\lambda]} = \left[K^{U} \right][\phi] - \left[M^{U} \right][\phi][\Lambda] \tag{15.75}$$

Therefore, the second necessary condition is

$$\frac{\partial L}{\partial[\lambda]} = \left[K^{U} \right][\phi] - \left[M^{U} \right][\phi][\Lambda] = [0] \tag{15.76}$$

Regarding the partial derivatives with $[\eta]$, we note that the derivative of only G is non-zero, which is obtained as

$$\frac{\partial G}{\partial \eta_{rs}} = \frac{\partial}{\partial \eta_{rs}} \left(\sum_{k=1}^{N} \sum_{j=1}^{m} \eta_{kj} \left([\Delta K] - [\Delta K]^{T} \right)_{kj} \right) \tag{15.77}$$

$$\frac{\partial G}{\partial \eta_{rs}} = \frac{\partial}{\partial \eta_{rs}} \left(\sum_{k=1}^{N} \sum_{j=1}^{m} \eta_{kj} \left(\Delta K_{kj} - \Delta K_{kj}^{T} \right) \right) \tag{15.78}$$

$$\frac{\partial G}{\partial \eta_{rs}} = \Delta K_{rs} - \Delta K_{rs}^{T} \tag{15.79}$$

$$\frac{\partial G}{\partial[\eta]} = [\Delta K] - [\Delta K]^{T} \tag{15.80}$$

Therefore, the third necessary condition is

$$\frac{\partial L}{\partial[\eta]} = [\Delta K] - [\Delta K]^{T} = [0] \tag{15.81}$$

Condition (15.81) is nothing but the symmetricity constraint of the change in the stiffness matrix. The corresponding Lagrange multipliers must also satisfy certain constraints to ensure they are unique. It can be shown that these constraints are

$$[\eta] = -[\eta]^{T} \tag{15.82}$$

Thus, we have the matrix equations (15.72), (15.76), and (15.82) that need to be solved to find the unknowns. We substitute $[\eta]$ from Eq. (15.82) into (15.72) and solve for $[\eta]$. Substituting that back into (15.82) and noting that $[\Delta K] = \left[K^{U} \right] - [K]$, we get

$$\left[K^{U} \right] = [K] - \left[M^{U} \right] \left([\lambda] [\phi]^{T} + [\phi] [\lambda]^{T} \right) \left[M^{U} \right] \tag{15.83}$$

Post-multiplying Eq. (15.83) by $[\phi]$ and making use of Eqs. (15.76) and (15.33), we obtain

$$\left[M^U\right][\phi][\Lambda] = [K][\phi] - \left[M^U\right][\lambda] - \left[M^U\right][\phi]\ [\lambda]^T\left[M^U\right][\phi] \tag{15.84}$$

Note that

$$[\lambda]^T\left[M^U\right][\phi] = [\phi]^T\left[M^U\right][\lambda] \tag{15.85}$$

and

$$\left([I]+[\phi]\ [\phi]^T\left[M^U\right]\right)^{-1} = [I] - \frac{1}{2}[\phi]\ [\phi]^T\left[M^U\right] \tag{15.86}$$

Making use of Eqs. (15.84), (15.85) and (15.86), $[\lambda]$ is obtained as

$$[\lambda] = \left[M^U\right]^{-1}[K][\phi] - \frac{1}{2}[\phi][\Lambda] - \frac{1}{2}[\phi][\phi]^T[K][\phi] \tag{15.87}$$

Substituting Eq. (15.87) into Eq. (15.83), we get the expression for the updated stiffness matrix as

$$\left[K^U\right] = [K] - [K][\phi]\ [\phi]^T\left[M^U\right] - \left[M^U\right][\phi]\ [\phi]^T[K] + \left[M^U\right][\phi]\ [\phi]^T[K][\phi]\ [\phi]^T\left[M^U\right] + \left[M^U\right][\phi][\Lambda]\ [\phi]^T\left[M^U\right] \tag{15.88}$$

15.13.3 ADVANTAGES AND LIMITATIONS OF THE DIRECT MATRIX UPDATING METHODS

Some of the advantages and limitations of the direct matrix updating methods are as follows.

- Direct matrix updating gives a closed-form solution for the updated matrices. Hence it is easy to implement and computationally not expensive.
- It is a one-step procedure and doesn't require iterations. Consequently, there is no issue of convergence.
- The method exactly reproduces the measured data used for updating. Thus, it gives an updated model with a perfect correlation with the measured test data.
- The method, however, alters the individual entries in the mass and stiffness matrices. Even the zero entries may alter and become non-zero after updating, so the method may not preserve the connectivity of different DOFs.
- The adjustments made to the individual entries in the mass and stiffness matrices are difficult to interpret physically or relate to the inaccuracies or changes in the structure.
- The method requires the incompleteness of the measured data to be addressed either by expanding the measured mode shapes or by reducing the FE model.
- Since the correlation with the test data is achieved by a numerical adjustment of the structural matrices, which may not be related to the modeling inaccuracies, the updated model cannot be used confidently to predict the effects of any structural modification in the design.

Example 15.1: Updating with the Direct Method

In this example, we update an undamped model of the simulated cantilever structure (500mm × 50mm × 3mm) we have considered in earlier chapters, using the direct method described in the previous section. The objective of this and other examples in this chapter is to present the basic steps to be followed and highlight any relevant issues. Since it is a simulated study, we use another FE model of the structure to generate 'experimental data' that, in practice, would be measured through a modal test on the structure.

  a. The experimental structure (shown in Figure 15.2) is simulated with the following properties.
   - The model has ten beam elements and 11 nodes.
   - The fixed end, in reality, may not have ideal fixed boundary conditions. For simulation, it is assumed that while the displacement DOF is fixed, the rotational DOF has some flexibility represented by a torsional spring of stiffness $K_t = 2,500\ N-m/rad$.

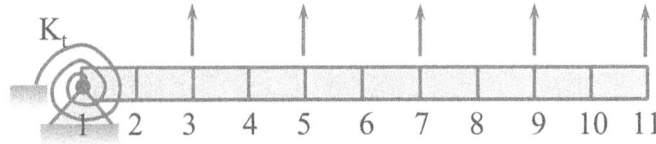

FIGURE 15.2 Simulated experimental structure.

TABLE 15.2

Comparison of the Measured and FE Model Natural Frequencies and MAC Values before and after Updating Using the Direct Method

		FE Model			
	Measured Natural	Before Updating		After Updating	
Mode No.	Frequency (Hz)	Natural Frequency (Hz)	MAC Value	Natural Frequency (Hz)	MAC Value
1	9.02	9.61	0.9999	9.02	1.0
2	56.64	60.27	0.9999	56.64	1.0
3	158.86	168.91	0.9998	158.86	1.0
4	311.90	331.42	0.9998	311.90	1.0
5	516.98	549.01	0.9998	549.00	0.9999
6	775.26	822.86	0.9983	822.85	0.9983

- Other properties are: $E = 1.8 \times 10^{11} \text{N/m}^2$, $\rho = 7,800 \text{ kg/m}^3$. Proportional viscous damping (with the damping matrix proportional to the stiffness matrix) is simulated with stiffness proportionality constant $\beta_k = 0.00006$.
- Transverse FRFs over 0–400 Hz frequency range are measured at 5 points, i.e., at alternate nodes of the model (shown by the arrows in Figure 15.2).

b. The initial FE model has ten finite elements with the node at the clamped end fixed. The other data are $E = 2 \times 10^{11} \text{N/m}^2$ (i.e. the nominal value) and $\rho = 7,800 \text{ kg/m}^3$.

c. Four measured eigenvalues and the corresponding eigenvectors are used for updating. Since the simulated experimental structure has proportional damping, the measured eigenvectors are real. (The measured eigenvectors, if complex, need to be first realized.)

d. Updating is carried out as follows.
- The order of each measured eigenvector is 5×1, while the DOF of the FE model is 22. The measured eigenvectors are expanded to the size of the FE model using the SEREP method to resolve the coordinate incompleteness. After the expansion, the measured eigenvector matrix $([\phi])$ is of order 22×4.
- The FE model mass matrix is updated using $[\phi]$ via Eq. (15.55).
- Using $[\phi]$, the updated FE model mass matrix, and the measured eigenvalue matrix $([\Lambda]$, which is a diagonal matrix of order 4×4), the FE model stiffness matrix is updated using Eq. (15.88).

e. Table 15.2 compares the measured and FE model natural frequencies and MAC values before and after updating. A comparison of modes 5 and 6 not used in updating is also shown. It is seen that the updated model exactly reproduces four measured natural frequencies used in updating. The MAC values of the corresponding eigenvectors are also 1.0, indicating a perfect correlation. However, no improvement in the predicted natural frequencies and MAC values of modes 5 and 6, the modes not used in updating, is seen.

f. Since the experimental data is simulated, we can also compare the changes in the mass and stiffness matrices due to updating. Figure 15.3 shows a comparison of $([M_X] - [M_A])$ (which is the actual error in the FE model mass matrix) and $([M_U] - [M_A])$ (which is the change in the FE model mass matrix due to updating). There was no error in the mass matrix, but the updating has changed many elements of the mass matrix. Figure 15.4 shows a comparison of $([K_X] - [K_A])$ and $([K_U] - [K_A])$, and it is observed that the adjustments in the stiffness matrix are also inconsistent with the errors in the stiffness matrix.

g. This example shows that the direct updating method reproduces the measured modal data used in updating by numerical adjustment of the mass and stiffness matrices, but the changes made are not consistent with the inaccuracies in the model. Since the modeling inaccuracies are not reduced, the updated model cannot be used with confidence to accurately predict the measured data not used in updating or the effects of any design modifications.

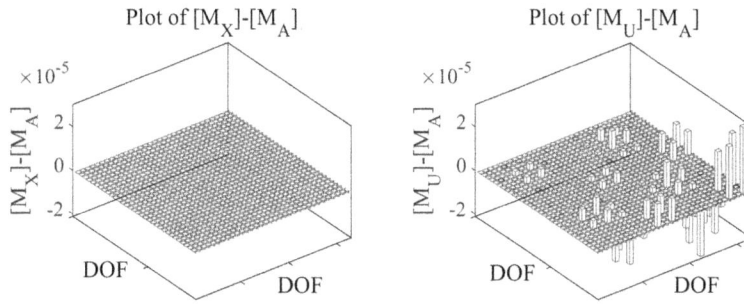

FIGURE 15.3 Comparison of $([M_X]-[M_A])$ and $([M_U]-[M_A])$.

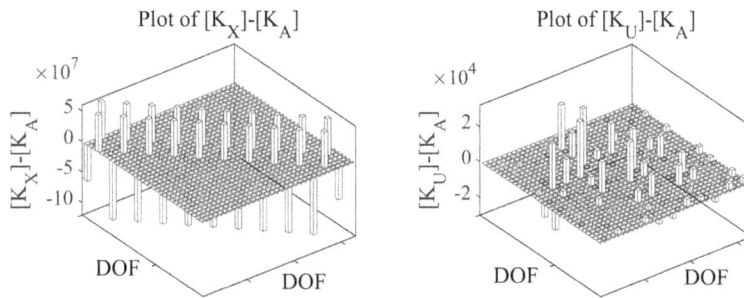

FIGURE 15.4 Comparison of $([K_X]-[K_A])$ and $([K_U]-[K_A])$.

15.14 ITERATIVE MODEL UPDATING USING MODAL DATA

In this method, the uncertain parameters of the FE model are updated to minimize the difference between the measured and FE model modal data. Thus, in this method, the structural matrices are not changed directly, as done in the direct method, but what is changed are the selected parameters which in turn change the structural matrices. The updating is based on linearizing the relationship between the modal data and updating parameters. Due to the approximate nature of the equations, the parameter estimates need to be revised iteratively until a prefixed convergence criterion is satisfied.

In this section, we present the formulation of the method to update the mass and stiffness parameters. The updating can be carried out either using only the eigenvalues or using both the eigenvalues and eigenvectors, but we present a general formulation using both these quantities. The measured eigenvalues and eigenvectors are assumed to be real or obtained by the realization of the complex modal data. The equations should be formulated with complex modal data if the damping parameters also need to be updated.

15.14.1 Basic Formulation

Let the measured eigenvalues and eigenvectors corresponding to 'm' correlated mode pairs (CMP)s are used for updating. Let each measured eigenvector has 'n' elements. Let there are q parameters to be updated and the vector of updating parameters is given by $\{p\} = \{p_1, \ p_2, ..., \ p_q\}^T$. All the symbols used in this method are shown in Table 15.3.

Analytical eigenvalues and eigenvectors are a nonlinear function of the updating parameters. Let $\{p\}^{k-1}$ be the estimate of the parameter vector at the end of the $(k-1)^{th}$ iteration. If $\{\Delta p\}^k$ represent the change in the parameter vector estimated in the kth iteration, then $\{p\}^{k-1}$ is updated to a new estimate $\{p\}^k$ at the end of the kth iteration

$$\{p\}^k = \{p\}^{k-1} + \{\Delta p\}^k \qquad (15.89)$$

TABLE 15.3

Symbols Used in the Iterative Method of Model Updating Using Modal Data

N: DOF of the analytical/FE model	m: Number of measured modes
n: Number of measured coordinates	q: Number of updating parameters
$\lambda_{X,i}$: i^{th} experimental eigenvalue	$\lambda_{A,i}$: i^{th} analytical eigenvalue
$\{\phi\}_{X,i}$: i^{th} experimental eigenvector	$\{\phi\}_{A,i}$: i^{th} analytical eigenvector
$\{p\}$: Parameter vector	$\{\Delta p\}$: Vector of changes in parameters
Suffixes 'X' and 'A': Experiment and analytical/FE model, respectively	Superscript 'k': Iteration number
$[M]_{N \times N}$: Analytical mass matrix	$[K]_{N \times N}$: Analytical stiffness matrix
$[S]$: Sensitivity matrix	$\{\Delta z\}$: Vector of difference between the experimental and analytical model eigendata

The updated analytical eigenvalue is obtained by linearizing the relationship between the eigenvalue and updating parameters. Therefore, the ith eigenvalue at the end of the kth iteration is

$$\lambda_{A,i}^{\ k} \approx \lambda_{A,i}^{\ k-1} + \sum_{j=1}^{q} \left(\frac{\partial \lambda_{A,i}}{\partial p_j} \bigg|_{\{p\}=\{p\}^{k-1}} \right) \Delta p_j^{\ k} \tag{15.90}$$

Equating the linearized estimate of the updated i^{th} eigenvalue (RHS of Eq. (15.90)) to the corresponding measured eigenvalue ($\lambda_{X,i}$), we get

$$\lambda_{X,i} = \lambda_{A,i}^{\ k-1} + \sum_{j=1}^{q} \left(\frac{\partial \lambda_{A,i}}{\partial p_j} \bigg|_{\{p\}=\{p\}^{k-1}} \right) \Delta p_j^{\ k} \tag{15.91}$$

Proceeding similarly as above gives the following equation for the ith eigenvector:

$$\{\phi\}_{X,i} = \{\phi\}_{A,i}^{\ k-1} + \sum_{j=1}^{q} \left(\frac{\partial \{\phi\}_{A,i}}{\partial p_j} \bigg|_{\{p\}=\{p\}^{k-1}} \right) \Delta p_j^{\ k} \tag{15.92}$$

For brevity, the reference to the iteration number is dropped in the equations below. Eqs. (15.91) and (15.92) can be written for all the m eigenvalues and eigenvectors and can be combined to obtain the following matrix equation:

$$\begin{Bmatrix} \lambda_{X,1} - \lambda_{A,1} \\ \{\phi\}_{X,1} - \{\phi\}_{A,1} \\ \vdots \\ \lambda_{X,m} - \lambda_{A,m} \\ \{\phi\}_{X,m} - \{\phi\}_{A,m} \end{Bmatrix} = \begin{bmatrix} \dfrac{\partial \lambda_{A,1}}{\partial p_1} & \dfrac{\partial \lambda_{A,1}}{\partial p_2} & \cdots & \dfrac{\partial \lambda_{A,1}}{\partial p_q} \\ \dfrac{\partial \{\phi\}_{A,1}}{\partial p_1} & \dfrac{\partial \{\phi\}_{A,1}}{\partial p_2} & \cdots & \dfrac{\partial \{\phi\}_{A,1}}{\partial p_q} \\ \vdots & \vdots & \cdots & \vdots \\ \dfrac{\partial \lambda_{A,m}}{\partial p_1} & \dfrac{\partial \lambda_{A,m}}{\partial p_2} & \cdots & \dfrac{\partial \lambda_{A,m}}{\partial p_q} \\ \dfrac{\partial \{\phi\}_{A,m}}{\partial p_1} & \dfrac{\partial \{\phi\}_{A,m}}{\partial p_2} & \cdots & \dfrac{\partial \{\phi\}_{A,m}}{\partial p_q} \end{bmatrix} \begin{Bmatrix} \Delta p_1 \\ \Delta p_2 \\ \vdots \\ \Delta p_q \end{Bmatrix} \tag{15.93}$$

The derivatives/sensitivities of the eigenvalues and eigenvectors appearing on the RHS of Eq. (15.93) can be calculated using the following expressions:

$$\frac{\partial \lambda_i}{\partial p_j} = \{\phi\}_i^T \left[\frac{\partial [K]}{\partial p_j} - \lambda_i \frac{\partial [M]}{\partial p_j} \right] \{\phi\}_i \tag{15.94}$$

$$\frac{\partial \{\phi\}_i}{\partial p_j} = -\frac{1}{2} \{\phi\}_i \{\phi\}_i^T \frac{\partial [M]}{\partial p_j} \{\phi\}_i + \sum_{r=1,\neq i}^{N} \{\phi\}_r \left(\frac{\{\phi\}_r^T \left[\frac{\partial [K]}{\partial p_j} - \lambda_i \frac{\partial [M]}{\partial p_j} \right] \{\phi\}_i}{\lambda_i - \lambda_r} \right) \tag{15.95}$$

Eq. (15.93) can be represented symbolically as

$$\underset{(1+n)m \times q}{[S]} \underset{q \times 1}{\{\Delta p\}} = \underset{(1+n)m \times 1}{\{\Delta z\}} \tag{15.96}$$

Eq. (15.96) represents simultaneous linear equations. Vector $\{\Delta p\}$ is the unknown and represents the parameter changes at the current iteration. The solution of the equations depends on whether the number of equations, which is $(1+n)m$, is more than, less than, or equal to q.

- If $(1+n)m < q$, then the equations are underdetermined and represent the situation where the amount of experimental data is less than the number of unknowns. In this case, theoretically, an infinite number of combinations of parameter changes exist to make the RHS vector zero. But these parameter changes may not correctly represent the parameter inaccuracies in the model.
- If $(1+n)m = q$, then the matrix $[S]$ is square, and if it has full rank, then a unique solution to Eq. (15.96) exists.
- If $(1+n)m > q$, (and assuming that the row rank of $[S]$ is more than its column rank) then the equations are overdetermined and represent a situation where the amount of experimental data is more than the number of unknowns. In this case, though there is no exact solution to the equations, it is the preferred case. We can find the least-squares solution by minimizing the residual error of the equations. The least-squares solution provides the estimate of parameters that gives the best possible correlation with the experimental data used in setting up the equations. The parameter estimates are expected to be physically meaningful (provided the choice of the parameters was appropriate). The over-determinacy of the equations also makes the parameter estimation process robust against any noise in the measured data.

Assuming overdetermined equations, and if $[S]$ is not rank-deficient, then pre-multiplying Eq. (15.96) by $[S]^T$ and then by the inverse of $[S]^T [S]$ gives the least-squares solution

$$\{\Delta p\} = \left([S]^T [S] \right)^{-1} [S]^T \{\Delta z\} \tag{15.97}$$

This method is also called the inverse eigen-sensitivity method (IESM), as it involves the inverse of the eigendata sensitivity matrix. The steps to perform model updating using this method are as follows:

1. Choose the measured data to be used and the parameters to be updated. Choose the initial values of the parameters $(\{p\}^k)$.
2. Build the FE model with current values of the updating parameters and get matrices $[M]$ and $[K]$.
3. Solve the eigenvalue problem based on $[M]$ and $[K]$. It gives $\lambda_{A,i}$ and $\{\phi\}_{A,i}$ for $i = 1, 2, .., m$.
4. Find the derivatives of the eigenvalues and eigenvectors using Eqs. (15.94) and (15.95).
5. Set up $[S]$ and $\{\Delta z\}$.
6. Obtain $\{\Delta p\}$ (representing $\{\Delta p\}^{k+1}$) using Eq. (15.97).
7. Obtain the updated values of the parameters: $\{p\}^{k+1} = \{p\}^k + \{\Delta p\}^{k+1}$.
8. Check the convergence criterion. It could be related to the tolerance in the correlation error, the parameter vector change over successive iterations, or some other suitable measure.
9. If the convergence criterion is not satisfied, then go to step 2. Otherwise, stop and obtain the updated mass and stiffness matrices using the latest parameter values.

15.14.2 Normalization of the Updating Equations

The condition of the sensitivity matrix in Eq. (15.96) affects the least-squares solution in Eq. (15.97).

a. The updating parameters with widely different orders of magnitude may adversely affect the condition of the sensitivity matrix. Normalization of the parameters can be used to improve the conditioning by defining the normalized parameters as

$$\Delta u_j^{\,k} = \frac{\Delta p_j^{\,k}}{p_j^{\,k-1}} \tag{15.98}$$

In Eq. (15.98) $p_j^{\,k-1}$ is the jth parameter value at the start of the kth iteration, $\Delta p_j^{\,k}$ is the change in the jth parameter estimated in the kth iteration and $\Delta u_j^{\,k}$ is the corresponding normalized value of the change. $\Delta u_j^{\,k}$ indicates a fractional correction or fractional change in the parameter.

b. It is noticed that the vector $\{\Delta z\}$ consists of the absolute error in eigenvalues and eigenvectors. The eigenvalue error is generally significantly more than the eigenvector error. As a result, the updating is influenced more by the eigenvalue error than the eigenvector error. For balancing the equations, the eigenvalue equation corresponding to each eigenvalue is divided by the corresponding eigenvalue.

Taking into account the modifications in (a) and (b), Eq. (15.93) can be written as

$$
\begin{Bmatrix}
\dfrac{\lambda_{X,1}-\lambda_{A,1}}{\lambda_{A,1}} \\[2mm]
\{\phi\}_{X,1}-\{\phi\}_{A,1} \\[2mm]
\cdots \\[2mm]
\dfrac{\lambda_{X,m}-\lambda_{A,m}}{\lambda_{A,m}} \\[2mm]
\{\phi\}_{X,m}-\{\phi\}_{A,m}
\end{Bmatrix}
=
\begin{bmatrix}
p_1 \cdot \dfrac{\partial \lambda_{A,1}}{\partial p_1} / \lambda_{A,1} & p_2 \cdot \dfrac{\partial \lambda_{A,1}}{\partial p_2} / \lambda_{A,1} & \cdots & p_q \cdot \dfrac{\partial \lambda_{A,1}}{\partial p_q} / \lambda_{A,1} \\[2mm]
p_1 \cdot \dfrac{\partial \{\phi\}_{A,1}}{\partial p_1} & p_2 \cdot \dfrac{\partial \{\phi\}_{A,1}}{\partial p_2} & \cdots & p_q \cdot \dfrac{\partial \{\phi\}_{A,1}}{\partial p_q} \\[2mm]
\cdots & \cdots & \cdots & \cdots \\[2mm]
p_1 \cdot \dfrac{\partial \lambda_{A,m}}{\partial p_1} / \lambda_{A,m} & p_2 \cdot \dfrac{\partial \lambda_{A,m}}{\partial p_2} / \lambda_{A,m} & \cdots & p_q \cdot \dfrac{\partial \lambda_{A,m}}{\partial p_q} / \lambda_{A,m} \\[2mm]
p_1 \cdot \dfrac{\partial \{\phi\}_{A,m}}{\partial p_1} & p_2 \cdot \dfrac{\partial \{\phi\}_{A,m}}{\partial p_2} & \cdots & p_q \cdot \dfrac{\partial \{\phi\}_{A,m}}{\partial p_q}
\end{bmatrix}
\begin{Bmatrix}
\Delta u_1 \\[2mm]
\Delta u_2 \\[2mm]
\cdots \\[2mm]
\cdots \\[2mm]
\Delta u_q
\end{Bmatrix}
\tag{15.99}
$$

Symbolically, Eq. (15.99) is written as

$$\underset{(1+n)m \times q}{[S']} \ \underset{q \times 1}{\{\Delta u\}} = \underset{(1+n)m \times 1}{\{\Delta z'\}} \tag{15.100}$$

Thus, Eq. (15.100) can be used in place of Eq. (15.96).

Let $\Delta u_j^{\,1}$, $\Delta u_j^{\,2}$,... $\Delta u_j^{\,k}$ be the fractional changes in the jth parameter at the end of first, second, and k^{th} iterations, respectively. Suppose $u_j^{\,k}$ is the cumulative change at the end of the kth iteration, defined as a fraction of the initial parameter value. Then $u_j^{\,k}$ can be calculated as

$$u_j^{\,k} = (1 + \Delta u_j^{\,1}) \cdot (1 + \Delta u_j^{\,2}) \ldots (1 + \Delta u_j^{\,k}) - 1 \tag{15.101}$$

15.14.3 Formulation Based on Minimization Problem

The solution to the linearized updating problem can also be derived through a minimization problem. From Eq. (15.96), we define the residual error vector as

$$\{e\} = \{\Delta z\} - [S]\{\Delta p\} \tag{15.102}$$

Let us also incorporate a weighting matrix $([W])$ in the formulation to choose a relative weighing of different data to reflect the relative importance of different data. One approach to choose the weighting factors could be to take them inversely proportional to the noise or the variance in the individual data. The objective function is defined as

$$J = \{e\}^T [W]\{e\} \tag{15.103}$$

Substituting Eq.(15.102) into Eq.(15.103) and simplifying, we get

$$J = \{\Delta z\}^T [W]\{\Delta z\} - \{\Delta z\}^T [W][S]\{\Delta p\} - \{\Delta p\}^T [S]^T [W]\{\Delta z\} + \{\Delta p\}^T [S]^T [W][S]\{\Delta p\} \tag{15.104}$$

J is a quadratic function of the unknown vector $\{\Delta p\}$. For the minimum value of J, its partial derivative with each parameter must be zero. We rewrite Eq. (15.104) as

$$J = \{\Delta z\}^T [W]\{\Delta z\} - \sum_{j=1}^{q}\left(\{\Delta z\}^T [W][S]\right)_{1j} \Delta p_{j1} - \sum_{j=1}^{q}\Delta p_{1j}{}^T \left([S]^T [W]\{\Delta z\}\right)_{j1} + \sum_{j=1}^{q}\Delta p_{1j}{}^T \sum_{i=1}^{q}\left([S]^T [W][S]\right)_{ji} \Delta p_{i1} \tag{15.105}$$

Taking the partial derivative with Δp_{r1} (which is the rth element of the column vector $\{\Delta p\}$, with subscript '1' indicating the only column in the vector), we get

$$\frac{\partial J}{\partial \Delta p_{r1}} = 0 - \left(\{\Delta z\}^T [W][S]\right)_{r1}^T - \left([S]^T [W]\{\Delta z\}\right)_{r1} + \sum_{i=1}^{q}\left([S]^T [W][S]\right)_{ri} \Delta p_{i1} + \sum_{j=1}^{q}\Delta p_{1j}{}^T \left([S]^T [W][S]\right)_{jr} \tag{15.106}$$

which yields

$$\frac{\partial J}{\partial\{\Delta p\}} = -[S]^T [W]\{\Delta z\} - [S]^T [W]\{\Delta z\} + [S]^T [W][S]\{\Delta p\} + [S]^T [W][S]\{\Delta p\} \tag{15.107}$$

Imposing the first-order necessary condition, $\dfrac{\partial J}{\partial\{\Delta p\}} = 0$, and simplifying, we obtain

$$\{\Delta p\} = \left([S]^T [W][S]\right)^{-1} [S]^T [W]\{\Delta z\} \tag{15.108}$$

We note that if we apply equal weighting to all the measured data, i.e. $[W] = \alpha[I]$, then Eq. (15.108) reduces to the solution (15.97).

15.14.4 ADVANTAGES OF ITERATIVE MODEL UPDATING USING MODAL DATA

- The method allows the choice of updating parameters. This flexibility gives control over the scope of the changes during updating. The user can judiciously choose the parameters to reflect those aspects of the model that are perceived to be in error. This way, only physically meaningful corrections in the model can be ensured.
- The method can be used with incomplete data without needing either the model reduction or mode shape expansion because the updating equation (Eq. 15.93) can be set up directly with incomplete data.

15.14.5 MODEL UPDATING USING UNCORRELATED MODES

It is noted that the inverse eigen-sensitivity method requires correlated pairs of experimental and FE model modes, as the difference between the corresponding eigendata is minimized by the method. It requires establishing correlated mode pairs, using some correlation index like MAC, and only those experimental modes that could be paired with FE model modes can be used for updating. However, it is possible that not all of the experimental modes are correlated with FE model modes, and there may be some uncorrelated experimental modes. Some of the situations that may lead to this state are listed below [68].

a. An experimental mode shape may have comparable MAC values for more than one FE model mode shape making it difficult to form a CMP reliably.
b. An experimental mode shape may have a relatively higher correlation with a particular FE model mode shape, however, the absolute MAC value for this pair might be small, indicating that this might not be a genuinely correlated pair.
c. Presence of close modes in a structure may make experimental modal analysis difficult and adversely affect the possibility of reliably establishing CMPs.
d. Structures that have higher levels of damping may also make modal analysis difficult and therefore may lead to a situation where CMPs cannot be reliably established.
e. Presence of excessive noise in the identified modes shapes may also make establishing correlation difficult. The excessive noise in mode shapes may be due to the excessive noise in the measured frequency response functions (FRFs) or may be due to the poor quality of the modal analysis of the measured FRFs.
f. Another possibility could be when the modeling error in the FE model is so high that mode shapes predicted by the FE model are too much different than those identified through a modal test on the structure resulting in many uncorrelated modes.

The consequence is that such uncorrelated modes cannot be used in updating even when they form valid known information about the structure. This is a disadvantage since it reduces the quantity of experimental data available for model updating.

Modak [68] proposed a method for model updating using uncorrelated modes in which an objective function is defined, allowing both correlated as well as uncorrelated modes to be included in the updating process. The method requires prescribing a correlation matrix between FE and experimental modes. Through this matrix, a list of experimental modes can be prescribed for every uncorrelated FE mode, among which lies a mode with which the FE mode is likely to correlate. The minimization of the objective function converges to the correct FE-experimental mode pairs, which otherwise are unknown at the start of the updating. The idea of model updating using uncorrelated modes is further extended in [69], in which the IESM is modified to include uncorrelated modes in the updating process.

Example 15.2: Iterative Model Updating Using Modal Data

In this example, we update an undamped model of the cantilever structure using the IESM described in sections 15.14.1 and 15.14.2.

a. The experimental data simulated in Example 15.1 is used as reference data.
b. The initial FE model to be updated has ten finite elements with $E = 2 \times 10^{11} N/m^2$ and $\rho = 7,800 \ kg/m^3$ and a torsional spring at the clamped end with stiffness $K_t = 4166.7 \ N - m/rad$.
c. Updating parameters and data for updating: The material property and the boundary condition are in error. We take E and K_t as updating parameters. We use 4 measured eigenvalues for updating. No weighting matrix is considered.
d. We build the FE model matrices $[M]$ and $[K]$ with the current values of the updating parameters and solve the eigenvalue problem to obtain $\lambda_{A,i}$ and $\{\phi\}_{A,i}$.
e. To set up the updating equations, the CMPs need to be known. To identify CMPs, we compute the MAC values (Eq. 15.2) between the measured and FE model eigenvectors. Figure 15.5 shows the 3D bar plot of the computed MAC matrix between the 4 measured and 6 FE model eigenvectors. Based on the highest MAC value in each row/column, we identify the CMPs as (1,1), (2,2), (3,3), and (4,4). We assume that the CMPs do not change during updating.
f. Updating is carried out using the following steps.
   - The derivatives of the eigenvalues and eigenvectors are found using Eqs. (15.94) and (15.95) and $[S]$ and $\{\Delta z\}$ are set up. We use fractional corrections to parameters (defined by Eq. 15.98) as unknowns and hence modify the matrix $[S]$ accordingly (as discussed in Section 15.14.2).
   - The least-squares solution of the updating equations (four equations and two unknowns) is found using the MATLAB® function 'lsqr()'. The solution yields $\{\Delta u\}$, from which $\{\Delta p\}$ is found.
   - The parameter vector is updated: $\{p\}^{k+1} = \{p\}^k + \{\Delta p\}^k$.
   - We check the convergence criterion. If the norm of $\{\Delta u\}$ is less than the tolerance (a value of 0.001 is used here), we stop and treat the FE model based on the updated parameter vector as the updated FE model. Otherwise, we repeat the above steps till the convergence criterion is satisfied.
g. Results: Figure 15.6 shows the convergence of the updating parameters over iterations. At convergence, the vector of cumulative fractional corrections to the parameters is $\{u\} = \{-0.1 \ -0.4\}^T$. It indicates a reduction of E and K_t by 10% and 40% respectively. The updated values of the parameters are $E = 1.8 \times 10^{11} N/m^2$ and $K_t = 2,500 \ N - m/rad$, which are identical to their values in the simulated experimental structure. Since the parameters in error are updated exactly, the eigenvalues and eigenvectors of the FE model also match exactly with the measured values.

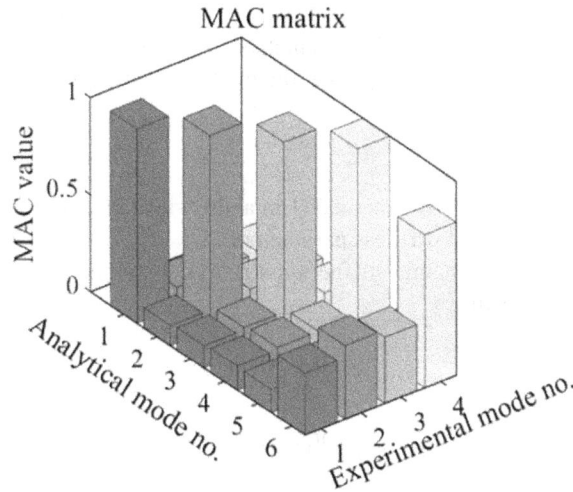

FIGURE 15.5 3D bar plot of the computed MAC matrix.

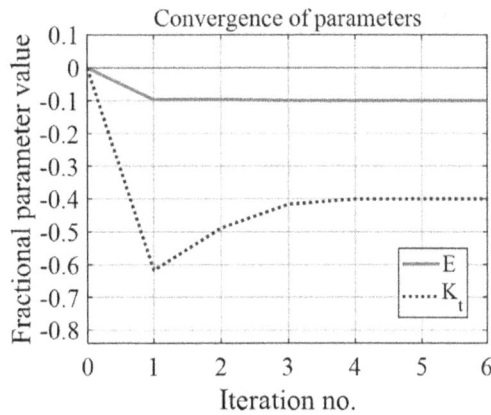

FIGURE 15.6 Convergence of the updating parameters over iterations.

15.15 ITERATIVE MODEL UPDATING USING FRF DATA

Frequency response functions (FRFs) provide another alternative as reference data for model updating. The FRFs and modal data are related, and theoretically, there is no particular advantage to preferring either. But the following considerations may influence the choice:

- The modal data is estimated from the curve-fitting of the FRFs. Curve fitting errors, if any, may adversely affect the estimated modal data accuracy and hence the updating process. It may be more of a concern in close modes and high-damping cases.
- An FRF contains the contribution of modes in the analysis frequency range and those outside it. In comparison, the modal data contains information about only the modes in the analysis frequency range.
- The use of FRF data requires the selection of frequency points for updating.
- Acceptable correlation criteria are often in terms of modal data correlation, like natural frequency and mode shape correlation. From this perspective, updating with modal data gives direct control over the desired level of correlation.

The FRF-based updating is formulated using either the equation or the output error. The equation error method is formulated by defining the error in the equation of motion in the frequency domain. As a result, the final equation contains the dynamic stiffness matrix. This offers the advantage that the final equation is related linearly to many commonly used updating parameters. However, the incompleteness of the experimental data needs to be addressed to implement the method.

The output error method is formulated by defining the error in the output in the frequency domain. As a result, the final equation contains the inverse of the dynamic stiffness matrix. Due to this, the final equation is related nonlinearly to the updating parameters. However, it has the advantage that the method can be used with incomplete experimental data.

15.15.1 RESPONSE FUNCTION METHOD

In this section, we look at an FRF-based updating method (Lin and Ewins [59]), which is like an equation error method. We present the method for updating mass and stiffness parameters.

The following identities relating to the dynamic stiffness matrix ($[Z]$) and the receptance matrix ($[\alpha]$) can be written for the analytical/FE model and test structure:

$$[Z_A][\alpha_A]=[I] \tag{15.109}$$

$$[Z_X][\alpha_X]=[I] \tag{15.110}$$

Substituting $[Z_X]=[Z_A]+[\Delta Z]$ into Eq. (15.110) and subtracting Eq. (15.109) from it, the following equation is obtained:

$$[\Delta Z][\alpha_X]=[Z_A]([\alpha_A]-[\alpha_X]) \tag{15.111}$$

Pre-multiplying Eq. (15.111) by $[\alpha_A]$ and using Eq. (15.109), we get

$$[\alpha_A][\Delta Z][\alpha_X]=[\alpha_A]-[\alpha_X] \tag{15.112}$$

Using only the ith column of the measured FRF matrix (represented by $\{\alpha_X\}_i$) in the RHS of (15.112) leads to

$$\underset{N\times N}{[\alpha_A]}\ \underset{N\times N}{[\Delta Z]}\ \underset{N\times 1}{\{\alpha_X\}_i}=\underset{N\times 1}{\{\alpha_A\}_i}-\underset{N\times 1}{\{\alpha_X\}_i} \tag{15.113}$$

Let $\{p\}=\{p_1,\ p_2,....,\ p_q\}^T$ be the vector of updating parameters. Linearizing $[\Delta Z]$ with the updating parameters gives

$$[\Delta Z]=\sum_{j=1}^{q}\frac{\partial[Z]}{\partial p_j}\Delta p_j \tag{15.114}$$

For an undamped model, $[Z]=[K]-\omega^2[M]$ and since $[K]$ and $[M]$ are related to the updating parameters, the derivatives $\dfrac{\partial[Z]}{\partial p_j}$ can be determined. Eq. (15.114) is substituted into (15.113), and the resulting equation is expressed at 'L' frequency points from the measured FRFs. All these equations can be combined and written in the following matrix form:

$$\underset{(N\times L)\times q}{[G]}\ \underset{q\times 1}{\{\Delta p\}}=\underset{(N\times L)\times 1}{\{\Delta\alpha\}} \tag{15.115}$$

We note from the LHS of Eq. (15.113) that the measured FRF matrix column $\{\alpha_X\}_i$ has to be of size N, the number of DOFs in the analytical model. However, due to incomplete experimental data, the experimental FRFs are not available at all N DOFs. One method to address the incompleteness is to substitute the unmeasured FRFs with the corresponding FE model FRFs.

Another issue with updating an undamped FE model using FRF data is that the FE model FRFs are real while the measured FRFs are complex. One strategy to deal with this is to use only the real part of the measured FRFs in the updating equations. The error due to this approximation can be minimized by choosing the frequency points by avoiding the resonance and antiresonance regions, as these regions are more affected by the system damping.

The steps followed for updating are similar to that for the iterative modal-data-based method. Eq. (15.115) is solved for $\{\Delta p\}$ by the method of least squares and then the updated parameter vector $\{p\}$ and the corresponding matrices $[K]$ and $[M]$ are computed. Eq. (15.115) is again set up, and the process is repeated until the convergence criterion is satisfied.

Updating of damping matrix

The RFM can also be used to update the damping matrix along with the mass and stiffness matrices. It can be done by incorporating the following changes in the formulation above.

- The dynamic stiffness matrix in Eq. (15.114) is now $[Z] = [K] - \omega^2[M] + i\omega[C]$.
- The experimental FRFs $(\{\alpha_X\}_i)$ are complex
- The analytical FRF matrix $([\alpha_A])$ corresponds to the damped FE model.
- The damping matrix parameters to be updated are added to the parameter vector $\{\Delta p\}$.

15.15.2 FRF-BASED UPDATING USING AN OUTPUT ERROR-LIKE FORMULATION

FRFs are nothing but the outputs of the system to unit amplitude harmonic forces. Along the lines of the formulation of the iterative method of model updating using modal data (in Section 15.14.1), a similar formulation based on the FRFs can be developed. If 'n' FRFs in the ith column of the FRF matrix are measured, then using a linearized relationship between the FRFs and the updating parameters, we can write

$$\{\alpha_X(\omega)\}_{n\times1} = \{\alpha_A(\omega)\}_{n\times1} + \sum_{j=1}^{q} \frac{\partial\{\alpha_A(\omega)\}}{\partial p_j}_{n\times1}\Delta p_j \tag{15.116}$$

Eq. (15.116) is expressed at 'L' frequency points, and all such equations are combined, yielding the following matrix equation:

$$\begin{Bmatrix} \{\alpha_X(\omega_1)\} - \{\alpha_A(\omega_1)\} \\ \{\alpha_X(\omega_2)\} - \{\alpha_A(\omega_2)\} \\ \vdots \\ \{\alpha_X(\omega_L)\} - \{\alpha_A(\omega_L)\} \end{Bmatrix} = \begin{bmatrix} \frac{\partial\{\alpha_A(\omega_1)\}}{\partial p_1} & \frac{\partial\{\alpha_A(\omega_1)\}}{\partial p_2} & \cdots & \frac{\partial\{\alpha_A(\omega_1)\}}{\partial p_q} \\ \frac{\partial\{\alpha_A(\omega_2)\}}{\partial p_1} & \frac{\partial\{\alpha_A(\omega_2)\}}{\partial p_2} & \cdots & \frac{\partial\{\alpha_A(\omega_2)\}}{\partial p_q} \\ \vdots & \vdots & \vdots & \vdots \\ \frac{\partial\{\alpha_A(\omega_L)\}}{\partial p_1} & \frac{\partial\{\alpha_A(\omega_L)\}}{\partial p_2} & \cdots & \frac{\partial\{\alpha_A(\omega_L)\}}{\partial p_q} \end{bmatrix} \begin{bmatrix} \Delta p_1 \\ \Delta p_2 \\ \vdots \\ \Delta p_q \end{bmatrix} \tag{15.117}$$

The derivative/sensitivity of the FRFs appearing on the RHS of Eq. (15.117) can be obtained from the derivative of the FRF matrix given below

$$\frac{\partial[\alpha_A]}{\partial p_j} = -[\alpha_A]\frac{\partial[Z_A]}{\partial p_j}[\alpha_A] \tag{15.118}$$

Eq. (15.117) can be symbolically represented as

$$[Q]_{(n\times L)\times q} \{\Delta p\}_{q\times1} = \{\Delta\alpha\}_{(n\times L)\times1} \tag{15.119}$$

Eq. (15.119) represents simultaneous linear equations and can be solved for $\{\Delta p\}$ by the least-squares method, and updating can be carried out iteratively, as discussed in sections 15.14.1 and 15.15.1. We note from Eq. (15.117) that the complete column of the FRF matrix is not required to set up this equation; hence, with the above formulation, the FRF-based updating can be carried out with the incomplete experimental data.

Example 15.3: Iterative Model Updating Using FRF Data

In this example, we update a damped model of our cantilever structure using the iterative FRF-based method (RFM) described in Section 15.15.1.

a. The experimental structure simulated in Example 15.1 is used. The structure is shown in Figure 15.2, and the details are given in Section 15.14(a).
b. The initial FE model to be updated has ten finite elements with $E = 2\times10^{11} N/m^2$ and $\rho = 7,800 \ kg/m^3$ and a torsional spring at the clamped end with stiffness $K_t = 4166.7 \ N-m/rad$. The DOF of the model is 22.
c. Updating parameters: The material property and the boundary condition are in error. We take E and K_t as updating parameters to address these errors. Since the objective is to update the damped model, we also choose

       parameters for updating the damping matrix. We take stiffness proportionality constant β_k also as an updating parameter with an initial value of 0.00001.

  d. Data for updating: We use the FRF data measured at 5 DOFs. Since the damping matrix is also updated, we use the measured complex FRFs to set up the updating equations. 8 frequency points, 2 points each near the four resonant peaks in the FRF, are used for updating.

  e. We build FE model matrices $[M]$ and $[K]$ with the current values of the updating parameters and compute the FE model FRF matrix $([\alpha_A]$, of order 22×22) at the 8 frequency points. For setting up updating equations, we also need $\{\alpha_X\}$ of order 22×1. Since $\{\alpha_X\}$ is of order 5×1, the incompleteness is resolved by substituting the unmeasured FRFs with the corresponding FE model FRFs.

  f. Updating is carried out using the following steps:

- Updating equation, Eq. (15.115), is set up by computing $[G]$ and $\{\Delta\alpha\}$. We use the fractional corrections to parameters (defined by Eq. 15.97) as unknowns and hence modify the matrix $[G]$ accordingly (as discussed in Section 15.14.2).
- The least-squares solution of the updating equations is found using the MATLAB function 'lsqr()'. The solution yields $\{\Delta u\}$, from which $\{\Delta p\}$ is found.
- The parameter vector is updated: $\{p\}^{\{k+1\}} = \{p\}^k + \{\Delta p\}^k$.
- We check the convergence criterion. If the norm of $\{\Delta u\}$ is less than the tolerance (a value of 0.001 is used here), we stop and obtain the updated parameter vector and the updated matrices. Otherwise, we repeat the above steps till the convergence criterion is satisfied.

  g. Results: Figure 15.7 shows the convergence of the updating parameters over iterations. At convergence, the vector of cumulative fractional corrections to the parameters is obtained as $\{u\} = \{-0.1 \ -0.4 \ +5.00\}^T$. It indicates a reduction of E and K_t by 10% and 40% respectively and an increase in β_k by 500.0%. The updated values of the parameters are $E = 1.8 \times 10^{11} N/m^2$, $K_t = 2,500 \ N-m/rad$ and $\beta_k = 0.00006$, which are identical to their values in the simulated experimental structure. Figures 15.8 and 15.9 show overlays of FRFs at node 11 before and after updating, respectively, showing that the updated model FRF matches exactly with the measured FRF.

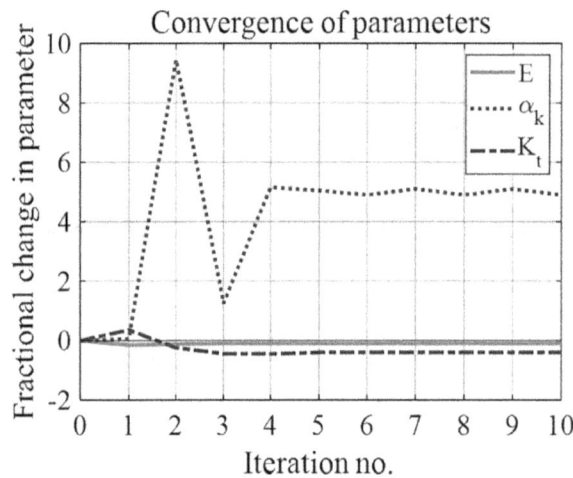

FIGURE 15.7   Convergence of the updating parameters over iterations.

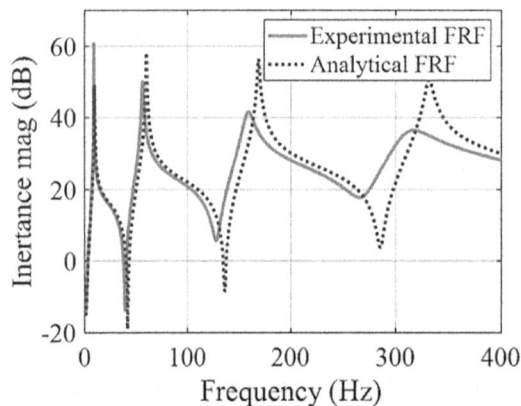

FIGURE 15.8   FRF overlay at node 11 before updating.

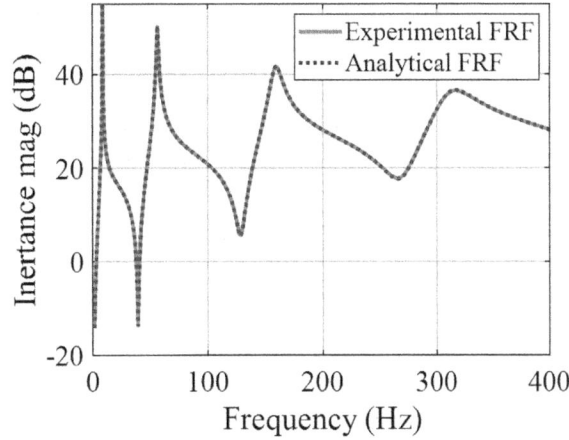

FIGURE 15.9   FRF overlay at node 11 after updating.

15.16   MODEL UPDATING USING NORMAL FRFs

15.16.1   UPDATING OF MASS AND STIFFNESS PARAMETERS

If only the mass and stiffness parameters are updated with the FRF-based updating, then one issue is that the measured and analytical FRFs are not consistent since the former are complex while the latter are real as the FE model is undamped. There could be two approaches to resolve this. One suggested approach is using only the real part of the measured FRFs. However, this approach is not correct theoretically since the damping in the structure affects not only the imaginary part but also the real part of the FRFs. The higher the damping, the higher the effect. The second approach is to include the damping in the FE model so that the analytical FRFs are also complex. This approach is also difficult to work on unless an accurate estimate of the damping matrix is available.

The problem of updating mass and stiffness parameters using FRF data can be formulated more consistently using the normal FRFs (Pradhan and Modak [87]). The method proposed is a modified version of the response function method (RFM) (described in Section 15.15.1) and is referred to as the normal response function method (NRFM). The normal FRFs are the FRFs of the structure if, hypothetically, the damping is removed from the structure or the structure doesn't have damping. In other words, the normal FRFs are the FRFs of the corresponding undamped structure. The normal FRFs of a structure can be computed from the measured complex FRFs using the following relation (Chen et al. [24]):

$$\left[\alpha_X^N\right]=\left[\alpha_X^R\right]+\left[\alpha_X^I\right]\left[\alpha_X^R\right]^{-1}\left[\alpha_X^I\right] \tag{15.120}$$

where $\left[\alpha_X^N\right]$ is the normal FRF matrix and $\left[\alpha_X^R\right]$ and $\left[\alpha_X^I\right]$ are the real and imaginary parts of the measured complex FRF matrix, respectively. The superscripts N, R, and I denote the normal and the real and imaginary parts, respectively.

The NRFM formulation proceeds along the lines of RFM but uses the normal FRFs and leads to the following equation:

$$\underset{N\times N}{\left[\alpha_A^N\right]}\ \underset{N\times N}{\left[\Delta Z_A^N\right]}\ \underset{N\times 1}{\left\{\alpha_X^N\right\}_i}=\underset{N\times 1}{\left\{\alpha_A^N\right\}_i}-\underset{N\times 1}{\left\{\alpha_X^N\right\}_i} \tag{15.121}$$

where $\left[\Delta Z_A^N\right]=[K]-\omega^2[M]$.

Setting up Eq. (15.121) needs the normal FRFs and should be estimated from the complex FRFs using Eq. (15.120). It requires the knowledge of the complete experimental FRF matrix, which for beam and plate-like structures includes both translational and rotational FRFs. However, from the studies conducted on the structures with beam members, it is observed that with a reasonable number of translational FRF measurements, and no rotational FRFs, the normal FRFs can be estimated accurately.

The square FRF matrix can be either measured (which requires measurement of the lower or upper triangular matrix, as the FRF matrix is symmetric) or regenerated by the modal model identified from the analysis of one measured row/column. The coordinate incompleteness in Eq. (15.121) is dealt with by replacing the unmeasured FRFs with the corresponding FE model FRFs. This equation is expressed at 'L' frequency points, and all such equations are combined and solved by the least-squares method.

Eq. (15.121) is based on the FRFs of the 'undamped' structure and hence provides a consistent formulation to update parameters of the undamped model (i.e., mass and stiffness parameters) using FRF data.

15.16.2 UPDATING OF DAMPING MATRIX PARAMETERS

The damping matrix parameters are updated separately after the mass and stiffness parameters have been updated using NRFM. The formulation of the method ([88]) is briefly described below.

Superscript 'c' denotes the complex version of the quantity, which also means the quantity is related to the damped system. The complex dynamic stiffness matrix $\left(\left[Z_X^c\right]\right)$ and FRF matrix $\left(\left[\alpha_X^c\right]\right)$ for the experimental structure are related by

$$\left[Z_X^c\right]\left[\alpha_X^c\right]=[I] \tag{15.122}$$

Similarly, $\left[Z_X^N\right]$ and $\left[\alpha_X^N\right]$, representing the corresponding quantities for the hypothetical undamped experimental structure are related by

$$\left[Z_X^N\right]\left[\alpha_X^N\right]=[I] \tag{15.123}$$

Subtracting Eq. (15.123) from Eq. (15.122), we get

$$\left[Z_X^c\right]\left[\alpha_X^c\right]-\left[Z_X^N\right]\left[\alpha_X^N\right]=[0] \tag{15.124}$$

Writing $\left[Z_X^c\right]=\left[Z_X^N\right]+i\left[Z_X^d\right]$, with $\left[Z_X^d\right]$ representing the damping part of the dynamic stiffness matrix, we get after simplification

$$\left[Z_X^N\right]\left(\left[\alpha_X^c\right]-\left[\alpha_X^N\right]\right)=-i\left[Z_X^d\right]\left[\alpha_X^c\right] \tag{15.125}$$

Substituting $\left[Z_X^N\right]=\left[\alpha_X^N\right]^{-1}$, and writing $\left[Z_X^d\right]=\left[Z_A^d\right]+\left[\Delta Z_A^d\right]$, we obtain

$$\left[\alpha_X^N\right]\left[\Delta Z_A^d\right]\left[\alpha_X^c\right]=i\left(\left[\alpha_X^c\right]-\left[\alpha_X^N\right]\right)-\left[\alpha_X^N\right]\left[Z_A^d\right]\left[\alpha_X^c\right] \tag{15.126}$$

Eq. (15.126) corresponding to the jth column of $\left[\alpha_X^c\right]$ can be written as

$$\left[\alpha_X^N\right]\left[\Delta Z_A^d\right]\{\alpha_X^c\}_j=i\left(\{\alpha_X^c\}_j-\{\alpha_X^N\}_j\right)-\left[\alpha_X^N\right]\left[Z_A^d\right]\{\alpha_X^c\}_j \tag{15.127}$$

Note that, in Eq. (15.127) $\left[Z_A^d\right]=\omega[C]$ and $\left[\Delta Z_A^d\right]=\omega[\Delta C]$. Thus, $\left[Z_A^d\right]$ depends on the damping matrix before the current iteration, while $\left[\Delta Z_A^d\right]$ depends on the change in the damping matrix. In this equation, the RHS is based on the experimental data and the current knowledge of the damping matrix and is constant for the current iteration. Representing it by vector $\{b\}_j$, we get

$$\left[\alpha_X^N\right]\left[\Delta Z_A^d\right]\{\alpha_X^c\}_j=\{b\}_j \tag{15.128}$$

Eq. (15.128) is the basic equation for updating the damping matrix involving the complex and normal FRFs. The method uses damping submatrix correction factors as updating parameters. The submatrices could be individual or groups of finite elements, and the coefficients (β_r) of these submatrices are the updating parameters. Thus,

$$[\Delta C]=\sum_{r=1}^{q}\beta_r[C]_r \tag{15.129}$$

Note that $\sum_{r=1}^{q}$ in Eq. (15.129) represents the assembly of finite elements groups and $[C]_r$ represents the damping matrix of the rth group at the current iteration. Eq. (15.128) is written at selected frequency points, and all such equations are combined and solved by the least-squares method.

Using the NRFM and the damping matrix updating approaches discussed above, a two-stage approach to updating is evolved ([89]) for mass, stiffness, and damping matrix updating. The mass and stiffness parameters are updated in the first stage using the normal FRFs, and the damping matrix parameters are updated in the second stage.